SPOT TESTS IN ORGANIC ANALYSIS

First English Edition, *1937* (translated by J. W. Matthews)

Second English Edition, *1939* (translated by J. W. Matthews)

Third English Edition, *1946* (translated by Ralph E. Oesper, on the basis of the translation by J. W. Matthews) [under the title *Qualitative Analysis by Spot Tests, Inorganic and Organic Applications*]

Fourth English Edition (in two volumes), *1954* (translated by Ralph E. Oesper) [under the title *Spot Tests, Vol. II, Organic Applications*]

Fifth English Edition, *1956* (translated by Ralph E. Oesper)

Sixth English Edition, *1960* (translated by Ralph E. Oesper)

Seventh English Edition, *1966* (translated by Ralph E. Oesper)

SPOT TESTS
IN
ORGANIC ANALYSIS

by

FRITZ FEIGL, Eng., D.Sc.

Laboratorio da Produção Mineral, Ministério da Agricultura,
Rio de Janeiro; Professor at the University of Brazil;
Member of the Austrian, Brazilian and Gotenburg Academies of Sciences

in collaboration with

Dr. VINZENZ ANGER

Research Laboratory, Lobachemie, Vienna, Austria

Translated by

RALPH E. OESPER, Ph. D.

Professor Emeritus, University of Cincinnati

Seventh English edition, completely revised and enlarged

ELSEVIER SCIENTIFIC PUBLISHING COMPANY
AMSTERDAM OXFORD NEW YORK
1975

ELSEVIER PUBLISHING COMPANY
335 JAN VAN GALENSTRAAT, P.O. BOX 211, AMSTERDAM

AMERICAN ELSEVIER PUBLISHING COMPANY, INC.
52 VANDERBILT AVENUE, NEW YORK, N.Y. 10017

ELSEVIER PUBLISHING COMPANY LIMITED
RIPPLESIDE COMMERCIAL ESTATE, BARKING, ESSEX

FIRST PUBLISHED 1966
FIRST REPRINT 1971
SECOND REPRINT 1975

ISBN 978 0 44440 209 7

LIBRARY OF CONGRESS CATALOG CARD NUMBER 65–13235

WITH 19 TABLES

Transferred to digital printing 2006

FOREWORD

In the Foreword to the first monograph on the use of spot tests in organic analysis (1954) the hope was expressed that the experiences and points of view yielded by the study of the chemistry of specific, sensitive, and selective reactions, which had played such an important role in the development of inorganic spot test analysis, would likewise be of much value in organic spot test analysis and so contribute to the growth and popularization of this branch of spot test analysis. The next ten years showed this hope to have been well founded. New spot tests in organic analysis followed in an almost continuous stream and these necessitated the issuing of enlarged editions in 1956 and 1960.

The present text brings this succession up to date so far as this is feasible in view of the fact that advances in the field are reported continuously. Through a critical revision of all the sections of the former text and by changing the arrangement of the material, the author has here presented a clearer picture of the application possibilities of organic spot test analysis, a field whose further growth and expansion may be expected with certainty.

The book was in danger of becoming too bulky as a result of the sheer mass of pertinent experimental material and also because of the corresponding growth of the didactic sections and paragraphs which presented the chemical bases of the tests and provided the related references to the literature. Accordingly, judicious pruning was indicated and abridgements were in order. For example, the chapter *Spot Test Techniques* was removed entirely since this topic was adequately covered in the companion volume *Spot Tests in Inorganic Analysis* (1958). Additional space was saved by omitting some of the Tables, by presenting certain structural formulae in a restricted form, and by employing other typographical simplifications. If despite these measures there has still been an unavoidable increase in the size of the volume the reason lies primarily in the fact that organic spot test analysis, which in the beginning was merely a supplement to inorganic spot test analysis, has now become a special field of qualitative organic analysis.

The text has been divided into seven chapters that are followed by a comprehensive Appendix of individual compounds and products examined, giving the limits of identification obtainable by the procedures. A complete author index is included as well as a subject index. The introductory chapter dealing with the development, present state and prospects of organic spot test analysis, has now been slightly enlarged to stress also the new guiding lines that have proved of use in the investigation of the field of qualitative organic analysis. Great expansions were necessary in other portions of the book that describe the making of spot tests and their trial on a sufficient number of organic materials. The first chapter of this kind logically takes up the so-called preliminary tests and includes 45 sections as opposed to 32 sections in the previous edition. The next chapters which were devoted to the detection of functional groups (109 sections instead of 70) and individual compounds (148 as against 133) were more sharply separated from each other by insertion of a chapter on the detection of characteristic compound types. It contains 74 sections. A new feature is a chapter with 54 sections describing the use of spot reactions for distinguishing isomers and for determining structures. The final chapter takes up the application of spot reactions in testing materials etc.; it has been enlarged to 131 sections in place of 111. In total the book now gives in 561 sections information about more than 900 tests as against 600 tests in 346 sections in the preceding edition.

It should be pointed out that for most of the studies included in the present edition there are now several tests available that differ in sensitivity and selectivity, a fortunate state of affairs that makes for greater reliability in the findings. The text is distinguished through the fact that numerous tests and remarks scattered throughout the book have been gleaned and collected from unpublished findings. Many valuable hints are sure to be found in these references, which number about 250.

The book has purposely been limited to description of spot tests and their chemical foundations. Accordingly no effort has been made to include references to the many publications that contain information about the use of spot tests in chromatography and concerning the adaptation of spot tests for colorimetric microdeterminations. It has been observed that these numerous and varied uses, especially in the biological sciences, have now become so commonplace and self-evident that in many instances it is no longer felt necessary to give proper credit to the original sources of these spot tests. The present text can well serve to provide the interested and conscientious reader with these fundamental references and facts.

As in the earlier editions, the author has stressed in this new edition all the features that are characteristic of spot test analysis. At first sight, spot test analysis will appear to many users of the book as an analytical technique that enables the operators to accomplish satisfactory semi-micro, micro-, and ultramicro tests with simple equipment and a minimum consumption of sample and time. However, from the scientific and practical points of view, it should not be forgotten that the characteristic features of spot test procedure are not the fundamental reason for spot test analysis and its development. The essential point is that the most feasible means are involved in its working technique. This is best illustrated by the fact that most spot tests are the end result of searches for chemical changes of the most diverse kinds that can be used in analysis. In this respect, and considering the high value of manual dexterity in the sciences, spot test analysis becomes experimental chemistry with analytical objectives. It widens and deepens our knowledge of chemical changes through its connections with other branches of chemistry. Accordingly, this text has been designed not only for professional analysts, organic chemists, biochemists, biologists, forensic chemists, etc., who are concerned with solving definite chemical analytical problems on the micro scale or who are interested in new reactions of organic compounds, but also for the body of chemistry teachers. The needs of organic chemistry laboratory courses have especially been kept in mind since those spot tests which can be carried out successfully with little trouble and expense can contribute much to the understanding of the reaction possibilities of organic compounds.

It is the author's hope that this book will serve to correct the erroneous opinion, widespread among non-analysts, that physical instrumentation is always superior to chemical methods for solving analytical problems. Each of the chapters presents instances of problems for which no solutions by physical means have previously been developed or for which the quick spot tests are often equal or superior to the expensive instrumental procedure. Accordingly, in the future as in the past, it may be expected that analysts, organic chemists, pharmaceutical chemists, etc. will resort to spot tests, not only when seeking solutions to problems of ordinary import but also when confronted with questions that arise in connection with paper-, thin layer-, and gas chromatography. So long as this is the case—and as shown by the current literature there is an unceasing search for new tests and a never-ending desire for the improvement of tests already described—qualitative analysis, which looks back on a proud past, will continue to preserve its value. It will aid in keeping classical analytical chemistry, of which it is an essential part, in the position it so rightfully occupies in pure and applied chemistry. Spot test analysis was founded and developed on this basis and not merely as a branch of microchemistry.

During my career as scientist and teacher I have had to overcome many hindrances, surmount many hurdles, and experience many upheavals. On the other hand, and as compensation, I have been fortunate in having the support of many friends, colleagues, and devoted coworkers. It is my duty and pleasure to express my thanks here first of all to Dr. Vinzenz Anger (Research Laboratory, Lobachemie, Vienna). He was my first coworker when I founded organic spot test analysis so many years ago and he has always maintained his active interest in this field. He has made notable contributions to its extension on his own. His experimental studies and literature investigations, together with his critical sense, have been of inestimable value and assistance to me during the compilation and writing of this text.

The translation has been the work of Dr. Ralph E. Oesper, emeritus professor of analytical chemistry at the University of Cincinnati. He has stood by me in many sad times and his encouragement has been invaluable in every respect. His interest in this undertaking has gone much beyond that of a mere translator: at considerable sacrifice of time and energy he has secured from many places, companies, and individuals in the United States specimens of compounds not available to me and which have been indispensable in trying out new tests. In this respect special thanks are due to Dr. Elizabeth Weisburger of the National Cancer Institute at Bethesda, Maryland, and to Dr. Keith W. Wheeler of the Richardson–Merrell Company of Cincinnati.

Finally mention should be made of the helpful support I received from Dr. Oswaldo Erichsen de Oliveira, Director of the Laboratorio da Produção Mineral and from the Conselho Nacional de Pesquisas who have provided me with all facilities necessary to pursue my researches.

Rio de Janeiro, Brazil   FRITZ FEIGL
December 1965

CONTENTS

CHAPTER 4. DETECTION OF STRUCTURES AND CERTAIN TYPES
OF ORGANIC COMPOUNDS 314

CHAPTER 6. APPLICATION OF SPOT TESTS IN THE DIFFERENTIA-
TION OF ISOMERS AND HOMOLOGOUS COMPOUNDS.

Chapter 1

Development, Present State and Prospects of Organic Spot Test Analysis

From the standpoint of a formal priority, the organic province of spot test analysis is older than its inorganic counterpart, since the earliest spot test was used for the detection of an organic compound. In 1859 HUGO SCHIFF reported that uric acid can be detected by placing a drop of an aqueous solution of this acid on filter paper which had been impregnated with silver carbonate. Silver precipitates and the finely devided metal produces a gray to black fleck which stands out prominently against the white paper. This test demonstrated the fundamental practicality of spot reactions on filter paper, and by actual trial SCHIFF also showed that this method of procedure permits the attainment of a considerable degree of sensitivity. At the then state of chemical analysis, this was truly a remarkable finding. It coincided with the beginnings of the classical studies by CHRISTIAN FRIEDRICH SCHÖNBEIN and FRIEDRICH GOPPELSROEDER, who in their "capillary analysis" demonstrated the great analytical significance of the spreading of liquids and solutions of organic and inorganic compounds with consequent localized fixation in filter paper of small amounts of the dissolved materials. The publications of these three pioneers obviously contain descriptions of effects which have been used to the widest extent in modern spot test analysis. Accordingly, it is logical to inquire why real attention was not given to spot reactions on filter paper until a much later date (around 1920), attention which might have led to the founding first of inorganic and then of organic spot test analysis. The correct answer to this question is readily revealed by a consideration of the history of the spot test detection of inorganic materials. In the beginning, chemical reactions were employed which respond not only to slight, often minimal, amounts of a particular material, but which in many cases also make possible its detection in the presence of other substances. Such tests were fairly uncommon as long as inorganic reagents predominated in inorganic analysis. This situation was not changed until the advent of organic reagents, especially those which give colored reaction products. Only then was it possible to work out numerous sensitive and unequivocal methods of detection and determination. The nickel–dimethylglyoxime reaction, introduced into qualitative and quantitative inorganic analysis by L. TSCHUGAEFF (1905) and O. BRUNCK (1914), has become the classic example of this type of progress even though P. CAZE-NEUVE, G. DENIGÈS, and M. ILLINSKY had previously described useful color reactions employing organic reagents. Consequently, the earliest efforts of the present author to use spot reactions for spot tests were directed particularly toward combining the great advantages of organic reagents with those of the Schönbein-Goppelsroeder capillary analysis. It was found that when drops of the test- and reagent solutions are brought together on filter paper, colored reaction products are fixed on the surface of the paper and produce flecks or rings within a circle wetted by water, and can thus be distinctly detected. The water circles which likewise appear when a dry reagent paper is spotted with a drop of an aqueous test solution, show that the products of chemical reactions may be separated from the reaction theater and fixed in the surface of the paper. This local accumulation accompanying spot reactions on filter paper is of much importance because it enhances the discernibility of colored reaction products, and especially if they are insoluble in aqueous media, or when colored water-soluble reaction products are adsorbed on filter paper. However, this type of spot tests, in which inorganic and organic reagents can be used, is obviously limited to reactions which do not require either strong heating or evaporation, or a highly alkaline or acidic reaction theater, since filter paper is attacked in such instances. The procedural aspect of spot test analysis received a very useful extension when it was found feasible to unite drops of the test- and reagent solutions on non-porous surfaces (spot plate, watch glass, micro test tube, etc.) where the limitations no longer apply. A decisive impulse to the development of inorganic spot test analysis came however from the logical but, at the time, novel idea that the restrictive preference for certain kinds of reactions and ways of carrying them out is a handicap. In every case, the sole objective is the discovery and employment of all possibilities through which the highest sensitivity and certainty can be attained with drop reactions. The proper appreciation of these goals, a matter of prime importance in spot test analysis, greatly extended the ways in which chemical reactions can be employed in analysis, and thus led to great advances. Such possibilities include: the wide-spread use of organic compounds as precipitation-, color-, and masking agents; the utilization of catalyzed and induced reactions; solid-body reactions at elevated temperatures and reactions in the gas phase through contact with suitable solid or dissolved reaction partners; reactions on the surface of a solid by contact with a dissolved partner; reactions which yield fluorescent products or those that quench fluorescence; interfacial effects (adsorption, capillarity, flotation). Last, but not least, was a proper consideration of the enormous influence of the test conditions on the course of the underlying reactions, the physical nature of the reaction products and the behavior of the accompanying material. In short, and using a terminology that has now been generally accepted, the conditioning of tests can in many cases bring about a distinct enhancement of the sensitivity and selectivity. Experience has shown that inorganic spot test analysis is the outstanding field of application of specific, selective, and sensitive tests with the goal of rapidly solving problems in qualitative microanalysis. It may be said that inorganic spot test analysis has initiated a hitherto nonexistent liberalism in the utilization of chemical reactions. This liberalism has benefited not only spot test analysis but likewise has had a lasting effect on the search for new procedures in inorganic qualitative and quantitative micro- and macroanalysis and qualitative organic analysis.

Spot tests will reveal as little as 0.001–10 micrograms (gamma $= \gamma = \mu g$) of material (solid or in drops of a solution). Those low "limits of detection" or "limits of identification" correspond, in so far as solutions are being examined, to great dilutions (dilution limit) or low concentrations. If these limits of detection or limits of identification can be attained with spot tests, it is proper to speak of micro tests.* It is evident that only larger amounts of materials will respond to less sensitive spot reactions. In general, only such spot reactions are used as are applicable to microanalytical or at least semi-microanalytical problems. The identification limit and dilution limit, as determined experimentally, are guides in this appraisal. These determinations are intimately related to the setting-up of the particular test conditions since the numerical values of the sensitivity are not constants of the fundamental reaction. Therefore it is not correct to speak of the sensitivity of a *reaction* as is the usual practice. Only a *test* is sensitive because the test conditions are likewise then taken into account. Strictly speaking, the same limitation applies to the terms "selectivity" and "specificity" since here too account must be taken of the very important matter of conditioning.

Spot test analysis in research or practice has brought into the foreground a very important problem, namely what determines the specificity, selectivity, and sensitivity of its procedures. In other words, what factors influence

---

* Many spot reactions are the equal of the crystal precipitations of the classical qualitative microanalysis with respect to quantity sensitivity (limit of identification), and in most cases they are superior to the latter with regard to concentration sensitivity (dilution limit). None the less, identifications by means of spot reactions are frequently called semimicro tests in Anglo-Saxon countries. This obviously incorrect designation originated in the fact that for many years microanalysis was identified with micromanipulation. However, the objective of qualitative microanalysis is the identification or detection of extremely small quantities of material, no matter what particular technique is employed. Precisely this point was very impressively demonstrated by spot reactions. Compare Chapter 1 in *Inorganic Spot Tests* regarding this point.

these characteristics in the positive and negative direction? The scientific treatment of the problem obviously is of fundamental importance not only to spot tests but also to all procedures of qualitative and quantitative inorganic and organic chemical analysis and constitutes the "chemistry of specific, selective and sensitive reactions". This field of experimental chemistry includes the consideration of the numerous relations of analytical chemistry to other areas of chemistry; pertinent studies suggested by spot test analysis have brought the importance of this field to the attention of chemists.*

It may be remembered that the searches for organic reagents and studies of their activity in relation to structural factors resulted in the preparation of new organic compounds and led to the isolation of new complex organometallic compounds with consequent enrichment and extension of the chemistry of the coordination compounds. The same benefit came to the chemical adsorption of dyestuffs on metal oxides or hydroxides with production of color lakes in which the chelate bondings are fundamental. The appreciation of the great microanalytical importance of fluorescence reactions, of catalyzed reactions, and of induced reactions has led to the discovery and employment of new accompanying effects which have become the subject of physical-chemical studies or have yielded pertinent leads. Observations and statements of new kinds of reactions obtained in analytical researches have in this way provided important material to other fields of chemistry.

Although directing advances in inorganic spot test analysis were obtained relatively soon, its systematic development was gradual and some time elapsed before its usefulness in qualitative semimicro- and microanalysis was extensively appreciated. Consequently, not much attention was given to the applicability of spot tests for organic compounds and the characteristic groups they contain until (since about 1930) the important foundations of inorganic spot testing had already been laid and the conditions of its further development recognized. The number of spot reactions, which can successfully be applied in qualitative organic analysis, has become large enough to justify the use of the term "organic spot test analysis", and all the more so because the advances in the last 10 years point to the deepening and widening of this field.

Organic spot testing likewise strives to attain microanalytical goals. Therefore, all possibilities must be explored so that the specificity, selectivity, and sensitivity of the particular test may be brought to a maximum. It must not be forgotten that when dealing with organic compounds, the objective of the analysis and the analytical utilization of chemical reactions should be considered from standpoints which differ from those adopted with respect to inorganic compounds. In qualitative inorganic analysis the aim

---

* See F. FEIGL, *Chemistry of Specific, Selective and Sensitive Reactions*, Academic Press, New York, 1949; F. FEIGL, *Research*, 4 (1950) 550.

is to detect metallic and non-metallic elements, and almost always it is possible to do this by chemical means. In contrast, the detection of elements has only an orienting value in qualitative organic analysis, because the important goal is the detection of particular compounds or the identification of characteristic groups in organic compounds, whose ultimate constituents are usually known. Chemical methods cannot contribute beyond a limited extent to the solution of these two problems, especially the former. The reason resides not only in the enormous number of organic compounds and the variety in their architecture. The decisive factors are that many organic compounds undergo chemical changes under conditions which cannot be realized in analytical work, and furthermore a uniform mode of reaction is encountered incomparably more often than with inorganic ions. Consequently, specificity and selectivity are much rarer in tests for organic compounds than in inorganic identifications and separation processes, such as those successfully used in the systematic qualitative inorganic scheme in the form of group precipitations or group solutions, play little or no part in qualitative organic analysis. Many tests for organic materials depend on the participation of certain groups in chemical reactions; but entirely apart from the fact that some important groups are not reactive, it must be kept in mind that the detection of groups gives information only about a certain region of the molecule of an organic compound. Therefore, reliable identifications of individual compounds by purely chemical tests are infrequent. As a rule, appeal must be made also to physical methods based on the determination of physical properties related to the structure and size of the organic molecule. Despite this limitation, chemical tests have a considerable practical importance in qualitative organic analysis and its many fields of application. Problems whose solution is facilitated by analytical procedures seldom involve totally unknown materials or artificial mixtures. The available information regarding origin, method of preparation, intended use *etc.*, as well as the color, form of aggregation, and so on, of the sample, almost invariably give clues as to the direction the examination should take. Frequently, the analyst is not asked to detect a particular compound, but is required merely to find out whether members of a particular class of compounds are present or absent. For these latter purposes, it is often sufficient to prove the presence of certain groups, or even of given elements. Accordingly, chemical methods can contribute a great deal to the solution of many of the problems encountered in the study of organic materials. Consequently, efforts must be made to apply all reaction possibilities of organic compounds, which may be valuable as analytical aids.

As mentioned above, the objectives and problems of qualitative inorganic analysis differ from those of qualitative organic analysis. Frequently, these fields also differ with respect to the reaction milieu and the mode of reaction.

The analytical employment of reactions of inorganic compounds involves primarily ionic reactions in aqueous solutions, but this is not true of purely organic compounds. Although water-soluble organic acids, bases, and salts and their corresponding reactive ions are known, the majority of organic compounds are non-ionogens of hydrophobic character, and the aqueous medium does not play the dominant role in their reactions that it does in inorganic analysis. Many organic compounds react only when dissolved in organic liquids, or when in the gas phase, in melts, and non-homogeneous systems. As a rule, such reactions proceed much slower than ionic reactions in aqueous solutions, they are incomplete, and frequently are accompanied by side reactions. Despite these handicaps, molecular reactions of organic compounds in the absence of water deserve every attention because, on occasion, such reactions provide an entirely satisfactory substitute for the lack of reactivity in water solutions. This holds especially for condensation and pyro-reactions in the regions 120–250° for which no counterparts are known in inorganic analysis. The same is true with respect to the release of characteristic products through ignition or sintering of organic compounds with inorganic or organic partners.

The use of reactions of organic compounds to obtain analytical objectives is still inconsiderable as compared with their widespread employment in synthetic organic chemistry. However procedures are known for the identification of many organic compounds and the groups contained in them which can be applied directly, or with proper modifications or additions, in organic spot test analysis. The importance of the special working techniques generally used in spot testing should not be overrated. The employment of drops of a solution is admittedly a very fundamental part of spot test analysis and superficially it may seem to be its most characteristic feature. However spot testing does not owe its development solely to its peculiar technique through which microanalytical goals may be reached but primarily to the efforts to find suitable subjects for this working technique. This search leads to the consideration of reaction possibilities that are already known and to the search for new reactions of organic compounds and characteristic parts of their molecules which can be used for detection or identification. Attention is therefore directed to the facts and findings of the "chemistry of specific, selective, and sensitive reactions", which render excellent service in this regard.

The use of spot tests in qualitative organic analysis can be discussed under six headings, each division having distinct objectives. Separate chapters will be devoted to these categories in this text They are:

1) The identification or detection of nonmetallic and metallic elements; the behavior on ignition and combustion; the proof of the basic or acidic

character; the possible redox activity of organic compounds; differentiation between aliphatic and aromatic compounds; study of the behavior toward reactive and nonreactive solvents.

2) The detection of certain reactive (so-called functional) groups included in the molecule of organic compounds.

3) The identification of characteristic compound types.

4) The identification or detection of individual organic compounds.

5) The differentiation of isomers and homologs of organic compounds and determination of constitution.

6) The use of procedures included in 1–5 for the solving of special analytical problems in tests of materials, *etc.*

The solution of the problems cited in 1) involves the use of "preliminary tests". The results of their procedures are extremely valuable adjuncts to the chemical examination of organic materials since with little expenditure of sample and time they yield important clues for the tests to be made subsequently. However the real province of qualitative organic analysis and hence also of organic spot test analysis is the detection of certain groups in organic compounds, the detection of characteristic types of compounds, and the identification of individual compounds. Chemical procedures for the solution of these problems are invariably based on the fact that organic compounds enter into chemical reactions not as entire individuals but through the action of certain atom groups. There are two ways of utilizing such reactions: If groups are present which react in such manner that addition (molecular) compounds, salts, condensation-, oxidation- or reduction products are formed and which because of color, fluorescence in daylight or u.v. light, solubility, *etc.* can serve for the identification of the starting material or the groups it contains, it is permissible to speak of "direct tests". In contrast, "indirect tests" employ the reactivity of certain groups to arrive at compounds which in their turn can be identified by salt-formation, condensation, *etc.* Many indirect tests employ operations that are commonly used in preparative organic chemistry for the tearing down, building up, and transformation of compounds. But instances are known in which the reactions of pyrolytic cleavage products of organic compounds make sensitive indirect tests possible.

It need hardly be stressed that any direct or indirect tests used in spot test analysis must be satisfactory with respect to sensitivity and reliability, and also must meet the requirement of being quickly conducted in a satisfactory fashion with amounts of material of the order of micrograms to milligrams. It is particularly essential that the preparative operations required in indirect tests be such as can be accomplished without elaborate apparatus

and loss of material. Consequently, in organic spot test analysis it is often
necessary to forego the use of reactions which lead to the desired goal when
carried out on a macro scale in preparative operations. Precisely as in inorganic spot test analysis, it by no means follows that analytical macro methods
can be used directly as a matter of course. The reverse process, namely the
adoption of the methods of spot testing on the macro scale is possible far
more often, though not always. This special position of the methods and
procedures of spot test analysis is based not merely on their special working
technique, but above all in accepting specificity, selectivity and sensitivity
as the prime analytical objective.

With respect to the analytical utility of direct and indirect tests, it seems
natural to assume that direct tests, which are more rapidly executed, are
also more sensitive, since they are not burdened with the supplementary
operations and the consequent inevitable losses of material involved in the
indirect tests. On the other hand, precisely because of the supplementary
steps, the indirect tests would be expected to have a lower sensitivity, but
instead a greater specificity or selectivity. All experience shows however that
these assumptions are unreliable generalizations, which need not be taken
seriously either in the use of known direct or indirect tests, or in the search
for new ones. Both types of tests are of equal value in spot test analysis,
provided they meet the requirement of sensitivity and reliability and merit
the same attention.

What was said concerning the past and present development of inorganic
spot test analysis* holds also for its organic counterpart. It is mostly a matter
of adapting previously known macroanalytical tests to spot testing, and of
finding new tests. Frequently these two objectives are closely related since
the knowledge of the chemistry of a test and appropriate modes of conditioning can lead to such extensive improvements that a practically new test
results. Adaptation and improvement of existing tests have not yet received
the attention they deserve, and much fruitful work along this line can be confidently expected. Color reactions, whose chemistry is known, should be
particularly considered in this connection, because such studies will lead to
an understanding of the details of the procedure, and also give a certain
measure of guidance relative to specificity and selectivity, points which still
need to be investigated. Furthermore, the books and journal literature
dealing with problems of organic analysis often contain descriptions of color
reactions between organic compounds (usually in the presence of concentrated acids and alkalies) which obviously were discovered empirically or
accidentally. When they are examined to determine their possible use in
spot test analysis, studies should also be undertaken to elucidate their chem-

* See Chapter I of F. FEIGL, *Spot Tests in Inorganic Analysis*, 5th edition, Elsevier,
Amsterdam, 1958.

ical background. Both endeavors contribute to the removal of the doubt, which still exists in many quarters, regarding the reliability of color reactions in qualitative organic analysis.

Very useful suggestions for the working out of new direct tests are obtained by considering studies dealing with the analytical employment of the ability of acidic and basic organic compounds to form salts with inorganic ions. On the basis of such investigations, which occupy a broad field of interest in the chemistry of specific, selective and sensitive reactions, it can now be stated with assurance that analytically useful effects of organic reagents can invariably be attributed to the presence and activity of certain groups in the molecule of the compound in question. For example, if dioximes, acyloinoximes, derivatives of α,α'-dipyridyl, and 8-hydroxyquinoline react sensitively and in selective fashion with certain metal ions to yield colored insoluble or soluble compounds, then this salt-formation, which occurs because of the presence of the respective functional groups:

$$
\begin{array}{l} -C=NOH \\ | \\ -C=NOH \end{array} \qquad -CH(OH)-C(NOH)-
$$

can always be applied conversely to the detection of the particular chelating groups in the molecule of the organic reagent. This is nothing more than an application of the longstanding principle (well known in inorganic analysis) that binary reactions can often be used analytically for either of the reactants.* Pertinent examples including also the formation of normal salts of acidic organic compounds will be found in the sections of this text dealing with practical applications.

It is readily apparent that the use of binary reactions between inorganic and organic compounds for the detection of either of the participants is not limited to the formation of normal and complex salts but must hold likewise for other types of reactions. This point is illustrated by the following example. A sensitive and specific test for hydrazine is based on its practically instantaneous condensation with salicylaldehyde in aqueous solution to give a precipitate of light yellow salicylaldazine:

$$
2 \ \text{(OH, CHO)} + H_2N-NH_2 \longrightarrow \text{(salicylaldazine)} + 2\,H_2O
$$

This condensation occures when an aqueous solution of salicylaldehyde is added to acidic solutions of hydrazine salts. The salicylaldazine shows an intense yellow green fluorescence in ultraviolet light, permitting the ready

---

* See in this respect the suggestions advanced by R. PALAUD, *Chim. Anal.*, 34 (1952) 194.

detection of even small quantities of this product. (The condensation products of *m*- and *p*-hydroxybenzaldehyde do not fluoresce.) As was to be expected, the formation of fluorescent salicylaldazine can be used conversely for the detection of salicylaldehyde and of compounds which split off hydrazine (*e.g.*, hydrazides of acids). The use of the condensation of hydrazine can be carried a step farther if consideration is given to the experiences of the chemistry of specific, selective and sensitive reactions relative to the activity of certain groups in organic reagents. In accord with the concept of group action, it could be expected that since all *o*-hydroxy-aldehydes and -ketones contain the same functional groups as salicylaldehyde, they too would necessarily condense with hydrazine and yield fluorescent aldazines. This was found to be the case; most of the fluorescence hues are the same and the sensitivities of these reactions are so satisfactory that it was possible then to develop a sensitive general spot test for *o*-hydroxyaldehydes and *o*-hydroxyketones, which can be successfully applied to even very complicated members of this class of compounds.

Proceeding from the dual analytical employment of inorganic–organic reactions in the wet way, a further step can be taken, namely the preparation of organic reagents may be expected to make possible tests for organic compounds involved in such preparation procedures. In other words, the sample being studied must be subjected to the conditions of the preparation of a particular organic reagent, which in turn can then be identified by means of inorganic ions. Here again there is an obvious close relation to analytical research directed toward the discovery of organic reagents for inorganic ions. Of course, the analytical employment of preparation procedures should not be limited to the production of organic reagents and their detection by an inorganic ion; the goal must be more general. The possibility of an analytical utilization exists whenever the main product of a synthesis or preparation procedure can be detected by means of an appropriate inorganic or organic reagent, or whenever this product is characterized per se by its color, fluorescence, solubility features, *etc*. Furthermore, it must not be forgotten that the detection of a characteristic side product may serve in some cases as a direct or indirect proof of the presence of a particular organic participant in the preparation process. Obviously, such cases require the recognition of the stoichiometric reaction underlying the preparation procedure. It is clear that the attainable yield of the particular organic product which is so important in preparative work is no longer the dominant factor. Instead, the analytical usefulness of the procedure depends solely on whether certain products of the synthesis or preparation procedure are formed at an adequate rate and in amounts sufficient to exceed the identification limits of their direct or indirect detectability. In such instances, low yields lose their deterrent effect and certain interesting aspects come into view, particularly when consideration is given not solely to reactions conducted in the wet way but also by melting and sintering.

In the first place, it may be expected that preparation procedures conducted on a small scale can still yield enough of the desired product to make its detection possible either directly or by macro-, semimicro- or micro tests. An excellent example, which demonstrates that it is worth carrying out syntheses with small amounts of material for analytical purposes, is provided in the classic preparation of alizarin (GRAEBE, LIEBERMANN, PERKIN, 1869). In this procedure, anthraquinone-2-sulfonic acid is fused with a caustic alkali in the presence of alkali salts of oxidizing acids:

This reaction succeeds even with the dry residue from a drop of a colorless 0.004% solution of anthraquinone-2-sulfonic acid and leads to the violet sodium alizarinate. Consequently, it is possible to detect 2 γ anthraquinone-sulfonic acid through this preparation of alizarin. It must not be expected that such simple macro→microtransformations will lead to the desired goal in all cases. However, when this is not possible, attempts can be made to depart from rigid adherence to complicated working directions to achieve simplifications that are in line with analytical techniques. Such modifications might include: changes of concentration and amounts of the reactants; changes of solvents; shortening of the reaction period; shifting of the reaction theater from solution into the vapor phase; conducting the reaction in melts or by means of solid body reactions, thus taking advantage of the higher reaction temperature, *etc.* The emancipation from rigid procedures may even go much further if it is kept in mind that the sole objective is to detect an organic compound through its participation in a selected reaction. Consequently, it is profitable to scan the literature with respect to reports on the formation of compounds which are easy to detect. Thus methods may be considered which have no value from the preparation standpoint because the yields are inadequate. If such methods of forming the desired compounds go fast enough, if they can be translated into the chemical analytical technique, and if small amounts of the resulting product can be reliably detected by a simple test, then a better analytical utilization can be secured in this novel fashion than by a familiar method of preparation which fails to meet adequately the conditions just set forth. Even though simplifications of the customary procedure usually lead to lower yields, this disadvantage can be overcome in large part when only analytical goals are being sought. Preparation procedures often necessitate the isolation of intermediate products and purification of the final product. Both of these steps entail unavoidable losses. However, such losses are not encountered and need not be considered if the test can be conducted in the reaction mixture without isolation of the intermediate products or without removal of the side products. This is often possible and deserves full consideration.

An excellent example of an indirect test accomplished through the synthesis of a characteristic organic compound displaying a sensitive reaction, is again provided through the formation of the fluorescent aldazines by condensation of hydrazine with *o*-hydroxyaldehydes. The latter can be prepared from phenols (with a free *ortho*-position) by the familiar Reimer-Tiemann reaction, which has been widely used in preparative organic chemistry since its discovery in 1875. In this procedure, phenols and chloroform are refluxed for a considerable time with strong caustic solutions. The following net reaction occurs:

$$\text{(phenol)}-\text{OK} + \text{CHCl}_3 + 3\,\text{KOH} \longrightarrow \text{(phenol)}\begin{array}{l}-\text{OK}\\-\text{CHO}\end{array} + 3\,\text{KCl} + 2\,\text{H}_2\text{O}$$

As a rule, the above formylation of phenols gives low yields of *o*-hydroxyaldehydes, and this fact, together with the relative complexity of the procedure, would seemingly present little promise of satisfactorily realizing the synthesis with only small quantities of a phenol. However, if chloroform is added to an alkaline solution of a phenol or to a solid alkali phenolate and the excess chloroform then evaporated, the quantity of *o*-hydroxyaldehyde produced is sufficient to be detected by the formation of the fluorescent aldazine. On the basis of this finding, it was possible to arrive at a test for phenols with a free *ortho*-position, since the reaction can be successfully conducted with as little as one drop of the solution being tested for phenol. It was found that the Reimer-Tiemann formylation occurs in a surprisingly rapid way in aqueous solution. Based on this fact and on the formation of the fluorescing salicylaldazine, spot tests for chloroform, bromoform, chloral and trichloroacetic acid were developed. The utilization of the Reimer-Tiemann formylation shows that the use of binary reactions is not restricted to the direct or indirect detection of either of the participants of ionic or condensation reactions, but may have a more general application.

The familiar Skraup synthesis of quinoline (1882) pointed to the possibility of a test for glycerol. This reaction, which can be expressed:

$$\text{(aniline)}\!-\!\text{NH}_2 + \begin{array}{l}\text{CH}_2\text{OH}\\\text{CHOH}\\\text{CH}_2\text{OH}\end{array} + [\text{O}] \xrightarrow[\text{H}_2\text{SO}_4]{\text{conc.}} \text{(quinoline)}\!-\!\text{N} + 4\,\text{H}_2\text{O}$$

is of no value as a means of detecting glycerol because no suitable specific test for the resulting quinoline is available. However, if the aniline is replaced by *o*-aminophenol, 8-hydroxyquinoline results:

$$
\underset{\text{OH}}{\text{(o-aminophenol)}} \ + \ \underset{\text{CH}_2\text{OH}}{\overset{\text{CH}_2\text{OH}}{\text{CHOH}}} \ + [\text{O}] \xrightarrow[\text{H}_2\text{SO}_4]{\text{conc.}} \underset{\text{OH}}{\text{(8-hydroxyquinoline)}} \ + \ 4\,\text{H}_2\text{O}
$$

This product (oxine) is a widely used precipitant for numerous colorless metal ions, yielding light-yellow inner-complex salts which, in most instances, give an intense yellow-green fluorescence in ultraviolet light. As anticipated, the oxine produced by the micro Skraup-synthesis can be detected through the production of fluorescent oxinates. This test is so sensitive that it reveals the formation of oxine when as little as one drop of a dilute solution of glycerol is warmed with *o*-aminophenol and concentrated sulfuric acid. After the oxine has been synthesized in this fashion, the fluorescent metal oxinate is produced by making the reaction mixture basic and adding a magnesium or alumininum salt. It should also be noted that oxine produces fluorescent salts not only with Mg^{2+}, Zn^{2+}, and Al^{3+} ions, but the same fluorescence effects are obtained as are displayed by the formula-pure oxinates when this reagent is adsorbed by coming into contact with the solid hydroxides or oxides of these metals. Derivatives of oxine, including its water-soluble sulfonic acids, which do not function as cation precipitants, are likewise chemically adsorbed to yield fluorescing systems. These adsorption effects can be utilized in sensitive spot tests for 8-hydroxyquinoline and its derivatives. They, like the fluorescent aldazines of *o*-hydroxyaldehydes, prove that the fluorescence depends on the presence and activity of particular groups in organic compounds, and hence can be called on for the detection of these groups. In this connection it should be noted that the Skraup synthesis with *m*-aminophenol leads to 7-hydroxyquinoline which fluoresces yellow-green in u.v. light. Thus a sensitive spot test has been developed for *m*-aminophenol (also *m*-nitrophenol after reduction to the amino compound) and distinguishes this substituted phenol from its isomers.

A striking example that demonstrates the use of a preparation of an organic reagent for the detection of a participant in the synthesis is found in a sensitive test for phenols which could be developed from the detection of cobalt by 1-nitroso-2-naphthol (ILLINSKY-KNORRE, 1885). The basis of this sensitive test is the production of a red-brown cobalt(III) inner-complex salt and the further fact that, in conformity with the functional group activity in organic compounds, other *ortho* nitrosated phenols behave analogously to the nitrosonaphthols. The chelate compounds may be formed directly by warming phenols, which have a free *ortho* position, with an acetic acid solution of sodium cobaltinitrite. The following successive reactions occur in the mixture:

$$[Co(NO_2)_6]^{3-} + 6\ H^+ \rightarrow Co^{3+} + 6\ HNO_2$$

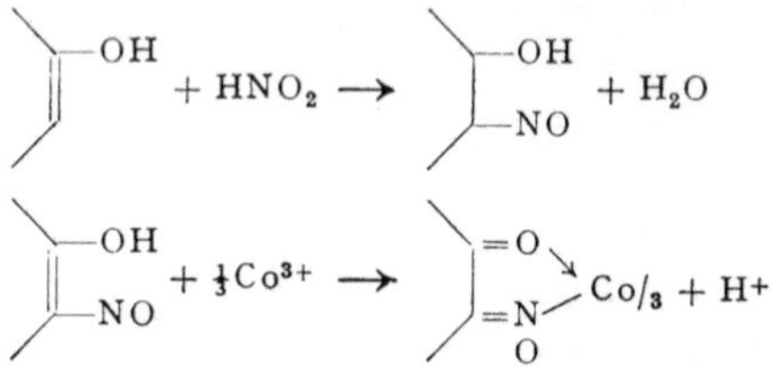

Accordingly, this procedure, which succeeds also with phenols having complicated structures, presents a synthesis which is conducted in the presence of a reagent for the reaction product. This is a matter of fundamental importance and has become the model of other tests described in this text.

An interesting example is provided by the tests which have grown out of the synthesis of diphenylcarbazide (SKINNER and RUHEMANN, 1881), through the condensation of phenylhydrazine and urea:

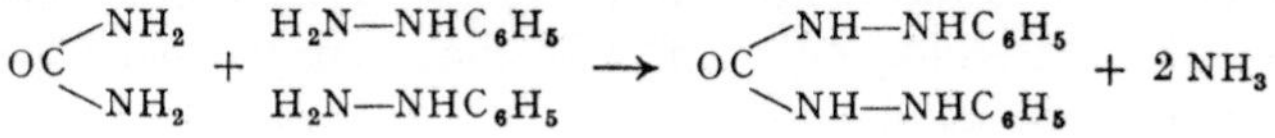

In qualitative inorganic analysis, diphenylcarbazide serves as a very sensitive reagent for quite a few metal ions, with which it yields colored inner-complex salts. Nickel diphenylcarbazide is produced with neutral and ammoniacal nickel solutions, and the violet salt, which is insoluble in water, dissolves readily in ether, chloroform, *etc.* Hence the occurrence of the condensation can be readily demonstrated through the formation of this highly colored salt. Consequently, there is the possibility of detecting either urea or phenylhydrazine. If unsymmetrical mono alkyl- and arylhydrazines are used in place of phenylhydrazine, the resulting derivatives of diphenylcarbazide react analogously to the parent compound because they too contain the same salt-forming groups. Accordingly, condensation with urea can be used as a general means of detecting monoalkyl- (aryl-) hydrazines. The reaction scheme of the diphenylcarbazide synthesis shows plainly which groups of the two reactants are involved in the condensation. It can therefore be anticipated that modified diphenylcarbazides will be produced by employing certain derivatives of urea. For instance, urethanes can also be detected by the synthesis of diphenylcarbazide. All of these syntheses, as well as the related tests for diphenylcarbazide through formation of the violet nickel salt, can be conducted within the bounds of spot test analysis.

It was pointed out on p. 11 that, when preparation procedures are being adapted to analytical purposes, the attainable yields are far less decisive than the rapid formation of products which can be detected with high sensitivity. In other words, when new organic tests are being sought, it is permissible to give serious consideration to reports of the formation of organic compounds and their chemical behavior which have no contemporary standing in preparation chemistry. A characteristic instance follows.

If mixtures of sodium formate and alkali salts of sulfonic acids are sintered or fused, the sulfonic group is replaced by the carboxyl group:

$$(R, Ar)SO_3Na + HCOONa \rightarrow (R, Ar)COOH + Na_2SO_3$$

This reaction, which was reported in 1870 by V. MEYER, has no value as a method for preparing carboxylic acids because the unavoidable losses due to thermal decomposition are too great. Nevertheless, it provides the basis of a characteristic indirect test for aliphatic and aromatic sulfonic acids, sulfones, and for sulfonamides through the formation of the heat-resistant alkali sulfite, which can be readily detected as a result of the liberation of SO_2 when treated with mineral acids. The entire procedure is simple, rapid, and can be accomplished within the technique of spot test analysis. This example emphasizes the important fact that it sometimes is profitable to modify a preparation procedure which is too complicated to have any analytical value and to transform it into a formation procedure which is readily accomplished and feasible from an analytical standpoint. A case in point is a test for aliphatic esters of fatty acids, which is based on a classic synthesis of acyloins (BOUVEAULT-BLANC, 1903). The usual procedure prescribes prolonged warming of an ethereal or benzene solution of the ester with metallic sodium to form the sodium salt of the enediol, which, after mechanical removal of the excess sodium, is saponified to the acyloin:

$$2\,R{-}COOR + 4\,Na^\circ \rightarrow \begin{matrix} R{-}C{-}ONa \\ \| \\ R{-}C{-}ONa \end{matrix} + 2\,RONa$$

$$\begin{matrix} R{-}C{-}ONa \\ \| \\ R{-}C{-}ONa \end{matrix} + 2\,H_2O \rightarrow \begin{matrix} R{-}CHOH \\ | \\ R{-}CO \end{matrix} + 2\,NaOH$$

At first glance it would appear improbable that this procedure, which demands considerable time and equipment, could be converted into an operation simple enough for analytical purposes and for spot test analysis in particular. Actually, the reactions can be accomplished with one drop of the ester and within one minute and still yield sufficient acyloin to be detectable by a sensitive color reaction for reducing compounds with 1,2-dinitrobenzene.

It may be safely predicted that the use of recently discovered tests for functional groups will be of much importance in the analytical employment of the formation of organic compounds. This point is well illustrated by the following example. The classic synthesis of 2-methylbenzimidazole (LADEN-BURG, 1875) is based on heating o-phenylenediamine with anhydrous acetic acid:

$$\text{(benzene ring)}\begin{array}{l}\text{—NH}_2\\ \text{—NH}_2\end{array} + \begin{array}{l}\text{O}\\ \text{C—CH}_3\\ \text{HO}\end{array} \rightarrow \text{(benzimidazole ring)}\text{C—CH}_3 + 2\,\text{H}_2\text{O}$$

The imidazoles contain a tertiary N atom and consequently give a positive response to a color reaction (OHKUMA, 1955) for tertiary amines. This occurs on warming the sample with a mixture of acetic anhydride and malonic or citric acid. Therefore, if o-phenylenediamine is added to this mixture that invariably contains anhydrous acetic acid, the reactions occur that are required both for the synthesis of the benzimidazole and for the color reaction. These facts served as the basis for the development of a selective test for o-phenylenediamine and its differentiation from its isomers. It should be pointed out that probably this procedure may be successfully extended to other *ortho*-diamines that form 2-methylimidazoles with glacial acetic acid. It has been found that other condensation reactions which lead to tertiary amines can be recognized through the Ohkuma color test. Selective detection of amides, amidines, and 1,2-dioxo compounds and specific identification of glyoxal will be found in this text.

The foregoing discussion has dealt with the chemical backgrounds of several new types of tests for functional groups and individual organic compounds. The aim has been to demonstrate that in the light of deliberations and experiences regarding the chemistry of specific, selective and sensitive reactions, it is possible to endow syntheses and modes of formation, which attracted little if any attention over a long period of years, with an actual analytical value. Valuable orienting ideas and interesting possibilities in the search for new tests for organic compounds are thus unfolded which can be of prime importance to the development of organic spot test analysis as will be shown in the remainder of this introductory chapter and likewise in the description of many new and improved tests contained in this book.

Primarily because of the hydrophobic character of many organic compounds, but also sometimes for other reasons, great importance attaches in qualitative organic analysis to reactions whose initiation and progress do not require the presence of water as an independent phase. In indirect tests which involve separate synthesis operations, primary reactions in organic solvents, reactions in the gas phase, fusion reactions, and the like are plausible. However, reactions of these kinds are also of great importance in direct tests for organic compounds. For example, salicylaldehyde and 8-hydroxyquinoline, which volatilize somewhat at room temperature and still more when gently warmed, can be detected within the technique of spot test analysis with high sensitivity through the formation of fluorescent products by the action of the vapors on hydrazine (or its salts) or metal hydroxides, respectively. Derivatives of the organic parent compounds contain the same functional groups but their vapor pressures are too low to permit their reaction in the gas phase. Consequently, the analogous mode of reaction, which likewise leads to fluorescent products, holds only for reactions in solutions. Therefore selective reactions may be the basis of specific tests when it is possible to transfer the reaction locale from the solution into the gas phase. This possibility should be kept in mind quite generally when considering the detection of organic compounds which are sublimable or volatile per se or which can be volatilized with water vapor. Pertinent examples will be found in this text.

Compounds which yield characteristic gaseous fission products when subjected to dry pyrolysis can sometimes be detected indirectly by an appropriate reaction, in the vapor phase. Unfortunately, it is impossible to make reliable predictions in such instances because the nature and course of thermal decompositions depend on the nature and constitution of the pyrolyzed material and the conditions of the heating. It is very likely that complicated processes are involved in many cases.* Nevertheless, definitive fission processes and fission products seem to be the rule in some classes of organic compounds. Familiar instances are: formation of hydrogen halides on combustion of compounds containing halogen, even though they contain little hydrogen; production of hydrogen cyanide in the pyrolysis of many nitrogenous organic compounds. In contrast, relatively few of the latter yield dicyanogen when they are thermally decomposed, and the evolution of ammonia is limited to the pyrolysis of primary and secondary aliphatic amines, urea derivatives, guanidine salts and some of its derivatives with free NH_2 groups. Similarly, the release of trimethylamine is characteristic for the pyrolysis of choline and betaine. When compounds containing carbon and oxygen are decomposed by heating, they invariably yield water in the form of superheated steam which is able to bring about hydrolytic splittings not realizable by the wet method. For example, when sulfoxylate compounds, oxymethylene compounds, salicin, and cellulose are subjected to dry heating, the resulting steam participates in the formation of hydrogen sulfide, formaldehyde, salicylaldehyde and furfuraldehyde together with acetaldehyde, as the case may be. These respective cleavage products may be detected easily in the vapor phase by sensitive tests and accordingly conclusions may be drawn as to the presence or absence of the initial materials. Pertinent

* Thermoanalytical studies of this kind instituted by Cl. DUVAL (*Inorganic Thermogravimetric Analysis*, Elsevier, Amsterdam, 1963) with inorganic and inorganic–organic compounds will doubtless provide much information of great importance and significance in clearing up pyrolytic decompositions involved in qualitative organic analysis.

instances that have been found recently include the formation of nitrous acid in the pyrolysis of organic compounds containing groups with N and O atoms (nitro- and nitroso compounds, oximes, *etc.*) and the production of volatile phenols through the pyrolysis of aromatics which contain O atoms in the nucleus or in a side chain. Since nitrous acid is readily detectable by the Griess reaction (red color with naphthylamine plus sulfanilic acid) and phenols are readily revealed by the Gibbs reaction (formation of blue in-dophenol with 2,6-dichloroquinone-4-chloroimide), these provide very useful preliminary tests.

Many interesting effects were observed in the pyrolysis of certain organic compounds in the presence of reactive inorganic compounds. As an example, it may be pointed out that all nitrogenous organic com-pounds yield nitrous acid when strongly heated in mixture with manganese dioxide. This finding has been made the basis of a reliable and exceptional-ly sensitive test for nitrogen. Another example is the formation of readily detectable sulfur dioxide when compounds containing a CH_2Hal group or alkyl esters of noncarboxylic acids are dry heated with sodium thiosulfate. This result is due to the fact that the initial product is a so-called Bunte's salt, which in turn undergoes a pyrolytic inner-molecular redox reaction. The pertinent partial reactions are:

$$(R,Ar)CH_2Hal + Na_2S_2O_3 \rightarrow NaHal + Na[(R,Ar)CH_2S_2O_3]$$
$$2\ Na[(R,Ar)CH_2S_2O_3] \rightarrow Na_2SO_4 + (R,Ar)CH_2\text{—}S\text{—}S\text{—}CH_2(Ar,R) + SO_2,\ and$$
$$R_2SO_4 + 2\ Na_2S_2O_3 \rightarrow Na_2SO_4 + 2\ Na[RS_2O_3]$$
$$2\ Na[RS_2O_3] \rightarrow Na_2SO_4 + RSSR + SO_2$$

The occurrence of the above reactions, detectable through the SO_2 formed, is the basis of selective spot tests for the methene-halide group and alkyl esters of noncarboxylic acids.

As a further illustration of what can be accomplished through the py-rolysis of organic compounds in the presence of inorganic materials, two effects will be cited here which are characteristic for non-volatile organic compounds: (1) the formation of gaseous hydrogen cyanide on ignition with mercuriamido chloride and (2) the production of gaseous nitrous acid on ignition with alkali nitrate. The first case involves the reaction of carbon (or carboniferous material) with this reagent; it may be written:

$$HgNH_2Cl + C \rightarrow Hg^0 + HCl + HCN$$

The second case involves the reaction series:

$$KNO_3 + C \rightarrow KNO_2 + CO$$
$$KNO_2 + H_2O \rightarrow KOH + HNO_2$$

Here the hydrolysis of the alkali nitrite is brought about by water pyrolytic-ally split out of organic compounds containing hydrogen and oxygen. This is the inorganic model for interesting hydrolytic effects that occur in the pyrolysis of organic compounds in the presence of materials that give up water, effects which have led to the discovery of new tests in organic spot test analysis, which will be discussed later. All of the above findings are of recent date. They have their prototype in the long known "lime test" for nitrogen, which is based on the fact that many nitrogenous organic materials yield ammonia when ignited with lime.

As already noted, the dry-heating of organic materials often results in volatile compounds which doubtless are hydrolysis products of the starting material or of its thermal cleavage products. In cases of this kind, the hydrolysis is brought about by the water split out of organic materials containing hydrogen and oxygen, which appears at the site of its production as quasi superheated steam. It was merely to be expected that suitable compounds would undergo a hydrolytic cleavage likewise under less drastic conditions than ignition. It actually was found that hydrolyses did occur when dry mixtures of certain organic compounds and an excess of water-donating compounds, either organic or inorganic, were heated, hydrolyses which otherwise could not be accomplished at all or only in the presence of acids or alkalis. Active water-donating compounds include: (1) Solid oxalic acid dihydrate at 80–105°; (2) Solid hydrates of metal sulfates (e.g. $MnSO_4.4 H_2O$ and $MnSO_4.H_2O$) at 100–200°; (3) Melts of hydrated oxalic acid at 120–160°; (4) Fused succinic and phthalic acid (through anhydride formation) above 220°.

The term "pyrohydrolysis" is proposed for hydrolyses that are brought about through contact with water that has been thermally released from solids or melts. It is easy to perceive the occurrence of such reactions if gaseous cleavage products result, which can be detected by appropriate reagent papers. Typical examples of pyrohydrolyses are:

Cleavage of acetic acid from O-acetyl compounds by heating (105°) with $H_2C_2O_4.2 H_2O$. – Cleavage of formaldehyde and acetaldehyde from choline hydrochloride by heating (180°) with $MnSO_4.4 H_2O$ or $MnSO_4.H_2O$. – Cleavage of salicylaldehyde from helicin by heating (105°) with $MnSO_4.4 H_2O$ or $MnSO_4.$ H_2O. – Cleavage of salicyl alcohol from salicin (with subsequent partial autoxidation to salicylaldehyde) on heating (150°) with $MnSO_4.H_2O$. – Cleavage of aniline from anilides of aliphatic carboxylic acids by heating (130°) with $MnSO_4.4 H_2O$. – Cleavage of hydrogen halides from aliphatic halogen compounds by heating (220–250°) with succinic acid or phthalic acid. – Cleavage of sulfurous acid from aromatic sulfonic acids by heating (230°) with succinic or phthalic acid. – Cleavage of methanol or ethanol (with subsequent partial autoxidation to formaldehyde or acetaldehyde) from N-methyl or N-ethyl compounds by heating (220°) with succinic or phthalic acid. – Cleavage of acetaldehyde (via initially produced ethylene glycol) by heating (220°) alkali salts of ethylenediaminetetraacetic acid with succinic or phthalic acid. – Cleavage of hydrogen sulfide from mercapto compounds by heating (130°) with $H_2C_2O_4.2\ H_2O$, $MnSO_4.4\ H_2O$ or by heating (200°) with succinic or phthalic acid.

Many of the above mentioned pyrohydrolyses can be accomplished with anhydrous ammonium oxalate as water donor, due to its transformation into oxamide when dry heated at 150–280°: $(COONH_4)_2 \rightarrow (CONH_2)_2 + 2\ H_2O$. Interesting examples of tests based on pyrohydrolyses are given in the later portions of this text. In this connection it may be emphasized that not only solids and liquids but also compounds in their gas phase may undergo pyrohydrolysis. This is shown by the fact that vapors of the chemically resistant carbon tetrachloride in contact with $MnSO_4.H_2O$ at 150–200° form hydrogen chloride: $CCl_4 + 2\ H_2O \rightarrow CO_2 + 4\ HCl$. Chloroform and other volatile organic liquids containing halogens behave similarly.

It has been found that tests for chloroform and carbon tetrachloride can be achieved with the ammonia generated when urea, biuret, or guanidine carbonate are heated The underlying reactions, which occur in the absence of any solvent, represent a special type of ammonolysis, which can fittingly be designated as *pyroammonolysis*, in analogy to pyrohydrolysis (*q.v.*) since they too occur at pyrolysis temperatures. Accordingly, aniline is formed from anilides, phenol from phenol esters, whereas hydrogen sulfide or hydrogen cyanide is formed from thioketones or chloroform.

$$(R,Ar)CONHC_6H_5 + NH_3 \rightarrow (R,Ar)CONH_2 + C_6H_5NH_2$$
$$(R,Ar)COOC_6H_5 + NH_3 \rightarrow (R,Ar)CONH_2 + C_6H_5OH$$
$$>C{=}S + NH_3 \rightarrow\ \ >C{=}NH + H_2S$$
$$CHCl_3 + 4\ NH_3 \rightarrow HCN + 3\ NH_4Cl$$

This book contains instances of the application of pyroammonolysis, and other examples will surely come to light when this new type of reaction is explored more thoroughly.

The pyrohydrolyses and pyroammonolyses used up till now for analytical purposes have been based mostly on dry-heating of the organic compounds in the presence of an excess of an inorganic or organic water or ammonia donor. However, slight amounts of organic compounds which yield water or ammonia when pyrolyzed are likewise capable of bringing about the hydrolysis or ammonolysis of other compounds which are stable when heated alone. Accordingly, such donors can be detected in this way provided characteristic products result from the heating of such mixtures. Examples are: the pyrolytic formation of H_2S from thiobarbituric acid (and thio-Michler's ketone) through the action of water or ammonia donors, and the formation of oxamide (detectable by a color reaction) through the action of ammonia donors on dimethyl oxalate, and the formation of formaldehyde and ammonia through the pyrolytic action of water donors on hexamine.

It should be pointed out here that the action of superheated water or ammonia in its own phase is not necessarily to be assumed whenever pyrolytic hydrolyses or ammonolyses occur. It is entirely conceivable that compounds which are capable of splitting off water or ammonia during pyrolysis may also be capable of reacting with heated H_2O- or NH_3 acceptors, which are in contact with such donors, even before the cleavage occurs. This would mean that in pyrolyses there may be also manifested a tendency or preparedness both on the part of the pyrolyzing and the accepting compounds, a characteristic which may possibly be applied in analysis and possibly also in preparation studies.

Many examples of redox reactions which may be designated as "pyrohydrogenolysis" are known. They are somewhat analogous to pyrohydrolysis and pyroammonolysis just described. This additional effect can be accomplished through molten benzoin (m.p. 135°) which acts as hydrogen donor through conversion into benzil:

$$C_6H_5CHOHCOC_6H_5 \rightarrow C_6H_5COCOC_6H_5 + 2\ H^0$$

It was found that aliphatic compounds of the general formula RX are split through taking up two hydrogen atoms when heated (at about 160°) with fused benzoin:

$$RX + 2\ H^0 \rightarrow RH + HX \qquad (X = Hal, SO_3H, NO_2)$$

The HX compounds noted here are volatile and may readily be detected in the gas phase by means of suitable reagent papers. Consequently good spot tests for the detection of aliphatically bound halogen, sulfonic-, sulfhydryl-, and nitro groups could be developed. It should be pointed out that arylmercuric nitrates, nitrates of organic bases, esters of nitric acid, and nitramines likewise split off nitrous acid when heated (160°) with fused benzoin and can be detected in this way through pyrohydrogenolysis.

In view of the pyrolytic behavior of benzoin, it may be expected that stoichiometrically defined redox reactions of organic compounds occur when they are sintered or fused with oxidants. For example, solid hexamine is oxidized when heated with manganese dioxide (140°) or with benzoyl peroxide (110–120°). However, and contrary to expectations, the underlying reactions do not lead to the same products. With manganese dioxide the oxidation proceeds:

$$(CH_2)_6N_4 + 12\ [O] \rightarrow 6\ CO_2 + 4\ NH_3$$

whereas the oxidation with benzoyl peroxide can be represented:

$$(CH_2)_6N_4 + 6\ [O] \rightarrow 6\ CH_2O + 2\ N_2$$

These redox reactions can serve as the basis for the spot test detection of benzoyl peroxide and hexamine.

Interesting reductive and oxidative cleavages are realizable in melts of sodium formate or benzoyl peroxide. When a mixture of sodium formate and hydroxide is heated to 205°, hydrogen is released:

$$HCOONa + NaOH \rightarrow Na_2CO_3 + 2\ H^0$$

Similarly, benzoyl peroxide (m.p. 103°) can serve as a source of oxygen:

$$(C_6H_5CO)_2O_2 \rightarrow (C_6H_5CO)_2O + O$$

It has been found that compounds with p-phenylenediamine- and p-nitroso-aniline structure produce p-phenylenediamine when fused with sodium formate–hydroxide. For example, the reductive cleavage of aniline yellow can be represented:

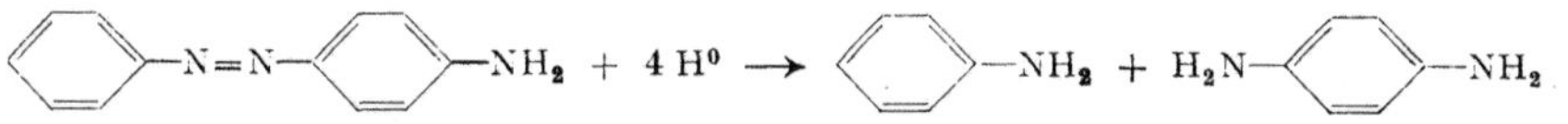

Since p-phenylenediamine is easily detected in the vapor phase by the blue color produced with aniline and persulfate, and since even complicated members of these classes of compounds are susceptible to smooth reductive cleavage, even in micro quantities, the process has considerable practical value in the testing of azo dyestuffs with p-phenylenediamine and p-nitroaniline structure.

Remarkable effects can be obtained with benzoyl peroxide fusions at 110–130°. For instance, O-ethyl and N-ethyl compounds undergo the oxidative cleavages:

$$R\!\!-\!\!OC_2H_5 + O \rightarrow ROH + CH_3CHO$$
$$>\!\!N\!\!-\!\!C_2H_5 + O \rightarrow\ >\!\!NH + CH_3CHO$$

O-Methyl and N-methyl groups in organic compounds behave similarly, but formaldehyde is formed instead of acetaldehyde. Here again, the resulting acetaldehyde can be detected in the vapor phase through the color reaction (blue) with sodium nitroprusside solution containing piperidine or morpholine. The detection of these groups has hitherto been rather difficult, but the fusion with benzoyl peroxide (within a minute or two) is quite simple when micro quantities are taken for examination.

From the analytical standpoint, it is interesting that hydrolytic cleavages, such as occur in pyrolyses of organic compounds, can likewise be accomplished in many instances by heating the material with concentrated sulfuric acid. At first glance, this effect is somewhat astonishing since concentrated sulfuric acid is ordinarily viewed as a classic dehydrant. However, it is easy to understand its action as a water donor, especially on warming, if it is remembered that the concentrated acid, which in fact is a hemihydrate, always contains some water which does not volatilize on heating but is brought to a state equivalent to superheated steam. The latter is able to accomplish hydrolyses which cannot be brought about by hot water or with H^+ and OH^- ions at all or but slowly and incompletely. Clear instances of this hydrolytic action of concentrated sulfuric acid are provided in the tests for monochloroacetic acid and phenoxyacetic acid, in which the sample is heated to 150–170° with concentrated sulfuric acid and the resulting formaldehyde then detected in the vapor phase. The reactions occur:

$$CH_2ClCOOH + H_2O \rightarrow CH_2OHCOOH + HCl$$
$$C_6H_5OCH_2COOH + H_2O \rightarrow CH_2OHCOOH + C_6H_5OH$$
$$CH_2OHCOOH \rightarrow CH_2O + CO + H_2O$$

Similarly, the production of formaldehyde when oxymethylene compounds are heated with concentrated sulfuric acid to 150–170° can be ascribed to the hydrolysis brought about by the concentrated acid or more correctly the water it contains:

$$\begin{array}{c} \diagdown -O \\ \diagdown -O \end{array}\!\!CH_2 + H_2O \longrightarrow \begin{array}{c} \diagdown -OH \\ \diagup -OH \end{array} + CH_2O$$

Concentrated sulfuric acid can function as a water donor also under milder conditions. For instance, the characteristic blue color reaction for diphenylamine with nitrate (nitrite) appears at once when *asym*-diphenylurea or diphenylguanidine is dissolved in concentrated sulfuric acid, warmed gently, and then treated with alkali nitrate (nitrite). This color reaction is not shown by *sym*-diphenylurea or *sym*-diphenylguanidine. Consequently, there is no doubt that the water present in the concentrated sulfuric acid accomplishes the cleavages leading to diphenylamine necessary for the color reaction:

$$OC\!\!\begin{array}{c} \diagup N(C_6H_5)_2 \\ \diagdown NH_2 \end{array} + H_2O \longrightarrow CO_2 + NH_3 + HN(C_6H_5)_2$$

$$HN\!\!=\!\!C\!\!\begin{array}{c} \diagup N(C_6H_5)_2 \\ \diagdown NH_2 \end{array} + 2\,H_2O \longrightarrow CO_2 + 2\,NH_3 + HN(C_6H_5)_2$$

The relationships are equally clear with respect to N-nitroso- and *p*-nitrosodiphenylamine. It is merely necessary to warm these compounds with concentrated sulfuric acid to cause them to split into diphenylamine and nitrous acid, which then yield the blue color:

$$\left.\begin{array}{c} ON\!-\!N\!\!\begin{array}{c} \diagup C_6H_5 \\ \diagdown C_6H_5 \end{array} \\[2ex] HN\!\!\begin{array}{c} \diagup C_6H_5 \\ \diagdown C_6H_4NO \end{array} \end{array}\right\} + H_2O \longrightarrow HN(C_6H_5)_2 + HNO_2$$

The effects just cited are the basis of a spot test for derivatives of diphenyl-amine. The role of concentrated sulfuric acid as water donor at room tem-perature is well illustrated by the fact that all O—NO_2, N—NO_2 and N—NO compounds behave like free nitric or nitrous acid when treated with the solution of diphenylamine in concentrated sulfuric acid. Obviously in all these cases there is first a hydrolysis which leads to these nitrogen acids.

The above examples demonstrate plainly that hydrolytic fissions must be regarded as essential partial reactions when color reactions occur in concen-trated sulfuric acid. Insights into the chemistry of many color reactions, that were discovered empirically, may be obtained in this manner as well as by taking account of the activity of concentrated sulfuric acid not only as a dehydrant but also as an oxidant through the change into sulfurous acid. Such new approaches obviously can yield valuable hints and leads in the search for new color reactions and tests.

A reaction in the gas phase can occur and be employed in analysis if a mixture of two solids, though stable at room temperature, produces a gas when warmed, because of the thermal decomposition of one of the solids and provided this gas then reacts with the other solid. A pertinent example is a mixture of molybdenum trioxide with organic compounds. On heating, the volatile organic decomposition products react with the solid molybdenum oxide to give molybdenum blue. This interpretation is supported by the fact that readily volatile organic compounds, such as alcohol, ether, *etc.*, on contact with hot molybdenum trioxide reduce it to molybdenum blue. Such production of the blue oxide can be applied within the technique of spot test analysis for the detection of traces of organic materials. This purpose can also be accomplished by means of potassium iodate. Even prolonged heating at 350° has no effect whatever on this salt. However, if a mixture of the iodate with organic materials is heated, there is rapid reduction to potassium iodide. This reduction can be disclosed by treating the sinter residue with a drop of acid, when iodine will be set free because of the familiar iodate–iodide reaction. Although there is no doubt that these two tests involve the action of gaseous decomposition products on a solid co-reactant, this action does not exclude the possibility that the redox reaction really begins as a solid–solid reaction, which perhaps proceeds to a considerable extent. The following example may show that this is actually the case. In inorganic spot test analysis, free (unbound) sulfuric acid is detected by its action on heated methylenedisalicylic acid. Red quinoidal formaurindicarboxylic acid is formed:

Concentrated sulfuric acid functions here as dehydrant and oxidant. Unexpectedly, it was found that the red formaurindicarboxylic acid is also formed when a mixture of sulfosalicylic acid (m.p. 120°) and methylenedisalicylic acid (m.p. 238°) is heated to 150°. (In this way, 5 γ sulfosalicylic acid can be detected by a drop reaction.) It is natural to suppose that the sulfosalicylic acid reacts by producing sulfur trioxide, and the latter then reacts with methylenedisalicylic acid in the same way as concentrated sulfuric acid. However, this is not the case. The red formaurindicarboxylic acid can be produced by heating the mixture of sulfosalicylic and methylenedisalicylic acids to only 100° or to 120°, *i.e.* to the melting point of sulfosalicylic acid. Consequently, in the sulfosalicylic acid–methylenedisalicylic acid system, the same reaction may occur either as a solid–solid reaction, as a solid–melt reaction, and as a solid–gas reaction.

Similar types of reaction are likewise possible in the system: citric acid (m.p. 152°) – urea (m.p. 132°). The ammonium salt of citrazinic acid, which has a blue fluorescence, is produced by heating this mixture to 150°. The fluorescence reaction, which will reveal as little as 2 γ citric acid, involves two partial reactions. First of all, biuret and ammonia are formed when urea is heated above its melting point:

$$2\ \underset{\underset{\displaystyle NH_2}{|}}{O}C-NH_2 \longrightarrow O=\underset{\underset{\displaystyle NH_2}{|}}{C}-NH-\underset{\underset{\displaystyle NH_2}{|}}{C}=O\ +\ NH_3$$

The superheated ammonia generated in this decomposition can react directly with citric acid and produce citrazinic acid by ammonolysis and condensation:

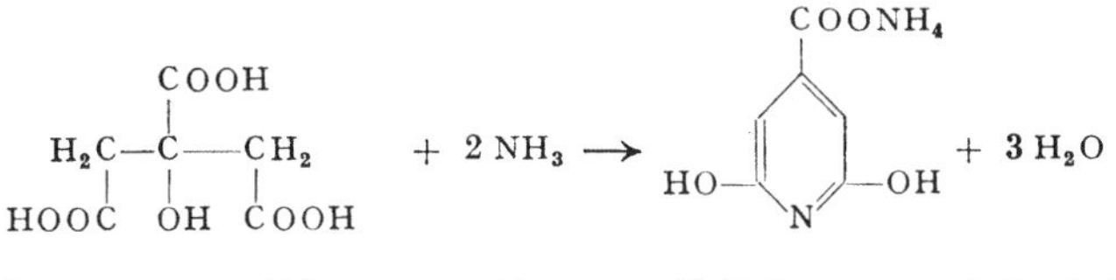

$$\underset{\underset{\displaystyle HOOC}{|}\ \ \underset{\displaystyle OH}{|}\ \ \underset{\displaystyle COOH}{|}}{H_2C-\underset{\underset{\displaystyle COOH}{|}}{C}-CH_2}\quad +\ 2\ NH_3\ \longrightarrow\ \text{[citrazinic acid ammonium salt]}\ +\ 3\ H_2O$$

This is shown as a solid–gas reaction, or, if it is assumed that the citric acid is melted, as a gas–melt reaction. It has now been found that the fluorescent citrazinic acid is likewise formed when the mixture of citric acid and urea is heated to only 110–120°, *i.e.*, below the melting point of either of the reactants. This means that urea functions as NH_3 donor in a solid body reaction before its transformation into biuret occurs. It was found, in addition, that the non-fusing alkali- and alkaline earth citrates likewise react under these same conditions to produce fluorescent alkali salts of citrazinic acid through a pyroammonolysis.

The above reactions in the sulfosalicylic acid–methylenedisalicylic acid and citric acid (citrate)–urea systems reveal two important facts. Firstly, they demonstrate the analytical usefulness of the participation of organic compounds in sintering and fusion reactions; secondly, they show that the respective chemical reaction can occur even at temperatures below the decomposition temperature of a thermally decomposable compound. This latter fact signifies that a readiness to react can be manifested by a low-melting compound even in temperature regions which heretofore have not been thought worth studying. Accordingly, it may be expected that organic compounds, which melt or decompose at fairly low temperatures, can by fusion or sintering be made to participate in reactions which are not realizable in the wet way. An excellent example of this possibility is found in the behavior of benzoin when fused with sulfur. The redox reaction

$$C_6H_5CHOHCOC_6H_5 + S^0 \rightarrow C_6H_5COCOC_6H_5 + H_2S$$

occurs at once in the vicinity of the melting point of benzoin (135°), and the resulting hydrogen sulfide may be detected by lead acetate paper. This finding has been made the basis not only of a specific test for elementary sulfur, but it was also the starting point for the finding that many fusible compounds containing a secondary alcohol group react analogously to benzoin and even when they are present in small amounts. Consequently, a spot test for the secondary alcohol group based on its transformation into a keto group, was worked out. In this connection it may be pointed out that the dehydrogenating action of molten sulfur (m.p. 113°) at 250° is so great that hydrogen sulfide is formed with all kinds of H-containing organic compounds. A rapid test for hydrogen is based on this long known fact.* New color tests that are described in this text belong in this category of fusion and sintering reactions.

An important chapter in complex chemistry deals with the formation and properties of purely organic molecular (addition) compounds, which doubtless deserve attention from the analytical standpoint. In this respect the analytical usefulness of the immediate production of intensely colored molecular compounds, in the form of solvates, when quinones or nitro compounds are fused with tetrabase (m.p. 90°) or when amines are melted with p-nitrophenetole (m.p. 60°) or duroquinone (m.p. 110°) is quite impressive. Minute amounts of quinones, amines, and C-nitro compounds can be detected in this simple manner. The behavior of a mixture of orange phenanthraquinone and colorless guanidine carbonate belongs in the category of melt- and sinter reactions which lead to colored products. A blue-violet product forms almost instantaneously when this mixture is heated to

---

* It is possible that the formation of hydrogen sulfide is based on the fact that, in the vicinity of 250°, pyrolysis of organic compounds occurs with formation of water; the latter then reacts with sulfur forming hydrogen sulfide. In line with this assumption, the water donors cited on p. 19 yield hydrogen sulfide when heated with elementary sulfur.

180°. This permits the sensitive and selective detection of phenanthraquin-one and guanidine.

Another interesting effect was observed when a mixture of colorless ox-amide (m.p. 419°) and thiobarbituric acid (m.p. 235°) was heated at 140–160°. A red product is formed rapidly and obviously a solid body reaction occurs at the beginning. This is the first time that a color reaction has been reported with the very resistant colorless oxamide. On the basis of this find-ing, it has been possible to develop an excellent spot test for oxamide and also for participants of those reactions which lead to the formation of oxam-ide, namely oxalic acid and its amides, anilides, and hydrazides. It is clear that the facts just cited may open interesting prospects for the development of new tests in organic analysis and especially in organic spot test analysis.

Characteristic examples in even the earliest contributions to spot test analysis indicated the possibilities of extensions of "fluorescence analysis". For a long time, the latter restricted itself primarily to registering chance fluorescences of organic and inorganic compounds, whereby data were obtained of importance in the testing of materials, determinations of origins, etc. The extension motivated by spot test analysis consisted essentially of including a stoichiometrically defined production of fluorescent products as a test for one of the compounds participating in the particular reaction. The mode of reaction is irrelevant here; the procedure may involve the formation of soluble or insoluble salts, condensation products, etc. It is essential that fluorescent products be formed through the reaction of non-fluorescent materials, or that there be a characteristic change of the fluores-cence hue in a fluorescing system. Such fluorescence reactions make direct tests possible, as shown by the transformation of m-aminophenol into fluorescing 7-hydroxyquinoline through the Skraup reaction and the forma-tion of fluorescing aldazines.

One method of employing fluorescence reactions for indirect tests is to produce, by suitable methods, non-fluorescent compounds, which can then in turn be detected by fluorescence reactions. The tests for o-hydroxyalde-hydes, glycerol, and citric acid, which were discussed above, demonstrate the usefulness of this procedure in organic spot test analysis. In general, direct and indirect tests based on fluorescence reactions go far toward satis-fying the requirements of selectivity and sensitivity. A limitation is imposed however by the fact that many organic compounds are fluorescent in their own right, or sometimes because of traces of impurities of unknown nature. Furthermore, non-fluorescent compounds can quench fluorescence by absorb-ing ultraviolet light. Such interferences are especially serious near the detection limits of fluorescence reactions. Some relief is afforded by compar-ison tests, by isolating the fluorescent compounds, and by observing the behavior of fluorescent materials when subjected to pH changes, which frequently are accompanied by characteristic alterations in the fluorescence or by the disappearance of the fluorescence.

Although not much evidence is available, it does appear that the fluorescence of a compound is sometimes dependent on its phase state and degree of dispersion. This point is illustrated by the metal oxinates and also by the condensation product of hydrazine with salicylaldehyde. The former fluoresce both in the solid state and when dissolved in organic liquids. In contrast, the aldazine of salicylaldehyde fluoresces only in the solid state, and not at all, or only to a minimum extent, when dissolved in ether, chloroform, *etc.* The dependence of fluorescence of compounds on their phase state and degree of dispersion merits consideration here because use of this relation may contribute to the enhancement of the specificity or selectivity of fluorescence reactions.

Far more significance attaches to the development of fluorescence resulting from the adsorption of non-fluorescent soluble or insoluble compounds. The following example shows that this can happen: hydrazine condenses with *p*-dimethylaminobenzaldehyde and yields a water-insoluble yellow aldazine which dissolves in acids with production of an orange quinoidal cation:

$$2 \ (CH_3)_2N-\!\!\!\big\langle\ \big\rangle\!\!\!-CHO \ + \ H_2N-NH_2 \ \longrightarrow$$

$$(CH_3)_2N-\!\!\!\big\langle\ \big\rangle\!\!\!-CH\!\!=\!\!N-N\!\!=\!\!CH-\!\!\!\big\langle\ \big\rangle\!\!\!-N(CH_3)_2 \ + \ H^+ \ \longrightarrow$$

$$(CH_3)_2N-\!\!\!\big\langle\ \big\rangle\!\!\!-CH\!\!=\!\!N-NH-CH\!\!=\!\!\big\langle\ \big\rangle\!\!\!=\!\!\overset{+}{N}(CH_3)_2$$

Neither the insoluble aldazine or its colored solution in acids is fluorescent. However, if a drop of the orange solution is placed on filter paper, a red (not orange) fleck is formed, which fluoresces luminously red in ultraviolet light, and assumes a blue-green fluorescence when spotted with alkali or ammonia (whereby aldazine base is regenerated). Obviously the explanation is that a fluorescence is produced through the adsorption on filter paper of non-fluorescent ions or molecules. The importance to organic spot test analysis of the discovery of fluorescence effects of this kind, in which filter paper functions somewhat as an active participant, hardly needs to be stressed. This phenomenon of "adsorption fluorescence" is not only of theoretical importance but it has analytical value. It served as the basis of a spot test for hydrazine in solutions diluted to $1:50,000,000$ and also of a spot test for the hydrazides of carboxylic acids.

It must be remembered that catalysis may also be included when considering the analytical employment of various reaction types. The familiar fact that the catalytic acceleration of chemical changes is often due to traces of materials which enter into intermediate reactions and which limit their activity to definite homogeneous and heterogeneous reaction systems, indicates that the detection of accelerations of reactions can be made the basis of specific and sensitive tests for the particular catalyst. The importance of catalysis reactions has long been overlooked by those seeking new tests, but due consideration of this field has led within a comparatively short time to many sensitive and specific tests in qualitative inorganic analysis and spot test analysis.* Thus there is no doubt that hastening of purely organic reactions by organic catalysts will similarly be of great value. A pertinent example is the catalysis by glycerol of the decomposition of oxalic acid, which allows the detection of glycerol, and another example is the acceleration of the immeasurably slow reaction $2 NaN_3 + I_2 \rightarrow 2 NaI + 3 N_2$ by traces of organic compounds containing CS- or SH groups. Despite the large number and variety of organic reactions, catalysis will be encountered in such cases relatively less often because the tendency of inorganic ions to participate in catalysis by virtue of polyvalence and change in valency is absent in the organic province. The analytical employment of catalyzed reactions presents the problem of detecting the acceleration of reactions which innately are sluggish or which in some cases proceed at an extremely low rate.

Color reactions present the following possibilities: the catalyzed reaction may (1) yield directly a colored product, or (2) a colorless product may be revealed by adding appropriate reagents. For instance, 1,2-dinitrobenzene gives a color reaction that is based on the fact that organic compounds, which serve as hydrogen donors in an alkaline medium, reduce this pale yellow nitro compound to the violet water-soluble alkali salt of the aciform of o-nitrophenylhydroxylamine:

$$\text{(benzene ring)}\begin{matrix}-NO_2 \\ -NO_2\end{matrix} + 4 H^0 + 2 OH^- \longrightarrow \text{(benzene ring)}\begin{matrix}=NO^- \\ =NO_2{}^-\end{matrix} + 3 H_2O$$

Among the hydrogen donors yielding this color reaction are the acyloins and benzoins, whose –CHOH—CO– group can be converted to the –CO—CO– group by the loss of hydrogen. Accordingly, it was anticipated that catalysts which lead to the formation of o-hydroxyketones as end- or intermediate products in suitable reaction systems, should be detectable through the color reaction with 1,2-dinitrobenzene. This has been found to be the case. For example, when benzaldehyde is warmed with an alkali cyanide, it undergoes the familiar benzoin condensation, the cyanide acting as catalyst:

$$2 C_6H_5CHO \rightarrow C_6H_5CHOHCOC_6H_5$$

* Compare P. W. West, *Anal. Chem.*, 23 (1951) 176.

When the goal is the actual preparation of benzoin, concentrated alkali cyanide solution is used, but in line with the catalytic nature of the condensation, even traces of cyanide are sufficient to produce within a few minutes enough benzoin to give a positive response to the color reaction with 1,2-dinitrobenzene. A sensitive test thus results for soluble, insoluble, and complex inorganic cyanides.

A second example is the reaction of formaldehyde with 1,2-dinitrobenzene:

$$2 CH_2O + \underset{\text{NO}_2}{\overset{\text{NO}_2}{\big|\big|}} + 4 OH^- \longrightarrow \underset{\text{NO}_2^-}{\overset{\text{NO}^-}{\big|}} + 3 H_2O + 2 HCOO^-$$

Under certain conditions, this redox reaction proceeds so sluggishly that no violet color becomes visible within three to five minutes. However, the color appears much sooner if o-diketones are present. The latter act as catalysts because they are promptly reduced by formaldehyde to o-hydroxyketones, which react at once with 1,2-dinitrobenzene to regenerate the diketone, which then again enters into the reaction with formaldehyde, and so on. p-Quinones and quinoneimides participate in similar partial reactions, and therefore a catalytic test for these compounds results. It is much more reliable and sensitive than any of the earlier tests.

The foregoing are model examples of the fact that catalytic reactions may be the basis of color tests for organic compounds that can be employed in organic spot test analysis. Although no comprehensive instances can be cited as yet, it may be assumed with certainty that a catalytic hastening of precipitation- and fluorescence reactions will be of use in analysis.

When catalysts are detected through their hastening of reactions, the prime reactants are always employed in considerable concentrations, and enough of the product is formed so that even insensitive tests, which need not be selective, are adequate. However, in another type of tests, sluggish reactants are invigorated by a suitable catalyst. Obviously, the sensitivity of such tests is not high. Nevertheless, this type of procedure can be expected to be feasible in qualitative organic spot test analysis, particularly if the reaction product of the catalyzed reaction can be readily detected even in small amounts through its self-color, insolubility, fluorescence, or by means of a supplementary reaction. The same holds for accelerations resulting from irradiation with daylight or ultraviolet rays (photocatalysis). FREYTAG* was the first to call attention to such tests, which can be conducted on paper as spot reactions and which are appropriately called "photoanalytical procedures". It seems that redox reactions and hydrolyses are especially susceptible to photoanalytical influences. For example, ultraviolet irradiation of

---

* F. FREYTAG, Z. Anal. Chem., **103** (1935) 334; **129** (1949) 366; **142** (1954) 12.

alkaline solutions of coumarin converts the non-fluorescing compound into an isomeric fluorescing product or enormously hastens this transformation. This finding has been made the basis of a specific and extremely sensitive test for coumarin, either in the solid, dissolved, or vapor state.

Photosensitive reactions constitute the basis of the detection of nitroso compounds with the yellow aqueous solution of sodium pentacyanoferroate:

$$Na_3[Fe(CN)_5NH_3] + ArNO \rightarrow Na_3[Fe(CN)_5ArNO] + NH_3.$$

The above exchange reaction, which ordinarily yields green products, does not proceed at all or at an exceedingly low rate if light is excluded. It may be assumed that the employment of photosensitive reactions in spot test analysis will not remain limited to the instance just noted, provided more attention is given to all the effects involved than has been the case up to the present[*]. It is entirely possible that many sluggish color reactions of organic compounds can be accelerated by illumination (particularly u.v. light).

The familiar discharge of the blue color of starch–iodine solutions on warming and the return of the color on cooling provides an excellent example of a thermochromic system. Another example is the behavior of the aqueous solution of the alkali salts of anthrahydroquinone. This solution is faintly red in the cold, but becomes deep red when heated. Such reversible color changes accompanying reactions of organic compounds involving thermal dissociation or changes in structure have been neglected for too many years. Recently, SAWICKI[**] and his associates have reported interesting examples which demonstrate that thermochromic reactions can lead to sensitive spot tests. It may be confidently expected that further examples of the use of thermochromic reactions in analysis will be developed before long.

As shown by the preceding statements, considerations of the development, present state, and prospects of organic spot test analysis deal primarily with the analytical utilization of chemical reactions. It is of great importance in this respect that organic compounds provide many more and more varied reaction possibilities than do inorganic compounds. This broadened horizon justifies an optimism regarding the further development of organic spot test analysis through the discovery of new tests. Consequently investigative studies in this field must direct special attention to the search for and the recognition of hitherto unknown reactions of organic compounds. Such investigations need not be confined to empirical and time-consuming trials and experiments since certain guide lines are available. Chemical reactions of organic compounds are invariably based on the activity of definite atom groups already present in the molecule or which can be introduced into it.

---

[*] Comp. the findings of R. PREUSSMANN, *Nature*, 201 (1964) 502, concerning the photolysis of nitrosamines.

[**] E. SAWICKI *et al.*, *Chemist-Analyst*, 48(1959)4, 68, 86.

If characteristic products of known composition result and if they can be properly placed in stoichiometric equations along with the starting compound, then a new kind of reaction behavior can be recognized and formulated. A finding of this kind obviously permits merely a statement that this particular reaction occurs but it does not give any information as to its extent. This limitation is important with regard to any projected application of newly discovered reactions to preparative purposes. The latter is likely to lead to success only if the reaction in question proceeds to a considerable extent and is not merely a side reaction delivering small yields. However, the latter are adequate for analytical purposes provided even small quantities of a rapidly formed characteristic reaction product are detectable. Such new kinds of reactions are most readily recognized if they yield gaseous reaction products for which sensitive and characteristic tests are available. Pertinent instances include the pyrohydrolyses, pyroammonolyses, and pyrohydrogenolyses discussed earlier. Additional examples, which deserve attention also from the preparative standpoint, are cited in what follows.

The fact that Raney nickel, because of its content of nickel hydride (Ni_2H), is a very powerful reductant has been learned from preparative organic chemistry. The preparation of the unstable hydrides is, however, a handicap with regard to their possible use in analysis, especially in spot test analysis. It has been found that Raney nickel *in statu nascendi* can be employed in their place. This is true for instance if the stable Raney nickel alloy (50% Ni, 50% Al) is allowed to react in an aqueous alkaline milieu. Atomic hydrogen is formed through solution of the aluminium and is partly retained as hydride by the metallic nickel. Interesting new kinds of reductions can be accomplished thus on warming.

For instance, aromatic sulfonic acids undergo the reaction:

$$ArSO_3Na + 6 Ni_2H \rightarrow ArH + NiS + NaOH + 2 H_2O + 11 Ni$$

This desulfurization can be detected through the production of hydrogen sulfide when the resulting nickel sulfide is dissolved by acids. Proof of this reaction is furnished by the finding that warming of an alkaline solution of sulfanilic acid with Raney nickel alloy yields aniline vapor:

$$H_2N-\langle\bigcirc\rangle-SO_3Na + 6 Ni_2H \rightarrow \langle\bigcirc\rangle-NH_2 + NiS + NaOH + 2 H_2O + 11 Ni$$

Such removal of sulfur by means of Raney nickel alloy makes possible a sensitive test for aromatic sulfonic acids and their derivatives, and also a demonstration of a sulfonation of aromatic compounds by concentrated sulfuric acid.

The vigorous reductive power of Raney nickel alloy in alkaline media is likewise exhibited by the finding that pyridine is hydrogenated rapidly and perhaps quantitatively to piperidine, which is readily detected in the gas phase by an appropriate color reaction. Pyridinesulfonic acid is likewise reduced to piperidine when warmed with Raney nickel alloy in alkaline surroundings, a reaction in which the hydrogenation is accompanied by a reductive splitting off and desulfurization of the SO_3Na group. Pyridine and pyridinesulfonic acids can be sensitively detected in this way within the technique of spot test analysis.

The above redox reactions in alkaline surroundings are doubtless due to the efficacy of the nickel hydride present in the Raney nickel since these reactions are not accomplished with nascent hydrogen (Devarda's alloy plus caustic alkali for instance). It was therefore to be expected that Raney nickel alloy would also be active in acidic media since nickel hydride must be produced transiently during the entire course of the solution of the alloy. The expectation was proven true by trials in which portions of a hydrochloric acid solution of methylene blue (a derivative of thiodiphenylamine) were warmed with Devarda's alloy and Raney nickel alloy respectively. In both cases the dye was decolorized by the nascent hydrogen but hydrogen sulfide was produced only in the portion heated with the Raney nickel alloy. Organic compounds carrying 2- or 4-valent sulfur behave like methylene blue in this respect and hence they can be detected in this manner by spot test analysis. It should be noted however that thiophene and its derivatives do not conform to this general behavior.

Interesting and in part unexpected pyrolytic reactions were found in the case of mercuric cyanide. This compound may be heated to 180° without decomposition or sublimation. In contrast, when mixed with certain organic compounds it yields hydrogen cyanide which can then be detected in the gas phase without difficulty. For instance, the following reaction occurs with alkali salts and esters of formic acid:

$$2\ HCOOX + Hg(CN)_2 \rightarrow Hg^0 + (COO)_2X_2 + 2\ HCN \qquad X = Na\ or\ R$$

This reaction can be employed for the detection of formates. Of more analytical importance is the finding that acidic compounds of all kinds, even those whose acidic groups do not respond to indicators, promptly yield hydrogen cyanide when dry-heated with mercuric cyanide. This holds true even for organic compounds whose molecules contain both acidic and basic groups, in which the basic character may be dominant. Consequently, quasi occult acidic groups can be detected by pyrolysis with mercuric cyanide, an accomplishment for which no procedure has been hitherto available.

Quite astounding is the finding that hydrogen cyanide is produced when organic compounds containing 2-valent sulfur are heated to 160° with mercuric cyanide. It is most probable that in such cases there is first a desulfurization with production of mercuric thiocyanate:

$$2 \; {>}C{=}S + 2 \; Hg(CN)_2 \rightarrow 2 \; Hg(CNS)_2 + {>}C{=}C{<} \qquad (1)$$

$$2 \; {>}C{-}S{-}C{<} + 2 \; Hg(CN)_2 \rightarrow 2 \; Hg(CNS)_2 + {>}C{-}C{<} \qquad (2)$$

The mercuric thiocyanate produced by (1) or (2) decomposes on heating to yield mercuric sulfide and volatile dicyanogen sulfide (3) which is saponified on contact with air and water (4):

$$Hg(CNS)_2 \rightarrow HgS + (CN)_2S \qquad (3)$$

$$(CN)_2S + H_2O + O \rightarrow 2 \; HCN + SO_2 \qquad (4)$$

A selective test for divalent sulfur in organic compounds could be based on the rapid occurrence of (1)–(4). The removal of sulfur by means of mercuric cyanide as per (1) and (2) may have interest for preparative purposes provided the yield of sulfur-free reaction products is satisfactory.

New kinds of pyrolytic reactions of hexamine were pointed out in connection with its oxidation by means of manganese dioxide and benzoyl peroxide (see p. 21). It is therefore quite understandable that the methenylation of NH_2 groups, which is well known in the case of Sørensen's formol titration, can also be accomplished by dry heating of pertinent compounds with hexamine. The occurrence of the reaction involved here

$$6 \; RNH_2 + (CH_2)_6N_4 \rightarrow 6 \; RN{=}CH_2 + 4 \; NH_3$$

is readily detected through the splitting out of ammonia and its action on acid–base indicator paper or Nessler's solution. It is rather remarkable and also of analytical significance that the methenylation of NH_2 groups in organic compounds also occurs when their molecules contain acidic SO_3H- or COOH groups. In such cases, e.g. amino acids, there is a concurrent salification of acidic groups by the strongly basic hexamine. The use of hexamine as pyrolytic reagent makes possible the discernment in simple fashion of occult amino groups in aliphatic and aromatic compounds, i.e. of such groups whose presence could not be established directly up till now. This new method can be employed in the examination of dyes to reveal free or salified NH_2 groups.

The pyrolytic effectiveness of hexamine with relation to NH_2 groups suggested that a study should be made of the behavior of alkylated amines. It was found that heating of hexamine (200°) with aliphatic or aromatic compounds which contain N-dimethyl or N-diethyl groups, produced vapors which gave the color reactions of formaldehyde. Since a direct cleavage of formaldehyde is excluded, obviously some other reaction path must be considered. A plausible assumption is methenylation with production of volatile methyl- or ethylformimine and saponification of the imines:

$$2\ RN(CH_3)_2 + (CH_2)_6N_4 \rightarrow 2\ RN{=}CH_2 + 4\ CH_2{=}NCH_3$$
$$2\ RN(C_2H_5)_2 + (CH_2)_6N_4 \rightarrow 2\ RN{=}CH_2 + 4\ CH_2{=}NC_2H_5$$
$$CH_2{=}NCH_3 + H_2O \rightarrow CH_2O + NH_2CH_3$$
$$CH_2{=}NC_2H_5 + H_2O \rightarrow CH_2O + NH_2C_2H_5$$

Accordingly, and in line with this assumption, pyrolysis with hexamine can bring about a transfer of N-dialkyl groups into the N-methylene group, a change that has hitherto not been known. It is noteworthy that mono-alkylamines do not react with hexamine in an analogous way as was demonstrated by comparison trials with mono- and diethylaniline. The detection of formaldehyde after saponification of the alkylformimine split off is best made with Tollens reagent. The procedure described here makes possible the detection of dialkylamine compounds in spot test analysis provided the melting points of the latter are in the vicinity of 200°. Despite this limitation, this method of detecting dialkylamines and of differentiating them from the monalkylamines and primary amines can be of analytical value. The previously unknown pyrolytic formation of alkylformimines may perhaps be of value in preparative organic chemistry.

As was pointed out above, the analytical utilization of the reactions of organic compounds always involved the formation of characteristic reaction products that can be sensitively detected. In this regard, it is of great significance that reliable tests are available for numerous gaseous or easily volatilized compounds. Among these are:

acetaldehyde	hydrogen halides
acetone	hydrogen sulfide
acetylene	hydroxylamine
acrolein	hydroxyquinoline (oxine)
ammonia	nitric acid
aniline	nitrous acid
benzaldehyde	phenol
carbon disulfide	phosgene
carbon oxysulfide	piperazine
carbon monoxide	piperidine
cyanogen bromide (chloride)	pyridine
dicyanogen	pyrrole
dicyanogen sulfide	salicylaldehyde
formaldehyde	salicylic acid
formic acid	sulfur dioxide
furfural	thiocyanic acid
halogens (Cl, Br, I)	thiophenol
hydrogen cyanide	volatile primary and secondary amines

Accordingly, if one of these compounds results from the binary reactions of organic compounds in the wet or pyrolytic way, this product must be brought into harmony with the stoichiometric formulation of the underlying reaction. Such success points to the possibility of detecting or identifying the participants in the reaction. Furthermore, the detection of reaction products in their gas phase may indicate the occurrence of hitherto unknown reactions. The analytical importance of such conclusions is obvious.

The examples cited in this chapter show the existence of new or for a long time little considered possibilities of using reactions of organic compounds for analytical purposes especially in spot test analysis. Of course it is not implied that the previous methods of employing organic reactions should be relegated to the rear. The classic methods of qualitative organic analysis and their improvement will continue to play an important role in the development of organic spot test analysis. Accordingly, it seems well to give here a general survey of the most important types of reaction which deserve attention in the employment of and search for new direct and indirect tests for organic compounds or the identification (detection) of characteristic groups contained in them.

(1) *Direct tests*
a) Precipitation and color reactions (including fluorescence reactions) with inorganic or organic participants in aqueous or organic media.
b) Reactions in the gas phase with dissolved or solid participants.
c) Reactions of solids with solutions.
d) Reactions of solids with gases.
e) Reactions of solids with melts, and fusion reactions.
f) Reactions of solids with solids, and sintering reactions.
g) Adsorption effects connected with change of color (fluorescence).
h) Catalysis reactions in which the material to be detected acts as catalyst.
i) Catalysis reactions in which the compound to be detected is brought to rapid reaction.

(2) *Indirect tests*
a) Degradation (thermal, hydrolytic, oxidative, reductive) with formation of cleavage products which are detectable by (1) (a–i).
b) Syntheses or modes of formation which lead to compounds readily detectable by (1) (a, g, h).
c) Ascertainment of the blocking of the reactivity of particular groups (masking of reactions) given in (1) (a).

The great majority of the direct and indirect tests used in organic spot test analysis are satisfactory with respect to quantity- and concentration sensitivity, whose characteristic values are the values of the identification limit and the dilution limit. As an average, the identification limits attained lie between 0.4–10 micrograms (gamma) of the sought material per drop (0.05 ml) and fractions of a milligram can still be detected by even the least sensitive tests. Frequently the sensitivity of spot tests can be improved (sometimes by powers of ten) through refining the technique (micro drops, working in capillaries, magnification, *etc.*). For many purposes it is sufficient to follow the procedures as described in this text, which does not prescribe micromanipulations. The directions are adequate for a refinement of the tests as well as for conducting them as test tube reactions on the macro scale. The necessary simple apparatus and manipulations in spot test analysis are discussed in detail in Chap. 2 of *Inorganic Spot Tests*.

The sensitivity attainable in the chemical detection of organic compounds, or of certain groups included in them is governed by the same considerations that hold for the detection of inorganic materials. The limits of identification and the dilution limits are not characteristic constants of the underlying reactions, even though they are often employed as numerical expressions of the sensitivity. They are dependent on the particular reaction conditions, the type of observation, the reaction period, the presence of other materials, *etc.* Therefore, the conditioning of tests has the same importance in qualitative organic analysis as in its inorganic counterpart with respect to attaining a maximum of sensitivity.

The detection of functional groups in organic compounds is further complicated by still another sensitivity-determining factor, namely the influence, sometimes important, exerted by the aliphatic or aromatic remainder of the molecule, in whole or in part, with respect to occurrence, speed, and extent of the chemical reactions of certain groups and also with respect to the discernibility of the characteristic reaction product through its color, solubility, volatility, fluorescence, *etc.* Definite reasons and reliable predictions concerning such factors which occur especially in aromatic compounds can be stated only in exceptional cases. Steric effects on the mobility and reactivity of functional groups often make themselves apparent. Furthermore, the reaction picture and the characteristics of reaction products may be extensively altered by chromophoric, hydrophobic and hydrophilic groups, which are not direct participants in the reaction in question. Such influences are evidenced by the fact that the attainable identification limits may vary within wide limits not only with respect to their absolute values but also in comparison with the values given by equimolar quantities of various compounds carrying the same functional groups. In fact, cases are known of compounds which do not respond at all to tests which in other instances are quite reliable for the particular functional groups. The earliest report of a test ordinarily covers only a few materials and so does not give a complete picture of the whole field of possible application. This wider view is obtained only after a comprehensive selection of cases has been tried. What has just been said about sensitivity applies equally to the selectivity of tests for functional groups, for characteristic types of compounds and for individual compounds.

As was pointed out above, the reactivity of certain groups in organic compounds is sometimes markedly influenced by the remainder of the molecule. In the case of isomers which differ solely in the position of like groups, this influence is sometimes so great that one isomer will fail completely to give certain reactions, or it may show new types of reaction. When this is true, the isomers can often be distinguished by simple spot tests. The latter can also be used for distinguishing between homologous compounds as well as for determining constitutions and structures. Pertinent examples, including the differentiation of isomers which contain different functional groups are compiled in Chap. 6.

In closing this present chapter, which deals with the development, present state and prospects of organic spot test analysis, the following statement is justified: There are now only rare cases of a total lack of information regarding the specificity, selectivity and sensitivity of direct and indirect tests for organic compounds or the groups they contain. In many cases the knowledge of the chemical background of a test provides valuable orientation concerning selectivity and provides valuable hints for improvements. On the other hand there have been as yet no systematic studies of the lowering of the sensitivity of tests by seemingly indifferent accompanying materials, an effect often encountered in inorganic analysis. The lack of such information is related to the fact that the enormous number of organic compounds makes it incomparably more difficult to review and actually try the tests than in the inorganic field. However, it may be anticipated that with growing interest in organic spot test analysis and its application to the detection of organic compounds in mixtures, new facts relative to specificity, selectivity and sensitivity will be uncovered. Although there are characteristic exceptions, chemical methods in general do not have the same reliability in organic analysis, and hence not in organic spot test analysis, as in inorganic analysis. Nevertheless, in view of past and present experiences, it may be expected that spot reactions will render excellent service in solving special problems encountered in the chemical testing of organic materials, and they will doubtless be found equal and sometimes superior to the corresponding macromethods. In this respect it may be emphasized that recently spot tests were elaborated for the solution of important problems of testing materials for which no macromethods were available hitherto. The ecomomy in material, time and labor accompanying the use of spot reactions is a special advantage. The expected growth of organic spot test analysis is all the more likely because there are now in use many tests which are not even realizable in classic microanalysis.

It is self-evident that the organic spot test analysis, which limits itself to chemical methods, cannot be nearly as serviceable as the classical qualitative organic macroanalysis, which does not hesitate to enlist physical methods. For the latter—infrared spectroscopy is an exception—chemical methods constitute a kind of preliminary examination whose objective is to provide information about preparative measures to be taken for the isolation of compounds and preparation of derivatives, which can be definitely identified by physical methods. Since the extension of organic spot test analysis requires the improvement of existing methods and the discovery of new tests, it seems certain that efforts along these lines will also enrich the chemical methods of classical qualitative and quantitative organic macro analysis, including chromatography. The experiences gained from organic spot test analysis in the examination of organic materials have already proven very useful supplements to the statements contained in the standard works on qualitative organic analysis. Some of these texts are:

N. CAMPBELL, *Qualitative Organic Chemistry*, McMillan, London, 1939.

N. D. CHERONIS and J. B. ENTRIKIN, *Semimicro Qualitative Organic Analysis*, 3rd ed., Interscience, New York, 1957.

N. D. CHERONIS, J. B. ENTRIKIN and E. M. HODNETT, *Identification of Organic Compounds*, Wiley-Interscience, New York, 1964.

H. T. CLARKE, *Handbook of Organic Analysis*, London, 1928, reprinted 1937.

D. DAVIDSON and D. PERLMAN, *A Guide to Qualitative Organic Analysis*, 2nd ed., Brooklyn College Bookstore, 1958.

E. H. HUNTRESS and S. P. MULLIKEN, *Manual of the Identification of Organic Compounds*, New York, 1941.

E. H. HUNTRESS, *The Preparation, Properties, Chemical Behavior, and Identification of Organic Chlorine Compounds*, Wiley, New York, 1948.

O. KAMM, *Qualitative Organic Analysis*, 2nd ed., New York, 1932.

R. P. LINSTEAD and B. C. L. WEEDON, *A Guide to Qualitative Organic Chemical Analysis*, Academic Press, New York, 1956.

S. M. McELVAIN, *The Characterisation of Organic Compounds*, 2nd ed., Macmillan, New York, 1953.

A. McGOOKIN, *Qualitative Organic Analysis and Scientific Method*, Reinhold, London, 1955.

H. MEYER, *Nachweis und Bestimmung organischer Verbindungen*, Julius Springer, Berlin, 1933.

H. MIDDLETON, *Systematic Qualitative Organic Analysis*, 2nd ed., Arnold and Co., London, 1943.

O. NEUNHOEFFER, *Analytische Trennung und Identifizierung organischer Substanzen*, W. de Gruyter, Berlin, 1960.

H. T. OPENSHAW, *Laboratory Manual of Qualitative Organic Analysis*, 3rd ed., University Press, Cambridge, 1955.

M. PESEZ and P. POIRIER, *Méthodes et Réactions de l'Analyse Organique*, *Vol. III*, Masson et Cie., Paris, 1954.

L. ROSENTHALER, *Der Nachweis organischer Verbindungen*, 2nd ed., Ferdinand Enke, Stuttgart, 1923.

F. SCHNEIDER, *The Microtechnique of Organic Qualitative Analysis*, Wiley, New York, 1946.

R. L. SHRINER, R. C. FUSON and D. Y. CURTIN, *The Systematic Identification of Organic Compounds*, 4th ed., Wiley, New York, 1956.

S. SIGGIA and H. T. STOLTEN, *Introduction to Modern Organic Analysis*, Interscience, New York, 1956.

F. J. SMITH and E. JONES, *A Scheme of Qualitative Organic Analysis*, London, 1953.

H. STAUDINGER, *Introduction à l'analyse organique qualitative*, Dunod, Paris, 1958.

S. VEIBEL, *The Identification of Organic Compounds*, 5th ed., G. E. C. Gad, Copenhagen, 1959.

A. I. VOGEL, *Elementary Practical Organic Chemistry: Part II, Qualitative Organic Analysis*, Longmans-Green, New York, 1957.

F. WILD, *Characterization of Organic Compounds*, Cambridge Univ. Press, Cambridge, 1962

When dealing with organic spot test analysis, it must always be remembered that it is and must be more than mere technique, description of procedures, and a collection of recipes. The purely manual aspect, which plays such a particularly important role in spot test analysis, deals with the conducting of specific, selective and sensitive tests. However, an exact knowledge of the chemistry of the reactions involved and the reasons for all the steps taken is indispensable to a real understanding of a procedure through which the specificity, selectivity and sensitivity can be brought to a maximum. In this way, analytical labors become experimental chemistry conducted on genuine scientific lines and thus come to stand in close relation to many provinces of chemistry. Spot test analysis, both inorganic and organic, owes its development and present rank to adherence to this thesis, which is also of very great didactic value and will be our guide in this text.

Chapter 2

Preliminary (Exploratory) Tests

General remarks

The objective of the so-called preliminary or exploratory tests is to provide reference points and guiding principles which have value with respect to the detection of individual compounds by characteristic reactions or to the chemical analysis of mixtures. This kind of examination is particularly useful in the analytical study of organic materials because of the great number and variety of compounds for whose detection there are no systematic schemes of analysis such as are available in inorganic analysis. The best that can be attained in the case of mixtures of organic compounds is a separation of certain individuals or members of certain types of compounds. Means to this end include the appraisal of solubility characteristics in acids, bases and organic liquids, distillation in air or steam, sublimation, adsorptive separations (chromatography). Such procedures, as preliminary analytical steps, are often demanding with respect to material and time; they are not always applicable, frequently they are not wholly reliable, and they inevitably involve loss of material. In addition, the number of chemical reactions of organic compounds that have analytical value is still quite limited and consequently relatively few specific and selective tests for such compounds are available. Therefore, every orientation or hint which can be drawn from preliminary studies conducted within the technique of spot test analysis and with little expenditure of time and material is valuable in an analytical study of organic samples.

Both negative and positive findings are informative in exploratory tests, whose most important objectives in qualitative organic analysis are to learn the presence of certain elements, or to discover whether acidic or basic compounds are at hand, or to uncover redox effects or other special characteristics. Orientation of this kind as well as conclusions derived from physical properties (color, odor, fusibility, behavior with respect to water and organic solvents, *etc.*) can indicate the presence of certain compounds and members of certain types of compounds. The results of such exploratory tests can often relieve the analyst of the need of making special tests which usually consume time and material. For instance, if a simple preliminary test has established the absence of nitrogen, there is obviously no need to test for NO_2, NO, NH_2, NOH, NH groups or alkaloids, *etc.* Similarly, if sulfur is absent, compounds containing SH, SO_3H, or CS groups cannot be present, and tests for such compounds are superfluous. The same is true of groups which contain nonmetals or metals, whose absence can be proved by simple preliminary tests. Consideration of the results of different preliminary tests on the same sample can yield important clues. If one preliminary test has proved the presence of nitrogen and another procedure has shown the sample to be neutral, the variety of nitrogenous compounds that need to be considered and tested for is reduced considerably. On the other hand, it is apparent that additional tests may be required if, for instance, preliminary tests have established the presence of nitrogen (sulfur, *etc.*) as well as redox or aromatic qualities. So many and such varied preliminary tests are available, that full consideration should always be given to their revelations in rationalizing subsequent analytical studies of the sample.

The analyst is not always expected to furnish exhaustive information about the kind and amount of all the constituents present in a given sample. Often it is sufficient to know whether certain elements, compounds, or members of certain classes of compounds are present or absent. Preliminary tests may then be decisive or they may provide reliable guidance in the choice of confirmatory tests, particularly if the intensity of the responses to the preliminary tests is taken into account. It should not be forgotten that sometimes the analytical examination may be confined to the use of sensitive tests to establish the absence of a particular item. Appropriate preliminary procedures can render excellent service in such instances, in fact they sometimes are all that is necessary. Therefore, the widely accepted view that the results of preliminary tests are merely diagnostic is incorrect. Rather, the preliminary tests can at times lead to results which are so decisive that further and different tests are redundant. An excellent example is the differentiation of artificial and fermentation vinegar through a simple and rapid preliminary test for nitrogen conducted on the evaporation residue from one drop. Only fermentation vinegar yields a positive response to this preliminary test, since it, in contrast to artificial vinegar, contains small amounts of amino acids.

The following sections contain descriptions of preliminary tests which can be carried out rapidly and with small amounts of materials by means of the spot test technique. It is earnestly advised that they be tried on the sample before proceeding to the tests for functional groups and for characteristic types of compounds discussed in Chap. 3 and 4. The results of the latter tests must always be brought into accord with findings yielded by the preliminary tests. This applies also to the tests for individual compounds discussed in Chap. 5. In chemical analysis, it is a cardinal rule that tests should be repeated and if possible confirmed by other tests having different sensitivities. This rule applies also to preliminary tests, which can be conducted within the spot test technique, not only with the respective original samples but also with sublimates, and evaporation residues from solutions in water or organic liquids. The relatively small consumption of time and material required by additional different preliminary tests will be more than compensated by the resulting greater certainty in the insights into the composition of the sample and through the wider fund of information that will serve in the choice of other analytical examinations.

There is no sharp boundary between chemical preliminary tests and tests for functional groups. The latter are invariably involved whenever organic compounds or their particular cleavage products take part in reactions. Preliminary tests—with the exception of non-chemical proofs and solubility tests—occupy no special position in this respect. However, the distinction employed in this text between preliminary tests and tests for functional groups is justified by the fact that the latter tests are invariably more selective.

1. Non-chemical proofs

Every chemical examination of organic substances should be preceded by gross observations as to the homogeneity of the material, its color, and odor. If the sample is solid, homogeneity may be determined to a certain extent by examining a fraction of a milligram with a magnifying glass or under a microscope. The heterogeneity of mixtures of colored and colorless constituents, which appear homogeneous to the naked eye, is often revealed under magnification. Similarly, amorphous and crystalline constituents, or different varieties of crystals can be distinguished by this means. If the optical examination indicates that a single substance is at hand, it is advantageous to attempt to determine the melting point of the sample, provided it melts without too much decomposition. Should the first portions melt at a temperature considerably lower than the rest of the sample, then there is a great probability that a mixture is at hand. If, however, the sample melts uniformly, *i.e.*, within a very narrow temperature range, a portion should be recrystallized and the determination repeated. If the rise in melting point is not greater than 1°, the material may be regarded as pure. Melting point tables of organic compounds[1] should then be consulted to discover what compounds have the observed melting point. There is no such simple method of so considerably limiting the number of possibilities when, because of extensive decomposition, it is not possible to determine a satisfactory melting point. Witn materials that cannot be melted but which nevertheless can be dissolved in water or organic liquids, there is a possibility of testing for homogeneity by constructing chromatograms. In these, the constituents of the solution are separated adsorptively.*

A test to discover whether the sample or one of its ingredients sublimes may be of great value in establishing the purity of the material or in obtaining pure products. This information can also be utilized in the subsequent chemical examination of the sample. Sublimation tests can be made with milligram amounts of the sample between two watch glasses. One serves as the heating vessel; the other (smaller) watch glass, placed on it, is cooled with moist filter paper and serves as condenser. With compounds which can be sublimed at water bath temperature, it is advisable to use a micro dish covered with a watch glass, whose convex side is cooled with a drop of water.[2] The sublimate collects on the concave side. As a rule, products obtained by sublimation are of highest purity. Decisive melting point determinations can be made on sublimates, and chemical tests, including spot tests, can be conducted with confidence on such products.** If one component of a mixture can be completely removed by sublimation, the residue can be subjected to chemical tests. The same holds for solutions which contain volatile and nonvolatile materials.***

Color examinations provide hints as to the nature of a sample since the majority of organic compounds are colorless in daylight. A yellow color, which persists when the solid is dissolved, indicates nitro-, nitroso- or azo compounds, quinones, quinonimides, o-diketones, aromatic polyhydroxyketones, certain thioketo compounds, dyes, etc. A red, blue, yellow or green

---

* A detailed discussion of chromatographic separation and analysis, which have acquired high importance in modern microanalysis, is beyond the scope of this text. See, for example, H. H. STRAIN, *Chromatographic Adsorption Analysis*, Interscience, New York, 1942; R. J. BLOCK, *Paper Chromatography*, Academic Press, New York, 1952; E. LEDERER and M. LEDERER, *Chromatography, A Review of Principles and Applications*, 2nd ed., Elsevier, Amsterdam, 1957; R. C. BRIMLEY and F. C. BARRETT, *Practical Chromatography*, Reinhold, New York, 1953; E. BAYER, *Gas Chromatography*, Elsevier, Amsterdam, 1961; E. STAHL, *Dünnschicht-Chromatographie*, Springer Verlag, Berlin, 1962; K. MACEK et al., *Bibliography of Paper Chromatography 1957–1960 and Survey of Applications*, Publ. House of the Cz. Akad. Nauk, Prague, 1962.

It should be noted that spot tests based on color reactions can be frequently and succesfully used for the identification of materials which have been adsorptively separated.

** Later chapters contain instances of sensitive tests involving the reaction of appropriate reagents with sublimable, *i.e.* gasifiable materials, or with pyrolytic cleavage products.

*** It must be noted that complete sublimation may be prevented or stopped because the sublimating material is coated with nonvolatile admixtures. In such cases, it is necessary to free the surface if a complete separation is required. (Compare N. D. CHERONIS and J. B. ENTRIKIN, *Semimicro Qualitative Organic Analysis*, 2nd. ed., Interscience, New York, 1957, Chap. 4).

color indicates that the solid may contain dyestuffs or certain organometallic compounds, especially normal or complex salts. If the sample is solid, it should always be tested with respect to its behavior on contact with water, organic liquids, acids and bases. A resistance of the color, the formation of colored solutions, changes in color or discharge of color, can sometimes be of great value, especially in conjunction with the findings yielded by other preliminary tests and tests for functional groups. It is well to examine the sample in ultraviolet light to determine possible fluorescence properties. Some organic compounds fluoresce strongly in daylight and/or in ultraviolet light.[3] While daylight fluorescence is relatively rare and gives valuable indications, no decisive significance can be attached, in general, to a fluorescence in ultraviolet light, because even slight admixtures or contaminants may be responsible for such effects. Conversely, in some cases, they partially quench an existing fluorescence. In contrast, great analytical value is attached to fluorescence of sublimed products.

If a sample fluoresces in daylight or ultraviolet light, a test should always be made to determine if the fluorescence persists on the addition of acid or base, or if it is quenched, or changed in color. Such trials may have diagnostic value. When making fluorescence tests it should be remembered that the state of aggregation of the sample can play a part. Not all compounds fluoresce in both the solid and dissolved state; sometimes the fluorescence is shown only by the solid (compare p. 28).

Instances are known in which fluorescence is produced by chemical adsorption of non-fluorescing components (compare detection of 8-hydroxyquinoline and certain synthetic fibers in Chaps. 5 and 7). Little attention has hitherto been given to the production of fluorescence by ultraviolet irradiation (compare test for coumarin on p. 480).

In connection with the test for color and fluorescence it should be noted that some compounds when dissolved in concentrated sulfuric, phosphoric, trichloroacetic, or perchloric acid, have a characteristic color or fluorescence which disappears when the solution is diluted with water. This phenomenon is known as halochromism or halofluorism,[4] a term also applied to color and fluorescence effects, whose production involves other materials present in the same solvent, in so far as the color or fluorescence disappears on dilution with water. The chemistry of these changes is not satisfactorily developed. Solutions of halochromic and halofluoric materials may contain colored solvates of the particular materials, and these loose molecular compounds dissociate into their components on dilution. Tests for the formation of such colored addition compounds, which are of particular significance in the study of steroids,[5] can be made with dust particles of the solid plus a drop or two of concentrated sulfuric acid, *etc.* on a spot plate, or in a micro test tube if warming is necessary.

Since the vast majority of organic compounds are odorless, the detection of an odor can be of aid. Among the compounds which exhibit characteristic odors even at room temperature are members of such classes of compounds as: alcohols and phenols; mercaptans; lower fatty acids and alicyclic monocarboxylic acids; naphthenic acids and their amides and esters; aldehydes; naphthol esters; amines; indole and indole derivatives; nitriles and isonitriles; coumarin; allyl compounds. The odor varies in intensity from case to case and also in character (attractive, repellent), which in itself is a guide of sorts.

The classification of odors in Table 1* of some organic compounds illustrates the range of this quality.

Sometimes the type of the odor depends on the quantity or concentration of the substance. A comparison of the odor with that of small amounts of the corresponding pure compound can serve for orientation purposes. Odor tests are best carried out by rubbing several milligrams of the solid material or a drop of a liquid on the palm of the hand to obtain a greater surface from which the evaporation is faster. When dealing with solutions, it is often useful to place a drop or two on filter paper and allow them to evaporate from there.[6] It should be observed whether an initial odor persists, disappears, or is replaced by another type of odor. It is advisable also to conduct odor tests with tiny amounts in a micro crucible and heat to 100–150°.

Extremely small quantities of certain compounds can be discerned by their odors. This is shown by Table 2.*

Since the discernment of odor is not an objective test, but involves the operator's olfactory capabilities, the figures in Table 2 have only a limited quantitative meaning and furnish only an approximate orientation. But even assuming that for the average individual the limits of detecting odor are at one thousand times the stated molecular values, there might always be dilutions at which the most sensitive chemical tests fail.** Accordingly, only in the case of materials which give off a very intensive smell is there a possibility of identifying by chemical tests those compounds whose presence has been indicated by odor tests. On the other hand, odor tests can sometimes furnish indications of the presence of slight contamination or admixtures in odorless materials. It should also be noted that some materials, which in fact are odorless when pure, ordinarily are assumed to possess a characteristic odor because they are usually contaminated with an odoriferous impurity. Familiar examples are acetamide and acetylene. Indole

---

* Taken in part from P. KARRER, *Organic Chemistry*, 4th English edition, Elsevier, Amsterdam, 1950, p. 933.

** Concerning the importance of olfactory tests in perfumery and cosmetics, compare the statements of F. EXNER and G. JELLINEK, *Z. Anal. Chem.*, 201 (1964) 278.

TABLE 1

Character of odor	Typical examples
1. Ethereal	ethyl acetate, ethanol, acetone, amyl acetate
2. Aromatic	
a) almond	nitrobenzene, benzaldehyde, benzonitrile
b) camphor	camphor, thymol, safrole, eugenol, carvacrol
c) lemon	citral, linalool acetate
3. Balsam	
a) floral	methyl anthranilate, terpineol, citronellol
b) lily	heliotropin, styrone
c) vanilla	vanillin, anisaldehyde
4. Musk	trinitro-*iso*-butyltoluene, ambrette musk, muscone
5. Garlic	ethyl sulfide
6. Cacodylic	cacodyl, trimethylamine
7. Empyreumatic	*iso*-butanol, aniline, cumidine, benzene, cresol, guaiacol
8. Rancid	valeric acid, caproic acid, methyl heptyl ketone, methyl nonyl ketone
9. Narcotic	pyridine, pulegone
10. Nauseating	skatole, indole

TABLE 2

Substance	Still detectable in I ml		Dilution
	Number of molecules	Micrograms	
Ionone	$16 \cdot 10^5$	$5 \cdot 10^{-10}$	$1 : 2 \cdot 10^{15}$
Ethyl disulphide	$15 \cdot 10^6$	$2 \cdot 10^{-9}$	$1 : 4.5 \cdot 10^{14}$
Skatole	$16 \cdot 10^6$	$3 \cdot 10^{-9}$	$1 : 3.3 \cdot 10^{14}$
Vanillin	$20 \cdot 10^6$	$5 \cdot 10^{-9}$	$1 : 2 \cdot 10^{14}$
Coumarin	$33 \cdot 10^6$	$9 \cdot 10^{-9}$	$1 : 1.1 \cdot 10^{14}$
Citral	$40 \cdot 10^6$	10^{-8}	$1 : 10^{14}$
Butyric acid	$69 \cdot 10^6$	10^{-8}	$1 : 10^{14}$
Guaiacol	$20 \cdot 10^7$	$4 \cdot 10^{-8}$	$1 : 2.5 \cdot 10^{13}$
Nitrobenzene	$32 \cdot 10^7$	$6 \cdot 10^{-8}$	$1 : 2 \cdot 10^{13}$
Heliotropin	$40 \cdot 10^7$	10^{-7}	$1 : 10^{13}$
Thymol	$15 \cdot 10^8$	$4 \cdot 10^{-7}$	$1 : 2.5 \cdot 10^{12}$
Pyridine	$31 \cdot 10^8$	$4 \cdot 10^{-7}$	$1 : 2.5 \cdot 10^{12}$
Safrole	$48 \cdot 10^8$	10^{-6}	$1 : 10^{12}$
Bornyl acetate	$14 \cdot 10^9$	$5 \cdot 10^{-6}$	$1 : 2 \cdot 10^{11}$
Methyl acetate	$16 \cdot 10^9$	$2 \cdot 10^{-6}$	$1 : 5 \cdot 10^{11}$
Carvone	$22 \cdot 10^9$	$5.5 \cdot 10^{-6}$	$1 : 1.8 \cdot 10^{11}$
Trimethylamine	$22 \cdot 10^9$	$2 \cdot 10^{-5}$	$1 : 5 \cdot 10^{10}$
Phenol	$26 \cdot 10^{10}$	$4 \cdot 10^{-5}$	$1 : 2.5 \cdot 10^{10}$
Menthol	$26 \cdot 10^{10}$	$7 \cdot 10^{-5}$	$1 : 1.4 \cdot 10^9$

has an unpleasant smell when impure, whereas very pure specimens have a flower-like odor and are used in perfumes.

1 For example, W. UTERMARK and W. SCHICKE (Editors), *Melting Point Tables of Organic Compounds*, 2nd revised ed., Interscience, New York, 1963.
2 Comp. L. ROSENTHALER, *Mikrochemie Ver. Mikrochim. Acta*, 35 (1950) 164.
3 See J. DE MENT, *Fluorochemistry*, Chemical Publ. Co., New York, 1945, Chap. XII regarding fluorescence colors of organic compounds.
4 Ch. DHÉRÉ and L. LASZT, *Compt. Rend.*, 226 (1948) 809.
5 Comp. M. PESEZ and P. POIRIER, *Méthodes et Réactions de l'Analyse Organique, Vol. III*, Chap. IX, Mason et Cie, Paris, 1954.
6 H. KUNZ-KRAUSE, *Apoth. Ztg.*, 31 (1916) 903.

2. Burning and pyrolysis tests

When organic and metallo-organic compounds burn in the air, the external picture is not the same in all cases, even though approximately like conditions are maintained with respect to the fineness of the powder, temperature reached, and rate of heating. The reason is that combustion involves numerous partial processes such as dehydration, thermal cleavages, reaction of water (that is split off at high temperatures and is therefore quasi superheated steam), redox reactions, *etc.* Various items of information can be gleaned by carefully noting the manner in which the sample burns.

The burning test is conveniently conducted by first placing 1–2 mg of the sample on an inverted porcelain crucible lid or in a small evaporating dish and applying a small flame beneath. It should be noted whether explosion or detonation occurs (nitro, nitroso, azo compounds or azides). If this is not the case, the trial is repeated with 5–20 mg and note taken of fusion, liberation of gases, *etc.* From time to time, the flame is applied directly to the material from above, so that it will ignite before it volatilizes. Should the substance carbonize, the flame should be increased and finally the sample is heated strongly.

Important guides given by the burning test are:[1]

Aromatic compounds	burn with a smoky flame
Lower aliphatic compounds	burn with an almost non-smoky flame
Compounds containing oxygen	burn with a bluish flame
Halogen compounds	burn with a smoky flame
Polyhalogen compounds	in general do not ignite until the flame is applied directly to the substance which then momentarily renders the flame of the burner smoky
Sugars and proteins	burn with characteristic odor

If nonvolatile organic substances are heated rapidly with limited access of air, a kind of dry distillation occurs. This may involve, as important partial phases: removal of water, hydrolysis, pyrolytic cleavage, condensation, oxidation, reduction and interreactions between pyrolytic cleavage products, including the action of free radicals. From the analytical standpoint it is important to note that products may result which are easily detected in the vapor phase through contact with suitable reagent papers. Such products include: acidic or alkaline vapors; hydrogen cyanide; dicyanogen; acetaldehyde; reducing gases; hydrogen sulfide. There has been as yet no systematic study of the behavior of organic compounds when subjected to dry heating. However, preliminary studies have revealed that valuable hints concerning the nature of the sample can often be secured from the positive or negative responses to tests for volatile compounds.

The test is made in a glass tube (4 cm $\times$ 0.5 cm) held in an asbestos support. About 1 or 2 mg of the sample is placed in the tube whose mouth is then covered with a disk of an appropriate moist reagent paper. The bottom of the tube is heated with a microflame until distinct charring sets in. This procedure will serve to detect the following decomposition products by means of reagent papers (freshly prepared if need be):

1. Volatile acids   Congo red paper (blued)
2. Volatile bases   Phenolphthalein paper (reddened)
3. Hydrogen cyanide   Copper acetate–benzidine acetate paper (blued, see p. 546)
4. Dicyanogen   Potassium cyanide–oxine paper (reddened, see p. 549)
5. Reducing vapors   Phosphomolybdic acid paper (blued, see p. 130)
6. Hydrogen sulfide   Lead acetate paper (blackened)
7. Acetaldehyde   Morpholine–sodium nitroprusside paper (blued, see p. 438)

Table 3 gives the results of tests 1–7 when applied to the volatile pyrolysis products from various classes of compounds. Less than 0.5 mg often suffices to detect characteristic pyrolysis products.

This compilation shows that a pyrolytic splitting out of hydrogen cyanide occurs surprisingly from so many nitrogenous organic compounds that perhaps the negative response of the test for cyanide ions becomes of diagnostic value. The same holds for the relatively seldom detection of dicyanogen and acetaldehyde in the pyrolytic products of organic compounds. Dicyanogen is split out of guanidine and many, though not all, of its open and cyclic derivatives, and also from purine derivatives.[2] Acetaldehyde can be detected in the gaseous pyrolysis products of compounds which contain the OC_2H_5, NC_2H_5, OCH_2CH_2O, NCH_2CH_2N, or NCH_2CH_2O groups.[3] The last three groups are converted into volatile ethyleneglycol by pyrolytically split out water (i.e. quasi-superheated steam) and accordingly indirectly yield the reactions of acetaldehyde.

As stated previously, pyrolytic cleavage of organic compounds may be accompanied or followed by other processes such as hydrolyses, redox reactions, condensations, *etc.* An interesting illustration of this multiple process is found in the behavior of many halogen-bearing organic compounds, which on pyrolysis yield not only hydrogen halide but surprisingly also, at the same time and in considerable quantity, free halogen.* That the production of the latter is due to a secondary reaction is shown by the fact that the dry heating of a mixture of a halogenated organic compound and sodium carbonate or magnesium oxide yields solely sodium or magnesium halide. Accordingly, the halogens arise obviously from the air-oxidation of primarily produced anhydrous hydrogen halides: $2 \text{ HHal} + \text{O} \rightarrow \text{H}_2\text{O} + 2 \text{ Hal}^0$.

Likewise of secondary nature is the above-mentioned production of acetaldehyde in the pyrolysis of O- and N-ethyl compounds. In these cases, and also with the corresponding methyl compounds, the water produced during the pyrolysis initially saponifies the compound to ethanol or methanol (pyrolytic hydrolysis). The hot ethanol (methanol) vapors are partly oxidized to the corresponding aldehyde by contact with air.

TABLE 3

Compound	H+	OH−	HCN	(CN)$_2$	Reducing vapors	H$_2$S	CH$_3$CHO
Thiourea	−	+	+	−	+	+	−
Uric acid	+	−	+	+	−	−	−
Nitroso-R salt	+	−	−	−	+	+	−
Saccharin	−	+	−	−	+	+	−
Rhodamine B	+	−	+	−	+	−	+
Barbituric acid	+	−	+	−	−	−	−
Cinchonine	−	−	−	−	+	−	−
Glucose	+	−	−	−	+	−	+
Aminophylline	−	+	+	+	−	−	−
D,L-Isoleucine	−	+	+	−	+	−	−
Alloxantin	+	−	+	+	−	−	−
Benzidine	−	+	+	−	+	−	−
Hydrazobenzene	−	+	+	−	+	−	−
Dimethylglyoxime	−	+	+	+	−	−	−
Biuret	−	+	−	−	−	−	−
Cystine	−	+	−	−	+	+	−
Guanidine carbonate	−	+	+	+	−	−	−
8-Hydroxyquinoline	−	−	+	−	+	−	−
6-Nitroquinoline	+	−	+	−	−	−	−
Polyvinyl alcohol	+	−	−	−	−	−	+
Xanthopterin	−	+	+	+	+	−	−
p,p'-Diaminodiphenyl sulfone	+	−	+	−	+	−	−

* Only halogen is released when H-free organic compounds are ignited (chloranil, hexachloroethane, *etc.*)

It is interesting that numerous derivatives of aniline yield the latter when they are subjected to dry heating.[4] It can be detected in the gas phase by yielding a yellow stain (formation of a Schiff base, compare p. 243) on contact with filter paper impregnated with p-dimethylaminobenzaldehyde. The sensitivity of this preliminary test for aniline derivatives is demonstrated by the finding that as little as 5 γ of acetanilide, benzanilide, hydrazobenzene, phenylhydrazine or phenylhydrazones give aniline when pyrolyzed.

The following pyrolytic cleavages are of considerable analytical importance: (1) Organic compounds which have groups containing both N and O, without exception split out nitrous acid. (2) Aromatic compounds which have O atoms in the nucleus or side chain yield phenols. Since nitrous acid, as well as volatile phenols, are sensitively detectable in the gas phase, characteristic tests can be based on this behavior (compare pp. 93, 141).

When appraising the detection of gaseous chemical cleavage products it should be noted that mixtures of compounds may behave differently than the separate components because of the occurrence of reactions between them on heating their mixtures. An example is the formation of acetaldehyde when a mixture of alkali or alkaline-earth salts of acetic acid and formic acid is heated. This is the basis of a test for the purity of formic acid and alkali formates (see Chap. 7). Another example is the rapid formation of nitrous acid, when a mixture of nitrosamines with carbohydrates, citric acid, tartaric acid, *etc.* is pyrolyzed. The formation of hydrogen sulfide when sulfonic acids or their alkali salts are strongly ignited is quite remarkable. However, it appears that this is *not* a general characteristic of these compounds since some are known which yield no hydrogen sulfide. Instances are: 8-hydroxyquinoline-5-(7)-sulfonic acid, 1-naphthylamine-3-sulfonic acid, 1-amino-2-naphthol-4-sulfonic acid.

Interesting effects are observed in the dry heating of organic compounds in mixture with infusible inorganic oxidants. Topochemical redox reactions with the organic participant or its pyrolysis products can occur with formation of characteristic pyrolysis products that are easily detected. Pertinent instances of this kind of effect will be found in this text.

1 N. D. CHERONIS and J. B. ENTRIKIN, *Semimicro Qualitative Organic Analysis*, Interscience, New York, 1947, p. 85.
2 F. FEIGL and Cl. COSTA NETO, *Mikrochim. Acta*, (1955-56) 969.
3 Unpublished studies, with C. STARK-MAYER.
4 Unpublished studies, with J. R. AMARAL.

3. Examination of ignition residues

Important conclusions can be drawn from the behavior of organic and metallo-organic compounds when the burning and pyrolytic processes are continued to the ignition stage with access of air. Purely organic materials are completely consumed and leave no residue. A rapid disappearance of carbon and tarry products indicates the presence of materials rich in oxygen and hydrogen. In contrast, stubborn persistence of carbon is an indication of the presence or formation of heat-resistant mineral substances, which in fused or sintered form envelop particles of carbon and shield them against complete combustion. If such an effect is encountered and it is desired to remove the unburned carbon completely, the cooled residue should be moistened with a few drops of strong hydrogen peroxide, taken to dryness, and the residue ignited once more. Repeated evaporation with several drops of concentrated nitric acid is even more effective, but the possible formation of alkali and alkaline-earth nitrates with destruction of carbonates, oxides and halides previously present may complicate or obscure the real picture of the ignition residue.[1]

It is often said that ignition of organic materials containing metals invariably produces a residue which consists entirely of the carbonate of the metal(s) in question. Tables 4 and 5 show that this is not a correct generalization. The tables also include the behavior of ignition residues toward water and dilute acetic acid, points that should be given attention.

TABLE 4

Salts of carboxylic acids, phenols, nitro compounds and oximes with:	Ignition residue		
	Composition	Solubility	
		Water	Dilute acetic acid
Alkali metals including thallium	Carbonate	+	+
Alkaline-earth metals	Carbonate (and oxide)	—(+)	+
Other bivalent metals	Carbonate and oxide	—	+
Aluminum and other ter- and quadri-valent metals	Oxide	—	—
Noble metals	Metal	—	—
Mercury and ammonium	No residue		

TABLE 5

Salts of sulfonic acids and mercaptans	Ignition residue		
	Composition	Solubility	
		Water	Dilute acetic acid
Alkali metals including thallium	Sulfate	+	+
Alkaline-earth metals	Sulfate	— or ±	— or ±
Other bivalent metals	Sulfate	+	+
Ter- and quadri-valent metals	Basic sulfate	—	±
Mercury and ammonium	No residue		

It is notable that despite the heat stability of alkali nitrates and nitrites, alkali salts of acidic nitro compounds leave no nitrate (nitrite) when ignited. This is due to the complete conversion into carbonate when alkali nitrates and nitrites are ignited with an excess of carboniferous material.[2]

Besides the compounds included in Tables 4 and 5, the following organometallic compounds yield carbonate-free residues, or residues which contain carbonate plus other heat-resistant metal salts:

Salts of organic bases with metal acids (molybdic, tungstic, phosphomolybdic acid, *etc.*)	metal acid anhydride
Salts of organic derivatives of arsenic and arsenous acid	metal arsenates
Stannous and stannic salts of acid organic compounds	stannic oxide
Salts of organic derivatives of antimonic acid	metal antimonates
Salts of organic derivatives of phosphoric acid	metal phosphates
Salts of halogen-bearing carboxylic acids, phenols, oximes, *etc.*	metal halides (plus carbonate)
Salts of fluorine-containing acid compounds	metal fluorides
Salts of organic acids with noble metals	metals
Addition compounds of ignition-resistant metal halides or sulfates with organic bases	metal halides or sulfates

By taking account of the quantity of the sample used for the ignition, it is easy to decide whether traces or considerable amounts of an ignition residue remain. The particular demands of the analysis determine whether the residue should be studied in detail.

Only in exceptional cases should complicated mixtures be looked for in the ignition residues left by organic or organometallic compounds. As a rule, the range of inorganic compounds that come into consideration is quite limited. The examination is facilitated still more by the fact that, in addition to oxides and carbonates, only sulfates, phosphates, arsenates and perhaps halides of metals can be present. Sometimes it is sufficient to establish the presence of certain basic compounds, using the term in its broadest sense, *i.e.* water-soluble and water-insoluble materials susceptible to attack by acids. The following procedures are suitable for this purpose; they can be carried out rapidly with minimal quantities of material. The absence of admixed inorganic compounds is assumed.

I. Test for alkali carbonate (including thallous carbonate) and alkaline-earth oxides.

The ignition residue is suspended in a few drops of water and centrifuged. Alkaline reaction toward phenolphthalein indicates a positive response.

II. Test for carbonates and oxides of alkaline-earth metals. (The test[3] utilizes the conversion of the oxides and carbonates into the respective heat-resistant nitrates, which show the color reaction with diphenylamine.)

The ignition residue is taken up in water, centrifuged, and the insoluble sediment washed with water until there is no basic response to phenolphthalein. The residue is dissolved in dilute nitric acid, a drop of the solution is transferred to a porcelain microcrucible and taken to dryness. After heating to 350–400° for 5 min., the cooled residue is treated with a drop of 1% solution of diphenylamine in concentrated sulfuric acid. A blue color (nitrate test) indicates a positive response.

III. Test for alumina and phosphates (arsenates) of bivalent metals. (The test utilizes "nickel dimethylglyoxime equilibrium solution", which reacts with insoluble basic compounds[4].)

If I and II have proved the absence of alkali carbonate and also of carbonates and oxides of bivalent metals, a colorless water-insoluble ignition residue may contain $BaSO_4$, $PbSO_4$, SiO_2, also alumina as well as phosphates or arsenates of aluminum and the bivalent metals. These latter compounds can be tested for by spotting the residue with a drop of "nickel dimethylglyoxime equilibrium solution" (for preparation see p. 109). The formation of red Ni-dimethylglyoxime indicates alumina, phosphate, *etc*. Alumina can be separated from phosphates, *etc*. by digestion with dilute hydrochloric acid and filtering. After washing the residue, it is ignited gently and then tested with the equilibrium solution. The evaporation residue of the filtrate is tested in the same way.

IV. Test for magnesium oxide and alkaline-earth carbonate. (The test utilizes the formation of fluorescing oxinates[5].)

The ignition residue is suspended in a few drops of water and a drop placed on a small disk of quantitative filter paper. The spotted paper is laid on a porcelain crucible containing solid 8-hydroxyquinoline ("oxine"). The volatilizing oxine produces metal oxinates which fluoresce intensely yellow-green in ultraviolet light. The fluorescent fleck is held over acetic acid vapors. If the fluorescence disappears, the presence of Mg- and/or Ca-oxinate is indicated. If it does not disappear, the presence of Al-oxinate, which is resistant to acetic acid vapors, is indicated.

V. Test for sulfates of alkali and alkaline-earth metals. (The test utilizes the conversion of sulfates into the respective sulfides by ignition with calcium oxalate; comp. p. 80.)

The ignition residue, which has been intimately mixed with several cg of *sulfate-free* calcium oxalate, is placed in a micro test tube and heated for about 1 min. with a free flame. The CO resulting from the decomposition: $CaC_2O_4 \rightarrow CaCO_3 + CO$ reduces the sulfate to sulfide; carbon is produced concurrently: $2\ CO \rightarrow CO_2 + C$. After cooling, several drops of dilute hydrochloric acid are added until the evolution of CO_2 stops, and then a piece of lead acetate paper is placed over the mouth of the test tube, which is gently warmed. A positive response is shown by the blackening of the reagent paper.

VI. Test for sulfates, excluding $BaSO_4$. (The test utilizes the realisation of the Wohlers effect *i.e.*, the formation of red-tinted barium sulfate in the presence of potassium permanganate.)

The ignition residue is warmed in a micro crucible with a drop of a saturated potassium permanganate solution containing barium chloride. After cooling, a drop of 20% hydroxylamine hydrochloride is added to destroy the excess permanganate. A positive response is indicated if a red or violet precipitate remains in the colorless liquid.

VII. Test for metal halides. (The test utilizes the liberation of free halogens or formation of chromyl chloride with chromic–sulfuric acid.)

The ignition residue is placed in a micro test tube together with several drops of chromic–sulfuric acid (for preparation see p. 65). A piece of filter paper moistened with potassium iodide or an alkaline solution of a Tl(I) salt is placed over the mouth of the tube, which is then gently heated with a micro flame. Development of a brown stain on the paper (iodine or

VIII. Test for water-soluble salts of magnesium or/and alkaline-earth metals. (The test utilizes the formation of insoluble carbonate with basic behavior.)

TlOOH) indicates the presence of metal halides.

The ignition residue is suspended in a few drops of water, gently warmed and centrifuged. A drop or two of the clear solution together with several cg of ammonium carbonate are evaporated in a micro test tube. The residue is heated to 120° for 20 min. and then spotted with a drop of Ni-dimethyl-glyoxime equilibrium solution (see p. 109). The formation of red Ni-dioxime indicates a positive response.

IX. Detection of barium sulfate in the absence of other sulfates. (The test is based on conversion into $BaCO_3$ and reprecipitation of $BaSO_4$ in the presence of potassium permanganate.)

The ignition residue is taken to dryness along with 1 drop of 10% sodium carbonate solution and the residue then gently ignited:

$BaSO_4 + Na_2CO_3 \rightarrow BaCO_3 + Na_2SO_4$

The cooled residue is treated successively with 1 drop of $KMnO_4$–$BaCl_2$ solution (see VI) and 1 drop of 20% hydroxylamine hydrochloride solution. A violet residue indicates a positive response (Wohlers effect).

X. Detection of barium and strontium sulfates. (The test utilizes the conversion into the respective chlorides followed by the formation[6] of the colored rhodizonate.)

The ignition residue is mixed with 0.5 g of ammonium chloride and 1 drop of water. After evaporation, the residue is heated over a free flame until no more ammonium chloride vapors are evolved. The cooled residue is spotted with 1 drop of a freshly prepared 0.2% solution of sodium rhodizonate. The formation of a red or red-brown product indicates a positive response.

XI. Test for barium carbonate in the presence of other carbonates. (The test utilizes the Wohlers effect mentioned in VI.)

The ignition residue is suspended in a drop of water and then treated with a drop of concentrated solution of potassium permanganate and a drop of sulfuric acid (1 : 1). A drop of 20% hydroxylamine hydrochloride is then introduced to destroy excess permanganate. A violet precipitate signals the presence of barium carbonate in the original ignition residue. Water-soluble salts of barium as well as barium phosphate and arsenate behave in the same manner as barium carbonate. The procedure permits the detection of barium carbonate in the presence of barium sulfate.

When ignition residues are to be systematically tested with respect to the metals present, the residue must be brought into solution by treatment with acid or by fusion with alkali bisulfate. In many instances, it is also possible to test ignition residues directly for certain metals by applying selective and sensitive tests. Pertinent information is given in the companion *Spot Tests in Inorganic Analysis* (1958).

1 Comp. L. SPALTER and M. BALLESTER, *Anal. Chem.*, 34 (1962) 1183: Simple qualitative micro test for organosilicon and other organometallic compounds.
2 Comp. F. FEIGL and A. SCHAEFFER, *Anal. Chim. Acta*, 7 (1952) 507.
3 Unpublished studies.
4 F. FEIGL and C. P. J. DA SILVA, *Ind. Eng. Chem., Anal. Ed.*, 14 (1942) 316.
5 F. FEIGL and G. B. HEISIG, *Anal. Chim. Acta*, 3 (1949) 561.
6 F. FEIGL and V. GENTIL, *Mikrochim. Acta*, (1954) 435.

DETECTION OF NONMETALLIC AND METALLIC ELEMENTS BOUND DIRECTLY OR INDIRECTLY TO CARBON

Direct binding to carbon atoms is limited almost exclusively to hydrogen and nonmetals. The latter may also be bound indirectly to carbon, *i.e.* through other nonmetallic atoms. Direct stable unions between metal atoms and carbon atoms are comparatively rare. They are limited to metal salts of acetylene and its homologs (metal acetylides and carbides), and to mercury, which is directly bound to carbon in numerous aliphatic and aromatic compounds. The great majority of the metallo-organic compounds, that are presented for analytical investigation, will be: normal and inner complex metal salts of organic acids and acidic compounds; chelate compounds of organic acids containing the metal atom as part of an anion or cation; salts of organic nitrogen bases of metal acids; addition products of metal salts of inorganic and organic acids with organic compounds; adsorption compounds of acid and basic metal oxyhydrates with basic and acid organic compounds, dyes and pigments.

The fundamental basis of nearly all tests for nonmetallic and metallic elements (H and O are the sole exceptions) is the destruction of the organic skeleton to obtain inorganic compounds which can then be identified by spot tests. The destruction, whereby carbon is completely oxidized to CO_2 in most cases, can be accomplished by pyrolysis (alone or along with inorganic oxidants), or in the wet way with oxidizing agents. The proper choice is determined by the nonmetallic or metallic elements that may be present. Pertinent instances, as well as examples of reductive destructions, are given in the following sections.

4. Carbon

A special test for carbon may be necessary to distinguish between inorganic and organic material and to recognize rapidly the presence of organometallic or organic compounds as contaminants or components of mixtures with inorganic compounds. The procedure often used for this purpose, namely to set fire to the sample in the air and then to note whether glowing occurs or whether tar is formed, is reliable only when dealing with larger amounts of organic substances which are not volatile or sublimable.

The classic test based on the detection of CO_2 formed when organic materials are burned with oxygen is more reliable. It is exceedingly sensitive when conducted on the microscale[1], but it must be remembered that this procedure requires complete exclusion of air containing carbon dioxide, and also that infusible inorganic carbonates yield CO_2 when they are heated strongly. The following tests avoid these limitations, but in the form given here they are admittedly less sensitive than the micro CO_2-test.

1 F. EMICH, *Z. Anal. Chem.*, 56 (1917) 1.

(1) Test by ashing with molybdenum trioxide[1]

If organic compounds, of any nature whatsoever, are mixed with bright yellow molybdenum trioxide and gradually heated (the highest temperature should be around 500°) they or their thermal decomposition products are oxidized with concurrent production of lower molybdenum oxides (so-called molybdenum blue). Accordingly, a kind of total combustion occurs, in which the heated MoO_3 acts as the oxidant. The volatile combustion products are carbon dioxide and water, along with nitrous oxides when nitrogenous organic compounds are involved. Organic materials containing halogens and sulfur appear to yield hydrogen halide and sulfur trioxide on oxidation with molybdic anhydride.

The total oxidation of elementary carbon by means of molybdenum trioxide can be represented: $C + 4 MoO_3 \rightarrow CO_2 + 2 Mo_2O_5$.

When the test described below is used, the complete absence of other compounds which are oxidizable by molybdenum trioxide must be insured. For example, anhydrous alkali sulfite or arsenite, under the conditions of the test, likewise reduce MoO_3 to molybdenum blue, which indicates that the redox action can also occur between solids. The impairment of the test by oxidizable inorganic compounds can be averted by evaporating the test material to dryness several times with strong hydrogen peroxide. Ammonium salts must not be present, since the ammonia gas, which they evolve when they are thermally decomposed, reacts with hot molybdenum trioxide, probably according to* $2 NH_3 + 6 MoO_3 \rightarrow 3 H_2O + 3 Mo_2O_5$.

The behavior of ammonia proves that MoO_3 can react also with organic compounds or their pyrolytic cleavage products via their gaseous phase.

Procedure. A hard glass tube (about 75×7 mm) closed at one end is used. A small quantity of the solid is introduced, or a drop of the test solution is evaporated in the tube at 110°. The tube is then half filled with finely powdered molybdenum trioxide and connected with a pump by suction tubing. After the air has been removed, the tube, clamped at an angle, is heated for 1–2 min. by a small flame in such manner that the upper portion of the molybdenum oxide is heated first and then the lower part. If carboniferous material is present, a blue zone appears at the point of contact with the light yellow molybdic anhydride. The size and color intensity of the blue zone varies according to the carbon content of the sample.

This ashing procedure with molybdenum trioxide revealed:

5 γ carbon in mannitol and saccharose	3 γ carbon in salicylic acid
1 γ carbon in urea	8 γ carbon in citric acid

* This effect is in accord with the well known fact that powdered ammonium molybdate turns blue when heated in a covered crucible. Prolonged heating in air of course converts the blue Mo_2O_5 to yellow MoO_3.

1 F. FEIGL and D. GOLDSTEIN, *Mikrochim. Acta*, (1956) 1317.

(2) Test through reduction of silver arsenate to elementary silver[1]

Brown-red silver arsenate remains unchanged when heated to temperatures up to 900°, whereas in mixture with organic compounds metallic silver is produced even below this temperature. The reduction is due to the occurrence of two processes. A topochemical reaction first occurs between Ag_3AsO_4 and pyrolysis products containing carbon:

$$2 Ag_3AsO_4 + C \rightarrow 2 Ag_3AsO_3 + CO_2 \qquad (1)$$

The silver arsenite produced in (1) is not stable when ignited, but undergoes an inner molecular redox reaction:

$$2 Ag_3AsO_3 \rightarrow 4 Ag^0 + 2 AgAsO_3 \qquad (2)$$

Summation of (1) and (2) gives the following schematic representation of the reduction of silver arsenate:

$$2 Ag_3AsO_4 + C \rightarrow 2 AgAsO_3 + CO_2 + 4 Ag^0 \qquad (3)$$

The occurrence of redox reaction (3) and hence the presence of organic materials, can be recognized directly by a dark color due to finely divided silver, if considerable quantities of the latter have been formed. Because of much greater sensitivity, it is better to detect metallic silver through its reaction with a hydrochloric acid solution of phosphomolybdic acid. Silver chloride and the highly tinctorial hydrosol of molybdenum blue result. Small amounts of the latter can be extracted from the aqueous phase by shaking with amyl alcohol.

Procedure. Several mg of silver arsenate are mixed in a micro test tube with the solid sample or with a drop of the solution to be tested. In the latter case, evaporate to dryness. The mixture is covered with about an equal quantity of silver arsenate and heated for several minutes over a microburner. After cooling, a drop or two of the solution of phosphomolybdic acid is added. If an organic substance was present, a blue color appears at once or within 1–2 min.

> *Reagents*: 1) Silver arsenate prepared by action of sodium arsenate with silver nitrate. The precipitate may be isolated by centrifuging. The product is washed thoroughly and then dried at 110°.
>
> 2) 3% solution of phosphomolybdic acid in 1 : 1 hydrochloric acid.

The procedure revealed 5 γ carbon in mannitol, saccharin, and urea.

1 F. Feigl, D. Goldstein and M. Steinhauser, unpublished studies.

(3) Test by heating with potassium iodate[1]

Potassium iodate undergoes no loss of weight even though kept for hours at temperatures up to 500°.[2] When heated to its melting point (560°) it evolves oxygen and leaves potassium iodide: $KIO_3 \rightarrow KI + 3\,O$. However, if a mixture of the powdered salt and nonvolatile organic substances is heated, potassium iodide is formed at 300–400°. The extent of the reduction depends on how much organic material is present and on the length of the reaction period. Obviously, hot potassium iodate in contact with solid or fused organic materials is capable of oxidizing their carbon and hydrogen to CO_2 and H_2O.* The iodide formed in this solid body reaction may be detected, after solution in water and acidification, by the familiar redox reaction:

$$5\,I^- + IO_3^- + 6\,H^+ \rightarrow 3\,H_2O + 3\,I_2$$

The reduction of potassium iodate when heated in contact with organic compounds permits a sensitive test for the latter in the absence of inorganic oxidizable materials, such as ammonium salts, alkali sulfite, arsenite, *etc.*

---

* When organic compounds containing sulfur or arsenic are heated with KIO_3 plus alkali carbonate, alkali sulfate or arsenate results. Other alkali halogenates as well as alkali perchlorate and periodate behave analogously, *i.e.* when heated below their decomposition points in contact with organic compounds, the corresponding halide results.

as well as considerable quantities of alkali cyanide. Ammonium salts and cyanides can be removed by evaporating the test material with dilute caustic hydroxide or hydrochloric acid. Other oxidizable inorganic compounds (sulfides, sulfites, arsenites, *etc.*) can be rendered harmless by evaporating the finely powdered sample several times with acidified (HCl) or alkaline hydrogen peroxide solution.

Procedure. A drop of the test solution is evaporated to dryness in a small Pyrex test tube (7 × 70 mm) or a tiny portion of the solid is taken. After mixing intimately with several mg of finely powdered potassium iodate, the mixture is covered with more of the salt, and kept for about five minutes in an oven heated to 300–400°. After cooling, the residue is taken up in sulfuric acid (1 : 2) and tested for iodine by adding starch solution or by shaking with chloroform. It is imperative to carry out a blank test on a like quantity of potassium iodate, since the latter frequently contains small amounts of iodide. The test tubes employed should be ignited briefly before the test. If starch solution is used to reveal the iodine, it should be added before making the solution acid, since the presence of traces of iodide is essential to the formation of the blue starch–iodine complex. Addition of several mg of thyodene indicator is recommended instead of starch solution.

The procedure gave positive results with the following amounts of carbon:

3 γ in benzoinoxime	2 γ in citric acid	1 γ in mannitol
3 γ in nitroalizarin	1.5 γ in saccharose	3 γ in salicylic acid
0.5 γ in sebacic acid	2 γ in sulfosalicylic acid	3 γ in theobromine
6 γ in thionalide		

Positive results were obtained also with: sodium acetate, nickel dimethylglyoxime, cobalt α-nitroso-β-naphthol, mercury succinate, Al-, Ca- and Mg-oxinates, zirconium mandelate, barium rhodizonate, calcium oxalate.

1 F. FEIGL and D. GOLDSTEIN, *Mikrochim. Acta*, (1956) 1317.
2 C. DUVAL, *Anal. Chim. Acta*, 15 (1956) 224.

(4) *Test through pyrolytic formation of hydrogen cyanide*

When sodium amide (m.p. 200°) is strongly ignited, it decomposes into its elements; if heated with carbon, sodium cyanide results.[1] Accordingly, sodium cyanide is formed if sodium amide is ignited with nonvolatile organic compounds. This fact, if combined with a sensitive test for cyanide, forms the basis of a good test for carbon,[2] which in contrast to Tests 1–3 is not affected by the presence of reducing or oxidizing materials.

If organic sulfur compounds are ignited with sodium amide, sodium thiocyanate results along with sodium cyanide. These same products likewise result if sulfur-free organic materials are mixed with inorganic sulfur compounds and then ignited along with sodium amide. Inorganic compounds containing carbon (carbonates, normal and complex cyanides, cyanates, thiocyanates) interfere, as do thiosulfates and sulfites.

It has been found[3] that sodium amide may be replaced by the readily available mercuriamido chloride; volatile hydrogen cyanide results. The net reaction may probably be written: $HgNH_2Cl + C \rightarrow Hg° + HCl + HCN$. In the procedure recommended here, the hydrogen cyanide formed, and hence indirectly the presence of organic materials, is detected through the color reaction with copper–benzidine acetate (see p. 546).

The use of mercuriamido chloride has the advantage over sodium amide in this procedure that, because of the production of mercuric sulfide, no thiocyanate is formed with organic sulfur-bearing compounds.

Procedure.   The test is conducted in a micro test tube fixed in an asbestos board support. A slight amount of the sample or the evaporation residue from a drop of the test solution is mixed with mercuriamido chloride or with a mixture of mercuric oxide and ammonium chloride. The open end of the test tube is covered with a disk of filter paper moistened with the reagent solution. The heating is begun at the upper end of the test tube and advanced downward. A blue stain appears on the paper if the response is positive.

Limit of identification: 1 γ carbon.

1 L. Wöhler, *Z. Elektrochem.*, 24 (1918) 261; see also J. Müller, *J. Prakt. Chem.*, 95 (1917) 53.
2 T. Momose *et al.*, *Chem. Pharm. Bull. (Japan)*, 6 (1958) 128.
3 A. Caldas and V. Gentil, *Talanta*, 2 (1959) 222.

5. Hydrogen

Test by pyrolysis with sulfur[1]

When nonvolatile organic compounds containing hydrogen are pyrolyzed in the presence of molten sulfur, hydrogen sulfide is produced no matter what other metallic or nonmetallic elements are present in the molecule. This dehydrogenation occurs rapidly even at 250°. Probably the withdrawal of hydrogen from the starting material or its pyrolytic cleavage products is not the only path to the formation of hydrogen sulfide. It must be remembered that water produced in pyrolyses reacts as quasi superheated steam to give H_2S with sulfur. Evidence in favor of this activity of pyrolytically split-out water is furnished by the fact that a mixture of sulfur and manganous sulfate monohydrate (which is dehydrated in the temperature range 150–200°) yields copious amounts of hydrogen sulfide on heating.

Procedure.   A micro test tube is used. A little of the solid or the evaporation residue from a drop of its solution is mixed with several cg of sulfur. The mouth of the tube is covered with a disk of lead acetate paper. The tube is placed in a glycerol bath that has been heated to 220–250°. If organic compounds containing hydrogen are present, a black or brown stain appears on the paper within about 2 min.

Hydrogen was detected in the following amounts of:

0.1	γ naphthalene	0.2 γ urea
0.5	γ anthracene	0.2 γ 2,4-dinitrochlorobenzene
0.1	γ phenol	0.2 γ dimethylglyoxime
0.2	γ naphtholsulfonic acid (1,4)	0.1 γ succinic acid
0.05	γ phenanthraquinone	0.5 γ phthalic acid
0.1	γ aniline	0.2 γ phenylarsonic acid
0.2	γ tetrabase	0.1 γ mannitol

No H_2S was yielded by 0.5 mg samples of the following hydrogen-free compounds: hexachlorobenzene, tetra(chloro-, bromo-)-p-benzoquinone (chloranil, bromanil).

1 F. FEIGL and E. JUNGREIS, *Mikrochim. Acta*, (1958) 812.

6. Halogens

(1) Test by conversion into copper halide (Beilstein test)[1]

Organic compounds which contain hydrogen and halogens (chlorine, bromine, iodine, or cyanogen and thiocyanate groups) are decomposed on ignition with production of the corresponding hydrogen halide. If the sample is mixed with copper oxide and then heated, copper halide is formed (along with CO_2 and H_2O) which gives a characteristic green or blue-green color to a non-luminous gas or alcohol flame. This effect is shown by all classes of organic compounds containing halogen, but is must be noted that certain halogen-free substances containing nitrogen also give a positive response; hydroxyquinoline, thiourea, and substituted pyridines as well as guanidine carbonate, mercaptobenzothiazole and salicylaldoxime, are pertinent instances.[2] It is assumed that this anomalous Beilstein test behavior results from the formation of HCN or HCNO during the thermal decomposition of these particular nitrogenous compounds and that their copper salts likewise color the flame. This explanation is not satisfactory however, since numerous organic nitrogen-bearing compounds yield HCN when pyrolyzed (compare p. 50), but they nevertheless do not respond positively to the Beilstein test,* which is the case with acetylacetone and other acidic organic compounds which form volatile complex Cu salts.

* It appears that the production of HCN is a necessary but not sufficient condition for the occurrence of the anomalous Beilstein test. Probably volatile copper cyanide is formed only when there is a splitting-off of HCN directly on the surface of CuO. The fact that HCN can be detected among the gaseous pyrolysis products when organic compounds are dry-heated is no guarantee that the above condition has been satisfied, because HCN can also be produced by hydrolysis of initially formed dicyanogen: $(CN)_2 + H_2O \rightarrow HCN + HCNO$. If this hydrolysis does not occur directly on CuO, but at some distance from it, then copper cyanide is not formed and hence an anomalous Beilstein indication is not observed. Recent studies[3] indicate that the anomalous Beilstein test is connected with the pyrolytic release of NH_3.

Organic compounds containing fluorine do not respond to the Beilstein test as copper fluoride is not volatile.

Procedure. A piece of copper wire, about 1 mm thick, is fused into a glass rod, and the end of the copper wire beaten out to form a spatula 2 to 3 mm wide. The spatula is heated in the oxidizing flame and thus coated with copper oxide. Such spatulas may be kept on hand. To test for halogen, a little of the powdered sample is placed on the spatula, or a drop of a solution (which should contain no inorganic halogen compound) is gently evaporated to dryness. The charged spatula is then heated fairly strongly in the nonluminous flame of a Bunsen burner (first in the blue inner zone and then in the lower part of the outer zone). A blue or green color appears in the flame, and lasts for a time varying with the halogen content. Sometimes only a momentary flash of color is seen. The yellow flame due to traces of sodium salts is seen almost always, particularly at the start. It may hide the green. This interference can be eliminated by viewing the flame through a cobalt glass.

A platinum spatula 2–3 mm wide fused to a glass rod is still better. A little copper oxide is mixed with the solid, liquid or dissolved sample and then heated on the tip of the spatula, which has been ignited beforehand.

Halogen was detected in the following amounts:

0.5 γ o-chloronitrobenzene	0.25 γ iodoeosin
0.25 γ eosin	0.3 γ cyanoacetic acid

Even smaller amounts of halogens may be detected in organic gases, liquids and solids by using an ingenious device which includes, as an essential part, a microflame issuing from a copper capilliary tube.[4] The anomalous Beilstein test can be avoided through a procedure described by JUREČEK[2].

1 F. BEILSTEIN, *Ber.*, 5 (1872) 620.
2 See: H. MEYER, *Analyse und Konstitutionsermittlung organischer Verbindungen*, I, 5th ed., Springer, Berlin, 1931, p. 132; M. JUREČEK and F. MUŽIK, *Collection Trav. Chim. Tchécoslov.*, 15 (1950) 236; J. VAN ALPHEN, *Rec. Trav. Chim.*, 52 (1933) 567.
3 Unpublished studies, with D. GOLDSTEIN.
4 H. JURÁNY, *Mikrochim. Acta*, (1955) 135.

(2) Test by conversion into free halogen[1]

If nonvolatile compounds containing chlorine, bromine, iodine are mixed with sodium carbonate and ignited, heat-resistant sodium halide remains. Warming with chromic–sulfuric acid mixture releases the halogen, whose detection in the gas phase accordingly will indicate the presence of organically bound halogen in the original material. Thio-Michler's ketone (4,4'-bis-dimethylamino-thiobenzophenone) serves as sensitive reagent; it is converted into a blue quinoidal compound by free halogens.

The basis of the color reaction[2] is that the brown-yellow thio-Michler's ketone (I) in its zwitter ion form (II), which is in isomeric equilibrium with the keto-form, is oxidized by halogen to the bivalent diquinoidic cation (III):

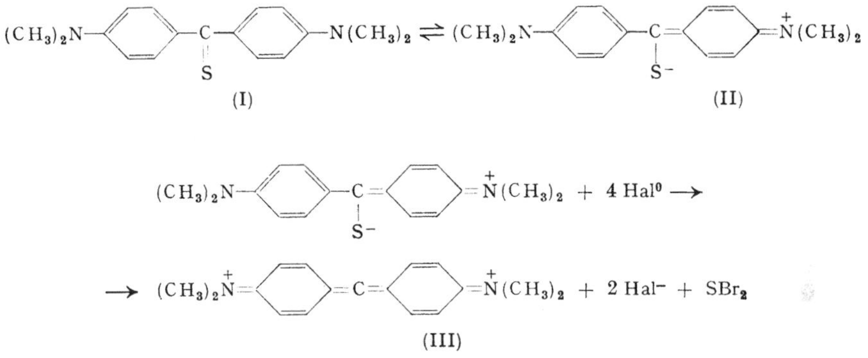

$$(CH_3)_2N-\overbrace{}-\underset{S}{\overset{|}{C}}-\overbrace{}-N(CH_3)_2 \rightleftharpoons (CH_3)_2N-\overbrace{}-\underset{S^-}{\overset{|}{C}}=\overbrace{}=\overset{+}{N}(CH_3)_2$$

(I)   (II)

$$(CH_3)_2N-\overbrace{}-\underset{S^-}{\overset{|}{C}}=\overbrace{}=\overset{+}{N}(CH_3)_2 + 4\ Hal^0 \longrightarrow$$

$$\longrightarrow (CH_3)_2\overset{+}{N}=\overbrace{}=C=\overbrace{}=\overset{+}{N}(CH_3)_2 + 2\ Hal^- + SBr_2$$

(III)

The procedure given below cannot be carried out as rapidly as the Beilstein test, but it has the advantage over the latter that it provides an absolutely reliable test for Cl, Br, and I.

Procedure. A little (not more than 10 mg) of magnesium carbonate is placed in a micro test tube; one drop of the test solution is added, and the solvent evaporated in a water bath. The residue is ignited, cooled, and 4 drops of chromic–sulfuric acid added. The mixture is placed in a boiling-water bath, and a piece of filter paper impregnated with the reagent is placed over the mouth of the tube. A blue fleck on the reagent paper indicates a positive response.

    Reagents: 1) Strips of filter paper are soaked in a 0.1% benzene solution of thio-Michler's ketone. The paper is then dried in the air and stored in the dark.

             2) Chromic–sulfuric acid. A saturated solution of potassium bichromate in conc. sulfuric acid is heated until the vapors do not affect the reagent paper described in (1). This preliminary treatment is essential to remove all traces of chloride possibly present in the bichromate or sulfuric acid.

The following amounts were detected:

5 γ chloranil   7 γ bromanil
5 γ 2-chloromandelic acid   8 γ 4-iodomandelic acid
5 γ 2-bromomandelic acid   10 γ 7-iodo-8-hydroxyquinoline-5-sulfonic acid

Another method for the detection of free halogens is based on the fact that they oxidize N,N-dimethyl-p-phenylenediamine to the so-called Wurster's red. The latter is a molecular compound of the diamine (I) and its quinoidal imine (II).

$$\text{H}_2\text{N}-\!\!\!\left\langle\overline{}\right\rangle\!\!\!-\text{N}(\text{CH}_3)_2 + 2\,\text{Hal}^\circ \longrightarrow \text{HN}\!\!=\!\!\!\left\langle\overline{}\right\rangle\!\!\!=\!\!\overset{+}{\text{N}}(\text{CH}_3)_2 + 2\,\text{Hal}^- + \text{H}^+$$

$$\text{(I)} \qquad\qquad\qquad\qquad\qquad \text{(II)}$$

Procedure.[3] As described before, but filter paper impregnated with a saturated aqueous solution of N,N-dimethyl-*p*-phenylenediamine is used as reagent. A positive response is indicated by the appearance of a reddish-violet fleck on the paper within 2–3 min.

The following amounts were detected:

0.5 γ hexachlorobenzene   0.5 γ pentachlorophenol

1.0 γ tetrachlorohydroquinone   5 γ *p*-nitrobenzyl chloride

5   γ 4-(bromo-, chloro-, iodo)-mandelic acid   5 γ chloranil

1 F. FEIGL, D. GOLDSTEIN and R. A. ROSELL, unpublished studies.
2 Cl. COSTA NETO, paper in press.
3 E. JUNGREIS (Jerusalem), private communication.

(3) Test through conversion into volatile hydrogen halides[1]

As pointed out in Test (2), if organic halogen compounds are ignited with sodium carbonate (or magnesium carbonate), the corresponding heat-resistant halides are left. When the latter are warmed with concentrated sulfuric acid, they form volatile hydrogen halides which can be detected through their action on filter paper impregnated with white silver ferrocyanide and moistened with ferric sulfate. The mixture is almost colorless but Prussian blue appears rapidly because of the formation of hydrated ferrocyanide ions from the silver salt:

$$\text{Ag}_4[\text{Fe(CN)}_6] + 4\,\text{Hal}^- \rightarrow \text{Fe(CN)}_6{}^{4-} + 4\,\text{AgHal}$$

Procedure. As described in Test 2, with the difference that concentrated sulfuric acid and $\text{Ag}_4[\text{Fe(CN)}_6]$-paper moistened with 0.1% ferric sulfate are used. A positive response is indicated by the formation of a blue fleck on the reagent paper.

Reagent paper: Silver ferrocyanide is precipitated by adding a small excess of silver nitrate to a neutral solution of potassium ferrocyanide. The precipitate is washed thoroughly with water and then suspended in concentrated ammonia. Quantitative filter paper is bathed in the ammoniacal suspension and dried in a blast of warm air. The ammonia escapes and the $\text{Ag}_4[\text{Fe(CN)}_6]$ remains in the pores of the paper. The reagent paper is stable if stored away from the air and light.

The following amounts were detected:

20 γ *p*-amino-α-chloroacetophenone   50 γ 2,6-dichlorobenzaldehyde

30 γ 4-(iodo, bromo, chloro)mandelic acid 100 γ 4-chloro-2-aminoanisole

25 γ dichlorofluorescein

1 Unpublished studies.

(4) Dectection of aliphatic bound halogens through pyrohydrolysis[1]

Aliphatic bound halogen in nonvolatile compounds can be detected by heating with succinic or phthalic acid to 200–230°. There is marked anhydration at this temperature of both acids and the superheated water formed accomplishes the pyrolytic hydrolysis (pyrohydrolysis):

$$RHal + H_2O \rightarrow ROH + HHal$$

The resulting hydrogen halide and hence indirectly the aliphatic bound halogen can be detected, as described in *(3)*, by the action of filter paper impregnated with silver ferrocyanide and moistened with ferric sulfate.

Procedure.   Several cg of succinic (phthalic) acid and a slight amount of the solid sample, or a drop of its solution in alcohol, ether, *etc.*, are placed in a micro test tube. The solvent is evaporated if need be. The mouth of the tube is covered with a disk of silver ferrocyanide paper moistened with 0.1% ferric sulfate solution. The test tube is dipped into a glycerol bath that has been heated to 200°, and the temperatures is raised to 230°. A positive result is indicated by the development of a more or less intense blue stain on the paper within $\frac{1}{2}$–3 min.

Limits of identification:

10 γ p-amino-α-chloroacetophenone, $p\text{-}NH_2\text{—}C_6H_4\text{—}CO\text{—}CH_2Cl$

40 γ 1,1,1-trichloro-2,2-bis(p-chlorophenyl)ethane (DDT), $CCl_3\text{—}CH(C_6H_4Cl)_2$

30 γ γ-1,2,3,4,5,6-hexachlorocyclohexane (Lindane), $C_6H_6Cl_6$

100 γ p-nitrobenzyl chloride, $O_2N\text{—}C_6H_4\text{—}CH_2Cl$

10 γ chloromycetin, $p\text{-}O_2N\text{—}C_6H_4\text{—}CHOH\text{—}CH(CH_2OH)\text{—}NH\text{—}CO\text{—}CHCl_2$

30 γ chloral-2-ethylquinoline,

$\text{—}CH(CH_3)\text{—}CHOH\text{—}CCl_3$

20 γ 1,1,1-trichloro-2-hydroxy-3-phenylaminopropane,

$C_6H_5\text{—}NH\text{—}CH_2\text{—}CHOH\text{—}CCl_3$

50 γ (o-methoxy-phenoxy)-ethyl bromide, $H_3C\text{—}O\text{—}C_6H_4\text{—}O\text{—}CH_2\text{—}CH_2Br$

20 γ N-trichloromethylmercaptotetrahydrophthalimide,

From among approximately 40 aromatic halogen compounds that were tested, only one, bromothymol blue (identification limit 25 γ) behaved as an aliphatic compound. It is significant that 2,4-dinitrochlorobenzene does not react although it contains a mobile Cl atom.

The above procedure cannot be applied in the presence of inorganic or organic salts of hydrogen halides or organic derivatives of hypohalogenous acids, *e.g.* chloramine T, 2,6-dichloroquinone-4-chloroimine, because, when

fused with succinic acid, these materials yield hydrogen halide or hypohalogenous acid. Likewise, organic compounds interfere if their pyrohydrolysis produces volatile acids which decompose $Ag_4[Fe(CN)_6]$. Such
compounds include thiol compounds, which give hydrogen sulfide, and
also phenoxyacetic acid and its derivatives, which hydrolyze to yield volatile
glycolic acid.

1 F. FEIGL, D. HAGUENAUER-CASTRO and E. JUNGREIS, *Talanta*, 1 (1958) 80.

7. Chlorine

(*1*) Detection through conversion into silver chloride[1]

Ignition of organic halogen compounds in the presence of manganese
dioxide results in oxidative decomposition with formation of hydrogen
halide, which reacts with MnO_2:

$$MnO_2 + 4\ HHal \rightarrow Mn^{2+} + 2\ Hal^- + 2\ Hal^0 + 2\ H_2O$$

Accordingly, half of the halogen is left in the ignition residue as water-soluble
manganese halide. When chloride is to be detected in the presence of bromide
and iodide ions, use can be made of the finding that the latter are completely
removed by warming with a mixture of permolybdic and sulfosalicylic acids.
The basis of this removal is that permolybdic acid oxidizes Br^- and I^- ions
to the corresponding halogens, which then unite with the sulfosalicylic acid
to yield bromo- and partly iodosulfosalicylic acid respectively. These water-
soluble products contain bromine or iodine in nonionogenic form, *i.e.* not
precipitable by silver ions. Since the chloride ions remain unchanged, they
can be detected through the precipitation of silver chloride. The procedure
given here therefore permits the detection of organically bound chlorine
without interference by organically bound bromine or iodine.

Procedure. A micro test tube is used. A drop of the sample is added to several cg of manganese dioxide and after evaporation to dryness is gradually brought
to red heat over a micro flame. The cooled mass is treated with several drops of
acetic acid, and then warmed and centrifuged. The clear centrifugate is transferred to a micro test tube and several drops of reagent and several cg of sulfosalicylic acid are introduced. The mixture is kept in the hot water bath for 5
min. If the solution remains clear after a drop of 5% silver nitrate solution is
added, the absence of chlorine has been established. If organically bound chlorine
was present in the sample, either a turbidity or a precipitate will appear.

Reagent: Solution of permolybdic acid; 5 ml of 30% hydrogen peroxide and
5 ml of concentrated acetic acid plus 2–3 drops of 0.5 M ammonium
molybdate.

The procedure revealed chlorine in:

8 γ 3,4-dichloroaniline	20 γ 2,4-dinitrochlorobenzene
16 γ p-chlorobenzophenone	16 γ p-chloroaniline
16 γ chloral hydrate	20 γ o- ,,
16 γ monochloroacetic acid	20 γ 1-chloronaphthalene
20 γ chlorotoluene	5 γ 5-chloro-8-hydroxyquinoline
20 γ chlorobenzene	

1 F. Feigl, E. Jungreis and V. Lipetz, *Anal. Chem.*, 36 (1964) 885.

(2) *Detection through selective oxidation of diphenylamine* [1]

If organic compounds containing halogen are heated with a mixture of potassium permanganate and concentrated sulfuric acid, oxidative decomposition occurs with production of free halogen. The latter, like other strong oxidants, converts diphenylamine dissolved in concentrated sulfuric acid into a blue quinoidal compound (see p. 301). If however a saturated solution of diphenylamine in ethyl acetate that contains some trichloroacetic acid is used, only chlorine yields a blue coloration. Bromine and iodine produce a yellow color that can be discharged by means of sodium thiosulfate.

Procedure. A small amount of the solid test material or a drop of its solution is placed in a micro test tube and evaporated if need be. A few mg of potassium permanganate and 2 drops of 6 N sulfuric acid are introduced and the mouth of the tube is covered with a disk of filter paper moistened with freshly prepared reagent solution. The tube is then immersed in boiling water. The appearance of a blue-green stain within 2–3 min. shows the presence of chlorine. If the quantity of chlorine is small, the fleck is grey-violet. The test is clearly distinct even in the presence of a one hundred-fold amount of bromine or iodine since the latter yield a yellowish stain that is characteristic of these two halogens. The latter fleck, in contrast to the chlorine stain, is easily reduced (made colorless) with a drop of 0.1 N sodium thiosulfate.

Reagent: 0.5 g of trichloroacetic acid is added to 10 ml of a saturated solution of diphenylamine in ethyl acetate. The freshly prepared solution is colorless and remains so for several hours, and then turns greenish-blue.

The test revealed the following amounts:

10 γ chloramine T	30 γ o- and p-chloroaniline
20 γ 2,4-dinitrochlorobenzene	40 γ monochloroacetic acid
10 γ chloral hydrate	25 γ 1-chloronaphthalene
20 γ 5,7-dichloro-8-hydroxyquinaldine	

A positive response was given by: 2,5- and 3,4-dichloroaniline; di- and trichloroacetic acid; o- and p-chlorobenzaldehyde; chlorobenzene; p-chlorobenzophenone; o-chlorobenzoic acid; chloroform; 2,7-dichlorofluorescein; 5-chloro-8-hydroxyquinaldine; p-chlorotoluene; dichloroacetic acid; chloranil; 3,4-dichloroaniline; adipic monomethyl ester monochloride.

1 L. Ben-Dor and E. Jungreis, *Mikrochim. Acta*, (1964) 100.

8. Bromine

(1) Detection through fluorescein following oxidative production of bromine[1]

If organic halogen-containing compounds are heated with chromic-sulfuric acid, wet combustion occurs and halogen is released. Bromine vapors produced in this fashion can be detected through the fluorescein test (conversion of yellow fluorescein into red eosin).[2] However this behavior is reliable only if iodine is absent since the latter produces red iodoeosin (erythrosine). The procedure given here, through which organically bound bromine can be detected in the presence of organically bound chlorine or iodine, is based on the conversion of halogens into halide ions and the latters' different behavior toward a mixture of hydrogen peroxide, acetic acid, and ammonium molybdate (permolybdic acid). The conversion can be accomplished by trapping the halogens in caustic alkali and reduction with hydrogen peroxide:

$$2 \text{ Hal}^0 + 2 \text{ NaOH} \rightarrow \text{NaHalO} + \text{NaHal} + \text{H}_2\text{O}$$
$$\text{NaHalO} + \text{H}_2\text{O}_2 \rightarrow \text{NaHal} + \text{H}_2\text{O} + \text{O}_2$$

Permolybdic acid does not affect chloride ions; bromide ions are converted to elementary bromine; iodine is converted via free iodine into iodic acid. Thus the conditions are provided under which the color reaction with fluorescein becomes unequivocal.

Procedure. A small amount of the solid test material or a drop of its solution is treated in a micro test tube with several drops of chromic–sulfuric acid (preparation see p. 65). The open end of the test tube is covered with a disk of filter paper moistened with 2 N sodium hydroxide. The test tube is kept for 5 min. in boiling water. The disk of reagent paper is then placed on a glass plate, spotted with several drops of permolybdic acid solution (for preparation see p. 68), and then dried with a current of cold air. If iodine is present a brown fleck will appear but it fades within a few minutes. The stain is then treated with a drop of 0.1% alcoholic solution of fluorescein. A positive response is indicated by the development of a red spot.

Bromine was detected in:

25 γ bromobenzene	60 γ p-bromoanisole
25 γ bromonaphthalene	25 γ p-bromoacetanilide
40 γ p-bromoaniline	40 γ N-bromosuccinimide

1 F. FEIGL, E. JUNGREIS and V. LIPETZ, *Chemist-Analyst*, 53(1964)9.
2 D. GANASSINI, *Chem. Zbl.*, 1904 I, 1172.

(2) Detection through p-aminophenol following oxidative formation of bromine[1]

Elementary bromine that has been produced during the oxidative decomposition of organic compounds by chromic–sulfuric acid can be detected directly and in minimal amounts through reaction with *p*-aminophenol. A blue water-soluble product results. The basis of the color reaction is that the initial oxidation product is *p*-quinone imine which reacts with unaltered aminophenol to form an indamine. (See detection of *p*-aminophenol, p. 517).

The color reaction is shown by bromine but not by chlorine. Larger amounts of iodine react with *p*-aminophenol to yield a brown color and therefore the procedure described here is reliable for the detection of bromine only if iodine is known to be absent (the test for iodine described on p. 72 should be applied).

Procedure. A little of the solid test material or a drop of its solution is united in a micro test tube with several drops of chromic–sulfuric acid and then heated in a boiling-water bath. The open end of the test tube is covered with a disk of filter paper moistened with a saturated alcoholic solution of *p*-aminophenol. A positive response is indicated by the development of a violet stain on the reagent paper within 1–3 min.

Bromine was detected in:

25 γ bromoanisole	25 γ bromobenzene
20 γ *p*-bromoacetanilide	25 γ α-bromonaphthalene
	20 γ *p*-bromoaniline

1 E. JUNGREIS and V. LIPETZ (Jerusalem), unpublished studies.

9. Iodine*

(1) Test by conversion into iodic acid[1]

When methyl or ethyl iodide is treated with bromine dissolved in glacial acetic acid, iodic acid is formed quickly. This is the basis of an iodometric determination of methoxy and ethoxy groups[2]. The behavior of these iodides is merely a special case illustrating the fact that iodic acid is formed by organic compounds, which contain iodine bound to carbon, when they are treated with bromine (or bromine water). The net reaction is

$$RI + 3\ Br_2 + 3\ H_2O \rightarrow RBr + 5\ HBr + HIO_3$$

Since this transformation proceeds rapidly and with small amounts, the proof of the production of iodic acid may serve as the basis of an indirect test for organically bound iodine.

The following partial reactions may explain the chemical foundation of the test. First of all, bromine acts to produce the equilibrium:

$$2\ RI + Br_2 \rightleftharpoons 2\ RBr + I_2 \tag{1}$$

This reaction then continues to a marked extent because of the removal of the iodine from the equilibrium through conversion into iodic acid:

* Non-volatile I-containing organic compounds when ignited split off I- and HI vapors which may be recognized by the bluing of starch paper (D. GOLDSTEIN, *Chemist-Analyst*, in press).

$$I_2 + 5 Br_2 + 6 H_2O \rightarrow 2 HIO_3 + 10 HBr \qquad (2)$$

If the iodic acid produced in (2) is to be detected by the redox reaction

$$HIO_3 + 5 HI \rightarrow 3 H_2O + 3 I_2 \qquad (3)$$

the excess bromine must be completely removed beforehand. This can be accomplished by adding sulfosalicylic acid, which is immediately brominated:

$$C_6H_3(OH)(COOH)(SO_3H) + Br_2 \rightarrow HBr + C_6H_2Br(OH)(COOH)(SO_3H) \qquad (4)$$

Although four successive reactions are involved, it is still possible to carry out a sensitive spot test for iodine in organic compounds based on these reactions.

Procedure A drop of the test solution (or a little of the solid) is mixed in a micro-test tube with a drop of a saturated solution of bromine in 5% potassium bromide and warmed gently. After cooling, solid sulfosalicylic acid is added until the color is discharged. A drop of water is added and the liquid shaken. (No bromine vapor may persist above the liquid.) A drop of 5% potassium iodide solution which contains thyodene indicator is then introduced. A more or less intense blue color appears.

This procedure gave positive results with the following in one drop:

0.05 γ methyl iodide   1 γ 4-iodomandelic acid
0.1 γ iodoform   0.05 γ 7-iodo-8-hydroxyquinoline-5-sulfonic acid
0.5 γ iodoeosin   0.5 γ diiodotyrosine.

The detection of iodine by converting it into iodic acid is not interfered with by other halogens. When using the test it is necessary to take account of the presence of oxidizing compounds which can liberate iodine from potassium iodide in acidified solutions. Their presence or absence can be established by stirring a drop of the test solution with acidified potassium iodide solution. If iodine is set free, a new test portion should be taken to dryness along with sulfurous acid and the evaporation residue then subjected to the above procedure. Only organic derivatives of arsenic and antimonic acid cannot be rendered harmless in this way.

Cyanoacetic acid and its derivatives must be absent because they form an oxidant when treated with bromine (see p. 551).

1 F. FEIGL, *Anal. Chem.*, 27 (1955) 1318.
2 F. VIEBOECK *et al.*, *Ber.*, 63 (1930) 2819, 3207.

(2) *Test by conversion into catalytically active iodide*[1]

If organic compounds containing iodine are ignited with sodium carbonate sodium iodide results, and it can be recognized because the aqueous solution of the ignition residue exhibits the reactions of the iodide ion. When magnesium carbonate is used in place of the sodium carbonate, magnesium iodide is formed. Soluble iodides can be detected with astounding sensitivity through the fact that they catalytically hasten the oxidation of colorless tetrabase (in acetic acid solution) by chloramine T. A blue color appears rapidly. The catalysis is due to the following effect: Chloramine T (sodium salt of p-toluenesulfonechloramide) hydrolyzes in aqueous solution:

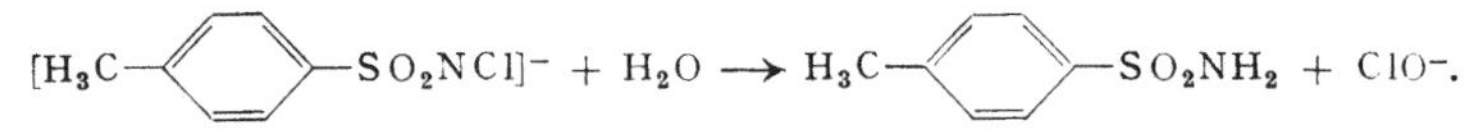

$$[H_3C-\langle\ \rangle-SO_2NCl]^- + H_2O \rightarrow H_3C-\langle\ \rangle-SO_2NH_2 + ClO^-. \quad (1)$$

However, the concentration of ClO^- ions in freshly prepared dilute solutions[2] of chloramine T (see p.126) is not sufficient to oxidize the acetate of tetrabase (N,N'-tetramethyldiaminodiphenylmethane) (I) to the blue quinoid compound (II):

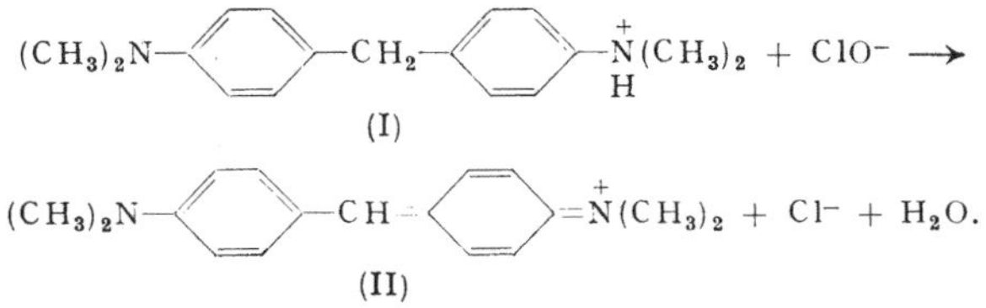

$$(CH_3)_2N-\langle\ \rangle-CH_2-\langle\ \rangle-\overset{+}{\underset{H}{N}}(CH_3)_2 + ClO^- \rightarrow$$

(I)

$$(2)$$

$$(CH_3)_2N-\langle\ \rangle-CH=\langle\ \rangle=\overset{+}{N}(CH_3)_2 + Cl^- + H_2O.$$

(II)

On the other hand, the ClO^- concentration is adequate to oxidize iodide ions:

$$ClO^- + 2\,I^- + 2\,H^+ \rightarrow Cl^- + H_2O + I_2 \quad (3)$$

The free iodine reacts, even in traces, with tetrabase to give a blue color:

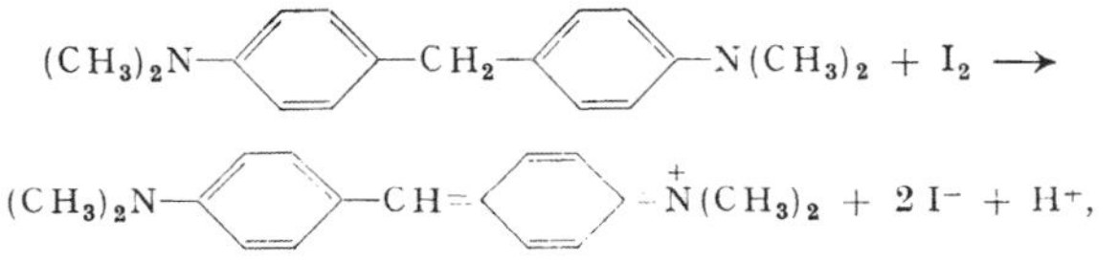

$$(CH_3)_2N-\langle\ \rangle-CH_2-\langle\ \rangle-N(CH_3)_2 + I_2 \rightarrow$$

$$(4)$$

$$(CH_3)_2N-\langle\ \rangle-CH=\langle\ \rangle-\overset{+}{N}(CH_3)_2 + 2\,I^- + H^+,$$

whereby the iodide ions are regenerated. The latter then react with the ClO^- again furnished by the hydrolysis equilibrium (1), and this further reaction (3) is followed by (4) once more, and so on. Addition of the rapid reactions (3) and (4) yields the redox reaction (2), which proceeds but sluggishly in the absence of a catalyst.

If the conditions stated here are maintained, the test for iodine in organic compounds is not only very sensitive, but also specific. Obviously, it is restricted to nonvolatile compounds.

Procedure. A slight amount of the solid or dissolved substance is stirred with several mg of magnesium carbonate in a hard glass micro centrifuge tube. If necessary, the mixture is taken to dryness, and then heated briefly with a micro flame. After cooling, 1–2 drops of water are added, stirred with a glass rod, and centrifuged. A drop of the clear liquid is placed on a spot plate and treated with 1 drop each of tetrabase and chloramine T solutions. Depending on the quantity of iodine, a more or less intense blue color appears at once or within several seconds. The color quickly turns to green and then to yellow.

Reagents: 1) Tetrabase solution. Several ml of 1 *N* acetic acid are warmed on the water bath along with excess tetrabase of high purity for several minutes. The warm suspension is filtered. On cooling, the solution desposits a little tetrabse, which need not be removed.

2) 0.015% solution of purest chloramine T in distilled water.

The following were detected:

0.02 γ *p*-iodomandelic acid	0.002 γ *p*-iodoacetanilide
0.005 γ iodoeosin	0.001 γ 2-amino-3,5-diiodobenzoic acid
0.005 γ rose Bengal	0.001 γ 3-iodobenzoic acid
0.02 γ 7-iodo-8-hydroxyquinoline-5-sulfonic acid	

1 F. FEIGL and E. JUNGREIS, *Z. Anal. Chem.*, 161 (1958) 87.
2 Comp. also E. BISHOP and V. J. JENNINGS, *Talanta*, 1 (1958) 127.

10. Fluorine

(1) Test by conversion into alkali fluoride

On heating nonvolatile organic compounds with metallic sodium or potassium, any fluorine present is converted into alkali fluoride.

Fluorides can be detected with high sensitivity through their reaction with "zirconium alizarinate". This test[1] (see *Inorganic Spot Tests*, Chap. 4) is based on the facts that in mineral acid solution, zirconium salts yield a red-violet color with alizarin and that this color changes to yellow (color of acid alizarin solution) on the addition of excess fluoride because of the production of complex zirconium hexafluoride ions. The reactions can be represented by:

$$Zr^{4+} + 4 \text{ Aliz}^- \rightarrow Zr(Aliz)_4$$
$$\underset{\text{red-violet}}{Zr(Aliz)_4} + 6 \text{ F}^- \rightarrow [ZrF_6]^{2-} + 4 \underset{\text{yellow}}{\text{Aliz}^-}$$

The zirconium alizarinate does not correspond precisely to an inner complex zirconium salt of alizarin; rather it is the hydrosol of a violet zirconium–alizarin lake, *i.e.*, a colored adsorption compound of hydrolysis products of aqueous zirconium salt solutions with alizarin[2] (compare p. 347).

Procedure.[3] A few mg of the sample are taken, or a drop of the solution is evaporated to dryness in an ignition tube. A piece of potassium the size of a pin head is added. The tube is heated, gently at first, and then strongly for 1 min. The molten metal thus comes into intimate contact with the sample. After cooling, 0.5 ml water is placed in the tube and then a few drops of a strongly acid (hydrochloric) red zirconium–alizarin solution. It is not necessary to filter off particles of carbon. The red color changes to yellow when the response is positive.

Alternatively, the sample can be ignited with calcium oxide or magnesium

oxide. The calcium (magnesium) fluoride in the residue can be tested with zirconium alizarinate (as just described).

   Reagent: An acid solution of zirconium chloride is prepared containing about
      0.5 mg Zr per ml. Several ml are treated with a slight excess of an
      alcoholic solution of alizarin. The excess alizarin may be recognized
      by extracting a portion of the solution with ether, which becomes
      yellow.

Fluorine was detected in:

200 γ acetofluoroglucose	100 γ β-fluoronaphthalene
200 γ lactosyl fluoride	100 γ o-fluorobenzoic acid

1 J. H. DE BOER, *Chem. Weekblad*, 21 (1934) 404.
2 F. FEIGL, *Chemistry of Specific, Selective and Sensitive Reactions*, Academic Press,
   New York, 1949, p. 560.
3 Unpublished studies, with D. GOLDSTEIN.

(2) *Test through chemical adsorption of hydrofluoric acid on glass* [1]

Vapors of hydrogen fluoride, released from metal fluorides by heating with
concentrated sulfuric acid, roughen glass surfaces as indicated by a more or
less intense frosting. Small quantities of hydrofluoric acid which do not produce a visible etching may nevertheless change the glass surface sufficiently
so that it is no longer wetted by concentrated sulfuric acid. The acid then
does not run off smoothly but collects in drops like water on an oily surface.
This non-wetting effect is due to a surface reaction between the silica of the
glass and hydrofluoric acid, the concentrated sulfuric acid functioning as
dehydrating agent (see *Inorganic Spot Tests*, Chap. 4):

$$[SiO_2]_x + 2\,HF \underset{+\,HO_2}{\overset{+\,H_2SO_4}{\rightleftharpoons}} [(SiO_2)_{x-1} \cdot OSiF_2] + H_2O$$

This schematic equation [2] is intended to show that the surface reaction
leads to a silicon–fluorine binding, in which the silicon atoms do not leave
their original phase association. Surface reactions without formation of
reaction products as independent phases are the essential feature of a
chemical adsorption.

   The non-wetting effect can be utilized for the detection of fluorine in
nonvolatile organic compounds. It is necessary to convert the fluorine
bound to carbon into inorganic bound fluorine. A reliable procedure is to
ignite the sample with lime to produce calcium fluoride. The hydrogen
fluoride is then released from this ignition residue by concentrated sulfuric
acid. The visibility of the non-wetting effect is improved by using chromic-
sulfuric acid [3].

   Procedure. [4] The solid test substance is mixed with several mg of calcium
oxide in a small pyrex test tube (capacity about 5 ml). Alternatively, a drop of

the test solution is added to the calcium oxide and taken to dryness. The mixture is then ignited, starting at the top and gradually advancing to the bottom of the tube. The cooled residue is treated with 0.5 ml of chromic–sulfuric acid and heated, either over a free flame or for 10 min. in a water bath. The presence of fluorine is revealed by the uneven wetting of the walls when the tube is shaken; the picture is reminiscent of an oily surface.

Reagent: Chromic–sulfuric acid. 1 g K_2CrO_4 in 100 ml conc. sulfuric acid.

This test revealed fluorine in:

50 γ N-trifluoroacetyl-*p*-toluidine   25 γ *m*-acetamidobenzotrifluoride
50 γ 4,4'-difluorobenzophenone   20 γ *m*-trifluoromethylbenzoic acid
13 γ α-hydroxy-α-trifluoromethylpropionic acid
40 γ tosyl-*m*-hydroxybenzotrifluoride

If the quantity of fluorine is not too small, the calcination with lime prescribed in the procedure is not essential. Instead, the oxidative decomposition with production of HF can be accomplished by warming with chromic-sulfuric acid [5].

1 B. FETKENHEUER, *Chem. Abstr.*, 17 (1923) 1398.
2 G. CANNERI and A. COZZI, *Anal. Chim. Acta*, 2 (1948) 321.
3 S. K. HAGEN, *Mikrochemie*, 15 (1934) 313.
4 Unpublished studies, with D. GOLDSTEIN.
5 O. DIMROTH and W. BOCKEMÜLLER, *Ber.*, 64 (1931) 521.

11. Oxygen*

Test through solvate formation with ferric thiocyanate [1]

Red ferric thiocyanate can be extracted from its aqueous solution by shaking with ether, amyl alcohol, and other oxygen-containing organic solvents. In contrast, solvents that are free from oxygen, such as benzene, toluene, carbon tetrachloride, *etc.*, are inactive. Since the solid salt shows this same solubility distinction toward these two classes of liquids, a test for oxygen in organic compounds can be based on this differential behavior. Obviously, the molecules of the oxygen-containing liquid are capable of forming stable solvates with $Fe(CNS)_3$ molecules through coordination on their O atoms. The molecules of liquid hydrocarbons and their halogenated derivatives lack these coordination-centers, and accordingly are inactive. This assumption is supported by the fact that such hydrocarbons and their halogenated derivatives acquire the ability to dissolve $Fe(CNS)_3$ when oxygen-containing materials are dissolved in or mixed with them. This effect is analogous to the solvate formation that occurs when iodine dissolves in organic solvents and produces a brown color. The violet iodine solutions, which probably contain no solvate, change toward brown when coordina-

* Comp. Sect. 34 concerning the behavior of O-containing aromatics.

tion-active organic compounds are introduced, and thereby cause the formation of iodine addition compounds which dissolve with a brown color[2].

It should be noted that sulfur- and nitrogen-containing compounds act
like the oxygenated materials relative to the solubility of ferric thiocyanate.
Acids and oxidizing compounds interfere with the test.

Procedure. Filter paper is plunged into a methanolic solution of $Fe(CNS)_3$
and then dried in air. The reagent paper must be freshly prepared each time.
Several drops of the solution to be tested are placed on the paper. A positive
response is indicated if a wine-red color appears. Solids should be dissolved
beforehand in hydrocarbons or their halogenated derivatives.

Reagent: Separate solutions of 1 g $FeCl_3$ and 1 g KCNS in 10 ml methanol are
united and allowed to stand for several hours before filtering off the
precipitate of potassium chloride.

In the absence of nitrogen and sulfur, the development of a red color in
the above procedure (when 20–50 mg of sample is used) indicates the presence of oxygen. On the other hand, a negative response is somewhat uncertain in its implications because no reaction is given by some oxygen-
bearing compounds, *e.g.*, high molecular ethers, nitro compounds, *etc.*

Another procedure, which makes better use of the solvation of the red
$Fe(CNS)_3$ and which also is successful when organic compounds are fused,
consists in stirring a drop of the test solution or melt with the solid salt.

Procedure.[3] A drop of the test solution is placed on a slide or in a micro test
tube and stirred with a thin glass rod, whose tip carries a little solid $Fe(CNS)_3$. If
oxygen-bearing compounds are present, the drop becomes light to dark red.
Solid samples can be melted beforehand on the slide or in the test tube. Since the
reagent may decompose during the melting, it is better to test a highly concentrated solution in an oxygen-free organic liquid. Chloroform is particularly
suited to this purpose. The stirring rod is charged with $Fe(CNS)_3$ by dipping it to
a depth of several millimeters into an ethereal solution of the reagent, and allowing
the solvent to evaporate into the air.

Reagent: Ethereal solution of iron(III) thiocyanate. 5 g of KCNS and 4 g of
$FeCl_3.6 H_2O$ are dissolved in 20 ml water each. The solutions are
united and the mixture is extracted with 30 ml ether. The ether solution will keep for several weeks if stored in the dark.

The *limit of identification* is 5–10 mg and accordingly this procedure is considerably more sensitive than the test on $Fe(CNS)_3$ paper. For example, diethyl
malonate and ethyl mandelate react promptly as do solutions of benzil and stearic
acid in chloroform, whereas no solvate formation is seen with these materials and
solutions on ferric thiocyanate paper.

1 D. DAVIDSON, *Ind. Eng. Chem., Anal. Ed.*, 12 (1940) 40.
2 F. FEIGL, *Chemistry of Specific, Selective and Sensitive Reactions*, Academic Press,
New York, 1949, p. 121.
3 J. GOERDELER and H. DOMGOERGEN, *Mikrochem. Ver. Mikrochim. Acta*, 40 (1953) 212.

12. Sulfur

(1) Test by fusion with alkali metals (Lassaigne's test)[1]

When nonvolatile organic sulfur compounds are heated with metallic sodium or potassium, alkali sulfide is formed along with carbon and other reduction products. This treatment involves redox reactions of a molten alkali metal with the solid organic compound and its solid, liquid or gaseous thermal decomposition products[2].

A sensitive test for the sulfide produced by the alkali metal fusion is provided by its catalytic effect on the reaction

$$2\,NaN_3 + I_2 \rightarrow 2\,NaI + 3\,N_2$$

This reaction is so sluggish that a yellow or brown solution of alkali azide and alkali polyiodide remains practically unchanged even after months of standing. The addition of soluble or insoluble sulfides initiates the reaction immediately, as can be seen from the fading or complete discharge of the color of the solution, or still better by the production of free nitrogen which escapes as small bubbles. Traces of sulfides can be detected in this way[3].

The detection of sulfides produced by the fusion of organic compounds with an alkali metal requires the preparation of an acetic acid solution of the fusion residue. Any loss of hydrogen sulfide from the acid solution can be prevented by the addition of cadmium acetate (formation of CdS).

Procedure.[4] A tiny portion of the sample is placed in a small hard glass tube with the end blown out to a bulb. If a solution is used, a microdrop is evaporated to dryness in the tube. A small piece of potassium (size of a pin head) is introduced on a glass rod. The tube is then carefully heated, starting from the open end, until the potassium melts and mixes with the sample. Finally, the tube is heated for a short time to redness, and plunged at once into a micro test tube containing 5 drops of water. Without removing the particles of glass and carbon, 1 drop of 20% cadmium acetate is added, followed by 2 drops of 20% acetic acid and, once cold, 1 or 2 drops of iodine–azide solution. If a sulfur compound was present, bubbles of nitrogen appear.

Reagent: Sodium azide–iodine solution. 3 g sodium azide in 100 ml 0.1 N iodine solution.

Sulfur was detected in:

0.3 γ thiourea	1 γ sulfosalicylic acid	1.2 γ H-acid
1.2 γ sulfanilic acid	1.2 γ potassium isatinsulfonate	

1 J. L. LASSAIGNE, *Ann.*, 48 (1843) 367.
2 See G. KAINZ and A. RESCH, *Mikrochemie Ver. Mikrochim. Acta*, 39 (1952) 75.
3 F. FEIGL, *Z. Anal. Chem.*, 74 (1928) 369; see also F. RASCHIG, *Ber.*, 48 (1915) 2088.
4 Unpublished studies, with L. BADIAN.

(2) Test by reduction to sulfide in an alcohol flame[1]

Sulfur in nonvolatile organic compounds can easily be converted into alkali sulfide by fusing with an alkali hydroxide in the reducing portion of an alcohol flame. The resulting sulfide is detected by the iodine–azide reaction. The following test is about ten times as sensitive as the Lassaigne test.

Procedure. The end of a piece of thin copper wire (0.1–0.2 mm) is cleaned with emery paper and washed with distilled water. The end of the wire is dipped into a saturated solution of caustic alkali in water and cautiously passed into the flame of an alcohol lamp to evaporate the water and to obtain a small bead of molten alkali at or near the end of the wire. If the bead does not form at the tip, the wire beyond the bead may be cut off. After breathing on the bead, it is touched to the substance to be tested, so that a little of the latter adheres to the alkali. Then the bead is introduced into the flame, moved slowly into the reducing zone and after a few seconds it is allowed to cool near the wick. The end of the wire is cut off, placed on a slide, treated with a drop of iodine–azide solution and observed under the microscope. If not less then 0.03 γ sulfur is present, the evolution of gas is general and quite vigorous, so that the positive result can be plainly distinguished from that given by the blank. The latter sometimes gives small gas bubbles at single scattered places on the wire and these obviously are due to sulfide inclusions in the copper.

Reagent: 1 g sodium azide in 50 ml 0.5 N iodine.

1 F. L. HAHN, *Ind. Eng. Chem., Anal. Ed.*, 17 (1945) 199.

(3) Test by formation of silver sulfide from volatile compounds[1]

The tests described under (1) and (2), which are based on the production of alkali sulfide, cannot be applied successfully to volatile organic compounds. If, however, sulfur-bearing materials are pyrolysed in small capillary tubes, whose inner walls have been silvered, the action of the hot decomposition products yields silver sulfide. Since the latter catalyzes the iodine–azide reaction just as soluble sulfides do, an extremely sensitive test for sulfur in easily volatile organic compounds becomes possible.

Procedure. Ordinary glass tubing is drawn down into 0.5–1 mm capillaries. These are filled with a 0.1 N silver nitrate solution containing alkali tartrate, ammonia and sodium hydroxide and allowed to stand overnight. After rinsing with water, the tubes are dried by warming and drawing air through them. The silvered tubes are then drawn out to 6 to 8 cm so that the ends are as thin as paper. The closed tubes can be kept indefinitely. A tube is prepared for use by breaking off the very fine point and allowing a droplet of the sample to enter. Solid samples are fused, or dissolved in the smallest possible volume of a solvent that has been tested to insure the absence of sulfur. The capillary is closed by touching the tip momentarily to the flame. A zone about 2 cm from the end is heated. When it becomes hot, the capillary bends so that the sample approaches the flame, volatilizes, and is reduced in the superheated zone. At this instant, a light snap is heard. After cooling, the tip is broken off and a drop of iodine–azide solution is allowed to enter. The reaction is observed under the microscope.

Limit of identification: 0.05 γ sulfur.

1 F. L. HAHN, *Ind. Eng. Chem., Anal. Ed.*, 17 (1945) 199.

(4) Test by conversion into hydrogen sulfide through pyrolytic reduction[1]

When sodium formate is heated above its melting point (250°) it decomposes: $2\ HCOONa \rightarrow 2\ H° + Na_2C_2O_4$. Consequently, sodium formate melt acts as hydrogen donor and therefore as a strong reducing agent.[2] If nonvolatile sulfur-bearing organic compounds are heated with sodium formate, hydrogen sulfide results which can be readily detected, even in traces, by the blackening of lead acetate paper. The procedure given here, though less sensitive than the iodine–azide tests (*1*) and (*2*), is reliable and fast.

Procedure. A micro test tube is used. A little of the solid or a drop of its solution is evaporated with a drop of 20% sodium formate solution. The mouth of the tube is then covered with a disk of lead acetate paper and the bottom of the tube heated over a microflame. If sulfur-bearing material is present, the paper develops a grey or black stain.

Sulfur was revealed in:

20 γ sulfanilic acid	5 γ 2,6-naphthylaminesulfonic acid
10 γ thiourea	5 γ Victoria violet
2.5 γ sulfosalicylic acid	5 γ taurine
0.5 γ phenylthiohydantoic acid	5 γ methylene blue
2.5 γ 3-acetamino-4-methoxy- benzenesulfinic acid	6 γ saccharin
	10 γ sulfanilamide
8 γ diphenylthiocarbazone	
4 γ 4,4'-diaminodiphenyl-2-sulfonic acid (sodium salt)	
5 γ anthraquinone-2-sulfonic acid (sodium salt)	

1 Unpublished studies, with Cl. COSTA NETO.
2 C. A. VOURNASOS, *Ber.*, 43 (1910) 2269.

(5) Test by pyrolysis with calcium oxalate[1]

When nonvolatile organic compounds containing sulfur are ignited with calcium oxalate, hydrogen sulfide results and can be readily detected through the blackening of lead acetate paper. It may be assumed that the carbon monoxide resulting from the reaction : $CaC_2O_4 \rightarrow CaO + CO_2 + CO$ reacts with sulfur to form volatile carbon oxysulfide. The latter then is hydrolytically decomposed by the steam arising during the combustion of organic compounds, or when it comes into contact with the moist reagent paper, with production of hydrogen sulfide:

$$S + CO \rightarrow COS \qquad COS + H_2O \rightarrow CO_2 + H_2S$$

It has not yet been definitely ascertained whether the formation of carbon oxysulfide requires the previous liberation of sulfur from organic compounds or whether there is a direct removal of sulfur by carbon monoxide at higher temperatures from organic compounds or their pyrolysis fission products. It is possible that in some cases sulfur dioxide formed by ignition reacts in the gaseous state:

$$3\ CO + SO_2 \rightarrow COS + 2\ CO_2$$

The production of hydrogen sulfide during the pyrolysis of organic sulfur compounds with calcium oxalate makes possible a reliable test for sulfur in organic compounds.

Procedure. A micro test tube is used. A tiny amount of the test material or a drop of its solution is mixed with several cg of calcium oxalate and taken to dryness if need be. The test tube is fastened in a perforated asbestos board and the mouth of the tube is covered with a disk of lead acetate paper. The bottom of the tube is heated to redness with a bare flame. The evolution of H_2S starts when the mixture assumes a dark hue, a change attributable to the liberation of carbon during the disproportionation: $2\ CO \rightarrow CO_2 + C$. A positive response is indicated by the development of a black or brown stain on the reagent paper.

The procedure was applied to the following compounds and sulfur was detected in:

5	γ sulfanilic acid	1	γ thiourea
2.5	γ sulfosalicylic acid	2	γ phenylthiohydantoic acid
5	γ diphenylthiocarbazone	1	γ 4,4'-diaminodiphenyl-2-sulfonic
2.5	γ anthraquinone-2-sulfonic acid		acid
3	γ Victoria violet	2.5	γ 2,6-naphthylaminesulfonic acid
2.5	γ methylene blue	0.5	γ taurine
8	γ N-phenylsulfanilamide	2	γ saccharin
3	γ benzenesulfinic acid (sodium salt)		
5	γ 3-acetamino-4-methoxybenzenesulfinic acid		
3	γ 1-amino-2-naphthol-4-sulfonic acid		
8	γ 1,2-naphthoquinone-4-sulfonic acid (sodium salt).		

1 D. GOLDSTEIN and E. LIBERGOTT, *Analytical Chemistry (Proceedings Feigl Anniversary Symposium, Birmingham)*, 1962, p. 70.

(6) *Test through conversion into sulfuric acid*[1]

Perchloric acid is a powerful oxidant in the vicinity of its boiling point (202°) and rapidly disintegrates the C-skeleton of organic compounds (formation of CO_2 and H_2O). In this respect, perchloric acid is far more effective than chromic–sulfuric acid or than persulfuric acid activated by Ag^+ ions (see p. 105). Organic compounds which contain nonvolatile nonmetals leave the respective oxygen acids when heated with perchloric acid. Since this treatment converts sulfur into sulfuric acid, a test for sulfur in organic compounds has been based on this behavior.

Procedure. A small amount of the test material or a drop of its solution is placed in a micro test tube. After evaporation to dryness if need be, 2 or 3 drops of concentrated perchloric acid are added and the test tube is immersed in a glycerol bath, preheated to around 150°, and whose temperature is gradually raised to 200°. After 5 min. at this temperature, the contents of the test tube are cooled and then 1 or 2 drops of water and 1 drop of 0.5% barium chloride solution are added. A white precipitate or turbidity indicates a positive response.

This procedure revealed sulfur in:

50 γ alizarinsulfonic acid	50 γ sulfanilic acid	50 γ dithizone
100 γ thio-Michlers ketone	100 γ thionalide	

1 Unpublished studies, with S. OFRI.

(7) Test by pyrolytic formation of ammonium sulfide [1]

When nonvolatile sulfur-bearing organic compounds are mixed with hexamine and ignited, hydrogen sulfide and ammonia are evolved. On contact with a solution of sodium nitroprusside, these gases produce the blue-violet color that is characteristic of alkali sulfides. (The chemistry of this color reaction is discussed in Chap. 1 of *Inorganic Spot Tests*). The action of hexamine may be due to its thermal decomposition, which brings about the pyrolytic reductive loss of hydrogen sulfide from sulfur-bearing compounds, the ammonia or sublimed hexamine providing the conditions for the occurrence of the sodium nitroprusside reaction.*

Procedure. A little of the test material or the evaporation residue of its solution is mixed with about 1 cg of hexamine in a micro test tube, which is fastened in a piece of asbestos board. The mouth of the test tube is covered with a piece of filter paper that has been moistened with a freshly prepared aqueous solution of sodium nitroprusside. The bottom of the test tube is heated to slight redness. A positive response is indicated by the development of a violet stain on the reagent paper.

Sulfur was revealed in:

100 γ benzenedisulfonic acid, sodium salt	400 γ Victoria violet
200 γ thiodiphenylamine	400 γ chloramine T
100 γ methylene blue	300 γ methyl orange

1 Unpublished studies, with E. LIBERGOTT.

(8) Detection of organically bound di- and tetravalent sulfur [1]

If organic compounds containing 2- or 4-valent sulfur are warmed with

---

* This assumption is supported by the finding that the purely inorganic compound sulfamic acid (NH_2SO_3H) yields hydrogen sulfide when heated with hexamine.

hydrochloric acid and Raney alloy (50% Ni, 50% Al) reductive cleavage occurs with evolution of hydrogen sulfide. The latter can be readily detected in the gas phase through the blackening of lead acetate paper. The Raney nickel alloy may play a double role as hydrogen donor during the reduction: a) nascent hydrogen is formed when the aluminium is dissolved out of the alloy and b) Raney nickel is produced (at least transiently) at the same time and because of its nickel hydride content is a stonger reductant than nascent hydrogen. It is probable that, depending on the nature of the organic sulfur compound under study, both reduction effects occur concurrently or in succession.

Among the compounds which contain 2- or 4-valent sulfur are:

thionic compounds $>$C$=$S group	thiocyanates —S—CN group
thiols $>$C—SH group	isothiocyanates —N$=$C$=$S group
sulfides $>$C—S—C$<$ group	sulfoxides $>$C—SO—C$<$ group
(open and cyclic thioethers)	sulfinic acids $>$C—SO$_2$H group
disulfides $>$C—S—S—C$<$ group	

No hydrogen sulfide was evolved when organic compounds containing 6-valent sulfur were warmed with acid and Raney alloy. Typical examples are: aromatic and aliphatic sulfonic acids and their derivatives, and also sulfones.

Procedure. A tiny amount of the test material or a drop of its solution is treated in a micro test tube with several mg of Raney alloy and 1 or 2 drops of hydrochloric acid (1 : 1). After the vigorous evolution of hydrogen has subsided, a piece of lead acetate paper is placed over the open end of the test tube. The tube is then placed in boiling water. A positive response is indicated by the development of a grey or dark stain on the reagent paper within 1–2 min.

The test revealed sulfur in:

2.5 γ dichlorothiourea	2.5 γ methylene blue
2.5 γ 1-naphthylthiourea	2.5 γ L-cystine
2.5 γ thiocarbanilide	3 γ phthalylsulphathiazole
5 γ dithiooxamide	1 γ benzenesulphinic acid, sodium salt
2.5 γ 3-acetamino-4-methoxybenzenesulphinic acid	

The procedure just described must be modified if selenium compounds are present since on warming with acid and Raney alloy they yield hydrogen selenide which likewise gives a dark stain (PbSe) on lead acetate paper. The test for selenium described on p. 88 can be applied to the sample and if the response is positive the operator is faced with the problem of detecting hydrogen sulfide in the presence of hydrogen selenide. Use can be made here of the fact that metal sulfides but not metal selenides catalytically accelerate the iodine–azide reaction (see p. 78). If the test material is treated as described in the above procedure, and if the open end of the test tube is covered with a piece of silver foil, and the resulting dark stain (Ag$_2$S and Ag$_2$Se) is spotted with a sodium azide–iodine solution, the formation of nitrogen bubbles is proof of the presence of silver sulfide and consequently of organic compounds containing 2- and/or 4-valent sulfur.

1 F. FEIGL and P. HAGUENAUER-CASTRO, *Talanta*, 9 (1962) 540.

(9) Detection of double bonded divalent sulfur (thionic compounds) [1]

If pertinent organic compounds containing a CS group are treated with potassium permanganate in neutral aqueous or acetone solution, an exchange of the S atom for an O atom takes place whereby sulfuric acid is formed:

$$\text{>C=S} + 4\,O + H_2O \rightarrow \text{>C=O} + H_2SO_4$$

When the redox reaction is carried out in the presence of BaCl$_2$, violet tinted BaSO$_4$ is formed. The color is resistant to reductants. This so-called Wohlers effect [2] (employed in inorganic spot test analyses for the detection of Ba^{2+} and SO$_4{}^{2-}$ ions) is due to the inclusion of KMnO$_4$ in the BaSO$_4$ lattice. [3] The realisation of the Wohlers effect permits the selective detection of thionic compounds which are soluble in water or acetone.

Thiols, sulfides, disulfides, sulfonic acids or thiocyanates* do not interfere.

The method described here has the limitation that materials oxidizable by permanganate must not be present in considerable amounts since in such cases the condition that the Wohlers effect occurs only when the precipitation of BaSO$_4$ takes place in the presence of much permanganate, is not satisfied.

Procedure. Two drops of a saturated aqueous solution of potassium permanganate containing 2 g barium chloride per 100 ml are placed in a micro test tube. A drop of the test solution is added and the tube is kept in boiling water for 5 min., whereby MnO$_2$ together with tinted BaSO$_4$ is formed. Care must be taken that the reaction mixture contains excess permanganate; therefore the solution may not be too concentrated. After cooling, the suspension is cleared and decolorized by adding a drop of 20% solution of hydroxylamine hydrochloride. A violet precipitate which can be collected by centrifugation indicates a positive response.

The following *limits of identification* were found for water-soluble (also very slightly soluble) compounds:

7	γ thiourea	200	γ diethylthiourea
50	γ ethylenethiourea	20	γ thiobarbituric acid
2.5	γ thioacetamide	2.5	γ rhodanine
10	γ sodium diethyldithiocarbamate	25	γ thiosemicarbazide.

* R. POHLOUDEK-FABINI (Greifswald), private communication.

Positive responses were also given by aqueous solutions of carbon disulfide (approx. 0.1% CS_2); sodium ethylxanthate; ammonium dithiocarbamate and 1-sulfanyl-2-thiourea (sulfathiourea).

Water-insoluble, acetone-soluble thionic compounds gave the following identification limits:

100 γ naphthylthiourea	10 γ dixanthogen
50 γ diisopropylthiourea	25 γ dibutylthiourea
25 γ phenylisothiocyanate	50 γ sym-diphenylthiourea
5 γ dimethylaminobenzylidenerhodanine	30 γ thio-Michlers ketone
1.8 γ dithiooxamide (rubeanic acid)	8 γ thiobenzanilide

It may be noted that aminobenzenesulfonic acids show the Wohlers effect. This is due to the fact that these acids, when treated with potassium permanganate, form sulfates, besides azo- and azoxy compounds.[4]

1 F. FEIGL, I. T. DE SOUZA CAMPOS and S. LADEIRO DALTO, *Anal. Chem.*, 36 (1964) 1657.
2 H. E. WOHLERS, *Z. Anorg. Chem.*, 59 (1908) 203.
3 H. YAGODA, *J. Ind. Toxicol.*, 26 (1964) 224.
4 H. LIMPRICHT, *Ber.*, 18 (1885) 1414.

13. Compounds with groups containing sulfur and oxygen

(1) Test through formation of sulfite by fusion with alkali formate[1]

If the alkali salts of benzene- and naphthalene-1-sulfonic acid are melted along with sodium formate they are converted to the corresponding carboxylic acids.[2] This exchange appears to be characteristic of all aromatic and aliphatic sulfonic acids. If the starting material is the free acid, the addition of alkali hydroxide brings about the following reaction on sintering:

$$(R,Ar)SO_3H + HCOONa + 2\ NaOH \rightarrow (R,Ar)COONa + Na_2SO_3 + 2\ H_2O$$

Because of unavoidable partial thermal decomposition, this reaction has no significance in preparative work. On the other hand, it permits a rapid test for sulfonic acids, since the resulting alkali sulfite, which is stable under the conditions of the test,* or the gaseous sulfurous acid released from the sulfite by acidification, can be detected even in small amounts. The reaction on ferri ferricyanide to produce Prussian blue is recommended.[3]

Sulfinic acids and sulfonamides react in the same manner as sulfonic acids when fused with sodium formate. Although no trials have been made as yet with sulfones, it is extremely likely that they too will produce alkali sulfite.

* A contributing factor may be the fact that when alkali salts of sulfonic acids are fused with an excess of sodium formate, the air oxidation of the resulting sulfite to sulfate is lessened by the thermal decompositions: $2\ HCOONa \rightarrow 2\ H° + (COONa)_2$ and $(COONa)_2 \rightarrow Na_2CO_3 + CO$.

Procedure.　The apparatus shown in Fig. 23, *Inorganic Spot Tests*, is used. A little of the solid or a drop of its aqueous solution is placed in the bulb and taken to dryness with a drop of an alkaline solution of sodium formate. The residue is heated over a bare flame for about 30 sec, *i.e.* until a grey tinge indicates that charring has begun. The cold mass is acidified with 1:1 sulfuric acid, and the apparatus is closed after a drop of the ferri ferricyanide reagent solution has been placed on the knob of the stopper. The suspended drop turns blue if sulfonic acids *etc.* were present. Even slight color changes become visible if the drop is wiped onto a spot plate.

> *Reagents:* 1) Alkaline sodium formate solution. 5 g sodium formate and 6 g sodium hydroxide dissolved in 100 ml water.
>
> 2) Ferri ferricyanide solution. 0.08 g anhydrous ferric chloride and 0.1 g potassium ferricyanide in 100 ml water.

The test revealed:

2.5 γ naphthalenedisulfonic acid (1,4; 1,5; 2,6)	1　γ sulfanilic acid
2.5 γ naphtholdisulfonic acid (2,6,8)	0.25 γ taurine.
1　γ chromotropic acid	
5　γ dihydroxynaphthalenesulfonic acid (2,3,6)	
2.5 γ naphthylaminesulfonic acid (1,3; 2,6)	
1　γ naphthylaminedisulfonic acid (2,3,6)	

Positive responses were given by H-acid, sulfosalicylic acid, Congo red, sulfapyridine, sulfamethazine, sulfonal, trional.

1 F. FEIGL, *Anal. Chem.*, 27 (1955) 1317.
2 V. MEYER, *Ber.*, 3 (1870) 112, 364.
3 G. B. HEISIG and A. LERNER, *Ind. Eng. Chem., Anal. Ed.*, 13 (1941) 843.

(2) Test through formation of sulfite by alkali fusion [1]

The classical fusion of aromatic sulfonic acids with alkali to produce phenols yields alkali sulfite and its detection provides the basis of a test for aromatic sulfonic acids. However, the production of sulfite by alkali fusion is not restricted to aromatic sulfonic acids; it is also formed from aliphatic sulfonic acids, and from amides of aromatic and aliphatic sulfonic acids. The reactions in the various alkali fusion processes are shown by equations (1)–(4) in which R may denote alkyl or aryl:

$$RSO_3H + 3\ NaOH \rightarrow RONa + Na_2SO_3 + 2\ H_2O \tag{1}$$

$$RSO_3Na + 2\ NaOH \rightarrow RONa + Na_2SO_3 + H_2O \tag{1a}$$

$$RSO_2NH_2 + 3\ NaOH \rightarrow RONa + Na_2SO_3 + H_2O + NH_3 \tag{2}$$

(N-substituted sulfonamides will evolve amines instead of ammonia)

$$RSO_2H + 3\ NaOH + O \rightarrow RONa + Na_2SO_3 + 2\ H_2O \tag{3}$$

$$\begin{matrix} R \\ R \end{matrix}{>}SO_2 + 4\ NaOH + O \rightarrow 2\ RONa + Na_2SO_3 + 2\ H_2O \tag{4}$$

Production of sulfite on alkali fusion is thus characteristic of compounds containing oxidized, *i.e.*, 4- and 6-valent sulfur. It is possible to differentiate such compounds, in combination with the alkali fusion, if solubility differences are taken into consideration. For instance, sulfonic acids and their alkali salts are soluble in water, whereas sulfonamides and sulfones do not dissolve in water or acids. According to (2) and (3), sulfonamides and sulfones differ in that only sulfonamides yield ammonia (or amine) when fused with alkali. (It should be noted that amides of carboxylic acids behave analogously to sulfonamides when fused with alkali.) Sulfinic acids can be recognized by their precipitability from mineral acid solutions by ferric chloride. Though this reaction is not very sensitive, it can serve to separate sulfinic from sulfonic acids.[2]

All the tests given in *Inorganic Spot Tests* for the detection of sulfur dioxide liberated by acids from alkali sulfites can be used to reveal the production of sulfite resulting from the alkali fusion. Especially good is the test which involves the formation of black Ni(IV) oxyhydrate from green Ni(II) hydroxide on contact with sulfur dioxide.[3] The reaction involves autoxidation of SO_2, which in turn causes the transformation of $Ni(OH)_2$ into $NiO(OH)_2$[4] which normally occurs only through strong oxidants. It is probable that when SO_2 comes in contact with $Ni(OH)_2$, the initial product is basic sulfite, whose cationic and anionic components are then oxidized by atmospheric oxygen together with or induced by the oxidation of sulfur dioxide:

$$2\ Ni(OH)_2 + SO_2 \rightarrow (NiOH)_2SO_3 + H_2O$$
$$(NiOH)_2SO_3 + O_2 \rightarrow NiO(OH)_2 + NiSO_4$$

The production of $NiO(OH)_2$ is easily discerned through a blackening or graying of the green $Ni(OH)_2$. Traces of the higher oxide can be detected by spotting with tetrabase; a blue color develops.

It should be noted that alkali sulfide is produced in the alkali fusion of organic compounds which contain bivalent sulfur (thiophenols, thioalcohols, thioethers, disulfides, thioketones). On acidification, this sulfide yields H_2S, which transforms the green $Ni(OH)_2$, to black NiS. The latter may be mistaken for black $NiO(OH)_2$. Therefore, it is well to make a preliminary alkali fusion; allow the mass to cool, add acid, and test the gas for H_2S with lead acetate paper. If the test fails, the procedure outlined below provides an indisputable test for oxidized sulfur in organic compounds. If the lead sulfide test was positive, the sample may be desulfurized (thiols, thioketones) by warming the aqueous solution or suspension with lead carbonate. The filtrate or centrifugate from the PbS–$PbCO_3$ mixture is evaporated and the test for oxidized sulfur is then made with the dry residue.

Procedure. A very little of the solid, or the evaporation residue from one drop of the test solution is heated with a grain of sodium hydroxide over a small flame until the mixture just melts. A hard glass tube, whose bulb has a capacity of about 3 ml, is used. After cooling, the melt is dissolved in 2 drops of water, 1 or 2 drops of concentrated hydrochloric acid are added (the acid reaction should be tested with litmus paper), and the walls of the tube washed with water. The rim of the tube is carefully wiped dry, and a strip of filter paper carrying green $Ni(OH)_2$ is laid over the mouth. The bulb is immersed in hot water for a few minutes to hasten the evolution of the sulfur dioxide. Sulfite is evidenced by the change from green to black or gray. When very small amounts are to be detected, it is best to spot the nickel hydroxide with a solution of tetrabase; a blue color is formed in the presence of a trace of $NiO(OH)_2$.

Reagents: 1) Nickel hydroxide paper, for preparation see p. 230.

2) Solution of tetrabase in chloroform (see p. 547).

The procedure revealed:

12 γ sulfanilic acid	6 γ potassium methionate
12 γ tartrazine	5 γ monopotassium benzenedisulfonate
6 γ sulfonal	3 γ monosodium β-naphthalenedisulfonate
10 γ trional	14 γ α-bromocamphorsulfonic acid
6 γ benzenesulfinic acid	12 γ benzaldehyde-o-sulfonic acid
5 γ camphorsulfonic acid	12 γ α-naphthalenesulfinic acid
	20 γ monosodium hydroxyquinolinedisulfonic acid

1 F. Feigl and A. Lenzer, *Mikrochim. Acta*, 1 (1937) 129.

2 See F. Feigl, *Chemistry of Specific, Selective and Sensitive Reactions*, Academic Press, New York, 1949, p. 289.

3 F. Feigl and E. Fraenckel, *Ber.*, 65 (1932) 545.

4 F. Haber and F. Bran, *Z. Physik. Chem.*, 35 (1900) 94; cf. W. Böttger and E. Thoma, *J. Prakt. Chem.*, [2], 147 (1936) 11.

14. Selenium

Test through wet oxidation to selenious acid [1]

Perchloric acid is not an oxidant in aqueous solution. It forms an azeotropic mixture (72.5% $HClO_4$) with water which boils at 170° and at this temperature is a powerful and rapid oxidant (see p. 81). Organic compounds containing selenium are destroyed by this reagent and leave selenious acid which can be detected by the redox reaction with hydrazine salts:

$$SeO_3{}^{2-} + NH_2NH_2 + 2 H^+ \rightarrow Se^0 + N_2 + 3 H_2O$$

The test described here is specific for selenium in organic compounds.

Procedure. A micro test tube is used. A small amount of the test material or a drop of its solution in alcohol is taken to dryness if need be. A drop of concentrated perchloric acid is added and the mixture heated for about 3 min. to 205° (glycerol bath) to bring about the oxidative destruction and to expel the excess perchloric acid. After cooling, the material in the test tube is treated with a drop of a saturated aqueous solution of hydrazine sulfate and then warmed in boiling water. A red precipitate or pink color indicates a positive response.

The following amounts were detected:

> 5 γ bis-(2-naphthyl) selenide
> 5 γ 2-nitrobenzeneseleninic acid
> 5 γ o-benzoylbenzeneseleninic acid
> 30 γ benzoylphenyl selenium tribromide
> 6 γ phenyl (m-nitrophenyl) selenide
> 5 γ phenyl (2,4-dinitrophenyl) diselenide
> 5 γ 2-nitro-4-methylphenyl selenocyanate
> 5 γ phenyl (2-nitro-4-methylphenyl) diselenide
> 5 γ benzyl (2-nitro-4-methylphenyl) diselenide
> 5 γ benzyl (2-nitro-4-chlorophenyl) diselenide
> 30 γ phenyl (2-nitro-4-chlorophenylthio) selenide
> 5 γ 2-nitro-4-methylbenzeneseleninic acid
> 6 γ phenylthio (2-nitro-4-methylphenyl) selenide

1 F. FEIGL, *Anal. Chim. Acta*, 24 (1961) 501.

15. Nitrogen

(1) *Test by ignition with calcium oxide (lime test)*

When organic compounds containing nitrogen and hydrogen are strongly heated with lime, ammonia results. Various kinds of nitrogen compounds but not all give this result though the yield of ammonia may vary. The paths by which ammonia is formed are not known. It is possible that calcium cyanamide and calcium cyanide are the initial products, and these then hydrolyze on heating with water to produce ammonia. The necessary water is provided in the lime tests by the combustion of the hydrogen almost always contained in the sample, and reacts at the site of its production as superheated steam, in other words in a particularly reactive state.

The ammonia produced in the lime test can be detected by indicator papers. Especially effective is filter paper moistened with silver nitrate–manganese nitrate solution.[1] On contact with ammonia, a gray or black fleck appears on the white paper. The stain is due to a mixture of free silver and manganese dioxide, resulting from the reaction:

$$2\ Ag^+ + Mn^{2+} + 4\ OH^- \rightarrow 2\ Ag^0 + MnO_2 + 2\ H_2O$$

The test can be intensified still more by utilizing the fact, employed in the detection of manganese (see *Inorganic Spot Tests*, Chap. 3), that when benzidine is oxidized by MnO_2, the highly colored meriquinoidal derivative of this base, namely benzidine blue, results. If the sample is heated with lime in a hard glass tube and a piece of filter paper moistened with the reagent solution is laid over the open end, it often happens that substances which contain no nitrogen also cause a darkening of the paper due to the formation of pyro-compounds during the incomplete combustion of the organic material. This interference with the test can be avoided if a little MnO_2 powder is mixed with the lime; the evolution of oxygen on heating to redness ensures complete combustion. Some of the ammonia may be oxidized in this case, but since the test for ammonia is very sensitive this loss is not important.

Procedure.[2]   A little of the sample is mixed in a small hard glass tube with an ignited mixture of lime and manganese dioxide (10 : 1). Alternatively, a drop of the test solution is evaporated to dryness in the tube. The open end of the tube is covered with filter paper moistened with the reagent solution, the glass stopper is put in place and the tube is slowly heated to redness. A black or gray stain appears on the paper, according to the amount of nitrogen present. The stain immediately turns blue on spotting with benzidine solution.

Reagents: 1) Manganese nitrate–silver nitrate solution. 2.87 g $Mn(NO_3)_2$ and 3.35 g $AgNO_3$ are dissolved in 80 ml water, and the mixture is diluted to 100 ml. Then dilute alkali is added, drop by drop, until a black precipitate is formed: this is filtered off. The reagent solution will keep if stored in a dark bottle.

2) Benzidine solution. 0.05 g benzidine base or hydrochloride dissolved in 10 ml strong acetic acid, diluted to 100 ml with water and filtered. Benzidine can be replaced by tetrabase.

Nitrogen was revealed in:

2 γ sulfanilic acid	1 γ *p*-nitrosophenol
2 γ 1-nitroso-2-naphthol	3 γ codeine hydrochloride

1 F. FEIGL, *Mikrochemie*, **13** (1933) 132.
2 Unpublished studies, with L. BADIAN.

(2) *Test through pyrolytic oxidation*[1]

If nitrogenous organic compounds, no matter what their type, are strongly heated along with excess MnO_2, nitrous oxides result. The latter can be detected in the gas phase by the Griess[2] reaction which involves the diazotization of sulfanilic acid (I) and the coupling of the resulting diazonium compound or its cation (II) with α-naphthylamine (III) to produce the red water-soluble azo dye (IV):

$$HO_3S—\langle\rangle—NH_2 + HNO_2 + H^+ \longrightarrow HO_3S—\langle\rangle—\overset{+}{N}{\equiv}N + 2\,H_2O$$

(I)   (II)

$$HO_3S-\!\!\!\bigcirc\!\!\!-\overset{+}{N}\!\!\equiv\!\!N + \bigcirc\!\!\!-NH_2 \longrightarrow HO_3S-\!\!\!\bigcirc\!\!\!-N\!\!=\!\!N-\!\!\!\bigcirc\!\!\!-\overset{+}{N}H_3$$

(III) (IV)

The general transformability of organically bound nitrogen into nitrous acid is surprising at first sight. A hint as to the partial reactions involved here is given by the fact that compounds possessing a group containing both N and O (consequently a structural element of nitrogen oxides) split off nitrous acid when dry heated. (This is the basis of the test for such compounds described in Sect. 16.) Accordingly, it may be assumed that nitrogenous oxygen-free pyrolysis products are converted into acidic nitrogen oxides by contact with manganese dioxide. It is possible that initially these acidic nitrogen oxides are fixed as nitrate or nitrite by the basic MnO. It is known that these manganese compounds behave analogously to other heavy metal nitrates (nitrites) in splitting out N_2O_4 or N_2O_3 when heated. The assumption that a pyrolytic oxidation occurs because of the atomic oxygen released by the MnO$_2$ (accompanied by a rupturing of the carbon framework) is strengthened by the finding that PbO_2[3], Pb_3O_4, Co_2O_3, Ni_2O_3, Mn_2O_3 behave in the same way as MnO_2, whereas CuO and reducible acidic metal oxides (CrO_3, WO_3, V_2O_5) are practically without effect. The behavior of silver and mercuric cyanide, copper tetrammine sulfate, magnesium nitride, and sodium azide is in conformity with an oxidation of oxygen-free nitrogenous pyrolysis products. On heating, these inorganic compounds split off only dicyanogen, ammonia, or nitrogen, but in mixture with MnO_2, their pyrolysis yields nitrous acid. Other inorganic nitrogen compounds, such as ammonium-, hydroxylamine- and hydrazine salts, alkali cyanides, alkali ferro- and ferricyanides, alkali thiocyanates, alkali nitrates and nitrites, exhibit the same behavior.

Another possible explanation of the transformation of organically bound nitrogen during pyrolysis with MnO_2 is that the initial products of this operation are NH radicals which then directly yield nitrous acid through reaction with higher metal oxides.

The formation of nitrous acid through pyrolytic oxidation occurs with even minimal quantities of nonvolatile nitrogenous organic compounds. A sensitive test for nitrogen has been based on this behavior.

It seems that the grouping N=CH has an influence on the response to the test here described. This is shown by the finding that a polyurethane resin with 4% N fails in the test.

Procedure. The test is conducted in a micro test tube which projects through a perforated asbestos sheet. A little of the solid is brought into the test tube, or a drop of the test solution is evaporated there and then intimately mixed with about 0.2 g of manganese dioxide or sesquioxide. The mouth of the test tube is covered with a disk of filter paper moistened with the Griess reagent, and the bottom of the tube is strongly heated with a microflame. In general it is sufficient to heat for 1–2 min. A positive response is indicated by the development of a pink or red circle on the colorless paper.

Reagents: 1) 1% solution of sulfanilic acid in 30% acetic acid.

2) 0.1% solution of 1-naphthylamine in 30% acetic acid.

Equal volumes of 1 and 2 are mixed just before using.

Instead of the unstable solution of 1-naphthylamine, the use of N-1-naphthylethylenediamine dihydrochloride is recommended[4].

About 150 nonvolatile compounds were examined by this procedure in amounts less than 0.5 mg. Though these test materials were taken from the most varied classes, a strong positive response was observed in every case. The *limits of identification* were determined with 10 appropriate representatives. In all instances the limit lay between 0.02 and 0.03 γ nitrogen.

A special treatment is necessary for volatile or sublimable organic compounds. Volatile bases can be transformed beforehand into the corresponding chlorides by evaporation with excess hydrochloric acid. This process, conducted in the micro test tube if need be, made it possible to detect nitrogen in 0.1 γ methylamine, 0.2 γ ethylamine, 0.1 γ ethylenediamine.

When dealing with volatile nitrogen compounds or compounds with low melting points, a drop of the solution is placed in the test tube, which is then half filled with MnO_2 or Mn_2O_3. The heating is begun just below the asbestos sheet, and the flame is gradually advanced toward the bottom of the test tube. In this way nitrogen was detected in 0.3 γ nitromethane (b.p. 101.2°), 0.3 γ nitroethane (b.p. 114°), and 0.4 γ ethyl carbamate (m.p. 50°).

The extreme sensitivity of this test makes it necessary to point out that ordinary glass test tubes and commercial manganese dioxide are not always nitrogen-free, so that a positive response may be obtained in a blank run. If manganese dioxide of adequate purity is not available, it is best to use Mn_2O_3 prepared by heating MnO_2 to 500–600°. The heating expels any nitrogen present as nitrate. When pure MnO_2 or Mn_2O_3 was used, and micro test tubes and centrifuge tubes of Pyrex glass were employed, the response in the blanks was negative. Sometimes, a slight but still distinct Griess reaction was observed when pure MnO_2 or Mn_2O_3 was ignited in ordinary glass tubes. This result is due to the presence of alkaline earth nitrate in the glass, which may participate in a solid body reaction with the manganese oxides. The resulting $Mn(NO_3)_2$ yields N_2O_3 on ignition.

1 F. FEIGL and J. R. AMARAL, *Anal. Chem.*, 30 (1958) 1148.

2 P. GRIESS, *Ber.*, 12 (1879) 427; *Z. Angew. Chem.*, 12 (1899) 666.

3 L. ROSENTHALER, *Z. Anal. Chem.*, 108 (1937) 25.

4 B. E. SALTZMAN, *Anal. Chem.*, 26 (1954) 498.

16. Compounds with groups containing nitrogen and oxygen

Test through pyrolytic cleavage of nitrous acid [1]

Organic compounds of all kinds containing groups that include both nitrogen and oxygen bonded together have a common characteristic: they split off nitrous acid when subjected to dry heating. Such compounds include: nitro- and nitroso compounds, oximes, hydroxamic acids, nitrites and nitrates, nitramines, azoxy compounds, amine oxides. The nitrous acid can be readily revealed by the Griess reaction (p. 90). It is probably formed by the initial production of N_2O_3 or N_2O_4 with collaboration of the atmospheric oxygen. However, it should be noted that, when organic materials containing hydrogen and oxygen are pyrolyzed, water results, which can function as superheated steam at the temperature and site of its production. Consequently, nitro- and nitroso compounds can experience a hydrolytic cleavage of nitrous acid, which in general cannot be realized in aqueous solution.

The test for pyrolytic generation of nitrous acid provides a good preliminary test for the compounds mentioned above. If the response is negative, no further tests for N- and O-containing groups are necessary.

The absence of alkali salts of phenols and/or carboxylic acids in the test material is a necessary condition for the validity of the test. In such cases the pyrolysis apparently follows another path. Alkali nitrite or nitrate are formed initially and subsequently reduced to N_2O or NH_3 (compare the remarks on p. 53). Accordingly, if these alkali salts are suspected, it is best to begin the procedure, by taking the sample to dryness with hydrochloric acid, or to eliminate any alkali salts by digestion with water.

The presence of sulfur along with groups containing nitrogen and oxygen is of interest. Some compounds of this type give the Griess reaction on pyrolysis, *e.g.* nitrosulfones, nitrothio acids, *etc.*; but others give no response, such as azo dyes containing NO_2 and SO_3H groups, or disulfides with NO_2 groups.

It appears that in such cases the pyrolytic production of SO_2 or H_2S has an influence in that the nitrous acid is destroyed if these volatile compounds are produced together. If nitrous acid is formed before the SO_2 or H_2S, or if the ratio $-NO_2/-SO_3H$ or $-S-S-$ is greater than 1, there is no interference. This assumption is supported by the finding that, if compounds with a group containing N and O are pyrolyzed along with excess sulfosalicylic acid, the Griess reaction inevitably fails.

Procedure. A micro test tube is supported in a perforated asbestos board. A little of the solid is used, or a drop of the test solution is evaporated to dryness in the test tube. The mouth of the reaction vessel is covered with a disk of filter paper moistened with Griess reagent. The bottom of the test tube is heated with a micro flame. It is often advisable, particularly when testing volatile or sublimable materials, to start the heating just below the asbestos support and gradually progress toward the bottom of the test tube. As a rule it is sufficient to heat for 1–2 min. If the result is positive, a red or pink circular stain appears on the colorless reagent paper.

Griess reagent. For preparation see Sect. 15.

Nitrous acid was given by the following compounds when tested in amounts less than 1 mg:

C-Nitro compounds: 1-nitronaphthalene (0.2 γ), *p*-nitrophenol, *p*-dinitrobenzene, *o*-nitrobenzoic acid, 4-hydroxy-3-nitrobenzoic acid, *p*-nitrobenzyl cyanide, *p*-nitroacetanilide, 3,5-dinitrobenzoic acid, 3-nitroalizarin, 2,4-dinitrobenzenesulfonic acid, 2,7-dinitrophenanthrenequinone, 5-nitrobarbituric acid, chloromycetin, nitromethane, nitroethane, nitropropane, 2,5-dimethyl-2,5-dinitrohexane (13 γ), tris(2-methyl-2-nitropropyl) phosphate, 5-methyl-5-nitro-2-hexanone, 3-methyl-5-nitro-2-hexanone, ethyl γ-methyl-γ-nitrovalerate.

N-Nitro compounds: nitroguanidine (0.1 γ).

C-Nitroso compounds: *p*-nitrosodimethylaniline, *p*-nitrosodiethylaniline, 2-nitroso-1-naphthol (0.3 γ), nitroso R-salt, 5-nitroso-8-hydroxyquinoline (10 γ).

N-Nitroso compounds: N-nitrosodibenzylamine, N-nitrosodicyclohexylamine, dinitrosopiperazine (0.4 γ), N-nitrosodiphenylamine (0.2 γ).

Hydroxamic acids: *p*-methoxybenzhydroxamic acid (0.2 γ), phenylacethydroxamic acid (0.2 γ).

Oximes: α-benzildioxime, diacetylmonoxime, 1,2-cycloheptanedionedioxime, camphoroxime, α-benzilmonoxime, 2-furaldoxime, dimethylglyoxime (0.4 γ), salicylaldoxime (5 γ).

Azoxy- and hydroxylamine compounds: azoxybenzene (15 γ), β-phenylhydroxylamine oxalate.

N-Oxides and nitriles: piperidine N-oxide, isobutyronitrile (0.3 γ).

The procedure given here permits the identification of groups containing N and O in organometallic normal and complex salts. The compounds tested successfully were: Ni- and Pd-dimethylglyoxime, Ni- and Pd-furildioxime, Cu- and Zn-salicylaldoxime, Cu- and Mo-benzoinoxime, Ca-picrolonate, PdCl₂-nitrosodiphenylamine.

1 F. FEIGL and J. R. AMARAL, *Mikrochim. Acta*, (1958) 337; also unpublished studies.

17. Sulfur and nitrogen (joint detection)

Test with calcium oxalate and sodium nitroprusside[1]

As shown on p. 80, hydrogen sulfide is invariably obtained if organic sulfur compounds are pyrolyzed in the presence of calcium oxalate. Calcium oxide is formed from the oxalate and yields ammonia when organic nitrogenous materials are ignited in its presence. (See the lime test described on p. 89.) Consequently, both hydrogen sulfide and ammonia should result when compounds containing both nitrogen and sulfur are ignited in the presence of calcium oxalate, a fact that can be verified by means of the violet coloration with sodium nitroprusside which is characteristic for alkali sulfides. However, this test gives a positive response only with some of the organic compounds under discussion here as shown by the following trials.

Procedure. A micro test tube is inserted through a perforated asbestos board. About 10 mg of the solid specimen is mixed with several cg of calcium oxalate and the mouth of the tube is covered with a piece of filter paper that has been moistened with a 5% solution of sodium nitroprusside. The bottom of the tube is heated to redness and the reagent paper is observed for the development of a violet stain.

The findings are given in the following tabular compilation; the detection limits (in γ) are stated in those instances in which the nitroprusside test was positive. A distinctly better limit was found when sodium oxalate was substituted for calcium oxalate. These latter figures are given in parentheses.

Positive Response	Negative Response
taurine, 100 (50)	sulfanilic acid
cysteine, 25 (25)	naphthionic acid
cystine, 300 (50)	1-amino-2-naphthol-4-sulfonic acid
sulfanilamide, 400 (100)	sulfapyridine
saccharin,	thiodiphenylamine
thiobarbituric acid, 1000 (250)	thiocarbanilide
chloramine T, 1000 (250)	thio-Michler's ketone
diphenylthiocarbazone, 1000 (500)	8-hydroxyquinoline-5-sulfonic acid
thiosemicarbazide, 1000 (150)	2,4-dinitro-1-naphthol-7-sulfonic acid
acetylthiourea, 1000 (150)	diphenylamine-4-sulfonic acid
	methyl orange.

The foregoing exhibit shows that the nitroprusside test fails invariably when the calcium oxalate ignition procedure is applied to compounds containing sulfur and nitrogen if the nitrogen is exclusively aromatically bound. A positive response to the color reaction is given by compounds containing aliphatically bound nitrogen (though aromatically bound nitrogen may also be present) and by amides of aromatic sulfonic acids. The behavior of sulfa drugs is notable; in accord with their general structure $H_2NC_6H_4SO_2NHR$ they are derivatives of sulfanilamide. Fourteen such compounds were tested and with the sole exception of sulfapyridine, which is a derivative of the aromatic aminopyridine, they all gave a positive response.

A plausible explanation for the divergent behavior of the compounds included in the above exhibit may be the assumption that the occurrence of the color reaction requires the concurrent or almost simultaneous pyrolytic production of H_2S and NH_3 so that ammonium sulfide can be formed. This condition is obviously not met with compounds in which the nitrogen is present solely in aromatic binding. Under such circumstances, there is rapid pyrolytic production of hydrogen sulfide in accord with the pattern shown by all sulfur-bearing compounds, whereas the formation of ammonia due to the action of the CaO does not occur at all or to an inadequate extent. The opposite to this condition is found in those cases in which there is approximately simultaneous generation of H_2S and NH_3 as may be true with sulfur-bearing compounds containing non-aromatic bonded nitrogen. This assumption is supported by the fact that the compounds included in the right-hand column of the tabulation, and which show no nitroprusside test when pyrolyzed with calcium oxalate, behave like the compounds in the left-hand column if the pyrolysis is conducted in the presence of ammonia-donors such as urea, biuret, and guanidine.

Therefore the procedure given above is not reliable in all cases for the joint detection of sulfur and nitrogen in organic compounds. It nonetheless provides a very much desired insight into the nature of the nitrogen bonding in such compounds, provided the presence of nitrogen and sulfur has been established by separate tests.

Since the test for ammonium sulfide by nitroprusside is very sensitive, the relatively high identification limits given in the tabulation indicate that a complete simultaneous splitting off of ammonia and hydrogen sulfide is never attained. It is possible that the carbon oxysulfide formed initially when sulfur-bearing compounds are pyrolyzed with calcium oxalate (see p. 80) reacts: $COS + NH_3 \rightarrow HCNO + H_2S$ and consequently the ammonium sulfide requisite to the color reaction does not form.

1 Unpublished studies, with E. LIBERGOTT.

18. Phosphorus

Test by ignition with lime (conversion to phosphate)

Nonvolatile organic compounds containing phosphorus leave heat-resistant tertiary calcium phosphate when ignited with calcium oxide. The phosphate ions obtained by dissolving the ignited residue in acid are precipitated as crystalline yellow ammonium phosphomolybdate by a nitric acid solution of ammonium molybdate:

$$PO_4^{3-} + 12\, MoO_4^{2-} + 3\, NH_4^+ + 24\, H^+ \rightarrow (NH_4)_3PO_4.12\, MoO_3 + 12\, H_2O$$

Traces of ammonium phosphomolybdate can be detected by spotting with benzidine and alkali acetate (see *Inorganic Spot Tests*, Chap 4).[1] An intense blue appears; it is due to a blue quinoidal oxidation product of benzidine ("benzidine blue") and a blue reduction product of hexavalent molybdenum ("molybdenum blue"). The color reaction is brought about as follows: NH_3 molecules of the $(NH_4)_3PO_4.12\ MoO_3$ are exchanged for benzidine molecules. However, benzidinium phosphomolybdate is stable only in mineral acid solution. In contact with alkali acetate or ammonia, the complexly bound molybdenum, which has higher oxidizing power than $MoO_4{}^{2-}$ ions, is reduced by benzidine with production of the two blue reaction products mentioned.

Procedure.   One drop of the test solution is placed on a few mg of calcium oxide in a platinum spoon and evaporated. Alternatively, a few grains of the powdered substance are mixed with lime. The spoon is heated, at first gently, and finally kept for some time at red heat. After cooling, two drops of 2 N nitric acid are added to dissolve the residue. A drop of ammonium molybdate solution is placed on quantitative filter paper followed by the contents of the spoon. After 1–2 min., a drop of benzidine solution is added. The paper is held over ammonia. A blue stain is formed, the intensity depending on the phosphate content.

Reagents: 1) Molybdate solution. 5 g ammonium molybdate is dissolved in 100 ml cold water and poured in 35 ml nitric acid (s. g. 1.2).

2) Benzidine solution. Comp. Sect. 15.

Phosphorus was detected by this procedure in the following quantities of organic compounds. Most of them are derivatives of phosphoric or thiophosphoric acid, and have acquired importance as insecticides[2].

1   γ 1-phenyl-1-chloroethylene-2-phosphonic acid

$$C_6H_5\diagdown\\ \qquad C=CH-P=O\ \ (OH)_2\\ Cl\diagup$$

2.5 γ O,O-diethyl-O-*p*-nitrophenyl thiophosphate (Parathion)

$$O_2N-C_6H_4-O-P=S\ (OC_2H_5)_2$$

5   γ octamethylpyrophosphoramide

$$(CH_3)_2N\diagdown\qquad\qquad\diagup N(CH_3)_2\\ \qquad\quad O=P-O-P=O\\ (CH_3)_2N\diagup\qquad\qquad\diagdown N(CH_3)_2$$

8   γ diethyl *p*-nitrophenyl phosphate (Para-oxon)

$$O_2N-C_6H_4-O-P=O\ (OC_2H_5)_2$$

3   γ tetraethyl dithiopyrophosphate (Sulfotepp)

$$C_2H_5O\diagdown\qquad\qquad\diagup OC_2H_5\\ \qquad\quad S=P-O-P=S\\ C_2H_5O\diagup\qquad\qquad\diagdown OC_2H_5$$

Organic compounds containing phosphorus, which are volatile or which sublime when heated, yield no or only insignificant amounts of calcium phosphate when carried through the above procedure. In such cases the test material must be oxidatively decomposed with conc. sulfuric acid to give phosphoric acid. A procedure of this kind permits the detection of phosphorus in microgram amounts of all kinds of compounds.[3]

1 F. FEIGL, Z. Anal. Chem., 61 (1922) 454; 74 (1928) 386; 77 (1929) 299.
2 Comp. S. A. HALL, Advances in Chemistry, Series 1 (1950) 150; S. A. HALL et al., Anal. Chem., 23 (1951) 1830; 23 (1951) 1866.
3 C. M. WELCH and Ph. W. WEST, Anal. Chem., 29 (1957) 874.

19. Arsenic

Test by ignition with lime (conversion to arsenate)

All nonvolatile organic compounds containing arsenic leave heat-resistant tertiary calcium arsenate when ignited with calcium oxide.

The arsenic(V) in calcium arsenate can be detected by procedures I–III, whose chemical bases are discussed in a–c.

(a) When heated above 300°, sodium formate yields hydrogen which in its nascent state transforms As(V) into arsenic hydrides[1]:

$$2 \text{ HCOONa} \rightarrow \text{Na}_2\text{C}_2\text{O}_4 + 2 \text{ H}^0$$

$$\text{Ca}_3(\text{AsO}_4)_2 + 16 \text{ H}^0 \rightarrow 2 \text{ AsH}_3 + 3 \text{ CaO} + 5 \text{ H}_2\text{O}$$

The gaseous arsine reacts with silver nitrate to give metallic silver[2]:

$$\text{AsH}_3 + 6 \text{ AgNO}_3 + 3 \text{ H}_2\text{O} \rightarrow 6 \text{ Ag}^0 + 6 \text{ HNO}_3 + \text{H}_3\text{AsO}_3$$

(b) Red-brown Ag_3AsO_4 is produced by spotting $\text{Ca}_3(\text{AsO}_4)_2$ with silver nitrate solution. The test is far less sensitive than the test based on the formation of AsH_3. It should also be noted that halogen-bearing organic compounds on ignition with lime yield calcium halides which react with silver nitrate, and that the resulting AgCl and AgBr darken on exposure to light (photo halide formation) which makes it difficult to detect small amounts of Ag_3AsO_4.

(c) Dilute sulfuric acid and alkali iodide react with calcium arsenate:

$$\text{AsO}_4^{3-} + 2 \text{ I}^- + 2 \text{ H}^+ \rightarrow \text{AsO}_3^{3-} + \text{H}_2\text{O} + \text{I}_2$$

The free iodine can be detected by the starch test. When applying the redox reaction with iodide, it should be remembered that antimony-bearing organic compounds leave antimony pentoxide or calcium antimonate after ignition with lime, and these products likewise set iodine free. The same is true when the ignition residue contains ferric oxide.

Procedure I.[3] A few grains of the powdered substance are placed in a micro test tube of hard glass, or alternatively a drop of the test solution is evaporated to dryness there. Several mg of ignited lime is mixed in, and the tube is heated gently at first and finally to redness for a few minutes. After cooling, a 10–15 fold excess of sodium formate is added, and the open end of the test tube is covered with a disk of filter paper moistened with 10% silver nitrate solution. The tube is heated until its contents turn slightly brown. A positive response is shown by the formation of a dark or brown circle on the paper.

Overheating must be avoided because the pyrolysis products of sodium formate can react with silver nitrate and produce faintly brown circles on the paper. In contrast to the metallic silver formed with calcium arsenate, these products are not resistant to 15% hydrogen peroxide.

Arsenic was revealed in:

5 γ chloroarsanilic acid   4 γ sodium p-hydroxyphenylarsonate
5 γ dinitroarsanilic acid   2 γ salvarsan
5 γ sodium acetylarsanilate

1 C. A. VOURNASOS, *Ber.*, **43** (1910) 2269.
2 M. GUTZEIT, *Pharm. Ztg.*, **24** (1879) 263.
3 E. JUNGREIS (Jerusalem), unpublished studies.

Procedure II. One drop of the test solution is added to about 1 mg CaO in a microcrucible and evaporated to dryness at 110°. Alternatively, a few grains of the powdered sample are mixed with the lime. The crucible is then heated strongly, and a drop of $AgNO_3$ is added to the cold residue. Depending on the arsenic content, red-brown Ag_3AsO_4 separates at once or after several minutes.

Reagent: 7% solution of silver nitrate in 6 N acetic acid.

This test revealed arsenic in 60 γ nitrophenylarsonic acid, 20 γ salvarsan.

Procedure III.[1] A mixture of the test substance and calcium oxide is prepared and ignited as described in Procedure II. The cooled ignition residue is treated with a drop of 1 : 1 sulfuric acid and the contents of the crucible cooled by setting it in cold water. A drop of cadmium iodide–starch solution* is then introduced. A blue color that appears within 30 sec. indicates arsenic.

Reagent: 5% solution of cadmium iodide in 1% solution of cadmium sulfate. Starch or solid thyodene indicator is added just before the test.

Arsenic was revealed in 1.4 γ phenylarsonic acid, 1.7 γ nitrophenylarsonic acid.

In the case of aliphatic and aromatic arsonic acids, which in accord with the general formula $(R,Ar)AsO(OH)_2$ contain quinquevalent arsenic, no calcination is needed to produce calcium arsenate. These acids, analogously to arsenic acid, react directly with acidified alkali iodide solutions:

$$(R,Ar)AsO(OH)_2 + 2\ HI \rightarrow (R,Ar)As(OH)_2 + I_2 + H_2O$$

with probable intermediate formation of R—AsI_2.[3] The attainable limits of detection are the same as those obtained by the calcination procedure.

* The use of cadmium iodide solution in place of the usual potassium iodide solution has the advantage that, because of the lower dissociation of CdI_2, the release of small quantities of iodine through dissolved oxygen and traces of N_2O_3 is much slower[2].

The tests described here for 5-valent organically bound arsenic of course cannot be applied in the presence of arsenic acid or arsenate. The addition of magnesia mixture to the ammoniacal solution of the sample will give a precipitate of $MgNH_4AsO_4$ if these are present. Organic derivatives of arsenic acid yield no precipitate.

Organic compounds of trivalent arsenic can be converted into arsenic acids by evaporation with hydrogen peroxide but the detection of quinque-valent arsenic compounds, formed in this manner, by the addition of acidified potassium iodide solution is not reliable. It has been found[4] that many organic compounds of the most varied kinds when evaporated with hydrogen peroxide leave products which, often even after heating to 200°, liberate iodine from acidified alkali iodide solutions.

The greater part of the organic arsenic compounds are free aliphatic or aromatic arsonic acids or their salts for which a reliable test in the wet way is described on p. 308.

1 Unpublished studies, with W. A. MANNHEIMER.
2 P. ARTHUR, T. E. MOORE and J. LAMBERT, *J. Am. Chem. Soc.*, 71 (1949) 3260.
3 J. GOLSE, *Bull. Soc. Pharm. Bordeaux*, 67 (1929) 84.
4 Unpublished studies, with R. MOSCOVICI.

20. Antimony

When organic compounds containing antimony are ignited they leave Sb_2O_5 and Sb_2O_4, which may readily be detected by (*1*) or (*2*) without interference from arsenic if this is present.

(*1*) Test with rhodamine B

Antimony pentoxide and tetroxide are dissolved by strong hydrochloric acid in the presence of excess alkali iodide. Iodine is set free and SbI_3 or $H[SbI_4]$ is formed. This complex acid gives a red-violet water-insoluble salt with the basic dyestuff rhodamine B. The chemistry of this sensitive test and the procedure, which has an *identification limit* of 0.6 γ antimony when conducted as spot reaction, is discussed on p. 677.

Organic arsenic compounds do not interfere. If organic bismuth or gold compounds are present, the test for antimony should be conducted according to (*2*), because under the conditions of the test there is production of $H[BiI_4]$ and $H[AuI_4]$, which behave like $H[SbI_4]$.

(*2*) Test with diphenylamine or diphenylbenzidine[1]

Diphenylamine or N,N-diphenylbenzidine, dissolved in concentrated sulfuric acid, is oxidized by antimony pentoxide to blue quinimine dyestuffs.

The chemistry of the color reaction is the same as that of the oxidation of these amines by nitric acid (see *Inorganic Spot Tests*, Chap. 4).

When testing for antimony in organic compounds, a small sample is ashed in a micro crucible and the residue is treated with a freshly prepared 1% solution of diphenylamine or diphenylbenzidine in concentrated sulfuric acid. A blue color appears at once or within several minutes if antimony is present. The *limit of identification* of this procedure is 5 γ antimony.

The test may not be applied directly if the ignition residue contains alkali or alkaline-earth nitrates even in small amounts. These compounds can be completely destroyed by evaporation with concentrated formic acid and subsequent ignition. Nitrates and nitrites are thus quantitatively converted into carbonates,[2] whereas Sb_2O_5 remains unaltered.

1 R. FRESENIUS, *Qualitative Analyse*, 19th ed., 1918, p. 336.
2 F. FEIGL and A. SCHAEFFER, *Anal. Chim. Acta*, 7 (1952) 507.

21. Silicon

Test by oxidative destruction with perchloric acid[1]

Organic compounds can be decomposed quickly and completely by warming with concentrated perchloric acid (see p. 88). Compounds which in addition to C, H, O, N, contain metals or nonvolatile nonmetals, produce residues containing the former as chlorides and the latter as water-soluble oxygen acids (compare test for Se, p. 88). Organic silicon compounds, which have recently attained much commercial importance, produce water-insoluble silicic acid when heated with concentrated perchloric acid. This fact serves as the basis for the test described here for organically bound silicon. The test is unequivocal in the absence of water-soluble alkaline-earth salts of organic sulfonic acids which yield water-insoluble alkaline-earth sulfates when oxidatively decomposed.

Procedure.[2] A small quantity of the solid test material or a drop of its solution is placed in a micro test tube and after evaporation (if need be) two drops of concentrated perchloric acid are added. The tube is immersed in a glycerol bath which has been preheated to 150° and the temperature is slowly (about 10 min.) raised to 210°. After cooling, about 1 ml of water is added and the liquid is carefully watched for the presence of a white residue. Small amounts of SiO_2 can be detected by shaking the water suspension with ether; the residue gathers in the H_2O–ether interface (flotation).

Limit of identification: about 100 γ silicon.

1 H. GILMAN, R. N. CLARK and R. WILEY, *J. Am. Chem. Soc.*, 68 (1946) 2728; 76 (1954) 918.
2 Unpublished studies, with S. OFRI and M. M. POLLAK.

22. Mercury

(1) Test by demasking of potassium ferrocyanide[1]

The tendency of mercuric ions to form water-soluble undissociated mercuric cyanide is so great that they are able to liberate ferrous ions from ferrocyanide ions. If this demasking occurs in the presence of α,α'-dipyridyl, a red color appears because of the formation of $[Fe(\alpha,\alpha'\text{-Dip})_3]^{2+}$ ions. (Compare *Inorganic Spot Tests*, Chap. 3 and 4.) The underlying reactions are:

$$Fe(CN)_6{}^{4-} + 3\ Hg^{2+} \rightarrow 3\ Hg(CN)_2 + Fe^{2+} \tag{1}$$

$$Fe^{2+} + 3\ \alpha,\alpha'\text{-Dip} \rightarrow [Fe(\alpha,\alpha'\text{-Dip})_3]^{2+} \tag{2}$$

Mercury salts of organic acids and likewise compounds in which mercury is bound to carbon are capable of demasking ferrocyanide ions.

A demasking analogous to that in (1) and (2) can be effected also by Ag^+ and Pd^{2+} ions (see *Inorganic Spot Tests*, Chap. 3). The test described here is consequently reliable only in the absence of organic silver and palladium compounds. In practice, only silver compounds are likely to be encountered and then but rarely.

Procedure.　A micro test tube is used. A drop of the test solution or a little of the solid is treated with one drop each of a 1% solution of potassium ferrocyanide, 2 N ammonia, and 1% alcoholic solution of α,α'-dipyridyl. The mixture is warmed on the water bath. Depending on the quantity of mercury present, a red or pink color will appear at once or within a few minutes.

Limit of Identification: 2 γ mercury.

1 F. FEIGL and A. CALDAS, *Anal. Chim. Acta*, 13 (1955) 526.

(2) Test by volatilization of mercury

Mercury compounds of all varieties when heated to dull red heat leave no residue or this contains no mercury. This behavior, which is unique among metal compounds, is due to the sublimation of mercury salts and to thermal decomposition with production of mercury vapor. The latter can be detected by the sensitive test with palladium chloride[1] in which the reaction

$$PdCl_2 + Hg^0 \rightarrow Pd^0 + HgCl_2$$

yields highly divided black free palladium. If filter paper, moistened with palladium chloride, comes into contact with mercury vapor, the brown-yellow reagent paper turns black (Procedure I).

Procedure II is based on the production of red cuprous tetraiodomercurate by the action of mercury vapor on insoluble white cuprous iodide:[2]

$$2\ Cu_2I_2 + Hg^0 \rightarrow Cu_2(HgI_4) + 2\ Cu^0$$

When organic mercury compounds are heated, mercury vapor is formed mostly through thermal dissociation. To detect this mercury vapor by means of $PdCl_2$ or Cu_2I_2, steps must be taken to prevent the formation of carbon monoxide and volatile tarry products when the organic material burns. CO likewise reduces $PdCl_2$ to the metal, and tarry products can collect on the reagent paper to form dark stains. Both of these interferences can be obviated if the ignition is made in mixture with cupric oxide.[3]

Certain organic mercury compounds produce some ethylene on ignition, and this cannot be prevented even by mixing the sample with copper oxide. Ethylene likewise reduces palladium chloride to free palladium.[4] Obviously, this would be no handicap in the detection of mercury vapor; rather it would be an advantage since more palladium chloride would be reduced under such circumstances. However, it is possible (though not yet proved) that some ethylene is produced when mercury-free compounds burn. In such cases, the reduction of $PdCl_2$ by C_2H_4 would simulate the presence of mercury vapor. Consequently, when in doubt, it is better to trap the mercury vapor initially on gold (wire, foil). Heating this amalgam will then drive the mercury out.

Procedure I.[5] A little of the test material, mixed with a few cg of CuO, is placed in a pyrex ignition tube (20 mm long, 10 mm diameter) and covered completely with CuO. The tube is held in a hole drilled through an asbestos board. The latter rests on a tripod. The mouth of the tube is covered with a disk of filter paper, moistened with 1% aqueous $PdCl_2$. The tube is heated cautiously by gradually raising a low flame, which finally is increased until the bottom of the tube is red hot. A dark stain on the paper indicates the presence of mercury. In case of doubt, the paper should be held over ammonia which discharges the yellowish brown color of the paper because of the formation of $[Pd(NH_3)_2]Cl_2$, the stain then becomes more obvious. Care must be taken to prevent the top of the tube from becoming hot; otherwise the $PdCl_2$ may be reduced by the paper itself.

When only traces of mercury are expected or if there is a chance that ethylene may be evolved, crushed gold leaf should be inserted in the upper part of the ignition tube, which may be provided with a constriction about 1/3 from the bottom to support the foil. The substance is pyrolyzed for about 2 min. but without warming the gold leaf, which therefore should be far enough above the asbestos board. The tube is then pushed down so that the gold leaf can now be heated with the full flame for a short time. The reagent paper should be applied to the mouth of the tube only during the second heating period.

A positive response was given by:

1.75 γ acetoxymercuryethyl acetate	2.5 γ bis(camphor-10)mercury
2.75 γ bis(3-chlorocamphor-10)mercury	2.5 γ camphor-10-mercury iodide
2.25 γ aceto-(2-chloromercuryethyl)mesidide	
2.0 γ bromomercuryethylpyridine bromomercurate	
5.0 γ iodomercuryethylpyridine iodomercurate	

These *identification limits* correspond (with one exception) to about 1γ Hg. The lower sensitivity in the case of iodomercuryethylpyridine iodomercurate may be attributed to the volatilization of a part of the mercury as HgI_2 which does not react with $PdCl_2$. Because of the formation of mercury halides, the test may be less sensitive, or even fail completely for small quantities of mercury, when larger quantities of organic and inorganic halogen compounds are present. In such cases it is better to use Procedure II, employing solid cuprous iodide which reacts not only with dissolved mercury salts,[6] but also with vapors of sublimable mercury halides. Red cuprous tetraiodomercurate results:

$$2\,Cu_2I_2 + HgHal_2 \rightarrow Cu_2[HgI_4] + Cu_2(Hal)_2$$

When organic mercury compounds are ground with dry Cu_2I_2 or with this salt moistened with water or alcohol, they react. A general test for organic mercurials probably could be developed from this observation.

Procedure II.* The method described in Procedure I is used for the destruction of the mercury-bearing test substance. Paper carrying a drop of Cu_2I_2 paste is used in place of $PdCl_2$ paper. A salmon or red color develops on the paper when exposed to mercury vapor or volatilized mercuric halides.

　　Reagent: Cuprous iodide paste. A solution of 5 g $CuSO_4.5\,H_2O$ in 75 ml water is mixed with a solution of 5 g Na_2SO_3 and 12 g KI in 75 ml water. The white precipitate is centrifuged or filtered, washed well with water, and stored moist. Just before the test, some of it is stirred to a thin slurry with water.

A positive response was given by amounts of the various test materials containing 4–8 γ mercury.

　　* After unpublished studies by H. E. FEIGL with partial modification of the procedure described by I. STONE (*loc. cit.*, ref. 2).

1 A. MERGET, *Compt. Rend.*, **73** (1871) 1356.
2 I. STONE, *Ind. Eng. Chem., Anal. Ed.*, **5** (1933) 220.
3 F. FEIGL, *J. Chem. Educ.*, **22** (1945) 344.
4 F. C. PHILIPS, *Z. Anorg. Chem.*, **6** (1894) 237.
5 G. SACHS, *Analyst*, **78** (1953) 185.
6 P. ARTMANN, *Z. Anal. Chem.*, **60** (1921) 81.

23. Metals

A direct detection of metallic elements in organic compounds is possible provided the test material is soluble in water to at least a slight extent, or if it can be appreciably decomposed by dilute acids or alkali hydroxides. When this is not the case, the test for metals must be preceded by destruction of the organic skeleton by ignition or by oxidizing acids. Sometimes, losses of material are inevitable when organometallic compounds are ignited. Certain materials (especially diazo-, nitro- and nitroso compounds) explode when heated. Others (particularly inner-complex salts) sublime to a considerable extent prior to their thermal oxidative fission. In some cases, complete ashing is difficult because carbon particles become coated with sintered metal carbonate or oxide which is formed during the ignition of the test material. The residue must be dissolved in dilute acids before undertaking the tests for metals. If the residues are not soluble in acids, fusion with sodium carbonate or alkali bisulfate is necessary. The difficulties which were pointed out above do not arise if the sample is oxidatively disintegrated by digestion with hot concentrated nitric acid (CARIUS) or sulfuric acid (KJEL-DAHL). If the action of these acids is prolonged sufficiently, water-clear solutions of the metal nitrates and sulfates, respectively, will be obtained. These procedures have the disadvantage that they introduce large amounts of concentrated acids, which must be eliminated before further examination.

A rapid process for the wet decomposition of nonvolatile organometallic (and purely organic) materials is to heat them with dilute sulfuric acid and potassium persulfate in the presence of silver salts[1]. The hydrolysis: $H_2S_2O_8 + H_2O \rightarrow 2 H_2SO_4 + O$ is catalyzed by silver ions[2], and the atomic oxygen can bring about the oxidative decomposition[3]. It is not certain whether the rupture of the organic skeleton proceeds so far in all cases that the entire carbon content is oxidized to carbon dioxide[*]. However, it appears that, with few exceptions, the disintegration is so extensive that the metals contained in organic compounds are converted to sulfates. Exceptions are: Co(III)-α-nitroso-β-naphthol, Sb(III)-pyrogallate and Cu-alizarin blue, which remain completely unaltered. As a rule, metal salts of thiols leave elementary sulfur when treated with persulfuric acid (activated by Ag+ ions).

The best method for the wet decomposition of organometallic compounds in spot test analysis is heating with perchloric acid[5] whose azeotropic mixture with water (72.5% $HClO_4$) boils at 203° and hence permits reaction at elevated temperatures. It is so powerful under these conditions that thiol compounds yield sulfate rather than free sulfur (as with persulfate). Furthermore, the metal salts noted above are completely disintegrated. Consequently the treatment with perchloric acid with subsequent heating at 210° (to remove the excess acid) can be highly recommended for the complete destruction of organometallic compounds. The pertinent procedure is given on

---

[*] Sometimes compounds may be formed that are resistant to activated persulfuric acid to a greater or lesser extent. An instance of this kind is the formation of maleic acid by warming hydroquinones and quinones with a sulfuric acid solution of silver sulfate and alkali persulfate[4].

p. 88. After treatment with perchloric acid and dilution, the resulting solution can be readily examined for metals. When organometallic compounds are oxidatively disintegrated, it should be remembered that water-insoluble inorganic metal salts are produced in some cases. For instance, organic silver compounds yield silver chloride, and organic compounds of tungsten give WO_3. Lead and alkaline-earth salts of organic sulfonic acids leave the corresponding insoluble sulfates. See p. 101, regarding the behavior of organic compounds containing silicon.

1 F. Feigl and A. Schaeffer, *Anal. Chim. Acta*, 4 (1950) 458.
2 H. Marshall, *Chem. News*, 83 (1901) 76; *Z. Anal. Chem.*, 43 (1904) 4181.
3 R. Kempf, *Ber.*, 38 (1905) 3963.
4 R. Kempf, *Ber.*, 39 (1906) 3717, 3726.
5 G. F. Smith, *Anal. Chim. Acta*, 8 (1953) 397.

24. Tests for basic or acidic behavior

The basic or acidic character of organic compounds is due to the presence and activity of certain elements in particular bonding forms which are responsible for the acceptance or delivery (consumption and production) of hydrogen ions (protons) or for the closely related process of the formation of salts. Salt-forming groups are situated, as a rule, at definite locales in the molecule of organic compounds. In some cases, compounds which are neutral in their normal structure can assume an acidic salt-forming character through tautomeric rearrangement or through the action of OH^- ions (keto–enol equilibrium, thioketo–thioenol equilibrium, nitro–nitrosyl acid equilibrium).

Practically all stable basic organic compounds are derivatives of ammonia, hydroxylamine or hydrazine.* The salts with strong mineral acids correspond to the ammonium-, hydroxylaminium-, and hydrazinium salts. With few exceptions, these salts are water-soluble. It is of practical importance that the hydrochlorides of all organic nitrogen bases are quite soluble in water if an excess of hydrochloric acid is present. The strength of organic nitrogen bases, with respect to the binding of H^+ ions or the delivery of OH^- ions, depends on the molecular remainder bound to the nitrogen atom. For example, NH_2- and NH groups bound to alkyl radicals are strongly basic, whereas they are weakly basic when bound to aryl radicals. If these groups are directly linked to a CO group, the products are neutral, acid-insoluble compounds (amides and anilides). Cyclic and noncyclic compounds containing NH groups adjacent to a CO or CS group function as acids. This is likewise

---

* A few compounds, such as pyrones, and the anthocyanidin pigments of some flowers, possess oxygen atoms with sufficient basicity to cause them to dissolve in dilute hydrochloric acid and to form iodomethylates (ethylates) by adding CH_3I or C_2H_5I.

true of sulfonamides which contain the SO_2NH_2- or SO_2NHR group.

Hydrogen atoms which can be split off as hydrogen ions, or which can be replaced by an equal number of metal equivalents to form salts, are essential to the acidic character of organic compounds. When hydrogen atoms are linked directly to carbon, they exhibit this faculty only in acetylene and its derivatives, which contain the —C≡CH group.* In general, acidic H atoms are contained in OH-, SH-, and NH groups, which are linked to carbon either directly of through an intermediate nonmetallic atom:

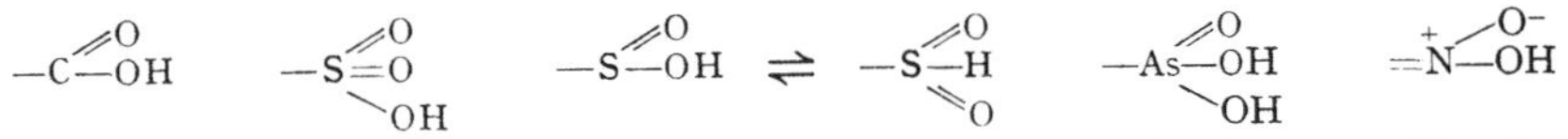

| Carboxylic | Sulfonic | Sulfinic | Arsonic | Nitroxylic |

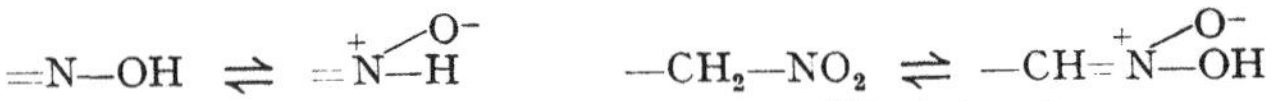

Oximic   Nitro (primary)

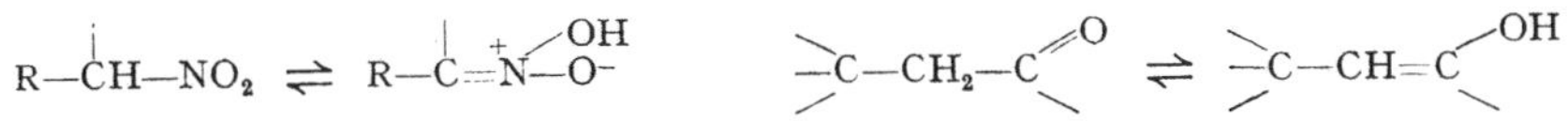

Nitro (secondary)   Enolic

Phenolic (enolic)   Thiophenolic   Thioenolic

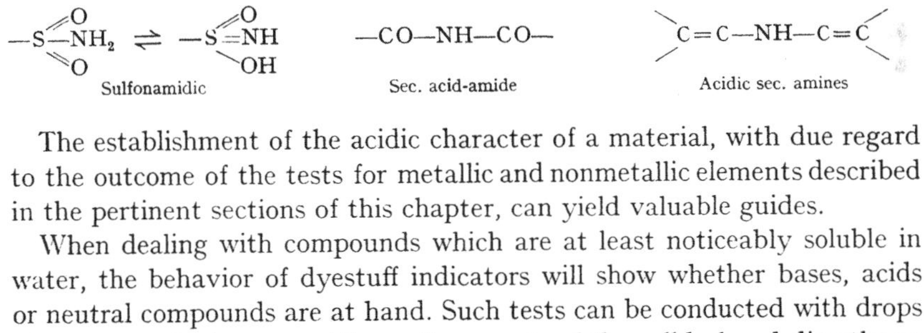

Sulfonamidic   Sec. acid-amide   Acidic sec. amines

The establishment of the acidic character of a material, with due regard to the outcome of the tests for metallic and nonmetallic elements described in the pertinent sections of this chapter, can yield valuable guides.

When dealing with compounds which are at least noticeably soluble in water, the behavior of dyestuff indicators will show whether bases, acids or neutral compounds are at hand. Such tests can be conducted with drops of aqueous solutions, or with small amounts of the solid placed directly on moistened indicator papers, or on spot plates with microdrops of suitable

* Mercury occupies a special position; it has a marked tendency to form a stable binding with carbon atoms and so it can take the place of hydrogen in many aliphatic and aromatic compounds (so-called mercuration). See in this connection F. C. WHIT-MORE, *Organic Compounds of Mercury*, The Chemical Catalog Co., New York, 1921. Concerning organometallic compounds in general, see E. G. ROCHOW, D. T. HURD and R. N. LEWIS, *The Chemistry of Organometallic Compounds*, Wiley, New York, 1957; H. C. KAUFMAN, *Handbook of Organometallic Compounds*, Van Nostrand, Princeton, 1961; G. E. COATES, *Organo-Metallic Compounds*, Methuen, London, 1961; H. ZEISS, *Organometallic Chemistry*, 1960.

indicator solutions. Congo red is recommended for acidic compounds; bromothymol blue or phenolphthalein for basic compounds. Four kinds of mixed indicators have been proposed[1] for classifying organic compounds, according to the pH, into the following classes: strong acids, intermediate acids, ampholytes, neutral compounds, weak bases, intermediate bases. The orientation in this respect is excellent, provided the directions for the preparation of the indicator solutions are followed.

An indication of salt-forming groups in water-insoluble organic compounds is given by their behavior with dilute hydrochloric acid or dilute caustic alkali. Basic compounds are thus dissolved as the chlorides, and acidic compounds as the corresponding alkali salts.* These tests can be accomplished with tiny amounts of the sample on a watch glass by adding a drop of acid or alkali, warming gently if necessary. It is well to carry out the treatment with acid or base in a micro-centrifugetube, and to segregate any undissolved material by centrifuging. Ammonia or dilute hydrochloric acid should then be added drop by drop to the clear liquid until a distinct alkaline or acid reaction is attained. If a precipitate or turbidity appears, it indicates the presence of water-insoluble basic or acidic compounds. When a water-insoluble material dissolves in hydrochloric acid as well as in alkali, an ampholyte is at hand, i.e., a compound whose molecule contains both basic and acidic salt-forming groups.

The foregoing tests for basic and acidic compounds are adequate for orientation purposes. Even γ quantities of water-soluble materials can produce plainly visible color changes of indicators; at least 0.2–2 mg of solid samples are required for securing distinct effects in the solubility tests in acid or alkali. Other micro and semimicro procedures for testing the basic or acidic character of organic compounds are given in the following sections.

---

* There are notable exceptions to this rule. R. ADAMS, *J. Am. Chem. Soc.*, **41** (1919) 247, showed that many substituted phenols are not soluble even in boiling 10% caustic alkali; he attributed this to the great water-insolubility of the particular phenols. The same is true of some sulfonamides.[2] Obviously the ability of bases and acids to dissolve in aqueous acid or alkali depends not alone on their acid or base strength but also on their ability to dissolve in water as an unionized molecule (compare ref. 1).

1 D. DAVIDSON, *J. Chem. Educ.*, **19** (1942) 221, 532.
2 C. S. MARVEL and F. E. SMITH, *J. Am. Chem. Soc.*, **45** (1923) 2696.

(1) Tests for basic compounds with nickel dimethylglyoxime- or zinc 8-hydroxyquinoline-equilibrium solutions[1]

When dimethylglyoxime (DH_2) is added to aqueous solutions of a nickel salt, the resulting red nickel dimethylglyoxime, being readily soluble in dilute mineral acids, is only partly precipitated because of the immediate establishment of the equilibrium:

$$Ni^{2+} + 2\ DH_2 \rightleftarrows Ni(DH)_2 + 2\ H^+$$

If the precipitate is removed by filtration, the filtrate is a saturated solution of nickel dimethylglyoxime and contains Ni^{2+} and H^+ ions at concentrations corresponding to the equilibrium. If this solution is placed in contact with H^+ ion-consuming (*i.e.* basic) materials, the equilibrium is disturbed by the removal of H^+ ions, and red nickel dimethylglyoxime precipitates to reestablish the equilibrium. Therefore, this test will reveal the presence of basic materials which are soluble or insoluble in water (Procedure I).

Another equilibrium solution responding to organic bases can be prepared by treating a solution of a zinc salt with excess 8-hydroxyquinoline (HOx) and filtering. Because of the establishment of the equilibrium

$$Zn^{2+} + 2\ HOx \rightleftarrows Zn(Ox)_2 + 2\ H^+$$

the light-yellow zinc oxinate is only partially precipitated. If the clear filtrate is brought into contact with organic bases, H^+ ions are withdrawn and zinc oxinate precipitates. Traces of this salt, which are too slight to be discernible as a turbidity, are clearly revealed by the strong yellow-green fluorescence in ultraviolet light (Procedure II).

Employment of these procedures, which are based on disturbing defined equilibria, assumes the absence of inorganic basic compounds and also of alkali and alkaline-earth salts of weak acids which react basic because of hydrolysis. Organic nitrogen bases can be liberated from aqueous solutions of their salts by alkalization with ammonia, and then extracted with ether, chloroform, *etc.* Ammonia should be used* because it is completely removed when the organic solvent is volatilized and the residue heated to 110°. When dealing with organic materials, which burn without residue or whose ignition residues contain no oxides or carbonates, a positive response to the tests described below is a strong indication of the presence of organic bases.

Procedure I. A drop of the aqueous test solution or the residue from the evaporation of a drop of a solution of nonvolatile bases in alcohol, ether, *etc.*, is mixed with a drop of equilibrium solution on a spot plate. If basic materials are present, a red crystalline precipitate appears, either immediately or after a transient yellow color, depending on the quantity of basic materials involved.

Reagent: Nickel dimethylglyoxime-equilibrium solution. 2.3 g nickel sulfate, dissolved in 300 ml water, is united with 2.8 g dimethylglyoxime dissolved in 300 ml alcohol. The precipitate is filtered off.

---

* When aqueous solutions of caustic alkali are shaken out with ether, the evaporation residue of the ethereal extract shows an alkaline reaction toward acid–base indicators. This is due to the distinct solubility of water in ether with consequent entrance of the alkali into the ether.

This procedure revealed the basic behavior of:

5 γ ethylenediamine	15 γ p-phenylenediamine
20 γ diethanolamine	10 γ benzidine
20 γ α-naphthylamine	15 γ tetrabase.

Procedure II. A drop of the test solution is taken to dryness in an Emich tube. The cooled residue is treated with a drop of the equilibrium solution and viewed under the quartz lamp. In the presence of nonvolatile organic bases, the residue exhibits a yellow-green fluorescence. If desired, the system may be diluted with water following the action of the equilibrium solution. Even traces of the resulting fluorescent zinc oxinate can then be seen.

> *Reagent:* Zinc oxinate-equilibrium solution. Equal volumes of 1% zinc chloride solution and 1% water–alcohol solution of 8-hydroxyquinoline are mixed. The precipitate is removed.

This procedure revealed the basic behavior of:

10 γ ethylenediamine	15 γ p-phenylenediamine
15 γ diethanolamine	5 γ benzidine
15 γ α-naphthylamine	10 γ tetrabase.

Sometimes, it is desired merely to establish the presence of basic compounds, without distinguishing between their organic or inorganic natures. In such cases it is sufficient to stir a drop of the equilibrium solution with a little of the test material on a spot plate, and to note whether a yellow-green fluorescence appears in ultraviolet light.

1 F. FEIGL, Cl. COSTA NETO and J. E. R. MARINS, unpublished studies. See also F. FEIGL and C. P. J. DA SILVA, *Ind. Eng. Chem., Anal. Ed.*, 14 (1942) 316; L. VELLUZ and M. PESEZ, *Ann. Pharm. Franç.*, 4 (1946) 10.

(2) Test for basic compounds through fixing of hydrogen chloride[1]

If strong organic nitrogen bases are taken to dryness with dilute hydrochloric acid, the respective chlorides are obtained. The latter show the same behavior as ammonium chloride, *i.e.*, they can be heated to the evaporation temperature (120°) of aqueous hydrochloric acid without undergoing decomposition. The water-soluble chloride in the evaporation residue can then be detected by adding silver nitrate.

Very weak bases do not seem to form a heat-stable union with hydrochloric acid; theobromine and α,α'-dipyridyl are typical examples. However, other factors are also responsible for the heat-stability of solid chlorides of organic bases, particularly the duration of the heating. This is shown by the relatively high identification limit of pyridine. Amphoteric compounds in which the basic character is dominant, glycine for instance, behave like the organic bases. Amphoteric compounds which produce zwitter ions (sulfanilic acid, naphthionic acid, *etc.*) and hence have the character of ammonium salts, do not bind hydrochloric acid, or the salts decompose with evolution of the acid, when kept at 120° for a short time.

The detection of organic bases through binding of hydrochloric acid cannot be applied in the presence of alkali- and ammonium salts of organic carboxylic- and sulfonic acids, phenols, amides and imides of acids, since they form alkali- or ammonium chloride when evaporated with hydrochloric acid. In such cases it is necessary to separate the bases beforehand by treating the sample with a solution of alkali hydroxide and then shaking out with ether. Acidic compounds remain in the water layer as alkali salts, while the organic bases pass into the ether.

The acetic acid solutions of organic bases can also be used for their detection. When such solutions are evaporated with hydrochloric acid, acetic acid is volatilized and the chlorides of the bases are left.

Procedure. One drop of the test solution or a tiny quantity of the solid (*e.g.*, the residue after evaporating a drop of an ether solution) is placed in a depression of a black spot plate. One drop of dilute hydrochloric acid is added, the mixture is taken to dryness, and the plate is kept for 2–5 min. in an oven heated to 120°. After cooling, one drop of 5% silver nitrate solution is stirred in. A precipitate or turbidity (sometimes only after several minutes) indicates the presence of basic compounds.

This procedure revealed:

2	γ ethylenediamine	5	γ glycine
2.5	γ benzidine	2.5	γ cysteine
5	γ α-naphthylamine	25	γ sulfanilamide
25	γ o-phenanthroline	5	γ sulfathiazole
50	γ caffeine	5	γ 8-hydroxyquinoline

No reaction was shown below 500 γ by: theobromine, pyridine, α,α'-dipyridyl.

Chlorides of nonvolatile bases, weaker than ammonia (as well as the amphoteric sulfanilic and naphthionic acids) behave like free acid when heated to 140° with sodium thiosulfate; this means that sulfur dioxide is formed which can be detected in the gas phase through the bluing of Congo paper moistened with hydrogen peroxide. In this way the presence of pertinent bases (in γ amounts) can be recognized. This test is not directly applicable in the presence of acids. Furthermore, compounds which contain a CH_2Hal group as well as alkyl esters of noncarboxylic acids must be absent, because compounds of this kind for other reasons release SO_2 when dry heated with sodium thiosulfate (see pp. 171 and 311).

1 Unpublished studies, with V. GENTIL.

(3) Detection of volatile organic bases

The general tests for basic compounds with equilibrium solutions (p. 109) become selective for those organic bases which volatilize on warming or which are evolved to a considerable extent with steam. Such bases, which usually are of medium strength, react in the vapor phase with the equilibrium solutions and precipitate red nickel dimethylglyoxime, or zinc oxinate which exhibits a yellow-green fluorescence. The bases in the present category include primary, secondary, and tertiary aliphatic amines with low carbon content which are volatile at room temperature; aniline; and also some heterocyclic bases, such as pyridine, quinoline, piperidine, nicotine, coniine.

When testing for volatile bases, it is best to start with aqueous solutions of their salts, which are readily prepared by digesting the test material with dilute mineral acids. The organic bases are liberated from these solutions (which need not be free of any acid-insoluble material) by warming with an excess of caustic alkali.

The test given here is not directly applicable when ammonium salts are present, since they yield ammonia. A separate test for ammonium salts should be made with Nessler solution. If this test is positive, an alkaline solution of bromine can be used to make the solution basic. The ammonia is thus oxidized ($2 \, NH_3 + 3 \, NaOBr \rightarrow N_2 + 3 \, H_2O + 3 \, NaBr$) as soon as it is set free.

Procedure.[1]   A drop of the acid test solution and a drop of 2 N sodium hydroxide are mixed in the bulb of the apparatus shown in Fig. 23, *Inorganic Spot Tests*. A drop of one of the equilibrium solutions (see test (*1*)) is placed on the knob of the stopper. The apparatus is closed and placed in boiling water for 1 or 2 min. If volatile bases are present, red nickel dimethylglyoxime precipitates in the hanging drop, or white-yellow zinc oxinate is formed which fluoresces in ultraviolet light.

This procedure gave positive results

(*a*) with nickel dimethylglyoxime equilibrium solution:

15 γ pyridine	10 γ ethanolamine
100 γ ethylenediamine	20 γ aniline
1 γ methylamine	

(*b*) with zinc hydroxyquinoline equilibrium solution:

5 γ pyridine	10 γ ethanolamine
2.5 γ ethylenediamine	20 γ aniline
0.1 γ methylamine	

1 F. Feigl, V. Gentil and D. Goldstein, unpublished studies.

(*4*) Tests for acidic compounds with iodide–iodate mixture[1]

Water-soluble acidic compounds, either inorganic or organic, liberate iodine from a solution containing both iodide and iodate ions:

$$5 \, I^- + IO_3^- + 6 \, H^+ \rightarrow 3 \, H_2O + 3 \, I_2 \qquad (1)$$

The occurrence of the reaction is revealed by a yellowing of the solution, or by the blue color which appears on the addition of starch solution.

If the reaction with iodide–iodate is carried out in a closed vessel, and with warming (Procedure I), it is possible to detect organic acids which are so weak that they do not affect dyestuff indicators, at all, or indecisively at most. Such acids (HAc) are usually not very soluble, and their reaction with iodide–iodate is due in only small measure to the fact that the equilibrium

$$HAc \rightleftarrows H^+ + Ac^- \tag{2}$$

is shifted toward the right by warming. The essential feature is that the H^+ ions consumed by reaction (1) are delivered by the dissociation equilibrium (2), which quickly re-establishes itself at the expense of the weak acid, which thus constantly replenishes the supply of H^+ ions.

Water-insoluble higher mono- or dibasic fatty acids, which melt at low temperatures, produce iodine in accord with (1) if a dry mixture of the acid and potassium iodide plus iodate is brought to the temperature of a boiling-water bath, or if the mixture is vigorously ground at room temperature (Procedure II). Under these conditions, topochemical reactions occur on the surface of the solid potassium iodide and iodate.

Procedure I.   The test is conducted in a micro test tube provided with a glass or rubber stopper. A drop of the alcoholic solution of the acid and one drop each of 2% iodide and 4% iodate solutions are introduced. The test tube is stoppered and held in boiling water for 1 min. and then cooled. A drop of starch solution or several mg of solid thyodene indicator is added and the mixture shaken. If acids are present, a blue to violet color appears.

This procedure established acid character in:

2 γ capric acid	2 γ sebacic acid	4 γ myristic acid
5 γ lauric acid	5 γ stearic acid	5 γ palmitic acid

If the highest sensitivity is not required, fractions of a milligram of the solid non-dissolved acids can be warmed directly with the freshly prepared iodide–iodate mixture. In this case, 3 or 4 drops of the starch solution should be added to the cooled solution to produce the blue iodine–starch complex. Solid thyodene indicator may be used if preferred.

Procedure II.   A drop of the alcoholic solution of the acid is evaporated to dryness in a micro test tube, several mg of potassium iodate containing potassium iodide added, and mixed intimately. When larger amounts of acid are involved, the production of free iodine is revealed directly by a yellow-brown color. Small quantities of liberated iodine are detected, as in Procedure I, by adding starch solution.

Reagent: Potassium iodate containing iodide. Finely powdered potassium iodate is heated to incipient fusion in a quartz crucible; the iodate is thus partially converted to iodide. The cooled mass is pulverized.

Sometimes the product is light yellow, but it gives neither an alkaline reaction nor does it respond to the starch test.

The following revealed acid character when carried through this procedure:

	m.p.		m.p.
5 γ capric acid	31°	10 γ palmitic acid	64°
10 γ lauric acid	44–48°	10 γ stearic acid	69.4°
5 γ myristic acid	58°	5 γ sebacic acid	133°

If carboxylic or sulfonic acids do not volatilize or sublime at 110–120°, they will remain when solutions of their ammonium salts are taken to dryness and the evaporation residue heated to 120°. Accordingly, if an organic mixture is to be tested for carboxylic or sulfonic acids, the sample can be digested with ammonia water, filtered if necessary, and a drop of the filtrate evaporated in a micro test tube. The residue is kept at 120° for 10–20 min., cooled, and then tested with iodide–iodate solution by Procedure I or II. An indication of the attainable identification limits is given by the fact that 5 γ sulfosalicylic and 5 γ benzoic acid were detected.

Those carboxylic acids, which volatilize to a marked degree when heated to 100°, or are volatile with steam, can be detected through the production of iodine when the acid vapors are brought into contact with iodide–iodate solution. Indicator papers can be used, if preferred. The test can be made on the respective alkali or alkaline-earth salts by warming the solid or a drop of its solution with concentrated phosphoric acid. Those acids whose vapors evolved at water bath temperature react immediately with iodide–iodate solution include: formic, acetic, propionic, lactic, salicylic. Considerable quantities of carbonates must not be present, since carbonic acid reacts noticeably on iodide–iodate solution.

1 F. FEIGL, V. GENTIL and J. E. R. MARINS, unpublished studies.

(5) Test through liberation of nitrous acid[1]

The sensitive Griess test for nitrous acid (p. 90) can be used conversely for the detection of acidic compounds. Thus, the colorless solution of sodium sulfanilate, sodium nitrite and α-naphthylamine becomes red or orange on adding a tiny quantity of acidic compounds because of the formation of nitrous acid.

Procedure. In a depression of a spot plate a small amount of the test material, or the evaporation residue of a drop, is treated with a drop each of the reagents (1) and (2). Depending on acidity and solubility of the sample, a red or orange color appears at once or after a few minutes.

Reagents: 1) 2.9 g sodium sulfanilate and 0.7 g sodium nitrite dissolved in 30 ml water.

2) 1.8 g α-naphthylamine dissolved in 40 ml ethanol or dioxane.

The *Limits of identification* are 25–40 γ.

The following results were obtained:

Immediate intense red color. All water-soluble acids including caproic acid.

Immediate orange to orange-red color. Caprylic, pelargonic, capric, palmitic, stearic, oleic, elaidic, erucic, cholic, aspartic, and glutamic acid.

Orange color within a few minutes. Lauric, myristic, and 12-hydroxyoleic acid.

α-Amino acids (the dicarboxylic aspartic and glutamic acids are exceptions) gave no positive response because they have the character of inner-complex ammonium salts; but after addition of 2–3 drops of formalin to liberate the COOH group they react positively. This was observed with glycine, alanine, cystine, valine, methionine, leucine, isoleucine, phenylalanine, tyrosine, histidine and asparagine.

1 Y. NOMURA, *Bull. Chem. Soc. Japan*, 32 (1959) 536.

(6) *Detection of organic acids through fixing of ammonia* [1]

When the ammonium salts of weak organic acids are heated to 120°, the respective acids remain. In contrast, the NH_4 salts of strong and moderately strong acids are stable or decompose to only a slight extent. However, ammonium formate is entirely decomposed even though formic acid is a strong acid. The reason for this divergent behavior is the volatility of formic acid. The formation of heat-stable ammonium salts can be readily detected by evaporating the test material to dryness with ammonia and then heating the residue briefly to 120°. If the residue gives a positive response on treatment with Nessler solution (red-brown precipitate of $HgI_2.HgNH_2I$) the presence of a strong or medium strong organic acid is indicated.

The direct application of the test requires the absence of free mineral acids and of ammonium salts of inorganic and organic acids. The latter can be converted into alkali salts by warming with alkali hydroxide solution; the ammonia is completely volatilized. If alkali salts of organic acids or basic solutions of alkali salts are presented for examination, the sample can be fumed with hydrochloric acid and then heated to 120° to remove the unused acid. The nonvolatile organic acids remain and are then converted to their ammonium salts by evaporation with ammonia. Salts of organic bases must not be present, because they are decomposed when evaporated with ammonia. The organic base is liberated and the respective ammonium salt is left on heating to 120°.

Procedure.   One drop of the test solution or a pinch of the solid is placed in a micro crucible and taken to dryness with a drop of concentrated ammonia. The residue is kept at 120° in an oven for a minute or two. After cooling, a drop of Nessler solution is added. A red-brown precipitate or a yellow color indicates the presence of strong or medium strong organic acids. The hydrochloric acid and ammonia used must be free from ammonium salts.

Reagent: Nessler solution: 50 g potassium iodide, dissolved in 35 ml water, is treated with saturated mercuric chloride solution until a slight precipitate persists. Then 400 ml of 9 *N* caustic alkali is added. The solution is diluted to 1000 ml, allowed to settle, and decanted.

This procedure revealed:

10 γ citric acid	10 γ phthalic acid	5 γ oxalic acid
5 γ tartaric acid	10 γ sulfanilic acid	10 γ mandelic acid
5 γ malonic acid	5 γ sulfosalicylic acid	
10 γ succinic acid	4 γ pyrocatecholdisulfonic acid	

No reaction was given by acetic, formic, aminoacetic, propionic acid.

1 F. FEIGL, V. GENTIL and J. E. R. MARINS, unpublished studies.

(7) *Detection of benzene-soluble acidic compounds*[1]

If benzene-soluble acidic compounds, such as higher monobasic fatty acids, aromatic monocarboxylic acids, phenols, imide compounds, *etc.* are treated, in mg quantities, with a colorless benzene solution of rhodamine B, a red color appears. This is intensified, and is produced with smaller amounts of the above compounds, if the system is shaken with an aqueous solution of uranyl salts. The red benzene solution fluoresces orange-red in ultraviolet light. The color (fluorescence) reaction is based on the formation of benzene-soluble salts of the basic dyestuff rhodamine B with acidic compounds or with complex uranyl carboxylic acids.

The chemistry is discussed in more detail on p. 704 in connection with the detection of higher fatty acids in paraffins, waxes, *etc.*

Procedure. A micro test tube is used. A little of the solid or a drop of its benzene solution is treated with five drops of a saturated benzene solution of rhodamine B and one drop of 1% uranyl acetate solution. The mixture is shaken. If acidic compounds are present, a red or pink color appears, the shade depending on the amount present. The red system fluoresces orange in ultraviolet light.

The procedure revealed:

0.5 γ benzoic acid	25	γ benzotriazole	0.5	γ pentachlorophenol
2.5 γ palmitic acid	20	γ benzoinoxime	1	γ salicylic acid
2.5 γ myristic acid	20	γ phenylglyoxal	1	γ salicylamide
2.5 γ stearic acid	0.1	γ picrolonic acid	5	γ salicylanilide.
5 γ *m*-nitrophenol	200	γ nitromethane		
6 γ *p*- ,,	25	γ acetoacetic ester		

A strong positive response to the color reaction was observed with 100 γ quantities of the following acids: mono-, di-, and tri-chloroacetic, phenoxy- and *p*-chlorophenoxyacetic, mercaptoacetic, anthranilic, N-methyl and N-phenylanthranilic, *p*-aminobenzoic, cinnamic, mandelic, lauric.

The procedure given here makes it possible to detect benzene-soluble acids rapidly in the presence of benzene-insoluble acids, for instance, benzoic acid in the presence of phthalic acid, and salicylic acid in the presence of sulfo-salicylic acid. The solid mixture (or evaporation residue) is digested with benzene and the resulting solution subjected to the procedure as described.

If a halogenoacetic acid is to be detected in the presence of benzene-insoluble acetic acid and halogen acids, a drop or two of the test solution is evaporated in a micro test tube. The residue is heated to 130° for several minutes. The cold residue is shaken with a drop of uranyl acetate solution and three or four drops of the benzene solution of rhodamine B. Halogeno-acetic acid is present if the benzene layer turns red and the color persists even after the addition of water.

1 F. FEIGL, V. GENTIL and D. GOLDSTEIN, unpublished studies.

25. Detection of occult acidic groups

Test by pyrolysis with mercuric cyanide[1]

There are numerous organic compounds possessing acidic hydrogen atoms (*i.e.* replaceable by metals) which however do not respond to the usual tests for acidic character described in previous sections or which give uncertain results. This is true of water-insoluble acidic compounds (oximes, dioximes, acyloins, *etc.*) and of compounds in which acidic OH- or SH groups are produced only after enolization of CO- or CS groups, and likewise of compounds containing acidic NH groups. The acidic character may be diminished or even entirely disappear in compounds that contain basic groups along with acidic OH-, COOH- or SO_3H groups (taurine, sulfanilic acid). Acid amines and aliphatic primary and secondary nitro compounds either undissolved or dissolved in neutral media show so little tendency to undergo transformation into the aci-forms that the acidic character does not make itself evident.

$$-CONH_2 \longrightarrow -C\overset{\nearrow OH}{\underset{\searrow NH}{}} \qquad -SO_2NH_2 \longrightarrow -SO\overset{\nearrow OH}{\underset{\searrow NH}{}} \qquad RCH_2NO_2 \longrightarrow RCH{=}NO_2H$$

The detection of acidic groups in colored compounds through color reactions is not feasible or is impeded. In the cases just cited in which obviously acidic groups cannot be detected by conventional methods, it is entirely proper to speak of occult acidic groups in analogy to the occult basic NH_2 groups discussed in the next section.

A possibility of detecting occult acidic groups, almost without exception, is based on pyrolysis with mercuric cyanide. This salt can be heated to 180°

without notable sublimation or decomposition, whereas when mixed with acidic compounds and heated it yields hydrogen cyanide:

$$2 \text{ HAc} + \text{Hg(CN)}_2 \rightarrow \text{HgAc}_2 + 2 \text{ HCN}$$

The above reaction occurs rapidly no matter what the strength of the acid (HAc). Even compounds in which the acidic nature is extremely slight (*e.g.* diphenylamine) undergo pyrolytic reaction with mercuric cyanide. The occurrence of this reaction and hence the presence of acidic groups, including occult acidic groups, can be revealed through the detection of the resultant hydrogen cyanide. The sensitive color reaction (blue color) on contact with copper acetate–benzidine acetate mixture is recommended (see p. 546).

The method given here for the revealing of occult acidic groups is not specific; other compounds that are not acidic likewise yield hydrogen cyanide when pyrolyzed with mercuric cyanide (see p. 147). A confirmation supporting indications of the presence of acidic groups is however frequently of value, especially if basic groups are likewise present as is so often the case in dyestuffs. Moreover, the verification of the presence of occult acidic groups may be of significance when determining structures or differentiating isomers (compare Chap. 6). Furthermore, information about the presence of occult acidic groups in compounds of known composition may be of interest with respect to their ability to yield metal salts. This can be important in studies dealing with the use of organic reagents in inorganic analysis.

The description of the procedure will be followed by a list of compounds whose acidic character is not revealed by conventional methods or in an uncertain manner if small quantities of the sample are taken for the tests.

Procedure.   A small amount of the solid or a drop of its solution is taken to dryness in a micro test tube. Several cg of mercuric cyanide are added and the tube is immersed to a depth of about 1/2 cm in a glycerol bath preheated to 120°. The open end of the tube is covered with a disk of filter paper moistened with a drop of the hydrogen cyanide reagent (see p. 546). The temperature of the bath is increased to a maximum of 170°. If acidic compounds are present, a blue stain will appear on the reagent paper.

A positive response was given by;

50 γ diphenylamine	25 γ diazoaminobenzene
20 γ thionalide	20 γ mercaptobenzothiazole
10 γ diphenylcarbazide	15 γ diphenylcarbazone
25 γ alizarin blue	50 γ 6-nitroquinoline
50 γ luminol (3-aminophthalic acid hydrazide)	

A pyrolytic evolution of hydrogen cyanide was observed with: acetamide, acetylacetone, *p*-aminosalicylic acid, aminocaproic acid, alizarin, benzoic acid hydrazide, benzoylphenylhydroxylamine, quinalizarin, dimethylglyoxime, fluorescein, furildioxime, isatin, nitroethane, 1-nitroso-2-naphthol, phthalimide, succinimide, sulfanylamide.

It is notable that the SO_3H groups in taurine and sulfanilic acid are *not* detectable by pyrolysis with mercuric cyanide. Obviously these amino acids are completely of the ammonium sulfate type. The acidic groups in these compounds, as well as all aminocarboxylic acids, can be set free by evaporation with paraformaldehyde (140°) whereby the NH_2 groups are methenylated (compare detection of taurine, p. 507).

1 F. Feigl, J. R. Amaral and E. Libergott, unpublished studies.

26. Detection of occult (non-basic) amino groups

Test by dry heating with hexamine[1]

Compounds which contain NH_2 groups bound to C atoms can be expected to show basic properties and respond accordingly to pertinent tests and also to the tests for primary amines as described in Chap. 3. However, this expectation is not realized invariably or it is difficult to verify in some cases. It is logical then to speak of "occult amino groups." The fact that the basic character of NH_2 groups is neutralized in salts of primary amines where there has been addition of acids has a counterpart in the salification of this group through the adjacent COOH- or SO_3H groups of the same molecule. In the aminosulfonic acids taurine and sulfanilic acid this salification is so extensive that these compounds are quasi-ammonium salts. This reduction in the basic character of the NH_2 group by acidic groups likewise occurs in other amino acids. Aromatic primary amines whose molecules contain a number of acidic groups have acid character. The conventional methods are in fact not capable of revealing the NH_2 groups in amides of carboxylic and sulfonic acids; they are amphoteric and can produce salts with either acids or strong bases (compare Sect. 24). Actually, the influence of the rest of the molecule may be so marked in amino compounds that tests which normally are reliable for primary amines fail completely. For example, tribromoaniline does not respond to tests that give positive results with the parent compound aniline. The same is true of aminopyridines. It is also proper to speak of occult amino groups when the compound being tested is sufficiently colored to hide the occurrence of color reactions. Many dyestuffs fall into this category.

The test for occult NH_2 groups described here is based on the finding that the pertinent compounds yield ammonia when dry-heated along with hexamine. The ammonia can be readily detected in the gas phase by Nessler reagent. These instances involve a methenylation of the NH_2 group by pyrolytic means:

$$6 -NH_2 + (CH_2)_6N_4 \rightarrow 6 -N{=}CH_2 + 4 NH_3$$

The above reaction occurs also with chlorides, sulfates *etc.* of primary aliphatic and aromatic amines, and likewise with compounds in which the NH_2 group is intramolecularly salified by COOH- or SO_3H groups. Cases in point include taurine, sulfanilic acid and α-aminocarboxylic acids. The same behavior is shown by NH_2 groups in amides of carboxylic and sulfonic acids and in compounds that have acidic character because they contain several acidic groups. In such instances, the distinctly basic hexamine combines with the particular acid or acidic groups with freeing of the previously occult NH_2 groups, which then are methenylated by the excess hexamine. The NH_2 groups of even colored compounds can be readily detected in this way. This is of value in the examination of dyes.

It is obvious that compounds in which NH_2 groups are revealed by conventional procedures can also be detected by pyrolysis with hexamine to yield ammonia. The advantage of the method here described resides in the detectability of occult amino groups within satisfactory identification limits.

Procedure. A small quantity of the solid test material or a drop of its solution is taken to dryness (if need be) in a micro test tube. Then a drop or two of a concentrated chloroform solution of hexamine is added and the mixture taken to dryness. The test tube is then immersed to a depth of 1 cm in a glycerol bath preheated to 120°. The mouth of the tube is covered with a disk of filter paper moistened with Nessler solution and the temperature of the bath is then raised to 160°. A positive response is signaled by the appearance of a brown stain on the reagent paper.

The test revealed:

20 γ tribromoaniline	20 γ trinitroaniline	5 γ aniline
2 γ benzamide	3 γ acetamide	2 γ aspartic acid
10 γ sulfanilic acid	25 γ taurine	5 γ benzidine
50 γ Victoria violet	10 γ Bismarck brown	5 γ biuret
20 γ chrysoidine	20 γ Diazine green	

Positive response was obtained with Congo red, 3-hydroxyanthranilic acid and benzidine sulfate.

1 F. FEIGL, E. JUNGREIS and L. BEN-DOR, *J. Israel Chem. Soc.*, 1 (1963) 351.

27. Detection of water-soluble salts of organic bases and alkali salts of organic acids[1]

It is easily possible to obtain soluble calcium salts from water-soluble salts of organic bases (also from difficultly soluble sulfates) and from alkali salts of nonvolatile organic acids. The first case requires nothing beyond warming with calcium carbonate:

$$2[BH]X + CaCO_3 \rightarrow Ca^{2+} + 2X^- + 2B + H_2O + CO_2 \qquad (1)$$

If alkali salts of carboxylic and sulfonic acids (also polynitrophenols) are involved, the material should be taken to dryness with hydrochloric acid and the residue heated for a short time at 110° to drive off the unused mineral acid. There remain, in addition to alkali chloride, the respective nonvolatile organic acids (HAc), which promptly react with calcium carbonate:

$$\text{Alk.Ac} + \text{HCl} \rightarrow \text{Alk.Cl} + \text{HAc} \tag{2}$$

$$2\,\text{HAc} + \text{CaCO}_3 \rightarrow \text{Ca}^{2+} + 2\,\text{Ac}^- + \text{H}_2\text{O} + \text{CO}_2 \tag{3}$$

The detection of calcium ions formed according to (1)–(3), and hence indirectly of the salts cited in the heading, can be accomplished by means of an alkaline solution of sodium rhodizonate. A violet basic calcium salt is produced (see *Inorganic Spot Tests*, Chap. 3) whereas $CaCO_3$ remains unaltered.

The test for salts of organic bases described in Procedure I assumes of course the absence of free acids, ammonium salts, and of metal salts which give colored precipitates with alkaline sodium rhodizonate. A preliminary test with this reagent is essential. Oxalates, phosphates, or arsenates of organic nitrogen bases cannot be detected, since the underlying acids yield insoluble calcium salts on treatment with $CaCO_3$ which do not react with sodium rhodizonate. Neutral potassium and sodium salts do not interfere, even in considerable amounts.

It should be pointed out with respect to the test described in Procedure II for water-soluble alkali salts of organic acids, that the organic acids left after fuming off with hydrochloric acid can obviously also be revealed by the sensitive iodide–iodate reaction (see p. 112). However, the latter is not reliable in the presence of iodine-consuming substances when only small quantities of acid are present. Furthermore, it is always an advantage to utilize tests of different sensitivities since useful conclusions can often be drawn from their occurrence and intensity. Procedure II should be preceded by a test for metal ions which react with sodium rhodizonate. If such are present, the sample should be warmed with sodium carbonate solution and the filtrate then tested.

Procedure I. (Salts of organic bases). A spot plate or micro crucible is used. A drop of the test solution is stirred into a thin slurry with the least possible amount of calcium carbonate. When the evolution of carbon dioxide has stopped, a drop of 0.5 N hydrochloric acid and a drop of freshly prepared 0.2% aqueous sodium rhodizonate solution are added in succession. A positive response is indicated by a more or less intense violet color.

This procedure revealed:

5 γ aniline sulfate	10 γ α-naphthylamine chloride
5 γ benzidine chloride	50 γ guanidine sulfate
5 γ ethylenediamine chloride	5 γ diethanolamine chloride.

Procedure II.   (Alkali salts of organic acids.) A drop of the test solution is taken to dryness, on a spot plate or in a micro crucible, along with a drop of 0.5 N sodium hydroxide. The residue is kept at 110° for several minutes. A drop of water is then added to the cooled residue and stirred with very little calcium carbonate. The remainder of the procedure is as in I.

Positive response was given by:

5 γ tartaric acid	4 γ adipic acid	5 γ sulfanilic acid
2.5 γ citric acid	5 γ malic acid	10 γ aminoethanesulfonic acid.

A test for salts of *nonvolatile* bases with mineral acids, which also succeeds in the presence of ammonium salts, is based on the fact (Procedure III) that the base is liberated on warming the salt with ammonium carbonate:

$$2 \text{ B.HX} + (\text{NH}_4)_2\text{CO}_3 \rightarrow 2 \text{ B} + 2 \text{ NH}_4\text{X} + \text{CO}_2 + \text{H}_2\text{O}$$

After removal of the excess carbonate, the freed base can be detected through its reaction with nickel dimethylglyoxime equilibrium solution (see p. 109). Bases can also be set free from their salts by evaporation with pure piperidine (formation of piperidine salts), a suitable procedure since excess piperidine is readily removed by heating to 120°.

Procedure III.   (Salts of non-volatile bases.) A micro test tube is used. A small quantity of the solid test material is treated with a drop of cold saturated aqueous solution of ammonium carbonate, or a drop of the aqueous test solution is mixed with several cg of ammonium carbonate. After taking to dryness, the residue is kept at 120° for about 10 min. (drying oven) and then a drop of the equilibrium solution is added. A positive response is indicated by the formation of red dimethylgyoxime. An indication of the sensitivity of the test is that 100 γ benzidine chloride can be detected.

Procedure III may not be employed in the presence of water-soluble magnesium or alkaline-earth salts of carboxylic or sulfonic acids since they yield the corresponding carbonates on treatment with ammonium carbonate and these carbonates being H^+ ion-acceptors react with the equilibrium solution. On the other hand, this behavior permits the detection of magnesium and alkaline-earth salts of organic acids provided salts of organic bases are absent.

1 F. FEIGL and V. GENTIL, *Mikrochim. Acta*, (1954) 435; and unpublished studies.

28. Detection of water-soluble salts of sulfonic and sulfinic acids[1]

All alkali and nearly all alkaline-earth salts of sulfonic and sulfinic acids are known to be water soluble. Ignition of these salts produces the respective heat-resistant sulfates. It is important to note that the presence of caustic alkali is necessary to obtain a complete transformation, since otherwise half of the sulfur is lost.

The resulting sulfates ($BaSO_4$ is an exception) can be detected by treatment with a concentrated solution of potassium permanganate containing barium chloride. The barium sulfate precipitated under these conditions is tinted violet and the color is resistant to all reagents that normally reduce permanganate. See p. 84 regarding this so-called Wohlers effect. The latter also permits the detection of free sulfonic or sulfinic acids after their conversion into the respective calcium salts, which is easily accomplished by treating the test solution or suspension with calcium carbonate followed by filtration or centrifugation.

Procedure. A drop of the test solution is made alkaline with a drop of 1% NaOH and evaporated in a micro crucible. The residue is ignited and the colorless ash is treated with a drop of a solution containing 7 g $KMnO_4$ and 2 g $BaCl_2$ per 100 ml. The suspension is kept in a boiling-water bath for 5 min. After cooling, the excess permanganate is destroyed by adding a drop of 20% hydroxylamine hydrochloride solution. A positive response is indicated by a violet residue.

The test revealed the following amounts of sulfonic and sulfinic acids in the form of their calcium salts:

30 γ taurine	15 γ anthraquinonesulfonic acid
25 γ sulfanilic acid	25 γ pyrocatecholdisulfonic acid
25 γ sulfosalicylic acid	

This procedure provides a means for differentiating between the calcium salts of organic sulfonic- and carboxylic acids (also arsonic acids) since on ignition the salts of the carboxylic acids and arsonic acids yield $CaCO_3$ or $Ca_3(AsO_4)_2$ respectively.

1 F. Feigl, I. T. de Souza Campos and S. Ladeiro Dalto, *Anal. Chem.*, 36 (1964) 1657.

29. Detection of water-insoluble bases and acids[1]

If slight amounts of water-insoluble bases or acids are to be detected in a given mixture of materials the following method may be used. The compounds under study are first converted into soluble chlorides or soluble barium salts followed by treatment with ammonium carbonate; thereby the corresponding bases or barium carbonate are formed which can be readily detected by means of nickel dimethylglyoxime equilibrium solution.

When applying this test, it is assumed that the test material contains neither water-soluble salts of bases or water soluble alkaline-earth or magnesium salts of organic or inorganic acids. A test to establish the absence of such materials should be made on the water extract of the sample. The latter should be evaporated with ammonium carbonate and the residue then spotted with the equilibrium solution. If red nickel dimethylglyoxime precipitates, the test material should first be extracted with water and procedure I or II then applied to the water-insoluble residue.

Procedure I. (Test for water-insoluble bases) The sample is extracted with water if need be and several mg of the insoluble residue are warmed in a micro test tube with about 0.5 ml of dilute hydrochloric acid and then centrifuged. The clear liquid contains the chlorides of the bases along with the unused hydrochloric acid. A drop or two (more if necessary) is evaporated in a micro test tube and several cg of ammonium carbonate and a drop of ethanol are added to the residue. The evaporation to dryness is repeated and the residue is kept for about 20 minutes at 130° (drying oven) to remove the excess ammonium carbonate completely. If the residue turns red when treated with a drop or two of nickel dimethylglyoxime equilibrium solution (for preparation see p. 109) it may be taken as proved that the test material contains free bases.

Less than 0.5 mg portions of the test material suffice for the tests described.

Procedure II. (Test for water-insoluble acids) A small amount of the sample is extracted with water if need be and a little of the undissolved material is treated with about 0.5 ml of ethanol and several cg of barium carbonate. The mixture is warmed for several minutes. The suspension is then centrifuged and the clear solution, which contains the water-soluble barium salt, is treated with ammonium carbonate as described in Procedure I.

The liberation of bases from salts by evaporation with ammonium carbonate followed by extraction with ether, chloroform, *etc.* can be employed to obtain solutions for further tests. The same holds for the solutions of barium salts of organic acids obtained by treatment with barium carbonate.

1 Unpublished studies, with E. LIBERGOTT.

30. Detection of compounds which react with acids

Together with other preliminary tests, it may be of interest to determine whether organic materials that react with hydrochloric acid are present in a sample. Included are all bases, compounds with basic and hence acid-binding groups, and also alkali and alkaline-earth salts of organic acids. The first two categories leave the corresponding hydrochlorides after evaporation with excess hydrochloric acid (120°); the last category leaves nonvolatile organic acids along with alkali or alkaline-earth chlorides. The occurrence of these reactions as well as hydrolytic fissions of esters, amides, and hydrazides, in which nonvolatile acids remain along with ammonium chloride or hydrazine chloride, can be ascertained quite simply. It is sufficient to mix the evaporation residue with mercuric cyanide and heat to 160°. Hydrogen cyanide is evolved and can be readily detected in the gas phase by the blue color it imparts to a copper–benzidine acetate solution (comp. p. 546). The positive outcome of this test is, however, proof of the formation of acidic compounds on evaporation with hydrochloric acid only if the original sample yields no hydrogen cyanide when heated with mercuric cyanide. This latter test is obligatory since there are many compounds which yield hydrogen cyanide when they are pyrolyzed in the presence of mercuric cyanide (see p. 147). It should also be remembered that hydrogen chloride is evolved when chlorides of weak bases are heated to 120°.

The evaporation trials in the presence of hydrochloric acid and leading to acidic residues require only minimal amounts of the test material. For instance, as little as 10 γ of benzidine base gave a positive response when treated in this fashion.

When the procedure is to be applied to mixtures containing large proportions of indifferent acid-stable materials it is advisable after the evaporation with hydrochloric acid to prepare a water–alcohol extract and to test the latter. This method will lessen the mechanical interference of the indifferent materials with the essential contact between the mercuric cyanide and the acidic compounds during the pyrolysis. If the acidic compounds are not removed by extraction, the negative response to the cyanide test does not provide a reliable proof of the absence of compounds that react with acids.

31. Orientation tests based on redox reactions

Relatively few organometallic and organic compounds enter stoichiometrically defined redox reactions conducted in the wet way. Such cases always involve the activity of certain groups in the molecule, on which the reaction begins, and the carbon skeleton may either be preserved or it may split in characteristic fashion at certain places. The test for redox effects exerted by organic materials consequently furnishes leads regarding the presence or absence of particular types of compounds. As a guiding principle, it may be taken that in such tests the overwhelming majority of all organic compounds prove to be indifferent, and that reducing actions are encountered much oftener than oxidation effects.

Oxidation effects of organic compounds

In inorganic analysis, the best criterion for oxidizing compounds is their ability to release iodine from an acidified solution of alkali iodide. This effect can of course be expected only from organic oxidizing agents which, as such, or in the form of reaction products, have a perceptible solubility in ḋilute acids. This can be established in the case of certain colored quinones and quinoneimide compounds, and organic derivatives of hydrogen peroxide (colorless) and also salts of organic bases with inorganic acids, which of themselves oxidize hydriodic acid (chlorates, chromates, molybdates, phosphomolybdates, *etc.*). The category of organic compounds, which release iodine from acidified solutions of alkali iodide, likewise includes: alkyl- and arylarsonic and stibonic acids, C- and N-nitroso compounds, alloxan, as well as haloamines of carboxylic and sulfonic acids.

The redox reaction between hydrogen peroxide and iodide ions goes too slowly in acid solution to serve as the basis of a general test for organic per-compounds, which undergo hydrolysis on contact with water. The oxidation of black lead sulfide to white lead sulfate, which can be conducted as a very sensitive spot test for hydrogen peroxide is very well suited to this purpose. However, since organic per-compounds are rarely present, no precise description of this test (see *Inorganic Spot Tests*, Chap. 4) will be given here. Small amounts of benzoyl peroxide (dibenzoyl peroxide), an organic derivative of hydrogen peroxide, do not respond either to acidified iodide or to lead sulfide. The detection of this per-compound is discussed on p. 487.

(1) Oxidation of thio-Michler's ketone or tetrabase[1]

Organic oxidants, which are active in neutral media, can be detected by means of yellow-brown thio-Michler's ketone (4,4'-bis-dimethylaminothiobenzophenone) or colorless tetrabase (tetramethyldiaminodiphenylmethane). Blue quinoidal compounds result when these tertiary bases are oxidized.

The chemistry of the oxidation of thio-Michler's ketone is outlined in Sect. 6 in connection with the test for free halogens. When tetrabase is oxidized, there is production of a diphenylmethane dye:

$$(CH_3)_2N-\langle\!\!\!\bigcirc\!\!\!\rangle-CH_2-\langle\!\!\!\bigcirc\!\!\!\rangle-N(CH_3)_2 + O \longrightarrow$$

$$(CH_3)_2N-\langle\!\!\!\bigcirc\!\!\!\rangle-CH=\langle\!\!\!\bigcirc\!\!\!\rangle=\overset{+}{N}(CH_3)_2 + OH^-$$

Compounds which react with these two reagents include: polyhalides of organic bases and their salts with hydrogen halides; benzoquinone and tetrahalogenated benzoquinone; peracids or their anhydrides; haloimides and amides of mono- and dicarboxylic acids, which may be regarded as derivatives of hypohalogenous acids.[2]

Chloramine B and chloramine T suprisingly react with thio-Michler's ketone, but not with tetrabase.

Percarboxylic acids, arsonic acids and amine oxides, which, like the majority of the compounds just listed, immediately release iodine from acidified iodide solutions, do not react with either thio-Michler's ketone or tetra-base. This is likewise true with nitro compounds.

Procedure. Reagent papers are used which are prepared by bathing quantitative filter paper in 0.1% benzene solution of thio-Michler's ketone or in 2.5% solution of tetrabase in ether. The dry papers are stable provided air is excluded. If a drop of the test solution is placed on the yellow or colorless reagent paper, a blue stain or ring appears after evaporation of the solvent, if necessary.

The behavior of the oxidants that have been tested in neutral surroundings, and the limits of identification of the test are:

Oxidant tested	Ident. limit (γ)		Solvent
	Tetrabase	Thioketone	
Chloramine B	–	0.04	water
Chloramine T	–	0.04	water
N-Bromosuccinimide	0.5	0.06	water
N-Chlorosuccinimide	0.5	0.06	water
Chloranil	0.12	0.12	ether or benzene
Bromanil	0.12	0.12	ether or benzene
Benzoyl peroxide	1.2	1.2	ether or benzene
2,6-Dibromoquinone-4-chloroimine	2.0	0.5	ether
2,6-Dichloroquinone-4-chloroimine	1.0	0.25	ether

1 F. FEIGL and R. A. ROSELL, *Z. Anal. Chem.*, 159 (1958) 335.
2 J. TSCHERNIAK, *Ber.*, 34 (1901) 4209.

(2) Oxidation of thallous hydroxide[1]

Organic oxidants that are active in strong alkaline solution can be detected through the brown precipitate of thallium(III) oxyhydrate which they produce in alkaline thallium(I) solutions: $Tl^+ + [O] + OH^- \rightarrow TlOOH$. Very few compounds fall into this class. Examples are: haloimides, such as N-(bromo-, chloro- or iodo-)-succinimide, which react even at room temperature, and also chloramine T and related compounds, which react on warning. When conducted as spot reactions on a spot plate or in a micro test tube, it is possible to detect 2.5 γ chloramine T or 3 γ N-bromosuccinimide by means of a 4% alkaline solution of thallous sulfate.

1 F. FEIGL and R. A. ROSELL, *Z. Anal. Chem.*, 159 (1958) 335.

(3) Detection of oxidizing compounds with N,N'-diphenylbenzidine[1]

The tests for oxidants described before are restricted to the examination of neutral or alkaline solutions of the sample. A test that can be carried out in strong sulfuric acid is based on the oxidation of N,N'-diphenylbenzidine (I) to the blue quinoidal compound (II). This transformation is brought about by organic compounds that act as O donors or H acceptors:

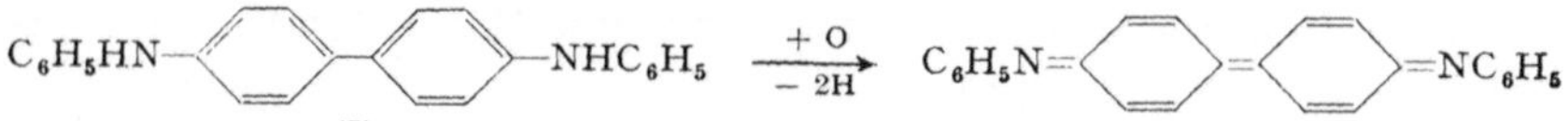

$$C_6H_5HN-\langle\!\!\!\!\bigcirc\!\!\!\!\rangle-\langle\!\!\!\!\bigcirc\!\!\!\!\rangle-NHC_6H_5 \quad \xrightarrow[-2H]{+O} \quad C_6H_5N=\langle\!\!\!\!\bigcirc\!\!\!\!\rangle=\langle\!\!\!\!\bigcirc\!\!\!\!\rangle=NC_6H_5$$

(I)   (II)

Pertinent compounds are:

Salts of nitrogen bases with HNO_3, H_2CrO_4, H_2MoO_4, etc.	O donors through liberation of oxidizing acids
Esters of nitric or nitrous acids	O donors through liberation of HNO_3 or HNO_2
Nitrosamines	O donors through liberation of HNO_2
N-Halogen compounds	O donors through formation of halogenous acids
Peroxides	O donors and H acceptors
Quinones[2]	H acceptors through formation of OH groups
C-Nitroso compounds[2]	H acceptors through formation of NHOH groups

The two last-mentioned classes of compounds react in general when warmed slightly. Surprisingly, benzoyl peroxide is practically without action. Organic derivatives of arsenic acid do not react.

The tests are carried out in a micro test tube by adding 1–2 drops of a solution of 10 mg N,N′-diphenylbenzidine in 100 ml concentrated sulfuric acid to the evaporation residue of a drop of the solution of the sample in water, alcohol or benzene.

Limits of identification: 0.2–30 γ.

1 Unpublished studies, with V. GENTIL.
2 V. ANGER, *Mikrochim. Acta*, (1959) 387; (1960) 58.

Reduction effects of organic compounds

Qualitative inorganic analysis utilizes a general test for reducing compounds which is based on their ability to discharge immediately the color of acid solutions of iodine (in alkali iodide), with addition of starch as indicator, if needed. Since the action of acidic iodine solutions on organic compounds ordinarily proceeds sluggishly and to not more than slight extent, no general preliminary test for reducing organic compounds can be based on this action. Suitable redox tests are described in the following tests (4)–(9).

(4) Reduction of periodic acid[1]

Periodic acid has a selective oxidizing action upon compounds with two hydroxyl groups attached to adjacent carbon atoms. Such compounds, dissolved in water or dioxane, give the reaction:

$$RCH(OH)CH(OH)R + HIO_4 \rightarrow 2\ RCHO + HIO_3 + H_2O \tag{1}$$

Compounds which contain a $\geqslant C(OH)$ group adjacent to a $>CO$ or $\geqslant C-NH_2$ group, and also compounds which contain a $-CO-CO-$ group, likewise react with periodic acid with production of iodic acid:

$$R_1CH(OH)CH(NH_2)R_2 + HIO_4 \rightarrow R_1CHO + R_2CHO + HIO_3 + NH_3 \tag{2}$$

$$R_1CH(OH)COR_2 + HIO_4 \rightarrow R_1CHO + R_2COOH + HIO_3 \tag{3}$$

$$R_1COCOR_2 + HIO_4 + H_2O \rightarrow R_1COOH + R_2COOH + HIO_3 \tag{4}$$

It is likely that redox reactions (2)–(4) do not proceed directly, but are preceded by more or less extensive reactions of water with the CNH_2- and CO groups contained in the starting compounds:

$$-\overset{|}{\underset{|}{C}}-NH_2 + H_2O \ \rightleftarrows\ -\overset{|}{\underset{|}{C}}(OH) + NH_3$$

$$\overset{\diagdown}{\underset{\diagup}{}}CO + H_2O \ \rightleftarrows\ \overset{\diagdown}{\underset{\diagup}{}}C\overset{OH}{\underset{OH}{\diagdown}}$$

These reactions accordingly produce 1,2-glycols, which then react with periodic acid as per (1).* The relative velocity of the oxidation by periodic acid is: 1,2-glycols $>$ α-hydroxyaldehydes $>$ α-hydroxyketones $>$ 1,2-diketones $>$ α-hydroxyacids.

The iodic acid resulting from reactions (1)–(4) can be detected by the precipitation of silver iodate, which is not soluble in dilute nitric acid. Since IO_4^- ions are not precipitable by Ag^+ ions, a solution of potassium periodate containing silver nitrate and nitric acid can be used as reagent.

Compounds, which under the conditions of the test hydrolyze to 1,2-glycols, α-hydroxyaldehydes, *etc.*, can also be detected by the behavior toward periodic acid. The detection of anhydrides of 1,2-glycols, the so-called epoxides,[3] whose cleavage proceeds:

$$\underset{\displaystyle O}{-HC\underline{\quad\quad}CH-} + H_2O \ \rightarrow\ \underset{\displaystyle OH\ \ OH}{-HC\underline{\quad\quad}CH-}$$

has practical importance. It may be assumed that all hydrolytic splittings are rapid and extensive when hydrolysis products react with periodic acid and are thus removed from the hydrolysis equilibrium.

When using this test it must be remembered that in addition to the hydroxy- and oxo-compounds noted above, thiols (mercaptans and thiophenols) as well as sulfinic acids and disulfides, likewise reduce periodic

---

* Rate studies show that an intermediate compound is formed between the 1,2-glycol and periodic acid before the carbon–carbon bond is broken. This intermediate is believed to be a cyclic ester of hydrated (tribasic) periodic acid.[2]

acid. The same is true of hydrazones, acid hydrazides, aldoximes and ketoximes, which are cleaved under the conditions of the test and yield alkyl-
hydrazine, hydrazine, or hydroxylamine, which in turn reduce periodic acid.

Procedure.[4] One drop of the (aqueous or dioxane) test solution is mixed
with one drop of the reagent solution in a depression of a black spot plate. A
positive response is signalled by the appearance (immediate or after 1–5 min.) of
a white or light yellow precipitate or turbidity.

Reagent: Mixture of 2 ml of conc. nitric acid, 2 ml of 10% silver nitrate solution and 25 ml of 2% potassium periodate solution. If a yellow precipitate develops after a time it may be filtered off.

The following *identification limits* were obtained by the periodic acid test:

50	γ maltose	5	γ ascorbic acid
50	γ tartaric acid	5	γ diacetyl
2.5	γ glycerol	2.5	γ picolinic acid hydrazide
2.5	γ mannitol	2.5	γ nicotinic acid hydrazide
5	γ galactose	2.5	γ isonicotinic acid hydrazide
5	γ glucose		

Phenylhydrazine, benzil, benzoin and dimethylglyoxime gave a positive response.

1 R. L. SHRINER and R. C. FUSON, *Identification of Organic Compounds*, 3rd ed.,
  Wiley, New York, 1948, p. 115: see also L. MALAPRADE, *Compt. Rend.*, 186 (1928)
  382; P. FLEURY and S. BOISSON, *Compt. Rend.*, 204 (1937) 1264.
2 Comp. G. HUGHES and T. P. NEVELL, *Trans. Faraday Soc.*, 44 (1948) 941.
3 R. FUCHS, R. C. WATERS and C. A. VAN DER WERF, *Anal. Chem.*, 24 (1952) 1514.
4 Unpublished studies, with V. GENTIL.

(5) *Reduction of phosphomolybdic acid*[1]

As pointed out (p. 97) in connection with the detection of phosphorus
in organic compounds, the Mo(VI) in the anion of the complex phosphomolybdic acid $H_3PO_4.12\ MoO_3.aq.$ is much more readily reduced to the so-called
molybdenum blue than in the normal molybdate anion. In mineral acid
solutions, this enhanced reactivity appears only toward strong reducing
agents. However, with proper buffering, weak reductants likewise produce
molybdenum blue and this fact provides a sensitive preliminary test for
reducing organic compounds. It is not necessary to establish a particular
pH in order to realize this redox reaction, but merely to treat the sample or
its acid solution with phosphomolybdic acid and then with excess ammonia.
The pH region essential to the production of molybdenum blue is thus trav-
ersed. Excess ammonia does not affect the molybdenum blue produced but
it decomposes the yellow phosphomolybdate ion, whose color is discharged:

$$[PO_4.12\ MoO_3]^{3-} + 24\ OH^- \rightarrow PO_4^{3-} + 12\ MoO_4^{2-} + 12\ H_2O$$

It is remarkable, and also a sign of the rapidity of the redox reaction, that the latter occurs even before the alkaline decomposition of the phosphomolybdate, which likewise is a very fast reaction.

Procedure. A drop of the neutral or acid test solution is mixed in a depression of a spot plate with a drop of 5% aqueous solution of phosphomolybdic acid and then a drop of concentrated ammonia is added. A blue to blue-green color indicates the presence of reducing materials.

Molybdenum blue was formed through this procedure by:

10	γ benzaldehyde	0.5	γ dimethylaniline
0.10	γ hydroxybenzaldehyde (o)	0.1	γ benzidine chloride
10	γ ,, (m,p)	0.05	γ tetrabase
0.05	γ phenol	0.1	γ diphenylamine
0.01	γ resorcinol	0.25	γ p-phenylenediamine
0.05	γ hydroquinone	0.5	γ p-methylaminophenol sulfate
0.05	γ pyrogallol	0.05	γ 8-hydroxyquinoline
2.5	γ aniline	0.5	γ dithiooxamide
0.5	γ K-ethyl xanthate	0.2	γ ascorbic acid
5	γ thiourea	0.1	γ uric acid
10	γ mercaptobenzothiazole	0.2	γ alloxantin
0.05	γ thiosemicarbazide	0.7	γ picolinic acid hydrazide
0.5	γ rhodizonic acid	0.8	γ isonicotinic acid hydrazide
0.1	γ naphthylamine (α or β)	1	γ Na-diethyl dithiocarbamate.

These data show the usefulness of this test for neutral, acidic and basic compounds which are oxidized by phosphomolybdic acid. A degree of selectivity can be attained if the test solution is made alkaline with dilute base and then extracted with ether. The ether is removed and its evaporation residue is then tested. This method will reveal bases and neutral compounds which are soluble in ether. The alkali salts of acidic oxidizable compounds are left in the water layer. They can be tested for by acidifying and then adding phosphomolybdic acid and ammonia. Selectivity can also be achieved if the test is restricted to compounds that are volatile with steam. The apparatus (Fig. 23, *Inorganic Spot Tests*) can be used for this purpose.

1 F. FEIGL, W. A. MANNHEIMER and L. VOKAČ, unpublished studies.

(6) *Reduction of Fehling, Tollens or Nessler reagents*

Fehling reagent intended for the detection of reducing compounds (particularly reducing sugars)[1] is prepared just prior to use by mixing equal volumes of copper sulfate solution (7.5 g $CuSO_4.5\ H_2O$ in 100 ml water) and an alkaline solution of sodium potassium tartrate (35 g tartrate plus 25 g potassium hydroxide in 100 ml water). The reagent contains complexly bound anionic copper, which is masked against precipitation by OH⁻

ions. However, the slight concentration of Cu^{2+} ions in Fehling solution is sufficient to react, on warming, with organic compounds which can function as strong reducing agents in alkaline solution. Copper(I) oxide (or oxyhydrate) is precipitated; its color (yellow to red) depends on the degree of dispersion and the particle size. When conducted as a drop reaction in a micro test tube it is possible to detect mg quantities of sugars (aldoses, ketoses), phenylhydrazine, and other derivatives of hydrazine.

A far more sensitive test of wider applicability is provided by a solution of silver oxide in ammonia. This so-called Tollens reagent[2] containing $[Ag(NH_3)_2]^+$ and OH^- ions reacts with reducing compounds, in the cold or on warming, to produce metallic silver as a black precipitate, which often forms an adherent mirror on the reaction vessel. This silver coating is favored if the vessel is cleaned beforehand with warm 10% sodium hydroxide and then rinsed with distilled water.

Tollens reagent oxidizes sugars, polyhydroxyphenols, hydroxycarboxylic acids, α-diketones, primary ketols, sulfinic acids, aminophenols, alkyl- and arylhydroxylamines, certain aromatic amines, hydrazo compounds, aldehydes, hydrazines, etc.[3]

Procedure.   The test is conducted in a micro test tube or in an Emich centrifuge tube. A drop of the test solution is mixed with a drop of the freshly prepared reagent and then placed for 15–30 sec in water warmed to 70–85°. Deposition of silver indicates reducing compounds.

    Reagent: Tollens solution. 1 ml of 10% silver nitrate solution is treated with 1 ml of 10% sodium hydroxide solution; the precipitate of silver oxide is dissolved by drop-wise addition of ammonia (1 : 1).*

The following *identification limits* indicate the sensitivity of the test:

0.1	γ formaldehyde	5	γ m-hydroxybenzaldehyde
1	γ galactose	0.25	γ isonicotinic acid hydrazide
0.5	γ glucose	1	γ benzoic acid hydrazide

Compounds which reduce in alkaline solution can be brought to reaction directly with silver oxide, instead of Tollens reagent, and the unused silver oxide dissolved by ammonia. This test[5] can be carried out as a sensitive spot reaction on filter paper.

The behavior of reducing compounds toward alkaline solutions of alkali iodide mercurate (Nessler's reagent, see p. 116) merits attention. Almost all aldehydes[6] precipitate finely divided mercury at room temperature, and aldoses and ketoses likewise react on warming.[7,8] The use of Nessler reagent in spot tests (plate or paper) is not limited to aliphatic aldehydes, but is capable of far wider application as shown by the following *limits of identification*:

---

* The Tollens solution should not be stored, because in time it becomes highly explosive through the formation of silver nitride.[4]

2.5 γ hydroquinone	1000 γ o-hydroxybenzaldehyde
5 γ pyrogallol	10 γ m- ,,
2 γ gallic acid	500 γ p- ,,
2.5 γ benzaldehyde	1 γ formic acid
10 γ nicotinic acid hydrazide	2 γ phenylhydrazine chloride.

1 H. FEHLING, *Ann.*, 72 (1849) 106; see also B. HERSTEIN, *J. Am. Chem. Soc.*, 32 (1910) 779.
2 B. TOLLENS, *Ber.*, 14 (1881) 1950; 15 (1882) 1635, 1828.
3 Comp. G. T. MORGAN and F. M. G. MICKLETHWAIT, *J. Soc. Chem. Ind.*, 21 (1902) 1375.
4 H. WALDMANN, *Chimia*, 13 (1959) 297.
5 F. FEIGL, *Chem. & Ind. (London)*, 16 (1938) 1161.
6 L. ROSENTHALER, *Pharm. Acta Helv.*, 29 (1954) 23.
7 T. H. LEE, *Chem. News*, 72 (1895) 153.
8 L. CRISMER, *Chemiker-Ztg.*, 4 (1889) 81.

(7) *Reduction of o-dinitrobenzene*

The colorless alcoholic-alkaline solution of 1,2-dinitrobenzene turns blue-violet on the addition of certain organic and inorganic reductants.[1] Only those organic compounds are effective which become reducing agents in alkaline solution through delivery of hydrogen. The 1,2-dinitrobenzene thus acts as acceptor for hydrogen, yielding, in cooperation with alkali hydroxide, blue-violet water-soluble alkali salts of a dibasic ortho-quinoidal nitrogen acid as shown by:[2]

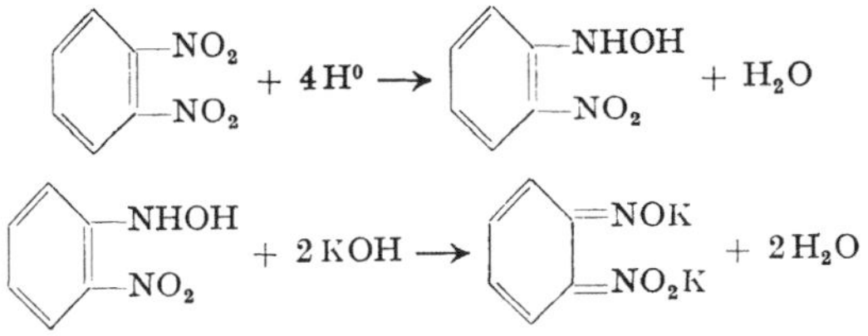

After observing and making analytical use of the color reaction of 1,2-dinitrobenzene with reducing sugars (aldoses),[3] the same behavior was found with respect to polyphenols and a series of other reducing organic compounds. This provides a selective preliminary spot test for certain classes of organic reductants.[4]

Procedure. A micro test tube is used. A drop of the aqueous or alcoholic test solution or a grain or two of the solid is treated in succession with one drop of 1% solution of 1,2-dinitrobenzene in alcohol and one drop of 0.5 N sodium hydroxide. The mixture is heated over a micro flame for not more than one minute at most. Depending on the amount of reducing substance, a more or less intense violet color appears.

This procedure revealed:

2.5	γ pyrocatechol	20	γ benzoic acid hydrazide
10	γ hydroquinone	25	γ nicotinic acid hydrazide
5	γ pyrogallol	5	γ glucose
10	γ tannic acid	10	γ lactose
10	γ ascorbic acid	10	γ maltose
5	γ phenylhydrazine chloride	5	γ l-sorbose

A positive response was given by: phenylhydroxylamine, benzoin, hydrazo-benzenes, adrenaline, glyceraldehyde, arabinose, rhamnose, xylose, α-galacturonic acid, xanthates.

This compilation shows that all of these compounds are reductants because of dehydrogenation. Since, with the exception of formaldehyde, other aldehydes react only in large amounts with 1,2-dinitrobenzene and then but weakly, the sensitive redox reaction with aldoses must be ascribed to the loss of the H-atoms (designated by*) of the $-\overset{*}{C}H(\overset{*}{O}H)CHO$ group.

1,2-Dinitrobenzene is readily reduced by hydroxylamine or hydrazine and their organic derivatives. The latter can be obtained by warming aromatic nitro-, nitroso-, azo-, azoxy compounds with zinc dust and ammonium chloride:

$$ArNO_2 \rightarrow ArNO \rightarrow ArNHOH \qquad ArAr_1N-NO \rightarrow ArAr_1N-NH_2$$

$$ArN=NAr \rightarrow ArNH-NHAr \qquad \underset{\overset{|}{O-}}{ArN^+=NAr} \rightarrow ArNH-NHAr$$

Accordingly, if the orginal sample does not react with an alcoholic alkaline solution of 1,2-dinitrobenzene, the alcoholic–aqueous solution can be warmed with zinc dust and ammonium chloride and the filtered (if need be) solution again tested with 1,2-dinitrobenzene. If a positive response is then obtained, the presence of nitrogen compound, reducible to aryl derivatives of hydroxylamine and hydrazine is indicated[5].

1 J. MEISENHEIMER, Ber., 36 (1903) 4174; J. MEISENHEIMER and E. PATZIG, Ber., 39 (1906) 2526; Ber., 50 (1919) 1161. Comp. also R. J. BLOCK and D. BOLLING, J. Biol. Chem., 129 (1939) 1.
2 R. KUHN and F. WEYGAND, Ber., 69 (1936) 1969.
3 P. K. BOSE, Z. Anal. Chem., 87 (1932) 110; Comp. M. PESEZ and P. POIRIER, Méthodes et Réactions de l'Analyse Organique, Vol. III, Masson et Cie, Paris, 1954, pp. 7, 10, 21.
4 F. FEIGL and L. VOKAČ, Mikrochim. Acta, (1955) 101.
5 M. PESEZ and P. POIRIER, loc. cit., Ref. 3, p. 22.

(8) Reduction of p-nitrosodimethylaniline[1]

The test for reducing compounds described under (7) is much more selective than those cited earlier, because it applies only to such compounds that function as hydrogen donors in strong alkaline solution. Still greater selectivity is shown in the behavior of p-nitrosodimethylaniline (I), because few hydrogen donors function for the redox reaction which leads to p-aminodimethylaniline (II):

$$(CH_3)_2N-\langle\bigcirc\rangle-NO + 4\,H^0 \rightarrow (CH_3)_2N-\langle\bigcirc\rangle-NH_2 + H_2O \qquad (1)$$

$$\qquad\quad (I) \qquad\qquad\qquad\qquad\qquad\qquad (II)$$

If this redox-reaction is conducted in neutral, slightly acid or slightly basic solution, in the presence of p-dimethylaminobenzaldehyde (III), the latter condenses immediately with the p-aminodimethylaniline (II) to give a colored water-soluble product which is the quinoidal protonized anion of the Schiff base (IV):

$$(CH_3)_2NC_6H_4NH_2 + OCHC_6H_4N(CH_3)_2 \rightarrow (CH_3)_2NC_6H_4N{=}CHC_6H_4N(CH_3)_2 + H_2O \qquad (2)$$

$$\quad (II) \qquad\qquad\quad (III) \qquad\qquad\qquad\qquad (IV)$$

The following compounds are capable of bringing about reaction (1) and the associated reaction (2): gallic acid, tannin, ascorbic acid, mercapto- and thiol compounds, and also organic derivatives of hydrazine. Such tiny amounts of these materials are needed that a selective preliminary test in the form of a spot reaction becomes possible.

The test described here is not applicable in the presence of compounds which yield colored condensation products with (I) or (III). Such compounds include primary aromatic amines and their salts, and also compounds with mobile CH_2 groups.[2] Therefore, the sample must be tested beforehand with regard to its behavior toward (I) and (III). This can be done by a spot reaction conducted on an alcoholic solution. If no change is seen, the color reaction with the mixture of the two dimethylamino compounds indicates the presence of one of the reducing compounds noted above.

Among the inorganic reducing agents, only hydrazine salts interfere; a red aldazine is formed with p-dimethylaminobenzaldehyde (compare *Inorganic Spot Tests*, Chapter 3).

The test may be made in a micro test tube, or better as a spot reaction on filter paper impregnated with both reagents. p-Dimethylaminobenzaldehyde can be replaced by p-dimethylaminocinnamaldehyde.

Procedure.   Filter paper (Whatman No. 120) is bathed in a solution of 0.2 g p-nitrosodimethylaniline and 1 g p-dimethylaminobenzaldehyde in 100 ml etha- nol. The impregnated paper keeps well in the dark. A drop of the test solution is placed on the yellow reagent paper. The solution may be aqueous (neutral, weakly alkaline or weakly acid). It is also permissible to test solutions in alcohol, ether, *etc.* The spotted paper is kept at 110° (drying oven) for 3–5 min. If con- siderable amounts of active reductants are present, the positive response is im- mediately discernible through the formation of a red or red-brown stain. If the paper is dipped into water, the yellow nitroso compound is leached out within several minutes, and the more or less intense red stain is left on the white background.

When strongly acidic solutions are to be tested, excess calcium carbonate is added, and a drop of the neutral filtrate or centrifugate is used.

A positive response was given by:

100 γ	gallic acid	2	γ benzoic acid hydrazide
5 γ	tannic acid	3	γ nicotinic acid hydrazide
1 γ	ascorbic acid	5	γ thionalide
1 γ	thioglycolic acid	5	γ mercaptobenzothiazole
2 γ	cysteine (as chloride)	10	γ toluene-2,3-dithiol
2 γ	semicarbazide	2	γ thiosalicylic acid
5 γ	thiosemicarbazide	7.5	γ thioglycolic acid anilide

As was to be expected, a response is given also by compounds which yield hydrolysis products with SH groups, namely xanthates, dialkyldithiocarbamates, salts of thiocarboxylic acids, and penicillin.

It should be noted that a reaction is given by those thioketo compounds which yield an SH group by enolization. Accordingly 5-thiobarbituric acid can be detected with an identification limit of 2.5 γ.

It may be pointed out with regard to the behavior of tannin and gallic acid, that the former reacts even at room temperature, though with less sensitivity, whereas 800 γ gallic acid is inactive. In these instances, the stains are not red but red-brown. This is due to the formation of brown oxidation products of these polyphenolic compounds along with the red Schiff base.

1 F. FEIGL and D. GOLDSTEIN, *Mikrochem. J.*, 1 (1957) 177.
2 P. EHRLICH and F. SACHS, *Ber.*, 32 (1899) 2341.

(9) Reduction of selenious acid

Many organic compounds of the most diverse classes are oxidized when warmed with selenious acid. Elementary selenium is formed. This reaction is of great importance in preparation procedures.[1] In contrast, very few organic compounds are known which react with this oxidant rapidly even at room temperature; hence a selective preliminary test becomes possible. Among the compounds in this class are: ascorbic acid (the sole compound containing only C, H, and O); acetylene[2]; thiourea, and almost all its derivatives; organic derivatives of hydrazine.

It is interesting that compounds with SH groups are not affected by selenious acid at room temperature, even though as derivatives of hydrogen sulfide they doubtless are far stronger reductants than thioketones. Since SH compounds are readily detected by (8), the positive or negative response to the test described here can give interesting hints regarding the nature of reducing compounds present.

Procedure. A drop of the test solution is brought to dryness on a spot plate. A drop of 10% solution of selenious acid in water or alcohol is then added. Depending on the quantity of oxidizable compound, red (sometimes black) selenium separates at once or within 5 min.

The following were detected:

10 γ thiourea	20 γ *sym-* or *asym-*diphenylthiourea
15 γ 1,3-diethylthiourea	50 γ 1,1,3,3-tetramethylthiourea
2.5 γ dithiooxamide	10 γ thiobarbituric acid
5 γ dithiotartaric acid	1 ʎ benzoic acid hydrazide
1 γ picolinic acid hydrazide	0.5 γ isonicotinic acid hydrazide

1 Comp. N. Rabjohn, *Organic Reactions*, Vol.V, Chap. 8, Wiley, New York, 1952.
2 A. Jouve, *Chem. Zbl.*, 1901, (I) 1389.

32. The Le Rosen test for aromatic compounds[1]

Concentrated sulfuric acid containing formaldehyde reacts with aromatic hydrocarbons, phenols, polyphenols, thiophene, *etc.* at room temperature, or on gentle warming. Red, violet, green precipitates or colorations appear. The same reactions occur if formaldehyde is added to solutions or suspensions of the organic compounds in concentrated sulfuric acid. The composition of the colored products is not known. The concentrated sulfuric acid, which is both an oxidant and dehydrant, probably brings about first a condensation of the aromatic compounds with formaldehyde, and then oxidizes the resulting diarylmethylene compounds to colored *p*-quinoidal products. Accordingly, benzene for example may undergo the reactions:

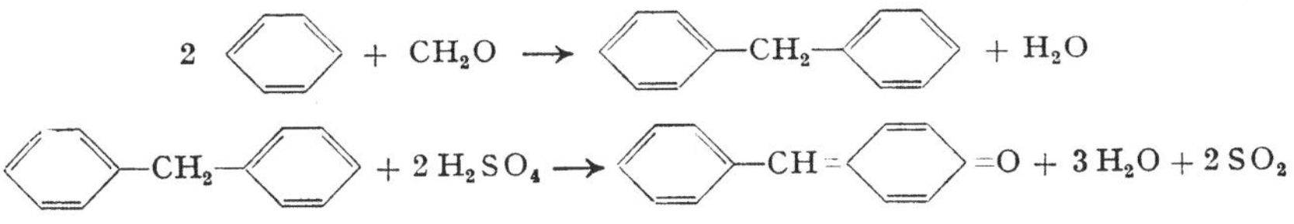

In accord with this explanation,[2] which can also apply to derivatives of benzene, the color reaction occurs with only those aromatic compounds which condense with formaldehyde to produce such methylene compounds as possess a free *para* position or a *para* OH group in order that an oxidation to colored *p*-quinoidal compounds can ensue through the action of the concentrated sulfuric acid. These conditions are satisfied in the case of pure aromatic hydrocarbons and also by phenols with a free *para* position. On the other hand, when dealing with substituted aromatic hydrocarbons and phenols, it must be kept in mind that the groups bound to the ring may retard or prevent the condensation with formaldehyde, or it may take place at positions where an oxidation to colored quinoidal products does not occur. A hindering of this kind is probably responsible for the fact that some aromatics do not react at all with formaldehyde–sulfuric acid, or only after long standing and warming. However, even small amounts of so many aromatics react so fast that the color test has considerable orienting value.

It should be kept in mind that some organic compounds dissolve in concentrated sulfuric acid with formation of unstable colored solvates (compare p. 156), and stable colored products may also result from the action of concentrated sulfuric acid. Therefore, the behavior of the test material toward concentrated sulfuric acid should be determined in advance. If a color appears, which is altered on addition of formaldehyde, it may be taken as an indication of the presence of reactive aromatics. When considerable amounts of the colored quinoidal compounds result, the color frequently becomes darker after a while. Possibly, this effect is due to a partial carbonization by the concentrated sulfuric acid.

The sample may be solid, liquid, gaseous, or dissolved in organic liquids. Procedure I will indisputably reveal aromatic hydrocarbons and phenols, which have a distinct vapor pressure at room temperature or when gently warmed. Procedure II is applicable to solid, liquid, or dissolved samples. In this case, however, in contrast to I, a negative result cannot be taken as absolute proof of the absence of aromatic compounds.

Procedure I.[3] (For volatile compounds.) A little of the solid or liquid sample is placed in the bulb of the apparatus (Fig. 23, *Inorganic Spot Tests*), or a drop of the ethereal test solution is evaporated there by brief immersion in warm water. A drop of the reagent solution is placed on the knob of the stopper; the apparatus is closed, and placed in hot water (60–80°). After 2–3 min., the drop on the knob is transferred (by wiping) to a depression of a white spot plate. A more or less intense color indicates the presence of volatile, reactive aromatic compounds.

If a solution in ether is used, and the solvent evaporated, considerable amounts of benzene, toluene, and phenol are carried along with the ether vapors. Such losses can be prevented by first placing one or two drops of concentrated sulfuric acid in the bulb of the apparatus and then adding the ethereal solution of the sample. Ether is soluble in concentrated sulfuric acid, and its vapor tension is thus reduced so much that only benzene, toluene, and the like vaporize at room temperature or on gentle warming.

Reagent: Formaldehyde–sulfuric acid. 0.2 ml of 37% formaldehyde plus 10 ml concentrated sulfuric acid. The reagent should be freshly prepared.

This procedure revealed, starting with ethereal solutions:

2 γ benzene	2 γ phenol	25 γ anthracene
1.5 γ toluene	4 γ o-cresol	2 γ thiophene
5 γ naphthalene	8 γ benzaldehyde	

Procedure II. (For nonvolatile compounds.) A drop of the non-aqueous test solution, or a little of the solid, is treated in a depression of a spot plate with a drop of the reagent solution. A positive reaction is signalled by the development of a color. A parallel trial with concentrated sulfuric acid must be conducted in all cases. If the latter yields a color, a drop of the reagent should be added and any change in the color noted. It is often advisable to conduct the test with diluted solutions, to establish the shade of the color better.

The following summary contains the findings with aromatic compounds of various classes.*

Anisole	red-violet	Methyl salicylate	red
Anthracene	yellow-green	Naphthalene	green
Benzaldehyde	red, † yellow-brown	Naphthoic acid	green,
Benzene	red		† slightly green
Benzildioxime	yellow	β-Naphthol	brown
Benzoinoxime	yellow	Naphthoresorcinol	green,† brown
Benzyl alcohol	red	Naphthylamine chloride	green
Benzyl mercaptan	orange	Nitronaphthalene	green-blue,
Carbazole	green		† red
Catechol	violet-red	Phenanthrene	green
Chlorobenzene	red	Phenol	red-violet
Cinnamic acid	brick red	Phenylarsonic acid	red
Cumene	red	Phenyl ether	violet
Dibenzyl	dark red	Phloroglucinol	brown-red
Diphenyl	blue-green	Pyrogallol	red,† yellow
Diphenylbenzidine	red	Resorcinol	red
Diphenylmethane	red	Salicylic acid	red
Ethylbenzene	red-tan	Stilbene	brown
Fluorene	green	Sulfosalicylic acid	
Furildioxime	brown,	(heating at 110°)	red
	† yellow	Tetralin	red
Gallic acid	yellow-green	Thionalide	green
Hexaethylbenzene	yellow	Thiophene	red-violet
Hydroquinone	black	Toluene	red
o-Hydroxydiphenyl	red	Triethylbenzene	orange
p- ,,	green	1,3,5-Triphenylbenzene	blue
Mandelic acid	violet-red	Triphenylmethane	red
Mesitylene	tan	o, m, p-Xylene	red

† in conc. H_2SO_4.

The following develop no color or there is no difference from the color of the solution in concentrated sulfuric acid: aminobenzoic acid, aniline, anthraquinone, azoxybenzene, benzidine, benzil, benzoic acid, benzophenone, salicylaldehyde, α,α'-dipyridyl, diphenylamine, 8-hydroxyquinoline, mercaptobenzothiazole, o-nitrobenzoic acid, o-nitrophenol, 6-nitroquinoline, pentachlorophenol, phthalic acid, picrolonic acid, salicylaldoxime.

The compilation reveals that diagnostic conclusions can be drawn both

* The statements are taken from the paper by LE ROSEN et al. (loc. cit.) and the experiments of D. GOLDSTEIN, Rio de Janeiro.

from the occurrence of a color reaction and from the actual shade obtained.

When examining mixtures, it is well to make a sublimation trial, and to subject the sublimate (if any) to the test. Since relatively few organic compounds are sublimable at normal pressure, a positive response to this color test considerably reduces the number of compounds which need be considered.

When using the Le Rosen test, it should be noted that higher alcohols (from propanol on) give a yellow, brown or red-brown color with the reagent.[4] The intensity depends on the amount and time of contact.

1 A. L. Le Rosen, R. T. Moravek and J. K. Carlton, *Anal. Chem.*, 24 (1952) 1335; Comp. G. Denigès, L. Chelle and A. Labat, *Précis de Chimie Analytique*, 7th ed., Maloine, Paris, 1930, p. 176.
2 Comp. also M. J. Rosen, *Anal. Chem.*, 27 (1955) 111.
3 Unpublished studies, with Cl. Costa Neto.
4 L. Rosenthaler, *Pharm. Acta Helv.*, 32 (1957) 440.

33. Sulfonation test for aromatics[1]

A test for aromatic sulfonic acids is described on p. 234; it is based on the formation of nickel sulfide when these compounds are treated with Raney nickel alloy and caustic alkali, and then with hydrochloric acid. Since the test is very sensitive it may be applied to detect even a slight sulfonation of aromatics by concentrated sulfuric acid. Aromatic hydrocarbons are readily sulfonated, but substitutions on the aromatic ring may impede this operation or even prevent it completely. The sulfonation test described here gives a positive response with aromatics which do not respond to the Le Rosen test (see Sect. 32) or to the chloranil test (see p. 144). It appears as though the converse may likewise be true. Therefore, indications of the presence or absence of certain aromatics can be obtained through three different kinds of tests.

The presence of aliphatic compounds does not impair the test described here. Such compounds are "kjeldahlized" by concentrated sulfuric acid and eventually resulting aliphatic sulfonic acids are resistant to Raney alloy and caustic alkali.

Procedure I. (Nonvolatile and nonsublimable aromatics.) A micro test tube is used. A small quantity of the solid test material or a drop of its solution in alcohol, ether, *etc.* is taken to dryness if necessary and 1 drop of concentrated sulfuric acid is added. The mixture is kept for 30 min. in a drying oven (140°) and then cooled. Concentrated caustic alkali solution is added and the basic solution is treated with several cg of Raney alloy and warmed for 5 min. in a boiling water bath. The system is then acidified with concentrated hydrochloric acid and the mouth of the test tube is covered with a piece of lead acetate paper and reheated in the water bath. A positive response is indicated by the development of a black or brown stain on the reagent paper.

Procedure II. (Low boiling or sublimable aromatics.) The test material or a drop of its solution is placed in a micro test tube and 1 drop of concentrated sulfuric acid is added. The upper end of the test tube is then sealed. After heating to 140° for 30 min., the constricted end of the test tube is cut off and the contents are heated briefly to expel any sulfur dioxide. The remainder of the test is conducted as described in Procedure I.

The following *identification limits* give an idea of the sensitivity of this test:

15 γ naphthalene   15 γ diphenyl   10 γ benzene   15 γ phenanthrene

Procedures I and II were applied to 0.5 mg specimens of the following compounds:

Positive response: anthracene, 7,8-benzoflavone, benzophenone, 1,4-diphenylbenzene, hydroquinone, α-naphthol, β-naphthol, pyrocatechol, pyrogallol, toluene.

Weaker response: alizarin, azobenzene, benzoic acid, carbazole, cinnamic acid, curcumin, diphenylamine, isoquinoline, α-nitroso-β-naphthol, quinaldine, quinine, quinoline, phenanthraquinone, tryptophan, tyrosine.

No response: anthraquinone, caffeine, chloranil, DDT, hexachlorobenzene, pentachlorophenol, phenolphthalein, phthalic acid, terephthalic acid, resorcinol, tribromoaniline.

1 Unpublished studies, with V. GENTIL.

34. Pyrolysis tests for aromatic compound constaining oxygen[1]

If phenolic compounds, including phenol esters and ethers, are strongly dry heated, phenol is split off. The tendency for this appears to be quite strong because non-phenolic aromatics, which have oxygen atoms in open or closed side chains likewise yield phenol under this treatment. It is likely that the formation of phenols is only a partial reaction during the pyrolysis and that phenol losses are inevitable because of combustion. Despite these limitations, a preliminary test for oxygen-containing aromatics can be based on the detection of pyrolytically produced phenols through their color reaction with 2,6-dichloroquinone-4-chloroimine (see p. 185).

The procedure described here is reliable in the absence of aliphatic or aromatic nitro compounds. Such compounds interfere through the pyrolytic splitting out of nitrous acid, which destroys the phenol reagent. It is easy to test beforehand for nitro compounds in a separate portion, since when dry heated they yield nitrous acid detectable by the sensitive Griess reaction (see p. 90). If the result is positive, another portion should be warmed with zinc and acetic acid, which easily converts NO_2 groups to NH_2 groups. Then the reaction residue is taken to dryness and subjected to the pyrolysis procedure. Aromatic mononitro compounds, *i.e.* nitrobenzene and nitronaphthalene yield phenol when pyrolyzed.

Procedure. About 0.5 mg of the sample is placed in a micro-test tube fixed in an asbestos support. The mouth of the tube (about 5 mm above the asbestos plate) is covered with a disk of filter paper impregnated with a saturated benzene solution of 2,6-dichloroquinone-4-chloroimine. The bottom of the test tube is heated, gently at first and then strongly. In most cases, this quasi dry distillation produces heavy brown or grey vapors. It is important that this "fog", which ascends with the heating, comes into contact with the reagent paper. This usually requires several minutes. The paper is then held over strong ammonia. If the response is positive, a blue spot appears which gradually fades on standing. The color can be restored by a renewed exposure to ammonia.*

The procedure was tested on the following phenolic compounds: α- and β-naphthol, resorcinol, pyrocatechol, *o*- and *p*-hydroxydiphenyl, hydroxy-hydroquinone, di-β-naphthol, 2,7-dihydroxynaphthalene, tetrahydroxy-anthraquinone, (2,7-), (2,6-), (1,4-) and (1,5-) naphtholsulfonic acids, ellagic acid, 4-amino-3-methylphenol, 1-amino-2-naphthol-4-sulfonic acid, 8-hydr-oxyquinoline-7-sulfonic acid, 8-hydroxyquinaldine, sulfosalicylic acid, tetra-chlorohydroquinone, 2-hydroxy-5-chlorobenzaldehyde, 2,4-dihydroxyaceto-phenone, *p*-hydroxybenzaldehyde, salicylaldoxime, pentachlorophenol, 2,4-dihydroxybenzaldehyde, 4-hydroxybenzophenone, orcinol, gallic acid, phe-nolphthalein, fluorescein, phenetole, veratrole, isoeugenol methyl ether, sa-frole, isosafrole, anethole, γ-phenoxybutyric acid, piperonal, rotenone, umbelliferone ethyl ether, narceine, emetine, *m*-phenetidine, 4-methoxy-benzaldehydesulfonic acid, anisic acid, narcotine, acetylsalicylic acid.

The following non phenolic compounds also yielded phenols when pyro-lyzed: benzoic, phthalic, naphthoic, mandelic, chloromandelic, *o*-nitroben-zoic, phenylanthranilic, phenylarsonic, *m*-nitrophenylarsonic acids, acetani-lide, benzanilide, α-acetylnaphthalide, *p*-chloro- and 3,4-dichlorobenzalde-hyde, quinone, anthraquinone, phenanthraquinone, cinnamic acid, hippuric acid, benzil, benzophenone, ninhydrin, diphenylmethylcarbinol, vulpinic acid, phenylacetic acid, β-phenylalanine, benzilmonoxime, acetophenone, xanthydrol, anthrone, 7,8-benzoflavone.

There appears to be a marked tendency toward the formation of phenol or volatile phenolic compounds during the pyrolysis of aromatics with oxygen atoms in substituents, and this is why phenyl compounds in which

* Blue-violet condensation products are formed immediately when the vapors of aromatic amines come into contact with the reagent paper. However, these products are decomposed by ammonia fumes and so do not interfere with the test for phenols, since then the blue only appears in ammoniacal medium.

oxygen atoms are bound to the benzene nucleus through metals or non-metals likewise split off phenol when pyrolyzed.

Pertinent examples include: phenylmercury acetate, phenylarsonic acid, m-tolylstibonic acid, triphenylarsinic oxide, nitrobenzene, nitronaphthol, but not dinitro compounds.

The range of applicability of the test is illustrated by the finding that the pyrolytic splitting off of phenol is readily detectable from amounts as small as 0.5 mg or less in the case of phenylurea, acetyldiphenylamine, antipyrine, and luminal, i.e. from compounds in which the phenyl groups and O atoms are far apart.

1 F. FEIGL and E. JUNGREIS, Anal. Chem., 31 (1959) 2099.

35. Ehrlich diazo test for compounds capable of coupling

The Ehrlich diazo test,[1] which is widely used in biological chemistry for detecting physiologically important substances, especially phenol- and imidazole derivatives can serve as a sensitive preliminary test for compounds capable of entering into coupling reactions. The reagent employed is a freshly prepared solution of diazobenzenesulfonic acid (II) obtained by diazotizing sulfanilic acid (I) with nitrous acid, in the presence of hydrochloric acid:

$$HO_3S-\langle\bigcirc\rangle-NH_2 + HNO_2 + H^+ \longrightarrow HO_3S-\langle\bigcirc\rangle-\overset{+}{N}\equiv N + 2\,H_2O$$

(I)   (II)

Compound (II) in the colorless solution couples instantly with phenols and aromatic amines to produce acid or basic azo dyestuffs:

$$HO_3S-\langle\bigcirc\rangle-\overset{+}{N}\equiv N + \langle\bigcirc\rangle-OH \longrightarrow HO_3S-\langle\bigcirc\rangle-N=N-\langle\bigcirc\rangle-OH + H^+$$

$$HO_3S-\langle\bigcirc\rangle-\overset{+}{N}\equiv N + \langle\bigcirc\rangle-NH_2 \longrightarrow HO_3S-\langle\bigcirc\rangle-N=N-\langle\bigcirc\rangle-NH_2 + H^+$$

If the para position of the phenol or amine is occupied, the coupling occurs in the position ortho to the OH or NH_2 group. The addition of sodium carbonate is necessary to bring about and complete the coupling. The following test is suitable for the rapid detection of compounds which couple with diazo compounds. Of primary interest in this connection are phenols, and primary, secondary, and tertiary aromatic amines.

In the case of imidazole derivatives, the coupling with diazotized sulfanilic acid occurs in the α position to the imide group of the imidazole ring:[2]

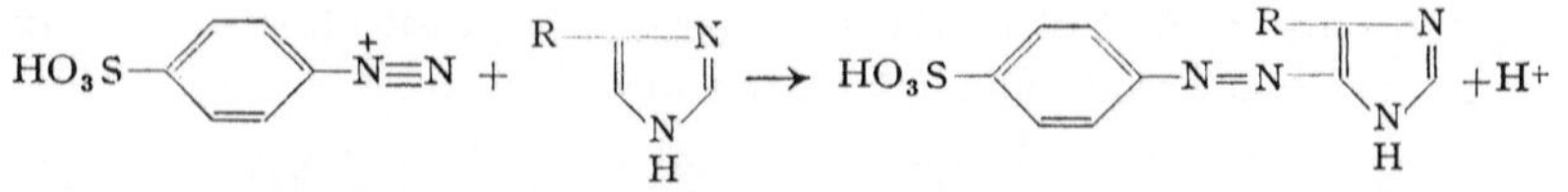

R = $CH_2CH(NH_2)COOH$ for histidine, and $CH_2CH_2(NH_2)$ for histamine.

Coupling likewise occurs with certain other nitrogen heterocycles such as thymine, thiamine (vitamin B_1 etc.)[3] and with primary aliphatic nitro compounds and with aldehyde arylhydrazones.

Procedure.[4]   One drop of the reagent and one drop of 0.5% alkali nitrite solution are mixed in a depression of a spot plate, and then one drop of the test solution is added with stirring. A drop of 10% $NaHCO_3$ solution is then introduced. The immediate production of a color is taken as a positive response. Some compounds are more sluggish and 1–2 min. waiting is not too long, but a blank test is essential in such cases. Likewise, a blank comparison test is always advisable when alcoholic solutions are being examined.

Reagent: 0.5% solution of sulfanilic acid in 2% hydrochloric acid.

The procedure revealed:

0.2 γ phenol	reddish yellow	
2 γ diethylaniline	reddish brown	
0.8 γ resorcinol	,,	,,
15 γ dimethylaniline	,,	,,
2 γ α-naphthol	,,	,,
5 γ β- ,,	yellow	
0.1 γ m-hydroxybenzaldehyde	brown-yellow	
50 γ p- ,,	orange-red	
5 γ p-hydroxydiphenyl	red	
1.5 γ tyrosine	,,	
5 γ aniline	greenish yellow	

1 P. EHRLICH, *Z. Klin. Med.*, 5 (1882) 285.
2 R. BURIAN, *Ber.*, 37 (1904) 696; H. PAULY, *Z. Physiol. Chem.*, 44 (1905) 159.
3 Comp. M. PESEZ and P. POIRIER, *Méthodes et Réactions de l'Analyse Organique,* Vol. III, Masson et Cie., Paris, 1954, p. 82.
4 Unpublished studies, with A. BONDI (Rehovoth).

36. Colmant chloranil test[1]

When warmed with potassium chlorate and concentrated hydrochloric acid, many aromatic compounds form chloranil (tetrachloro-*p*-benzoquinone) as the stable end product.[2] The necessary oxidation, oxidative cleavage and chlorination, are accomplished by the gaseous reaction products formed by warming chlorate with concentrated hydrochloric acid:

$$HClO_3 + 5\ HCl \rightarrow 3\ H_2O + 3\ Cl_2 \tag{1}$$

$$4\ HClO_3 \rightarrow 4\ ClO_2 + O_2 + 2\ H_2O \tag{2}$$

Since chloric acid is completely destroyed by (1) and (2), any chloranil that may have been formed can be detected by adding potassium iodide (liberation of iodine) or still more conclusively (after extraction with ether) by means of tetrabase which yields a blue quinoidal compound (see p. 447).

Procedure.[3] A micro test tube is used. A drop of the test solution or a pinch of the solid is warmed with one drop of saturated potassium chlorate solution and one drop of concentrated hydrochloric acid. After no more chlorine is evolved, the solution is cooled, 2 or 3 drops of water added, and shaken with 5–10 drops of ether. A drop or two of the ether layer is placed on filter paper and spotted with one drop of a 1% ethereal solution of tetrabase. After the ether has evaporated (air blast) a blue stain is left indicating the presence of aromatic compounds that are oxidatively decomposed to form chloranil.

The following compounds were tested in mg quantities. The *identification limits* were determined in those cases in which the response was especially strong.

I. Strong response:

acetanilide; benzidine (0.5 γ); p-chloroaniline (0.5 γ); coumarin; diphenylamine (0.25 γ); emetine; p-hydroxybenzaldehyde (0.25 γ); m-hydroxybenzaldehyde; 8-hydroxyquinoline; p-hydroxydiphenyl; α-naphthol (10 γ); β-naphthol (10 γ); isatin (1 γ); methylene blue (1 γ); mercaptobenzothiazole (10 γ); naphthylamine (5 γ); naphthylamine hydrochloride; p-nitrophenol (0.5 γ); papaverine hydrochloride; pentachlorophenol; phenolphthalein (1 γ); phenylurea (10 γ); picric acid (5 γ); quinone; tryptophan; salicin (2.5 γ); salicylic acid (1 γ); sulfamethazine; sulfanilic acid (0.5 γ); sulfapyridine; sulfathiazole; sulfosalicylic acid (1 γ).

II. Weaker response:

azobenzene; ephedrine; isoquinoline; Congo red; nitrobenzene; o-nitrobenzoic acid; quinoline; tetrabase; toluene.

III. No response:

alizarin; anthraquinone; anthracene; benzoic, chromotropic and ellagic acid; curcumin; hexachlorobenzene; morin; morphine chloride; naphthalene; α-nitroso-β-naphthol; phthalic acid; pyrogallol; pyrocatechol; quinalizarin; quinine chloride; resorcinol; saccharin; Sudan III.

These findings show that many but by no means all aromatic compounds give the chloranil test. However, the results are useful if they are correlated with the findings of other tests for aromatics. It is remarkable that polyphenols, hydroxyanthraquinones and flavanols, and also such simply constructed compounds as benzoic acid and phthalic acid, yield no chloranil. Therefore, certain groups and structural factors are responsible for the formation of chloranil. This is shown plainly in the behavior of naphthol- and naphthylamine-sulfonic acids, which can be differentiated with the aid of the chloranil test (see Chap. 7).

1 R. L. P. COLMANT, *Chem. Zbl.*, (1931), I, 3705.
2 Comp. P. KARRER, *Organic Chemistry*, 4th Engl. ed., Elsevier, Amsterdam, 1950, p. 580.
3 F. FEIGL, V. GENTIL and J. E. R. MARINS, *Anal. Chim. Acta*, 13 (1955) 210.

37. Fuming-off test with concentrated nitric acid[1]

Characteristic alterations may ensue on warming nonvolatile organic compounds with concentrated nitric acid. Of interest are: nitration, oxidation of particular groups, oxidative cleavage, deamination as shown in the following assembly:

1. Aromatic hydrocarbons are nitrated.

2. Phenols are converted to mono- and dinitrophenols, which are more acidic than the initial compounds. Aci-nitro compounds are formed if the nitro group enters *para* to an OH group.

3. Thiophenols, mercaptans, sulfides, and disulfides are oxidized to sulfonic acids. Nitration may occur in addition.

4. Aromatic amines are deaminated and nitrated, *i.e.*, acidic nitrophenols are produced.

5. Certain cyclic nitrogen bases, and also alkaloids, are split with production of carboxylic acids.*

6. Aliphatic polyhydroxy compounds are converted into the corresponding dicarboxylic acids by oxidation of the terminal $-CH_2OH$ and $-CHO$ groups.**

7. Glycosides, which are esters of phenols and alcohols with various sugars, are split hydrolytically and the sugars are then oxidized as per 6. Any liberated phenol is nitrated.

* Familiar instances include: oxidative cleavage of: pyrrole to oxalic acid, piperidine to aminobutyric acid, narcotine to opianic acid, nicotine and alkaloids of the pyridine series to pyridinecarboxylic acids.

** Cyclohexanol and cyclohexanone are transformed into the noncyclic dibasic adipic acid by treatment with strong nitric acid. This behavior can be employed for analytical purposes (see p. 414).

1 Unpublished studies, with V. GENTIL.

38. Fusion test with benzoin[1]

Fused benzoin is a very energetic hydrogen donor with respect to suitable acceptors. Benzil results. It should be remembered that this transformation can occur not only from the hydroxyketone form but also from the isomeric more acidic reductone form of benzoin:

$$C_6H_5CHOHCOC_6H_5 \rightarrow C_6H_5COCOC_6H_5 + 2 H^0$$

$$C_6H_5C(OH){=}C(OH)C_6H_5 \rightarrow C_6H_5COCOC_6H_5 + 2 H^0$$

The pyrolytic generation of hydrogen can bring about the cleavage of certain compounds with acceptance of hydrogen, an action that may be called pyrohydrogenolysis. The following cleavages come into consideration:

$RX + 2 H^0 \qquad \rightarrow RH + HX$ (X = Hal, SO_3H, NO_2, SH; R = aliphatic radical)

$RN{-}NO_2 + 2 H^0 \rightarrow RNH + HNO_2$

$RONO_2 + 2 H^0 \quad \rightarrow ROH + HNO_2$

$B.HNO_3 + 2 H^0 \quad \rightarrow B + H_2O + HNO_2$ (B = organic base)

$(R,Ar)CNS + 2 H^0 \rightarrow (R,Ar)CN + H_2S$

$HgArX' + 2 H^0 \quad \rightarrow Hg + ArH + HX'$ (X' = CH_3COO, Cl, CN)

These redox reactions produce volatile acidic hydrogen compounds which can be readily detected in the gas phase by suitable reagent papers. This makes possible a preliminary test for compounds which undergo a pyrohydrogenolysis. Benzoin (m.p. 135°) is melted in a micro test tube along with a minimal quantity of the test material, the test tube being immersed in a glycerol bath which has been preheated to 120–160°. The mouth of the test tube is covered with a piece of indicator paper or reagent paper suitable for the detection of H_2S, SO_2, HNO_2, or HCN.

The fusion test assumes the absence of compounds which yield halogen hydride, SO_2, etc. when heated to 180°. Although such compounds will be encountered rarely, a blank test should be made invariably.

Compare the pertinent statements in Chap. 3 regarding the procedure for the fusion test with benzoin for the detection of functional groups included in the above list and the identification limits obtained.

It should be pointed out that some organic compounds split off volatile bases when melted with benzoin. Included are: hydrazides of carboxylic and sulfonic acids, which lose NH_3, and carbanilide and phenylhydrazides of carboxylic acids which yield aniline. This behavior too makes possible a preliminary test, but of course it is characteristic only if the test material does not yield ammonia or aniline when heated without benzoin.

1 F. Feigl, *Angew. Chem.*, 73 (1961) 656.

39. Sinter test with mercuric cyanide[1]

Mercuric cyanide can be heated in a micro test tube to 170° with no signs of sublimation or decomposition being shown by the cyanide test (bluing of a copper–benzidine acetate solution). In contrast, there is rapid generation of hydrogen cyanide if a mixture of mercuric cyanide and certain compounds is heated. Examples are:

1. Metal formates and esters of formic acid (see p. 453)
2. Acidic compounds of all kinds, including those which show no acidic behavior when subjected to the procedure given in Sect. 24 (oximes, phenols, thiols, *etc.*)
3. Salts of organic bases with inorganic or organic acids
4. Amphoteric compounds (*e.g.* hydroxyquinoline, amino acids, benza⁻midine)
5. Compounds in which acidic OH groups are developed through enolization of CO or NO groups (*e.g.* acetylacetone, acetoacetic ester, acid amide hydrazides, primary and secondary aliphatic nitro compounds)
6. Compounds with acidic imide groups
7. Compounds which contain 2- or 4-valent sulfur (see p. 83)
8. Compounds, which in addition to acid groups contain several basic groups and consequently take on a basic character.

As shown by 1–8, the compounds involved here are mostly those with occult acid groups, whose detection by pyrolysis with mercuric cyanide was mentioned in Sect. 25. The test described here is suitable for compounds whose acidic character is extremely weak. This suitability is demonstrated by the positive response to the sinter test given by ammonium salts and by diazoaminobenzene (C_6H_5—N=N—NH—C_6H_5). The aminosulfonic acids taurine and sulfanilic acid do not react with $Hg(CN)_2$. These amino acids behave as absolutely neutral inner ammonium salts (compare p. 368).

Procedure.   The test is conducted in a micro test tube. A little of the test material, if need be the evaporation residue from a drop of the solution, is mixed with several cg of mercuric cyanide. The mouth of the tube is covered with a piece of filter paper moistened with cyanide reagent solution (for preparation see p. 546). The test tube is immersed to about half its length in a glycerol bath preheated to 120°. The temperature can be raised to 160° if necessary. A positive response is shown by the development of a blue stain on the reagent paper within 3 min. at most.

The procedure was checked with numerous compounds of the kinds listed above and in amounts below 0.5 mg.

It should be noted that compounds which lose water at 170° should not be present since hydrogen cyanide will then be formed through pyrohydrolysis: $Hg(CN)_2 + H_2O \rightarrow Hg(OH)CN + HCN$. A test for compounds of this kind is given in the next section.

---

1 Unpublished studies, with E. LIBERGOTT.

40. Detection of compounds which split off water or ammonia when heated to 190°[1]

The presence of such compounds can be revealed by the fact that they split off hydrogen sulfide when heated to 190° along with thio-Michler's ketone. The reason for this result is that the pyrolytically released water (as quasi superheated steam) or ammonia brings about a hydrolysis or ammonolysis of the thioketone:

$$(CH_3)_2N-\langle\ \rangle-\underset{\underset{S}{\|}}{C}-\langle\ \rangle-N(CH_3)_2 + H_2O\ [NH_3]\ \longrightarrow$$

$$(CH_3)_2N-\langle\ \rangle-\underset{\underset{O(NH)}{\|}}{C}-\langle\ \rangle-N(CH_3)_2 + H_2S$$

Thiobarbituric acid, which likewise contains a CS group, may be used instead of the thio-Michler's ketone as acceptor for the water or ammonia produced by the pyrolysis. As a rule, heating for 2–5 min. is sufficient to reveal the hydrogen sulfide in the gas phase through the blackening of lead acetate paper. The upper temperature given here should not be exceeded since temperatures above 200° result in a slight but still detectable production of hydrogen sulfide from these thioketo compounds.

Compounds, which yield ammonia when dry heated, include primary alkyl amines, polyamines and β-amino acids. They undergo condensation:

$$2\ RNH_2 \rightarrow RNHR + NH_3$$

Urea and guanidine as well as their derivatives with free NH_2 groups are subject to analogous condensations.

Water is also split off by aliphatic di- and tricarboxylic acids which yield anhydrides or whose hydroxyl groups are dehydrated. In this connection it is worth noting that succinic acid (m.p. 185–187°) and phthalic acid (m.p. 230°), which are known to form their respective anhydrides when their melts are heated, do this far below their respective melting points if they are sintered with thio-Michler's ketone or thiobarbituric acid.

This category includes also cyclic secondary alcohols, which split out water when subjected to dry heating and yield unsaturated ring compounds (e.g. cyclohexanol, p. 416) Accordingly, these are instances of topochemical reactions that occur or at least begin at the contact sites of two solids.

Compounds which split off water, include aliphatic and aromatic acid amides and aldoximes; cyanides result:

$$(R,Ar)CONH_2 \rightarrow (R,Ar)CN + H_2O*$$

* This cleavage at temperatures up to 200° appears to be characteristic for acid amides; the amides referred to later did not give any ammonia (that could be detected by Nessler reagent) through formation of acid imides.

Procedure. A small amount of the solid or a drop of its solution is placed in a micro test tube and taken to dryness if necessary. A slight quantity of thio-Michler's ketone or thiobarbituric acid is added and then several drops of highest purity benzene or chloroform to insure intimate mixing. The solvent is removed and the test tube then placed in a bath which has been preheated to 120°. A piece of lead acetate paper is placed across the mouth of the test tube. The temperature of the bath is then raised to 180°. A positive response is shown by the development of a brown or black stain on the reagent paper within at most 2–3 min. after the temperature has reached 180–190°.

The following amounts were detected:

50 γ spermidine phosphate	2.5 γ nicotinamide	10 γ citric acid
10 γ cyclohexylamine	5 γ mandelamide	2.5 γ sulfosalicylic acid
5 γ taurine	5 γ succinic acid	10 γ phthalic acid
5 γ glycine	2 γ malonic acid	2.5 γ mandelic acid
1.5 γ urea (allylurea)	5 γ maleic acid	1.5 γ glutaric acid
0.3 γ acetamide	2.5 γ adipic acid	50 γ mucic acid
2.5 γ benzamide	1 γ tartaric acid	50 γ spermine phosphate

1 F. FEIGL and J. R. AMARAL, *Mikrochim. Acta*, (1960) 816.

41. Detection of compounds which yield ammonia when pyrolyzed[1]

If nitrogenous compounds, which yield ammonia when heated to 160°, are brought to this temperature after being mixed with dimethyl oxalate (m.p. 54°; b.p. 163.3°), oxamide results:

$$\begin{array}{l} \text{COOCH}_3 \\ | \\ \text{COOCH}_3 \end{array} + 2\,\text{NH}_3 \longrightarrow \begin{array}{l} \text{CONH}_2 \\ | \\ \text{CONH}_2 \end{array} + 2\,\text{CH}_3\text{OH}$$

Oxamide can be almost specifically detected through the formation of a deep red product when dry heated with thiobarbituric acid (see p. 535). The formation of oxamide as shown in the above equation and the production of the colored material can be accomplished in one operation and thus permit the detection of compounds which yield ammonia on pyrolysis.

Pertinent examples are: (1) primary and secondary aliphatic amines, (2) β-aminocarboxylic acids, (3) ammonium salts of organic acids, (4) amides of aliphatic carboxylic acids, (5) urea and its derivatives with free NH_2 groups, (6) guanidine and its derivatives with free NH_2 groups.

It should be noted that ammonium salts of volatile inorganic acids give the oxamide reaction if heated to 160° with dimethyl oxalate and thiobarbituric acid. Accordingly, the chlorides and nitrates of primary and secondary aliphatic amines likewise give a positive response.

Procedure. The test is conducted in a micro test tube. A little of the solid or the evaporation residue of its solution is united with several cg dimethyl oxalate and thiobarbituric acid, and heated in a glycerol bath at 130–160°. A positive response is indicated by the formation of a red product within 1–3 min.

As an indication of the sensitivity of this test, the following quantities of ammonia-donors (oxamide formers) were detected:

15 γ ethylenediamine chloride	20 γ azelaic acid diamide
5 γ α-alanine	10 γ methylurea
5 γ asparagine	17 γ *asym*-diphenylurea
15 γ nicotinamide	25 γ phenylbiguanide
10 γ acrylamide	25 γ *sym*-diphenylguanidine

This procedure revealed the following in about 0.5 mg amounts: benzylamine, spermidine and spermine phosphate, ethylenediamine chloride, α-alanine, *l*-asparagine, *l*-histidine, mandelamide, azelaic acid diamide, urea, acetylurea, methylurea, allylurea, phenylurea, biuret, guanidine, diphenylguanidine, benzoguanidine, *sym*-triphenylguanidine*, 1-phenylsemicarbazide (170°), semicarbazide chloride.

* This compound according to the symmetric formulation should not react. Probably the following equilibrium must be assumed:

$$C_6H_5N=C\diagup^{NHC_6H_5}_{\diagdown NHC_6H_5} \rightleftharpoons C_6H_5N=C\diagup^{N(C_6H_5)_2}_{\diagdown NH_2}.$$

1 F. FEIGL and J. R. AMARAL, *Mikrochim. Acta*, (1960) 816.

42. Detection of compounds which yield ammonia when heated with caustic alkali and Devarda alloy[1]

After the presence of nitrogen in organic compounds has been established through appropriate tests (see Sect. 15), further information may be secured by examining the behavior of the material when it is warmed with caustic alkali and Devarda alloy (50% copper, 45% aluminium, 5% zinc). The nascent hydrogen formed through its reaction with alkali leads to reductive cleavage of ammonia from the following classes of compounds:

1. Nitrates of alcohols and organic bases
2. Nitramides
3. Nitrosamines
4. Oximes of aldehydes and ketones
5. Hydroxamic acids
6. Alkyl and aryl hydrazides
7. Hydrazides of carboxylic and sulfonic acids
8. Aliphatic nitro and aromatic polynitro compounds

The ammonia produced can be detected in its gas phase by the color change of acid–base indicator paper or through the Nessler test.

When the preliminary test described here is used, it must be remembered that ammonium salts and amides of most carboxylic acids yield ammonia when warmed with alkali. Accordingly, it is advisable to warm the sample first with alkali alone until any evolution of ammonia has ceased and then introduce the Devarda alloy.

Procedure. A micro test tube is used. A little of the solid or a drop of its solution is treated with 1–2 drops of 5 N alkali. Several cg of Devarda alloy are added, the mouth of the tube is covered with a piece of indicator paper, and the tube then placed in boiling water. A positive response is revealed by a color change on the indicator paper or by the development of a yellow or brown stain on the Nessler reagent paper.

Amounts of 0.5 mg or less suffice for this test.

1 Unpublished studies.

43. Detection of compounds which split off water when heated to 180°[1]

The detection of such compounds can be revealed by the fact that they split off formaldehyde and ammonia when heated to 180° along with hexamine (hexamethylenetetramine). The reason for this result is that the pyrolytically released water brings about the hydrolysis

$$(CH_2)_6N_4 + 6\,H_2O \rightarrow 6\,CH_2O + 4\,NH_3.$$

The above pyrohydrolysis can be detected by the reaction of Nessler's reagent with both gaseous products formed (formation of black metallic mercury and brown $HgNH_2I.HgI_2$).

Compounds which split off water are mentioned in Sect. 40.

Procedure. A small amount of the solid or a drop of its solution is placed in a micro test tube and taken to dryness if necessary. Several cg of hexamine are added and intimately mixed. The tube is then placed in a glycerol bath which has been preheated to 110°. A piece of qualitative filter paper moistened with Nessler's reagent is placed on the mouth of the tube. The temperature of the bath is then raised to 180°. A positive response is shown by the development of a black or brown fleck on the reagent paper within 1 or 2 min.

No more than 100 γ are necessary for the test.

1 F. FEIGL and J. R. AMARAL, *Mikrochim. Acta*, (1960) 816.

44. Compounds containing reactive $>$CH$_2$ and —NH$_2$ groups

Test with sodium 1,2-naphthoquinone-4-sulfonate[1]

The yellow neutral or faintly alkaline solution of 1,2-naphthoquinone-4--sulfonic acid reacts with compounds which contain two removable hydrogen atoms attached to one carbon atom or nitrogen atom. Deeply colored *para*-quinoid condensation products result. In the case of primary aromatic amines, the reaction can be represented:

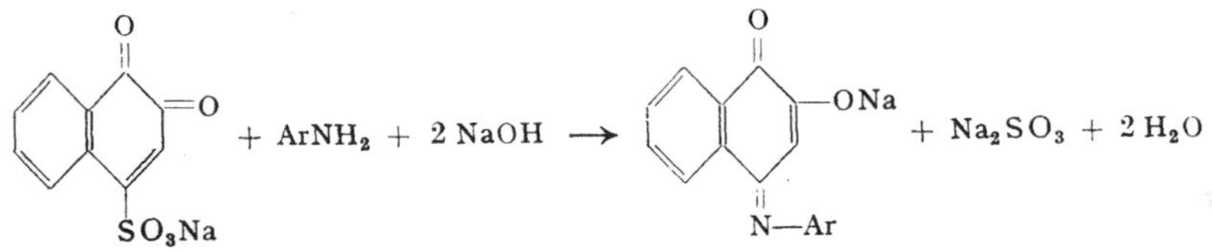

The colored products belong to the class of indophenol dyes. Sulfonamides behave like primary aromatic amines[2]. In alkaline solution, 1,2-naphthoquinone-4-sulfonic acid also reacts with α-aminocarboxylic acids[3] and likewise with guanidine[4] (but not with its derivatives containing a free NH$_2$ group) to yield red *p*-quinoidal compounds.

p-Quinoid compounds of similar structure containing a $>$=C group instead of a $>$=N– group are produced in the case of an active CH$_2$ group. Compounds which contain a CH$_2$Hal group (Hal = Cl, Br, I, CN) belong in this category. An active CH$_2$ group can also be formed through isomerization; pertinent examples are resorcinol, phloroglucinol, indole, pyrrole and nitromethane.

The test with sodium 1,2-naphthoquinone-4-sulfonate described below may only be applied for orientation purposes, since the reactivity of the CH$_2$ and NH$_2$ groups is influenced by the other groups in the molecule. For example, in contrast to aniline, its derivatives with strongly negative groups, *e.g.*, trinitroaniline and tribromoaniline, will not condense in this way. Negative groups in the *ortho* and *para* position particularly cause much interference with the reaction; negative groups in the *meta* position are less harmful. 1-Aminoanthraquinone, di-N-nitrosopiperazine and N-ethylpiperidine show no reactivity.

Procedure.[5]   A little of the solid or a drop of its solution is placed in a micro crucible along with two drops of a saturated solution of sodium 1,2-naphthoquinone-4-sulfonate in 50% alcohol. The mixture is made faintly alkaline with diluted sodium hydroxide.

Limits of identification and colors are:

0.6 γ piperidine (red)	0.6 γ pyrrole (violet)
1.2 γ malonic ester (dark violet)	0.6 γ rhodanine (dark blue-violet)
1.2 γ ethyl acetoacetate (blood red-violet)	12 γ dibenzoylmethane (red-violet)
0.6 γ thiocyanatoacetic acid (red-brown)	0.6 γ benzylamine (green)
6 γ *m*-nitroaniline (yellow-brown)	0.12 γ aniline (brick red)
0.6 γ semicarbazide (orange-red)	0.6 γ indole (emerald green)

0.6 γ β-naphthylamine (yellow-red to red-brown)
1.2 γ β-naphthoquinoline iodomethylate (dark violet)

Color reactions are shown by[6]: resorcinol (violet); phloroglucinol (brown-violet); trinitrotoluene (brown-red); acetylacetone (brown-red); pyruvic ester (blue-green); nitromethane (violet); phenylmethylpyrazolone (green-blue); and compounds with a cyanomethyl group (violet).

The reaction of sodium 1,2-naphthoquinone-4-sulfonate with compounds containing reactive CH_2 and NH_2 groups can also be accomplished by contact of vapors of such compounds with filter paper moistened with the reagent. If need be, steam can be employed as carrier gas. Aniline, tolidine, piperidine, piperazine, pyrrole, indole, benzylamine, anisidine, and morpholine can be detected in this way. The selectivity of the test is distinctly enhanced by transferring the reaction into the gas phase.

1 P. EHRLICH and C. A. HERTER, *Z. Physiol. Chem.*, 41 (1904) 329.
2 E. G. SCHMIDT, *J. Biol. Chem.*, 122 (1938) 787; F. J. BANDELIN, *Science*, 106 (1947) 426.
3 O. FOLIN, *J. Biol. Chem.*, 51 (1922) 377, 393.
4 M. X. SULLIVAN and W. C. HESS, *J. Am. Chem. Soc.*, 58 (1936) 47.
5 F. FEIGL and O. FREHDEN, *Mikrochemie*, 16 (1934) 79.
6 F. SACHS and M. CRAVER, *Ber.*, 38 (1905) 3685.

45. Solubility tests

The behavior of solid, liquid, and gaseous materials and mixtures toward water, dilute acids and alkalis, as well as non-aqueous solvents, is of analytical importance because: 1) such tests give valuable guides regarding the preparation of suitable reaction milieus in which to carry out precipitation and color reactions; 2) the solubility behavior *per se* can be characteristic for certain compounds or types of compounds.

Ordinarily, a distinction is made between inert solvents and those which are chemically active (reaction solvents). This practice is justified with respect to the residues obtained when the solvent is removed by evaporation or displaced by the addition of another solvent. In the case of inert solvents, the dissolved material is always unchanged, whereas, with reactive solvents the residue consists of products of the dissolution process. If, however, the solution process as such and the state existing in solution are considered, then the distinction between indifferent and chemically active can no longer be maintained. In every case, addition reactions produce solvates, *i.e.*, compounds made up of molecules of the solvent plus molecules (or ions) of the now dissolved material, which originally was present as solid, liquid, or gas. It is in accord with the chemical nature of solution that the presence and activity of certain groups in the solvent and solute are responsible for the occurrence and extent of these solution processes.* Therefore, it obviously may be expected that useful conclusions concerning the molecular structure and the presence of certain groups can be drawn from the behavior of an organic material toward selected solvents. This is true to the greatest extent for the behavior toward reaction solvents, as is shown in the section dealing with the detection of the acidic and basic behavior of organic materials. However, here also there are notable exceptions to the rule that basic compounds are soluble in acids and acidic compounds in alkali hydroxides. Several characteristic instances demonstrate that in organic compounds the point of attack by acids and bases can be blocked as well as activated by the remainder of the molecule. For example, 8-hydroxyquinoline is easily soluble in dilute mineral acids, in contrast to its halogenated derivatives. Phenol does not dissolve in aqueous ammonia or in alkali bicarbonate, but this fact does not permit generalization because the behavior of nitroso- and nitrophenols, hydroxyanthraquinones, and hydroxyaldehydes shows clearly that certain compounds which are throughly phenolic in constitution may be soluble in aqueous ammonia and alkali bicarbonate. Certain phenols, in opposition to the general rule, are not soluble in caustic alkalis,[1] for instance the phenolic azo dyestuff Sudan III (tetrazobenzene β-naphthol).

With regard to divergences from general solubility rules involving reacting solvents, it is very instructive to examine the behavior of alizarin blue (I) which contains not only the characteristic functional groups of alizarin (II) but also of 8-hydroxyquinoline (III).

(I)   (II)   (III)

Contrary to expectation, this dyestuff, which is an exceedingly sensitive and specific reagent for copper,[2] is soluble neither in alkalis as are (II) and (III), nor in dilute acids as is (III). Consequently, the results of these

---

* See the very instructive discussion of the analytical significance of this topic in S. M. McElvain, *The Characterization of Organic Compounds*, Macmillan, New York, 1945, Chap. II; N. D. Cheronis and J. B. Entrkiin, *Semimicro Qualitative Organic Analysis*, New York, 1947, Chap. 4.

particular solubility tests would not justify any statement about the presence of acidic phenolic groups or of the basic nitrogen atom in the dyestuff.

Cold concentrated sulfuric acid dissolves unsaturated hydrocarbons, polyalkylated aromatic hydrocarbons and many oxygen-containing organic compounds. Sulfonation, esterification, oxidation, removal of water, isomerization, polymerization, hydrolytic splitting, or formation of addition compounds may occur when concentrated sulfuric acid acts; the nature of the action depends on the kind of compound involved. In the case of organic compounds that contain oxygen, the result is often the production of addition compounds soluble in concentrated sulfuric acid. Sometimes, the solvates of oxygen-containing compounds in concentrated sulfuric acid are colored or the solutions show a deeper color than the solute itself. Examples are unsaturated ketones[3] and phenazines (condensation products of o-phenylenediamine with 1,2-diketones). This effect is known as halochromy or halochromism. In general, the addition products with concentrated sulfuric acid are decomposed on dilution with water and the initial compounds come out of solution. Consequently, the close relation between reactive and inert solvents is shown particularly well in the behavior of concentrated sulfuric acid.* However, there is no universal rule that all organic compounds behave in this manner. For example, many aromatic hydrocarbons (benzene, toluene, etc.), and their halogen derivatives, as well as diaryl ethers are not soluble in concentrated sulfuric acid. Solubility in concentrated sulfuric acid may therefore be merely indicative of the presence of oxygen-containing organic compounds. It sometimes serves as a guide in the separation and isolation of a particular organic compound.

Syrupy (85%) phosphoric acid has been recommended as an inert solvent for differentiating water-insoluble organic oxygen compounds.[5] It dissolves alcohols, aldehydes, methyl ketones, cyclic ketones and esters of less than 9 C atoms, but there are numerous exceptions in these groups. Some olefins, amylene for example, are also soluble in syrupy phosphoric acid, which behaves like concentrated sulfuric acid with respect to removal of water from organic compounds but, in contrast, it never acts as oxidant.**

Concentrated aqueous trichloroacetic acid (10 parts crystalline acid + 1 part water) has, in analogy to concentrated sulfuric acid, been found to be

---

* The production of color or the change in color that accompanies solution in concentrated sulfuric acid in many cases, is probably due in large measure to the fact that the concentrated acid can function as dehydrant as well as strong oxidant. Furthermore, small impurities in concentrated sulfuric acid may play an important role.[4]

** When syrupy phosphoric acid is heated to 300° a liquid is obtained which contains 44 wt.% ortho- and 39 wt.% pyrophosphoric acid besides tri- and tetrapyrophosphoric acids.[6] Through addition of $SnCl_2$ or K_2CrO_4 to this "300° strong phosphoric acid", powerful reducing or oxidizing systems are obtained from which very interesting applications to qualitative analysis may be expected.[7]

an excellent solvent for organic compounds.[8] Among the organic liquids which are miscible are: ether, higher aliphatic aldehydes and alcohols, benzene*, chloroform, cumene, cyclohexane, o-dichlorobenzene, diisopropyl ether, higher fatty acids, fats and ethereal oils, carbon disulfide, carbon tetrachloride, toluene, trichloroethylene, 1,2,4-trichlorobenzene, xylene. The solid organic compounds which dissolve in trichloroacetic acid include: p-aminobenzoic acid, p-aminosalicylic acid, anthranilic acid, benzoic acid, benzonaphthol, cetyl alcohol, cholesterol, coumarin, iodoform, ionone, naphthalene, naphthols, phenacetin, phenol, salicylic acid, stearin, sulfonal, veronal.

In addition to its value as a solvent, trichloroacetic acid has the further analytical value that certain color reactions, which occur with concentrated sulfuric acid alone or on the addition of suitable reagents, also succeed with concentrated trichloroacetic acid or its melt (m.p. 56°). Such reactions can be conducted as spot reactions on filter paper, which are not feasible with concentrated sulfuric acid because the latter carbonizes the paper. The solubility of trichloroacetic acid in ether and benzene provides still another advantage, because it makes possible the impregnation of filter paper with mixtures of the acid and ether- or benzene-soluble reagents. Interesting prospects are thus opened for additional spot reactions and their application in paper chromatography.[9]

Since there are marked departures from solubility rules even with respect to both chemically active and inert inorganic solvents, it is not strange that this is still more true of the behavior of organic compounds toward inert organic solvents. Consequently, the results of preliminary tests of the behavior of organic materials toward such solvents must always be accepted with caution. Insolubility or slight solubility must especially not be taken as a sure indication of the presence or absence of certain groups or members of particular types of compounds. These reservations must be kept in view when applying the following statements, which are intended solely to give an orientation concerning the solubility behavior of the most important classes of organic compounds, and which in addition furnish the basis for separations of the constituents of organic mixtures.** Before making solubility tests, it is well to make the preliminary tests described in the previous sections, and to take the findings into account along with the results of the solubility tests.

* Benzine and petroleum ether are not miscible with trichloroacetic acid solution; consequently these liquids can be distinguished from benzene through this behavior.

** Consult H. STAUDINGER, *Anleitung zur organischen qualitativen Analyse*, 6th edition, Springer Verlag, Berlin 1955, concerning the data indispensable to such separations. For a new micro method which gives information about the behavior of organic compounds towards volatile organic solvents, see J. A. JAEKER and F. SCHNEIDER, *Mikrochim. Acta*, (1959) 801.

I. Compounds soluble in water, insoluble in ether

Only carbon, hydrogen, oxygen present:	Dibasic and polybasic acids
	Hydroxy acids
	Polyhydroxyphenols
	Simple carbohydrates
Metals present:	Salts of acids and phenols
	Miscellaneous metallic compounds
Nitrogen present:	Ammonium salts
	Amine salts of organic acids
	Amino acids
	Amides
	Amines (aliphatic-aromatic, polyamines, oxyamines)
	Aminophenols
	Nitro acids
	Nitrophenols
	Semicarbazides
	Semicarbazones
	Ureas
Halogens present:	Halo acids
	Halo alcohols, aldehydes, *etc.*
	Acyl halides (by hydrolysis)
Sulfur present:	Sulfonic acids
	Mercaptans
Nitrogen and halogen present:	Amine salts of halogen acids
Nitrogen and sulfur present:	Aminosulfonic acids
	Bisulfates of weak bases
	Cyanosulfonic acids
	Nitrosulfonic acids

II. Compounds soluble in water and ether

Only carbon, hydrogen, oxygen present:	Carboxylic acids
	Alcohols
	Aldehydes and ketones
	Anhydrides
	Esters
	Ethers
	Polyhydroxyphenols
Nitrogen present:	Amides
	Amines
	Amino acids and aminophenols
	Nitro acids and nitrophenols
Halogens present:	Halo acids and phenols
Sulfur present:	Heterocyclic hydroxy sulfur compounds
	Mercaptans
	Thiophenols

III. Compounds insoluble in water, soluble in ether

Only carbon and hydrogen present:	Liquid and solid aliphatic and aromatic hydrocarbons

Only carbon, hydrogen, oxygen present:	Esters and lactones
	Higher lactones, alcohols, and aldehydes
	Ethers
	Higher fatty acids
	Aromatic carboxylic acids
	Phenols and enols
Nitrogen present:	Nitro compounds
	Nitrophenols and arom. nitrocarboxylic acids
	Esters of nitrous acid
	Aromatic bases
	Hydrazine derivatives
	Acid amides, imides, anilides
	Aromatic amino acids
	Azo compounds
Halogens present:	Aliphatic and aromatic halogen compounds
	Halogen-substituted alcohols, aromatic aldehydes, phenols, carboxylic acids, esters
	Halo-nitro compounds
Sulfur present:	Thiol compounds
	Thioacid amides and anilides
	Thio ethers
	Esters of aliphatic and aromatic sulfonic acids
	Disulfides

IV. Compounds insoluble in water and ether

Only carbon, hydrogen, oxygen present:	Aliphatic, alicyclic and certain aromatic poly-carboxylic acids (gallic, phthalic acid, *etc.*)
	Hydroxyanthraquinones
	High molecular ketones (anthraquinones, *etc.*)
	Polymerization products of aldehydes
Nitrogen present:	Aminoanthraquinones
	High molecular aromatic bases (benzidine and benzidine derivatives)
	Acid amides and anilides
	Phenylhydrazones of aldehydes and ketones
	Salts of organic bases
Sulfur present:	Sulfones
	Polymeric thioaldehydes
	Sulfamide derivatives of secondary amines

V. Compounds insoluble in water, soluble in 10% hydrochloric acid

Amines
Amino acids
Aryl substituted hydrazines
N-dialkyl amines
Amphoteric compounds

VI. Compounds insoluble in water, soluble in 10% sodium hydroxide and in 10% sodium bicarbonate solutions

Only carbon, hydrogen, oxygen present:	Acids and anhydrides
	Hydroxyanthraquinones

Nitrogen present: Aromatic aminoacids
                                 Nitro acids
                                 Cyano acids
                                 Polynitrophenols
Halogens present: Halo acids
                                 Polyhalophenols
Sulfur present: Sulfonic acids
                                 Sulfinic acids
                                 Mercaptans
Nitrogen and sulfur present: Nitrothiophenols
                                 Sulfates of weak bases
                                 Sulfonamides
Sulfur and halogens present: Sulfonhalides

*VII. Compounds insoluble in water and 10% sodium bicarbonate
solution; soluble in 10% sodium hydroxide solution*

Only carbon, hydrogen, oxygen   Phenols
   present: Enols
Nitrogen present: Amino acids
                                 Nitrophenols
                                 Amides
                                 Aminophenols
                                 Cyanophenols
                                 Aromatic N-monoalkylamines
                                 N-substituted hydroxylamines
                                 Primary and secondary nitroparaffins
                                 Aromatic trinitrohydrocarbons
                                 Oximes
                                 Ureides
Halogens present: Halophenols
Sulfur present: Mercaptans
                                 Thiophenols
Nitrogen and halogen present: Polynitro halogenated aromatic hydrocarbons
                                 Substituted phenols
Nitrogen and sulfur present: Alkylsulfonamides
                                 Arylsulfonamides
                                 Aminothiophenols
                                 Aminosulfonic acids
                                 Thioamides

VIII. Compounds insoluble in all previously employed solvents

Nitrogen present: Anilides and toluidides
                                 Amides
                                 Nitroarylamines
                                 Nitrohydrocarbons
                                 Diarylamines
                                 Azo-, hydrazo-, and azoxy-compounds
                                 Dinitrophenylhydrazines
                                 Nitriles
                                 Aminophenols

Sulfur present:

Sulfides
Sulfones
N-dialkyl sulfonamides
Thio esters
Thiourea derivatives

IX. Compounds soluble in concentrated sulfuric acid,
insoluble in all other solvents previously tried

Alcohols
Aldehydes and ketones
Esters
Ethers
Unsaturated hydrocarbons
Anhydrides

X. Compounds containing no nitrogen or sulfur, insoluble
in all other solvents tried previously

Hydrocarbons
Halogen derivatives of hydrocarbons

The differentiation between soluble, difficulty soluble, and insoluble in the behavior of organic compounds towards the solvents cited in I–IX is not sharp. Accordingly, certain types of compounds are included in various solubility classes.

When selecting suitable solvents it is very worth-while to take into account the statements regarding the solubility of organic compounds, a subject on which much accurate information is available.*

* See for example N. A. LANGE, *Handbook of Chemistry*, 10th ed., New York, 1963; *Handbook of Chemistry and Physics*, 44th ed., Cleveland, 1962; A. and E. ROSE, *Condensed Chemical Dictionary*, 6th ed., New York, 1961; W. UTERMARK and W. SCHICKE, *Melting Point Tables of Organic Compounds*, 2nd ed., New York, 1963.

1 R. ADAMS, *J. Am. Chem. Soc.*, 41 (1919) 247.
2 F. FEIGL and A. CALDAS, *Anal. Chim. Acta*, 8 (1953) 117, 339.
3 G. REDDELIEN, *Ber.*, 45 (1912) 2904.
4 L. ROSENTHALER, *Pharm. Ztg.*, 104 (1957) 73.
5 R. L. SHRINER and R. C. FUSON, *Identification of Organic Compounds*, 2nd ed., Wiley, New York 1940, p. 23.
6 S. OHASHI and H. SUGATANI, *Bull. Chem. Soc. Japan*, 30 (1957) 864.
7 T. KIBA et al., *Bull. Chem. Soc. Japan*, 30 (1957) 482.
8 L. ROSENTHALER, *Pharm. Ztg.*, 100 (1955) 475.
9 O. E. SCHULTZ and R. GMELIN, *Arch. Pharm.*, 287 (1954) 344.

Chapter 3

Detection of Characteristic Functional Groups in Organic Compounds

General remarks

The tremendous number of organic compounds would present a vast chaos of materials if there was no possibility of arranging them into classes whose members can be characterized with respect to their structural constitution, since they contain certain identical groups of atoms within the framework of their molecules. Accordingly, one of the most important tasks of qualitative organic analysis is to recognize such typical groups unfailingly. When attempting this task, it must be kept in mind that in many cases there are distinct relations between atom groups and the chemical and physical properties of an organic compound. The same group whose presence constitutes the basis for the classification of an organic compound is often the place in the molecule where chemical reactions are likely to occur. It has become customary to designate such groups as functional groups. Obviously, an organic molecule may include not merely one but two or even more different functional groups.

Comparatively few characteristic functional groups can be detected directly by the rapid action of suitable reagents which produce compounds having an appropriate insolubility, a distinctive color in day- or ultraviolet light, or other properties which can be the basis for identification by spot reactions. Usually the sample must be subjected to condensation, oxidation, *etc.* to arrive at products* which, in their turn, can be detected by appropriate reagents. In spot test analysis the preliminary operations on the starting material, which are necessary for the production of characteristic and sensitive reacting compounds, must be such as can be successfully conducted on a micro- or semimicro scale without too great a loss of the prime material and with little expenditure of time. Because of these requirements, it is by no means possible to employ in spot test analysis all of the tests that are of use in qualitative organic macroanalysis. On the other hand many spot tests cannot be translated to the technique of macroanalysis.

---

* This type of production is fundamentally different from that employed in classic qualitative organic macroanalysis. The latter has as its objective the preparation and isolation of derivatives, which can be characterized by their melting points, *etc.*

The objectives of the examination for functional groups in organic compounds correspond somewhat to the testing for certain cationic or anionic constituents in inorganic compounds. However, the literature of inorganic analysis contains much observational material concerning the influence on and impairment of tests by accompanying materials, but as yet such records exist to only a limited extent with respect to organic analysis. This lack or paucity of information must be kept in mind when applying the tests described in this chapter. When testing for functional groups in organic compounds, account must be taken not only of the behavior of organic accompanying materials toward the reagents being employed, but also of a special circumstance that plays no part in inorganic analysis. In contrast to ions, functional groups are not independent participants; instead they represent only the active portions of the molecular species participating in the reaction. Consequently, the remainder of the molecule has an influence on the solubility of the particular compound and its reaction products in either water or organic liquids, on the speed and extent of the reactions, and also on the color of the reaction products in daylight or ultraviolet light. In isolated cases the influence of the remainder of the organic molecule on the reactivity of functional groups may be so extensive that certain compounds do not respond at all to tests for these groups, even though experience with other compounds containing these groups has shown the tests to be effective. As a rule, these deviations from the general pattern are only a matter of degree. Accordingly, the limits of identification vary widely at times when compounds containing the same functional group are tested under like conditions. A study of the table following Chap. 7 is enlightening in this respect.

Tests for functional groups do not necessarily demand that the test material be in solution. Sometimes fusion or sintering reactions can be made with the solid sample. In such cases, because of the higher temperature and absence of dilution, reactions may sometimes be realized which do not occur at all in solution. Spot test analysis also makes use of the reactions of gaseous thermal fission products, which are brought into contact with suitable reagents, and thus reveal the presence or absence of particular functional groups. Reactions in the vapor phase deserve special attention when testing for functional groups in compounds which are volatile at room temperature or on gentle warming. If this is the case, a test which is merely selective when conducted in solution, can be rendered specific for an individual volatile or sublimable compound.

Because of their importance as medicinals, technical materials, stabilizers, *etc.*, organometallic compounds are often the subject of qualitative organic analysis. Sometimes it is possible to test them directly for functional groups. The more reliable method, however, is first to decompose the material in such a manner that the organic constituent remains intact and can be isolated. Although no general procedure can be advocated, this goal is often achieved by treating organometallic compounds with acids, hydrogen sulfide, alkali hydroxide and the like, and then taking up the liberated organic compounds in solvents which are not miscible with water.

It is not good practice to start the chemical examination of organic materials with tests for typical functional groups. Rather, as in qualitative inorganic analysis, a series of preliminary tests should be made which often afford valuable clues as to which groups are likely to be present or absent. Directions for carrying out such exploratory tests, whose negative response may be of higher significance than the positive response, have been given in Chap. 2. On the other hand, tests to establish the presence of functional groups should invariably precede the testing for characteristic types of compounds and individual compounds (Chap. 4 and 5).

When testing for functional groups it is always advisable (particularly for beginners) to carry out "model tests". In these, trials are made with a compound which contains the functional group in question and at various concentrations. The operator thus becomes acquainted with the typical reaction picture. If more than one test is available, the trials should not be limited to a single selected test. Such comparative tests are no great handicap in spot test analysis since but little time and material are consumed and the same conditions apply to repetitions of a procedure. In addition the intensity of the responses yield valuable information.

It should be noted regarding chemical tests that the numerical values of the attainable identification limits are not constants for the underlying reactions; rather they often are greatly dependent on the reaction conditions. Under circumstances which seemingly are identical, almost all tests have a concentration region in which the test sometimes succeeds and sometimes fails. (This is the "region of uncertain reaction", a term coined by F. EMICH.) In addition, the color of a compound or reagent can impair the discernment of a color or the observance of precipitation of a reaction product. When minimal amounts are taken for a test, it is therefore best in all cases to make a blank test, *i.e.* in the absence of the material to be detected, but of course under otherwise identical conditions. The results of the test can then be compared with this standard.

A. GROUPS CONTAINING CARBON AND HYDROGEN

This section will include tests for hydrocarbon radicals that are characteristic for particular types of compounds. Pertinent tests for acetylene, naphthalene, anthracene, and phenanthrene are described in Chap. 5. Chap. 2 should be consulted with regard to the detection of aromatics.

1. O- and N-Methyl compounds (—OCH₃ and >NCH₃ groups)

Test through oxidative cleavage with molten benzoyl peroxide[1]

Compounds containing the above groups are oxidatively split when heated to 120° with benzoyl peroxide (m.p. 103°):

$$>C{-}OCH_3 + (C_6H_5CO)_2O_2 \rightarrow\ >C{-}OH + (C_6H_5CO)_2O + CH_2O \qquad (1)$$

$$>NCH_3 + (C_6H_5CO)_2O_2 \rightarrow\ >NH + (C_6H_5CO)_2O + CH_2O \qquad (2)$$

The gaseous formaldehyde produced in (*1*) or (*2*) can be detected with a solution of chromotropic acid in concentrated sulfuric acid; a violet color appears. The chemistry of this very sensitive color reaction is discussed on p. 434. Accordingly, the fusion reaction with benzoyl peroxide permits the reliable detection of methoxy and N-methyl groups in the absence of compounds which give the formaldehyde reaction under the conditions of the test. Such materials include hexamine and its salts as well as Schiff bases of formaldehyde.

If O- and N-methyl compounds are dissolved in ether, they may readily be separated by shaking with dilute mineral acid. The N-methyl compounds pass into the water layer as salts, and may then be extracted from this solvent with ether after the bases are liberated by alkali. The evaporation residues of such ethereal solutions can then be subjected to the benzoyl peroxide fusion.

Procedure.   The apparatus shown in Fig. 24, *Inorganic Spot Tests*, is used. A little of the sample or a drop of its solution in benzene, ether or chloroform, is placed in the bulb. Two drops of a 10% solution of benzoyl peroxide in benzene are added. The solvent is evaporated. The knob of the stopper is charged with chromotropic acid solution and the stopper put in place. The apparatus is immersed in a glycerol bath that has been heated to 120–130°. A positive result is given if the hanging drop becomes violet within several minutes. If care is used, the fusion may be accomplished without fulmination.

Reagent: Chromotropic acid–sulfuric acid mixture. Several mg of pure chromotropic acid, or better its sodium salt, are stirred with 2 ml of concentrated sulfuric acid. The reagent must be freshly prepared at not too long intervals.

A positive result was obtained with the following compounds:

O-Methyl compounds: anisole, veratrole, *p*-methoxybenzhydrol, *p,p*′-dimethoxybenzhydrol, methyl cellulose, codeine, brucine, papaverine.

N-Methyl compounds: choline chloride, *m*-dimethylaminophenol, N-methyldiphenylamine, antipyrine, pyramidone, caffeine, theobromine, theophylline, pilocarpine, methyl orange, methyl red, methyl violet, malachite green.

The following amounts were detected:

20 γ pyramidone	20 γ brucine	100 γ antipyrine
40 γ caffeine	40 γ codeine	10 γ methyl orange

1 F. FEIGL and E. SILVA, *Analyst*, 82 (1957) 582.

2. Ethylene derivatives (—CH₂—CH₂— group)

Test through pyrolytic formation of acetaldehyde [1]

When compounds which contain the —CH₂—CH₂— group linked with O, N, S or Hal atoms are heated with fused 95% zinc chloride at 230–250°, acetaldehyde is formed which can be detected in the gas phase by the bluing of a mixture of sodium nitroprussiate and morpholine (see p. 438). The formation of acetaldehyde is due to the fact that zinc chloride acts as water donor for the pyrohydrolytic cleavage of ethylene derivatives to glycol (*1*), and as water acceptor[2] in the dehydration of the glycol (*2*):

$$X-CH_2-CH_2-X + 2\ H_2O \rightarrow HO-CH_2-CH_2-OH + 2\ XH \qquad (1)$$

$(X = OR; N<; SH; SR; Hal; etc.)$

$$HO-CH_2-CH_2-OH \rightarrow CH_3-CHO + H_2O \qquad (2)$$

It is remarkable that even cyclic ethylene derivatives yield acetaldehyde when heated with zinc chloride at 250°. Pertinent examples are morpholine (b.p. 128°) and dioxane (b.p. 101°). These compounds most likely do not evaporate because of the formation of addition compounds with zinc chloride which in turn are pyrohydrolyzed by the excess of this salt.

Piperazine and ethylenediamine and its derivatives do not react; when dry heated they form pyrrole (see p. 393 and 507).

Procedure. A small amount of the sample is mixed in a micro test tube with some cg of anhydrous zinc chloride. The open end of the tube is covered with a piece of filter paper moistened with the reagent solution (for preparation see p.439). The tube is heated to 250°. A positive result is indicated by the appearance of a blue stain on the reagent paper. A glycerol bath should not be used, because acrolein may develop which behaves like acetaldehyde.

The procedure revealed:

2 γ morpholine	5 γ "Endoxan"	2 γ glycol
4 γ 2,4,6-triethyleneamino-1,3,5-triazine		

The following gave a positive response: diethanolamine, triethanolamine, EDTA, ethylenechlorohydrin, β,β'-dichlorodiethyl sulfide, β,β'-dichlorodiethyl-methylamine, dioxane.

1 F. FEIGL and V. ANGER, unpublished studies.
2 A. WIRTZ, *Ann.*, 108 (1858) 68.

3. Vinyl compounds (—CH=CH₂ group)

Test by conversion into primary bromides[1]

If pertinent compounds are treated with a concentrated solution of bromine in carbon tetrachloride, bromine is added on the double bond:

$$-CH=CH_2 + Br_2 \rightarrow -CHBr-CH_2Br$$

The resulting primary bromide can be detected by the procedure described on p. 171, in which heating with sodium thiosulfate to 180° yields sulfur dioxide. The reaction with bromine, the evaporation of the solvent and excess bromine, as well as the heating with sodium thiosulfate, can all be conducted in the same micro test tube, in which a tiny amount of the solid test material has been placed initially. If a solution of the latter is presented for testing, the solvent can of course be removed in this same test tube.

The test revealed: 10 γ quinine.

1 F. FEIGL and V. ANGER, unpublished studies.

4. Propenyl compounds (—CH=CH—CH₃ group)

(1) Test by formation of acetaldehyde through the action of benzoyl peroxide[1]

Compounds which contain the propenyl group, which is isomeric with the allyl group, produce acetaldehyde when oxidatively cleaved by gentle heating with benzoyl peroxide:

$$CH_3-CH=CH-R + 2 (C_6H_5CO)_2O_2 \rightarrow CH_3CHO + RCHO + 2 (C_6H_5CO)_2O$$

The acetaldehyde may be detected in the vapor phase by the blue color it produces on contact with a solution of sodium nitroprusside containing morpholine (compare p. 438).

The test given here may not be used if acetaldehyde or compounds which yield acetaldehyde under the prescribed conditions are present. Such materials include ethanol, O- and N-ethyl compounds (see p. 165). A separation from ethylated nitrogen bases is readily secured by shaking the solution of the test substance in benzene, ether, *etc.* with dilute hydrochloric acid. The bases then pass into the water layer as chlorides.

Compounds which contain an allyl or vinyl group do not form acetaldehyde on treatment with benzoyl peroxide.

Procedure. A drop of the test solution is united in a micro test tube with a drop of 5% solution of benzoyl peroxide in benzene. The mouth of the tube is covered with a disk of filter paper moistened with sodium nitroprusside–morpholine solution (for preparation see p. 439). The tube is dipped into boiling water. If propenyl compounds are present, the paper turns blue.

Ethanol should not be used as solvent when very slight amounts of propenyl compounds are sought; there is distinct oxidation of ethanol to acetaldehyde by benzoyl peroxide.

The following were detected:

100 γ isosafrole	60 γ anethole	60 γ isoeugenol
40 γ anol	50 γ Crotamiton	

The rather volatile propenyl compounds just mentioned give off vapors at water bath temperature (or even at room temperature) which themselves show the color reaction without the reaction of benzoyl peroxide. This is probably due to the fact that air-oxidative cleavage occurs when the vapors come into contact with the reagent. This behavior seems to be characteristic for volatile propenyl compounds and can be employed for their detection in mg amounts.

1 F. Feigl and E. Silva, *Analyst*, 82 (1957) 582.

(2) Test through iodination[1]

Propenyl compounds add iodine when they are taken to dryness with a carbon disulfide solution of iodine:

$$R—CH=CH—CH_3 + 2 I \rightarrow R—CHI—CHI—CH_3$$

Accordingly, through the above reaction followed by the removal of the unconsumed iodine, there is produced from the iodine-free compound a material that contains iodine which then responds to the test for iodine outlined on p. 71. The allyl group, which is isomeric with the propenyl group, is not iodinated under the conditions prescribed here, and consequently the isomeric compounds of these two classes can be differentiated.

Procedure. A tiny portion of the test material or 1 drop of its ethereal solution is mixed on a watch glass with 3 drops of a 2% solution of iodine in carbon disulfide and allowed to stand for 5 min. at room temperature. The solvent is then removed in a drying oven and the residue is kept at 140° for about 30 min. After cooling, 3 drops of strong bromine water are added and thorough mixing is insured by blowing air through the liquid through a pipette. Several cg of sulfosalicylic acid are added to consume the unused bromine (decolorization). Here again it is well to remove all traces of bromine from the watch glass by aeration through a pipette. Several drops of zinc–iodide solution (Trommsdorf reagent–Merck) are added. A deep to light blue appears, the depth of the color depending on the quantity of propenyl compound present.

If the directions are followed faithfully, a blank remains completely colorless for 1–2 min.

The test revealed:

   2.5 γ isoeugenol   5 γ isosafrole   10 γ anethole
   8 γ anol   12 γ Crotamiton

1 F. FEIGL and E. LIBERGOTT, Z. Anal. Chem., 192 (1963) 91.

5. Allyl compounds ($>$C—CH=C$<$ group)

Test with phloroglucinol–hydrochloric acid

Certain allyl compounds, such as allyl disulfide and allyl chloride, and also arylallyl compounds, such as eugenol or cinnamic alcohol, react with phloroglucinol–hydrochloric acid to yield red-violet products[1]. The reaction is based on the condensation to trimethine dyes with participation of the atmospheric oxygen, intermediate peroxides being formed. Allyl compounds without aryl groups may perhaps condense with two molecules of phloroglucinol to yield compound (I); arylallyl compounds condense analogously with one molecule of phloroglucinol with formation of trimethine dyes of the general structure (II)[2]. Arylvinyl aldehydes yield the same condensation products with phloroglucinol.

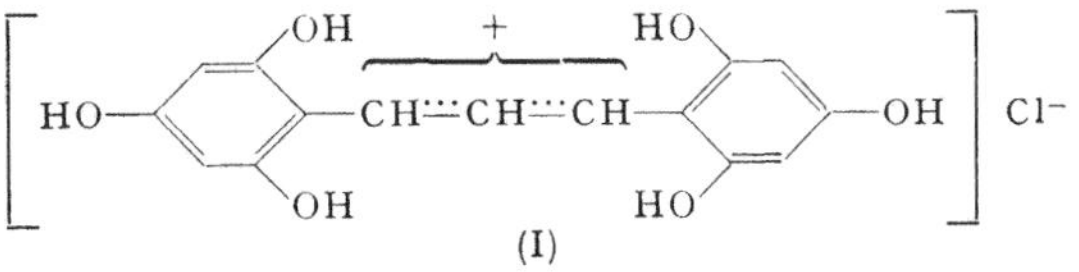

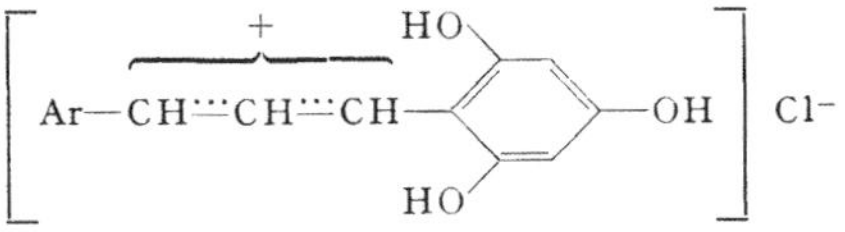

Aromatic aldehydes must be absent since they give the same color reaction with phloroglucinol. The test with naphthoresorcinol given on p. 206 should be employed to test for aromatic aldehydes or compounds that split off such compounds under the action of acids, since all other tests for aromatic aldehydes likewise respond to certain allyl compounds.

Procedure.[3] A drop of the test solution is treated in a depression of a spot plate with a drop of the reagent. A pink to red-violet color indicates the presence of the pertinent allyl compounds.

Reagent: Saturated solution of phloroglucinol in conc. hydrochloric acid.

Limits of identification:
8 γ allyl disulfide 10 γ eugenol methyl ether 10 γ estragole 10 γ coniferyl alcohol

Allyl chloride, allylamine, safrole and cinnamic alcohol gave a positive response.

No reaction was shown by allyl ether and thioether, nor by alkylallyl compounds such as diallylbarbituric acid (Dial).

1 K. KOBERT, Z. Anal. Chem., 46 (1907) 711.
2 V. ANGER, Analytical Chemistry 1962 (Proceedings Feigl Anniversary Symposium, Birmingham), 61.
3 Y. HASHIMOTO (Kyoto), private communication.

6. Benzal compounds ($C_6H_5CH=$ group)

Test by pyrolytic oxidative cleavage[1]

Pertinent compounds when dry-heated (150–180°) with lead dioxide (also MnO_2) split off benzaldehyde; e.g. cinnamic acid reacts in the following manner:

$$C_6H_5CH=CHCOOH + 2 [O] \rightarrow C_6H_5CHO + CHOCOOH$$

The occurrence of the above oxidative cleavage and therefore the presence of benzal compounds can be detected through a test for the benzaldehyde vapors evolved. The color reaction of aromatic aldehydes with thiobarbituric acid (see p. 205) is appropriate.

The test described here is highly selective but not very sensitive, probably due to the further oxidation of the benzaldehyde formed to benzoic acid. Agreeing with this assumption, only slight traces of formaldehyde, or none at all, can be detected when styrene is pyrolysed with lead dioxide. Probably the formaldehyde primarily formed is oxidized to formic acid, which combines with lead oxide to yield lead formate which is stable under the test conditions.

Procedure. In a micro test tube a drop of the benzene test solution is added to a few mg of lead dioxide. After evaporation in a water bath, the tube is immersed in a glycerol bath heated to 150–180° and a piece of filter paper moistened with a suspension of thiobarbituric acid in syrupy phosphoric acid is placed on the open end of the tube. A positive response is indicated by the appearance of a yellow fleck on the reagent paper within 5–10 min. If only small amounts of the yellow reaction product are formed, recognition is somewhat difficult because of the transparency of the reagent paper, in which case it is advisable to place a piece of dry filter paper below the reagent paper.

The following amounts were detected:

50 γ stilbene	*500 γ styrene
100 γ cinnamic acid	50 γ ω-nitrostyrene
100 γ methyl cinnamate	50 γ dibenzylideneacetone
100 γ p-hydroxycinnamic acid	50 γ benzylideneacetophenone

* The low sensitivity for styrene is probably due to its partial polymerization.

1 F. FEIGL, L. BEN-DOR and R. COHEN, Mikrochim. Acta, (1964) 1181.

B. ORGANIC HALOGEN COMPOUNDS

If the presence of halogen has been established by the preliminary tests described in Chap. 2, the following tests will reveal the particular bonding modes of the halogen. In this connection, the procedure for distinguishing between aliphatic- and aromatic-bound halogen should be consulted (see p. 67).

7. Primary halogenoalkyls (—CH₂Hal group)

Test through pyrolysis with sodium thiosulfate[1]

Compounds containing the halomethyl group react with hydrated or anhydrous sodium thiosulfate to give a so-called Bunte[2] salt, which in turn yield the corresponding disulfides and sulfur dioxide when pyrolyzed:[3]

$$RHal + Na_2S_2O_3 \rightarrow NaHal + NaRS_2O_3 \tag{1}$$

$$2\,NaRS_2O_3 \rightarrow RSSR + Na_2SO_4 + SO_2 \tag{2}$$

Reactions (1) and (2) can be realized by dry-heating and the resulting sulfur dioxide detected by means of color reactions on contact with appropriate reagent papers.

The test is not applicable in the presence of compounds which react with sodium thiosulfate to give sulfur dioxide when dry heated. Cases in point are: acidic compounds of all kinds; salts of weak bases with strong acids (see p. 111); alkyl esters of non-carboxylic acids (see p. 311).

Attention must be given to the fact that certain compounds containing $CHal_2$ and $CHal_3$ groups are hydrolyzable whereby the respective hydrogen halides are formed. (Pertinent examples are chloromycetin, chloral, iodoform.) The water necessary for such hydrolysis can be split off from hydrated sodium thiosulfate*. The action of acids already present or formed by hydrolysis takes place at 140°, whereas for the pyrolysis of Bunte's salt a temperature of 180° is necessary. This permits a clear differentiation.

It must be remembered that sodium thiosulfate may react with hydrogen halide pyrolytically split off from organic compounds, whereby sulfur dioxide is formed. An example is the behavior of 1,2-dibromocyclohexane which is converted to bromocyclohexene.

Procedure. A small amount of the solid or a drop of its solution is mixed in a micro test tube with several cg of sodium thiosulfate and the test tube is then placed in a glycerol bath preheated to 110° and then further heated at 180°. The thiosulfate first melts in its water of crystallization and after loss of water

---

* Anhydrous sodium thiosulfate is also reactive due to the action of inevitable traces of water which are regenerated when hydrogen halides react with thiosulfate.

resolidifies. The mouth of the test tube is covered with a disk of filter paper moistened with ferri ferricyanide solution (see p. 86) or a disk of Congo paper moistened with hydrogen peroxide. A positive response is signalled by the appearance of blue stains on the reagent papers.

The following were detected:

10 γ α-bromoacetophenone   100 γ epichlorohydrin
 5 γ p-amino-α-chloroacetophenone   10 γ chloroacetamide

A positive response was obtained with: dichlorodiethyl ether, ethylhexyl bromide, dichloroisoprene, benzyl chloride, ethyl monochloroacetate.

The isomeric p-bromoacetophenone and α-bromoacetophenone, which both melt at 50°, can be readily distinguished by the test given here, because only the latter gives a positive response due to its CH_2Br group.

1 F. FEIGL, V. ANGER and D. GOLDSTEIN, *Mikrochim. Acta*, (1960) 231.
2 H. BUNTE, *Ber.*, 7 (1874) 646.
3 R. B. WAGNER and H. D. ZOOK, *Synthetic Organic Chemistry*, Wiley, New York, 1953, p. 797.

8. Polyhalogen compounds (—CHal$_3$- and —CHHal$_2$ groups)

Test with pyridine and alkali hydroxide

A red color is produced when chloroform, bromoform, iodoform, or chloral is warmed briefly with pyridine and aqueous caustic alkali[1]. Analogous color reactions were observed with other polyhalogen compounds, and employed for detection and in colorimetric determinations.[2] Only those halogen compounds are reactive which have at least two halogen atoms bound to one carbon atom. The red water-soluble reaction products may be Schiff bases of glutaconic aldehyde. They are formed by opening the pyridine ring after addition of the polyhalogen compounds to the cyclic nitrogen atom (compare tests for pyridine and its derivatives, p. 384). In the case of chloroform, the reaction scheme might be:

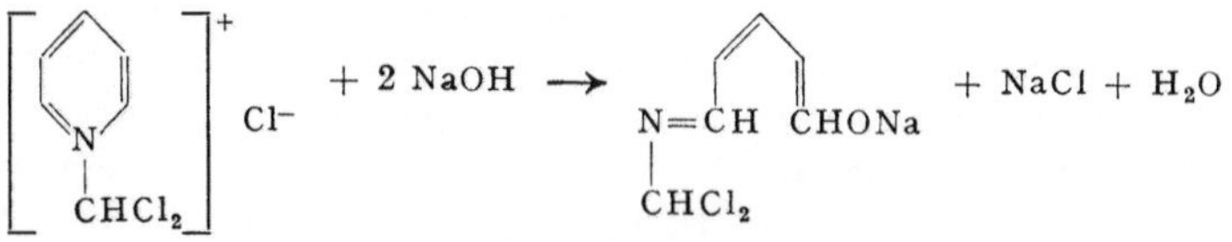

This scheme, which could apply to other polyhalogen compounds, is supported by the finding that addition of acetic acid to the red solution turns the color to yellow, and subsequent addition of primary aromatic amines, for instances benzidine, results in the precipitation of the violet Schiff base of glutaconic aldehyde obtained in the pyridine test.

Not enough cases have been investigated to permit any statement regarding the selectivity of the test. Basically it may be expected that not only polyhalogen but also monohalogen compounds with mobile halogen atoms are capable of addition to pyridine, and thus susceptible to a subsequent cleavage to give colored derivatives of glutaconic aldehyde.

Procedure.[3] One drop of the test solution (acetone is recommended for water-insoluble compounds) is mixed in a micro test tube with two drops of pyridine, one drop of 5 N sodium hydroxide, and warmed in boiling water. A positive response is indicated by the appearance of a red or pink color in the pyridine layer within a few seconds or minutes. More prolonged heating causes the color to fade or to turn brown or yellow. If a little benzidine chloride is added and the mixture acidified with dilute acetic acid, a violet color or precipitate is obtained.

The following amounts were detected:

1 γ chloroform	0.5 γ chloral hydrate
5 γ bromoform	0.5 γ trichloroacetic acid
50 γ iodoform	

A strong response was given by: chloromycetin (see p. 650), trichloroethylene, tribromoethanol, technical carbon tetrachloride. Carefully purified CCl_4 does not respond[4].

No reaction was given by: hexachloroethane and DDT (p,p'-dichlorodiphenyl-trichloroethane).

1 K. Fujiwara, *Chem. Abstr.*, 11 (1917) 3201.
2 Comp. M. Pesez and P. Poirier, *Méthodes et Réactions de l'Analyse Organique*, Vol III, Masson et Cie., Paris, 1954, pp. 111 and 130.
3 Unpublished studies, with V. Gentil.
4 L. Jenšoský, Ref. *Z. Anal. Chem.*, 152 (1956) 305.

9. Organic iodine compounds

Test through catalysis of the formation of Prussian blue from ferric salts, arsenious acid and ferricyanide[1]

A mixture of ferric salts, ferricyanide, and arsenious acid remains unchanged since there is no reduction to the ferrous condition that would result in the formation of Prussian blue. However, this reduction is promoted by iodine, even when organically bound, and Prussian blue appears.

Iodide ions, of course, also act as catalyst in the redox reaction Fe(III) + As(III) → Fe(II) + As(V) necessary for the formation of Prussian blue.

Procedure. A drop of the test solution is treated in a spot plate or on filter paper with several drops of reagent. A blue color appears if iodine compounds are present.

Reagents: (A) Solution of 2.7 g ferric chloride in 100 ml of 2 N HCl
(B) 5 g of potassium ferricyanide in 100 ml water

(C) 5 g of NaAsO$_2$ is dissolved in 30 ml N NaOH and mixed (vigorous stirring) with 65 ml of 2 N HCl

(A), (B) and (C) are mixed prior to use in the ratio 5:5:1.

Limit of identification: 0.002 γ thyroxine, triiodothyronine, and diiodotyrosine were detected on the chromatogram.

The sensitivity is distinctly less on the spot plate.

1 R. GMELIN and A. I. VIRTANEN, *Acta Chem. Scand.*, 13 (1959) 1469.

C. OXYGEN COMPOUNDS
(ORGANIC DERIVATIVES OF WATER)

This section will deal with tests for groups that in addition to oxygen contain only hydrogen.

10. Primary, secondary and tertiary alcohols

(1) Test with vanadium oxinate[1]

The black-green vanadium compound (I) of 8-hydroxyquinoline (oxine) can, be precipitated from acetic acid solution. In ethanol it yields a red solution, whereas in solvents such as benzene, toluene, trichloroethylene, that are not miscible with water, its solutions are grey-green. On addition of alcohol, these green solutions become red. If an aqueous suspension of this phenol ester of orthovanadic acid[2] is treated with alcohol and then shaken with benzene, *etc.*, the second layer is not green but red. These findings indicate that the formation-tendency and stability of the red solvates of (I)* are greater than those of the grey-green solvate. Probably these effects involve alcoholates, in which, as shown in (II), the OH group of the ester acid takes over the binding of the alcohol molecules. This idea is supported by the fact that the orange esters of H$_2$MoO$_4$ and H$_2$WO$_4$, as in (III), which contain no free OH groups, are not soluble in alcohol and organic liquids.

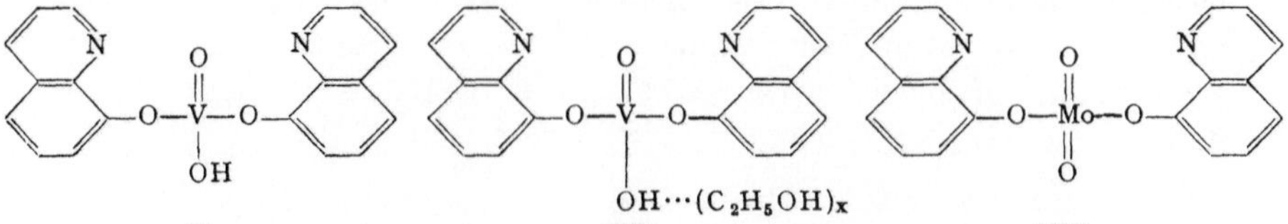

(I)   (II)   (III)

The color change of the grey-green benzene solution of vanadium oxinate

* Comp. also R. MONTEQUI, *Anales Real. Soc. Espan. Fis. Quim.*, 60 (1964) 277.

is brought about by primary, secondary and tertiary alcohols, and also by some compounds containing alcoholic OH groups provided certain conditions are met. One of these requirements appears to be a slight solubility in benzene, toluene, *etc.* Accordingly, glycerol and esters of lactic acid bring about the color change, while sugars are ineffective. No effect is obtained with compounds which, in addition to alcohol groups, also contain carboxyl-, phenol groups, or basic nitrogen atoms. This is shown by the behavior of lactic-, tartaric-, citric-, and mandelic acid, and choline chloride.

The formation of red solvates of vanadium oxinate makes possible the detection of lower (water-soluble) and higher (benzene-soluble) alcohols. It is best to use a benzene solution of vanadium oxinate as reagent.

Procedure.[3] One drop of the test solution (in water, benzene, or toluene) is treated in a micro test tube with four drops of the grey-green reagent solution. The mixture is heated, with intermittent shaking, in a water bath at 60°. A change to red after 2–8 min. is easily visible.

Reagent: Benzene solution of vanadium oxinate. One ml of a solution that contains 1 mg of vanadium is treated with 1 ml of 2.5% solution of 8-hydroxyquinoline in 6% acetic acid. The mixture is shaken with 30 ml benzene. The reagent solution will keep for about one day.

The procedure revealed:

20 γ methanol	5 γ isoamyl alcohol	100 γ glycol
20 γ ethanol	6.5 γ cetyl alcohol	500 γ glycerol
20 γ n-butanol	5 γ benzyl alcohol	5 γ menthol
20 γ isopropanol	10 γ cyclohexanol	5 γ borneol
20 γ 2,4-dichlorophenoxyethanol		10 γ chloromycetin
10 γ 3-nitro-4-hydroxybenzyl alcohol		3 γ salicyl alcohol (violet)

The following gave a positive response: hexanol, octanol, methylcyclohexanol, benzoin (hot, yellow; on cooling, red), terpineol, terpenes, vitamin A, vitamin D_2, cinchonine, sitosterol, tropine, atropine, the unsaponifiable residues of carnauba wax.

Xanthydrol, which is soluble in benzene, showed no reaction.

When using the test, it should be noted that phenols, ketones, ethers, thiols and amines give a grey-green, green or yellow color under the prescribed conditions.[4]

1 F. BUSCARONS, J. L. MARINS and J. CLAVER, *Anal. Chim. Acta*, 3 (1949) 310, 417.
2 M. BORREL and R. PARIS, *Anal. Chim. Acta*, 4 (1950) 267.
3 F. FEIGL and C. STARK, *Mikrochim. Acta*, (1955) 996.
4 A. J. BLAIR and D. A. PANTONY, *Anal. Chim. Acta*, 13 (1955) 1.

(2) Test with nitratocerate anions[1]

The conversion of $[Ce(NO_3)_6]^{2-}$ into $[Ce(OR)(NO_3)_5]^{2-}$ is accompanied by a color change (yellow to red) which can be used for the detection of primary, secondary and tertiary alcohols. One ml of a solution containing 40 g $(NH_4)_2[Ce(NO_3)_6]$ in 100 ml 2 N nitric acid is diluted with 2 ml water or dioxane. To this is added a drop of a solution prepared by dissolving the material to be tested in as little water or dioxane as possible (*Idn. Limit*: about 400 γ). Aliphatic bases interfere by forming precipitates; oxidizable compounds (phenols and aromatic amines) by giving colored products.

1 F. R. DUKE and G. F. SMITH, *Ind. Eng. Chem. Anal. Ed.*, 12 (1940) 201.

11. Primary and secondary alcohols

Test by conversion into xanthates [1]

Primary and secondary alcohols react rapidly with carbon disulfide at room temperature, and in the presence of alkali hydroxides to give water-soluble alkali-alkyl xanthates:

$$ROH + CS_2 + NaOH \rightarrow CS(OR)(SNa) + H_2O$$

Alkali-alkyl xanthates form a violet product with molybdates in strongly acidic solutions (comp. the sensitive Mo-test, *Inorganic Spot Tests*, Chap. 3). The colored product, whose composition is $MoO_3.2\ CS(OR)SH$,[2] is soluble in organic liquids that are not miscible with water. Primary and secondary alcohols can be detected by conversion into the respective xanthates followed by the molybdate reaction.

The xanthate–molybdate test for alcohols is not sensitive because the condensation of alcohols or alcoholates to alkali-alkyl xanthate is not quantitative. Furthermore, when xanthate solutions are acidified even in the presence of molybdate, there is an unavoidable partial decomposition of xanthates with regeneration of carbon disulfide and the respective alcohols.

Esters react similarly to alcohols since, under the conditions of the test, they are partially saponified to alcohols. Compounds containing the CH_2COCH_2 group also form orange-red products[3] which on treatment with molybdate give brown precipitates insoluble in chloroform and so interfere with the detection of small amounts of alcohols by the xanthate test.

Procedure.[4]   A drop of the test solution (in ether, if possible) is placed in a small test tube along with a drop of carbon disulfide and a few cg of powdered sodium hydroxide. The mixture is shaken for about 5 min. One or 2 drops of 1% ammonium molybdate solution are then added, and as soon as the alkali has dissolved, the solution is carefully acidified with 2 N sulfuric acid and shaken with 2 drops of chloroform. When primary or secondary alcohols are present, the chloroform layer is violet.

The following amounts were detected:

1 mg ethanol	0.5 mg isoamyl alcohol
1 mg methanol	1.0 mg allyl alcohol
1 mg propanol	0.5 mg cyclohexanol
1 mg butanol or isobutanol	0.1 mg phenylethyl alcohol

1 G. Dragendorff, *Die gerichtliche chemische Ermittlung von Giften*, 1895, p. 111.
2 M. Siewert, *Z. Anal. Chem.*, 60 (1927) 464; J. Koppel, *Chemiker-Ztg.*, 43 (1919) 777.
3 H. Apitzsch, *Ber.*, 38 (1905) 2895.
4 Unpublished studies, with R. Zappert.

12. Secondary alcohols

Test by heating with sulfur[1]

If nonvolatile compounds which contain secondary reducible alcohol groups are fused for a short time with sulfur (m.p. 119°), hydrogen sulfide results:

$$>CHOH + S^0 \rightarrow >CO + H_2S$$

This redox reaction seems to be especially realizable with compounds that melt at temperatures between 120–180° or whose boiling points lie above this temperature range. The procedure described here for the detection of the CHOH group is not very sensitive but it has some value if used in conjunction with tests described in Sect. 10 and 11.

Fatty acids with long chains, *e.g.*, palmitic-, stearic-, oleic acid, as well as fats and waxes likewise yield hydrogen sulfide when heated with sulfur, a result that probably is related to a dehydrogenation of —CH_2—CH_2— to —CH=CH— groups. However, these reactions are sluggish even when much sulfur is used, while secondary alcohols rapidly give hydrogen sulfide when heated with even slight amounts of sulfur.

Procedure. A micro test tube is used. A little of the solid or a drop of its solution in alcohol, ether, *etc.*, is treated with a drop of a 2% solution of sulfur in carbon disulfide. The mixture is brought to dryness by brief heating. The mouth of the tube is covered with a disk of lead acetate paper and the tube is placed in a glycerol bath previously heated to 150°. If necessary, the temperature is raised to 180°. If secondary alcohols are present, a black or brown stain (lead sulfide) appears on the paper within two or three minutes.

The procedure revealed:

10 γ benzoin (135–140°)	20 γ benzoin oxime (150–160°)
50 γ furoin (135–150°)	20 γ xanthydrol (155°)
2 γ codeine (154–160°)	50 γ chloromycetin (175°)
200 γ ephedrine (180°)	200 γ menthol (200°)
40 γ borneol (200–210°)	

Cyclohexanol (180°), cinchonine (180–190°), morphine (180–190°), adrenaline (170°), 9-hydroxyfluorene (160°), tropine (150–155°), also give a positive response. The same is true with 25 γ cholesterol (180°) and 100 γ cholic acid (180°).

1 F. FEIGL, V. GENTIL and C. STARK-MAYER, *Mikrochim. Acta*, (1957) 342.

13. Enols

Test with bromine and potassium iodide

Ketones in which the CO group is adjacent to a CH_2 group have acid character because in solution there is an equilibrium between the keto form and a tautomeric unsaturated alcohol form, the so-called enol form. In the case of ethyl acetoacetate, the keto–enol equilibrium is:

$$CH_3-CO-CH_2-COOC_2H_5 \rightleftarrows CH_3-C=CH-COOC_2H_5$$
$$\overset{|}{OH}$$

All enols take up bromine instantaneously with intermediate formation of enol dibromides, which form labile α-bromoketones on elimination of hydrogen bromide. The bromoketones oxidize hydrogen iodide and liberate iodine, the enols being regenerated.[1] Enols can therefore be detected by treating the sample with excess bromine, removing the unused bromine, and adding potassium iodide to the acid solution; iodine is liberated.

In the case of the enol form of ethyl acetoacetate, the equations relating to the bromination and the liberation of iodine are:

$$CH_3C(OH)=CHCOOC_2H_5 + Br_2 \rightarrow CH_3C(OH)BrCHBrCOOC_2H_5 \tag{1}$$

$$CH_3C(OH)BrCHBrCOOC_2H_5 \rightarrow CH_3COCHBrCOOC_2H_5 + HBr \tag{2}$$

$$CH_3COCHBrCOOC_2H_5 + 2 HI \rightarrow CH_3COCH_2COOC_2H_5 + HBr + I_2 \tag{3}$$

After the α-bromoketone has been formed (*2*), the excess bromine is removed with sulfosalicylic acid (see p. 72), whereby the color is discharged immediately. Reactions (*1*) and (*2*) are very rapid whereas (*3*) is slow.

Procedure.[2]   Saturated bromine water is added to a drop of the test solution until the yellow color persists. The solution is then decolorized with a saturated solution of sulfosalicylic acid. A few drops of 5% potassium iodide solution and starch solution are added. A blue color indicates the presence of enols.

A positive reaction was given by:
60 γ acetoacetic ester   40 γ malonic ester   100 γ benzoylacetic ester.

1 K. H. MEYER, *Ber.*, 47 (1914) 835.
2 Unpublished studies, with O. FREHDEN.

14. Phenols

(1) Test with nitrous acid [1]

Many phenols react with nitrous acid to give p-nitroso derivatives (I) which by reacting in the isomeric p-quinoid oxime form (II) condense with the excess phenol in the presence of concentrated sulfuric acid to yield the intensely colored indophenols (III): [2]

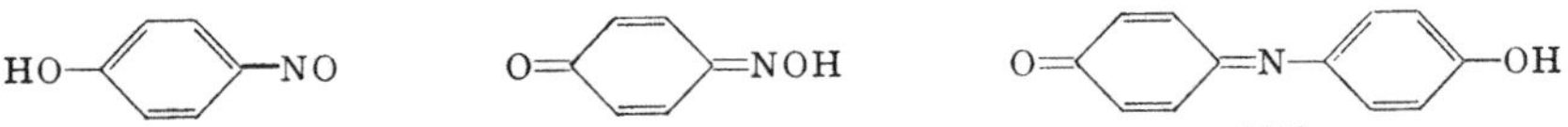

(I)   (II)   (III)

Para-substituted phenols and nitrophenols do not react; phenol ethers and thiophene give an intense phenol reaction.

Procedure. [3] A drop of the test solution (in ether) is allowed to evaporate to dryness in a micro crucible, then treated with a drop of the freshly prepared reagent, and left for a few minutes. The mixture is then cautiously diluted with a drop of water. Sometimes the color deepens. After cooling, the mixture is made alkaline with 4 N sodium hydroxide, whereupon a further color change often results.

Reagent: 1% solution of sodium nitrite in conc. sulfuric acid.

Limits of identification are given in Table 6.

1 C. Liebermann, *Ber.*, 7 (1874) 248, 806, 1098.
2 H. Decker and B. Solonina, *Ber.*, 35 (1902) 3217.
3 Unpublished studies, with R. Zappert.

(2) Test with 5-nitroso-8-hydroxyquinoline [1]

The *Liebermann reaction*, described in (1) is based on the conversion of a portion of the phenol present (with free *para*-position) into p-nitrosophenol by action of nitrous acid and condensation of the resulting nitrosophenol (in its oxime form) with unchanged phenol to produce a colored indophenol. An alternative method to obtain indophenol dyes is to use an appropriate p-nitrosophenol directly as the reagent. A light yellow solution of 5-nitroso-8-hydroxyquinoline (I) in concentrated sulfuric acid is suitable. Its oxime form (II) reacts with phenols. The condensation product of the oxime with phenol is shown in the indophenol compound (III). Other phenols yield analogous indophenol dyes.

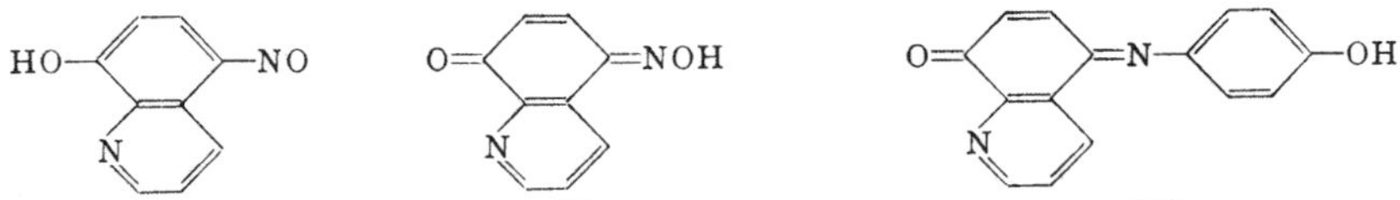

(I)   (II)   (III)

Procedure. A drop of the alcoholic or aqueous alkaline solution of the test material is evaporated to dryness in a micro crucible, and a drop of the reagent

TABLE 6

PHENOLS

Name	Test with	
	Liebermann's reagent	Millon's reagent
Phenol	1 γ Blue–red–green	1 γ Red
Resorcinol	5 γ Red–blue	0.5 γ ,,
Pyrocatechol	5 γ Green–red–blood-red	5 γ ,,
Hydroquinone	10 γ Green–red	10 γ ,,
Orcinol	5 γ Yellow–red–purple	5 γ ,,
Phloroglucinol	10 γ Blood-red	5 γ ,,
Pyrogallol	10 γ Violet–brown	5 γ ,,
Thymol	5 γ Green–red–blue	
p-Nitrophenol	No reaction	2 γ ,,
Salicylaldehyde	2 γ Red–light green	5 γ ,,
m-Hydroxybenzaldehyde		10 γ ,,
p- ,,	No reaction	1 γ ,,
Protocatechuic aldehyde	,,	4 γ ,,
Vanillin	,,	4 γ ,,
Methyl salicylate		1 γ ,,
Phenyl salicylate	4 γ Green–red–blue	
p-Hydroxybenzoic acid	No reaction	2 γ ,,
Methyl p-hydroxybenzoate	,,	1 γ ,,
α-Naphthol	Green	1 γ ,,
β- ,,	Dark green	1 γ ,,
o-Hydroxyquinoline	No reaction	0.5 γ ,,
m-Hydroxycinnamic acid		5 γ ,,

is added to the cold residue. The indophenol dyestuff is formed on gentle warming.

Reagent: 1% solution of 5-nitroso-8-hydroxyquinoline in conc. sulfuric acid.

This procedure revealed:

1 γ phenol	dark brown		7 γ pyrogallol	black
5 γ o-cresol	,, ,,		4 γ pyrocatechol	greenish black
10 γ α-naphthol	,, ,,		5 γ o-nitrophenol	green-yellow
2 γ resorcinol	red-violet		5 γ xylenol	violet

1 Unpublished studies, with T. ØSTERUD.

(3) Test with nitrous acid and mercuric nitrate [1]

Solutions of nitrous acid containing mercuric nitrate react with phenols, either in the cold or on slight warming, producing red colors or yellow precipitates, which dissolve in nitric acid to form red solutions. The reaction probably depends on the formation of a nitro compound, which then reacts with the phenol. Both aniline and phenol ethers likewise show this reaction, since they produce phenol on boiling with nitrous acid.

Di-*o*- and di-*m*-substituted phenols,[2] such as picric acid, do not react; neither do hydroxyanthraquinones. This *Millon test* is especially recommended for *p*-substituted phenols, which do not respond to Tests *1* and *2*.

Procedure.[3] A drop of the aqueous, alcoholic or ethereal solution is mixed in a micro crucible with a drop of the reagent solution and left for a few minutes. If no change occurs, the mixture is briefly heated to boiling. A red color forms if phenols are present.

Reagent: One part mercury is dissolved in one part fuming nitric acid and diluted with 2 parts water.

Limits of identification are given in Table 6.

1 E. Millon, *Compt. Rend.*, 28 (1849) 40; A. Almén, *Z. Anal. Chem.*, 17 (1878) 107; P. C. Plugge, *Arch. Pharm.*, 9 (1890) 9.
2 W. Vaubel, *Z. Angew. Chem.*, 13 (1900) 1127.
3 Unpublished studies, with R. Zappert.

(4) Tests through formylation[1]

A selective and sensitive test for *o*-hydroxyaldehydes is described on p. 342. It is based on the condensation of these compounds with hydrazine to give water-insoluble aldazines, which fluoresce yellow-green (sometimes orange) in ultraviolet light. In two relatively simple steps it is possible to arrive at *o*-hydroxyaldehydes by starting with phenols. Accordingly, sensitive tests for phenols have been developed from these syntheses in combination with the aldazine reaction.

The classic method for the *o*-formylation of phenols consists in protracted refluxing of the alkaline phenolate solution with chloroform (Reimer-Tiemann method[2]). The net reaction[3] can be written:

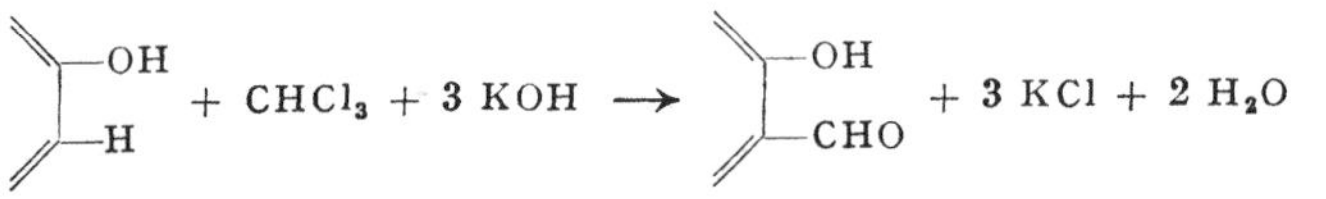

$$\text{(phenol)}-\text{OH} + CHCl_3 + 3\ KOH \longrightarrow \text{(phenol)}-\text{OH, }-CHO + 3\ KCl + 2\ H_2O \qquad (1)$$

More recently, hexamine has been found to be a reagent for *o*-formylation (Duff method[4]). The action of hexamine resides in its condensing, in its trialkylamine form[5], with phenols:

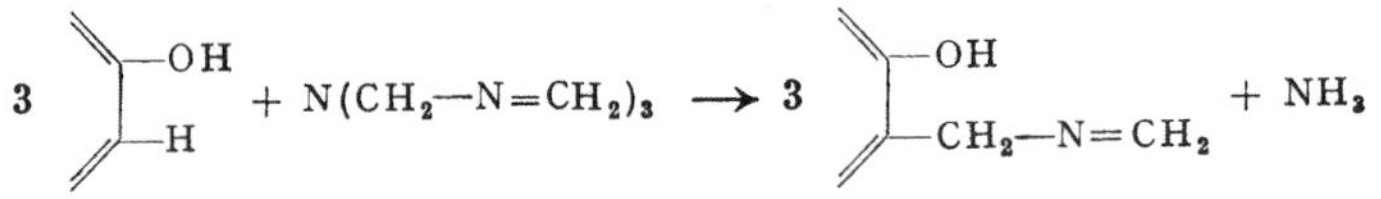

$$3\ \text{(phenol)}-\text{OH} + N(CH_2-N=CH_2)_3 \longrightarrow 3\ \text{(phenol)}-\text{OH, }-CH_2-N=CH_2 + NH_3 \qquad (2)$$

The condensation product, in its isomeric form as the Schiff base of monomethylamine, is saponified by acids to give *o*-hydroxyaldehydes:

$$\text{(structure)}\begin{matrix}-OH\\-CH=NCH_3\end{matrix} + H_2O + H^+ \longrightarrow \text{(structure)}\begin{matrix}-OH\\-CHO\end{matrix} + \overset{+}{N}H_3CH_3 \qquad (3)$$

Reactions (2) and (3) are accomplished in simple fashion by heating phenols with a mixture of hexamine and oxalic acid dihydrate. The yield appears to be better than those obtained when o-hydroxyaldehydes are produced by (1).
The formylations outlined here are used in Procedures I and II.

Procedure I. A drop of the alkaline test solution is evaporated to dryness in a micro crucible. The residue is treated with 10–20 drops of chloroform and taken to dryness in an oven. This treatment with chloroform is repeated once or twice. The residue is taken up in a drop of 6 N acetic acid, a drop of hydrazine solution is added, and the contents of the crucible transferred to filter paper. If phenols were present, a yellow to orange fluorescing fleck is seen under ultraviolet light. To avoid a false conclusion due to the self-fluorescence of phenols in acetic acid solution, it is well to conduct a blank with a drop of water in place of the hydrazine solution. In contrast to the fluorescence of the acid-stable aldazines, the fluorescence of phenols can usually be made to disappear by bathing the fleck in alcohol or 2 N hydrochloric acid.

Reagent: Hydrazine solution. 10 g hydrazine sulfate boiled with 10 g sodium
        acetate in 100 ml water. Filter after cooling.

Limits of identification are given in Table 7.

The formylation of phenols also occurs when their aqueous alkaline solutions are briefly warmed with chloroform or chloral. Therefore the procedures described in Chap. 5 for the detection of these compounds may be employed for the detection of phenols. A higher sensitivity is obtained (*i.e.* 5 γ phenol).

Procedure II. A micro test tube is used. Several cg of the oxalic acid–hexamine mixture and 1 drop of the alcohol or ether test solution are mixed and the solvent removed. It is also permissible to begin with a tiny portion of the sample. The test tube is placed in a glycerol bath previously heated to 150°. The temperature is raised to 160° and kept there for 1–2 min. After cooling, the reaction mass is taken up in 1 drop of hydrazine sulfate solution and the suspension is shaken if necessary after adding 1 drop of water. The liquid is then placed on filter paper and dried briefly. When examined under ultraviolet light, the stains exhibit a blue-green (sometimes orange) fluorescence if phenols were present.

Reagents: 1) A freshly prepared mixture of 1.3 g of crystallized oxalic acid and
        1.4 g of hexamine (hexamethylenetetramine).
        2) Hydrazine solution (see Procedure I).

Limits of identification are given in Table 7.

The salicylaldehyde methylimine, formed by heating the phenol with the mixture of hexamine and oxalic acid, fluoresces itself intensively yellow-green.

TABLE 7

PHENOLS

Name	Fluorescence color	Limit of identification (γ)	
		Proc. I	Proc. II
Phenol	Yellow	12	0.25
α-Naphthol	,,	7	0.25
β- ,,	,,	6	0.5
Di-β-naphthol	No reaction	–	–
o-Hydroxydiphenyl	Orange	20	5
p- ,,	,,	2.5	1
2,7-Dihydroxynaphthalene	Yellow	50	15
Naphthoresorcinol	No reaction	–	–
H-acid	,,	–	–
Resorcinol	Yellow	3	0.25
Salicylic acid	,,	5	2
m-Hydroxybenzoic acid	,,	10	1
Acetylsalicylic acid	,,	5	1
Phenyl salicylate	,,	5	
2,4-Dinitroresorcinol	No reaction	–	–

The *Limits of identification* are:[6]

1 γ phenol	0.5 γ 3,4-dimethylphenol	0.5 γ α-naphthol
	1 γ 3,5- ,,	5 γ β- ,,

A positive response is given by: 2,3- and 2,5-dimethylphenol.
No response is given by: 2,4- and 2,6-dimethylphenol.

1 Unpublished studies, with W. A. MANNHEIMER; F. FEIGL and E. JUNGREIS, *Analyst*, 83 (1958) 666.
2 K. REIMER and F. TIEMANN, *Ber.*, 9 (1876) 824; 10 (1877) 213.
3 For details of the course of the reaction and literature comp. L. F. and M. FIESER, *Organic Chemistry*, 2nd ed., Reinhold, Boston, 1950, pp. 720–724.
4 J. C. DUFF, *J. Chem. Soc.*, (1941) 547.
5 G. LÖSEKANN, *Chemiker-Ztg.*, 14 (1890) 1409.
6 V. ANGER and M. M. POLLAK, unpublished studies.

(5) *Test by conversion to complex cobaltic salts of o-nitrosophenols*[1]

Compounds which have a free position *ortho* to a phenolic OH group can be converted by nitrous acid into o-nitrosophenols. As shown on p. 394, the latter can be detected by the formation of the brown cobaltic chelate compounds. It is not necessary to conduct a separate nitrosation and reaction of the isolated nitrosophenols with cobalt salts in acetic acid solution. It is perfectly feasible to start with phenols (solid or dissolved) and to arrive at the cobalt chelate compounds by warming with an acetic acid solution of sodium cobaltinitrite. The following succession of reactions occurs:

$$[Co(NO_2)_6]^{3-} + 6\,H^+ \rightarrow Co^{3+} + 6\,HNO_2 \tag{1}$$

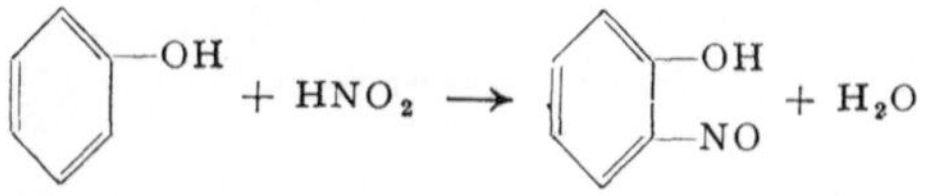

$$\text{(2)}$$

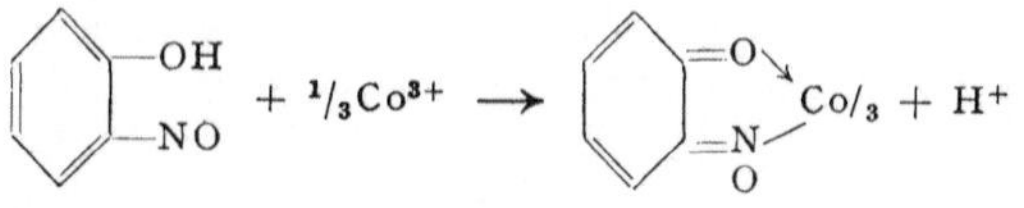

$$\text{(3)}$$

The *o*-nitrosation of phenols seemingly is favored by the occurrence of (*3*). This is shown by the finding that phenol, which yields almost exclusively *p*-nitrosophenol with nitrous acid, gives notable amounts of the cobalt salt of *o*-nitrosophenol when an acetic acid solution of cobaltinitrite is employed as nitrosating agent. On the other hand, the groups included in the phenol also exert an influence. For example, reactions (*2*) and (*3*) cannot be realized in the case of salicylic acid or the three isomeric hydroxybenzaldehydes.

When the test described here is used, the absence of primary aromatic amines should be assured. Under the conditions specified by the procedure, these amines exchange the NH_2 for an OH group and then of course show the phenol reaction. Amines which contain no hydrophilic groups can be removed by extracting the alkaline solution or suspension with ether; the phenols remain in the water layer, which can be tested directly.

Procedure. One drop of the test solution or a very little of the solid is treated in a micro test tube with a drop of the reagent and a drop of glacial acetic acid. A blank test is set up in a second test tube with a drop of water. The tubes are heated over a free flame until the blank has acquired a pink color. A positive response is given by a brown to yellow color or brown precipitate. The cobalt chelate compounds are insoluble in water and soluble or extractable in chloroform if the phenols in question do not contain hydrophilic groups.

Reagent: Freshly prepared 5% aqueous solution of sodium cobaltinitrite.

The procedure revealed:

1 γ α-naphthol	0.5 γ resorcinol
5 γ 1,4-naphtholsulfonic acid	2　γ ellagic acid
1 γ sulfosalicylic acid	0.5 γ morin
5 γ chromotropic acid	0.5 γ 2,4-dihydroxybenzaldehyde

A positive response was given by: pyrogallol, *o*-, *m*-, *p*-cresol, eugenol, hydroquinone, pyrocatecholdisulfonic acid, *p*-hydroxydiphenyl, adrenaline, gallic acid, tropaeolin 0, morphine, thymol, arbutin, *o*-hydroxyacetophenone, stovarsol, salicylic acid, phenyl salicylate, naphthoresorcinol, *β*-naphthol, antipyrine.

1 F. Feigl, *Anal. Chem.*, 27 (1955) 1315.

(6) Detection of volatile phenols with 2,6-dichloroquinone-4-chloroimine[1]

Many phenols condense with 2,6-dichloroquinone-4-chloroimine to give colored indophenols[2]. For example, phenol reacts:

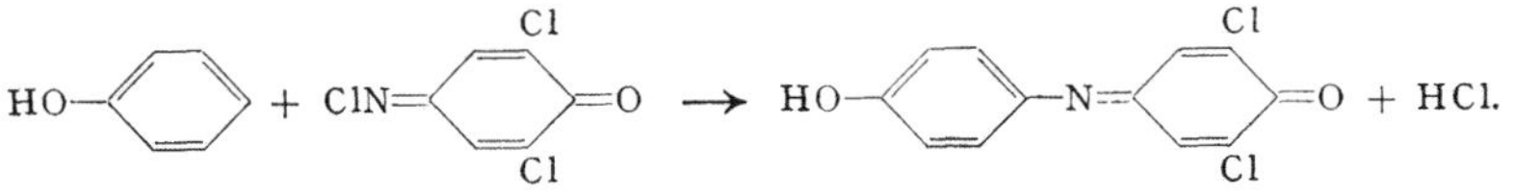

Indophenols of this type are acid–base indicators. The indophenols themselves are brown to yellow, their alkali and ammonium salts blue. The condensation requires that the phenol have a free *para* position* and that the benzene ring does not have CHO, NO_2, NO, or COOH groups in *ortho* position to the phenolic OH group[3].

It is interesting that, in contrast to the parent compound, salicylaldoxime and salicylamide but not salicylaldazine do give the indophenol reaction.

m-Dihydroxybenzenes with no CHO, NO_2 *etc.* groups give red-brown rather than the usual blue condensation products with the reagent.

The indophenol reaction is trustworthy for the detection of volatile phenols because their vapors yield yellowish brown products with the reagent, which turn blue when exposed to ammonia vapors. The test can also be accomplished as a solid body reaction if solid phenols are rubbed with 2,6-dichloroquinone-4-chloroimine.

Procedure. A little of the test material or a drop of its solution is placed in a micro test tube and taken to dryness if need be. The mouth of the tube is covered with a disk of filter paper impregnated with a saturated benzene solution of 2,6-dichloroquinone-4-chloroimine. A stable diluted (faintly yellow) solution in $CHCl_3$ can be used with advantage. The tube is placed in a glycerol bath that has been heated to 150°. After several minutes, the reagent paper is held in ammonia vapors. If the response is positive, the paper develops a blue stain which is not stable due to the loss of ammonia from the phenolate. The blue color reappears when the spot is held over ammonia.

The procedure revealed:

0.3 γ phenol	2 γ (o, m)-cresol	0.5 γ α-naphthol
0.5 γ thymol	2 γ isoeugenol	1 γ pyrocatechol

Resorcinol yields a violet color in this test (1 γ).

* The groups OCH_3, CHOH and SO_3H seem to be eliminated in the course of the reaction. The H atom of a vinyl group in *para*-position behaves like a reactive phenol.

1 F. FEIGL and E. JUNGREIS, *Anal. Chem.*, 31 (1959) 2099, 2101.
2 H. D. GIBBS, *J. Biol. Chem.*, 72 (1927) 649.
3 Comp. S. BOHDANETZKÝ, *Anal. Abstr.*, 2 (1955) 2473; and also F. FEIGL, V. ANGER and H. MITTERMANN, *Talanta*, 11 (1964) 662.

(7) *Test with p-nitrosophenol* [1]

Certain phenols condense in concentrated hydrochloric acid with *p*-nitroso-phenol to yield red quinoidal indophenols (I). When the red solution is made basic with ammonia it turns blue with production of the salt (II):

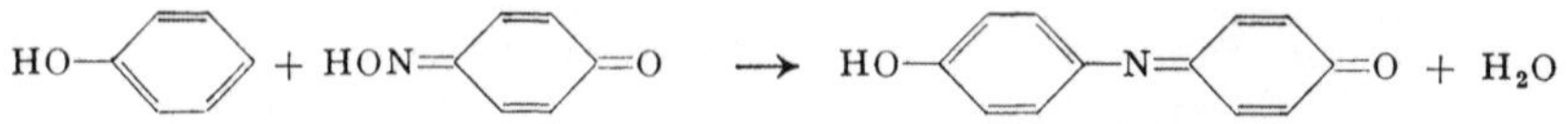

(I)

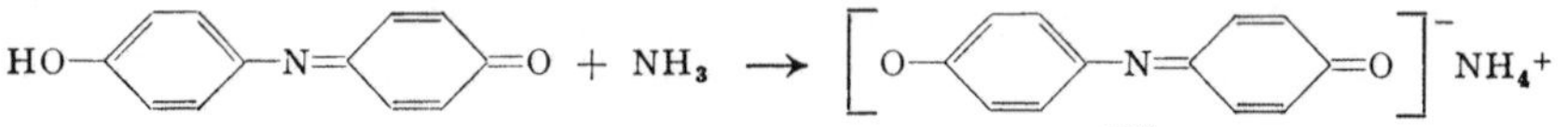

(II)

The condensation occurs with the hydrogen *para* to the phenolic hydrogen; the terminal hydrogen of a *para* vinyl group reacts in the same way with resultant lengthening of the *para*-substituent:

These reactive hydrogen atoms may not be substituted. The hydrogen atoms of the phenol molecule that do not participate in the condensation may be substituted by hydrocarbon groups, halogens, alkoxy groups and oxymethyl groups. Substituents with doubly bound oxygen, such as SO_3H, CHO, COOH groups, as well as nitrogen in the ring or as substituent prevent the occurrence of the reaction. Additional hydroxyl groups attached to the benzene ring do not of themselves impair the reaction but they influence the color of the alkali salt of the indophenols produced, which in these instances show only green shades instead of blue.

Procedure.   A slight quantity of the solid test material or the evaporation residue from a drop of its alcoholic or ethereal solution is treated in a depression of a spot plate with a drop of a 0.05% solution of *p*-nitrosophenol in concentrated hydrochloric acid. If phenols are present, a red color appears which goes over into a blue or grey-green on the addition of concentrated ammonia water.

Limits of identification and colors with ammonia:

5.0 γ phenol	blue-green	1.0 γ guaiacol	blue
0.5 γ o-cresol	,,	0.5 γ isoeugenol	,,
1.0 γ m- ,,	,,	0.5 γ pyrocatechol	green-grey
0.5 γ thymol	,,	0.5 γ resorcinol	,,
0.5 γ carvacrol	,,	0.5 γ orcinol	,,
0.5 γ α-naphthol	,,	2.5 γ pyrogallol	grey
2.0 γ o-hydroxydiphenyl	,,	1.0 γ phloroglucinol	orange-grey
0.5 γ orthoeugenol	,,		

A positive response is also given by hydroxyhydroquinone.

No reaction was given by: p-cresol, β-naphthol, p-hydroxydiphenyl, hydroquinone, hydroquinone monomethyl ether, phenol-p-sulfonic acid, 1-naphthol-2-sulfonic acid, o-aminophenol, o-, m-, and p-nitrophenol, o-, m- and p-hydroxybenzaldehyde, vanillin, orthovanillin, salicylamide, gallic acid, eugenol, 8-hydroxyquinoline, pyridoxine, theophylline, antipyrine.

1 V. ANGER and S. OFRI, *Z. Anal. Chem.*, 200 (1964) 217.

(8) *Test with pentacyano iron complex salts and hydroxylamine employing a granular ion exchange resin* [1]

Phenols condense with hydroxylamine in alkaline solution and yield aminophenols, which unite with pentacyano iron complex salts to form green to blue compounds[2] (see p. 248). Indophenols often are produced as by-products, probably through reaction of the resulting p-aminophenol with additional phenol and atmospheric oxygen.[3]

If the reaction is carried out with a pentacyano iron complex salt adsorbed on an ion exchange resin, very low identification limits are obtained.

Procedure. A few grains of the resin and a drop of the test solution are placed in a depression of a spot plate. After about 15 min., a drop of a 3% hydroxylamine hydrochloride solution and a drop of 5% sodium carbonate are introduced. The mixture on the spot plate is then heated for a few minutes with an infrared lamp. If the response is positive, the grains turn yellow-green or blue-green depending on the quantity of phenol present. Aromatic amines must be absent.

Preparation of the resin grains: About 0.1 g of Diaion SA-201 in the hydroxide form is treated with 1 ml of 2% sodium carbonate solution containing 1% sodium nitroprusside which has been aged either by prolonged storing, or by irradiation in ultraviolet light for around 15 min. The suspension is stirred until the grains change from light yellow to faint dull green, and then the excess reagent is removed and the grains washed with water. The resin grains are ready for use after their color has changed to light yellowish brown.

Limits of identification:

0.1 γ phenol	0.4 γ o-ethylphenol	4.2 γ 2,4-dimethylpheno
0.5 γ o-cresol	0.5 γ o-chlorophenol	0.5 γ 2,6- ,,
0.6 γ m- ,,	2.3 γ p- ,,	1.2 γ 3,4- ,,
1.2 γ p- ,,	2.3 γ pyrocatechol	1.9 γ p-hydroxydiphenyl
1 γ thymol	0.4 γ 2-bromo-m-cresol	0.5 γ 2-chloro-m-cresol
1.5 γ eugenol	1 γ 8-hydroxyquinoline	1.6 γ 4- ,, -m- ,,
1 γ carvacrol	1.2 γ salicylaldehyde	0.4 γ 2-methoxy-m-cresol
4 γ vanillin	12 γ 1-bromo-2-naphthol	1 γ m-hydroxybenzaldehyde
0.25 γ α-naphthol	0.4 γ salicylamide	1.9 γ p- ,,
4.2 γ β- ,,	1.9 γ salicylic acid hydrazide	1.1 γ o-hydroxyacetophenone
4 γ salicylic acid	1 γ m-hydroxybenzoic acid	8 γ p- ,,
0.2 γ guaiacol	2 γ p- ,, ,,	1.3 γ methyl salicylate
0.6 γ resorcinol	1 γ methylhydroquinone	1 γ phenol-m-sulfonic acid
1.1 γ hydroquinone	4.2 γ o-methylpyrocatechol	2 γ ,, -p- ,,

1.8 γ creosol (2-methoxy-4-methylphenol)   0.9 γ phenol-o-sulfonic acid
8.4 γ p-hydroxybenzoic acid methyl ester   0.5 γ salicylalhydrazone
1.8 γ p-hydroxybenzoic acid hydrazide

No reaction was observed with 3,5-dimethylphenol, nitrophenols, orcinol, pyrogallol, gallic acid, 2-hydroxy-3-methylbenzoic acid, phloroglucinol, chromotropic acid, G-acid, H-acid, rutin.

1 A. Tsuji and H. Kakihana, Mikrochim. Acta, (1962) 479.
2 S. Ohkuma, J. Pharm. Soc. Japan, 72 (1949) 872. Comp. V. Anger, Mikrochim. Acta, 2 (1938) 5.
3 T. Itai and S. Kamiya, Japan Analyst, 7 (1958) 616.

15. Ethers (R—O—R group)

Test through formation of peroxide[1]

If a drop of a volatile ether or a drop of a mixture of chloroform and ether (up to 800 : 1) is placed in a micro test tube, and the latter then plunged to $\frac{3}{4}$ of its length in a glycerol bath heated to 230°, and the mouth of the test tube covered with a disk of filter paper moistened with a freshly prepared mixture of equal parts of saturated copper acetate and benzidine chloride solutions, a blue color which constantly deepens will make its appearance after 1–2 min. The blue product is known as benzidine blue; it is a meriquinoid oxidation product of benzidine (see p. 546). The following series of reactions are involved in this test which is specific for ether; obviously they must proceed rapidly:

$$\text{ether} + \text{O (air)} \rightarrow \text{ether peroxide} \qquad (1)$$

$$\text{ether peroxide} + \text{Cu(II)} \rightarrow \text{Cu(II) peroxide} + \text{ether} \qquad (2)$$

$$\text{Cu(II) peroxide} + \text{benzidine} \rightarrow \text{Cu(II)} + \text{benzidine blue} \qquad (3)$$

Summation of the redox reactions (1)–(3), which repeat themselves again and again, yields reaction (4) which per se proceeds only at an immeasurably slow rate:

$$\text{benzidine} + \text{O (air)} \rightarrow \text{benzidine blue} \qquad (4)$$

Accordingly, the interposing of the rapid intermediate reactions (1–3) catalytically hastens the oxidation of benzidine to benzidine blue, where the autoxidation of the ether to a peroxide constitutes the initial reaction.

As little as 40 γ of ether in one drop of an ether–chloroform mixture can be detected through the production of benzidine blue.

The following redox reactions, which are accompanied by a color change, may also be used for the detection of ether because of the formation of peroxide. Appropriately impregnated filter papers containing reductants should be used:

(a) liberation of iodine from potassium iodide;

(b) formation of the blue oxidation product of thio-Michler's ketone (see p. 65);

(c) conversion of black lead sulfide to white lead sulfate.

It should be pointed out that the autoxidation of ether results in the formation of acidic materials, probably acetic acid, which react with acid–base indicator papers (*Limit of identification*: 20 γ diethyl ether). Positive responses were also given by dioxane and tetrahydrofuran; vapors of diisopropylether do not react. It seems therefore that a —CH$_2$—O—CH$_2$— group is responsible for the autoxidation.

The tests given here are successful in solutions which contain ether along with 800 times as much chloroform or carbon tetrachloride. The tests fail in mixtures containing ether mixed with about 7 times as much (by volume) benzene, carbon disulfide, petroleum ether. However, a positive response can be obtained by adding the same volume of chloroform.

1 F. FEIGL, J. R. AMARAL and D. HAGUENAUER-CASTRO, *Mikrochim. Acta*, (1960) 821.

16. Ethoxy compounds (—OC$_2$H$_5$ group)

Test by formation of acetaldehyde [1]

Acetaldehyde is produced if ethoxy compounds are treated with a solution of alkali bichromate in sulfuric acid. Probably, the initial step is hydrolysis to ethanol, followed by oxidation to acetaldehyde:

$$ROC_2H_5 + H_2O \rightarrow ROH + C_2H_5OH$$

$$C_2H_5OH + O \rightarrow CH_3CHO + H_2O$$

This assumption is supported by the fact that many ethoxy compounds are promptly oxidized in the wet way to yield acetaldehyde, whereas the direct oxidation with benzoyl peroxide (see p. 165) does not succeed in all cases.

The acetaldehyde in the vapor phase can be detected by the bluing of sodium nitroprusside–morpholine solution (see p. 438).

The behavior of N-ethyl compounds toward chromic acid is not uniform; many compounds are not attacked. Consequently, it is advisable when testing for ethoxy compounds to remove any basic N-ethyl compounds by shaking the chloroform test solution with dilute hydrochloric acid (formation of water-soluble chlorides).

Procedure. A drop of the test solution* or better a small amount of the not-dissolved sample is treated in a micro test tube with a drop of acidified bichromate solution. The mouth of the tube is covered with a piece of filter paper moistened with a drop of sodium nitroprusside–morpholine solution (for preparation see p. 439). The test tube is placed in boiling water. If ethoxy compounds are present, a more or less intense blue fleck appears.

Reagent: 1 g potassium bichromate is dissolved in 60 ml water and poured into 7.5 ml concentrated sulfuric acid (caution).

The procedure revealed:

30 γ phenetole	20 γ diethyl oxalate
50 γ phenacetin	200 γ ethylmorphine chloride dihydrate

A positive response was obtained with: ethyl benzoate, ethyl *p*-aminobenzoate (anesthesine), ethyl butyrate, ethyl lactate, diethyl phthalate, ethyl cellulose, and with the ethyl esters of phosphoric and thiophosphoric acids given on p. 97.

It must be noted that ethyl urethane and 3-ethoxy-4-methoxybenzyl cyanide form no acetaldehyde when treated with chromic acid, whereas this occurs when they are fused with benzoyl peroxide. Probably these ethoxy-compounds are not hydrolyzed to ethanol which is essential for the oxidation into acetaldehyde in the wet way.

* Alcohol or ether may not be used as solvent because these compounds produce acetaldehyde under the prescribed conditions.

1 F. FEIGL and E. SILVA, *Analyst*, 82 (1957) 582.

17. Alkyl phenyl ethers

Test by splitting to phenols [1]

Phenyl ethers can be dealkylated, with volatilization of phenols, by heating to 150° with a mixture of potassium iodide and oxalic acid dihydrate. The oxalic acid releases hydrogen iodide which in gaseous form reacts:

$$ArOR + HI \rightarrow RI + ArOH$$

At the same time, the water vapor given off by the hydrated acid facilitates the volatilization of the phenols, which can be detected in the gas phase by the indophenol reaction (p. 185) with 2,6-dichloroquinone-4-chloroimine. The alkyl iodide or hydrogen iodide formed do not interfere with the test.

The procedure may not be applied directly if phenols are present. In such cases the latter must be removed beforehand as water-soluble alkali phenolates by the process outlined on p. 216.

Procedure. A little of the sample or a drop of its solution in alcohol, ether, *etc.* is placed in a micro test tube and evaporated to dryness, if need be, along with several cg of a freshly prepared 1:1 mixture of KI and $H_2C_2O_4.2\ H_2O$. The tube is closed with a disk of filter paper impregnated with a saturated benzene solution of 2,6-dichloroquinone-4-chloroimine and heated in a glycerol bath at 150°. After a few minutes the paper is held in ammonia vapors. A blue stain indicates a positive response.

The test revealed:

0.5 γ veratrole	25 γ anethole	10 γ α,β-diphenoxyethane
1 γ phenetole	10 γ phenoxyacetic acid	5 γ o-methoxybenzoic acid
1 γ anisole	20 γ m-phenetidine	2.5 γ guaiacol benzoate
2 γ isoeugenol	10 γ m-ethoxyphenoxyacetic acid	

1 F. FEIGL and E. JUNGREIS, *Anal. Chem.*, 31 (1959) 2099, 2101.

18. Carbonyl compounds (>CO group)

(1) Test by interaction with bisulfite

Approximately neutral solutions of aldehydes and aliphatic methyl ketones combine with sodium bisulfite to form well crystallized water-soluble products[1] known as "aldehyde bisulfite" and "ketone bisulfite", respectively. They are alkali salts of α-hydroxysulfonic acids:[2]

$$\begin{array}{c} R \\ \diagdown \\ \qquad C=O \ + \ NaHSO_3 \ \longrightarrow \\ H \diagup \end{array} \qquad \begin{array}{c} R \diagdown \quad \diagup OH \\ \quad C \\ H \diagup \quad \diagdown SO_3Na \end{array}$$

$$\begin{array}{c} R \\ \diagdown \\ \qquad C=O \ + \ NaHSO_3 \ \longrightarrow \\ H_3C \diagup \end{array} \qquad \begin{array}{c} R \diagdown \quad \diagup OH \\ \quad C \\ H_3C \diagup \quad \diagdown SO_3Na \end{array}$$

An excess of aldehyde or methyl ketone rapidly consumes the bisulfite and converts it into compounds which, unlike free bisulfite, no longer react with iodine.[3] The familiar redox reaction:

$$SO_3{}^{2-} + I_2 + H_2O \rightarrow SO_4{}^{2-} + 2\ I^- + 2\ H^+$$

may thus be masked through the formation of aldehyde- or ketone bisulfite. Therefore, when a neutral test solution of aldehyde or methyl ketone is mixed with a dilute solution of bisulfite followed by a blue starch or thyodene–iodine solution, and the color is not discharged, it indicates the presence of aldehydes or ketones. Obviously, no other substance which will consume iodine may be present.

Procedure.[4] A drop of the alcoholic or aqueous test solution is mixed with a drop of approximately 0.001 N sodium bisulfite. If the original solution is alcoholic it is advisable to add 4 or 5 drops of water. After about 5 min., a drop of a 1% starch or thyodene solution faintly blued with iodine, is added. If the blue color remains, the test is positive, indicating the presence of aldehyde or ketone.

The dilutions should be so adjusted that 10 drops of the bisulfite solution require 11 or 12 drops of the iodine solution to give a permanent blue.

The following amounts were revealed:

0.05	γ formaldehyde	1	γ m-nitrobenzaldehyde
0.5	γ acetaldehyde	1	γ p-aminobenzaldehyde
5	γ oenanthal	20	γ vanillin
5	γ furfural	15	γ anisaldehyde
2	γ benzaldehyde	500	γ glucose
4	γ benzaldehyde-o-sulfonic acid	500	γ fructose
4	γ salicylaldehyde	500	γ lactose
1	γ m-hydroxybenzaldehyde	50	γ acetone
10	γ p- ,,	20	γ methyl ethyl ketone
4	γ o-nitrobenzaldehyde	20	γ acetophenone

No reaction was given by: saccharose, benzophenone, benzil, ethanol, purified dioxane (boiled with hydrochloric acid and distilled over sodium).

1 C. BERTAGNINI, *Compt. Rend.*, 35 (1852) 800.
2 F. RASCHIG and L. PRAHL, *Ann.*, 448 (1926) 265; see also G. SCHROETER, *Ber.*, 59 (1926) 2341; 61 (1928) 1616.
3 A. KURTENACKER, *Z. Anal. Chem.*, 64 (1924) 56.
4 Unpublished studies, with R. ZAPPERT.

(2) Test with azobenzenephenylhydrazinesulfonic acid [1]

The aqueous strongly acidic solution of azobenzenephenylhydrazinesulfonic acid (I) reacts with aldehydes giving deep red or blue solutions. Hereby salts of hydrazones of azobenzenephenylhydrazine (II or IIa) are formed [2]:

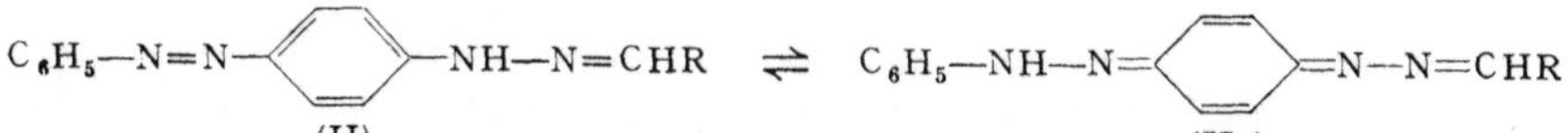

This color reaction may be applied to the detection of aldehydes. It is carried out in strongly acid solution, with heating, and the product is extracted with chloroform in the presence of alcohol. The color of the product differs for aromatic and aliphatic aldehydes; namely, red and blue. Aromatic and aliphatic aldehydes may be differentiated by this difference.

Ketones react similarly to aldehydes, but much less readily. Esters, alcohols, phenols, amines, amides, quinones, chloral do not react.

Procedure. A drop of the test solution is mixed with about 7 drops of reagent solution and 4 drops of concentrated sulfuric acid in a test tube. The tube is then placed in boiling water for 30 sec. and allowed to cool. A few drops of alcohol are added and enough chloroform to form a lower layer. About 5 drops of concentrated hydrochloric acid are added and the mixture shaken vigorously. In the presence of aldehydes (ketones) the chloroform layer is colored.

Reagent: 0.2% aqueous solution of azobenzenephenylhydrazine sulfonic acid.

The following were detected by this test:

0.25 γ	acetaldehyde	1 γ	protocatechuicaldehyde
0.25 γ	formaldehyde	0.5 γ	*m*-hydroxybenzaldehyde
0.36 γ	paraldehyde	4.8 γ	*o*-phthalaldehyde
0.16 γ	oenanthal	0.16 γ	vanillin
4.5 γ	glyceraldehyde	1 γ	piperonal
2 γ	phenylacetaldehyde	0.5 γ	2-hydroxy-1-naphthaldehyde
0.2 γ	benzaldehyde	1.2 γ	*p*-dimethylaminobenzaldehyde
0.35 γ	anisaldehyde	0.2 γ	acrolein
0.35 γ	salicylaldehyde	0.1 γ	crotonaldehyde
0.11 γ	cinnamaldehyde	0.35 γ	furfural
0.86 γ	benzaldehyde-*o*-sulfonic acid	40 γ	acetone
0.5 γ	*o*-nitrobenzaldehyde	130 γ	acetophenone
0.2 γ	*p*- ,,	25 γ	cyclohexanone
0.25 γ	3-nitrosalicylaldehyde	12 γ	benzoin
6.8 γ	5- ,,	55 γ	epichlorohydrin

1 Unpublished studies, with G. Frank.

2 J. Tröger and O. Müller, *J. Prakt. Chem.*, 78 (1908) 371.

19. Aromatic and α,β-unsaturated carbonyl compounds

Test with p-nitrophenylhydrazine[1]

Compounds containing the —C=C—C=O group condense with *p*-nitrophenylhydrazine to yield Schiff bases (I), which react with alkali hydroxide through prototropic rearrangement to form blue or violet alkali salts of *p*-quinoidal nitronic acids (II):

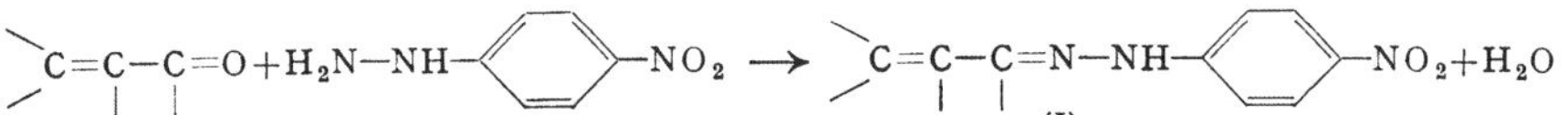

(I) (1)

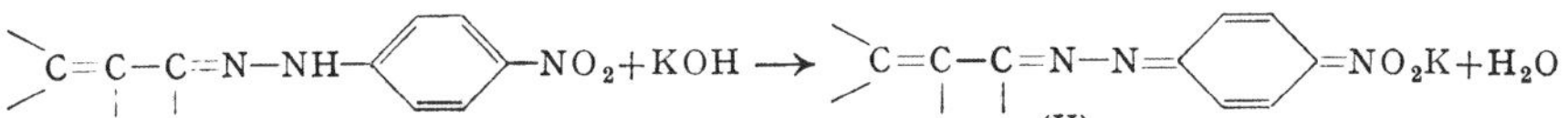

(II) (2)

As shown by (II), the intense color quality of the alkali salt is due to the development of a chain of continous conjugated double bonds.

Reactions (1) and (2) proceed very rapidly in solutions of isopropanol. It is noteworthy that the color differences that are still very apparent at 10 γ carbonyl compounds no longer are seen at the identification limits that are lower by many powers of 10. The colors are almost always pink or orange in the latter case.

The tendency of carbonyl compounds to give colored condensation products of type (II) is extremely strong. This is evident not only from the marked sensitivity of the color reactions but also from the finding that oximes and anils of the carbonyl compounds act similarly to the parent compounds; obviously the NOH- and NC_6H_5 groups are displaced by the condensation with nitrophenylhydrazine.

It should be noted that 1,2-dicarbonyl compounds react with p-nitrophenylhydrazine in a manner analogous to that of the aromatic and α, β-unsaturated carbonyl compounds (comp. p. 327).

Procedure. A drop of the test solution and 1 drop of the p-nitrophenyl-hydrazine solution are placed in a micro test tube and heated to boiling for 30 sec. Three drops of isopropanolic potassium hydroxide are then added. If the response is positive, an orange to red-violet color appears.

It is imperative to run a blank test since it likewise frequently gives a pale coloration.

> *Reagents:* 1) 2 mg of p-nitrophenylhydrazine are dissolved in 10 ml of isopropanol and then 5 drops of conc. hydrochloric acid are added.
> 2) Saturated solution of KOH in isopropanol–water (95:5).

Limits of identification:

0.005	γ acrolein	0.05	γ opianic acid
0.1	γ crotonaldehyde	0.0001	γ p-dimethylaminobenzaldehyde
0.00001	γ glutaconic aldehyde	0.01	γ o-nitrobenzaldehyde
0.01	γ benzaldehyde	0.0001	γ p- ,,
0.01	γ hydrobenzamide	0.05	γ 2,4-dinitrobenzaldehyde
0.005	γ benzalazine	0.001	γ furfural
0.01	γ benzalphenylhydrazone	0.1	γ β-indolylaldehyde
0.05	γ benzalsemicarbazone	0.05	γ pyridine-2-aldehyde
0.05	γ cinnamaldehyde	0.01	γ ,, -4- ,,
0.0001	γ o-phthalaldehyde	0.01	γ 6-methylpyridine-2-aldehyde
0.001	γ salicylaldehyde	0.05	γ acetophenone
0.01	γ salicylalhydrazone	0.1	γ ω-chloroacetophenone
0.01	γ salicylaldoxime	0.0001	γ benzalacetone
0.05	γ salicylazine	0.05	γ benzophenone
0.01	γ m-hydroxybenzaldehyde	0.05	γ deoxybenzoin
0.01	γ anisaldehyde	0.1	γ benzoin
0.0001	γ vanillin	1	γ benzoin oxime

1 F. FEIGL, V. ANGER and G. FISCHER, *Mikrochim. Acta*, (1962) 878.

20. Aldehydes (–CHO group)

(1) Test with fuchsin–sulfurous acid [1]

Sulfurous acid decolorizes *p*-fuchsin (I) by destroying the quinoid structure with production of the N-sulfinic acid of the leuco sulfonic acid (II). The resulting colorless solution turns violet to blue on the addition of aliphatic or aromatic aldehydes. They restore the quinoid structure, and consequently the color, by combining with the sulfurous acid which has reacted with the dye. On the addition of aldehyde, the initial product is the colorless compound (III). The second step involves the loss of the sulfonic acid group attached to the carbon atom and the pink quinoid dye (IV) is formed. Analogous changes occur with other triphenylmethane dyes.[2]

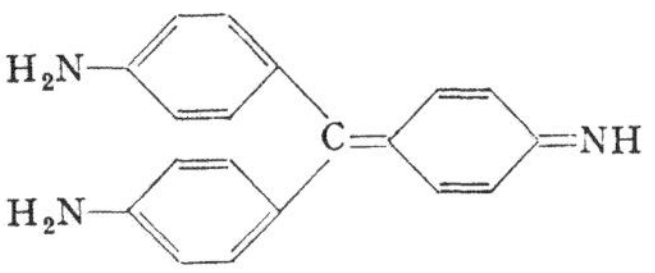

(I)
red

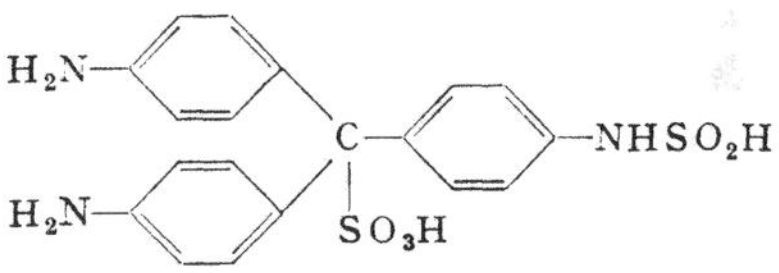

(II)
colorless

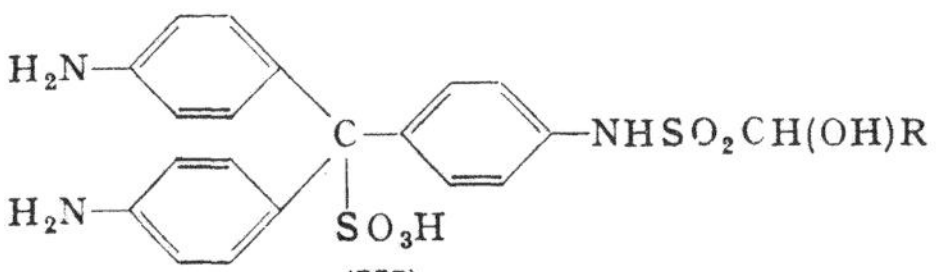

(III)
colorless

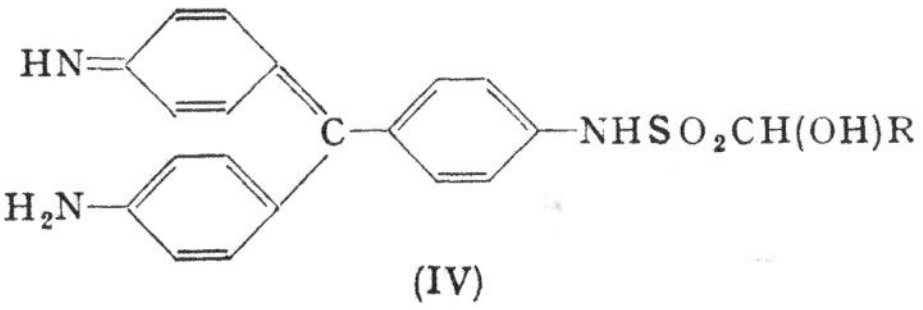

(IV)
pink

The reaction between aldehydes and solutions of fuchsin or malachite green, which have been decolorized by sulfite, can be carried out as a drop reaction on a spot plate, or on filter paper.

Procedure.[3] A drop of the alcoholic or aqueous solution is treated with a drop of sulfurous acid and a drop of fuchsin reagent. The violet to blue color appears in 2–30 min. depending on the amount of aldehyde present.

Reagents: 1) Fuchsin–sulfurous acid. SO_2 is passed through a 0.1% solution of fuchsin until the color is discharged.*

      4) Dilute solution of sulfurous acid (freshly prepared).

The following were detected:

1 γ formaldehyde	100 γ salicylaldehyde
4 γ acetaldehyde	50 γ *m*-hydroxybenzaldehyde
20 γ furfural	1000 γ *p*- ,,
30 γ benzaldehyde	1000 γ anisaldehyde
40 γ nitrobenzaldehyde (*o, m*)	8 γ cinnamaldehyde

* See W. C. Tobie, *Ind. Eng. Chem., Anal. Ed.*, 14 (1942) 405 regarding an improved reagent which also reveals free aldehyde groups in certain aldoses.

Oenanthal gives a slight reaction; vanillin, ethylvanillin hydrochloride, 2,4-dihydroxybenzaldehyde, chloral hydrate, p-amino- and p-dimethylaminobenzaldehyde give no response.

1 H. Schiff, Ann., 140 (1866) 93; B. v. Bitto, Z. Anal. Chem., 36 (1897) 373; E. Votocek, Ber., 40 (1907) 414; L. Rosenthaler and G. Vegezzi, Mitt. Lebensmittelunters. Hyg. (Eidg. Gesundheitsamt Bern), 45 (1954) 178.
2 H. Wieland and G. Scheuing, Ber., 54 (1921) 2527.
3 Unpublished studies, with R. Zappert.

(2) Test through conversion into ferric salt of hydroxamic acids [1]

Benzenesulfohydroxamic acid (I) is split by alkalis into benzenesulfinic acid (II) and the nitrosyl radical (III):

$$C_6H_5-SO_2NHOH + KOH \rightarrow C_6H_5-SO_2K + NOH + H_2O$$
$$\text{(I)} \qquad\qquad\qquad \text{(II)} \qquad\quad \text{(III)}$$

If this decomposition occurs in the presence of an aldehyde, the nitrosyl group attaches itself to the aldehyde and a hydroxamic acid results:

$$R-CHO + NOH \longrightarrow R-CH{\overset{\displaystyle /OH}{\underset{\displaystyle \backslash NO}{}}} \longrightarrow R-CONHOH$$

(Alkali salts of hydroxamic acids are formed in alkaline solutions.)
Hydroxamic acids produce violet inner-complex compounds (comp. p. 212) with ferric ions in acid solution. This conversion into iron(III) salts of hydroxamic acids serves as the basis of a spot test for aldehydes.

Procedure.[2] A drop of the test solution is treated with a drop of a 1% alcoholic solution of benzenesulfohydroxamic acid and then a drop of 1 N sodium hydroxide is added. The mixture is allowed to stand for 5 min., and is then acidified by adding a drop of 2 N hydrochloric acid. A drop of a 1% solution of ferric chloride is added. The appearance of a red-violet color signals the presence of aldehydes.

Toluenesulfohydroxamic acid may also be used as the reagent.

Limits of identification:

2 γ formaldehyde	8 γ cinnamaldehyde	100 γ salicylaldehyde
5 γ acetaldehyde	20 γ p-nitrobenzaldehyde	50 γ m-hydroxybenzaldehyde
12 γ furfural	10 γ o- ,,	
15 γ benzaldehyde	20 γ m- ,,	

The following gave no reaction: p-hydroxybenzaldehyde, vanillin, p-aminobenzaldehyde, p-dimethylaminobenzaldehyde.

1 A. Angeli, Gazz. Chim. Ital., 26 II (1896) 17; Chem. Ztg., 20 (1896) 176; E. Rimini, Gazz. Chim. Ital., 31 II (1901) 84.
2 Unpublished studies, with R. Zappert.

(3) Test with o-dianisidine [1]

Primary aromatic amines condense with aliphatic and aromatic aldehydes in acetic acid solution. The condensation products, which are colored in many cases, are known as Schiff bases. o-Dianisidine (I) is especially suitable for the formation of colored Schiff bases (II):

$$H_2N-\langle\rangle-\langle\rangle-NH_2 \xrightarrow{2\,RCHO} RHC=N-\langle\rangle-\langle\rangle-N=CHR + 2H_2O$$

(I)   R = alkyl or aryl   (II)

Ketones do not generally interfere as they are rather inactive. However, large amounts of certain ketones may give yellow, green or brown colorations, especially on heating, and so interfere with the detection of small amounts of aldehydes, or the ketone may be mistaken for an aldehyde. Formation of ketimides is probably responsible for these colors.

Alicyclic compounds, such as pinenes, camphenes and others, especially in high concentrations, give brown colors with the reagent. They can interfere with the test when the sample is a mixture (essential oils, etc.). The brown products are probably addition compounds.

Procedure. A drop of the sample is mixed with 3 or 4 drops of reagent in a micro crucible. A light color usually appears even in the cold, and is intensified by heating. The reaction may also be carried out on filter paper, by treating a drop of the sample solution with the reagent solution. The color can be deepened by warming over a micro flame.

Reagent: Saturated solution of pure o-dianisidine in glacial acetic acid. 2,7-Diaminofluorene may be used in place of o-dianisidine; it forms even brighter colors with some aldehydes.

Limits of identification are:

50	γ formaldehyde	0.05	γ cinnamaldehyde
30	γ acetaldehyde	5	γ salicylaldehyde
4	γ paraldehyde	4	γ m-hydroxybenzaldehyde
20	γ propionaldehyde	5	γ p- ,,
9	γ oenanthal	10	γ 2-hydroxy-1-naphthaldehyde
200	γ decylaldehyde	2	γ anisaldehyde
0.1	γ acrolein	4	γ piperonal
8	γ crotonaldehyde	70	γ opianic acid
0.1	γ citral	3	γ benzaldehyde-o-sulfonic acid
10	γ citronellal	0.4	γ p-aminobenzaldehyde
40	γ bromal	0.2	γ p-dimethylaminobenzaldehyde
3	γ benzaldehyde	5	γ o-nitrobenzaldehyde
5	γ tolualdehyde	1	γ p- ,,
3	γ cumic aldehyde	0.02	γ furfural

1 R. WASICKY and O. FREHDEN, *Mikrochim. Acta*, 1 (1937) 55.

TABLE 8

ALDEHYDES AND ALDEHYDE DERIVATIVES

Name	Color		Limit of identification, γ
	In acid solution	In neutral solution	
Formaldehyde	Green, Black	Black	0.02
Acetaldehyde	,, ,,	,,	0.01
Propionaldehyde	Brown, Black	Brown, Black	35, 10
Oenanthal	Green, Black	Violet, Black	0.3
Phenylacetaldehyde	,, ,,	Violet-black	10, 5
Glyceraldehyde	Violet, Black	Greenish, Black	12, 8
Acrolein	Green, Black	Red, Black	0.25
Crotonaldehyde	Light brown	Brown, Black	0.12
Benzaldehyde	Yellow	,, ,,	3.5
Benzaldehyde-o-sulfonic acid	Green	Red-violet, Black	1.6
o-Phthalaldehyde	Yellow	Black	1, 4
o-Hydroxybenzaldehyde	Orange-yellow	Yellow, Black	0.7
m- ,,	Yellow	Violet, Black	0.5
o-Nitrobenzaldehyde	Violet	,, ,,	0.6
p- ,,	Orange-red	Red-brown, Black	0.8
Anisaldehyde	Light yellow	Darker	5
Protocatechuic aldehyde	Orange	Greenish brown	1.2
Vanillin	Yellow	Brown, Black	2
Cinnamaldehyde	Orange	Violet, Black	0.2
Salicylaldehyde bisulfite	,,	Yellow, Black	0.8
4-Nitrosalicylaldehyde bisulfite	,,	Brown, Black	0.15
Anisaldehyde bisulfite	Light yellow	Brown	6.5
Cinnamaldehyde bisulfite	Yellow	Black	0.8
Furfural bisulfite	,,	Brown, Black	0.7
Acetaldehyde ammonia	Green, Black	Green, Black	0.06
Hexamethylenetetramine	Violet, Black	Violet, Black	50
Mandelonitrile	Yellow, Brown	Dark brown	4
Propionaldehyde cyanohydrin	Violet	,, ,,	34, 10

(4) Test by catalytic acceleration of the oxidation of p-phenylenediamine by hydrogen peroxide[1]

p-Phenylenediamine is oxidized by hydrogen peroxide, in acid or neutral solution, to give a black quinoidal compound known as Bandrowski's base:[2]

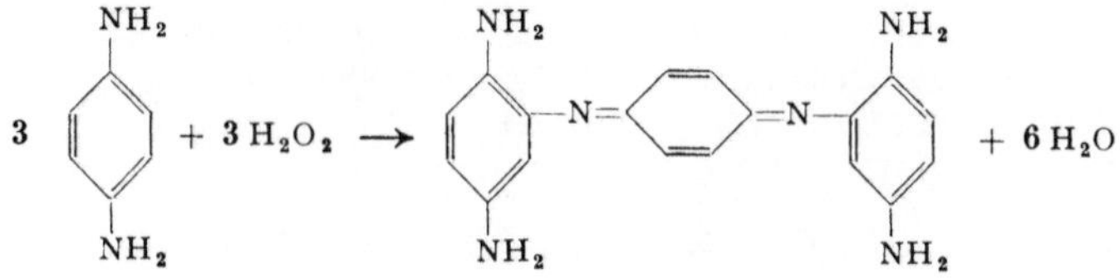

The oxidation is appreciably accelerated by aldehydes.[3] This action permits the detection of aldehydes, if the test is carried out in certain concentrations of reagent and acid, at which the rate of the uncatalyzed reaction is lowered, while the peroxidase action of the aldehydes is not appreciably affected.

In neutral solution, all aldehydes yield a black color or precipitate (with other preceding transitory colors) which lasts a little longer when aromatic aldehydes are involved. In acid solution, the aliphatic aldehydes behave in the same way but most aromatic aldehydes form a yellow precipitate or color that persists for some time. This difference in behavior is useful in distinguishing between aliphatic and aromatic aldehydes.

Cyanohydrins, aldehyde ammonia, and aldehyde bisulfite compounds behave like the aldehydes. Aliphatic and aromatic oximes are less reactive; ketones have no catalytic effect.

Procedure. A drop of the reagent, 2 drops of 2 N acetic acid and 2 drops of 3% hydrogen peroxide are mixed with a drop of the test solution on a spot plate. A color appears at once when large amounts of aldehyde are present, or after a short time with lesser quantities. It is advisable always to carry out a blank test on a drop of water, and further to carry out a parallel test omitting the acetic acid since some aldehydes react more rapidly in acid solution and others more rapidly in neutral solution. A yellow color in acid medium indicates aromatic aldehydes.

Reagent: 2% alcoholic solution of *p*-phenylenediamine (freshly prepared).

Limits of identification are given in Table 8.

1 Unpublished studies, with G. FRANK.
2 E. v. BANDROWSKI, *Ber.*, **27** (1894) 480; see also J. J. RITTER and G. H. SCHMITZ, *J. Am. Chem. Soc.*, **51** (1929) 1587.
3 G. WOKER, *Ber.*, **47** (1914) 1024. See also G. WOKER, *Die Katalyse* (Vols. 27/28 of *Die Chemische Analyse*), Enke, Stuttgart, 1931, p. 232.

(5) Test with indole[1]

In hydrochloric acid solution, indole reacts with aldehydes to yield red products. The reaction involves the condensation of the aldehyde with two molecules of indole, the initial condensation product being oxidized by oxygen from the air:

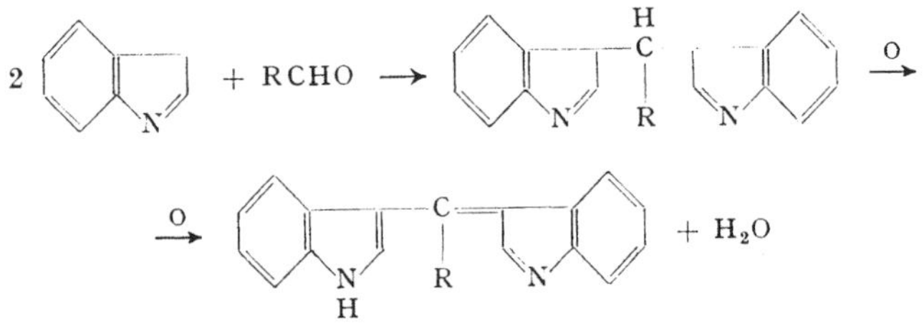

This assumption is supported by the finding that formic acid, which may also be regarded as an aldehyde, gives a particularly distinct reaction with indole. The colored quinoidal product obviously is formed directly with indole in the case of formic acid, *i.e.* without oxidation.

Procedure. A drop of the test solution is treated in a micro test tube with 5 drops of indole reagent and the mixture is kept in a boiling water bath until a color appears. A pink to orange color indicates the presence of aldehydes. It is essential to make a blank test and to do this at the same time as the prime test. If the aldehyde is dissolved in an organic liquid, the latter should be tested with regard to its possible aldehyde content. Alcohols should be avoided as solvent for this reason.

Reagent: Several platelets of purest indole (recrystallized from benzine or petroleum ether if possible) are suspended in 3 ml of pure conc. hydrochloric acid. The reagent must be freshly prepared and should not be used if it is more than 15 min. old. Specimens of indole which yield a pink color when heated on the water bath with hydrochloric acid alone are not suitable for use as reagent in this test.

Limits of identification:

0.2	γ formaldehyde	0.1	γ salicylaldehyde
1	γ acetaldehyde	0.1	γ anisaldehyde
10	γ chloral hydrate	0.0001	γ vanillin
10	γ isovaleraldehyde	0.5	γ p-dimethylaminobenzaldehyde
0.05	γ crotonaldehyde	0.01	γ cinnamaldehyde
20	γ glucose	0.05	γ furfural
10	γ glyoxal	1	γ paraldehyde
0.5	γ benzaldehyde	0.05	γ p-hydroxybenzaldazine
0.5	γ o-nitrobenzaldehyde	0.005	γ formic acid
0.1	γ p- ,,		

1 V. ANGER and G. FISCHER, *Mikrochim. Acta*, (1960) 592.

(6) Test with 1,2-dianilinoethane (for water-soluble aldehydes) [1]

1,2-Dianilinoethane (N,N'-diphenylethylenediamine) reacts (in methanolic solutions) with most aliphatic and aromatic aldehydes:

$$
\begin{array}{c}
H_2C-NHC_6H_5 \\
| \\
H_2C-NHC_6H_5
\end{array}
+ \; OCH(R,Ar) \longrightarrow
\begin{array}{c}
H_2C-NC_6H_5 \\
| \quad\;\; \diagdown \\
| \qquad CH(R,Ar) + H_2O \\
| \quad\;\; \diagup \\
H_2C-NC_6H_5
\end{array}
$$

The colorless condensation products (2-substituted 1,3-diphenyltetrahydroimidazoles) are not soluble and separate in a very pure form. Their characteristic melting points can serve to identify the initial aldehydes. This condensation is characteristic for aldehydes, chloral and glucose being exceptions. Ketones do not react.

Since the aldehyde condensation products are quite soluble in methanol, the precipitation sensitivity of the condensation reaction in this solvent is low for a general test for aldehydes in spot test analysis. However, water-soluble aldehydes react not only with the methanolic solution of the reagent but also with the aqueous solution of its chloride and the precipitation sensitivity attained in the latter case is adequate for spot tests.

Procedure.   One drop of the neutral or weakly acid test solution is treated in a micro test tube with a drop of reagent. Depending on the quantity of water-soluble aldehyde, a white precipitate or turbidity appears at once. When small amounts are present, the turbidity becomes clearly visible only after several minutes and is best confirmed by comparison with a blank.

Reagent: 1 g of dianilinoethane is dissolved in 100 ml glacial acetic acid and made up to 250 ml with water.

This procedure revealed:

0.2 γ formaldehyde (1 γ hexamine)	23 γ acetaldehyde
15 γ furfural	29 γ butyraldehyde

1 H. W. Wanzlick and W. Loechel, *Ber.*, 86 (1953) 1463.

(7) *Other tests for aldehydes*

(*a*) The test for sugars (see p. 338) by reducing silver oxide is a special application of a general test for aldehydes.[1] This reaction may be used to differentiate aldehydes from ketones. Among the aliphatic aldehydes, the reactivity decreases as the number of carbon atoms increases. Benzaldehyde and aromatic aldehydes react quite slowly.[2] Organic thiol compounds interfere because of the formation of black silver sulfide.

(*b*) One drop of the test solution is shaken with 1 ml of ammoniacal fuchsin solution. If a violet color appears within 1 min., aldehyde is present.[3] Aromatic aldehydes, such as vanillin and salicylaldehyde, which are but slowly soluble in water, also react. (The fuchsin solution contains 0.05 g of rosaniline base in 100 ml water, treated with 2 ml conc. ammonia, boiled for 5 sec, cooled, and made up to 200 ml with CO_2-free water.)

1 E. Tollens, *Ber.*, 15 (1882) 1635.
2 W. Ponndorf, *Ber.*, 64 (1931) 1913.
3 A. B. Wang, *Chem. Abstr.*, 26 (1932) 1920.

21. Aromatic aldehydes

(1) *Test with hydrogen sulfide and sodium pentacyanoammineferroate* [1]

The light yellow aqueous solution of sodium pentacyanoammineferroate $Na_3[Fe(CN)_5NH_3]$ gives a deep blue with thioketones, and with aromatic aldehydes in the presence of hydrogen sulfide. The reaction with these types of aldehydes probably involves the intermediate formation of thioaldehydes. When alone, these thioaldehydes polymerize readily, but in the presence of pentacyanoammineferroate, they react [in the monomolecular form initially produced in (*1*)], with replacement of ammonia as shown in (*2*):

$$ArCHO + H_2S \rightarrow ArCHS + H_2O \qquad (1)$$

$$Na_3[Fe(CN)_5NH_3] + ArCHS \rightarrow Na_3[Fe(CN)_5(ArCHS)] + NH_3 \qquad (2)$$

Procedure. A drop of 1% solution of sodium pentacyanoammineferroate and a drop of ammonium sulfide solution (free from polysulfides) are mixed in a micro crucible. A drop of the aqueous or alcoholic test solution is added, and the mixture is neutralized with dilute acetic acid. A blue to green color forms according to the amount of reactive aldehyde present.

An excess of acetic acid is to be avoided since a bluish turbidity may appear even in a blank. The strength of the acetic acid should therefore correspond to the sulfide solution used.

The procedure revealed:

1 γ benzaldehyde	1 γ hydroxybenzaldehyde (*o*-, *m*- or *p*-)
2 γ anisaldehyde	3 γ *o*-benzaldehydesulfonic acid
1 γ furfural	4 γ *o*-nitrobenzaldehyde
1 γ vanillin	1 γ *m*- ,,
2 γ cinnamaldehyde	2 γ *p*- ,,
1 γ piperonal	1 γ protocatechuic aldehyde
1 γ *o*-phthalaldehyde	

1 H. W. Schwechten, *Ber.*, **65** (1932) 1734. Comp. also F. Feigl, V. Anger and R. Zappert, *Mikrochemie*, **15** (1934) 192.

(2) *Test with phenylhydrazine hydrochloride and p-nitrosophenol*[1]

Aldehyde phenylhydrazones condense with *p*-nitrosophenol in concentrated sulfuric or hydrochloric acid to give deep red-brown or black-blue products. (See p. 281 regarding the chemistry of this reaction.) It has been found that the same colored compounds result if aromatic aldehydes, phenylhydrazine, and *p*-nitrosophenol are brought together in concentrated hydrochloric acid. Purely aliphatic and α, β-unsaturated aliphatic aldehydes show no reaction. This behavior provides the basis of a sensitive and selective test for aromatic aldehydes. Red-brown compounds are formed in almost all instances. With *p*-hydroxy- and *p*-alkoxyaryl aldehydes, the initial products are blue-green that change to brown. Such derivatives of aromatic aldehydes as Schiff bases, oximes, hydrazones, azines, *etc.* that can be hydrolyzed by hydrochloric acid behave like the parent aldehydes.

Procedure. A slight amount of the solid test material or the evaporation residue of a drop of the test solution is treated in a depression of a spot plate with 1 drop of each of the reagent solutions. A positive response is indicated by the appearance of a color (usually red-brown). Larger amounts of *p*-hydroxy- or *p*-alkoxyaldehydes yield black products. It is advisable to carry out not only a blank test but also two additional tests using 1 drop of each of the reagent solutions separately. The reason is that certain phenols and aldehyde phenyl- hydrazones give color reactions with *p*-nitrosophenol, and certain strongly un- saturated aliphatic and pyridine-aldehydes react with phenylhydrazine to pro- duce colored products.

Reagents: 1) Saturated solution of phenylhydrazine hydrochloride in conc. hydrochloric acid.

2) 0.05% solution of *p*-nitrosophenol in conc. hydrochloric acid (freshly prepared).

Limits of identification:

0.5	γ benzaldehyde		1	γ orthovanillin
0.5	γ benzalazine		0.1	γ vanillin
1	γ cinnamaldehyde		0.1	γ anisaldehyde
0.2	γ *o*-nitrobenzaldehyde		0.1	γ piperonal
0.2	γ *p*- ,,		0.2	γ *p*-dimethylaminobenzaldehyde
1	γ salicylaldehyde		0.05	γ furfural
5	γ salicylideneaniline		5	γ 6-methylpyridine-2-aldehyde
1	γ salicylalhydrazone		0.5	γ pyridine-3-aldehyde
1	γ salicylaldazine		1	γ ,, -2- ,,

No color reaction was given by: acetaldehyde, isovaleraldehyde, oenanthal, chloral hydrate, glyoxal, glucose, crotonaldehyde, citral, glutaconic aldehyde (red color with phenylhydrazine hydrochloride alone), opianic acid, β-indolyl- aldehyde.

1 V. ANGER and S. OFRI, *Z. Anal Chem.*, 203 (1964) 424.

(3) Test with N,N-dimethyl-p-phenylenediamine [1]

Aromatic aldehydes condense with N,N-dimethyl-*p*-phenylenediamine:[2,3]

$$\text{ArCHO} + \text{NH}_2\text{C}_6\text{H}_4\text{N(CH}_3)_2 \rightarrow \text{ArCH}=\text{NC}_6\text{H}_4\text{N(CH}_3)_2 + \text{H}_2\text{O}$$

The resulting yellow Schiff base forms salts, or more correctly undergoes protonation to yield an intense red color. This may be formulated:

$$\text{ArCH}=\text{NC}_6\text{H}_4\text{N(CH}_3)_2 + \text{H}^+ \rightarrow [\text{ArCH}=\text{NHC}_6\text{H}_4\text{N(CH}_3)_2]^+$$

The color reaction is best accomplished if a solution of the aldehyde is added to a salt of this diamine. Most favorable results are secured by heating the aldehyde with the solid oxalate or hydrochloride of the diamine.

Aliphatic saturated aldehydes yield colorless Schiff bases only and do not interfere with this test. This fact has been confirmed by trials with acetal- dehyde, glyoxal and crotonaldehyde. Unsaturated aldehydes that have more than one double bond conjugated with the carboxyl group might be expected to yield a less pronounced positive test.[4] Ketones do not interfere; pertinent instances are acetone, benzophenone, benzil, 2,3-butanedione.

Procedure. Several mg of the solid oxalate or chloride of N,N-dimethyl-p-phenylenediamine are added to a drop of the alcoholic or ethereal test solution in a micro test tube and then warmed in a preheated water bath. A red color that develops during the evaporation of the solvent indicates a positive response.

Limits of identification:

0.3 γ { benzaldehyde, salicylaldehyde, p-hydroxybenzaldehyde, anisaldehyde, piperonal, p-dimethylaminobenzaldehyde

100   γ pyridine-2-aldehyde

0.05 γ furfural

Furfural probably reacts with N,N-dimethyl-p-phenylenediamine in the same manner as with aniline, *i.e.* the furan ring is cleaved with formation of a dianilic compound. (See Chap. 5, Sect. 34.)

1 F. FEIGL, L. BEN-DOR and E. JUNGREIS, *Chemist-Analyst*, 52 (1963) 113.
2 E. CALM, *Ber.*, 17 (1894) 2938.
3 F. J. MOORE and R. D. GALE, *J. Am. Chem. Soc.*, 30 (1908) 399.
4 S. HÜNIG, J. UTTERMANN and G. ERLEMANN, *Ber.*, 88 (1955) 708; S. HÜNIG and J. UTTERMANN, *Ber.*, 88 (1955) 1201, 1485.

(4) *Test with phloroglucinol–hydrochloric acid*[1]

Phloroglucinol condenses with aryl aldehydes in conc. hydrochloric acid to yield orange-pink monomethine chloride:

$$Ar\text{—}CHO + \text{(HO—} \bigcirc \text{—OH)} + HCl \longrightarrow \left[Ar\text{—}CH\text{—} \text{(HO—} \bigcirc \text{—OH)} \right]^{+} Cl^{-} + H_2O$$

Most of these colored compounds are stable. Unstable colors that disappear rapidly are given by vinyl aldehydes and certain derivatives of benzaldehyde, such as o-nitrobenzaldehyde and o-phthalaldehyde. No color reaction whatsoever is given by p-nitrobenzaldehyde. It should be noted that allyl compounds likewise give color reactions with phloroglucinol–hydrochloric acid (see p. 169) but compounds of the type of eugenol or isoeugenol seemingly respond under the conditions described here only in dependence on the respective aldehyde content. Glutaconic aldehyde reacts positively and sensitively; it yields a red-violet color.

Procedure. A drop of the test solution is mixed with a drop of a saturated solution of phloroglucinol in conc. hydrochloric acid in a depression of a spot plate. An orange-pink color appears if aromatic aldehydes are present.

Limits of identification:

20   γ benzaldehyde   1 γ salicylaldazine

20	γ benzalazine	0.5	γ orthovanillin
0.05	γ cinnamaldehyde	0.1	γ vanillin
0.5	γ salicylaldehyde	0.1	γ anisaldehyde
0.5	γ salicylalhydrazone	0.1	γ piperonal
1	γ salicylideneaniline	0.01	γ furfural (turns green)

The following give a red-violet coloration:

0.01 γ glutaconic aldehyde	50 γ pyridine-3-aldehyde
2 γ p-dimethylaminobenzaldehyde (disappears quickly)	

Unstable colors were obtained with crotonaldehyde, citral, o-nitrobenzaldehyde, o-phthalaldehyde, 2,3-naphthalenedialdehyde. The following gave no reaction: p-nitrobenzaldehyde, 6-methylpyridine-2-aldehyde, pyridine-4-aldehyde, opianic acid, aliphatic aldehydes, ketones.

1 V. ANGER and S. OFRI, Z. Anal. Chem., 203 (1964) 425.

22. Aromatic and α,β-unsaturated aldehydes

(1) Test with thiobarbituric acid and phosphoric acid [1]

Aromatic aldehydes react with thiobarbituric acid in hydrochloric acid solution to give orange condensation products: [2]

$$\text{ArCHO} + \begin{matrix} \text{O C—NH} \\ | \quad | \\ \text{H}_2\text{C} \quad \text{C S} \\ | \quad | \\ \text{O C—NH} \end{matrix} \longrightarrow \begin{matrix} \text{O C—NH} \\ | \quad | \\ \text{ArCH=C} \quad \text{C S} \\ | \quad | \\ \text{O C—NH} \end{matrix} + \text{H}_2\text{O}$$

The condensation proceeds quicker and with greater sensitivity on warming with thiobarbituric acid and syrupy phosphoric acid. Such α,β-unsaturated aldehydes as cinnamaldehyde, pyridine aldehydes and citral react in the same manner as aromatic aldehydes. Some allyl compounds behave like aromatic aldehydes.

Procedure. A small quantity of the test material or the evaporation residue from a drop of the ethereal or benzene test solution is treated in a micro test tube with several cg of thiobarbituric acid and 1–2 drops of syrupy phosphoric acid. The tube is heated to 120–140° in a glycerol bath. A positive response is indicated by the appearance of a yellow or orange colored product within 1–3 min.

The following were revealed:

5	γ benzaldehyde	5 γ o-nitrobenzaldehyde
2.5	γ o-hydroxybenzaldehyde	5 γ 2-hydroxy-5-chlorobenzaldehyde
5	γ m- ,,	25 γ p-chlorobenzaldehyde
0.5	γ p- ,,	10 γ m,o-benzaldehydesulfonic acid
0.1	γ anisaldehyde	50 γ citral
0.5	γ cinnamaldehyde	10 γ crotonaldehyde
25	γ p-dimethylaminobenzaldehyde	10 γ acrolein

1 F. FEIGL and E. LIBERGOTT, *Anal. Chem.*, 36 (1964) 132.
2 A. W. DOX and G. P. PLAISANCE, *J. Am. Chem. Soc.*, 38 (1916) 2156.

(2) Test with naphthoresorcinol[1]

In hydrochloric acid solution, naphthoresorcinol yields condensation products with vinyl- and aryl aldehydes. These products may show all shades between yellow-green and blue; mostly they are orange however. The mechanism is analogous to that of the color reaction of aromatic aldehydes with phloroglucinol (see p. 204). Allyl compounds do not interfere.

Opianic acid, aliphatic aldehydes and ketones show no reaction.

Procedure. A drop of the alcoholic test solution is mixed in a depression of a spot plate with a drop of a 0.1% solution of naphthoresorcinol in methanol and several drops of concentrated hydrochloric acid. A color appears if vinyl- or aromatic aldehydes are present.

Limits of identification and colors:

			0.01	γ furfural	orange
0.5	γ crotonaldehyde	yellow-green	1	γ benzaldehyde	yellow
6	γ citral	,, ,,	1	γ benzalazine	,,
6	γ glutaconic aldehyde	blue-green	0.2	γ o-phthalaldehyde	,,
0.5	γ p-dimethylamino-benzaldehyde	yellow-green	2	γ 6-methylpyridine-2-aldehyde	,,
0.5	γ o-nitrobenzaldehyde	,, ,,	2	γ pyridine-2-aldehyde	,,
0.5	γ p- ,,	,, ,,	1	γ ,, -4- ,,	,,
0.5	γ salicylaldehyde	orange	100	γ ,, -3- ,,	red
0.5	γ salicylalhydrazone	.,	0.2	γ anisaldehyde	,,
1	γ salicylideneaniline	,.	0.1	γ vanillin	red-violet
1	γ orthovanillin	,,	0.1	γ coniferylaldehyde	,, ,,
0.05	γ 2,3-naphthalenedial-dehyde	,,	0.5	γ piperonal	,, ,,
			20	γ β-indolylaldehyde	,, ,,

1 V. ANGER and S. OFRI, *Z. Anal. Chem.*, 203 (1964) 425.

23. Aliphatic ketones (—CH$_2$—CO— group)

(1) Test with m-dinitrobenzene and alkali

Aromatic *m*-dinitro compounds react with ketones and alkali to yield violet products.[1] The mechanism of the color reaction and the constitution of the resulting reaction products are not known. Since neither aldehydes nor diketones give this color reaction, the latter can serve as the basis of a selective test for aliphatic monoketones.

Procedure.[2] A drop of a 1% solution of *m*-dinitrobenzene in alcohol is mixed in a micro test tube with a drop of a saturated methanolic solution of potassium hydroxide and a drop of the test solution. The tube is then warmed to 70–80° in a water bath for 2–3 min. A violet-red color appears if ketones are present.

Limits of identification:
20 γ acetone   20 γ methyl ethyl ketone 200 γ methyl isobutyl ketone
5 γ acetophenone   10 γ cyclohexanone

1 L. JANOWSKY, *Ber.*, 24 (1891) 971.
2 V. ANGER and S. OFRI, *Z. Anal. Chem.*, 206 (1964) 186.

(2) Test by conversion into 1,2-diketones [1]

The conversion of aliphatic 1,2-diketones into red or orange 1,2-dioxime nickel salts as described (p. 328) can also be used for the detection of oxomethylene compounds because the CH_2 group adjacent to the CO group can be oxidized to a CO group. Selenium dioxide (selenious acid) is a suitable oxidant for this transformation [2] whereby elementary selenium is formed.

The test is not completely decisive for the CH_2CO group, since the o-dioxo compounds, which finally enter the test reaction sometimes originate in other groupings when oxidized by selenious acid. Instances are:

(a) Compounds containing the —CHOH—CH_2— group.

(b) Compounds containing the —CO—CHOH— group, which can be oxidized, just as the oxomethylene groups, to the dioxo group.

TABLE 9

Name and formula	Oxidation product	Color	Limit of identification, γ
Acetoin, $CH_3CHOHCOCH_3$	$CH_3COCOCH_3$	Red	0.5
Acetone, CH_3COCH_3	CH_3COCHO	,,	50
Methyl ethyl ketone, $C_2H_5COCH_3$	C_2H_5COCHO	Brown-red	25
Cyclopentanone, 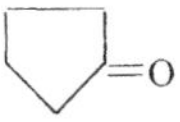	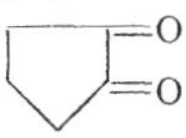	Orange	5
Cyclohexanone, 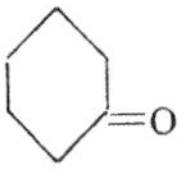	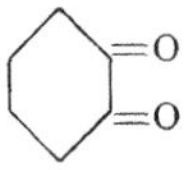	Red	2.5
Phenylacetaldehyde, $C_6H_5CH_2CHO$	C_6H_5COCHO	Orange	5
α-Hydrindone, 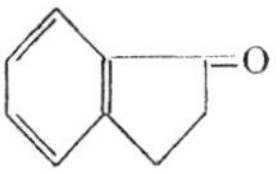	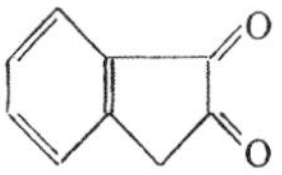	,,	10
α-Tetralone, 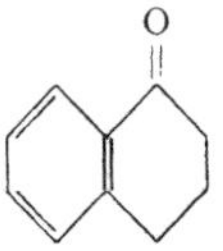	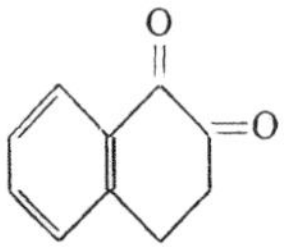	,,	10

(c) Unsaturated hydrocarbons which are oxidized to the corresponding α,β-unsaturated ketones[3] which in turn may be still further oxidized as in (b) to diketones.

Procedure. One drop of the alcoholic test solution is placed in a capillary tube along with several grains of selenious acid. The charged capillary is fused shut and heated for 20 min. at 150–170°. When cool, the reaction mass is transferred to a centrifuge tube, and warmed briefly with 2 drops of a solution of 2 g hydroxylamine-HCl plus 1 g sodium acetate in 2 ml water. The oxime is formed and the excess selenious acid is reduced to selenium. A little animal charcoal is added and the suspension is centrifuged. One drop of the clear supernatant solution is spotted on filter paper with a drop of 5% nickel acetate solution. A yellow or red fleck appears if the response is positive.

Limits of identification are given in Table 9.

1 M. Ishidate, *Mikrochim. Acta*, 3 (1938) 283.
2 H. L. Riley, J. F. Morley and N. A. C. Friend, *J. Chem. Soc.*, (1932) 1875.
3 E. Schwenk and E. Borgwardt, *Ber.*, 65 (1932) 1609; G. Dupont, J. Allard and R. Dulou, *Bull. Soc. Chim.*, [4], 53 (1933) 599; W. Zacharewicz, *Chem. Abstr.*, 30 (1936) 8191.

24. Methyl ketones (CH₃—CO— group)

(1) Test with sodium nitroprusside[1]

When acetone is treated with sodium nitroprusside it gives an intense red-yellow color that changes to pink-violet on acidifying with acetic acid. The basis of the color reaction is that the NO of the nitroprusside reacts with acetone to give isonitrosoacetone, which remains in the complex anion. At the same time, the iron(III) is reduced to iron(II):[2]

$$[\text{Fe(CN)}_5\text{NO}]^{2-} + \text{CH}_3\text{COCH}_3 + 2\,\text{OH}^- \rightarrow [\text{Fe(CN)}_5\text{ON}=\text{CHCOCH}_3]^{4-} + 2\,\text{H}_2\text{O}$$

Other methyl ketones and compounds, which contain an enolizable CO group give analogous color reactions, while ketones which lack methyl or methylene groups bound to CO groups are not active in this respect.

The nitroprusside reaction was long regarded as a characteristic test for methyl ketones. This belief is unfounded since the reaction occurs with all compounds which from the beginning contain an activated methylene group, or in which an active methylene group can arise through the shifting of a hydrogen atom.[3] This is shown by the finding that compounds which have no ketone character also yield colors with alkaline nitroprusside. Pertinent examples are: indene (I), pyrrole (II), indole (III), resorcinol (IV).

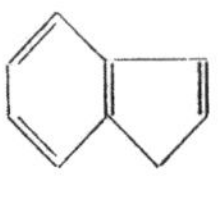

The color reaction is therefore due to the isonitrosation of CH_2 groups as shown by the fact that, in addition to (I)–(IV), the following also react with alkaline nitroprusside: cyclopentanone, cyclohexanone, orcinol, hexyl-resorcinol, phloroglucinol, malonic diethyl ester, phenylacetic ethyl ester, phenylisocrotonic ethyl ester, desoxybenzoin, hydantoin, cyanoacetic ethyl ester, vitamin C, digitoxin, k-Strophanthin, pyrrole- and indole-acetic acid.

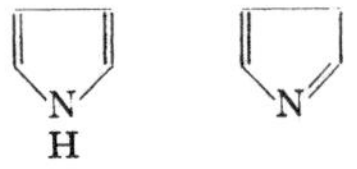

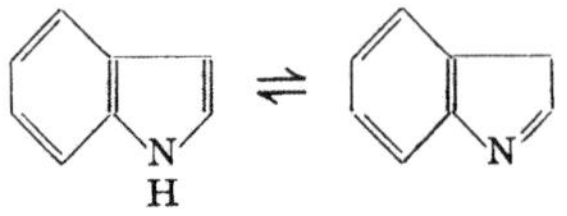

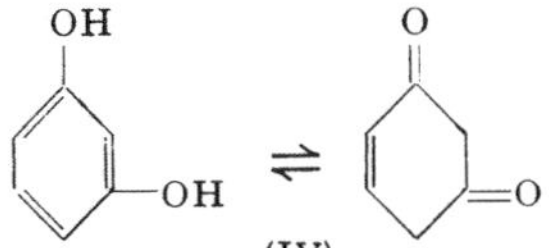

It is remarkable that cinnamic aldehyde, $C_6H_5CH=CHCHO$, apparently is the only aldehyde which gives the color reaction described here and hence can be characterized by this unique behavior.[4]

Procedure.[5] A drop of the aqueous or alcoholic test solution is mixed in a micro crucible with a drop of 5% sodium nitroprusside solution and a drop of 30% sodium hydroxide solution. After a short time, when a slight color usually has developed, 1 or 2 drops glacial acetic acid are added. A red or blue color indicates the presence of a methyl ketone.

The following amounts were detected:

1 γ acetophenone	blue	4 γ acetoacetic ester	orange
10 γ acetone	pink	10 γ diacetyl	pale pink
10 γ methyl ethyl ketone	,,	15 γ pyruvic acid, salt	red
10 γ 2-methylheptan-6-one	brown-violet	acid	violet
10 γ methyl stearyl ketone	red	15 γ propanone-1,1-di-	
2 γ acetylacetone	purple	carboxylic acid	,,

1 E. LEGAL, *Jahresber. Fortschr. Chem. u. verwandter Theile anderer Wissenschaften*, (1883) 1648; B. v. BITTO, *Ann.*, 267 (1892) 372; G. DENIGÈS, *Bull. Soc. Chim.*, [3], 15 (1896) 1058 and 17 (1897) 381.
2 L. CAMBI, *Chem. Zbl.*, (1913) I, 1756; (1914) II, 1100.
3 Personal communication by S. OHKUMA, Tokyo.
4 Comp. M. PESEZ and P. POIRIER, *Méthodes et Réactions de l'Analyse Organique*, Vol. III, Masson, Paris, 1954, p. 40.
5 Unpublished studies, with R. ZAPPERT.

(2) *Test by conversion into indigo*[1]

Indigo is formed by the action of o-nitrobenzaldehyde (I) on acetone in alkaline solution. o-Nitrophenylacetonylcarbinol (II) is formed and loses a

molecule of acetic acid and probably yields o-nitrostyrene (III) as intermediate, which by intramolecular condensation is converted into indolone
(IV), which polymerizes to indigo (V):*

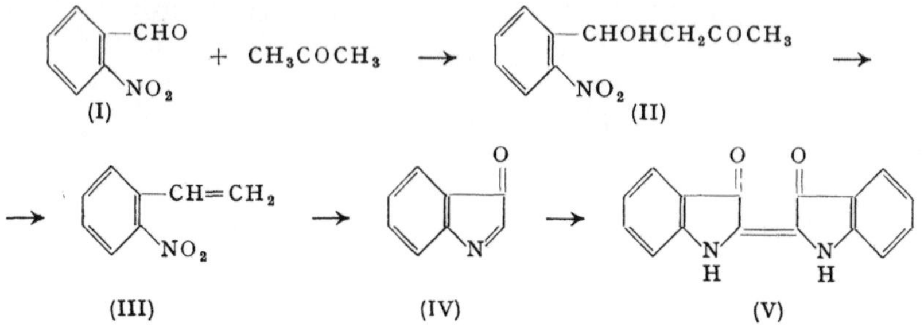

o-Nitrobenzaldehyde similarly forms indigo with all substances containing
the CH_3CO grouping, so that the formation of the dye may be applied as a
test for this group joined to a hydrogen atom or to a carbon atom that does
not carry groups which exert excessive hindrance. Since nitrobenzaldehyde
in alkaline solution is used as reagent in the indigo test, compounds in which
the CH_3CO group is set free by hydrolysis can likewise give the indigo
test.[2] Among these are: (a) halogen derivatives such as ethylidene chloride,
2,2-dibromopropane, $etc.$; (b) acetals such as acetaldehyde alcoholate acetal;
(c) oximes such as acetoxime, acetophenone oxime, $etc.$; (d) bisulfite addition
products such as acetaldehyde sodium bisulfite, acetone sodium bisulfite, $etc.$

Procedure.[3]   A drop of the test solution is treated in a micro test tube with
a drop of an alkaline solution of o-nitrobenzaldehyde, and gently warmed in a
water bath. The cooled mixture is extracted with chloroform. A blue color in the
chloroform layer indicates the presence of a methyl ketone. Alcoholic solutions
sometimes produce a red instead of a blue color in the chloroform layer and
accordingly should be avoided.

Reagent: Saturated solution of o-nitrobenzaldehyde in 2 N sodium hydroxide.

The following amounts were detected:

100 γ acetone	50 γ acetophenone	300 γ acetoacetic ester
150 γ methyl ethyl ketone	200 γ acetylacetone	100 γ acetaldehyde
150 γ 2-methylheptan-6-one	40 γ diacetyl	

* Another reaction path has been suggested by J. TANANESCU and A. BACIU, *Bull.
Soc. Chim. France*, [5], 4 (1937) 1673.

1 A. v. BAEYER and V. DREWSEN, *Ber.*, 15 (1882) 2856; F. PENZOLDT, *Z. Anal. Chem.*,
    24 (1885) 149. Comp. also R. J. LE FEVRE and J. PEARSON, *J. Chem. Soc.*, (1932) 2807.
2 J. KAMLET, *Ind. Eng. Chem., Anal. Ed.*, 16 (1944) 362.
3 F. FEIGL, R. ZAPPERT and J. V. SANCHEZ, *Mikrochemie*, 17 (1935) 169.

25. Acetonyl compounds (CH₃—CO—CH₂— group)

Detection through thermochromic reaction with 2-hydroxy-1-naphthaldehyde[1]

When methanolic solutions of pertinent compounds are treated with 2-hydroxy-1-naphthaldehyde and then subjected to a jet of hydrogen chloride gas, a brilliant green color appears. On warming, the color turns to yellow, pink or pale green but the original green is restored on cooling. This reversible color change permits a sensitive and selective test for the acetonyl group.

The thermochromic test is based on the following reactions:

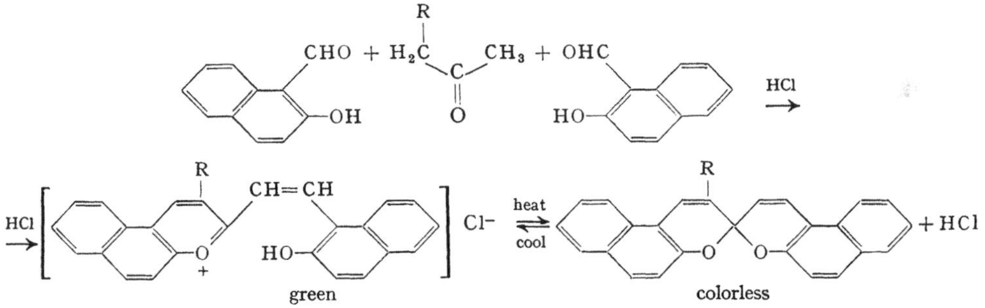

Procedure. To 0.1 ml of the methanolic test solution add 15 ml of a 0.5% methanolic solution of 2-hydroxy-1-naphthaldehyde. Gently stir the mixture for 30 sec. by means of a jet of hydrogen chloride gas. If aliphatic ketones are present at considerable concentrations, the mixture turns dark blue-green. Add 0.5 ml of *o*-dichlorobenzene and boil the mixture vigorously until all of the methanol but none of the *o*-dichlorobenzene is distilled off. It is best to heat the sides of the test tube first so that the removal is accomplished quickly and sharply. Cool the residual liquid. A positive response is indicated by a brilliant green color that fades and disappears completely on heating and then reappears on cooling. A blank test is advisable.

Limits of identification:

1 γ acetone	1 γ pentanone	1 γ heptadecanone
0.3 γ methyl ethyl ketone	0.2 γ octanone	1 γ nonadecanone
0.2 γ hexanone	1 γ undecanone	100 γ phenylpropanone
0.2 γ heptanone	1 γ tridecanone	1 γ acetylacetone

In addition to the preceding, a positive response is given also by:

4 γ cyclobutanone	0.2 γ cyclopentanone	5 γ cyclohexanone

No reaction was shown by: formaldehyde, diacetyl, 3-pentanone, 3-heptanone, 4-heptanone, 1,3-diphenylpropanone, acetophenone, 2-acetylnaphthalide, 4-acetylbiphenyl, acetaldehyde, oenanthal, cyclohexanol.

1 E. SAWICKI, *Chemist-Analyst*, 48 (1959) 4.

26. Carboxylic acids (—COOH group)

Test through conversion into hydroxamic acids [1]

Hydroxamic acids containing the CONHOH group react in neutral or weakly alkaline solutions with ferric ions giving red or bluish red inner-complex salts:

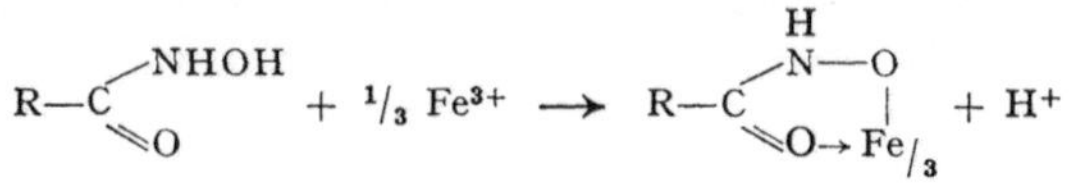

$$R-C\underset{\diagdown O}{\overset{\diagup NHOH}{}} + {}^1/_3 \ Fe^{3+} \longrightarrow R-C\underset{\diagdown O \to Fe/_3}{\overset{\diagup N-O}{\overset{H}{}}} + H^+$$

In strongly or weakly acidic solutions the colored inner-complex cations $[Fe(RCONHO)]^{2+}$ or $[Fe(RCONHO)_2]^+$ respectively are formed. [2]

Carboxylic acids and its derivatives which are convertible into hydroxamic acids can be detected by this ferric hydroxamate test.

Carboxylic acids cannot be converted into hydroxamic acids by direct action with hydroxylamine. The acid chloride must be formed first by the action of thionyl chloride and it then readily gives the alkali salt of hydroxamic acid on treatment with hydroxylamine and alkali. The underlying reactions are:

$$RCOOH + SOCl_2 \to RCOCl + SO_2 + HCl$$

$$RCOCl + NH_2OH + 2\,NaOH \to RCO(NHONa) + NaCl + 2\,H_2O$$

Procedure. A drop of the test solution is evaporated to dryness in a micro crucible, or particles of the solid are treated there with 2 drops of thionyl chloride. The mixture is evaporated, almost to dryness, to convert the carboxylic acid into its chloride. Two drops of a saturated alcoholic solution of hydroxylamine hydrochloride are then added and drops of alcoholic alkali until the liquid is basic to litmus paper. Reaction takes place on reheating. The mixture is acidified with a few drops of 0.5 N hydrochloric acid, and treated with a 1% aqueous solution of ferric chloride. The color change ranges from brown-red to dark violet.

Limits of identification:

100 γ sodium acetate	20 γ stearic acid	11 γ succinic acid
12 γ monochloroacetic acid	33 γ crotonic acid	11 γ tricarballylic acid
11 γ dichloroacetic acid	11 γ oleic acid	11 γ phenylacetic acid
15 γ glycine	15 γ ricinoleic acid	33 γ cinnamic acid
16 γ palmitic acid		12 γ anthranilic acid

Positive responses are also given by citric acid and thioacetic acid.

1 F. FEIGL, V. ANGER and O. FREHDEN, *Mikrochemie*, 15 (1934) 12, 18; comp. D. DAVIDSON, *J. Chem. Educ.*, 17 (1940) 81.
2 Comp. G. AKSNES, *Acta Chem. Scand.*, 11 (1957) 710.

27. Aliphatic and aromatic polycarboxylic acids and aralkyl monocarboxylic acids

Test through pyrolytic release of hydrogen chloride from sodium chloride[1]

If alkali halides are heated to 140–160° with the above acids, the respective hydrogen halides are split off:

$$—COOH + MeHal \rightarrow —COOMe + HHal$$

Aliphatic monocarboxylic acids, sulfonic acids, hydroxamic acids and aci-nitro compounds do not show this effect which is readily discerned with indicator paper. Since there is no relationship with melting point and dissociation constant of the pertinent organic acids in aqueous solution, it appears that the active acids are much stronger in their melts or in the vicinity of their melting points than when dissolved in water, and that this enhancement of the acidic nature is entirely absent or is present to only a much lesser extent in the case of inactive acids.

The reactions with fused or solid organic acids which result in the liberation of hydrogen halides occur on the surface of the solid alkali halide. Even though this is a handicap because of the limitation of the reaction site, the liberation of hydrogen chloride from sodium chloride is nevertheless successful with such slight amounts of the active acids, and it occurs so rapidly that a test for these acids can be based on this effect.

Procedure. A little of the solid or a drop of its solution is mixed in a micro test tube with several cg of sodium chloride and evaporated if need be. After removing any liquid, the test tube is placed in a glycerol bath previously heated to 150°, and the mouth of the tube is covered with a disk of moistened indicator paper. The temperature is raised to 160°. The change in the color of the paper requires 2 min. at most if active acids are present.

Limits of identification: 10 γ pimelic or mandelic acid, 20 γ adipic acid, 100 γ benzoic acid.

The following acids yielded hydrogen chloride: oxalic, malonic, succinic, malic, tartaric, azelaic, citric, phenylacetic, cinnamic, salicylic, phthalic, hemipinic, naphthalic.

No liberation of hydrogen chloride was observed from the following acids: caprylic, capric, lauric, myristic, palmitic, stearic, sulfanilic, 1,5-naphtholsulfonic, anthraquinone-1-sulfonic, anthraquinone-1,5- and -2,6-disulfonic acids.

It should be noted, that the pyrolytic reaction of active carboxylic acids with alkali halides has a counterpart in the formation of sulfur dioxide on heating with sodium thiosulfate at temperatures up to 140°:

$$2 —COOH + Na_2S_2O_3 \rightarrow 2 —COONa + H_2O + S + SO_2$$

All compounds having acid character, and likewise the chlorides of weak bases, are capable of this reaction (see p. 111). Microanalytical detection limits are achieved through the sensitivity of the test for the SO_2 formed.

1 F. FEIGL and C. STARK-MAYER, *Talanta*, 1 (1958) 252. Comp. PING-YUAN YEH and coworkers, *Anal. Chem.*, 34 (1962) 990.

28. Esters of carboxylic acids (—COO(R, Ar) group)

Test through conversion into hydroxamic acid[1]

Esters of carboxylic acids can be converted into alkali salts of hydroxamic acids on treatment with hydroxylamine hydrochloride and alkali hydroxide:

$$(R, Ar)COOR_1 + NH_2OH + NaOH \rightarrow (R,Ar)CO(NHONa) + R_1OH + H_2O$$

The hydroxamic acid released by acidification can be identified by the color reaction with ferric chloride (see p. 212). Lactones, which may be regarded as inner esters, react similarly to esters.

Procedure.   A drop of the ethereal solution of the ester is treated in a porcelain micro crucible with a drop of saturated alcoholic hydroxylamine hydrochloride solution and a micro drop of saturated alcoholic caustic potash. The mixture is heated over a micro flame until a slight bubbling occurs. After cooling, the mixture is acidified with 0.5 N hydrochloric acid, and a drop of 1% ferric chloride solution is added. A more or less intense violet color appears.

Limits of identification are:

10 γ ethyl formate	10 γ ethyl stearate	3 γ diethyl malonate
10 γ ethyl urethan	10 γ cetyl palmitate	3 γ propyl benzoate
2 γ ethyl acetate	10 γ glyceryl oleate	3 γ methyl salicylate
10 γ vinyl acetate	3 γ diethyl oxalate	5 γ phenyl salicylate
2 γ phenyl acetate	3 γ dimethyl oxalate	10 γ potassium ethyl xanthate
10 γ glycine ethyl ester hydrochloride		10 γ ethyl orthoformate
10 γ methyl naphthalene-1-dithiocarboxylate		5 γ coumarin

1 F. FEIGL, V. ANGER and O. FREHDEN, *Mikrochemie*, 15 (1934) 12, 18; comp. D. DAVIDSON, *J. Chem. Educ.*, 17 (1940) 81.

29. Alkyl esters of carboxylic acids (—COOR group)

Detection by reaction with sodium and 1,2-dinitrobenzene[1]

Aliphatic α-hydroxyketones (acyloins) may be prepared from aliphatic fatty acids by prolonged heating of ethereal or benzene solutions of their esters with metallic sodium.[2] There is intermediate formation of α-diketones and the light yellow benzene-insoluble sodium salts of enediols (1) which can be saponified to acyloins (2):

$$2\,R\!-\!COOR' + 4\,Na \;\longrightarrow\; \begin{array}{l} R\!-\!C\!-\!ONa \\ \quad\| \\ R\!-\!C\!-\!ONa \end{array} + 2\,R'ONa \qquad (1)$$

$$\begin{array}{l} R\!-\!C\!-\!ONa \\ \quad\| \\ R\!-\!C\!-\!ONa \end{array} + 2\,H_2O \;\longrightarrow\; \begin{array}{l} R\!-\!CHOH \\ \quad| \\ R\!-\!C\,O \end{array} + 2\,NaOH \qquad (2)$$

Acyloins may be readily detected through the fact that in alcoholic alkaline solution they reduce 1,2-dinitrobenzene to yield the violet water-soluble alkali salts of an *o*-quinoid nitrolnitroic acid (comp. p. 330). Consequently, it is fundamentally possible, by means of the synthesis of acyloins, to arrive at a method of detecting the aliphatic fatty acid esters participating in the synthesis. This objective is reached in surprisingly simple and rapid fashion through the realization of (*1*) in the presence of 1,2-dinitrobenzene. If the reaction mixture is treated with water, without removing the excess sodium, the violet color characteristic of acyloins appears at once.

It is noteworthy, and in itself characteristic, that when metallic sodium comes into contact with a benzene or ethereal solution of an ester, which also contains 1,2-dinitrobenzene, a red product appears almost immediately. Possibly, there is direct production here of the anhydrous salt of the *aci*-form of nitrosonitrobenzene as shown in (*3*):

$$\qquad (3)$$

On the addition of water, the salt dissolves to give a violet solution. Possibly the red product is a benzene-insoluble addition compound of the sodium salt of enediol and 1,2-dinitrobenzene, which forms acyloin when saponified, and the latter may react with the nitro compound more quickly than the hydrogen produced by the reaction of metallic sodium and water.

The prodecure given here for the detection of aliphatic esters should not be applied in the presence of α-diketones, since they too give rise to sodium salts of enediols, which then yield a violet color on reaction with 1,2-dinitrobenzene.

A much greater restriction of the test is imposed by the fact that only benzene, toluene, or chloroform can serve as solvent for the ester. Alcohol and impure ether react with sodium to give hydrogen, with consequent reduction of the 1,2-dinitrobenzene to the colored orthoquinoid compound.

Procedure. The test is conducted in a depression of a spot plate. A piece of metallic sodium the size of a small seed is pressed into a disk with a glass rod. A drop of the benzene solution of the ester is added and then a drop of a 2.5% benzene solution of 1,2-dinitrobenzene. The system is stirred with a fine glass rod and after 1–2 min. a drop of water is introduced. If esters are present, a deep to pale violet appears, the depth of the color depending on the quantity involved. A comparison blank is advisable when small amounts of ester are suspected.

The procedure revealed:

     5 γ ethyl acetate   10 γ butyl acetate   5 γ ethyl benzoate

A positive response was obtained also with:[*] propyl acetate, n-butyl formate, diethyl malonate, diethyl allylmalonate, diethyl oxalate, ethyl phenylacetate, diethyl phthalate, ethyl cinnamate.

[*] According to results obtained by Angelita BARCELÓN, University of Cincinnati.

1 Unpublished studies, with Cl. COSTA NETO.
2 S. M. MCELVAIN in: *Organic Reactions, Vol. IV*, Wiley, New York, 1948, p. 256.

30. Phenyl esters of carboxylic acids (—COOAr group)

Test through pyrohydrolytic splitting out of phenol[1]

The saponification of pertinent esters with acids or alkalis

$$(R,Ar)—COOAr' + H_2O \rightarrow (R,Ar)—COOH + Ar'OH$$

which normally requires long warming, can be quickly accomplished by sintering or fusing with oxalic acid dihydrate, which begins to lose water at 105° and continues to do so in the melt up to 160°. The water released in this temperature interval not only brings about the hydrolysis (pyrohydrolysis) of the ester, but also facilitates the volatilization of the resulting phenols as a quasi steam distillation. Since the phenols are readily detected by the indophenol reaction (p. 185), a very convenient test for phenol esters results. Phenols that are volatile with steam must be absent or removed. This is readily accomplished by shaking the sample (suspended in alkali) with ether. The alkali phenolates remain in the water layer while the phenol esters are taken up in the ether. The test is made on the ethereal solution.

Triphenyl phosphate is not saponified by pyrohydrolysis with hydrated oxalic acid; it yields phenol through pyroammonolysis (see p. 566).

Procedure. A little of the sample or a drop of its solution is mixed with several cg of oxalic acid hydrate in a micro test tube and taken to dryness if necessary. The test tube is placed in a glycerol bath previously heated to 150°, and its mouth is covered with a disk of filter paper impregnated with a saturated benzene solution of 2,6-dichloroquinone-4-chloroimine. After several minutes the paper is held in ammonia vapours. A more or less intense blue appears, which rapidly loses its color or turns to a dirty violet. The original blue is restored by ammonia vapors.

The procedure revealed 10–20 γ of the following esters: phenyl acetate, phenyl benzoate, phenyl anthranilate, diphenyl carbonate, phenyl salicylate, phenyl stearate, diphenyl phthalate.

1 F. FEIGL and E. JUNGREIS, *Anal. Chem.*, 31 (1959) 2099, 2101.

31. Alkyl and aryl acetates (CH_3—COO— group)

Test by pyrohydrolytic release of acetic acid [1]

If alkyl or aryl esters of acetic acid are heated to 110° with oxalic acid dihydrate, the water of crystallization brings about the saponification:

$$CH_3COO(R,Ar) + H_2O \rightarrow CH_3COOH + (R,Ar)OH$$

The acetic acid resulting from the pyrohydrolysis volatilizes at the reaction temperature and can be detected with indicator paper. The pyrohydrolysis can be accomplished also by heating with $MnSO_4.4\,H_2O$ at 120–140°. This is a fine proof that water of crystallization driven off at elevated temperatures is active without the intervention of H^+ or OH^- ions.

Since N-acetyl compounds are not altered by heating to these temperatures with $H_2C_2O_4.2\,H_2O$ or $MnSO_4.4\,H_2O$, the test described here affords a reliable method for detecting O-acetyl compounds, provided that compounds which split off volatile acids through pyrohydrolysis are absent, *i.e. o*-formyl compounds, or salts of acetic and formic acids with ammonia and organic bases.

Procedure. Several cg of pulverized oxalic acid dihydrate are mixed in a micro test tube with a little of the solid or a drop of its solution in alcohol or ether. The solvent is evaporated. A disk of acid–base indicator paper is placed over the mouth of the tube and the latter is then heated in a glycerol bath previously brought to 110°. A positive response is readily observed through the rapid change in the color of the indicator paper.

The test revealed the following amounts:

10 γ acetylsalicylic acid	50 γ acetylcellulose
100 γ methyl acetylsalicylate	50 γ acetoxyacetanilide

1 Unpublished studies, with E. JUNGREIS.

32. Anhydrides and imides of carboxylic acids

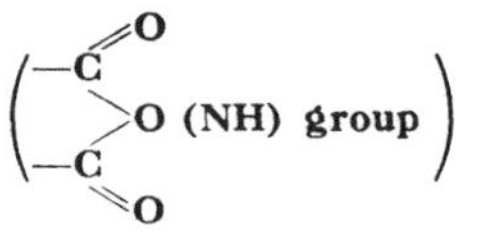

$$\left(\begin{array}{c} -C \diagup^{O} \\ {\diagdown} O \text{ (NH) } \textbf{group} \\ -C \diagdown_{O} \end{array} \right)$$

Test through conversion into hydroxamic acid [1]

Anhydrides and imides of mono- and dicarboxylic acids react directly with hydroxylamine to give hydroxamic acids:

$$((R, Ar)CO)_2O + NH_2OH \rightarrow (R, Ar)-CONHOH + (R, Ar)COOH$$

$$((R, Ar)CO)_2NH + NH_2OH \rightarrow (R, Ar)-CONHOH + (R, Ar)CONH_2$$

The colored iron(III) hydroxamate is formed if ferric ions are present (comp. p. 212).

It is remarkable that penicillin shows the same behavior.[2]

Procedure.   A drop of the ethereal solution of the anhydride is mixed in a micro crucible with 1 or 2 drops of the freshly prepared reagent solution and evaporated to dryness over a micro flame. A few drops of water are added. A violet or pink color is formed according to the amount of anhydride present.

> *Reagent:* A 0.5% alcoholic solution of ferric chloride is acidified with a few drops of concentrated hydrochloric acid and saturated (warm) with hydroxylamine hydrochloride.

The *limits of identification* and the respective colors are:

5 γ acetic anhydride	violet	5 γ succinic anhydride	red-brown
5 γ phthalic anhydride	,,	10 γ thapsic anhydride	pink
5 γ phthalimide	,,	6 γ benzoic anhydride	,,
10 γ camphoric anhydride	liiac	10 γ m-nitrophthalic anhydride	,,
5 γ hemipinic anhydride	,,		

1 F. FEIGL, V. ANGER and O. FREHDEN, *Mikrochemie*, 15 (1934) 12, 18; comp. D. DAVIDSON, *J. Chem. Educ.*, 17 (1940) 81.
2 J. H. FORD, *Anal. Chem.*, 19 (1947) 1004.

33. Carboxylic acid chlorides (—COCl group)

Test with pyridine-4-aldehyde benzothiazol-2-ylhydrazone[1]

Carboxylic acid chlorides react with pyridine-4-aldehyde benzothiazol-2-ylhydrazone on filter paper giving a purple color, which fades in a few minutes. This behavior is characteristic for carboxylic acid chlorides.

Procedure.   A drop of the test solution (dissolved in acetone, methyl ethyl ketone, chloroform, carbon tetrachloride, tetrahydrofuran or ether) is placed on filter paper previously impregnated with an acetone solution of 0.05% pyridine-4-aldehyde benzothiazol-2-ylhydrazone and dried. A positive response is shown by a purple color which fades within several minutes.

Limits of identification:

0.15 γ acetyl chloride	0.25 γ benzoyl chloride
0.1 γ chloroacetyl chloride	0.5 γ p-nitrobenzoyl chloride
0.5 γ dichloroacetyl chloride	0.2 γ 2-ethoxybenzoyl chloride
0.2 γ succinyl chloride	0.2 γ cinnamoyl chloride
0.1 γ phenylacetyl chloride	0.1 γ 1-naphthoyl chloride
0.06 γ nonanoyl chloride	0.2 γ 4-phenylazobenzoyl chloride
0.3 γ 1-furoyl chloride	

A color reaction is also shown by 6 γ 2-bromoacetophenone and, with blue color, by 0.05 γ cyanuric chloride. Trichloroacetyl chloride and 3,5-dinitrobenzoyl chloride give a yellow color in the spot test.

If the color forms in 5–15 min. alkyl or arylsulfonyl chlorides or anhydrides are present.

1 E. SAWICKI, D. F. BENDER, T. R. HAUSER, R. M. WILSON JR. and J. E. MEEKER, *Anal. Chem.*, 35 (1963) 1481, 1485.

D. SULFUR AND SELENIUM COMPOUNDS

If the presence of sulfur or selenium has been established by the preliminary tests described in Chap. 2, the following procedures can be employed to detect certain bonding types and oxidation states of these two elements. See p. 94 regarding tests for groups that contain sulfur along with nitrogen.

34. Thioketones and thiols ($\ge$CS and $\ge$CSH groups)

Test by catalytic acceleration of the iodine–azide reaction[1]

The redox reaction:

$$2\,NaN_3 + I_2 \rightarrow 2\,NaI + 3\,N_2$$

which proceeds extremely slowly, is catalyzed not only by inorganic sulfides, thiosulfates and thiocyanates (see *Inorganic Spot Tests*, Chap. 4) but also by solid or dissolved organic compounds containing the group C=S or C—SH. Organic sulfur derivatives such as thioethers (R—S—R), disulfides (R—S——S—R) [with the exception of diacyldisulfides (R—CO—S—S—CO—R)], sulfones (R—SO$_2$—R), sulfinic acids (R—SO$_2$H) and sulfonic acids (R—SO$_3$H) or the salts of these acids, exert no, or at the most very slight effect on the reaction. The catalytic hastening of the iodine–azide reaction is therefore characteristic for thioketones and mercaptans (thiols) and permits their sensitive detection.

The mechanism of the catalytic effect of mercaptans and thioketones on the iodine–azide reaction has not been completely elucidated. In all likelihood the same explanation will not be valid for all compounds. The action of mercaptans is probably analogous to the catalytic activity of hydrogen sulfide and soluble metal sulfides. It possibly involves the formation of a reactive labile intermediate compound, RSI (*1*), which reacts with sodium azide, as shown in (*2*). Summation of (*1*) and (*2*) gives the equation of the uncatalyzed reaction, in which the catalyst no longer appears:

$$RSNa + I_2 \rightarrow R\text{–}S\text{–}I + NaI \quad \ldots \ldots \ldots \text{rapid } (1)$$

$$R\text{–}S\text{–}I + 2\,NaN_3 \rightarrow RSNa + NaI + 3\,N_2 . \quad . \quad , \quad . \quad . \text{rapid } (2)$$

$$2\,NaN_3 + I_2 \rightarrow 2\,NaI + 3N_2 \quad \ldots \ldots \ldots \text{rapid } (1 + 2)$$

Thioketones probably form an analogous labile intermediate product containing iodine, $\diagdown\!\!\!\underset{\diagup}{C}\!\!<^{SI}_{I}$, which can react with sodium azide with re-generation of the catalytically active thioketones:

$$\diagdown\!\!\!\underset{\diagup}{C}\!\!<^{SI}_{I} + 2\,NaN_3 \rightarrow \diagdown\!\!\!\underset{\diagup}{C}{=}S + 2\,NaI + 3\,N_3$$

Another type of intermediate compound may be formed, as discussed in more detail (see p. 489) in the test for carbon disulfide.

When drawing conclusions from the results of the iodine–azide reaction regarding the type of binding of bivalent sulfur in organic compounds, it is necessary to keep the following points in mind. An immediate distinct production of nitrogen (spontaneous reaction) can be taken as certain proof of the presence of $\geqslant$CS- of $\geqslant$C—SH groups, provided not too small samples are being tested. Likewise, complete lack of reaction is a sure indication of the absence of such groups. However, between these extremes, instances occur in which a reaction begins only after several minutes or sometimes hours and furthermore usually continues quite slowly.[2] Such cases may involve slow hydrolytic cleavages being undergone by the test substance, with formation of SH- or SNa-containing products which, of course, react promptly with iodine–azide solution. It should also be noted that the reagent solution is slightly alkaline, because of its content of sodium azide, and this basicity favors hydrolytic splitting. Such hydrolysis may be responsible for the fact that disulfides,[3] particularly those which are but slightly soluble in water (*e.g.* cystine), and likewise thioethers and sulfur-bearing ring compounds, sometimes give a distinct iodine–azide reaction after they have been in contact with the reagent solution.

Inorganic compounds containing bivalent sulfur (sulfides, thiosulfates) must be absent.

Procedure. A drop of a solution of the test substance in water or an organic solvent* is mixed on a watch glass with a drop of iodine–azide solution, and observed for evolution of bubbles of nitrogen. Either a solid or liquid compound (ater removal of the solvent) may be tested. A positive reaction is then especially easy to see, even with very small amounts of substance.

Reagent: Solution of 3 g sodium azide in 100 ml 0.1 N iodine solution. The mixture is stable for months.

---

* Carbon disulfide should not be used as the solvent, since as a thioketo compound it reacts with the iodine–azide solution (see p. 488).

TABLE 10

THIOKETONES AND MERCAPTANS

Name and formula	Limit of identification and dilution limit	Solvent	No. of drops per ml
Thioacetic acid, CH_3—CO—SH	0.0003 γ 1 : 100,000,000	Acetone + water	30
Methyl methylxanthate, CH_3—O—CS—S—CH_3	0.03 γ 1 : 1,000,000	Acetone + water	30
Potassium ethylxanthate, C_2H_5—O—CS—SK	0.04 γ 1 : 1,000,000	Water	25
Rhodanine,  HN———=O S=_S/	0.003 γ 1 : 10,000,000	Acetone + water	35
3-Amino-4-phenyl-5-thiotriazole, 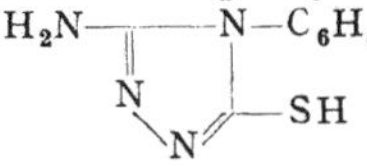 H_2N——N—C_6H_5 N_N/—SH	0.03 γ 1 : 1,000,000	Acetone + water	30
Allyl isothiocyanate (allyl mustard oil), CH_2=CH—CH_2—N=C=S	15 γ 1 : 2,000	Acetone + water	35
Phenyl isothiocyanate (phenyl mustard oil), C_6H_5—N=C=S	0.25 γ 1 : 100,000	Acetone + water	40
p,p'-Dimethoxythiobenzophenone, CH_3—O—C_6H_4—CS—C_6H_4—O—CH_3	0.1 γ 1 : 200,000	Acetone + water	50
4-Thio-α-naphthoflavone, 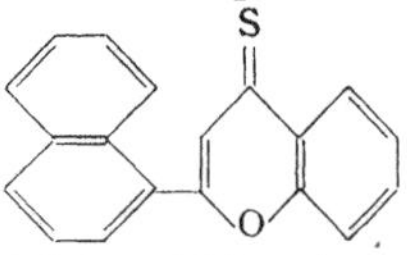	0.02 γ 1 : 1,000,000	Acetone + water	55
Diphenylthiocarbonate, C_6H_5—O—CS—O—C_6H_5	1 γ 1 : 20,000	Acetone + water	50
Di-p-tolyltrithiocarbonate, CH_3—C_6H_4—S—CS—S—C_6H_4—CH_3	0.1 γ 1 : 200,000	Acetone + water	50
Methyl α-naphthylcarbodithionate, —CS—S—CH_3 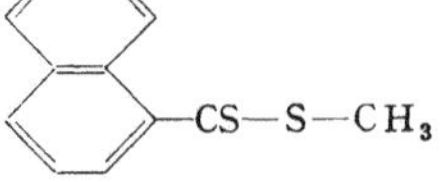	0.12 γ 1 : 100,000	Methanol	80
Thiourea, NH_2—CS—NH_2	0.005 γ 1 : 10,000,000	Water	20
$sym.$-Diphenylthiourea, C_6H_5—NH—CS—NH—C_6H_5	0.6 γ 1 : 200,000	Alcohol + water	80
Diphenylthiocarbazone (dithizone), C_6H_5—N=N—CS—NH—NH—C_6H_5	2.5 γ 1 : 10,000	Acetone + water	40
Dithiooxamide (rubeanic acid), NH_2—CS—CS—NH_2	0.03 γ 1 : 1,000,000	Acetone + water	30
Thioglycolic acid, HS—CH_2—CO—OH	0.05 γ 1 : 1,000,000	Water	20

Thioketones and thiols[2] may be differentiated by the fact that the latter are easily oxidized to the corresponding disulfides by iodine:

$$2 \ (R,Ar)—SH + I_2 → (R,Ar)—S—S—(R,Ar) + 2\ HI$$

which do not react with iodine–azide solution. Thioketones are not altered by iodine. Consequently, when a sample gives a positive reaction, a fresh sample should be warmed briefly with an excess of alcoholic iodine solution containing sodium acetate and, after cooling, again tested with iodine–azide solution. A positive reaction then indicates the presence of thioketones.

Limits of identification of thioketones and mercaptans through the catalysis of the iodine–azide reaction are given in Table 10. Since the drop size for the organic solutions is not constant the number of drops per ml have been determined in each instance to obtain the volume requisite for the calculation of the dilution limit.

1 F. FEIGL, *Mikrochemie*, 15 (1934) 1. Comp. also E. CHARGAFF, *Dissertation*, Vienna, 1928.
2 Comp. W. AWE, *Mikrochemie Ver. Mikrochim. Acta*, (1951) 574.
3 E. FRIEDMANN, *J. Prakt. Chem.*, [2], 146 (1936) 179.

35. Thiols (—SH group)

(1) Test by precipitation of cuprous salts[1]

Thiol compounds may react with cupric ions in various ways. Sometimes water-insoluble, mostly dark colored, cupric salts are produced. This is true of 8-mercaptoquinoline, dithiocarbamates and also of compounds which contain adjacent SH groups, *e.g.*, rubeanic acid. Another mode of reaction, which may occur in strong ammoniacal solution, leads to the production of black copper sulfide (*e.g.* cysteine). Along with this reaction, and sometimes predominating[2] in the case of some mercapto compounds, there is an initial redox reaction:

$$2 \ RS^- + 2\ Cu^{2+} → Cu_2^{2+} + RS—SR \tag{1}$$

Subsequently, the cuprous ions may react with unused mercaptan:

$$Cu_2^{2+} + 2\ RS^- → Cu_2(RS)_2 \tag{2}$$

The cuprous salts of mercaptans are yellow, orange-yellow or yellow-brown; they are insoluble in water, dilute acids or ammonia. If an acetic acid or ammoniacal cuprous salt solution is used, the redox reaction (*1*), through which the mercaptan is consumed in the production of disulfide, may be by-passed and immediate precipitation of the cuprous mercaptan may occur.* It is advisable to use acetic acid–cuprous salt solutions as reagents for mercaptans, because interfering side reactions (particularly the formation of copper sulfide) are ordinarily thus repressed. Since this is not invariably the case, the following procedure provides an absolutely reliable test for mercaptans only when a yellow or brown precipitate or turbidity appears.

Procedure.   A drop of the acetic acid or ammoniacal test solution is treated with one drop of the reagent on a spot plate. If mercaptans are present, a yellow or brown precipitate or coloration results.

Reagents: 1) 1.5 g cupric chloride and 3 g ammonium chloride are dissolved in a little water. The solution is treated with 3 ml concentrated ammonia and made up to 50 ml with water.

2) 20% hydroxylamine hydrochloride solution in water.

Equal volumes of (1) and (2) are mixed just before the test.

The procedure gave precipitates with:

yellow	*yellow-brown*
2.5 γ thioglycolic acid anilide	1 γ potassium ethyl xanthate
5 γ mercaptobenzothiazole	0.5 γ dithiooxamide
2.5 γ mercaptophenylthiadiazolone	

1 Unpublished studies, with D. GOLDSTEIN.
2 Compare F. FEIGL, *Chemistry of Specific, Selective and Sensitive Reactions*, Academic Press, New York, 1947, p. 236.

(2) Test for primary and secondary thiols by hydrolysis to hydrogen sulfide [1]

If primary or secondary thiols are heated with concentrated ammonia at boiling water temperature, hydrogen sulfide results:**

$$R—CH_2—SH + H_2O \rightarrow R—CH_2OH + H_2S$$

$$\frac{R}{R}{>}CH—SH + H_2O \rightarrow \frac{R}{R}{>}CH—OH + H_2S$$

Since tertiary thiols and thioketones do not undergo this hydrolysis under these conditions, it is possible to base a sharp distinction between the latter types of compounds and primary and secondary thiols on this reaction.

Procedure.   A small amount of the test material or a drop of its solution is treated in a micro test tube with a drop of concentrated ammonia. The mouth

* There are exceptions: Thioglycolic acid added to an acid solution of copper salts gives a yellow precipitate whereas when added to a blue ammoniacal solution only decolorization occurs. This indicates the formation of a soluble complex cupro compound with the metal in the anionic constituent; the compound is very stable as shown by the fact that its solution does not react with acetylene. Compare the test for acetylene, p. 404.

** N. KHARASCH has made interesting contributions to the clarification of the precise mechanism of the reactions (*Organic Sulfur Compounds, Vol. I*, Pergamon, New York, 1961).

of the tube is covered with lead acetate paper and the tube is placed in a boiling water bath. A black stain with a metallic lustre indicates a positive response.

The test revealed:

2 γ thionalide	100 γ cysteine
1 γ thioglycolic acid anilide	100 γ thiomalic acid

The following tertiary thiols and thioketones gave a negative response: mercaptobenzothiazole, thiourea, diphenylthiocarbazone, diphenylthiourea, sodium diethylthiocarbamate.

1 Unpublished studies, with E. JUNGREIS.

(3) Test for aliphatic thiols through pyrohydrogenolysis [1]

If appropriate non-volatile compounds are heated in a micro test tube with fused benzoin (m.p. 135°) at 150–200°, hydrogen sulfide (detectable by the blackening of lead acetate paper) is split off:

$$RSH + C_6H_5CHOHCOC_6H_5 \rightarrow C_6H_5COCOC_6H_5 + RH + H_2S$$

Aromatic thiols as well as aliphatic and aromatic disulfides do not react with benzoin. The above pyrohydrogenolysis is characteristic for aliphatic thiols provided that compounds which split off H_2S when heated alone are absent. For this reason the behavior of the test material on dry heating must first be examined.

The following amounts were revealed: 100 γ cysteine, 120 γ thionalide.

It seems that aliphatic CS-compounds which can react in a tautomeric C—SH (sulfhydryl) form, split off H_2S when fused with benzoin. An example is the behavior of thiourea.

1 Unpublished studies, with R. COHEN.

36. Disulfides (—S—S— group)

(1) Test through conversion into thiols [1]

Dialkyl- and diaryl disulfides are reduced to the corresponding thiols (mercaptans or thiophenols) by brief treatment in alcoholic hydrochloric acid solution with Raney nickel alloy. The nickel hydride (Raney nickel) formed from the alloy is the active agent here. Since thiols are readily revealed through the catalytic acceleration of the iodine–azide reaction, (see p. 219) this reduction can serve as the basis for a sensitive test for disulfides. Diselenides behave in the same way as disulfides (see p. 237). Thioethers (also thiophene), sulfoxides, sulfones, sulfinic- and sulfonic acids do not impair the detection. On the other hand, compounds that show the iodine–azide reaction (thiols, thioketones, thionic acids as well as inorganic sulfides, thiosulfates, thiocyanides) do interfere. The test described here for disulfides is nonequivocal if the test material does not react with iodine–azide. Disulfides can be separated from many reactive inorganic sulfur compounds by extraction with ether.

Procedure. A drop of the test solution is mixed in a micro test tube with 6 drops of ethanol, 2 drops of 5% alcoholic hydrochloric acid and several mg of Raney nickel alloy. The suspension is kept in boiling water for around 30 sec. and then cooled in a stream of running water. Then several drops of iodine–azide solution are added. (For preparation see p. 220). A positive response is indicated by the immediate evolution of nitrogen bubbles.

Limits of identification:

0.005 γ cystine
0.1 γ bis-(p-dimethylaminophenyl) disulfide
0.01 γ bis-(o-nitrophenyl) disulfide
0.2 γ bis-(p-nitrophenyl) disulfide

1 V. ANGER and G. FISCHER, *Mikrochim. Acta*, (1962) 501.

(2) Test by cleavage with alkali sulfite[1]

When disulfides are treated with alkali sulfite the respective thiols and sulfenylsulfites (Bunte salts) are formed:[2]

$$(R,Ar)S—S(Ar,R) + SO_3{}^{2-} \rightarrow (R,Ar)S^- + (R,Ar)SSO_3{}^-$$

Both cleavage products in contrast to the sulfite reduce ammonium molybdate in strong sulfuric acid to give molybdenum blue*. The test described here is based on this behavior, and is decisive in the absence of thiol compounds. Although hitherto few pertinent compounds have been examined, it seems that the above cleavage is characteristic of both aliphatic and aromatic disulfides.

A blank test with the sample without treatment with alkali sulfite is advisable.

Procedure. A drop of the neutral or slightly basic test solution is evaporated to dryness in a test tube with a drop of 10% sodium sulfite solution. After cooling, a drop of a solution of some cg of ammonium molybdate in 50% sulfuric acid is added. A positive response is indicated by the appearance of a blue color immediately or within a few minutes.

Limit of identification: 5 γ of cystine, homocystine, dixanthogen, diphenyl disulfide, bis-(p-nitrophenyl) disulfide.

* This action was first observed with alkali thiosulfates[3].

1 Unpublished studies, with R. COHEN.
2 Comp. A. J. PARKER and N. KHARASCH, *Chem. Rev.*, 50 (1950) 612.
3 E. E. POZZI-ESCOT, *Bull. Soc. Chim. France*, 13 (1913) 401.

37. Xanthates $\left(S{=}C\diagdown_{OR}^{S^-}\ \text{group}\right)$

Test with molybdic acid [1]

The xanthate group occurs in the relatively stable water-soluble alkali salts and in the water-insoluble heavy metal salts of the monoalkylester acid (I), which is derived from dithiocarbonic acid (II)

$$S\,C\diagup_{OR}^{SH} \qquad S\,C\diagup_{OH}^{SH} \qquad R{=}CH_3,\ C_2H_5,\ C_6H_5CH_2,\ \text{cellulose } etc.$$

(I)   (II)

Neither (I) nor (II) can be isolated because the water-soluble salts are decomposed by dilute acids. For instance:

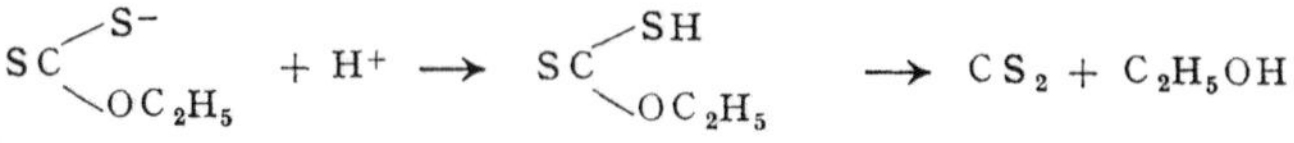

$$S\,C\diagup_{OC_2H_5}^{S^-} + H^+ \longrightarrow S\,C\diagup_{OC_2H_5}^{SH} \longrightarrow CS_2 + C_2H_5OH$$

By virtue of their thioketo and thiol groups, the water-soluble and water-insoluble xanthates catalytically accelerate the sodium azide–iodine reaction (compare p. 219). The soluble xanthates react with cuprous or cupric salts to yield yellow cuprous xanthates, which are insoluble in water and dilute acids. When cupric salts are used, the initial step produces cuprous ions and the disulfide of the particular monoalkylester acid. Neither the iodine–azide reaction nor the formation of cuprous salt is characteristic for xanthates. On the other hand, the behavior of xanthates toward acid solutions of alkali molybdate is unequivocal. The resulting plum-blue precipitate dissolves in ether, chloroform, etc. to give a violet color.[2] The composition of the precipitate corresponds to an addition compound of one molecule MoO_3 and two molecules of ester acid. For instance, sodium ethyl xanthate gives $MoO_3.2[CS(SH)OC_2H_5]$. It is notable that hereby the instable monoesters are stabilized through the coordinative binding to MoO_3.

Procedure. A micro test tube is used. A little of the solid, or a drop of solution of its alkali salt, is treated in succession with one drop each of 20% ammonium molybdate solution and 0.5 N hydrochloric acid and then shaken with several drops of chloroform. Depending on the quantity of xanthate present, the chloroform acquires a more or less intense violet color.

Limit of identification: 1 γ potassium ethylxanthate.

1 F. J. SMITH and E. JONES, *A Scheme of Qualitative Organic Analysis*, London, 1953, p. 200.
2 S. MALOWAN, *Z. Anorg. Allgem. Chem.*, 108 (1914) 73; *Z. Anal. Chem.*, 79 (1929) 202; J. KOPPEL, *Chemiker Ztg.*, 43 (1919) 777.

38. Aromatic sulfinic acids (ArSO₂H group)

(1) Test through reaction with o-dinitrobenzene[1]

A method for preparing nitroaryl sulfones is based on the reaction of alkali arylsulfinates with o-dinitrobenzene[2]:

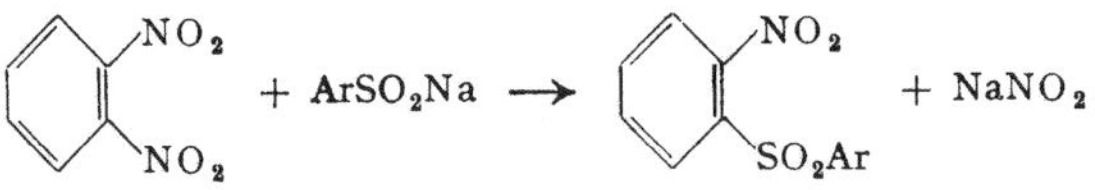

The occurrence of the above reaction and hence the presence of aryl sulfinates is revealed by the fact that the reaction mixture shows a positive response to the familiar Griess test for nitrous acid. The nitrite formation is accomplished most quickly by using glycerol as solvent and heating to 120°. *p*-Dinitrobenzene behaves in the same fashion as *o*-dinitrobenzene. *m*-Dinitrobenzene gives a very faint reaction and this probably can be attributed to a trace of the *ortho* isomer contaminating the preparation used for the trials.

Procedure. A drop of the test solution (containing alkali or alkaline-earth salts of the sulfinic acid) is taken to dryness in a micro test tube, and then several mg of *o*-dinitrobenzene and a drop of glycerol are added. The test tube is then kept at 120° for 5 min. in a glycerol bath. The cooled mixture is then treated with 2 drops of Griess reagent and gently warmed (water bath). A red or pink color signals a positive response. It is advisable to shake the mixture with several drops of chloroform to remove any unconsumed *o*-dinitrobenzene.

Griess Reagent: For preparation see p. 90.

The following were detected:
5 γ toluenesulfinic acid   10 γ benzenesulfinic acid
12 γ naphthalenesulfinic acid   20 γ 3-acetamino-4-methoxybenzenesulfinic acid

In case free acids are present, it is necessary to convert them into their calcium salts which behave in the same manner as the alkali salts. To this end, the aqueous test solution is warmed with excess calcium carbonate, filtered, and 1 drop of the filtrate is taken for the test. Another method is to add a minimal amount of calcium carbonate to a drop of the aqueous test solution in a micro test tube; after evaporation the procedure outlined above is employed. It is well to make a blank test with *o*-dinitrobenzene because the latter is denitrosated to trace extents with formation of nitrite.

1 F. FEIGL, D. HAGUENAUER-CASTRO and E. LIBERGOTT, *Mikrochim. Acta*, (1961) 595.
2 H. LANDERS, US patent 1,936,721; compare *Chem. Abstr.*, 28 (1934) 1049.

(2) Test through reductive cleavage with Raney nickel alloy[1]

If benzenesulfinic acid is treated with hydrochloric acid and zinc or Devarda's alloy, the nascent hydrogen produces thiophenol:

$$C_6H_5SO_2H + 4 H^0 \rightarrow C_6H_5SH + 2 H_2O$$

The thiophenol (b.p. 104°) is readily detected since it volatilizes on warming and gives a yellow stain of $Pb(SC_6H_5)_2$ on lead acetate paper (*Idn.Lim.* 15 γ).

If Raney nickel alloy is used instead of zinc or Devarda's alloy, hydrogen sulfide results and is detectable by the darkening of lead acetate paper. In other words, the reduction proceeds distinctly farther. The reason is that Raney nickel alloy on treatment with acid not only produces nascent hydrogen but furthermore Raney nickel which contains the hydride Ni_2H. The latter is a more powerful and faster reductant than nascent hydrogen; its action with benzenesulfinic acid may be:

$$C_6H_5SO_2H + 6 Ni_2H \rightarrow C_6H_6 + 12 Ni^0 + 2 H_2O + H_2S$$

Arylsulfinic acids behave in the same way as benzenesulfinic acid toward the reductants mentioned. The reduction with Raney nickel alloy is recommended for the test since a higher sensitivity can be attained in the detection of hydrogen sulfide as compared with thiophenol.

Arylsulfonic acids are not affected when warmed in acid media with Raney nickel alloy or Devarda's alloy. Therefore the following procedure permits an excellent differentiation between arylsulfinic and arylsulfonic acids or their alkali salts.

Procedure. A slight quantity of the solid test material or a drop of its solution together with a few mg of Raney nickel alloy are treated in a micro test tube with several drops of dilute hydrochloric or sulfuric acid. After the vigorous evolution of hydrogen has subsided, a piece of lead acetate paper is placed over the mouth of the tube, whose contents are then gently warmed in a water bath, and then the temperature is raised. A grey or black stain due to lead sulfide appears on the paper if the response is positive. In case no more than slight amounts are suspected, it is advisable to run a blank with the Raney nickel alloy alone, since at times this material contains slight amounts of sulfides which may react with acids. However, this comparison is required only when quantities of sulfinic acid below approximately 10 γ are to be detected.

The test revealed:

4 γ benzenesulfinic acid   5 γ toluenesulfinic acid   6 γ naphthalenesulfinic acid

1 F. FEIGL, *Anal. Chem.*, 33 (1961) 1118.

(3) Test through reaction with mercuric chloride [1]

If the alkali salts of benzenesulfinic acid are warmed with mercuric chloride there is rapid production of phenylmercury chloride and sulfur dioxide: [2]

$$C_6H_5SO_2Na + HgCl_2 \rightarrow C_6H_5HgCl + NaCl + SO_2$$

Since sulfur dioxide is readily detected through the bluing of Congo paper moistened with hydrogen peroxide the above reaction can serve as the basis of a test for benzenesulfinic acid, and because of similar behavior, also of a general test for arylsulfinic acids.

Procedure. A drop of the aqueous test solution is taken to dryness in a micro test tube, or a small portion of the solid test material is used. Several mg of mercuric chloride and a drop of alcohol are introduced and the tube is then placed in a water bath that has been heated to 80°. The mouth of the tube is covered with a disk of Congo paper that has been moistened with 5% hydrogen peroxide. After the alcohol has evaporated, the temperature is raised to 100°. A positive response is indicated by the formation of a blue fleck on the red indicator paper.

The test revealed:

0.25 γ naphthalenesulfinic acid	0.25 γ benzenesulfinic acid (Na salt)
2.5 γ 3-acetamino-4-methoxy- benzenesulfinic acid	2.5 γ toluenesulfinic acid

1 F. Feigl, D. Haguenauer-Castro and E. Libergott, *Mikrochim. Acta*, (1961) 595.
2 K. Peters, *Ber.*, 38 (1905) 2570.

39. Sulfones ($>SO_2$ group)

Test through pyrolytic release of sulfur dioxide[1]

When sulfones are dry heated SO_2 is given off. This gas when obtained by such thermal decomposition should not be detected by tests based on reduction reactions since the pyrolysis of organic compounds frequently yields reducing gaseous cleavage products (see Chap. 2). The sensitive test for SO_2, based on the fact that its autoxidation induces the oxidation of green $Ni(OH)_2$ to black $NiO(OH)_2$, may be employed.

The procedure given here for the detection of all kinds of sulfones through the sulfur dioxide evolved cannot be applied in the presence of thioketones, mercapto and thiol compounds, sulfides and disulfides, because they give off H_2S when ignited and the latter forms black NiS on contact with green $Ni(OH)_2$. The presence of such compounds can be revealed beforehand by heating a little of the sample in a micro test tube whose mouth is covered with moist lead acetate paper. If there is little or no blackening (formation of PbS), a positive response to the test for SO_2 with $Ni(OH)_2$ may be taken as positive proof of the presence of sulfones provided that aromatic sulfonic acids which split off sulfur dioxide when ignited are absent.

Procedure. A little of the solid or a drop of the test solution is placed in a micro test tube and evaporated if necessary. The mouth of the tube is covered with a disk of nickel hydroxide paper. The tube is heated over a micro flame. A black or gray fleck appears on the green paper if the response is positive.

If much sulfur dioxide is formed, the initial black stain disappears and only a gray ring remains at its edge. In such cases it is well to repeat the procedure with less of the sample.

Reagent: Nickel hydroxide paper. Strips of filter paper are placed in a 30% solution of $NiSO_4.6\ H_2O$ in concentrated ammonia. The dried strips are then bathed for several minutes in 1 N sodium hydroxide to obtain a homogeneous precipitation of $Ni(OH)_2$ in the pores of the paper. The paper, which should be washed with water, must not be allowed to dry. It keeps for months if stored over moist cotton wool.

The procedure revealed:

5 γ sulfonal	5 γ benzyl ethyl sulfone
20 γ dicetyl sulfone	10 γ 4,4'-diaminodiphenyl sulfone

A positive response was obtained also from trional, tetronal, di-*n*-butyl sulfone, di-*p*-nitrobenzyl sulfone, di-*p*-methoxyphenyl sulfone, di-2-phenylethyl sulfone.

1 Unpublished studies, with Cl. COSTA NETO.

40. Aryl- and aralkyl sulfones

Test through desulfonation with Raney nickel alloy[1]

If aryl and aralkyl sulfones are warmed with caustic alkali and Raney nickel alloy, the Raney nickel produced brings about their desulfonation in a manner analogous to that occurring with aryl and aralkyl sulfonic acids (see p. 233). As in the latter instances, the active material is the nickel hydride present in the Raney nickel:

$$R—SO_2—R' + 6\ Ni_2H \rightarrow RH + R'H + 2\ H_2O + 11\ Ni^0 + NiS$$
$$(R\ and\ R' = aryl\ or\ aralkyl\ radicals)$$

The key compound is the nickel sulfide produced in the above redox reaction; it reacts with dilute mineral acids to yield hydrogen sulfide that is readily detectable. It thus provides the basis for the detection of aryl- and aralkyl sulfones and also their differentiation from aliphatic sulfones since the latter are not desulfurized by Raney nickel.

The test described here is not applicable in the presence of sulfonic and sulfinic acids nor of sulfamic acids and their derivatives.

Procedure.   As described in Sect. 43 for the detection of aromatic sulfonic acids.

The following were detected:

25 γ benzyl ethyl sulfone	25 γ di-*p*-methoxyphenyl sulfone
2.5 γ di-*p*-nitrobenzyl sulfone	2.5 γ di-*p*-diaminodiphenyl sulfone

The two latter compounds on reductive cleavage with Raney nickel yield not only nickel sulfide but also aniline or *p*-methylaniline respectively. These products are volatile with steam and can be detected in the gas phase with sodium 1,2-naphthoquinone-4-sulfonate (see p. 153).

1 F. FEIGL, *Anal. Chem.*, 33 (1961) 1121.

41. Sulfonic acids (—SO₃H group)

(1) Test by conversion into the iron(III) salt of acetohydroxamic acid[1]

Sulfonic acids can be converted into sulfohydroxamic acids, which react with acetaldehyde in the presence of alkali:

$$(R,Ar)SO_2NHOH + CH_3CHO \rightarrow CH_3CONHOH + (R,Ar)SO_2H$$

giving acetohydroxamic acid and sulfinic acids. Both of these products react with ferric chloride in weak acid solution to give red soluble ferric complexes of hydroxamic acid (see p. 212), and an orange-red insoluble iron(III) salt of the sulfinic acid, respectively.[2]

The necessary conversion of the sulfonic acid to the sulfohydroxamic acid requires a preliminary preparation of the sulfonyl chloride, which then can be converted to the sulfohydroxamic acid by reaction with hydroxylamine. Thionyl chloride is used to prepare the sulfonyl chloride. The sulfohydroxamic acid reacts with acetaldehyde to give acetohydroxamic acid and sulfinic acid, both of which react with ferric chloride:

$$RSO_3H + SOCl_2 \rightarrow RSO_2Cl + SO_2 + HCl$$

$$RSO_2Cl + NH_2OH \rightarrow RSO_2(NHOH) + HCl$$

$$RSO_2(NHOH) + CH_3CHO \rightarrow CH_3CO(NHOH) + RSO_2H$$

$$3\ CH_3CO(NHOH) + FeCl_3 \rightarrow Fe[CH_3CO(NHO)]_3 + 3\ HCl$$

$$3\ RSO_2H + FeCl_3 \rightarrow Fe(RSO_2)_3 + 3\ HCl$$

$$(R = alkyl\ or\ aryl)$$

In spite of the lengthy series of reactions, the procedure for carrying out a spot test is relatively simple.

Procedure. A little of the solid or the evaporation residue from a drop of solution is fumed in a micro crucible with a few drops of thionyl chloride. The product is treated with 2 drops of a saturated alcoholic solution of hydroxylamine-HCl and a drop of acetaldehyde, then made slightly alkaline with 5% sodium carbonate solution. After some minutes, the mixture is acidified with 0.5 N hydrochloric acid. If a drop of a dilute aqueous solution of ferric chloride is added a brown to violet color or precipitate appears.

When testing the salt of a sulfonic acid, it should be evaporated with hydrochloric acid before the treatment with thionyl chloride. Many dyes containing an SO_3H group whose color interferes with the recognition of the color change can easily be decomposed with a few drops of bromine water.

The following amounts were detected:

15 γ H-acid	12 γ benzenesulfonyl chloride
30 γ methyl orange	20 γ naphthalene-β-sulfonic acid
25 γ sulfanilic acid	10 γ potassium hydrogen p-benzenedisulfonate

1 F. FEIGL and V. ANGER, *Mikrochemie*, 15 (1934) 23.
2 Comp. F. FEIGL, *Chemistry of Specific, Selective and Sensitive Reactions*, Academic Press, New York, 1949, p. 289.

(2) Test through pyrohydrolysis [1]

If succinic acid is heated above its melting point (189°), succinic anhydride is formed:

$$(CH_2COOH)_2 \rightarrow (CH_2CO)_2O + H_2O$$

The water lost in this dehydration is released as superheated steam, and if liberated in contact with certain organic compounds it is capable of accomplishing hydrolytic cleavings (pyrohydrolyses) which are not realizable by the wet way. Among these reactions are the hydrolyses

$$(R,Ar)SO_3H + H_2O \rightarrow (R,Ar)H + H_2SO_4$$

followed by the reduction of the sulfuric acid formed to sulfurous acid. Fused phthalic acid (m.p. 230°) likewise gives off water with formation of the anhydride, and accordingly can be used in place of succinic acid. The participation of sulfonic acids in pyrohydrolyses and hence their detection can be established through the resulting sulfurous acid, which precipitates Prussian blue from ferri ferricyanide solution.

When using this test, care must be taken to insure the absence of compounds which split off volatile reducing compounds when pyrohydrolyzed. Thio compounds are particularly to be noted in this connection because their SH groups are readily exchanged for OH groups.[2] Their presence can be established easily by the blackening of lead acetate paper held above the succinic (phthalic) acid melt.

Procedure. The test is made in a micro test tube. A little of the solid or a drop of its solution is united with several cg of succinic or phthalic acid and the solvent removed if need be. The test tube is placed into a glycerol bath previously heated to 200°. A disk of filter paper moistened with ferri ferricyanide solution is placed over the mouth of the test tube, and the temperature of the bath is raised to 250°. If the response is positive, a blue stain appears on the paper.

Reagent: 0.08 g of anhydrous $FeCl_3$ and 0.1 g of $K_3Fe(CN)_6$ are dissolved in 100 ml water.

The following amounts were detected:

10 γ 1-naphthol-4-sulfonic acid	15 γ benzenesulfonic acid
5 γ 1-naphthylamine-4-sulfonic acid	10 γ sulfosalicylic acid
1 γ taurine	20 γ Congo red

1 F. FEIGL, D. HAGUENAUER-CASTRO and E. JUNGREIS, *Talanta*, 1 (1958) 80.
2 F. FEIGL, *Angew. Chem.*, 70 (1958) 166.

42. Aliphatic sulfonic acids (R—SO₃H group)

Test through pyrohydrogenolytic formation of sulfurous acid[1]

If aliphatic sulfonic acids are heated to 160–170° along with benzoin, the redox reaction:

$$R—SO_3H + C_6H_5CHOHCOC_6H_5 \rightarrow RH + C_6H_5COCOC_6H_5 + H_2O + SO_2$$

occurs. When alkali salts are to be examined, previous evaporation with hydrogen chloride is necessary. The evaporation can be carried out in a micro test tube followed by heating (140°) to expel completely the hydrochloric acid. The sulfurous acid formed by the above pyrohydrogenolysis can be detected through the bluing of a Congo paper moistened with hydrogen peroxide.

Many aromatic sulfonic acids behave like aliphatic sulfonic acids. Therefore the absence of the former is essential. A test for aromatic sulfonic acids is described in the next section.

Procedure. A small amount of the test material or the residue which remains after transformation of alkali salt into acid is mixed in a micro test tube with several mg of benzoin. The tube is immersed about 0.5 cm deep in a glycerol bath that is at 140°. The temperature is then brought to 170°. A disk of Congo paper moistened with diluted hydrogen peroxide is laid across the mouth of the tube. The formation of a blue stain on the indicator paper within 5 min. signals a positive response. It is advisable to run a blank test.

15 to 20 γ of the following compounds were detected:

> butane-1-sulfonic acid (Na salt)
> 2-methylprop-2-ene-1-sulfonic acid (Na salt)
> 3-methylbutane-1-sulfonic acid (Na salt)

1 Unpublished studies, with A. DEL'ACQUA.

43. Aromatic sulfonic acids (Ar—SO₃H group)

Test through reductive removal of sulfur[1]

If arylsulfonic acids are warmed with Raney nickel alloy (50% Ni, 50% Al) in the presence of caustic alkali, neither SO_3^{2-} nor S^{2-} ions can be detected in the resulting solution. However, the residual alloy contains nickel sulfide that can be revealed through the formation of hydrogen sulfide when warmed with dilute mineral acid. Since Devarda's alloy is without action in this case, the effective reductant cannot be nascent hydrogen, and it may be assumed that the reductive desulfurization is due to the nickel hydride in the Raney nickel formed from the alloy through the action of the caustic alkali. The reduction may be represented simply:

$$ArSO_3Na + 6\ Ni_2H \rightarrow ArH + NiS + NaOH + 2\ H_2O + 11\ Ni^0$$

The ease with which the sulfur is removed from arylsulfonic acids is truly astounding. It is in accord with the desulfurization effects through Raney nickel previously observed.[2, 3, 4]

The test described here requires the absence of thioketones, thiol compounds, disulfides, and sulfides since they too react with Raney nickel to yield nickel sulfide. The presence of these interfering substances (with the exception of thioethers and disulfides) can be established through their catalytic hastening of the iodine–azide reaction (comp. p. 219). A preliminary test of this kind is necessary however only in the rare instance when these compounds are soluble in dilute alkali, as is true of sulfonic acids.

Aliphatic saturated sulfonic acids yield little or no nickel sulfide on warming with caustic alkali and Raney nickel alloy. This fact was established for taurine, 1,10-decanedisulfonic acid, camphorsulfonic acid, sodium 3-methyl-1-butanesulfonate, sodium 2-propanesulfonate, sodium 1-butanesulfonate. On the other hand, sodium 2-propene-1-sulfonate and benzylsulfonic acid react in the same manner as purely aromatic sulfonic acids. Therefore it appears that carbon double bonds adjacent to the $SO_3(H,Na)$ group activate the reducibility by Raney nickel, and that OH groups in aliphatic sulfonic acids have a similar effect though to a much lesser degree. This was observed in the cases of $ClCH_2CHOHCH_2SO_3Na$ and $HOCH_2CH_2SO_3Na$, which in amounts over 1000 γ yielded nickel sulfide on treatment with caustic alkali and Raney nickel alloy.

Procedure. A little of the solid test material or 1–2 drops of its solution is placed in a micro test tube along with several mg of Raney nickel alloy and 1 drop of 5% caustic alkali. The mixture is warmed in a water bath. If the evolution of hydrogen becomes too turbulent, the warming should be interrupted occasionally. The warming is continued for about 3 min.; the contents of the tube are allowed to cool, and then several drops of hydrochloric acid (1:1) are added. After the hydrogen ceases to come off, a piece of lead acetate paper is laid over the mouth of the tube, which is then warmed in the water bath. A brown or grey stain signals a positive response.

The test revealed:

1	γ anthraquinone-2,6-disulfonic acid	5	γ sulfosalicylic acid
1	γ ,, -1,5- ,, ,,	2.5	γ sulfanilic acid
5	γ 8-hydroxyquinoline-7-sulfonic acid	5	γ metanilic acid
5	γ naphthalene-1,5-disulfonic acid	1	γ saccharin
2.5	γ 1-amino-2-naphthol-4-sulfonic acid	2.5	γ methyl orange
2.5	γ 1-naphthylamine-3,6,8-trisulfonic acid	2	γ H-acid
2.5	γ 1-anthraquinonesulfonic acid (K salt)	5	γ chromotropic acid (Na salt)

1 F. Feigl, *Anal. Chem.* 33 (1961) 1119.
2 J. Bougault and coworkers, *Compt. Rend.*, 208 (1939) 657.
3 A. Trofonoff and coworkers, *Compt. Rend. Acad. Bulgare Sci.*, 7 (1954) 1.
4 L. Granatelli, *Anal. Chem.*, 31 (1959) 434.

44. Esters and N-organylamides of sulfuric acid

Detection through alkaline saponification[1]

Compounds in this category include alkyl sulfuric acids and their salts, alkyl sulfates, and alkylated (arylated) monoamides of sulfuric acid; the latter are derivatives of sulfamic acid (see p. 306).

The solutions of these compounds contain no $SO_4{}^{2-}$ ions precipitable by Ba^{2+} ions. However, extensive saponification (hydrolysis) of these compounds is brought about by evaporation of their solutions in the presence of caustic alkali:

$$SO_2(OR)OK + KOH \rightarrow K_2SO_4 + ROH$$

$$SO_2(OR)_2 + 2 KOH \rightarrow K_2SO_4 + 2 ROH$$

$$SO_2(NHR)OK + KOH \rightarrow K_2SO_4 + NH_2R$$

The sulfate ions resulting from the hydrolysis can be detected by precipitation as barium sulfate in the presence of permanganate. A violet product results whose color is stable against the action of reducing agents. See p. 84 regarding this so-called Wohlers effect.

Procedure.   A drop of the acetone solution of the test material is taken to dryness in a micro test tube along with a drop of 5% caustic alkali. The evaporation residue is dissolved in a drop of dilute hydrochloric acid and a drop of a solution of 7 g of potassium permanganate and 2 g of barium chloride per 100 ml water is added. The mixture is heated briefly, cooled, and then a drop of a 20% solution of hydroxylamine hydrochloride is added to destroy the excess permanganate and the manganese dioxide that may have been produced. A violet residue or suspension indicates a positive response.

Limits of identification:

20 γ N-cyclohexylsulfamic acid (Na salt)   10 γ dimethyl sulfate
40 γ N-phenylsulfamic acid (NH₄ salt)   10 γ diethyl sulfate
25 γ N,N-dicyclohexylsulfamic acid (Na salt)

1 Unpublished studies, with J. T. DE SOUZA CAMPOS and S. LADEIRA DALTO.

45. Sulfoxylate compounds (—CH₂—O—SONa group)

Test through pyrolytic release of hydrogen sulfide[1]

Dry heating of formaldehyde sodium sulfoxylate (Rongalite) produces hydrogen sulfide and formaldehyde:[2]

$$2 \; H_2C \overset{\displaystyle{-OH}}{\underset{\displaystyle{-O-SONa}}{\Big<}} \quad \longrightarrow \quad 2\, CH_2O + Na_2SO_4 + H_2S \qquad (1)$$

This effect is the basis of a specific test for Rongalite described in *Inorganic Spot Tests*, Chap. 4. Organic derivatives of Rongalite, which contain the —CH₂OSONa group, and which are easily formed by condensing primary aromatic amines with Rongalite in alkali carbonate solution, likewise evolve hydrogen sulfide when they are heated to 250–280°. The following partial reactions may occur:

$$RHN—CH_2—O—SONa \rightarrow RN{=}CH_2 + NaHSO_2 \qquad (2)$$

$$2\, NaHSO_2 \rightarrow H_2S + Na_2SO_4 \qquad (3)$$

However, it should be remembered that when organic compounds are pyrolyzed, the water produced acts locally as superheated steam, and effectuates the hydrolysis:

$$RHN—CH_2—O—SONa + H_2O \rightarrow RNH_2 + CH_2(OH)(O—SONa) \qquad (4)$$

which is followed by the thermal decomposition represented in (1).

The detection of sulfoxylate compounds through the detection of pyrolytically evolved hydrogen sulfide is specific under the conditions prescribed, provided primary thiol compounds which likewise yield H₂S on dry heating are absent.

Procedure. The test is conducted in a micro test tube. A little of the solid is introduced and the mouth of the tube is covered with a piece of moist lead acetate paper. The test tube is placed in a glycerol bath previously heated to 250° and the temperature is then raised to 270–280°. A positive response is signalled by a black or brown stain on the white paper.

The limits of identification cannot be determined because when solutions of sulfoxylate compounds are taken to dryness there is oxidation of the —CH₂SO₂Na group to the —CH_2SO_3Na group; the latter does not yield hydrogen sulfide on pyrolysis. However, about 0.3 mg of the following compounds gave a distinct response: rongalite, aldarsone, neosalvarsan, sulfoxone sodium.

1 F. FEIGL and E. SILVA, *Analyst*, 82 (1957) 582.
2 G. PANIZZON, *Melliand's Textilber.*, 12 (1931) 119.

46. Diselenides (—Se—Se— group)

Test through conversion into selenol[1]

A test for disulfides is described in Sect. 36 based on their reduction by Raney nickel alloy to thiols which like hydrogen sulfide and metal sulfides accelerate the iodine–azide reaction catalytically. The diselenides behave in the same manner as the disulfides. Hence (R,Ar)SeH compounds must result which catalyze the reaction between sodium azide and iodine that of itself proceeds immeasurably slowly. This effect is therefore very remarkable since neither hydrogen selenide nor metal selenides are catalysts under these conditions.

Procedure. As described on p. 225 for the detection of disulfides. The test revealed: 1 γ dibenzyldiselenide.

1 V. ANGER and G. FISCHER, *Mikrochim. Acta*, (1962) 501.

47. Seleninic acids (—SeO₂H group)

Test through pyrolysis with hydroxylamine sulfate[1]

In acid solution, selenious acid reacts with hydroxylamine:

$$H_2SeO_3 + NH_2OH \rightarrow Se^0 + 2 H_2O + HNO_2 \qquad (1)$$

Organic derivatives of selenious acid give *no* analogous reaction in the wet way. However if these acids are mixed with hydroxylamine sulfate and heated to 130–140°, elementary selenium is produced as well as nitrous acid, which is readily detected through the positive response to the Griess color reaction. Since an analogous reaction does not occur with hydroxylamine hydrochloride, it may be assumed that the sulfuric acid contained in the sulfate initially displaces selenious acid from the organic seleninic acids:

$$Ar—SeO_2H + H_2SO_4 \rightarrow Ar—SO_3H + H_2SeO_3 \qquad (2)$$

and that the redox reaction (*1*) follows reaction (*2*). Addition of (*2*) and (*1*) gives:

$$Ar—SeO_2H + NH_2OH.H_2SO_4 \rightarrow Se^0 + Ar—SO_3H + 2 H_2O + HNO_2 \qquad (3)$$

The realization of (3) permits a selective detection of organic seleninic acids but this test, however, is not very sensitive because the reaction temperature must not exceed 140°. Above this temperature, hydroxylamine reacts with the sulfuric acid to which it is bound

$$NH_2OH + 2 H_2SO_4 \rightarrow 3 H_2O + 2 SO_2 + HNO_2$$

with production of nitrous acid, the sulfuric acid functioning as oxidant.*

Procedure. A small amount of the test material or if need be the evaporation residue from a drop of its solution is mixed in a micro test tube with several cg of hydroxylamine sulfate. A disk of filter paper moistened with Griess reagent is laid over the mouth of the tube and the latter is then immersed in a glycerol bath at 140°. A positive response is indicated if a red stain appears on the reagent paper within 1–2 min.

Limits of identification:
      100 γ benzeneseleninic acid   120 γ tolueneseleninic acid

* Hitherto, it was apparently not known that N_2O_3 is always formed when hydroxylamine is oxidized in acidic medium.

1 Unpublished studies, with E. LIBERGOTT.

E. NITROGEN COMPOUNDS

If the preliminary tests described in Chap. 2 have shown the presence of nitrogen, perhaps also of sulfur, the following tests will furnish information as to which of the numerous possible kinds of combination of nitrogen is present.

48. Primary, secondary and methyl-bearing tertiary alkyl-, aryl- and acyl amines

Test by fusion with fluorescein chloride[1]

When these amines are fused with colorless fluorescein chloride (I) and anhydrous zinc chloride, red water-soluble rhodamine dyes result.[2] They have the general structures (II) and (III).

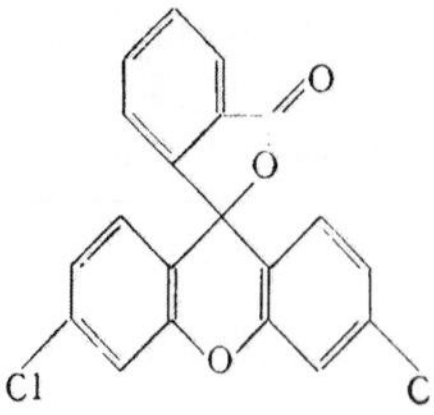

(I) 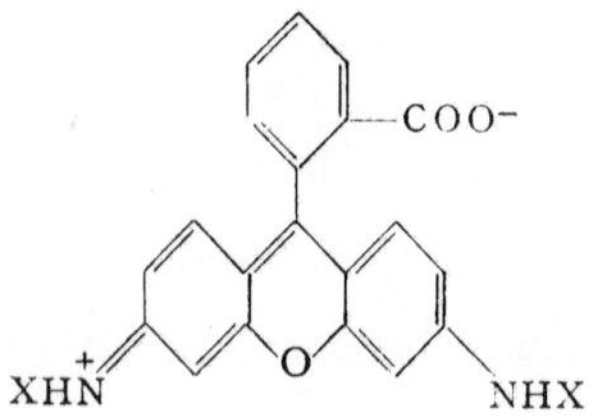   (II)   (III)

(X = R or Ar)

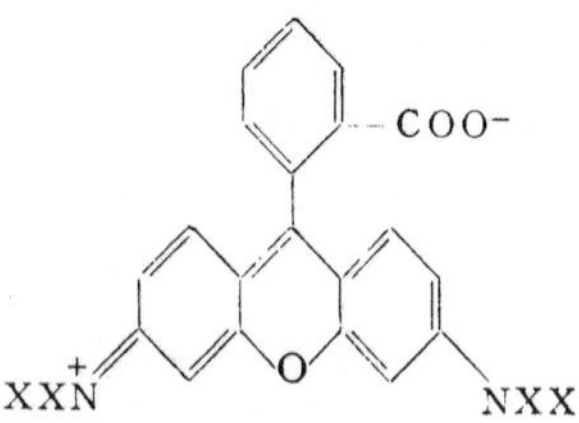

In addition to this *para* quinonoid ammonium salt formulation given, the rhodamine dyes can also be written as *ortho* quinonoid oxonium salts or as carbonium salts with the respective ions IV and V.[3,4,5]

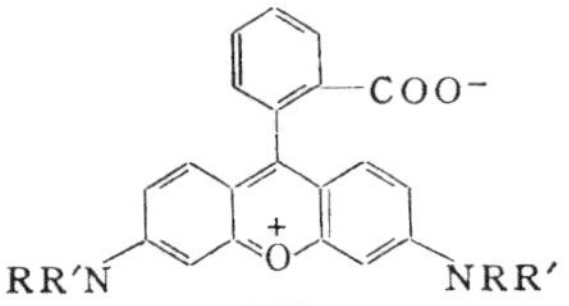

(IV) 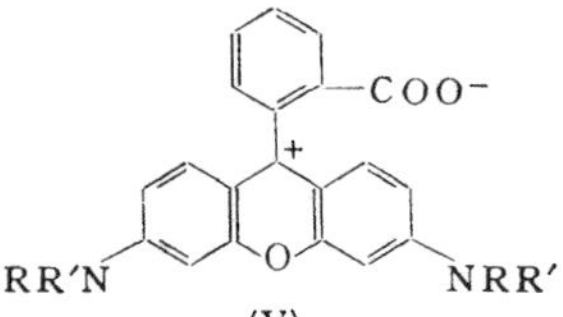   (V)

The rhodamine dyes of the types (II) and (III) (X=R) exhibit yellow-green fluorescence in aqueous solution or an orange fluorescence, respectively, both in daylight and ultraviolet light. In contrast, rhodamine dyes of the types (II) and (III) in which X=Ar are not fluorescent.

As cyclic secondary aliphatic bases, pyrrole bases form rhodamine dyes which fluoresce blue in ultraviolet light.

Accordingly, after the fusion has been made and the product taken up in water, a red color indicates the presence of amines in general. Additional information can be obtained by examining the product in ultraviolet light.

Procedure.   A drop of the test solution, containing hydrochloric acid, is evaporated to dryness in a test tube, and the residue is mixed with a little fluorescein chloride and twice the bulk of anhydrous zinc chloride. The mixture is then heated in an air bath (iron crucible or aluminum block) to 250–260° until all the zinc chloride has melted. When cool, the melt is dissolved in 10% alcoholic hydrochloric acid and examined in both daylight and ultraviolet light.

The following amounts were detected:

Primary aliphatic amines (yellow-green fluorescence)

4 γ methylamine chloride	10 γ ethylamine chloride
10 γ benzylamine	20 γ glycine ethyl ester

Secondary amines (orange fluorescence)

4 γ diethylamine	4 γ piperidine	20 γ aceturic acid ethyl ester

Arylamines (red to violet; no fluorescence)

5 γ aniline	5 γ p-nitroaniline	4 γ hydrazobenzene
5 γ 1-naphthylamine	4 γ p-chloroaniline	8 γ diphenylamine
7 γ 2- ,,	10 γ H-acid	4 γ acetanilide
4 γ aminoanthraquinone	1 γ p-phenylenediamine	2 γ Michler's ketone
4 γ benzidine	4 γ benzenesulfo-p-nitroanilide	
2 γ o-aminobenzaldehyde	4 γ p-dimethylaminobenzaldehyde	
8 γ benzylidene-p-nitroaniline		

Pyrrole bases (blue fluorescence)

12 γ indole	30 γ carbazole
20 γ acetindoxyl	40 γ pyrrole

1 F. FEIGL, V. ANGER and R. ZAPPERT, *Mikrochemie*, 16 (1934) 67, 70.
2 P. FRIEDLAENDER, *Fortschritte der Teerfarbenfabrikation*, Vol. II, 1891, pp. 79, 81; Vol. III, 1896, p. 174.
3 P. KARRER, *Organic Chemistry*, 4th Engl. ed., Elsevier, Amsterdam, 1950, p. 634.
4 L. F. FIESER and M. FIESER, *Organic Chemistry*, 2nd ed., Reinhold., New York, 1950, p. 905.
5 R. KUHN, *Naturwissenschaften*, 20 (1932) 622.

49. Primary, secondary and tertiary amines

(1) Test with 2,4-dinitrochlorobenzene [1]

These amine bases even in small amounts condense with the practically colorless 2,4-dinitrochlorobenzene to give yellow products [2,3] which are in the case of primary and secondary amines N-substituted 2,4-dinitroanilines. The following reaction may be assumed for primary amines:

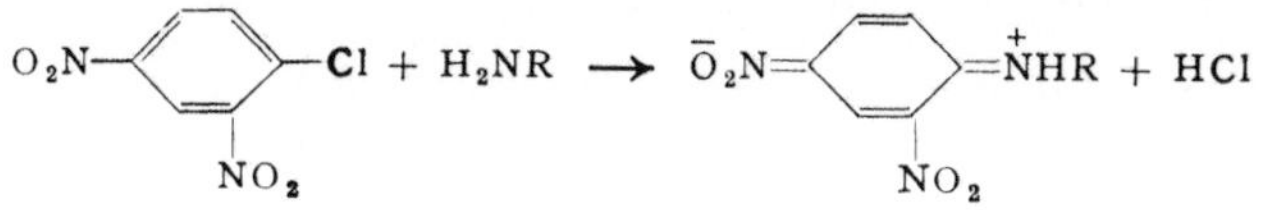

$$O_2N-\langle\ \rangle-Cl + H_2NR \longrightarrow \bar{O}_2N=\langle\ \rangle=\overset{+}{N}HR + HCl$$

$$NO_2 \qquad\qquad NO_2$$

Secondary amines may give analogous reactions. Tertiary amines also form colored products of unknown constitution.

The assumption that quinoidal zwitterion condensation products are formed is supported by the intense yellow color, the solubility in chloroform, and the resistance of the products, once they have been formed, to dilute acid. Hydrochloric acid is formed in the reactions involving primary and secondary amines. Noncyclic tertiary amines may perhaps yield alkyl (aryl) halide instead of hydrochloric acid. Pyridine and its derivatives (β- and γ-hydroxypyridine are exceptions) likewise give yellow products with the reagent but in these cases derivatives of glutaconaldehyde are formed [4].

When using 2,4-dinitrochlorobenzene as reagent, it is rather important that the condensation occurs only with free bases in the absence of acid or alkali. Aliphatic and aromatic amino acids do not react. Accordingly this test permits the detection of amines in the presence of aminocarboxylic- and aminosulfonic acids.

Procedure.   A drop of the ethereal test solution is treated in a depression of a spot plate with a drop of a 1% ethereal solution of 2,4-dinitrochlorobenzene. After the ether has evaporated, a residual yellow (brown) color or ring indicates a positive response.

Limits of identification:

1.6 γ aniline	6 γ ethylenediamine	5 γ N-ethylcarbazole
1.6 γ benzylamine	1.6 γ diphenylamine	1.7 γ 1-naphthylamine
3.3 γ piperidine	1.7 γ diethylaniline	5 γ isoquinoline
0.2 γ tetrabase	0.4 γ p-dimethylaminobenzaldehyde	

The following likewise gave a positive response: 2-naphthylamine, methyl-aniline, o- and p-phenylenediamines, p-chloroaniline, hydrazobenzene, indole.

A negative response was obtained with the following aminocarboxylic and aminosulfonic acids: glycine, β-alanine, asparagine, aspartic acid, anthranilic acid, sulfanilic acid, amino-p-toluenesulfonic acid, 4-aminonaphthalene-1-sulfonic acid, 1-amino-2-naphthol-4-sulfonic acid, 4-aminoazobenzene-4'-sulfonic acid.

When testing salts of volatile bases, a drop of the test solution is treated in a micro test tube with 1 drop of a 20% solution of sodium hydroxide. The mouth of the test tube is covered with a disk of filter paper moistened with a 1% ethereal solution of 2,4-dinitrochlorobenzene. The contents of the tube are then taken to dryness in an oven at 130°. If the response is positive, a yellow circle appears on the paper.

Positive results were observed with 500 γ of the following salts: ethylene-diamine dichloride, ephedrine chloride, piperazine dichloride, aniline chloride, p-aminodimethylaniline sulfate.

The test fails with 1-aminoanthraquinone and 4-amino-1,2-naphthoquinone.

1 Unpublished studies, with J. R. AMARAL.
2 F. J. SMITH and E. JONES, *A Scheme of Qualitative Organic Analysis*, London, 1953, p. 110.
3 Staff of Hopkins and Williams Research Laboratory, *Organic Reagents for Organic Analysis*, Brooklyn, 1947, p. 50.
4 Comp. P. KARRER and K. HELLER, *Helv. Chim. Acta*, 21 (1938) 463; R. KUHN and J. LÖW, *Ber.*, 72 (1939) 1457.

(2) Test with fused potassium thiocyanate[1]

When fused potassium thiocyanate (m.p. 173–179°) is brought in contact with solid ammonium salts, the following topochemical reaction takes place:

$$KCNS + NH_4X \rightarrow KX + NH_4CNS \quad (X = Cl, Br, I, \tfrac{1}{2} SO_4, etc.)$$

On further heating, the ammonium thiocyanate (m.p. 149°) undergoes a series of transformations depending on temperature, rate of heating, *etc.*[2] One transformation occurring in the fusion is the familiar rearrangement of ammonium thiocyanate into the isomeric thiourea which, in the imide form, decomposes on heating into hydrogen sulfide and cyanamide:[3]

$$NH_4CNS \rightleftharpoons S{=}C\!\!\begin{smallmatrix}\nearrow NH_2\\[2pt]\searrow NH_2\end{smallmatrix} \rightleftharpoons HS{-}C\!\!\begin{smallmatrix}\nearrow NH_2\\[2pt]\searrow NH\end{smallmatrix} \rightarrow H_2S + NH_2CN$$

Salts of aliphatic and aromatic amines, which are actually substituted ammonium salts, react in an analogous fashion. With primary and secondary amines the corresponding thiocyanates are formed first and remain in equilibrium with the respective substituted thioureas:

$$RNH_3CNS \rightleftarrows S{=}C{\overset{NH_2}{\underset{NHR}{}}} \quad \text{and} \quad R_1R_2NH_2CNS \rightleftarrows S{=}C{\overset{NH_2}{\underset{NR_1R_2}{}}}$$

Both of these substituted thioureas, probably in the isomeric imide forms, decompose at 200–250° with evolution of hydrogen sulfide to yield alkyl or aryl cyanamides. Salts of tertiary amines also split off hydrogen sulfide when heated with fused potassium thiocyanate, but the mechanism of the reaction is certainly different. It must be noted that cyclic nitrogen bases, in contrast to secondary and tertiary amines, react but faintly.

The facts that the reaction of salts of amines with fused potassium thiocyanate occurs almost instantaneously and that the hydrogen sulfide formed is easily detectable provide the basis of a test for aliphatic and aromatic amines. Free bases must be converted into their chlorides before carrying out the test. This is accomplished by evaporation with an excess of dilute hydrochloric acid*. The potassium thiocyanate must be well dried since the moist salt itself yields hydrogen sulfide when heated.

When the test is used, attention must be given to the fact that organic compounds which split off water when heated interfere (compare Chap. 2).

Procedure.   One drop of a solution of the amine in alcohol, ether or other solvent, and one drop of 1:10 hydrochloric acid are evaporated to dryness in a micro test tube at 110°. The dry residue is mixed with an excess of well dried potassium thiocyanate and heated in a bath to about 200–250°. A piece of filter paper moistened with a drop of 10% lead acetate solution is held over the mouth of the test tube. The presence of amines is revealed by an almost immediate blackening of the paper.

Reagent: Potassium thiocyanate (pulverized and dried at 110°).

The test revealed:

50 γ aniline	50 γ diphenylamine	50 γ histidine chloride
15 γ benzidine	15 γ ethylenediamine	50 γ monomethylaniline
25 γ benzylamine	5 γ dimethylaniline	50 γ nitrosomethylaniline
5 γ tetrabase		

A faint reaction was shown by the following cyclic nitrogen bases in amounts of 2000 γ: 8-hydroxyquinoline, 6-nitroquinoline, 1-naphthoquinoline, pyridine, acridine.

* In this connection it should be noted that when organic bases are taken to dryness with hydrochloric acid and the excess acid then expelled by heating to 110°, the resulting chloride is not always stable but may split into its components. A pertinent example is the chloride of α,α′-dipyridyl which is completely volatilized when brought to a temperature of 110°.

1 F. Feigl and H. E. Feigl, *Mikrochim. Acta*, (1954) 85.
2 W. Gluud, K. Keller and W. Klempt, *Z. Angew. Chem.*, 39 (1926) 1071.
3 H. Krall, *J. Chem. Soc.*, 103 (1913) 1383.

50. Primary and secondary amines

Test with p-dimethylaminobenzaldehyde[1]

These amines and their salts condense with p-dimethylaminobenzaldehyde to yield colored Schiff bases:

$$RNH_2 + OCH\!\!-\!\!\langle \bigcirc \rangle\!\!-\!\!N(CH_3)_2 \longrightarrow RN{=}CH\!\!-\!\!\langle \bigcirc \rangle\!\!-\!\!N(CH_3)_2 + H_2O$$

With monoamines the Schiff base is yellow, whereas di- and polyamines yield orange products in some cases. Probably more than one NH_2 group condenses in these instances.

The fact that the color of the Schiff bases becomes more intensive by the action of diluted acids is due to the protonation of the $—N{=}CH—$ to the $—\overset{+}{N}H{=}CH—$ group.

Procedure.[2] A drop of a saturated benzene solution of p-dimethylaminobenzaldehyde is placed on filter paper, which is then spotted with a drop of an ethereal solution of the amine or of an aqueous solution of its salt buffered with sodium acetate. The filter paper is kept in an oven at 100° for about 3 min. or held over vapors of acetic acid. A colored stain indicates a positive response.

The condensation with p-dimethylaminobenzaldehyde can also be accomplished with the molten aldehyde (m.p. 73°). Salts of the respective amines also react in this way, for example the infusible sulfate, chloride and acetate of benzidine.[3]

p-Dimethylaminocinnamaldehyde (solution in methanol plus trichloroacetic acid) can be advantageously employed in place of p-dimethylaminobenzaldehyde. The resulting Schiff bases of primary amines are blue, those of secondary amines are purple.[4]

Limits of identification:

0.2 γ aniline	2.2 γ aniline sulfate	0.2 γ p-phenylenediamine
1　γ benzylamine	0.6 γ diphenylamine	0.5 γ o-　　　,,
0.02 γ benzidine	1.7 γ 2-aminopyridine	1.7 γ 1-aminoanthraquinone
1　γ 1-naphthylamine	1　γ ethylenediamine dichloride	

4-Amino-1,2-naphthoquinone gave a positive response to the test.

1 S. N. Chakravarti and M. B. Roy, *Analyst*, 62 (1937) 603.
2 J. R. Amaral (Rio de Janeiro), unpublished studies.
3 Unpublished studies, with J. R. Amaral.
4 S. Sakai and coworkers, *Japan Analyst*, 9 (1960) 862.

51. Primary and secondary alkylamines

Test with carbon disulfide[1]

The dithiocarbamates of the corresponding bases are formed almost instantly at room temperature by the action of carbon disulfide on solid or dissolved primary and secondary aliphatic amines:[2]

$$\text{CS}_2 + 2\ \text{NH}_2\text{R} \longrightarrow \text{SC}\begin{smallmatrix}\diagup\text{SH.NH}_2\text{R}\\[2pt]\diagdown\text{NHR}\end{smallmatrix} \quad \text{and} \quad \text{CS}_2 + 2\ \text{NHR}_1\text{R}_2 \longrightarrow \text{SC}\begin{smallmatrix}\diagup\text{SH.NHR}_1\text{R}_2\\[2pt]\diagdown\text{NR}_1\text{R}_2\end{smallmatrix}$$

The dithiocarbamates formed by these addition reactions may be detected after removing the excess carbon disulfide (b.p. 46°). The iodine–azide test (see p. 219) may be applied because the dithiocarbamate contains the SH and CS groups necessary for the catalysis of the NaN_3–I_2 reaction. Alternatively, the dithiocarbamate may be revealed by the formation of silver sulfide when silver nitrate is added to the evaporation residue.

Tertiary aliphatic bases do not react. Aromatic amines, which likewise may react with carbon disulfide with production of thiourea derivatives[3]

$$2\ \text{C}_6\text{H}_5\text{NH}_2 + \text{CS}_2 \rightarrow \text{C}_6\text{H}_5\text{NH}{-}\text{CS}{-}\text{NHC}_6\text{H}_5 + \text{H}_2\text{S},$$

do not react under the mild conditions employed here. Accordingly, primary and secondary aliphatic amines may be detected in the presence of aromatic amines.

When salts of the bases are present, use may be made of the fact that triethylamine, which as a tertiary amine does not react with carbon disulfide, is so strong a base that it is able to liberate primary and secondary amines from solutions of their salts. It is merely necessary to add carbon disulfide and an excess of triethylamine, or a prepared mixture of these reagents, to the sample and proceed as prescribed. In the presence of $\text{N}(\text{C}_2\text{H}_5)_3$, dithiocarbamates are formed, *e.g.* $\text{RHNCSSH.N}(\text{C}_2\text{H}_5)_3$, or $\text{R}_1\text{R}_2\text{NCSSH-}$ $\text{N}(\text{C}_2\text{H}_5)_3$ in the case of a primary or secondary aliphatic amine.

The test with sodium azide–iodine cannot be used in the presence of mercaptans, thioketones and thiol compounds, which catalyze the sodium azide–iodine reaction.

Procedure. A drop of the alcoholic test solution, or of the free base, is mixed with a few drops of a 1:1 alcohol–carbon disulfide mixture in a micro crucible. After about 5 min., the excess carbon disulfide is volatilized. A few drops of an iodine–azide solution (3 g NaN_3 in 100 ml 0.1 N iodine) or of a 1% solution of silver nitrate in dilute nitric acid, is added. The mixture is observed for evolution of nitrogen or for a blackening due to silver sulfide.

The detection of dithiocarbamates may also be conducted on filter paper. The base and carbon disulfide (if necessary along with triethylamine) are mixed on filter paper and then spotted with acidified silver nitrate solution. About the same limits of identification are obtained.

Limits of identification are given in Table 11.

TABLE 11

PRIMARY AND SECONDARY ALIPHATIC AMINES

Name	Limits of identification	
------	with iodine–azide	with silver nitrate
Ethylamine	6.5 γ	3 γ
Propylamine	13 γ	10 γ
Isobutylamine	10 γ	8 γ
Dipropylamine	140 γ	125 γ
Di-isoamylamine	100 γ	115 γ
Heptylamine	12 γ	7 γ
Diethanolamine	1 γ	0.6 γ
Benzylamine	15 γ	8.5 γ
Piperidine	4.5 γ	–
Phenylhydrazine	250 γ	200 γ
Glycine ester hydrochloride	100 γ	70 γ
Leucylglycine	35 γ	35 γ
Leucylglycylglycine	30 γ	35 γ
Arginylglycine	25 γ	30 γ

1 Unpublished studies, with G. FRANK.
2 A. W. HOFMANN, *Ber.*, 1 (1868) 25, 169.
3 A. W. HOFMANN, *Ann.*, 57 (1846) 266; 70 (1849) 144.

52. Primary and secondary arylamines and secondary vinylamines

Test with glutaconic aldehyde[1]

Acid solutions of primary aromatic amines give colored products with pyridine irradiated with ultraviolet light.[2] The reaction depends on the hydrolysis of pyridine in ultraviolet light to give the ammonium salt of glutaconic aldehyde enol, as shown in (*1*). This forms condensation products of the Schiff base type[3] with primary aromatic amines (*2*):

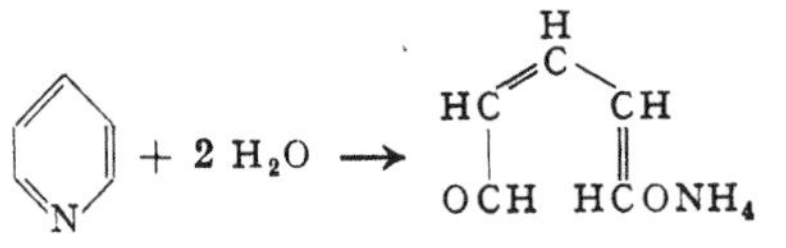

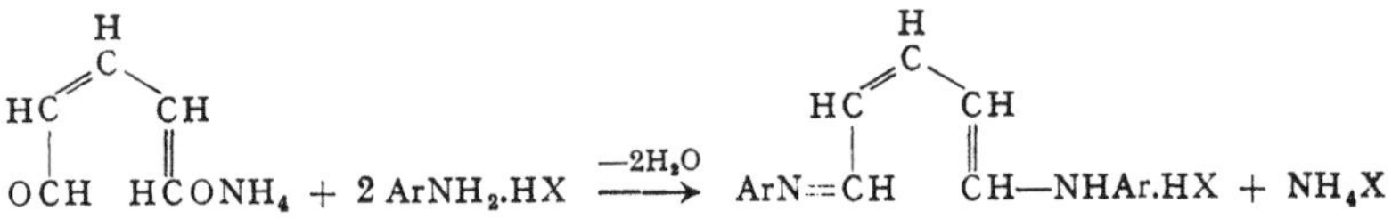

Since the condensation (2) proceeds even with very small amounts of primary amines, it may constitute the basis of a test for the latter. The colored products belong to the class of polymethine dyes.

Free glutaconic aldehyde is very unstable; it polymerizes readily, and even its yellow, water-soluble ammonium salt (formed by ultraviolet light on paper impregnated with pyridine) decomposes in a few days. It is therefore best to conduct the test with the aid[4] of 4-pyridylpyridinium dichloride (I) which reacts with alkali to give the alkali enolate of glutaconic aldehyde (II) and 4-aminopyridine (III):

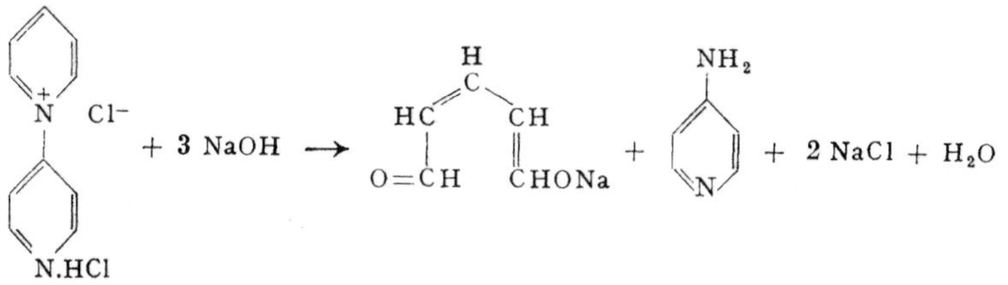

These polymethine dyes formed by (2) may also be formulated symmetrically:

$$[Ar\!-\!NH\!-\!CH\!=\!CH\!-\!\overset{+}{C}H\!-\!CH\!=\!CH\!-\!NH\!-\!Ar]X^-$$

This concept explains why secondary aromatic amines and likewise secondary vinylamines (primary vinylamines are not capable of existence) react positively.

Procedure. A drop of a 1% pyridylpyridinium dichloride aqueous solution is placed in a depression of a spot plate and followed by a drop of the test solution, a drop of 1 N sodium hydroxide, and a drop of concentrated hydrochloric acid. A positive response is indicated by an orange, red or violet color.

Limits of identification:

0.05 γ	aniline	red-brown	1	γ m-nitroaniline	bluish red
0.1 γ	benzidine	,,	1	γ o- ,,	rose-red
1 γ	1,8-naphthylenediamine	,,	0.3	γ p- ,,	,,
0.5 γ	anthranilic acid	orange	1	γ m-bromoaniline	red
0.1 γ	sulfanilic acid	,,	1	γ p-aminodiphenylamine	,,
0.5 γ	sulfanilamide	,,	0.05	γ 2,4-diaminopyridine	,,
2 γ	K-acid	,,	0.1	γ 2-naphthylamine	,,
0.05 γ	diphenylamine	,,	0.01	γ 1- ,,	scarlet
0.05 γ	diphenylbenzidine	,,	0 1	γ p-phenylenediamine	violet
1 γ	4-oxo-1,4-dihydropyridine	,,			

No reaction was shown by: 2-aminopyridine, 4-aminopyridine, dimethylaniline, aminotetrazole, acetanilide.

1 F. FEIGL, V. ANGER and R. ZAPPERT, *Mikrochemie*, 16 (1934) 74; comp. V. ANGER and S. OFRI, *Mikrochim. Acta*, (1964) 770.
2 H. FREYTAG and W. NEUDERT, *J. Prakt. Chem.*, [2], 135 (1932) 15.
3 F. FEIGL and V. ANGER, *J. Prakt. Chem.*, [2], 139 (1934) 180.
4 E. KOENIGS and H. GREINER, *Ber.*, 64 (1931) 1049.

53. Salts of primary amines

Test through pyrolysis with hexamine[1]

If a dry mixture of such salts with excess hexamine (a monoacid base) is heated to 160°, the corresponding amine is set free and reacts with hexamine whereby ammonia is lost:

$$(R,Ar)NH_2.HX + (CH_2)_6N_4 \rightarrow (CH_2)_6N_4.HX + (R,Ar)NH_2 \qquad (1)$$

$$12\,(R,Ar)NH_2 + (CH_2)_6N_4 \rightarrow 6 \begin{array}{c}(R,Ar)NH \\ \diagdown \\ CH_2 + 4\,NH_3 \\ \diagup \\ (R,Ar)NH\end{array} \qquad (2)$$

The realization of reactions (1) and (2) makes possible, through the detection of the resulting ammonia (indicator change or Nessler reaction), a proof of the presence of salts of primary alkyl- and arylamines, assuming the absence of ammonium salts which likewise split off ammonia when dry-heated with hexamine. Alkaline Nessler solution is a convenient means for the direct detection of ammonium salts. If the latter are present they must be removed prior to the pyrolysis reaction. This elimination is readily accomplished by adding alkali hypobromite (bromine water decolorized with alkali hydroxide), acidifying with dilute hydrochloric acid, and evaporating to dryness.

Procedure.   The mixture of a small amount of the dry salt and several cg of hexamine is placed in a micro test tube whose mouth is covered with a disk of acid–base indicator paper or a piece of filter paper moistened with Nessler solution. The test tube is immersed to a depth of about 1 cm in a glycerol bath preheated to 120° and the temperature is then raised to 140°. A positive response is indicated by a color change in the indicator paper or by the development of a brown fleck on the filter paper.

The test revealed 10 γ amounts of:

aniline chloride (sulfate)   ethylenediamine dichloride   2-aminopyridine chloride
ethylamine chloride   arginine chloride

Water-insoluble salts of primary amines also react with hexamine; for example a positive response was obtained when benzidine sulfate was pyrolyzed with hexamine.

1 Unpublished studies, with E. LIBERGOTT.

54. Aromatic primary amines

(1) Test with sodium pentacyanoaquoferriate [1]

A green or blue color is obtained when primary aromatic amines are mixed with the light yellow sodium carbonate solution of sodium pentacyanoaquoferriate:

$$Na_2[Fe(CN)_5H_2O] + ArNH_2 \rightarrow Na_2[Fe(CN)_5ArNH_2] + H_2O$$

A specific test for primary arylamines can be based on this exchange reaction.

Procedure. A drop of the test solution is mixed with a drop of the reagent solution on a spot plate or in a micro crucible. If necessary, a little dilute sodium carbonate solution is added. A more or less intense green or blue color appears at once or after a few minutes.

Reagent: A 1% solution of sodium pentacyanoammineferriate is treated first with bromine water until the color is violet, and then with sodium pentacyanoammineferriate until the color is yellow. 2% Na_2CO_3 solution is added.

Limits of identification:

0.5 γ aniline	green	0.1 γ p-phenylenediamine		blue	
0.2 γ o-toluidine	,,	1 γ 2,5-diaminoanisole		,,	
1 γ o-aminobiphenyl	,,	1 γ m-toluylenediamine		brown-green	
2 γ naphthylamine (α,β)	,,	10 γ 1,8-naphthylenediamine		,,	
1 γ o-phenylenediamine	,,	1 γ o-aminophenol		,,	
2 γ m- ,,	,,	1 γ benzidine		blue-green	
5 γ m-bromoaniline	,,	1 γ 8-aminoquinoline		,,	
50 γ m-nitroaniline	,,	10 γ m-aminophenol		grey	
10 γ p- ,,	,,	0.1 γ p- ,,		greenish blue	
0.5 γ p-anisidine	,,	100 γ aminotetrazole		violet-pink	
1 γ 2,4-diaminophenol	,,	5 γ 4-aminophenazone		pink	
10 γ anthranilic acid	,,				
10 γ 2-aminopyridine	,,				

Primary arylamines with strong acidic groups give no or very faint reactions as exemplified by the behavior of o-nitroaniline, 2,4-dinitroaniline, p-aminobenzoic acid, aminoterephthalic acid, sulfanilic acid, arsanilic acid.

1 V. ANGER, *Mikrochim. Acta*, 2 (1937) 3.

(2) Other tests for primary aryl amines

The fact that the hydrogen atoms of the NH_2 group are readily replaceable forms the basis of the following tests.[1-3] They may be carried out as spot reactions (in micro crucibles) with heating:

(a) Condensation with aromatic nitroso compounds to form azo dyes.

Reagent: Saturated solution of 5-nitroso-8-hydroxyquinoline or 10% nitrosodimethylaniline in glacial acetic acid (*Idn. Limit:* 1-100 γ).

(b) Condensation with furfural to give violet Schiff bases.

Reagent: Solution of 10 drops of furfural in 10 ml glacial acetic acid (*Idn. Limit:* 0.1–200 γ).

Secondary aromatic amines, also aliphatic amines and amino acids, react in the same manner.

(c) Condensation with chloranil to form blue, red, or brown products.

Reagent: Saturated solution of chloranil in dioxan (*Idn. Limit:* 0.2–200 γ). Secondary amines react similarly; amino acids also react. Phenols give red to violet colorations.

(d) Primary amines react with carbon disulfide to form symmetrical dialkylated thioureas and liberate hydrogen sulfide. The latter can be detected with lead acetate paper (*Idn. Limit:* 1 to 5 γ amine).

(e) The following prodecure has been recommended. One drop of the ethereal or acetone solution of the sample is placed on filter paper. After evaporation of the solvent the paper is exposed, successively, for 5 min. each, to the vapors of concentrated hydrochloric acid and ethyl nitrite (from C_2H_5OH, HCl and solid $NaNO_2$). Then the paper is spotted with an ethereal solution of resorcinol, held over ammonia, and finally steamed. An orange, red, or violet color indicates the presence of diazotizable NH_2 groups.

Test (a) is recommended especially.

In the absence of other compounds with reactive NH_2- or CH_2 groups, an excellent spot test for primary arylamines can be based on their condensation with sodium 1,2-naphthoquinone-4-sulfonate (comp. p. 153).

1 O. FREHDEN and L. GOLDSCHMIDT, *Mikrochim. Acta,* 1 (1937) 338.
2 O. FREHDEN and K. FUERST, *Mikrochim. Acta,* 3 (1938) 197.
3 S. J. BURNISTROV, *Chem. Abstr.,* 34 (1940) 2287.

55. Secondary amines

Test through conversion into thiazine dyes [1]

A synthesis of methylene blue is based on the interaction of thiodiphenyl-amine, bromine and dimethylamine [2] in the presence of bromine:

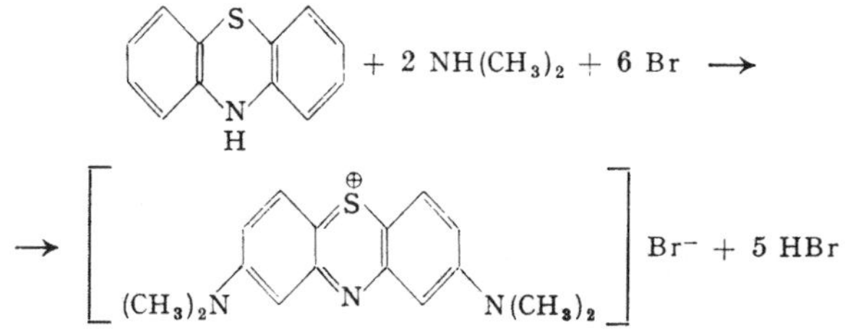

Other dialkylamines as well as secondary arylamines and aralkylamines react analogously, leading to the respective thiazine dyes. This behavior permits a selective test for secondary amines.

Procedure. A drop of a 0.1% methanolic solution of thiodiphenylamine is shaken in a test tube with a drop of a 2% methanolic solution of bromine and then a drop of the test solution (as near neutral as possible, pH 8) is added. A blue color signals the presence of secondary amines. The color is rendered more visible by extraction with a drop of ether.

Identification limits:

0.5 γ dimethylamine	1	γ monobenzylaniline
1 γ morpholine	0.5	γ monomethylaniline
10 γ piperidine	1	γ diethylamine hydrochloride
5 γ pyrrolidine	10	γ cetylphenylamine
10 γ diphenylamine	0.5	γ piperazine hexahydrate
10 γ diethanolamine	0.5	γ N-methylpiperazine
10 γ di-*n*-butylaniline	1	γ hexamethyleneimine
2 γ monoethylaniline		

1 H. Bröll and G. Fischer, *Mikrochim. Acta*, (1962) 250.
2 F. Kehrmann, *Ber.*, 49 (1916) 3832.

56. Aliphatic secondary amines

(1) Test through formation of copper dithiocarbamates [1]

Secondary aliphatic or cyclic amines react with carbon disulfide and ammoniacal copper sulfate solution to produce brown water-insoluble copper salts of the respective dithiocarbamic acids:

$$NHRR_1 + CS_2 + {}^1/_2\ Cu^{2+} + NH_3 \longrightarrow SC{\overset{S-Cu/_2}{\underset{NRR_1}{}}} + NH_4^+$$

These copper salts give brown solutions in benzene, chloroform, *etc.* and consequently the occurrence of the above reaction can be discerned even with solutions of the secondary bases that are too dilute to yield a precipitate of the copper salt. The test described here is based on this finding.

Primary and tertiary aliphatic amines, aromatic amines, and also secondary aromatic–aliphatic amines do not react under the specified conditions. Basic $>$NH compounds, which also contain hydrophilic groups, do not form benzene-soluble copper dithiocarbamates. Pertinent examples are: proline, diethanolamine, adrenaline, spermine, spermidine, piperazine. This divergent behavior may be of value for differentiating basic $>$NH groups.

Procedure. A micro test tube is used. One drop of the acid test solution plus a drop of 5% copper sulfate solution is made basic with ammonia. The blue solution, which may be turbid because of the precipitation of insoluble bases, is shaken with two drops of a mixture of 1 volume of carbon disulfide and 3 volumes of benzene. If the reaction is positive, the benzene layer turns brown or yellow.

The test revealed:

2 γ ephedrine   5 γ pervitin   0.2 γ piperidine   10 γ coniine (α-n-propylpiperidine)

A positive response is given by small amounts of morpholine, emetine, and after acid or alkaline hydrolysis also by piperine (see p. 642).

1 Unpublished studies, with L. HAINBERGER.

(2) Test with sodium nitroprusside and acetaldehyde[1]

Secondary aliphatic amines containing the $-CH_2NHCH_2-$ group form blue-violet compounds when they react with sodium nitroprusside and certain aldehydes (acetaldehyde is most suitable) in solutions made alkaline with sodium carbonate. Primary and tertiary amines do not interfere but the activity of the $>$NH group appears to be affected by other groups in the molecule as shown by the different limits of identification obtained with the test.

Procedure. A drop of the test solution is mixed with a drop of the reagent solution in a micro crucible or on a spot plate. The mixture is made basic with a 2% solution of sodium carbonate. A blue to violet color indicates a positive response.

Reagent: 1% solution of sodium nitroprusside to which 10% by volume of acetaldehyde is added. The mixture must be freshly prepared.

The test revealed:

4 γ diethylamine	5 γ piperidine	0.5 γ pyrrolidine
1 γ L-proline	80 γ spermine	2 γ di-*iso*-amylamine
10 γ adrenalone	70 γ spermidine	100 γ diethanolamine
10 γ adrenaline	10 γ 1-methylamino-3-phenylpropane	

Diisopropylamine does not react.

1 F. FEIGL and V. ANGER, *Mikrochim. Acta*, 1 (1937) 138.

57. Tertiary amines

Test with citric acid and acetic anhydride[1]

When distinctly basic tertiary amines are heated with a solution of citric acid (or *cis*-aconitic anhydride) in acetic anhydride, a red to violet color develops.[2] The color reaction appears to be selective for tertiary amines.

Salts of tertiary amines as well as quaternary ammonium salts—with the exception of sulfates and phosphates—behave similarly. Salts of potassium, rubidium, cesium, strontium and barium behave like quaternary ammonium salts. Salts of alkali and alkaline-earth metals with organic acids react in the same manner.[3] The chemistry of the color test seems to be complicated and has not been elucidated.

Procedure. A micro test tube (alkali-free glass) is used. A little of the tertiary amine or a drop of its alcoholic solution is united with a drop of a solution containing 2 g of citric acid dissolved in 100 ml of acetic anhydride. The mixture is warmed (80°) in a water bath. A red to purple color appears (5–10 min.) when the response is positive. A blank test is advisable.

The test may also be applied to the evaporation residue of the test solution, taken to dryness with dilute hydrochloric acid.

The procedure revealed:

7 γ N-ethylpiperidine (red)   2 γ pyridine (dark violet)
4 γ p-dimethylaminobenzaldehyde (red)

A positive response was given by the following tertiary amines or their salts: trimethylamine, triethanolamine, methylephedrine, tribenzylamine, hordenine, procaine, narceine, creatine, atropine, strychnine, brucine, codeine, scopolamine, caffeine, histidine, antipyrine, quinine, cinchonine, pilocarpine, berberine, veratrine, dimethylaniline, α-picoline, nicotinic acid, 2,6-lutidine, pyridine-2,6-dicarboxylic acid, nicotinic and isonicotinic acid hydrazide, 2-aminopyridine, 4-nitro-α-picoline N-oxide, quinoline, 3- and 8-hydroxyquinoline, 2-amino-4,6-dihydroxypyrimidine, 2-aminothiazole, vitamin B_1, acrinol.

α,α'-Dipyridyl and α,α'-phenanthroline do not react.[3]

1 S. Ohkuma, *J. Pharm. Soc. Japan*, 75 (1955) 1124.
2 P. Kalnin, *Helv. Chim. Acta*, 11 (1928) 977.
3 V. Anger, unpublished studies.

58. Ethylamines ($>$NC$_2$H$_5$ group)

Test through oxidative cleavage with molten benzoyl peroxide[1]

When heated to 120–130° with benzoyl peroxide, N-mono and N-diethyl compounds form acetaldehyde, for instance:

$$>NC_2H_5 + (C_6H_5CO)_2O_2 \rightarrow \, >NH + (C_6H_5CO)_2O + CH_3CHO$$

Consequently this oxidative splitting is analogous to that of O- and N-methyl compounds described on p. 165.

The gaseous acetaldehyde can be detected by the blue color it gives with a solution of sodium nitroprusside containing morpholine (see p. 438). This thus constitutes an indirect test for N-ethyl compounds.

With mild heating some ethoxy compounds are distinctly resistant to fused benzoyl peroxide. The cleavage of acetaldehyde is detectable only at higher temperature but then there is danger of explosive reaction.

Propenyl compounds likewise yield acetaldehyde when melted with benzoyl peroxide (see p. 167).

To detect N-ethyl compounds with certainty it is best to start with an aqueous mineral acid solution of the sample and to extract this with ether after it has been made alkaline. The bases are thus obtained in ethereal solutions which can then be tested.

Procedure. A little of the solid or the evaporation residue of a drop of a solution is treated in a micro test tube with 2 drops of a 10% solution of benzoyl peroxide in benzene and the mixture taken to dryness. The open end of the test tube is covered with a piece of filter paper moistened with sodium nitroprusside–morpholine reagent. The tube is placed in a glycerol bath at 120–130°. Depending on the amount of N-ethyl compounds present, a more or less intense blue stain appears.

Reagents: 1) 20% aqueous solution of morpholine.

2) 5% aqueous solution of sodium nitroprusside.

Equal volumes of (1) and (2) are united just before use.

A positive response was given by the following compounds:

Mono- and di-ethylaniline, N-ethyl-4-aminocarbazole, brilliant green, ethyl orange, coelestine blue, rhodamine B, procaine chloride, procaine penicillin G, coramine (N,N-diethylnicotinamide), quinacrine chloride [3-chloro-7-methoxy-9-(1-methyl-4-diethylaminobutylamino)-acridine dihydrochloride dihydrate], chloroquine phosphate [7-chloro-4-(4-diethylamino-1-methylbutylamino)-quinoline diphosphate].

An indication of the sensitivity of the test is given by the detection of 25 γ brilliant green, 50 γ rhodamine B, 15 γ ethyl orange.

The test is especially useful for the rapid chemical differentiation of N-methyl and N-ethyl compounds in the examination of dyestuffs (see Chap. 7).

1 FEIGL and E. SILVA, *Analyst*, 82 (1957) 582.

59. Dialkylarylamines (Ar—NRR group)

Test through pyrolysis with hexamine [1]

If compounds in this category are mixed with hexamine and heated to 180°, vapors are evolved which give the color reaction of formaldehyde with chromotropic acid (comp. p. 434). Since a direct production of this aldehyde from the oxygen-free components is impossible, it may be assumed that the pyrolysis yields volatile reaction products that are saponified to formaldehyde. The following reactions are proposed for the N-dimethyl and N-diethyl compounds:

$$2 \text{ X---N(CH}_3)_2 + (\text{CH}_2)_6\text{N}_4 \rightarrow 2 \text{ X---NCH}_2 + 4 \text{ CH}_2\text{=NCH}_3 \tag{1}$$

$$2 \text{ X---N(C}_2\text{H}_5)_2 + (\text{CH}_2)_6\text{N}_4 \rightarrow 2 \text{ X---NCH}_2 + 4 \text{ CH}_2\text{=NC}_2\text{H}_5 \tag{2}$$

$$\text{CH}_2\text{=NCH}_3 + \text{H}_2\text{O} \rightarrow \text{CH}_2\text{O} + \text{CH}_3\text{NH}_2 \tag{3}$$

$$\text{CH}_2\text{=NC}_2\text{H}_5 + \text{H}_2\text{O} \rightarrow \text{CH}_2\text{O} + \text{C}_2\text{H}_5\text{NH}_2 \tag{4}$$

In line with this assumption, the pyrolytic primary reactions yield two different Schiff bases of formaldehyde from which the volatile alkylform-imines yield formaldehyde when hydrolyzed as represented in (3) and (4). It is advisable to use Tollens reagent for the detection of this aldehyde.

Previous experience indicates that rapid accomplishment of the pyrolysis reaction requires that the N-dialkyl compounds involved melt at the reaction temperature or be brought near to their respective melting points.

Procedure. A small amount of the test material (or if need be the evaporation residue from a drop of its solution) is placed in the bulb of the gas evolution apparatus shown in Fig. 23, *Inorganic Spot Tests*, and a drop of a saturated solution of hexamine in chloroform is added. The solvent is driven off. A drop of Tollens reagent (preparation see p. 132) is placed on the glass knob, the apparatus is closed, and then immersed for about 1–3 min. in a glycerol bath that has been heated to 180°. A positive response is indicated by the development of a silver mirror that persists when the glass knob is plunged into a small dish filled with ammonia water. This latter treatment is essential in order to remove any black silver oxide deposited by the Tollens reagent.

The test revealed:

100 γ *p*-dimethylaminobenzaldehyde   100 γ *p,p'*-dimethylaminobenzalazine
100 γ dimethylaminobenzylidene   50 γ *p*-dimethylaminophenylarsonic
    rhodanine   acid

1 F. FEIGL, E. LIBERGOTT and L. BEN-DOR, unpublished studies.

60. Amides of carboxylic and sulfonic acids

(1) Detection through pyrolytic formation of oxamide[1]

If compounds of these classes are heated to 190–200° with hydrated oxalic acid, oxamide results along with the pertinent (perhaps partially decomposed) carboxylic- or sulfonic acids. The formation of the heat-resistant oxamide (m.p. 419°) may be accounted for as follows. Oxalic acid dihydrate loses its water of crystallization between 105° and 160° and the anhydrous acid sublimes on further heating. The liberated water (in the form of superheated steam) accomplishes the pyrohydrolyses:

$$\text{XSO}_2\text{NH}_2 + \text{H}_2\text{O} \rightarrow \text{XSO}_3\text{H} + \text{NH}_3 \qquad \text{X = aliphatic or} \tag{1}$$

$$\text{XCONH}_2 + \text{H}_2\text{O} \rightarrow \text{XCOOH} + \text{NH}_3 \qquad \text{aromatic radical} \tag{2}$$

$$XSO_2NHR + H_2O \rightarrow XSO_3H + NH_2R \qquad X = \text{aliphatic or} \qquad (3)$$

$$XCONHR + H_2O \rightarrow XCOOH + NH_2R \qquad \text{aromatic radical} \qquad (4)$$

The ammonia produced in (1) and (2) yields ammonium oxalate with oxalic acid and this product loses water to give oxamide. The primary aliphatic amines formed in (3) and (4) condense:

$$2\,NH_2R \rightarrow RNHR + NH_3$$

and thus produce the ammonia essential to the formation of oxamide.

Oxamide can be specifically detected through the red alcohol-soluble product which it yields when heated with thiobarbituric acid (see p. 536). The realization of the above pyrohydrolyses and the subsequent formation of oxamide accordingly permit the detection of the amides of aromatic carb-oxylic- or sulfonic acids. Obviously, compounds which yield ammonia directly when dry-heated must be absent, since they too would yield oxamide via NH_4 oxalate with oxalic acid. The presence of such interfering compounds can be established as described on p. 258 by heating with dimethyl oxalate and thiobarbituric acid to 125° to accomplish the oxamide reaction. If the response to this test is negative, or if it reveals the presence of no more than slight amounts of oxamide-formers, the test described here is decisive provided the absence of open and closed ureides is established. These compounds are pyrohydrolysed by oxalic acid with the formation of urea, which acts as NH_3 donor towards dimethyl oxalate.

Procedure. A small amount of the solid or a drop of its solution in a micro test tube is mixed with several mg of oxalic acid dihydrate and the tube is placed in a glycerol bath that has been preheated to 110°. The temperature is increased to 200° to remove the excess oxalic acid. (Compare with a like amount of oxalic acid dihydrate; no residue must remain in the cooled tube). After the reaction mixture has cooled, several mg of thiobarbituric acid are added and the contents of the tube rubbed with a glass rod. The mixture is then heated to 120–140° in the glycerol bath. If the response is positive, a red product appears.

The rest revealed:

5 γ benzamide	5 γ salicylamide	30 γ sulfapyridine
5 γ nicotinamide	20 γ sulfanilamide	10 γ sulfanilurea

1 Unpublished studies, with D. HAGENAUER-CASTRO.

(2) Differentiation of amides of carboxylic and sulfonic acids [1]

Amides of carboxylic and sulfonic acids have acidic character that reveals itself by their solubility in alkalis and also by their splitting off hydrogen cyanide when they are pyrolyzed with mercuric cyanide (comp. p. 546). The reason for the acidic character lies in the fact that in solutions and also in the solid state these materials are present in tautomeric equilibrium:

$$X—CONH_2 \rightleftharpoons X—C{\overset{OH}{\underset{NH}{}}} \qquad X—SO_2NH_2 \rightleftharpoons X—SO{\overset{OH}{\underset{NH}{}}}$$

In amides of sulfonic acids, the baso–aci equilibrium lies so far toward the side of the (OH-containing) aci-form that the amount of the (NH$_2$-containing) baso-form is not sufficient to permit the production of ammonium salt by hydrolysis with caustic alkali or of course still further to ammonia. The latter occurs with the less acidic amides of carboxylic acids:

$$X—CONH_2 + KOH \rightarrow X—COOK + NH_3$$

This difference in behavior can be used to distinguish the carboxylic from the sulfonic amides since the production of ammonia is readily detected. It is not yet clear whether this procedure can be reliably applied to all carboxylic amides because the nature of the radical X influences the position of the tautomer equilibrium.

The test described here involves saponification at 130–150° with a glycerol solution of caustic potash; the absence or prior removal of ammonium salts is essential. With regard to this elimination it is pertinent to point out that many amides of carboxylic acids are soluble in chloroform.

Procedure.   A few tiny granules of the solid are treated in a micro test tube with 2 drops of a 5% solution of KOH in glycerol. The mouth of the tube is covered with a disk of filter paper moistened with Nessler reagent and the tube is placed in a glycerol bath at 150°. A positive response is indicated by the appearance of a brown or yellow stain on the reagent paper.

The following were detected:

25 γ lauric acid amide	10 γ benzamide	25 γ salicylamide
15 γ azelaic acid diamide	10 γ nicotinamide	

1 Unpublished studies, with E. LIBERGOTT.

61. Amides and imides of carboxylic acids (—CO—NH— group)

(1) Test through acid saponification[1]

Most amides and imides of carboxylic acids are saponifiable by mineral acids:

$$RCONH_2 + H_2O + H^+ \rightarrow RCOOH + NH_4^+ \tag{1}$$

$$RCONHCOR + 2\,H_2O + H^+ \rightarrow 2\,RCOOH + NH_4^+ \tag{2}$$

$$RCONHR_1 + H_2O + H^+ \rightarrow RCOOH + NH_3^+R_1 \tag{3}$$

When even small amounts of the test material are taken to dryness only once with concentrated hydrochloric acid these hydrolyses proceed so far that ammonium chloride and nonvolatile carboxylic acids can be detected in the evaporation residue. This constitutes the basis of selective tests, and even of a differentiation of the organic starting materials.

Nessler reagent is used to detect the ammonium chloride in the residue from the hydrochloric acid evaporation. This procedure is reliable in the case of water-insoluble amides and imides, since if need be any ammonium salts are easily removed by digestion with water. If the presence of water-soluble amides (urea, guanidine and its derivatives with free NH_2 groups, saccharin, *etc.*) has to be considered, the absence of ammonium salts is essential. Accordingly, if the Nessler test is positive when applied to the original sample, it is advisable to test the residue from the hydrochloric acid evaporation for carboxylic acids. This is readily done by means of potassium iodide and iodate (see p. 112). The reaction with uranyl acetate and rhodamine B is recommended for benzene-soluble carboxylic acids (see p. 116). Both of these tests can of course be applied only in the absence of or after removal of organic carboxylic and sulfonic acids or their salts.

Certain amides of aromatic acids are not saponifiable by evaporation with concentrated hydrochloric acid. An instance is the amide of triphenylacetic acid, and likewise benzamide substituted in the 2,6-position. Such amides (and of course all other amides) can be converted into the respective carboxylic acids by means of nitrous acid (KNO_2+conc. HCl) at room temperature:[2]

$$RCONH_2 + HNO_2 \rightarrow RCOOH + H_2O + N_2$$

Procedure. A little of the solid or 1 or 2 drops of the solution is taken to dryness in a micro test tube along with a drop or two of concentrated hydrochloric acid. The excess acid is driven off by heating to 120°. To detect any resulting ammonium chloride, a drop of 1 N caustic alkali is added to the residue and the mouth of the tube is covered with a piece of filter paper moistened with Nessler reagent. Qualitative filter paper must be used because almost all quantitative papers contain ammonium salts. The development of a brown or yellow stain on gentle warming shows that hydrolyses (1) or (2) have taken place.

To detect acids in the evaporation residue, one drop each of 2% potassium iodide and 5% potassium iodate are added along with a little starch or thyodene solution. A blue color indicates the presence of acids. The production of benzene-soluble carboxylic acids is revealed by means of a drop of 1% uranyl acetate solution and 4 drops of a saturated solution of rhodamine B in benzene. A red color in the benzene layer is obtained on shaking.

Evaporation with hydrochloric acid is not necessary when testing for benzene-soluble carboxylic acids produced by saponifying the amides with nitrous acid. In this case it is sufficient to add several cg of KNO_2 and 1–3 drops of concentrated HCl to the sample. The mixture is warmed after about 10 min. and then shaken with several drops of benzene. The benzene solution is taken for the carboxylic acid test with rhodamine B.

The *Nessler reaction* revealed:

2 γ benzamide	5 γ salicylamide	0.5 γ succinimide
10 γ saccharin	5 γ phthalimide	40 γ gallamine blue
25 γ urea	10 γ benzilic acid amide	
5 γ guanidine	10 γ glycolic acid amide	

The *iodide–iodate* reaction revealed:

2 γ benzamide   2 γ benzilic acid amide

The reaction with *rhodamine B* revealed:

10 γ benzamide   1 γ benzilic acid amide

1 Unpublished studies, with L. HAINBERGER.
2 R. H. A. PLUMMER, *J. Chem. Soc.*, 127 (1925) 2651.

(2) Test through transformation into imidazoles[1]

When the solution of an amide containing a large excess of α-alanine (or other α-aminocarboxylic acid) is evaporated, a product remains which shows the color reaction of tertiary amines, described on p. 252. It may be assumed, therefore, that an imidazole is formed according to

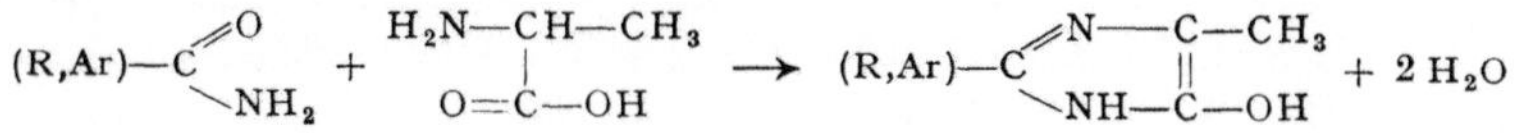

$$(R,Ar)-C\overset{\diagup O}{\diagdown NH_2} + \overset{H_2N-CH-CH_3}{\underset{O=C-OH}{|}} \longrightarrow (R,Ar)-C\overset{\diagup N\text{---}C\text{---}CH_3}{\diagdown NH\text{---}C\text{---}OH} + 2\,H_2O$$

The above condensation, combined with the color test, permits the detection of amides provided that tertiary amines or their salts are absent.

Procedure.   In a micro test tube a drop of the diluted alcoholic test solution is mixed with a drop of 10% aqueous α-alanine solution. After 15 min. the mixture is evaporated to dryness. The remainder of the procedure is the same as described in Sect. 57.

Limits of identification:

10 γ formamide	5 γ phenylacetamide	5 γ benzamide
8 γ acetamide	3 γ 3,5-dinitro-*o*-toluamide	

The following compounds also gave positive responses: urea, adipamide, asparagine.

1 Unpublished studies, with S. YARIV.

62. Amides of aliphatic carboxylic acids (R—CONH₂ group)

Test through heating with dimethyl oxalate[1]

If amides of aliphatic carboxylic acids are heated to 130–160° after being mixed with dimethyl oxalate (m.p. 51°) and thiobarbituric acid, the reaction characteristic of oxamide occurs. A red product, soluble in water or alcohol, results (see p. 536). The reactions for the formation of oxamide may be:

$$RCONH_2 + \begin{matrix} COOCH_3 \\ | \\ COOCH_3 \end{matrix} \rightarrow RCOOCH_3 + \begin{matrix} COOCH_3 \\ | \\ CONH_2 \end{matrix}$$

$$2 \begin{matrix} COOCH_3 \\ | \\ CONH_2 \end{matrix} \rightarrow \begin{matrix} COOCH_3 \\ | \\ COOCH_3 \end{matrix} + \begin{matrix} CONH_2 \\ | \\ CONH_2 \end{matrix}$$

These reactions do not occur with amides of aromatic carboxylic acids and hence a differentiation becomes possible. Amides of heterocylic carboxylic acids behave like aliphatic amides. The test is restricted in that compounds which yield ammonia when they are thermally decomposed may not be present because oxamide is also formed in such cases.

Procedure.   As given on p. 255.

The test revealed:

10 γ acrylamide	20 γ azelaic acid diamide
15 γ nicotinamide	10 γ chloroacetamide

1 Unpublished studies, with D. HAGUENAUER-CASTRO.

63. N-Arylamides (RCONHAr group)

Detection through nitrosation and coupling with α-naphthol[1]

Compounds which contain this group, where R may denote hydrogen, alkyl or aryl, methoxy or ethoxy and also NH_2, can be nitrosated by nitrous acid (*1*). The resulting N-nitroso compounds are saponified in alkaline solution, a fact first established in the case of nitrosoacetanilide.[2] Diazotates are formed (*2*) and can couple with α-naphthol to give red-violet water-soluble azo dyes (*3*). The following reactions:

$$Ar—NH—COR + HNO_2 \rightarrow Ar—N(NO)—COR + H_2O \qquad (1)$$

$$Ar—N(NO)—COR + 2\ KOH \rightarrow Ar—N{=}N{-}OK + RCOOK + H_2O \qquad (2)$$

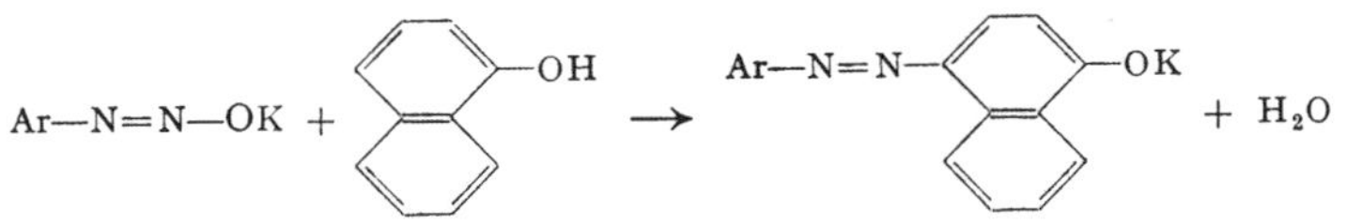

$$Ar—N{=}N—OK + \text{(naphthol—OH)} \rightarrow Ar—N{=}N—\text{(naphthyl—OK)} + H_2O \qquad (3)$$

can be realized within the technique of spot test analysis.

Procedure.   One drop of the test solution is mixed in a micro test tube or on a spot plate with a drop of 10% sodium nitrite solution and 1:1 hydrochloric acid and after a minute a drop of 10 *N* caustic alkali and a pinch of urea are added.* (See note on p. 260.) The liquid is shaken (stirred with a glass rod), a drop of 0.1% alcoholic solution of naphthol is introduced and the mixture then warmed. Depending on the quantity of acylide, *etc.* present, the liquid turns red or orange.

The following amounts were detected:

0.5 γ acetanilide	1 γ acetylsulfathiazole
0.5 γ acetylphenetidine	0.1 γ monophenylurea
1 γ p-acetylaminophenol	5 γ phenylurethan
5 γ 3-ureidoisonicotinic acid	1 γ acetylarsanilic acid

A positive response was also obtained with acetylnaphthalide, stovarsol, and monotolylurea.

* A solution of alkali nitrite that has been acidified with hydrochloric acid and then made basic gives a yellow color on the addition of a solution of α-naphthol. This effect is due to the production of significant amounts of nitrosyl chloride or chlorine when nitrite solution is acidified with hydrochloric acid, with consequent production of alkali hypochlorite when alkali is added. The hypochlorite oxidizes α-naphthol to a colored quinone. This interference is completely avoided by adding urea which has no effect on alkaline solutions of N-nitroso compounds.

1 F. FEIGL, *Anal. Chem.*, 27 (1955) 1315.
2 E. BAMBERGER, *Ber.*, 27 (1894) 915; 30 (1897) 366.

64. Anilides of carboxylic acids ((R,Ar)CONHC$_6$H$_5$ group)

Test through pyroammonolytic release of aniline[1]

If anilides of aliphatic and aromatic carboxylic acids are heated with guanidine carbonate to 250°, aniline is produced. The guanidine carbonate yields ammonia (comp. Sect. 67) which on contact with solid or fused anilides brings about the reaction:

$$(R,Ar)CONHC_6H_5 + NH_3 \rightarrow (R,Ar)CONH_2 + C_6H_5NH_2$$

The aniline can be detected in the gas phase through the formation of the yellow Schiff base with p-dimethylaminobenzaldehyde (see p. 243).

The above reaction is a fine example of an ammonolysis which is not realizable in the wet way but solely by the action of ammonia vapors released pyrolytically. Therefore the term *pyroammonolysis* is appropriate for this type of reaction.

The test described here cannot be applied in the presence of salts of volatile aromatic amines which form colored Schiff bases with p-dimethylamino-benzaldehyde.

Procedure. A drop of the test solution or a small amount of the solid is mixed in a micro test tube with several cg of guanidine carbonate and the solvent is removed by evaporation. A disk of filter paper moistened with an ethereal solution of p-dimethylaminobenzaldehyde is placed over the mouth of the tube. The latter is immersed in a glycerol bath that has been preheated to 180°. The temperature is raised to 250°. A positive response is indicated by the development of a yellow stain on the reagent paper within several minutes.

The test revealed:

1 γ formanilide	2 γ p-tolylurea	1.7 γ cyanoacetanilide
0.6 γ carbanilide	6 γ benzanilide	2 γ salicylanilide
2 γ oxanilide	20 γ phenylurea	20 γ phenylurethan

1 Unpublished studies, with J. R. AMARAL.

65. Anilides of aliphatic carboxylic acids (RCONH—C₆H₅ group)

Test through pyrohydrolytic release of aniline[1]

If anilides of aliphatic carboxylic acids are heated to 130° with MnSO₄.4 H₂O (which loses its water between 100 and 200°), the resulting steam brings about the hydrolysis:

$$RCONHC_6H_5 + H_2O \rightarrow RCOOH + C_6H_5NH_2$$

The aniline is carried along with the steam, so that it can be detected in the gas phase by contact with p-dimethylaminobenzaldehyde (see p. 243). A yellow Schiff base results. These effects provide a convenient means of distinguishing between anilides of aliphatic and aromatic carboxylic acids, since the latter are not affected in the following procedure.

Procedure. A drop of the test solution in ether is placed in a micro test tube along with several cg of MnSO₄.4 H₂O and taken to dryness. The mouth of the tube is covered with a disk of filter paper impregnated with a saturated benzene solution of p-dimethylaminobenzaldehyde. The test tube is placed for some minutes in a glycerol bath previously heated to 130°. A positive response is indicated by the rapid development of a yellow stain on the reagent paper.

The test revealed:

25 γ n-butyranilide	25 γ propionanilide	10 γ carbanilide
25 γ adipic acid anilide	10 γ formanilide	20 γ oxanilide
25 γ p-bromoacetanilide	10 γ acetanilide	

1 Unpublished studies, with J. R. AMARAL.

66. Amidines $\left(-C\diagup^{NH}_{\diagdown NH_2} \text{ group}\right)$

Test through transformation into imidazoles[1]

Amidines and their salts condense easily with α-hydroxyketones to imidazoles[2]. With benzoin the reaction is:

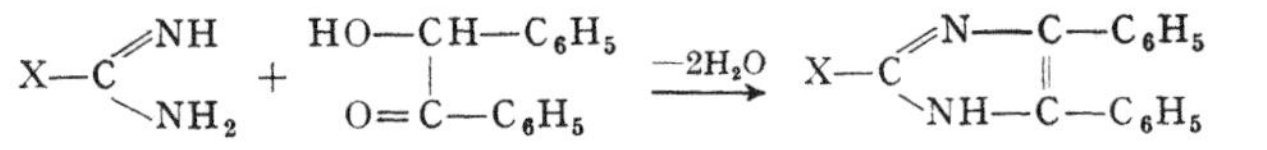

$$X-C\diagup^{NH}_{\diagdown NH_2} + \overset{HO-CH-C_6H_5}{\underset{O=C-C_6H_5}{|}} \xrightarrow{-2H_2O} X-C\diagup^{N-C-C_6H_5}_{\diagdown NH-C-C_6H_5} \qquad X=H, R, Ar$$

The imidazoles contain a cyclic bonded tertiary nitrogen atom and there-fore can be detected by the color reaction of tertiary amines with a mixture of acetic anhydride and citric acid, as outlined in Sect. 57. The test described here is specific for amidines provided that tertiary amines or their salts are absent.

Procedure.   In a micro test tube a few grains of benzoin are added to a drop of the test solution and after 10 min. the mixture is evaporated to dryness. The remainder of the procedure is the same as described on p. 258.

The following were detected:

5 γ formamidine acetate	7.5 γ acetamidine chloride
4 γ benzamidine chloride	7.5 γ guanidine chloride
5 γ dicyanodiamide	10 γ arginine

Positive responses were obtained with dicyanodiamidine sulfate and bigua-nide. Sulfaguanidine and streptomycin sulfate do not react.

1 Unpublished studies, with S. YARIV.
2 Comp. *V. Richter's Organic Chemistry, Vol. IV*, Elsevier, New York, 1947, p. 125.

67. Guanidino compounds $\left(\text{HN} = \text{C} \begin{smallmatrix} \diagup \text{NH}_2 \\ \diagdown \text{NH}- \end{smallmatrix} \text{ group} \right)$

(1) Test through pyrolytic release of ammonia [1]

Dry heating of guanidine yields biguanide and ammonia:

$$2 \ \text{HN} = \text{C} \begin{smallmatrix} \diagup \text{NH}_2 \\ \diagdown \text{NH}_2 \end{smallmatrix} \rightarrow \begin{smallmatrix} \text{HN} \diagdown \\ \text{H}_2\text{N} \diagup \end{smallmatrix} \text{C} - \text{NH} - \text{C} \begin{smallmatrix} \diagup \text{NH} \\ \diagdown \text{NH}_2 \end{smallmatrix} + \text{NH}_3 \qquad (1)$$

An analogous pyrolytic splitting out of ammonia also occurs when salts of guanidine or of its derivatives containing the group shown above are heated to about 250°. If biguanide or cyclic derivatives of guanidine are heated no ammonia is split off. In analogy to *(1)*, the loss of ammonia by salts of guanidine with free NH_2 groups may involve the condensation:

$$2 \ \text{HN} = \text{C} \begin{smallmatrix} \diagup \overset{+}{\text{N}}\text{H}_3 \\ \diagdown \text{NH}_2 \end{smallmatrix} \rightarrow \begin{smallmatrix} \text{H}_2\overset{+}{\text{N}} \diagdown \\ \text{H}_2\text{N} \diagup \end{smallmatrix} \text{C} - \text{NH} - \text{C} \begin{smallmatrix} \diagup \overset{+}{\text{N}}\text{H}_2 \\ \diagdown \text{NH}_2 \end{smallmatrix} + \text{NH}_3 \qquad (2)$$

The ammonia vapors may be detected by Nessler reagent.

The release of ammonia from water-soluble chlorides of these compounds is selective. No ammonia is given off when salts of primary, secondary, and tertiary amines are heated to 250°. This is also true of salts of amino acids, with the exception of salts of arginine, which is a derivative of guanidine. Guanidineacetic acid and methylguanidineacetic acid (creatine) yield no ammonia when heated because they lose water and form the lactams; also glycocyamidine and creatinine, which as cyclic derivatives of guanidine possess no free NH_2 groups available for condensation.

The test is not applicable in the presence of β-amino acids, urea, thiourea, biuret and other urea derivatives with free NH_2 groups, since these compounds likewise yield ammonia when heated to 250°.

Procedure. A little of the solid is placed in a micro test tube. If preferred, a drop of the hydrochloric acid solution may be evaporated there. All of the moisture is removed by heating in a glycerol bath (180°) for about 10 min. The mouth of the test tube is covered with a disk of filter paper moistened with a drop of Nessler reagent. The tube is then brought to 250° in the glycerol bath. The evolution of ammonia is revealed by the development of a brown or yellow stain at once or within several minutes.

This procedure revealed:

2 γ guanidine carbonate (chloride)	5 γ arginine chloride
5 γ nitroguanidine chloride	100 γ streptomycin sulfate
1 γ dicyanodiamidine	0.3 γ dicyandiamide

A positive response was given by: streptomycin, streptidine, synthalin. It is notable that dry guanidine and some of its derivatives split off $(CN)_2$ when heated. For the detection of $(Cn)_2$ see p. 549.

1 F. FEIGL and Cl. COSTA NETO, *Mikrochim. Acta*, (1955) 969.

(2) Test with diacetyl[1]

Numerous guanidino compounds with the group

$$\begin{array}{c} HN\diagdown \\ C-N \diagup \\ H_2N\diagup \end{array}$$

give an orange color if warmed with diacetyl and alkali hydroxide or lime. This color reaction is not given by the parent guanidine. The active material is not diacetyl itself, but an unstable aldol resulting from the action of alkali (see test for diacetyl, p. 445). Probably this product condenses with the free NH_2 group of the respective guanidino compounds. This hypothesis is supported by the behavior of creatine (I) and creatinine (II):

$$\begin{array}{cc}
\begin{array}{c} CH_2-N \diagup^{CH_3} \\ | \diagdown C=NH \\ COOH | \\ NH_2 \end{array} &
\begin{array}{c} CH_2-N \diagup^{CH_3} \\ | \diagdown C=NH \\ CO-NH \end{array} \\
(I) & (II)
\end{array}$$

Only (I), which has a free NH_2 group, but not the lactam (II), gives a red-orange color when warmed with diacetyl and caustic alkali.[2]

The reactivity of the NH_2 group in guanidyl compounds is dependent on the rest of the molecule. This is shown by the finding that methylguanidine, nitro- and aminoguanidine give no red color if warmed with diacetyl and alkali hydroxide*. Although the color reaction does not occur with all guanidino compounds, a positive response is of high diagnostic value in the examination of guanidyl derivatives.

Procedure.[3]   A drop of a 0.1% aqueous solution of diacetyl is placed in a micro test tube and followed by a drop of the test solution. A pinch (tip of spatula) of calcium oxide is then added. The test tube is immersed in a boiling water bath. The lime is tinted red and then the entire solution develops this color if a guanidino derivative is present.

The procedure revealed:

1 γ creatine hydrate	20 γ biguanide sulfate	10 γ dicyanodiamide
1 γ arginine chloride	50 γ guanylurea sulfate	

The following gave a positive response: streptomycin, p-chloroaniline biguanide, thioguanylurea, phenylbiguanide.

No response was observed with: sulfaguanidine, sym-diphenylguanidine (melaniline), melamine, guanine chloride, creatinine, benzoguanamine, acetoguanamine, guanidine salts.

---

* This lack of reactivity is probably the result of the fact that the velocity of the transformation of the active aldol into inactive dimethylbenzoquinone (see p. 446) is far greater than the speed of the condensation of the aldol with the NH_2 group in these guanidino compounds.

1 A. HARDEN and D. NORRIS, *J. Physiol.*, 42 (1911) 332.
2 G. ST. WALPOLE, *J. Physiol.*, 42 (1911) 301.
3 Unpublished studies, with C. STARK-MAYER.

68. Nitriles (cyanides) (—C≡N group)

Detection by conversion to oxamide[1]

If such compounds are heated to temperatures up to 180° with hydrated oxalic acid, which melts and splits off its water of crystallization, pyrohydrolysis occurs, whereby initially amides (*1*) are formed; these then undergo pyroacidolysis to give oxamide (*2*):

$$(R,Ar)CN + H_2O \rightarrow (R,Ar)CONH_2 \tag{1}$$

$$2\,(R,Ar)CONH_2 + (COOH)_2 \rightarrow 2\,(R,Ar)COOH + (CONH_2)_2 \tag{2}$$

After the excess of the sublimable anhydrous oxalic acid has been removed by heating to 200°, the test for oxamide (see p. 535) can be made. This test is based on the formation of an orange product on sintering with thiobarbituric acid.

Procedure.   A small amount of the sample or a drop of its solution is placed in a micro test tube together with some mg of hydrated oxalic acid. The tube is immersed as deeply as possible in a glycerol bath preheated to 200°. After 5 min. the tube is cut 2 cm below its open end and the heating is repeated, in order to eliminate any oxalic acid. After cooling some mg of thiobarbituric acid are added and the mixture is heated to 140–160° in a glycerol bath. A positive response is indicated by the formation of an orange product.

The behavior of 20 aliphatic and aromatic C- and N-cyano compounds was examined and limits of identification in the range 2–20 γ were obtained.

The test may not be applied directly in the presence of compounds which yield oxamide when treated with oxalic acid under the conditions of the test. Such interfering materials include: ammonium salts, amides of acids, amino acids and ureides. In most cases, the interfering materials can be removed by taking advantage of the solubility of cyanides in benzene.

1 Unpublished studies, with J. R. AMARAL.

69. Aliphatic nitriles (R—C≡N group)

(1) Test through pyrolytic release of hydrogen cyanide[1]

If aliphatic nitriles (cyanides) are heated to 250°, considerable amounts of hydrogen cyanide are split off; this is not true of aromatic cyanides nor of other nitrogenous organic compounds. The reaction is aided by the presence of calcium oxide and calcium carbonate; it then occurs even at 150°. This surprising result is due perhaps to an innate tendency of aliphatic cyanides containing the —CH_2CN group to release HCN when pyrolyzed, and the latter is then initially bound as calcium cyanide. The residual material or radical then yields water (as quasi superheated steam) which in turn brings about the hydrolysis (pyrohydrolysis)*:

$$Ca(CN)_2 + H_2O \rightarrow CaO + 2\,HCN$$

The volatile hydrogen cyanide formed is easily detectable by the color reaction with benzidine acetate and copper acetate described on p. 546.

Procedure. A little of the solid test material or the evaporation residue from a drop of its solution is mixed in a micro test tube with a few cg of a 1:1 mixture of calcium carbonate and oxide. The tube is placed in a glycerol bath

* The formation of HCN by pyrohydrolysis can be clearly demonstrated by the following experiment: KCN is fused in a Pyrex test tube until copper acetate–benzidine acetate paper is slightly blued. If the cooled salt in the test tube is then mixed with an excess of $MnSO_4.H_2O$ and the mixture taken to 200°, much HCN is formed. The steam given off from the hydrated manganese sulfate, which loses its water in the temperature range[2] 150–200°, hydrolyzes the alkali cyanide with production of HCN.

that has previously been brought to 250°, and the mouth of the tube is covered with a disk of filter paper moistened with copper(II) acetate–benzidine acetate solution. A positive response is indicated by the development of a blue stain after 3–4 min.

The test revealed:

5 γ 2,6-dichlorobenzyl cyanide   2 γ 3-ethoxy-4-methoxybenzyl cyanide

2 γ *p*-nitrobenzyl cyanide   50 γ α-cyano-N-methylaminocinnamic acid

25 γ cyanoacetic acid

1 F. FEIGL, V. GENTIL and E. JUNGREIS, *Mikrochim. Acta*, (1959) 47.
2 Comp. Cl. DUVAL, *Inorganic Thermogravimetric Analysis*, Elsevier, Amsterdam, 1963, p. 316.

(2) Test through dry-heating with manganese dioxide[1]

If a mixture of an aliphatic cyanide with excess of manganese dioxide is heated at 130–140°, volatile hydrogen cyanide is formed. The chemical background of this effect is not yet clear. Possibly in the case of primary and secondary nitriles the following transformations take place:

$$-CH_2CN + O \rightarrow -CHO + HCN$$

$$>CHCN + O \rightarrow >CO + HCN$$

The production of hydrogen cyanide does not occur in the case of aromatic cyanides or of N-cyano compounds, even when large amounts are present. Therefore a distinction is possible due to this different behavior. For the detection of the hydrogen cyanide formed the above mentioned color test with copper acetate–benzidine acetate is used (see p. 546).

Procedure.   A drop of the solution of the sample is mixed in a micro test tube with some cg of manganese dioxide and the solvent is evaporated. Then the tube, covered with a piece of the reagent paper (for preparation see p. 547), is placed in a glycerol bath preheated to 120°, and the temperature is raised to 130–140°. A positive result is indicated by the formation of a blue stain on the reagent paper within 1–3 min.

When ethereal solutions of the test material are under examination, the ether vapor which remains in the test tube after evaporation must be removed by blowing out with a pipet before starting the dry heating, to avoid autoxidation of ether (see detection of ether, p. 188).

Ten aliphatic cyanides were examined. *Limits of identification* between 2.5 γ and 150 γ were obtained.

1 Unpublished studies, with J. R. AMARAL.

70. Compounds with —C≡N and $>$C=N— groups

Test through pyrolysis in molten sulfur [1]

When aliphatic and aromatic cyanides (nitriles) as well as compounds with C=N— groups in open or closed chains are heated with molten sulfur, they produce not only hydrogen sulfide, as do all organic compounds containing hydrogen (see p. 62), but also thiocyanic acid. The latter may be detected by filter paper moistened with acidified ferric salt solution and held in the vapors arising from the pyrolyzed material. The chemical basis of this cleavage of thiocyanic acid has not yet been experimentally explored. It seems plausible to assume a reaction of the pyrolytically split off hydrogen sulfide with thiocyanogen produced in the same way:

$$H_2S + (CNS)_2 \rightarrow 2\ HCNS + S$$

Various paths are conceivable for the formation of thiocyanogen during pyrolysis with sulfur. One would be the cleavage of C≡N or C=N— groups as CNS or its dimer, namely dithiocyanogen.

Procedure. A little of the solid or the evaporation residue of a drop of its solution is mixed in a micro tube with several cg of sulfur. The tube, fixed in an asbestos support, is covered with a disk of filter paper moistened with acidified ferric nitrate solution. The heating over a micro flame should start at the upper part of the tube and gradually proceed to the bottom. Aliphatic cyanides react quickly; aromatic cyanides and compounds containing a C=N— group require stronger heating. Depending on the quantity of cyanide *etc.* present, a more or less intense red stain develops on the paper.

The following amounts of cyanides were detected:

3 γ 2,6-dichlorobenzyl cyanide	1 γ o-cyanoacetanilide
1 γ 3-ethoxy-4-methoxybenzyl cyanide	2 γ cyanoacetic acid
3 γ p-nitrobenzyl cyanide	20 γ 3-cyanoquinoline
2 γ α-cyano-N-methylaminocinnamic acid	20 γ 4-chlorobenzonitrile

The method revealed the following compounds containing the C=N- group:

10 γ uric acid   40 γ pyrazole   25 γ sulfathiazole   25 γ oxazoline

A positive response was given by: 10 γ theobromine, and 5 γ xanthine, caffeine, theophylline, hypoxanthine.

As the above findings indicate, compounds containing the C=N—group in open or closed chains cannot be detected by this procedure if cyanides are also present. When this problem arises, the cyanides must be saponified by warming with alkali hydroxide:

$$(R,Ar)CN + NaOH + H_2O \rightarrow (R,Ar)COONa + NH_3$$

which is accomplished quickly and completely. A convenient procedure is evaporating with dilute alkali solution and to test the residue with respect to its behavior when pyrolyzed in the presence of molten sulfur.

If basic compounds containing C=N— groups are precipitated as phosphomolybdates and then these salts are pyrolyzed in the presence of sulfur, thiocyanic acid is split off. This rather remarkable finding may be of value in the separation of cyanides.

Thiocyanic acid is likewise obtained from the pyrolysis of azo compounds in molten sulfur. Accordingly, the test given here is applicable not only to C≡N and C=N—, but also to C—N=N— groups in organic compounds. Azo compounds can be detected and, if necessary, destroyed as described on p. 283.

1 F. FEIGL, V. GENTIL and E. JUNGREIS, *Mikrochim. Acta*, (1959) 47.

71. N-Cyano compounds (>N—CN group)

Test through reductive release of hydrogen cyanide[1]

Aliphatic and aromatic cyanides (nitriles) are reduced by nascent hydrogen in an acid medium to salts of the corresponding primary amines:

$$(R,Ar)CN + 4 H^0 + H^+ \rightarrow (R,Ar)CH_2\overset{+}{N}H_3$$

In contrast to the behavior of C-cyano compounds, the reduction of N-cyano compounds leads to the release of hydrogen cyanide:

$$>N—CN + 2 H^0 + H^+ \rightarrow >\overset{+}{N}H_2 + HCN$$

Accordingly, detection of the resulting volatile hydrogen cyanide through the color test with copper acetate–benzidine acetate makes possible a specific test for the >N—CN group in such compounds. The test is not sensitive because the hydrogen cyanide formed is reduced by nascent hydrogen to methylamine.

Procedure. The test is conducted in a micro test tube (about 6 cm long). A little of the solid or a drop of its solution is treated with several cg of granulated zinc and a few drops of 1:5 hydrochloric acid. A disk of filter paper moistened with freshly prepared reagent solution (see p. 547) is placed over the mouth of the test tube. The tube is held in boiling water. A blue stain appears if the response is positive.

The procedure revealed:

80 γ cyanamide	100 γ dicyanodiamide
200 γ potassium cyanurea	200 γ dibenzylcyanamide

1 F. FEIGL and V. GENTIL, *Mikrochim. Acta*, (1959) 44.

72. Hydrazine compounds

Test with glutaconic aldehyde[1]

Practically all organic derivatives of hydrazine (*sym*-diaryl derivatives are an exception) react with glutaconic aldehyde (formed according to p. 246) to yield red to orange condensation products. The mechanism of the reactions is not uniform since it depends on the type of the particular hydrazine compound involved. It is notable that aldehydrazones of acylated hydrazine only are detectable by the test described here.

Procedure. A drop of the test solution is placed in a depression of a spot plate and treated in succession with 1 drop each of a 1% aqueous solution of pyridylpyridinium chloride, 1 N sodium hydroxide, and concentrated hydrochloric acid. This order of adding these reagents must be followed implicitly. A positive response is indicated by an orange or red color.

Limits of identification:

0.01 γ salicylalhydrazone	red	0.05	γ *p*-nitrophenylhydrazine	red-brown
0.1 γ phenylhydrazine	,,	6	γ *o*- ,,	orange
0.5 γ *m*-nitrophenylhydrazine	,,	0.05	γ ethylphenylhydrazine	,,
0.2 γ 2,4-dinitrophenylhydrazine	,,	1	γ semicarbazide hydrochloride	,,
0.1 γ gluconic acid hydrazide	,,	0.2	γ benzoic acid hydrazide	,,
0.1 γ glucosazone	,,	1	γ salicylic acid hydrazide	,,
0.1 γ benzalphenylhydrazone	,,	0.1	γ isonicotinic acid hydrazide	,,
0.05 γ salicylalphenylhydrazone	,,	0.05	γ diphenylcarbazide	,,
0.01 γ orthovanillinphenylhydrazone	,,	0.01	γ diphenylcarbazone	,,
0.01 γ vanillinphenylhydrazone	,,	10	γ benzalsemicarbazone	,,
0.01 γ cinnamalphenylhydrazone	,,	1	γ benzalbenzhydrazone	,,
		0.1	γ benzal-*m*-nitrophenylhydrazone	,,

No color reaction was seen with: benzal-*p*-nitrophenylhydrazone; dioxytartaric acid hydrazone; isatin-*p*-nitrophenylhydrazone; hydrazobenzene.

Aminoguanidine, being a hydrazine compound, gives a yellow color.

1 V. ANGER and S. OFRI, *Mikrochim. Acta*, (1964) 627.

73. Alkyl-, aryl- and acyl-hydrazines ($>$N—NH$_2$ group)

(1) Test with sodium pentacyanoammineferroate[1]

Hydrazines containing a free NH$_2$ group react similarly to nitroso compounds (see Sect. 88) and α,β-unsaturated and aromatic aldehydes (see p. 201) on treatment with a solution of the light yellow sodium pentacyanoammineferroate, Na$_3$[Fe(CN)$_5$NH$_3$]. Deeply colored soluble compounds are formed probably by the replacement of the NH_3 in the prussiate by the hydrazine; for instance:

$$Na_3[Fe(CN)_5NH_3] + NH_2NHR \rightarrow NH_3 + Na_3[Fe(CN)_5NH_2NHR]$$

Procedure.   Several drops of 1% sodium pentacyanoammineferroate solution are added to one drop of a neutral aqueous or alkaline solution of the test substance in a micro crucible. The mixture soon turns deep red to violet. A few materials turn yellow when the solution is made alkaline with 2 N alkali.

Limits of identification are:

3	γ 8-hydroxynaphthyl-1-hydrazine-3,6-disulfonic acid	violet
7	γ 1,1-diphenylhydrazine	,,
0.1	γ thiosemicarbazide	,,
0.1	γ benzhydrazide	,,
0.1	γ salicylic acid hydrazide	,,
0.6	γ 1,1-diethylhydrazine	dark cherry
0.1	γ aminoguanidine	red-violet
0.5	γ phenylhydrazine	red
0.5	γ acetylhydrazine	,,
0.3	γ semicarbazide	,,
1	γ o-nitrophenylhydrazine	brown-red
0.4	γ m- ,,	cherry red
0.5	γ p- ,,	blue-violet
0.2	γ nicotinic acid hydrazide	brown-violet
3	γ 1-methyl-1-phenylhydrazine	blue-red

1 F. FEIGL, V. ANGER and O. FREHDEN, *Mikrochemie*, 15 (1934) 184.

(2) *Test by formation of colored hydrazones with aldehydic azo dyestuffs*[1]

Certain water-soluble azo dyes, formed by coupling diazotized p-aminobenzaldehyde with naphthol- or aminonaphtholsulfonic acids,[2] by virtue of their free aldehyde group, can be condensed to hydrazones with acyl- and aryl-hydrazines in weak acetic acid solution. The color is thus deepened or changed. For example, the dye (I), easily produced by coupling diazotized p-aminobenzaldehyde and croceic acid (2-hydroxynaphthalene-8-sulfonic acid), is yellow in aqueous solution. The addition of an acyl- or aryl-hydrazine to this solution produces a characteristic color change due to the formation of the respective hydrazone. For example, with phenylhydrazine, the product is the violet hydrazone (II):

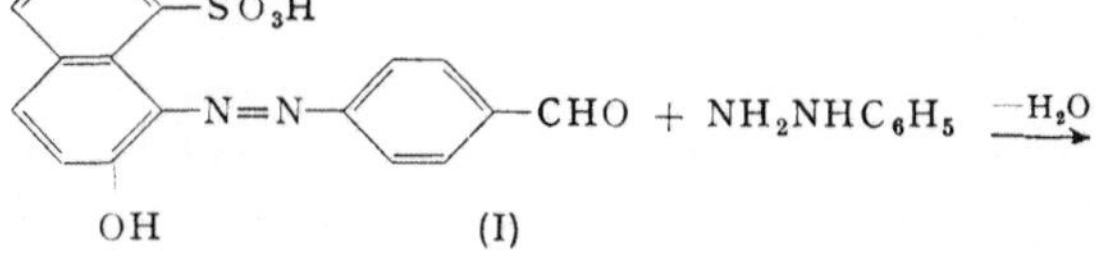

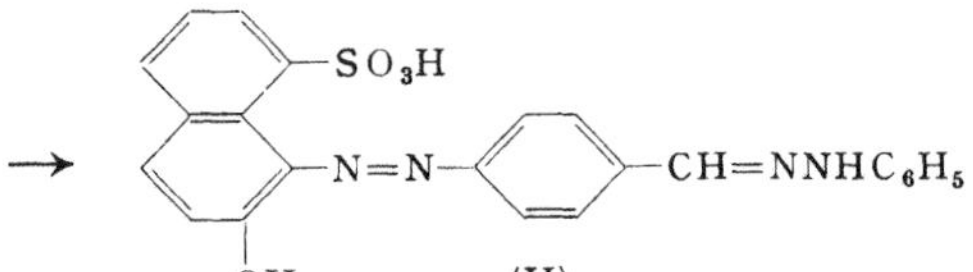

(II)

Procedure. A drop of a solution of the dyestuff in dilute acetic acid is mixed with a drop of 10% sodium acetate solution and a drop of the aqueous or alcoholic test solution is added. When large amounts of a hydrazine are present, the color change is almost instantaneous; smaller amounts require up to 15 min. A blank is advisable.

Reagent: p-Aminobenzaldehyde chloride is suspended in dilute hydrochloric acid and diazotized with the calculated amount of nitrite (1 mole $NaNO_2$ to 1 mole amine). The diazonium solution is added to the equivalent amount of croceic acid dissolved in excess sodium carbonate and the mixture finally acidified with acetic acid. The solution of the resulting azo dyestuff is stable.

Chromazone red B, which is commercially available, can likewise be recommended. It is a coupling product of diazotized p-aminobenzaldehyde and chromotropic acid.

Limits of identification:	*Colors, changing from yellow to:*
1 γ phenylhydrazine	violet
2 γ 1-methyl-1-phenylhydrazine	brown-red
1 γ o-nitrophenylhydrazine	,,
1 γ m- ,,	,,
5 γ benzoic acid hydrazide	orange
1 γ 1,1-diphenylhydrazine	red-brown

1 F. Feigl, V. Anger and R. Zappert, *Mikrochemie*, 15 (1934) 190.
2 P. Friedländer, *Fortschritte der Teerfarbenfabrikation*, Vol. IV, 1899, p. 705.

74. Mono- and *asym*-di-arylhydrazines $\left(\begin{smallmatrix} (Ar,R,H) \\ \\ Ar \end{smallmatrix} \!\!\diagdown_{\diagup} N—NH_2 \text{ group} \right)$

Test through condensation with pyridine-2-aldehyde[1]

A test for pyridinealdehydes is described on p. 524. It is based on the finding that addition of phenylhydrazine chloride or of an acetic acid solution of phenylhydrazine to an aqueous solution of one of the isomeric pyridinealdehydes produces a yellow color resulting from the formation of the corresponding phenylhydrazone. Salts of other arylhydrazines with a free amino group behave in like fashion:

$$C_5H_4NCHO + H_2N—NHAr \rightarrow C_5H_4NCH=N—NHAr + H_2O$$

The yellow color is discharged on the addition of pyridine or ammonia and reappears on acidification. This indicates that the color is due to salts of the hydrazone. The protonation involved is shown in formulas (I) and (II) of the phenylhydrazone of pyridine-2-aldehyde:

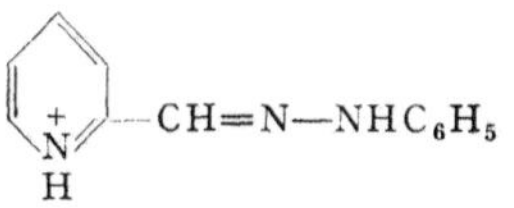

(I) 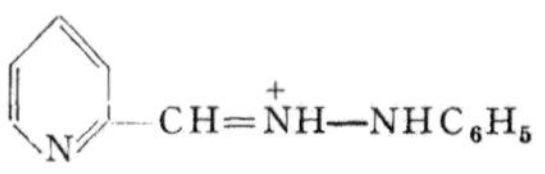   (II)

Analogous formulas hold for the phenyl- and arylhydrazones of the isomeric pyridine-3- and 4-aldehydes.

Procedure.   A drop of the neutral or weakly acid test solution is treated in a depression of a spot plate or in a micro test tube with a drop of a 10% aqueous solution of pyridine-2-aldehyde. A positive response is indicated by the appearance of a yellow color. If mono- or dinitrophenylhydrazines are presented for testing, their yellow mineral acid solutions should be decolorized beforehand by adding several mg of Devarda's alloy (reduction of the NO_2 groups to NH_2 groups). The reduced solution should then be neutralized.

The test revealed:

0.5 γ phenylhydrazine	5 γ *asym*-diphenylhydrazine
5 γ naphthylhydrazine	5 γ nitrophenylhydrazine (*o,m,p*)
1 γ *asym*-methylphenylhydrazine	5 γ 2,4-dinitrophenylhydrazine

The above procedure will also reveal arylhydrazines and osazones after they have been saponified by brief warming with dilute hydrochloric acid.

The following were detected:

2 γ benzophenone phenylhydrazone	1 γ arabinose osazone
30 γ dihydroxytartaric acid osazone (after reductive cleavage)	

1 F. FEIGL and L. BEN-DOR, *Talanta*, 10 (1963) 1111; and unpublished studies.

75. Monoarylhydrazines (Ar—NHNH$_2$ group)

Test with selenious acid and 1-naphthylamine[1]

Arylhydrazines are oxidized by selenious acid to diazonium salts, which can be coupled with aromatic amines (1-naphthylamine is the most suitable) to form bright red to red-violet azo dyestuffs.[2] Small amounts of arylhydrazines (and also arylhydrazones and osazones) may be detected by this reaction. The following equations exemplify the reactions in the test for the parent phenylhydrazine in acid solution:

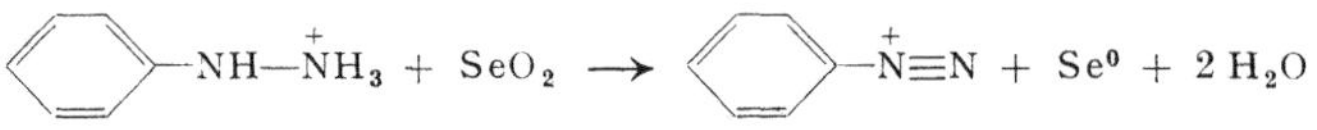

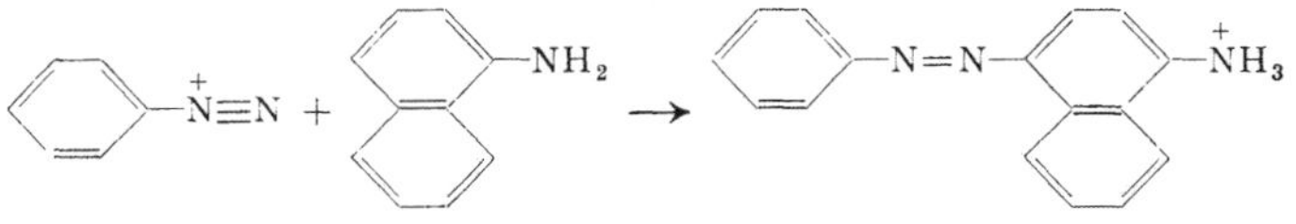

$$\langle\!\!\!\!\bigcirc\!\!\!\!\rangle\!-NH-\overset{+}{N}H_3 + SeO_2 \longrightarrow \langle\!\!\!\!\bigcirc\!\!\!\!\rangle\!-\overset{+}{N}\!\equiv\!N + Se^0 + 2\,H_2O$$

$$\langle\!\!\!\!\bigcirc\!\!\!\!\rangle\!-\overset{+}{N}\!\equiv\!N + \bigcirc\!\!\!\!\bigcirc\!-NH_2 \longrightarrow \langle\!\!\!\!\bigcirc\!\!\!\!\rangle\!-N\!=\!N\!-\!\bigcirc\!\!\!\!\bigcirc\!-\overset{+}{N}H_3$$

Procedure. A drop of the test solution is mixed with a drop of dilute hydrochloric acid, in a micro crucible or on a spot plate, and a small amount of selenious acid is added. When large amounts of arylhydrazine are involved, the mixture is allowed to stand 1 or 2 min. to permit the red selenium to separate. A drop of an acetic acid solution of 1-naphthylamine is added and a few crystals of sodium acetate. When primary arylhydrazines are present, red to violet dyes are formed. The color can be intensified by the addition of a drop of hydrochloric acid.

> *Reagent:* 0.3 g of 1-naphthylamine is dissolved by boiling in 70 ml water and 30 ml glacial acetic acid. After cooling the solution is filtered. It should be kept in the dark.

Limits of identification:

0.06 γ m-nitrophenylhydrazine	0.03 γ naphthylhydrazine
0.06 γ p- ,,	0.08 γ p-bromophenylhydrazine
0.04 γ phenylhydrazine	

1 F. FEIGL and V. DEMANT, *Mikrochim. Acta*, 1 (1937) 134.
2 J. J. POSTOWSKY and coworkers, *Ber.*, 69 (1936) 1913.

76. Monoarylhydrazones ($>$C$=$NNHAr group)

Test through hydrolytic release of arylhydrazine[1]

The test for arylhydrazines described in Sect. 75 can also be applied for the detection of arylhydrazones and osazones because these compounds yield the arylhydrazine chloride when saponified with hydrochloric acid.

Procedure. A drop of the test solution or preferably a crystal of the solid is taken almost to dryness in a micro crucible along with a few drops of concentrated hydrochloric acid. A small grain of selenious acid is added. The further treatment is the same as described in Sect. 75.

Limits of identification:

0.1 γ levulinic acid phenylhydrazone   0.1 γ glucosazone
0.1 γ propionaldehyde phenylhydrazone   0.7 γ dihydroxytartaric acid
0.7 γ glucuronic acid phenylhydrazone   osazone
2   γ cinnamoylformic acid phenylhydrazone
0.1 γ α-pyridylbenzoyl phenylhydrazone

1 F. FEIGL and V. DEMANT, *Mikrochim. Acta*, 1 (1937) 134.

77. Phenylhydrazones

Test through hydrolytic cleavage to phenylhydrazine [1]

Dilute acids readily hydrolyze these compounds and yield acid solutions of phenylhydrazine salts. When these solutions are alkalized with pyridine, the phenylhydrazine is set free and can be detected by its condensation with pyridine-2-aldehyde to yield the yellow hydrazone of this aldehyde (comp. p. 386).

Procedure. A drop of the aqueous or alcoholic test solution and a drop of 2 N hydrochloric acid are placed in a micro test tube and heated for a few minutes in a boiling water bath. After cooling, a drop of a 1% aqueous solution of pyridine-2-aldehyde is added and followed by a drop of pyridine. A positive response is indicated by the appearance of a yellow color.

The test revealed:

2 γ phenylhydrazone of benzophenone   1 γ arabinose osazone

1 F. FEIGL and L. BEN-DOR, *Talanta*, 10 (1963) 111.

78. Hydrazides of carboxylic acids (—CO—NHNH₂ group)

All acid hydrazides, including the cyclic hydrazide luminol which does not respond to any of the reactions cited in this section, show acidic character. Accordingly, they respond to the test given on p. 147 for acid groups that is based on the cleavage of hydrogen cyanide when pyrolyzed with mercuric cyanide. (*Idn. Limit:* 5 γ luminol)

(1) Test by hydrolytic cleavage and formation of salicylaldazine [1]

When the hydrazides of carboxylic acids are digested with warm concentrated hydrochloric acid, they are split with production of anions of the particular carboxylic acids and hydrazinium ions. However, the reaction is slow. In contrast, the cleavage is rapid and practically complete if these materials are warmed with dilute alkali hydroxide:

$$(R,Ar)CONH—NH_2 + H_2O \rightarrow (R,Ar)COO^- + NH_2—\overset{+}{N}H_3$$

In weakly acid aqueous solution, hydrazine (or hydrazinium ion) condenses rapidly and quantitatively with salicylaldehyde to produce light yellow, insoluble salicylaldazine:

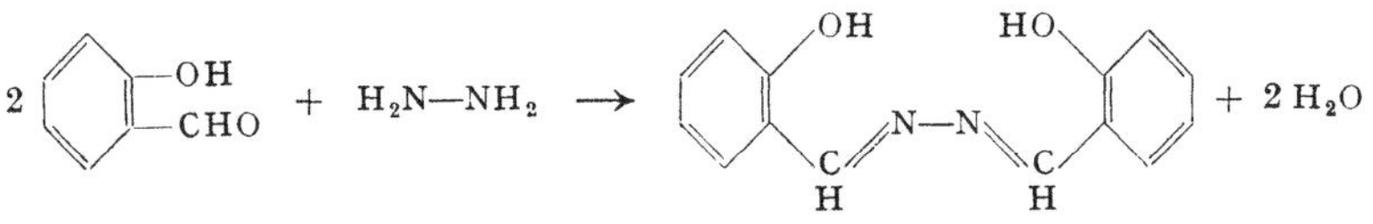

Solid salicylaldazine fluoresces orange-yellow in ultraviolet light, and may be detected with great sensitivity through this behavior (see p. 341).

The formation of salicylaldazine from acid hydrazides is best achieved by warming the test material with an alkaline solution of salicylaldehyde and then acidifying with acetic acid. In this way it is possible to detect acid hydrazides which have a free terminal NH_2 group. The test is reliable provided the sample contains no hydrazine itself or hydrazones of ketones and aldehydes which yield hydrazine on warming with alkali hydroxide.

Procedure: A drop of the test solution or a little of the solid is mixed in a micro test tube with a drop of saturated aqueous salicylaldehyde solution and a drop of 1 N alkali hydroxide. The mixture is warmed in the water bath for 5–20 min., cooled, and then treated with a drop of 3 N acetic acid. The solution is poured on filter paper and after 1-2 min. the paper is examined under ultraviolet light. An orange-yellow to yellow-green fluorescence indicates the presence of an acid hydrazide. The fluorescent fleck persists when the paper is bathed in 6 N acetic acid.

The following amounts were detected:

2.5 γ semicarbazide	0.15 γ isonicotinic acid hydrazide	
1 γ benzoic acid hydrazide	0.15 γ oxalyldihydrazide	
0.5 γ picolinic acid hydrazide	1 γ p-nitrobenzoic acid hydrazide	
0.5 γ nicotinic acid hydrazide	5 γ salicylic acid hydrazide	

1 Unpublished studies, with W. A. MANNHEIMER.

(2) Test through direct condensation with salicylaldehyde[1]

Like hydrazine and its derivatives with a free NH_2 group, the hydrazides of carboxylic acids, since they possess the group —$CONHNH_2$, can condense with aldehydes and ketones. With salicylaldehyde the reaction is:

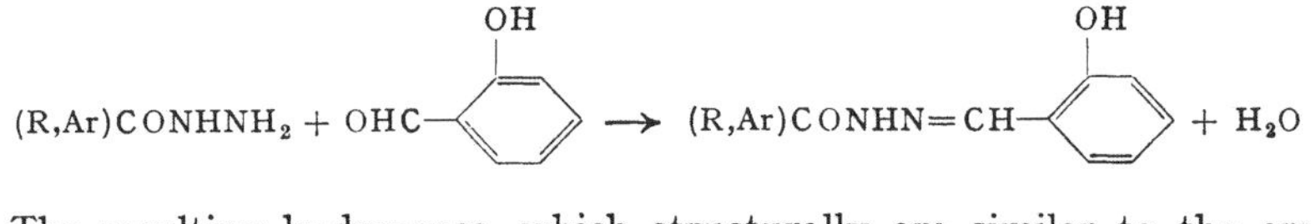

The resulting hydrazones, which structurally are similar to the orange-yellow fluorescing water-insoluble salicylaldazine (compare Test *1*), likewise are not soluble in water and fluoresce strongly (yellow-green to blue-green) in ultraviolet light. Because of the phenolic OH group, the above condensation products dissolve in ammonia, as do salicylaldehyde and salicylaldazine, and give yellow solutions. The behavior toward dilute acids (in which salicylaldazine is not soluble) seems to depend on the nature of R or Ar. Some hydrazones dissolve in dilute acids to only a slight extent or not at all (*e.g.* the product formed by salicylhydrazide); others are dissolved by acids without decomposition (*e.g.* the products formed by the three isomeric pyridinecarboxylic acid hydrazides), whereas some are decomposed by acids (*e.g.* the condensation product of benzoic hydrazide with salicylaldehyde). The realization of (*1*) and the action of acids on the hydrazones formed in this reaction provide the possibility of a rupture of hydrazine, which condenses with salicylaldehyde to produce the acid-stable fluorescing salicylaldazine.

A neutral or weakly ammoniacal milieu is required to make the direct condensation with salicylaldehyde into a general test for acid hydrazides. Since ammoniacal solutions of salicylaldehyde have a yellow-green fluorescence, it is impossible to detect directly the production of small quantities of ammonia-soluble condensation products which exhibit blue-green or yellow-green fluorescence. This impairment of the test is averted if a drop of the mixture of acid hydrazide and salicylaldehyde is treated with a little ammonia and then brought on filter paper. Under these conditions, the fluorescence due to the ammoniacal salicylaldehyde disappears within a few minutes, while the fluorescence of the condensation product of salicylaldehyde with the acid hydrazide persists. The intensity of this latter fluorescence gradually decreases and it completely disappears when small quantities are involved, but it is immediately restored if the fleck is held over ammonia. This is an instance of an effect sometimes encountered in spot testing, namely that a spot reaction on paper is particularly sensitive and reliable and hence the paper functions as if it were an active participant in the reaction. In the present case, the course of events is that ammoniacal solutions of salicylaldehyde are brought on paper and through loss of ammonia from the phenolate leave salicylaldehyde behind, which then either volatilizes from the paper or, because of the high state of division, is oxidized rapidly (by the air) to salicylic acid. This reasoning is supported by the fact that a drop of a saturated aqueous solution of salicylaldehyde placed on filter paper and treated with ammonia immediately yields a yellow-green fluorescent fleck, which disappears rapidly. If 3–4 min. elapse before the ammonia is brought into play, no fluorescence develops.

Procedure. One drop of the acid or neutral test solution is mixed with a drop of the solution of salicylaldehyde and transferred to a filter paper. After

3-5 min., the paper is held over ammonia, and the fleck is examined at once under ultraviolet light. A fleck which fluoresces blue-green or yellow-green indicates the presence of an acid hydrazide. The fluorescence disappears after several minutes, but can be regenerated by a fresh exposure to ammonia.

Reagent: 9 ml of saturated aqueous solution of salicylaldehyde are acidified with 1 ml glacial acetic acid just before using.

The procedure gave a positive response with:

1 γ semicarbazide	2 γ isonicotinic acid hydrazide
1 γ benzoic acid hydrazide	0.5 γ salicylic acid hydrazide
2 γ picolinic acid hydrazide	0.5 γ oxalyldihydrazide
1 γ nicotinic acid hydrazide	

The cyclic 3-aminophthalic acid hydrazide (luminol) does not react; *p*-nitro-benzoic acid hydrazide gives no response, probably because of a quenching effect by the NO_2 group.

1 F. FEIGL, H. E. FEIGL and W. A. MANNHEIMER, unpublished studies.

(3) Test with p-dimethylaminobenzaldehyde

When the hydrazide of a carboxylic acid is saponified with alkali hydroxide (compare Test *1*), and the reaction mixture then acidified with acetic acid, the resulting hydrazine can be detected by adding an excess of *p*-dimethylaminobenzaldehyde. An orange to yellow color appears, according to the quantity of hydrazine present.[1] The same reaction also occurs if a neutral or weakly acid solution of the hydrazide is warmed with an acid solution of the aldehyde. Obviously, the hydrazide is partially saponified, and the hydrazine is constantly removed from the rapidly established hydrolysis equilibrium by reaction with the aldehyde. The color reaction involves a condensation of aldehyde and hydrazine analogous to that in Test *1*, but differs from the latter in that the resulting aldazine (II) because of its basic nature dissolves at once in acids with production of the quinoidal cation (III):

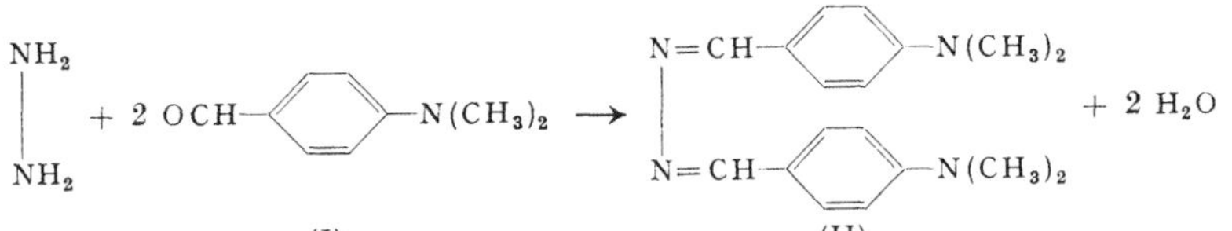

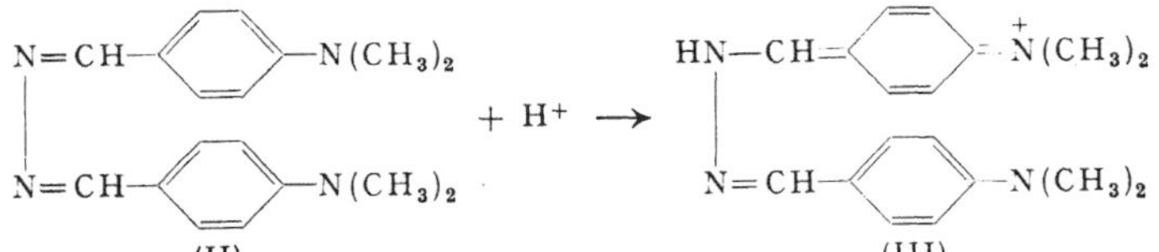

The formation of (III) can be detected with high sensitivity if a drop of the orange to yellow solution, which displays no fluorescence, is placed on filter paper. A red to pink fleck remains and fluoresces with an intense red to salmon hue in ultraviolet light.[2] This effect, which can be observed in very dilute almost colorless solutions, probably involves the formation of a fluorescing adsorbate of (III) with the paper whereby, in accord with the constitution of (III), consideration must be given to both an exchange- as well as an addition adsorption.[3] The red to salmon fluorescing fleck on the filter paper takes on a blue-green hue when treated with ammonia; the red fluorescence is restored by acids. These changes of the fluorescence color, which can be repeated at will, probably arise from a transition of adsorbed (III) into adsorbed (II) and vice versa.

Aromatic amines condense with p-dimethylaminobenzaldehyde to form Schiff bases, which are yellow to orange. Highly concentrated solutions of the latter leave green fluorescent flecks on filter paper, which disappear when bathed in ammoniacal alcohol and do not return on acidification. Since the blue-green fluorescing flecks of the Schiff base (II) remain on filter paper and the red fluorescence returns when spotted with acid, it is possible to detect acid hydrazides in the presence of salts of aromatic amines.

The following test requires an excess of p-dimethylaminobenzaldehyde. The quantity of reagent prescribed provides sufficient excess for testing solutions containing not more than 1 per cent hydrazide. If greater concentrations are at hand, the aldehyde then condenses with the unsaponified acid hydrazide at its free NH_2 group analogous to the action in Test 2. This condensation yields a yellow fluorescing product.

Procedure.[2]   A drop of the test solution is placed in a micro test tube along with a drop of the reagent solution and the mixture is warmed in the water bath for 5–10 min. A drop of water is added to the cold solution and a drop of the latter is then placed on filter paper. The paper is bathed in 1:250 hydrochloric acid and viewed in ultraviolet light. A red to salmon fluorescence indicates the presence of an acid hydrazide. When bathed in 1:10 ammonia, the fleck turns blue-green.

Reagent: 0.4 g of p-dimethylaminobenzaldehyde dissolved in 20 ml alcohol plus 2 ml concentrated hydrochloric acid.

The following amounts were detected:

0.5 γ semicarbazide	0.25 γ picolinic acid hydrazide
0.05 γ benzoic acid hydrazide	0.1 γ nicotinic acid hydrazide
0.1 γ p-nitrobenzoic acid hydrazide	0.05 γ isonicotinic acid hydrazide

The cyclic hydrazide luminol shows no reaction.

1 M. PESEZ and A. PETIT, *Bull. Soc. Chim. France,* (1947) 122.
2 F. FEIGL and W. A. MANNHEIMER, *Mikrochemie Ver. Mikrochim. Acta,* 40 (1953) 355.
3 F. FEIGL, *Specific, Selective and Sensitive Reactions,* Academic Press, New York, 1949, p. 530.

(4) Test through pyrolytic reaction with benzoin[1]

Molten benzoin (m.p. 135°) acts as a hydrogen donor when converted into benzil and therefore can cause reductive cleavage (pyrohydrogenolysis) of certain organic compounds (comp. p. 147). The following reaction occurs with acid hydrazides at temperatures up to 180°.

$$(R,Ar)CONHNH_2 + C_6H_5CHOHCOC_6H_5 \rightarrow (R,Ar)CONH_2 + C_6H_5COCOC_6H_5 + NH_3$$

The occurrence of this reaction and consequently the presence of acid hydrazides can be detected through the positive response of the resulting ammonia. The procedure assumes the absence of compounds that yield ammonia on heating to 180°. Consequently, a preliminary test along these lines is necessary.

Procedure. A small quantity of the solid test material or the evaporation residue from a drop of its solution is united in a micro test tube with several cg of benzoin. The mouth of the tube is covered with a piece of moist acid–base indicator paper. The tube is immersed to about half its length in a glycerol bath that has been preheated to 160°. If the temperature is increased to 180°, there will be an indicator color change almost immediately or after several minutes.

The test revealed:

50 γ benzoic acid hydrazide	80 γ 5-methoxysalicylic acid hydrazide
60 γ m-nitrobenzoic acid hydrazide	80 γ nicotinic acid hydrazide
80 γ p-methoxybenzoic acid hydrazide	80 γ naphthalene-1-acetic acid hydrazide

A positive response was given by the following: isonicotinic acid hydrazide, p-chlorobenzoic acid hydrazide, p-nitrobenzoic acid hydrazide, 2-hydroxy-3-naphthoic acid hydrazide.

1 F. FEIGL and D. HAGUENAUER-CASTRO, *Microchem. J.*, 6 (1962) 172.

(5) Test with 2,6-dichloroquinone-4-chloroimine[1]

Aliphatic and aromatic acid hydrazides react with Gibbs' reagent and ammonia to give brown-violet condensation products of unknown constitution. This reaction makes possible a spot test for acid hydrazides. Phenols yield a blue color (see p. 185).

Procedure. A drop of the test solution is placed on filter paper followed by a drop of an 0.1% solution of 2,6-dichloroquinone-4-chloroimine in chloroform. After the fleck has dried it is fumed with ammonia and a brown-violet stain appears.

Limits of identification:

1 γ semicarbazide hydrochloride	1 γ salicylic acid hydrazide
0.1 γ thiosemicarbazide	1 γ isonicotinic acid hydrazide
0.5 γ benzoic acid hydrazide	

1 V. ANGER and S. OFRI, *Z. Anal. Chem.*, 204 (1964) 265.

79. Acylphenylhydrazides (—CO—NH—NH—C$_6$H$_5$ group)

Test through pyrohydrolytic release of phenylhydrazine [1]

In acid solution, phenylhydrazine is oxidized by arsenic acid; phenol results and can be detected in the gas phase by the indophenol reaction (see p. 185). This is the basis of a test for phenylhydrazine described on p. 554. Phenylhydrazones and osazones, which are saponified in the wet way, behave in the same manner as phenylhydrazine.

Compounds containing the above group do not split out phenylhydrazine in the wet way and accordingly do not give a positive response to its tests. However, a pyrohydrolysis of acylphenylhydrazides occurs if the sample is heated to 160° with hydrated oxalic acid, whose water of crystallization is liberated as quasi superheated steam. When this pyrohydrolysis is conducted in the presence of arsenic acid, the phenylhydrazine is oxidized and yields phenol vapors that may be detected by the indophenol reaction as noted above. A preliminary separation from volatile phenols, phenol esters, phenylhydrazones, and osazones is necessary.

Procedure.   The solid sample or a drop of its aqueous solution is mixed in a micro test tube with several cg of oxalic acid hydrate and one cg of arsenic acid and taken to dryness if need be. The test tube is then placed in a glycerol bath previously heated to 160°. The mouth of the tube is covered with a disk of filter paper moistened with a saturated benzene solution of 2,6-dichloro-quinone-4-chloroimine. A positive response is shown by the development of a brown spot which turns blue when exposed to ammonia vapors.

The following amounts were detected:

10 γ diphenylcarbazone	20 γ phenylthiosemicarbazide
10 γ diphenylthiocarbazone	15 γ 1,4-diphenylsemicarbazide
20 γ phenylthiosemicarbazide	

1 F. Feigl and E. Jungreis, *Talanta*, 1 (1958) 367.

80. Aldehyde hydrazones (—CH=N—NH— group)

(1) Test with 2,6-dichloroquinone-4-chloroimine [1]

The alcoholic solution of aldehyde hydrazones remains practically unaltered upon addition of Gibbs' reagent; however when the mixture is evaporated a colored (mostly blue) product remains. No alkalization is necessary to obtain this effect.

The following explanation is presented: a condensation occurs on the active hydrogen atom of the CH group, whereby a solution of (I) is formed. By evaporation a molecular rearrangment occurs whereby (II) is formed. This transformation obviously occurs in the absence of a solvent.

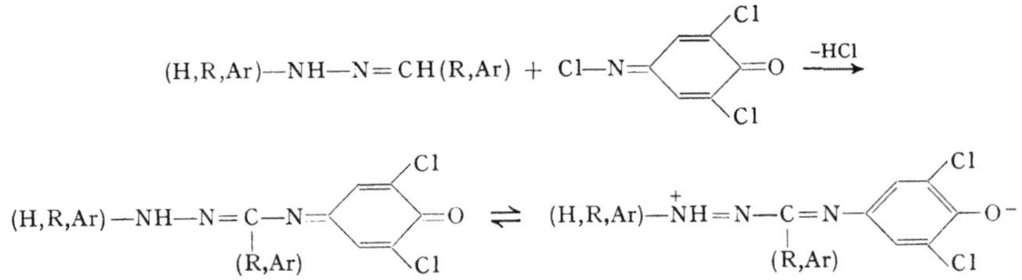

$$(H,R,Ar)-NH-N=CH(R,Ar) + Cl-N{=}\!\!\bigcirc\!\!{=}O \xrightarrow{-HCl}$$

(with Cl substituents on the ring)

$$(H,R,Ar)-NH-N=\underset{\underset{(R,Ar)}{|}}{C}-N{=}\!\!\bigcirc\!\!{=}O \rightleftarrows (H,R,Ar)-\overset{+}{N}H=N-\underset{\underset{(R,Ar)}{|}}{C}=N{-}\!\!\bigcirc\!\!{-}O^-$$

(I)   (II)
(dissolved, colorless)   (solid, colored)

Procedure. A drop of the alcoholic test solution is treated in a depression of a dry spot plate with a drop of an 0.1% solution of 2,6-dichloroquinone-4-chloroimine in chloroform. If hydrazones are present, the residue from the evaporation appears as a colored (usually blue) ring.

Limits of identification:

5 γ benzalphenylhydrazone	blue	5 γ salicylalhydrazone brown
0.5 γ salicylalphenylhydrazone	,,	1 γ vanillinphenyl-
1 γ orthovanillinphenylhydrazone	,,	hydrazone green-brown
5 γ benzal-*m*-nitrophenylhydrazone	,,	0.5 γ cinnamalphenyl-
5 γ ,, -*p*- ,, ,,	,,	hydrazone blue-green

Benzal-*o*-nitrophenylhydrazone does not react. No reaction was obtained with ketophenylhydrazones, benzalazine, semicarbazones, and other acylaldehyde hydrazones and arylhydrazines.

1 V. ANGER and S. OFRI, *Z. Anal. Chem.*, 204 (1964) 264.

(2) Test with p-nitrosophenol[1]

Aldehyde phenylhydrazones react with *p*-nitrosophenol in concentrated hydrochloric acid with production of brown to blue-black compounds. In contrast to the analogous blue or red reaction products of the nitrosophenol with phenols, no color change occurs toward blue on alkalization with ammonia. The mechanism of this color reaction obviously rests on the condensation of the nitrosophenol with the active CH (or CH$_2$ group) of the aldehyde phenylhydrazones since ketone phenylhydrazones do not react. The reaction product may have one of the tautomeric structures:

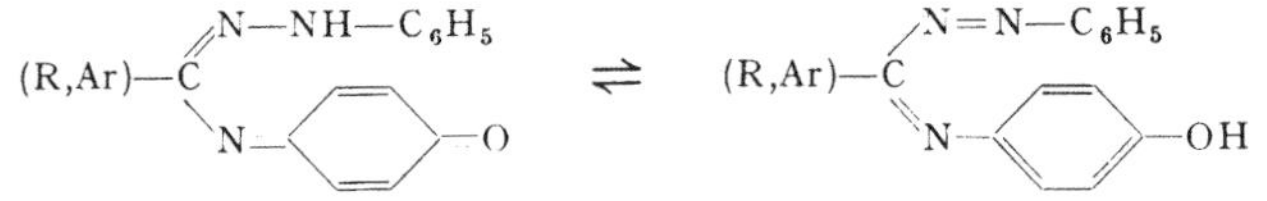

$$(R,Ar)-C\overset{\nearrow N-NH-C_6H_5}{\underset{\searrow N-\bigcirc-O}{}} \rightleftarrows (R,Ar)-C\overset{\nearrow N=N-C_6H_5}{\underset{\searrow N-\bigcirc-OH}{}}$$

The following ionogenic formulation can also be used:

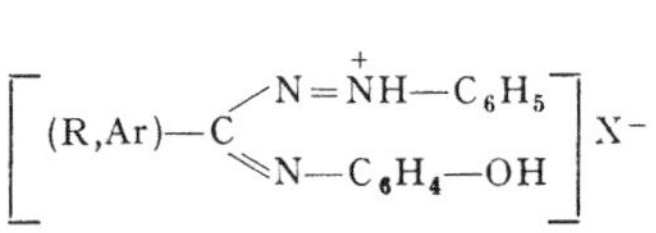

$$\left[(R,Ar)-C\overset{\nearrow N=\overset{+}{N}H-C_6H_5}{\underset{\searrow N-C_6H_4-OH}{}}\right] X^-$$

Procedure. A small amount of the solid test substance or the evaporation residue from a drop of the test solution is treated in a depression of a spot plate with several drops of a freshly prepared 0.05% solution of p-nitrosophenol in conc. hydrochloric acid. A positive response is indicated by the appearance of a brown or blue-black color.

Limits of identification:

1 γ benzalphenylhydrazone	brown	0.1 γ vanillinphenylhydrazone
1 γ salicylalphenylhydrazone	,,	blue-black
50 γ glucosazone	,,	
1 γ cinnamalphenylhydrazone	,,	
1 γ orthovanillinphenylhydrazone	,,	

Positive results were obtained from:

anisaldehydephenylhydrazone   blue-black   furfuralphenylhydrazone brown
piperonalphenylhydrazone   ,,   ,,

No reaction was obtained with salicylalhydrazone, nitrophenylhydrazones, dihydroxytartaric acid osazone.

1 V. ANGER and S. OFRI, *Z. Anal. Chem.*, 200 (1964) 217.

81. Hydrazo compounds (—NH—NH— group)

Test through pyrolysis with hexamine[1]

If compounds that contain the above group are heated to 150° along with hexamine, ammonia is split off. The following condensation probably occurs:

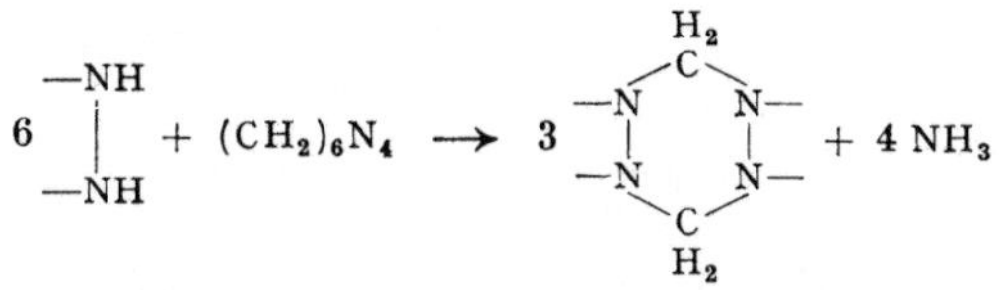

$$6 \begin{array}{c} -NH \\ | \\ -NH \end{array} + (CH_2)_6N_4 \longrightarrow 3 \begin{array}{c} H_2 \\ -N \overset{C}{\diagup} N- \\ | \quad | \\ -N \underset{C}{\diagdown} N- \\ H_2 \end{array} + 4\,NH_3 \qquad (1)$$

This assumption is supported by the finding[2] that an analogous condensation occurs when formaldehyde acts on hydrazobenzene whereby dimethylene tetraphenyltetrazine results:

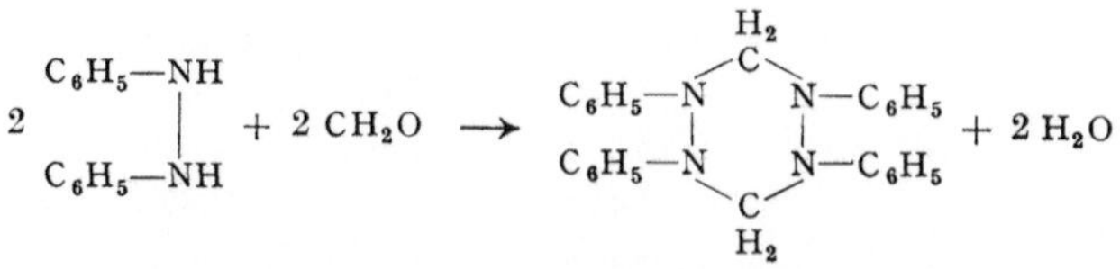

$$2 \begin{array}{c} C_6H_5-NH \\ | \\ C_6H_5-NH \end{array} + 2\,CH_2O \longrightarrow \begin{array}{c} H_2 \\ C_6H_5-N \overset{C}{\diagup} N-C_6H_5 \\ | \quad | \\ C_6H_5-N \underset{C}{\diagdown} N-C_6H_5 \\ H_2 \end{array} + 2\,H_2O \qquad (2)$$

The realization of (1) makes possible the detection of hydrazo compounds through the detection of the ammonia formed, provided compounds with NH_2 groups are absent since the latter likewise yield ammonia when pyrolyzed in the presence of hexamine (comp. p. 119). The interference due to primary alkyl- and arylamines (but not that occasioned by acid amides) can be dealt with successfully by digesting the test material with dilute sulfuric acid and then extracting with ether. The amines are quantitatively retained as sulfates in the water layer, while the neutral or weakly basic hydrazo compounds pass into the ether layer. The test is carried out with the ethereal solution.

Procedure. A tiny amount of the solid or a drop of its ethereal solution is united in a micro test tube with several cg of hexamine (or several drops of its chloroform solution). After any necessary evaporation of the solvents, the test tube is placed in a glycerol bath that has been preheated to 150° and the mouth of the tube is covered with a piece of qualitative filter paper moistened with Nessler reagent. A positive response is indicated by the development of a brown stain on the reagent paper.

The procedure revealed:

        5 γ diphenylcarbazone   5 γ diphenylcarbazide

        5 γ diphenylthiocarbazone

1 F. FEIGL, E. JUNGREIS and L. BEN-DOR, unpublished studies.
2 C. A. BISCHOFF, *Ber.*, 31 (1898) 3244.

82. Azo compounds (—N=N— group)

Test by reduction to primary aromatic amines

Compounds containing a —N=N— group bound to aromatic (usually substituted) radicals, are rapidly cleaved by warming with zinc and hydrochloric acid, and thus yield 2 molecules of primary aromatic amines (as chlorides):

$$RN=NR_1 + 4\ H^0 + 2\ H^+ \rightarrow R\overset{+}{N}H_3 + R_1\overset{+}{N}H_3$$

This reduction, and hence the presence of azo compounds, can be detected through the fact that primary aromatic amines are readily detectable through their condensation with *p*-dimethylaminobenzaldehyde, to yield colored quinoidal cations of Schiff bases (see p. 243) or colored quinoidal condensation products with sodium 1,2-naphthoquinone-4-sulfonate (see p. 153).

The detection of azo groups is especially important when testing dyes. The pertinent procedure is given in Chap. 7.

Azoxy compounds show the same behavior toward zinc and hydrochloric acid as azo compounds.

83. Azoxy compounds ($-N=\overset{+}{N}-\overset{-}{O}$ group)

Test by conversion into hydroxydiazo compounds[1]

If aromatic azoxy compounds are allowed to stand in contact with concentrated sulfuric acid or are gently warmed with it, they are converted into the isomeric azohydroxy (hydroxydiazo) compounds. For example:

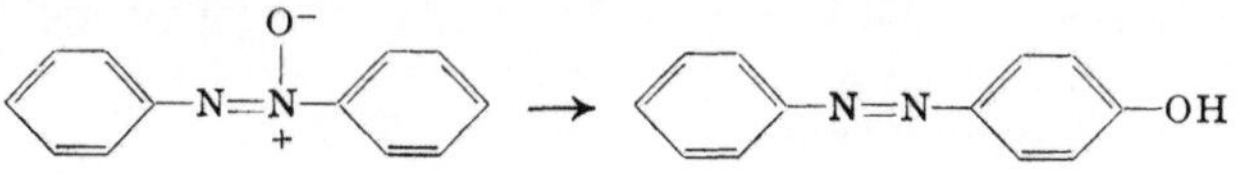

This so-called Wallach transformation[2] is accompanied by a color change because the hydroxydiazo compounds in sulfuric acid solution are not so highly colored as the isomeric azoxy compounds. The rearrangement apparently occurs via less stable, intensely colored intermediate compounds, probably of quinoidal character. These products, which are usually deep red or red-brown, appear within 3 sec when the temperature is raised to 95°. The color deepens within 30 sec and then fades somewhat because of the formation of the hydroxydiazo compounds. Polynitroazoxy compounds present an exception; their sulfuric acid solutions require longer or more intense heating to develop the optimum color intensification. The darkening observed when such solutions are boiled cannot be mistaken for charring (of a compound not containing an azo group) since the solution rapidly loses its color on further heating. Azo compounds do not exhibit this behavior; the color of their concentrated sulfuric acid solutions remains unchanged even on boiling. On the other hand, aromatic compounds containing iodine directly attached to the nucleus (*p*-iodophenol, *o*-iodoaniline, *etc.*) show a similar behavior, but the deepening of the color is slower.

Procedure. A drop of the test solution (in a volatile solvent) is evaporated to dryness on a watch glass. A drop of concentrated sulfuric acid is added and the temperature raised to 95°. Comparison is made with an unheated blank.

The behavior of a number of azoxy compounds is given in Table 12.

1 P. H. GORE and G. K. HUGHES, *Anal. Chim. Acta*, 5 (1951) 357.
2 O. WALLACH and L. BELLI, *Ber.*, 13 (1880) 525.

84. Arylhydroxylamines (Ar—NHOH group)

Test with sodium pentacyanoammineferroate

Aromatic hydroxylamine compounds yield deep red-violet complexes with prussic salts[1]:

TABLE 12

BEHAVIOR OF AZOXY COMPOUNDS WITH CONCENTRATED
SULFURIC ACID

Compound	Color
Azoxybenzene	lemon[a], very deep red[b, c]
3,3'-Dichloroazoxybenzene	lemon[a], orange-red[b], deep orange-red[c]
4,4'- ,, ,,	yellow[a], orange-red[b], bright blood red[c]
α-4-Bromoazoxybenzene ⎫ β-4- ,, ,, ⎭	orange[a], deep red-brown[b], intensely dark red-brown[c]
2,4,6,2',4',6'-Hexabromoazoxybenzene	pale brown[a], yellow-brown[b], deep rose-red[c]
3,3'-Dinitroazoxybenzene	yellow[a,b,c], very light brown[d]
3,5,3',5'-Tetranitroazoxybenzene	pale yellow[a,b], light yellow-brown[c], pale red-brown[d]
α-4-Nitroazoxybenzene	greenish yellow[a,b], orange red[c]
β-4- ,, ,,	pale yellow[a], orange-red[b], deep blood red[c]
2,2'-Dimethylazoxybenzene	orange-brown[a], chocolate brown[b], intensely dark brown[c]
4,4'- ,, ,,	orange[a], intensely dark blood red[b], deep red-brown[c]
2,2'-Dimethoxyazoxybenzene	yellow-brown[a], olive green → very dark royal blue[b], deep bluish purple[c]
4,4'- ,, ,,	yellow-brown[a], deep red-brown[b], deep chocolate brown[c]

[a] initial; [b] 3 sec., 95°; [c] 30 sec., 95°; [d] boiled.

$$\text{ArNHOH} + \text{Na}_3[\text{Fe(CN)}_5\text{NH}_3] \rightarrow \text{Na}_3[\text{Fe(CN)}_5\text{ArNHOH)}] + \text{NH}_3$$

This reaction can be employed for the detection of arylhydroxylamines.

Procedure.[2]   A drop of the test solution is mixed with a drop of a 1% solution of sodium pentacyanoammineferroate in a depression of a spot plate. A red-violet color indicates a positive response.

Limit of identification: 0.2 γ phenylhydroxylamine.

The color is independent of the pH. Nitrosobenzene as well as its homologs and halogen derivatives react in the same manner. The test can be supplemented by a negative response of tests for nitroso groups (see p. 290).

1 H. W. Schwechten, *Ber.*, 65 (1932) 1734.
2 V. Anger, *Mikrochim. Acta*, (1962) 94.

85. Oximes and hydroxamic acids (=NOH group)

Tests by splitting off hydroxylamine and oxidation to nitrous acid

If aldoximes and ketoximes, or hydroxamic acids derived from carboxylic and sulfonic acids, are warmed with concentrated hydrochloric acid, the NOH group is split off to give hydroxylamine hydrochloride (*1, 1a, 1b*). The hydroxylamine can be oxidized to nitrous acid by iodine in acetic acid solution (*2*).[1] If the oxidation occurs in the presence of sulfanilic acid, the latter is diazotized by the nitrous acid (*3*). After the excess iodine is removed by means of thiosulfate (*4*), the *p*-diazoniumbenzenesulfonic acid (I) can be coupled with 1-naphthylamine (II) to produce the red azo dye 4-(*p*-sulfophenylazo)-1-naphthylamine (III) (*5*). The reactions are:

$$\text{>C=NOH} + H_2O + HCl \rightarrow \text{>C=O} + NH_2OH.HCl \qquad (1)$$

$$RC(OH)NOH + H_2O + HCl \rightarrow RCOOH + NH_2OH.HCl \qquad (1a)$$

$$RSO_2(OH)NOH + H_2O + HCl \rightarrow RSO_3H + NH_2OH.HCl \qquad (1b)$$

$$NH_2OH + 2 I_2 + H_2O \rightarrow HNO_2 + 4 HI \qquad (2)$$

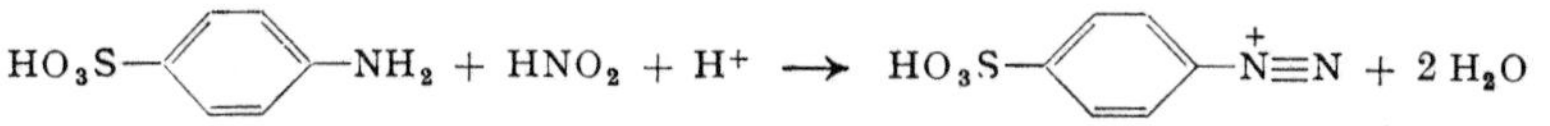

(3)

$$I_2 + 2 Na_2S_2O_3 \rightarrow 2 NaI + Na_2S_4O_6 \qquad (4)$$

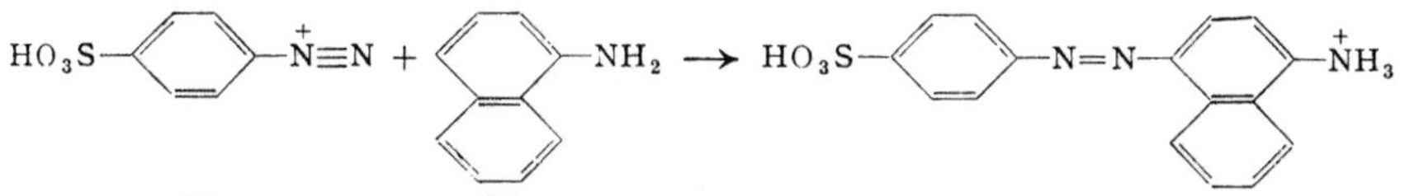

(5)

(I)   (II)   (III)

The reactions (*1*)–(*5*), of which (*3*) and (*5*) constitute the basis of the familiar and sensitive Griess test for nitrite, proceed so rapidly and completely that they can serve as the basis of a specific test (Procedure I) for small amounts of aldoximes, ketoximes, and hydroxamic acids.

Procedure I.[2] A little of the sample (solutions should be evaporated to dryness) is heated in a micro crucible with 3 drops concentrated hydrochloric acid until the volume is reduced to about one-fifth of the original. A few mg solid sodium acetate, 1 or 2 drops of sulfanilic acid solution, and a drop of iodine solution are added in succession. After 2 or 3 min. any excess free iodine is removed by 0.1 *N* sodium thiosulfate, and then a drop of 1-naphthylamine solution is added. A more or less intense red color appears, according to the amount of oxime or hydroxamic acid present.

Reagents: 1) Iodine solution. 1.3 g iodine in 100 ml glacial acetic acid.

       2) Sulfanilic acid solution. 10 g sulfanilic acid in 750 ml water plus 250 ml glacial acetic acid.

3) 1-Naphthylamine solution. 3 g base in 700 ml water plus 300 ml glacial acetic acid.

Limits of identification are:

0.08 γ acetone oxime	0.1 γ cyclohexanedione dioxime
0.03 γ dimethyl oxime	0.5 γ isonitrosocamphor
0.05 γ benzil dioxime	0.5 γ α-nitroso-β-naphthol
0.1 γ benzoin oxime	0.5 γ β- ,, -α- ,,
0.2 γ benzohydroxamic acid	0.9 γ benzenesulfohydroxamic acid
0.4 γ desoxybenzoin oxime	8 γ isonitrosomethyl ethyl ketone
0.6 γ benzylbenzoin oxime	7 γ isonitrosoacetophenone
6 γ o-methoxybenzaldoxime	
0.2 γ benzylidenebutan-2-one oxime	

In acid solution hydroxylamine is converted to nitrous acid by most oxidants ($NH_2OH + 2\,O \rightarrow H_2O + HNO_2$) as indicated by the positive response to the Griess test. This result, as well as the finding that the hydrolytic cleavages (*1*), (*1a*) and (*1b*) proceed to a notable extent even with brief warming, provide the basis of a test for oximes and hydroxamic acids. It is well to use bromine water as the oxidant because excess bromine is easily eliminated by means of sulfosalicylic acid (comp. p. 72). The test conducted by Procedure II is likewise specific but less sensitive than the test outlined in Procedure I.

Procedure II.[3] A small quantity of the solid or a drop of its solution is mixed in a micro test tube with a drop of hydrochloric acid (1:1) and heated for about 3 min. in boiling water. After cooling, bromine water is added, drop by drop, until the solution acquires a yellow color. A drop of a saturated aqueous solution of sulfosalicylic acid is added to discharge the color, and then a drop of Griess reagent (equal volumes of reagents 2 and 3 as described in Procedure I) is introduced. The mixture is then warmed in boiling water. A positive response is indicated by the development of a red color.

The *limits of identification* range from 1–10 γ for oximes and hydroxamic acids.

1 J. BLOM, *Ber.*, 59 (1926) 121; comp. also F. RASCHIG, *Schwefel- und Stickstoffstudien*, Leipzig, 1924, p. 183.
2 F. FEIGL and V. DEMANT, *Mikrochim. Acta*, 1 (1937) 132.
3 F FEIGL and E. SILVA, *Analyst*, 82 (1957) 583.

86. Aliphatic oximes ($R_2C{=}NOH$ group)

Test with benzoyl peroxide[1]

When oximes are heated to 120–130° with benzoyl peroxide (m.p. 103°), nitrous acid is produced only with aliphatic oximes, whereas aromatic oximes yield no more than traces of this product. Consequently, the fusion reaction may be used to distinguish aliphatic from aromatic oximes. The reaction may be:

$$\mathord{>}C{=}NOH + 2\ (C_6H_5CO)_2O_2 \rightarrow \mathord{>}C{=}O + 2\ (C_6H_5CO)_2O + HNO_2$$

The procedure and reagents are those described in Sect. 87 (1).

The following amounts were revealed:

10 γ dimethylglyoxime   20 γ cycloheptanedione dioxime   40 γ camphor oxime

A positive response was obtained with: phenylglyoxaldoxime, furfuraldoxime, 1,2-cyclohexanedione dioxime. No reaction was given by: salicylaldoxime, α-benzil monoxime, α-benzil dioxime, benzoin oxime.

Nitro compounds must be absent (see the next section).

1 F. FEIGL and E. SILVA, *Analyst*, 82 (1957) 585.

87. Hydroxamic acids (—CO—NHOH group)

(1) Test with fused benzoyl peroxide[1]

If aliphatic or aromatic hydroxamic acids are heated to 120–130° with benzoyl peroxide, nitrous acid is split out within several minutes:

$$(R,Ar)CONHOH + 2\ (C_6H_5CO)_2O_2 \rightarrow (R,Ar)COOH + 2\ (C_6H_5CO)_2O + HNO_2$$

The resulting nitrous acid can be detected in the gas phase by the Griess test.

Procedure.   A micro test tube is used. A little of the solid or the evaporation residue of a drop of its solution in ether or benzene is treated with a drop of benzoyl peroxide solution and taken to dryness. The mouth of the test tube is covered with filter paper moistened with nitrite reagent solution and the tube is immersed in a glycerol bath previously heated to 120–130°. A positive response is indicated if a pink or red stain appears within 3–10 min.

Reagents: 1) 10% solution of benzoyl peroxide in benzene.
2) Griess reagent (for preparation see p. 92).

The procedure revealed 30–40 γ benzohydroxamic acid, salicylhydroxamic acid, phenylacetohydroxamic acid or *p*-methoxybenzohydroxamic acid.

Since many aliphatic and aromatic nitro compounds likewise yield nitrous acid when heated with benzoyl peroxide, the fusion test is not applicable when such compounds are present.

1 F. FEIGL and E. SILVA, *Analyst*, 82 (1957) 586.

(2) Test through oxidation with iodine in acetic acid solution[1]

Hydroxylamine is obtained from hydroxamic acids by hydrolysis much more easily than from oximes. This is shown by the fact that nitrous acid is formed immediately when iodine acts in acetate buffered acetic acid solution:

$$(R,Ar)CONHOH + 2 I_2 + 2 H_2O \rightarrow (R,Ar)COOH + 4 HI + HNO_2$$

Oximes do not react under these conditions.

Procedure. As described in Sect. 85, Procedure I, with the difference that the fuming off with hydrochloric acid is omitted.

The following were revealed:

5 γ acetohydroxamic acid   0.25 γ p-methoxybenzhydroxamic acid

0.1 γ phenylacetohydroxamic acid   0.25 γ benzenesulfohydroxamic acid

1 Unpublished studies, with E. LIBERGOTT.

Nitroso compounds

Nitroso compounds may be divided into two classes: (*a*) the N-nitroso compounds (nitrosamines) in which the NO group is attached to nitrogen as in $\begin{smallmatrix} R \\ R_1 \end{smallmatrix}$N—NO (R and R_1 = alkyl or aryl radical) and (*b*) the C-nitroso compounds in which the NO group is attached to a carbon atom in an alkyl or aryl radical. The second class also includes the *iso*-nitroso compounds in which the NO group is enolized to the NOH group.

An indication of the presence of C-nitroso compounds is furnished by the fact that frequently they are green, or if colorless in the crystalline state (*e.g.* nitrosobenzene) they assume a green or blue color when fused or vaporized. Furthermore, freshly prepared solutions in benzene, alcohol, ether, *etc.* turn blue on standing and especially when warmed. Nitroso compounds, which form salts because of basic groups in the molecule, show this color change only after they are liberated from their salts by caustic alkali. In contrast, N-nitroso compounds are yellow liquids or solids, which show no color change when solidified or liquefied.

There is a fundamental difference between the two classes of nitroso compounds in their behavior toward hydrazoic acid. Even when warmed with sodium azide and dilute hydrochloric acid C-nitroso compounds are unaltered, whereas N-nitroso compounds are quickly denitrosated even at room temperature:[1]

$$>\!N\!-\!NO + HN_3 \rightarrow >\!NH + N_2O + N_2$$

Alkyl nitrites are likewise denitrosated by hydrazoic acid:

$$>\!C\!-\!ONO + HN_3 \rightarrow >\!C\!-\!OH + N_2O + N_2$$

These effects, together with the following tests, may have value in characterizing C-nitroso compounds.

1 F. FEIGL, *Anal. Chem.*, 27 (1955) 1315.

88. C-Nitroso compounds (—NO group)

(1) Test with phenol and sulfuric acid (Liebermann test) [1]

When warmed with phenol and concentrated sulfuric acid, aliphatic nitroso and iso-nitroso compounds and aromatic nitroso compounds give a red
color, which turns blue with alkalis. The sulfuric acid probably saponifies
nitroso compounds with production of nitrous acid, which nitrosates the
phenols in the unoccupied p-position. The resulting nitrosophenols condense
in their isomeric NOH (oxime) form with excess phenol to produce quinoid
indophenol dyes (compare tests (1) and (2) for phenol, p. 179).

The following test is characteristic for nitroso compounds, provided alkyl
nitrites are absent, since the latter, as derivatives of nitrous acid, likewise
show the Liebermann reaction.

Procedure.[2]   A little of the powdered substance is melted in a micro crucible
along with a particle of phenol. After cooling, a few drops of pure concentrated
sulfuric acid are added. The sample turns dark cherry-red. After diluting with a
little water, the solution is made alkaline with a few drops of 4 N sodium hydroxide. The solution turns deep blue. A solution of the sample in ether may be
used instead of the solid.

Limits of identification are given in Table 13.

1 C. LIEBERMANN, *Ber.*, 7 (1874) 247, 287, 806, 1098.
2 F. FEIGL, V. ANGER and O. FREHDEN, *Mikrochemie*, 15 (1934) 181, 183.

(2) Test with sodium pentacyanoammineferroate [1]

Nitroso compounds form brightly colored complex products with the prussic salt sodium pentacyanoammineferroate $Na_3[Fe(CN)_5NH_3]$. The mechanism of the color reaction [2] differs with regard to the production of the green
or violet materials. The green products result from the exchange of the
ammonia molecule of the prussic salt for a molecule of nitrosamine or
nitrosophenol:

$$Na_3[Fe(CN)_5NH_3] + HOC_6H_4NO \rightarrow NH_3 + Na_3[Fe(CN)_5(HOC_6H_4NO)]$$

The violet products result from the reduction of the nitroso compound to
the corresponding arylhydroxylamine by the complex Fe(II) salt and subsequent substitution of the ammonia by the arylhydroxylamine: [3]

$$ArNO + 2 Na_3[Fe(CN)_5NH_3] + 2 H_2O \rightleftharpoons ArNHOH + 2 Na_2[Fe(CN)_5NH_3] + 2 NaOH$$
$$ArNHOH + Na_3[Fe(CN)_5NH_3] \rightarrow Na_3[Fe(CN)_5ArNHOH] + NH_3$$

It is noteworthy that the entrance of the ArNO into the anion of the prussic
salt, with expulsion of the NH_3, occurs only in the light. In the dark, the
exchange proceeds at an immeasurably slow rate. This is one of the very
rare examples of the direct application in spot test analysis of a photo

reaction. Other instances may be found in the test for primary aromatic amines (p. 248) and for coumarin (p. 479).

Aromatic thioaldehydes and a few thioketones[4] also react with prussic salts to give a blue color; certain aromatic hydrazines give a red or violet color. This result probably likewise involves an exchange reaction with NH_3 and H_2O molecules of the prussic salts. Interference by hydrazines may be prevented by adding a few drops of formaldehyde: formaldehydehydrazone is formed which does not react with prussic salts.

The following compounds give blue or blue-green products with sodium pentacyanoammineferroate[5]: thiourea, phenylthiourea, formamidine disulfide, thiouracil. A reddish color is produced with isonicotinic acid (in contrast to nicotinic acid and picolinic acid, which do not react).[6]

It is noteworthy that pyridine (in contrast to picoline) hinders the color reaction of nitroso compounds with pentacyanoammineferroate; this masking is probably due to the formation of a pentacyanopyridineferroate.

As shown in the following table quinonedioxime behaves like a nitroso compound. This is because it reacts in its tautomeric form as p-nitrosophenylhydroxylamine (see p. 447).

Procedure. A drop of the test solution or a small amount of the sample is mixed on a spot plate with several drops of a freshly prepared 1% solution of sodium pentacyanoammineferroate. An intense green color, or more rarely, a violet color appears.

Limits of identification are given in Table 13.

TABLE 13

NITROSO COMPOUNDS

Name	Limit of identification (γ) and color	
	Test with phenol+H_2SO_4	Test with $Na_3[Fe(CN)_5NH_3]$
p-Nitrosodimethylaniline	0.5 pink–yellow–green	0.15 emerald green
p-Nitrosophenol	0.4 red–blue	0.15 dark green
α-Nitroso-β-naphthol	0.5 red–blue	1 olive green
β- ,, -α- ,,	0.6 red–green	1 olive green
Tetrahydro-β-nitroso-α-naphthol	0.5 red–green	—
Isonitrosoacetylacetone	1 dark red–green-yellow	2.5 brown lilac
Isonitrosoacetophenone	1 brick red–yellow	3 green
Nitrosobenzene	1 red–blue	0.2 violet
Quinonedioxime		0.8 blue

1 F. FEIGL, V. ANGER and O. FREHDEN, *Mikrochemie*, 15 (1934) 181, 184.
2 V. ANGER, *Mikrochim. Acta*, (1962) 94.
3 Comp. O. BAUDISCH, *Ber.*, 54 (1921) 413.
4 H. W. SCHWECHTEN, *Ber.*, 65 (1932) 1734.
5 R. R. FEARON, *Analyst*, 71 (1946) 562.
6 E. F. G. HERINGTON, *Analyst*, 78 (1953) 175.

(3) Test with N,N'-diphenylbenzidine[1]

As stated on p. 127, N,N'-diphenylbenzidine in strong sulfuric acid solution is converted into a blue p-quinoidal compound by nitrates, nitrites, and other strong oxidants. The same product is also obtained by the action of nitroso compounds. However, in this case the result is not due to the hydrolytic splitting out of oxidizing nitrous acid, as might be expected. Instead, the diphenylbenzidine is oxidized here because it functions as hydrogen donor and the NO group acts as hydrogen acceptor through conversion into the NHOH group:

$$ArNO + C_6H_5NHC_6H_4.C_6H_4NHC_6H_5 \longrightarrow ArNHOH + C_6H_5N{=}\langle\;\rangle{=}\langle\;\rangle{=}NC_6H_5$$

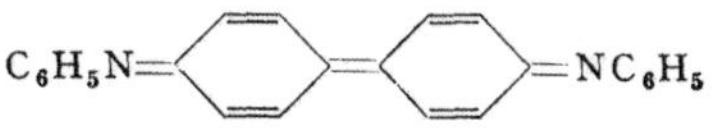

Unlike the nitroso compounds, the iso-nitroso compounds, which are really oximes, do not react to form hydroxylamine derivatives. The test given here is not applicable in the presence of strong oxidants, such as nitrates, nitrites, halogenates, esters of nitric and nitrous acids, organic compounds containing active halogen (for instance chloramine, etc.), as well as a number of quinones[2].

Procedure. A micro test tube is used. A small quantity of the solid or a drop of its solution is treated with about 0.5 ml of a solution of 10 mg N,N'-diphenylbenzidine in 10 ml 85% sulfuric acid (1 vol. of water plus 3 vol. concentrated sulfuric acid). If the response is positive, a more or less intense blue color develops at once, or after brief warming in a boiling water bath.

The test revealed:

1 γ nitrosobenzene	10 γ β-nitroso-α-naphthol
0.2 γ p-nitrosophenol	20 γ α- ,, -β- ,,
0.4 γ 5-nitroso-8-hydroxyquinoline	2 γ nitrosoantipyrine
30 γ p-nitrosophenylhydroxylamine	1 γ p-nitroso-N-dimethylaniline

1 V. ANGER, *Mikrochim. Acta*, (1960) 827.
2 Comp. V. ANGER, *Mikrochim. Acta*, (1959) 386.

89. Nitrosamines ($>$N—NO group)

Test by hydrolytic release of nitrous acid[1]

The denitrosation of N-nitroso compounds (nitrosamines) may be regarded as the result of the constant removal of nitrous acid from the equilibrium, in an acid medium,

$$>N{-}NO + H_2O \rightleftharpoons >NH + HNO_2$$

through the irreversible redox reaction

$$HNO_2 + HN_3 \rightarrow H_2O + N_2O + N_2$$

In harmony with this viewpoint is the fact that N-nitroso compounds, at suitable pH values, show the reactions of nitrous acid which is not true at all or only to a slight extent with C-nitroso compounds because these do not undergo an adequate primary hydrolytic cleavage.

The nitrous acid resulting from nitrosamines can be detected by means of the Griess reagent (Procedure I). Certain N-nitroso compounds give this color reaction even on slight warming with the acetic acid solution of sulfanilic acid and 1-naphthylamine. For the general test, it is best to warm the sample plus the reagent solution in the presence of strong hydrochloric acid. The chemistry of this Griess test is discussed on p. 90.

The nitrous acid furnished by the hydrolysis equilibrium may also be caused to react with sulfamic acid:

$$HNO_2 + NH_2SO_3H \rightarrow H_2SO_4 + H_2O + N_2$$

This redox reaction is the inorganic counterpart of the familiar exchange of the NH_2 groups bound to C atoms for OH groups brought about by nitrous acid. When carried out in the presence of Ba^{2+} ions, barium sulfate precipitates. This is the basis of a macro test and also of a gravimetric procedure for determining nitrite.[2] Procedure II makes use of this finding for the detection of nitrosamimes.

In both I and II, the hydrolytic splitting of N-nitroso compounds can be accomplished in the wet way by acid. In conformity with its stoichiometric representation, this hydrolysis should also be realizable in the absence of acid. Actually, if dry mixtures of nitrosamines with hydrated manganese sulfate are heated the sulfate loses its water and the resulting superheated steam accomplishes the hydrolysis to yield volatile nitrous acid, which can then be detected in the vapors by the Griess reagent (Procedure III).

O-Nitroso compounds, *i.e.* esters of nitrous acid, behave like N-nitroso compounds with respect to splitting off nitrous acid.

Procedure I. One drop of the test solution is treated in a micro test tube with a drop of freshly prepared Griess reagent and one drop of 1:1 hydrochloric acid and the mixture is warmed in a water bath. If nitrosamines are present, a more or less intense red-violet develops within several minutes.

This procedure revealed:

10 γ N-nitrosodibenzylamine	0.4 γ N-nitrosomethylurea
9 γ N-nitrosodicyclohexylamine	4 γ N,N'-dinitrosopiperazine
1 γ N-nitrosodiphenylamine	1 γ N-nitrosoacetanilide

Procedure II. A drop of the aqueous, alcoholic, *etc.* test solution and a drop of the reagent solution are mixed in a micro test tube and heated gently in warm water if necessary. A precipitate or turbidity appears if nitrosamines are present.

Reagent: 5 g $BaCl_2.2 H_2O$ and 5 g sulfamic acid (amidosulfonic acid) are dissolved in 100 ml of a mixture of equal volumes of water and dioxane. Any precipitate should be filtered off before use.

The procedure revealed:

    10 γ N-nitrosodiphenylamine   10 γ N-nitrosomethylurea

Procedure III. A micro test tube is used. One drop of the test solution or a little of the solid is mixed with several cg of $MnSO_4$.aq. and taken to dryness. The mouth of the test tube is covered with a disk of filter paper moistened with Griess reagent. The heating over a micro flame is continued to strong caramelization. In the presence of nitrosamines, a red-violet stain appears on the colorless paper.

The procedure revealed:

    10 γ N-nitrosodicyclohexylamine   5 γ N-nitrosodiphenylamine
    15 γ N,N'-dinitrosopiperazine

1 F. FEIGL and Cl. COSTA NETO, *Anal. Chem.*, 28 (1956) 1311.
2 P. BAUMGARTEN and J. MARGGRAFF, *Ber.*, 63 (1930) 1019.

90. Aliphatic nitrosamines

Test through denitrosation [1]

Aliphatic nitrosamines are denitrosated by hydrazoic acid to yield secondary amines (see p. 292) which react in characteristic manner with carbon disulfide and ammoniacal copper solution. Brown copper dithiocarbamates result which give brown solutions in benzene. These salts are formed directly from the components as outlined in Sect. 50.

$$CS_2 + NHRR_1 + {}^1/_2Cu^{2+} + NH_3 \rightarrow S C \overset{\displaystyle S-Cu/_2}{\underset{\displaystyle NRR_2}{}} + NH_4^+$$

The procedure given here is of course reliable only in the absence of secondary aliphatic amines.

Procedure. The test is conducted in a micro test tube. The solid or the evaporation residue from a solution is treated with a little solid sodium azide and several drops of dilute hydrochloric acid. The mixture is gently warmed until no more gas is evolved. One drop of 5% solution of copper sulfate is then introduced and an excess of ammonia is added. The blue solution is shaken with 2 drops of a mixture prepared from 3 parts benzene and 1 part carbon disulfide. A positive response is signalled by the benzene layer becoming brown or yellow.

The procedure revealed:

    1 γ N-nitrosodibenzylamine   10 γ N,N'-dinitrosopiperazine
    500 γ N-nitrosodicyclohexylamine

1 Unpublished studies, with L. HAINBERGER.

91. Nitro compounds (NO₂ group)

Test by fusion with tetrabase or diphenylamine[1]

If nitro compounds are added to molten tetrabase (m.p. 91°) or molten diphenylamine (m.p. 53°), or if their solid mixtures with these bases are heated to 100°, colored (usually orange-red) melts are obtained. These color reactions are based on the formation of molecular compounds between nitro compounds and tetrabase or diphenylamine[2]:

$$R-\overset{+}{\underset{O^-}{N}}=O\rightarrow \overset{CH_3}{\underset{CH_3}{N}}-C_6H_4CH_2C_6H_4-\overset{CH_3}{\underset{CH_3}{N}} \qquad \overset{C_6H_5}{\underset{C_6H_5}{HN}}\leftarrow O=\overset{+}{\underset{O^-}{N}}-R$$

The addition compounds are soluble in the molten bases (solvation). The postulate of solvate formation is supported by the finding that with some nitro compounds the colored melts become colorless when cooled or take on the color of the dry mixture. The color reappears on heating. This signifies that the molecular compounds in these instances are stable only in the form of their solvates in melts and dissociate into their components on cooling.

Mono- and polynitro compounds behave alike with respect to the formation of molecular compounds and their solvation in the melts of these bases. In general the products obtained from polynitro compounds are more intensely colored and more stable than the mono compounds.

p-Quinoid compounds interfere with the test because they yield red solvates of the respective molecular compounds when fused with tetrabase or diphenylamine. Nearly all of the melts with quinoid compounds lose their color on cooling.

Procedure. The test is conducted in a micro conical tube (Emich centrifuge tube). One drop of the ethereal or benzene solution of the sample, and one drop of a 5% solution of tetrabase or diphenylamine in benzene are introduced and the tube then dipped into boiling water. After the solvent has evaporated, a melt remains in the narrow part of the tube; it is more or less yellow depending on the quantity of nitro compound present.

The procedure revealed:

50	γ nitromethane	0.2 γ *p*-nitrophenetole
0.5	γ tetranitromethane	1 γ *o*-nitrobenzaldehyde
0.8	γ nitrobenzene	3 γ *o*-nitrobenzoic acid
1	γ (*p, o*)-dinitrobenzene	0.1 γ 2,4-dinitrochlorobenzene
0.3	γ *p*-nitrobenzonitrile	2.5 γ 3-nitro-4-hydroxybenzyl alcohol
2	γ *p*-nitroaniline	0.2 γ 2,4-dinitrophenol
2	γ (*p, m*)-nitrophenol	0.1 γ 2,4-dinitroresorcinol
1	γ 1-nitronaphthalene	5 γ chloromycetin

1 F. FEIGL, Cl. COSTA NETO and C. STARK-MAYER, unpublished studies.
2 Comp. H. BEILSTEIN, *Handb. Org. Chemie*, Bd 12 (1924) 174; 13 (1930) 242.

92. Primary aliphatic nitro compounds ($R-CH_2-NO_2$ group)

(1) Test through coupling with fast blue salt B[1]

Primary nitro compounds in alkaline solution, and hence in their aci-form, couple with diazonium salts to give colored (red or orange) condensation products[2,3]:

$$RCH=NO_2K + ArN=NCl + KOH \rightarrow R-C\begin{smallmatrix}NO_2K \\ N=NAr\end{smallmatrix} + KCl + H_2O.$$

In the absence of other compounds which couple with diazonium salts (especially phenols and aromatic amines), this reaction is characteristic for primary nitroparaffins. Secondary nitroparaffins and also aromatic nitro compounds do not react.

It is best to use a stable diazonium salt for the condensation. Well suited is fast blue salt B (tetrazotized di-o-anisidine = I). It reacts with alkaline solutions of primary nitroparaffins to yield mono- or di-coupling products (II or III):

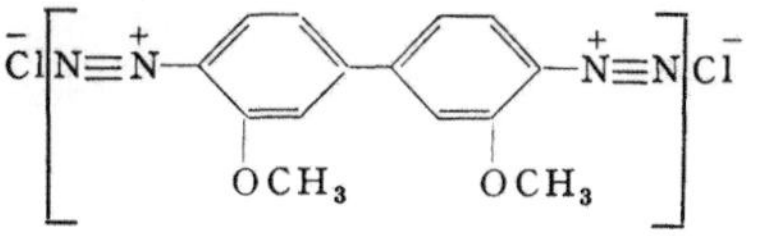

(I) 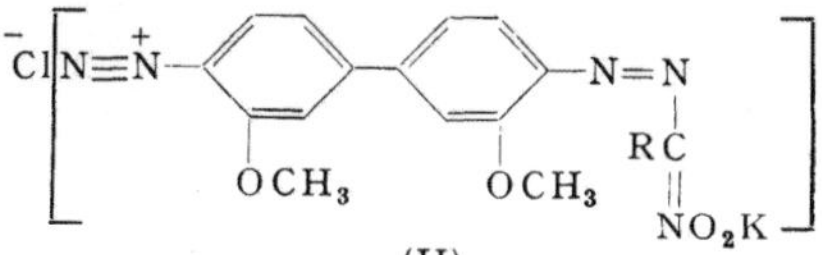 (II)

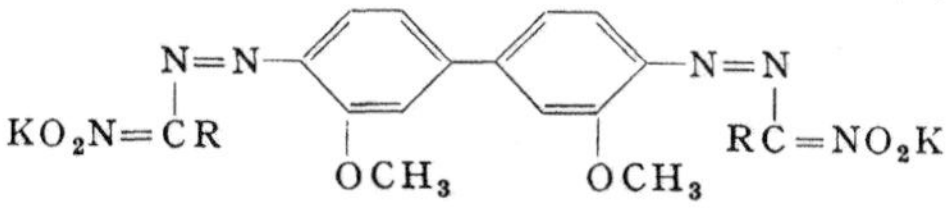

(III)

The postulation of the formation of different condensation products (II) or (III) is supported by the finding that the reaction picture differs when the test is made on a spot plate (or micro test tube) as opposed to filter paper. An orange color results in the first case, whereas a red stain or ring results on paper. The color reaction on filter paper is considerably more sensitive.

Procedure. One drop of the freshly prepared reagent solution is placed on filter paper (S & S, Black Ribbon) followed by 1 drop each of the alcoholic test solution and 0.5 N sodium hydroxide. The appearance of an orange fleck or ring, which turns red after a few minutes, indicates that a primary nitroparaffin was present originally.

Reagent: Fast blue salt B is suspended in ethanol, shaken thoroughly, and then centrifuged. The reagent solution does not keep.

The procedure revealed: 0.5 γ nitromethane, nitroethane, nitropropane.

1 F. FEIGL and D. GOLDSTEIN, Z. Anal. Chem., 158 (1957) 427.
2 V. MEYER and G. AMBÜHL, Ber., 8 (1875) 751.
3 F. TURBA, R. HAUL and G. UHLEN, Z. Angew. Chem., 61 (1949) 74.

(2) Test through pyrolytic–reductive cleavage to nitrous acid[1]

When aliphatic nitro compounds are heated to 160° with benzoin (m.p. 137°), nitrous acid is split off. Benzoin is converted to benzil and functions as hydrogen donor:

$$RNO_2 + C_6H_5CHOHCOC_6H_5 \rightarrow RH + C_6H_5COCOC_6H_5 + HNO_2$$

It is essential that the given temperature not be exceeded because aromatic polynitro compounds as well as nitroamines react with benzoin at temperatures above 190° to produce nitrous acid. Hexogen (see p. 684) is an exception. Since the nitrous acid split off in the above redox reaction can be detected with high sensitivity, through the Griess test, aliphatic nitro compounds can be differentiated from aromatic nitro compounds through fusion with benzoin.

Procedure. A slight quantity of the test substance, dissolved in or mixed with benzene, is brought together with several cg of benzoin in a micro test tube. The solvent is volatilized and the test tube is immersed to a depth of around 0.5 cm in a glycerol bath preheated to 130°. The mouth of the tube is covered with a disk of filter paper moistened with Griess reagent and the temperature is increased to 160°. A positive response is indicated by the appearance of a red or pink fleck on the reagent paper.

The *Limits of Identification* are in the range of 10–20 γ.

It is notable that low boiling aliphatic nitro compounds such as nitroethane (b.p. 114°) can be detected by this procedure. This finding is due to the fact that vapors of the nitro compound remain in the test tube when the heating occurs and therefore can react with the fused benzoin.

1 Unpublished studies, with E. LIBERGOTT.

93. Aromatic nitro compounds

(1) Test with sodium pentacyanoammineferroate after electrolytic reduction to nitroso compounds[1]

The color reaction of nitroso compounds with sodium pentacyanoammineferroate (p. 290) can also be applied to the detection of nitro compounds, if the nitro group is reduced to the nitroso group. This reduction can be carried out electrolytically in a drop of the test solution. The electrolysis is accomplished in neutral or alkaline solution, in the presence of the reagent, and between a nickel and a lead electrode.[2]

Procedure. A drop of the alcoholic or aqueous test solution is mixed in a micro crucible with a drop of a freshly prepared solution of 1% sodium pentacyanoammineferroate and a drop of 4 N alkali. If the addition of alkali causes a coloration, a drop of 4% sodium sulfate solution should be used instead as the electrolyte.

The current, drawn from a flashlight battery, is then passed for at least 10 min. or up to 30 min. for small amounts of test substances. The liquid becomes colored during the electrolysis. The color deepens on standing; it usually is green, more seldom violet.

A blank test with sodium sulfate shows no change in the pale yellow color, but there is a slight deepening of the yellow shade with alkali.

A study with mono- and polynitro aromatic compounds gave the following *limits of identification:*

15	γ p-chloronitrobenzene	green	8	γ m-nitrobenzoic acid	lilac
0.4	γ (o,p)-nitrophenol	,,	1.5	γ nitrobenzene	violet
3	γ (o,m,p)-nitrocinnamic acid	,,	3	γ m-chloronitrobenzene	,,
			3	γ m-dinitrobenzene	brown-violet
4	γ (o,m)-nitroaniline	,,	8	γ p-nitrotoluene	violet-red
0.25	γ p- ,,	,,	8	γ picric acid	olive green
0.5	γ (o,p)-nitrophenyl-hydrazine	,,	4	γ (o,m,p)-nitrobenz-aldehyde	,,
1.5	γ m-nitrophenyl-hydrazine	,,	0.3	γ 1,5-nitronaphthyl-amine	dark brown
2.5	γ p-nitrobenzoic acid	,,	5	γ 6-nitroquinoline	light green

1 F. FEIGL, V. ANGER and O. FREHDEN, *Mikrochemie*, 15 (1934) 183.
2 O. DIEFFENBACH, D.R.P. 192,519; *Chem. Abstr.*, 2 (1908) 1765.

(2) Test through reduction to phenylhydroxylamine derivatives or to nitroso-phenols and complex formation with pentacyanoammineferroate

If aromatic nitro compounds are reduced with metallic zinc in weakly acid solution arylhydroxylamines result. The weakly acid milieu can be obtained by means of ammonium chloride or calcium chloride. The aryl-hydroxylamines formed react with pentacyanoammineferroate to yield violet complexes. It should be pointed out that the reduction stops at the nitroso stage in the case of p-nitrophenols and p-nitrophenylamines.

If enough sample is available, it is well to reduce the nitro to nitroso groups by warming with calcium chloride and metallic zinc.[1] (Probably other reductants may also be used.)

Procedure. A few mg of the sample is dissolved in 3 ml of hot alcohol in a test tube and treated with 6 drops of 10% $CaCl_2$ solution. About 50 mg of zinc dust is added and the suspension is heated to strong boiling in a water bath. The cooled filtrate or centrifugate is treated with 1 drop of 1% solution of $Na_3[Fe(CN)_5NH_3]$. The color becomes purple, blue, or green if nitro compounds were present.

The procedure revealed: (o,m,p)-chloronitrobenzene, nitrobenzene, m-dinitro-benzene, (o,p)-nitrotoluene, (o,p)-nitrophenol, picric acid, (o,m,p)-nitrobenz-aldehyde, (m,p)-nitrobenzoic acid, (o,m,p)-nitrocinnamic acid, (o,m,p)-nitroani-line, 1,5-nitronaphthylamine, (o,m,p)-nitrophenylhydrazine, o-nitroquinoline, 3-nitro-4-aminoanisole, 2-amino-4-nitrophenetole, 4-nitroveratrole, 4-nitro-5-aminoveratrole, 3-nitro-4-acetaminoveratrole, p-nitroacetanilide, 1-nitro-2-naphthol, 4-nitropyridine-N-oxide, 4-nitroquinoline-N-oxide, 6-methoxy-8-ni-troquinoline, 5-nitrofurfural semicarbazone, 2,4-dinitrochlorobenzene, (2,4-, 3,5-)-dinitrotoluene, 2,4-dinitrophenol, 2,4-dinitrophenylhydrazine, 2,4-dinitro-1-naphthol-7-sulfonic acid (sodium salt), 1,3,5-trinitrobenzene, 2,4,6-trinitrotoluene.

Chloromycetin can likewise be detected by this method. Therefore, it is likely that the procedure for the detection of this compound given on p. 650 can also be used for the detection of other nitro compounds.

1 S. Ohkuma, *J. Japan. Chem.*, 4 (1950) 622.

94. Aromatic polynitro compounds

Test through saponification to nitrite[1]

Aliphatic and aromatic mono- and polynitro compounds exhibit character-istic differences in their behavior toward alkali hydroxide. Whereas aliphatic nitro compounds are saponified even in the cold, aromatic mononitro com-pounds, with a few exceptions* remain unaltered after prolonged boiling with even concentrated alkali. On the other hand, aromatic polynitro com-pounds produce nitrite after brief warming with concentrated alkali. Accor-dingly, at least one of the nitro groups in the molecule reacts:

$$Ar—NO_2 + 2 KOH \rightarrow Ar—OK + KNO_2 + H_2O$$

This difference in the behavior of alkali hydroxide toward mono- and poly-nitro compounds (with the nitro groups in the *ortho* and *para* positions) is in accord with the rule, which applies throughout the entire aromatic series, namely that negative substituents always activate and render mobile groups in the *ortho* and *para* position. Strangely enough, *meta*-dinitro compounds likewise react under the conditions given here.

The alkali nitrite produced by the saponification of aromatic polynitro compounds yield nitrous acid when acidified and the latter can be detected in the gas phase by the very sensitive Griess test.

If aryl and alkyl nitrites and aliphatic nitro compounds are absent, the following procedure is characteristic for aromatic polynitro compounds. No

* For example 9-nitrophenanthrene and its naturally occurring derivatives. Surpris-ingly these compounds give the color reaction with diphenylbenzidine described in Sect. 95 (according to V. Anger, Vienna).

nitrite is formed when C- and N-nitroso compounds are warmed with alkali solutions; N,N'-dinitrosopiperazine is an exception.

Procedure.[2] A tiny particle of the solid or a drop of its alcoholic solution is taken to dryness in a micro test tube along with 1 drop of 1 N alkali. The cooled residue is treated with 1 drop of each of the reagent solutions. Depending on the amount of polynitro compound present, a more or less intense red color appears.

Reagents: 1) 0.5% solution of sulfanilic acid in 1:1 acetic acid.
          2) 0.3% solution of 1-naphthylamine in 1:1 acetic acid.

Limits of identification:

0.5 γ dinitrobenzenes (*o,p*)
1   γ 2,4-dinitrochlorobenzene
1   γ 2,4-dinitrophenol

1   γ 2,4-dinitroresorcinol
0.5 γ hexanitrodiphenylamine
0.1 γ picric acid

A positive response was given by:

m-dinitrobenzene
dinitritoluenes (2,4; 2,6)
1-chloro-4-bromodinitrobenzene
2,4-dinitroaniline
dinitrobenzoic acids (2,4; 3,5)
dinitroxylenes (1,2,4,6; 1,2,3,4)
dinitro-*o*-cresols (3,5; 3,4)

2,4-dinitro-1-naphthol-7-sulfonic acid
dinitronaphthalenes (1,5; 1,8)
2,4-dinitrophenylhydrazine
trinitrobenzene (1,3,5)
trinitrotoluene (2,4,6)
trinitro-*m*-cresol (2,4,6)

1 P. K. Bose, *Analyst*, 56 (1931) 504.
2 Unpublished studies, with V. Gentil.

95. Nitrates, nitrites, nitroamines and nitrosamines
(—ONO$_2$, —ONO, >NNO$_2$, and >NNO groups)

Test through oxidation of diphenylamine or N,N'-diphenylbenzidine[1]

Among the organic nitrates are the salts of nitric acid with organic nitrogen bases (B) which as B.HNO$_3$ represent organic derivatives of ammonium nitrate. As solids or in solution they contain the NO$_3^-$ anion, and with few exceptions are water-soluble. The best known exception is nitron nitrate.[2] The —O—NO$_2$ group occurs in non-ionogenic form in the alkyl nitrates (RO—NO$_2$), which are alcohol esters of nitric acid. These esters of the lower C numbers are sweet-smelling, mobile liquids, which explode violently when heated. The nitric esters of polyhydric alcohols (nitroglycerine, nitrocellulose, nitromannitol, *etc.*) are solids if the parent alcohols are solid.

Salts of the nitrous acids with organic bases cannot be isolated. The nitrous esters RO—NO are isomeric with the corresponding C-nitro compounds R—NO$_2$, but differ from these stable materials in the ease with which they are saponified to nitrous acid and the respective alcohol. Therefore, nitrous esters are easily detected and distinguished from nitrates by adding a drop of the Griess reagent to a drop of the test solution; when gently warmed a red color appears (*Idn. Lim.* 0.2 γ amyl nitrite).

Nitroamines, as N-nitro compounds, are derivatives of primary or secondary amines in which an H atom is substituted by the NO_2 group. The pertinent compounds may also be considered as amines of nitric acid.

The present test for all organic nitrates, nitrites and nitroamines is based on the fact that they oxidize diphenylamine and diphenylbenzidine (dissolved in concentrated sulfuric acid) to intensely blue quinoidal compounds.[3] This oxidation of diphenylamine or diphenylbenzidine is brought about by the nitric or nitrous acid produced on saponification of alkyl nitrates, nitrites or nitroamines:

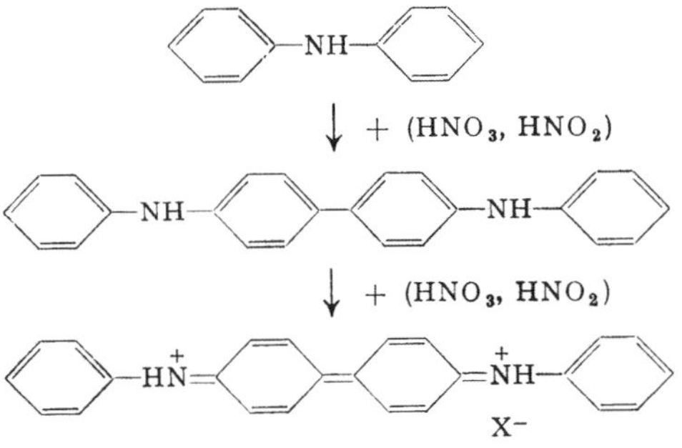

(I) Diphenylamine (colorless)

$\downarrow$ + (HNO$_3$, HNO$_2$)

(II) N,N-diphenylbenzidine (colorless)

$\downarrow$ + (HNO$_3$, HNO$_2$)

(III) Quinoidimonium salt (II) (blue)
X = $^1/_2$ SO$_4^{2-}$, Cl$^-$, *etc.*

This series of reactions shows that the use of diphenylbenzidine instead of diphenylamine has the advantage of making more effective use of the oxidizing action of the nitric or nitrous acid, and consequently the sensitivity of the test for nitrates, nitrites and nitroamines is raised.[4]

Certain aliphatic nitro compounds are resistant to diphenylamine (and probably diphenylbenzidine).[5] A negative result is obtained with RCH_2NO_2, $R_2C(NO_2)_2$ and $RC(NO_2)_3$ when R = alkyl, and $C(NO_2)_4$. The result is positive with $RCH(NO_2)R'$ and $RC(NO_2)R_2''$ when R = alkyl, R' = alkyl, NO_2 or halogen and R'' = alkyl or halogen.

The test described here assumes the absence of N-nitroso compounds (nitrosamines) since through cleavage of nitrous acid they likewise react with diphenylamine or diphenylbenzidine.

Procedure. About 0.5 ml of the reagent solution is placed on a spot plate and a little of the test material (fragment of a solid, a drop of a solution or suspension) is added. A blue ring is formed; the intensity of the color depends on the nitrate or nitrite content.

Reagent: Several crystals of diphenylamine or diphenylbenzidine are covered with concentrated sulfuric acid. A little water is added. When solution is complete, more concentrated acid is added. About 1 mg of the solute should be present in 10 ml of the finished reagent solution.

	(a) with diphenylamine	(b) with diphenylbenzidine
Limit of Identification:	0.5 γ nitric acid	0.07 γ nitric acid

A differentiation of nitroamines from nitro- and nitroso compounds is based on the fact that a 0.2% solution of diphenylamine in 8% phosphoric acid does not react with nitroamines, whereas nitrates and nitrites give a blue color.[6]

1 Unpublished studies.
2 Regarding water-insoluble nitrates, comp. F. FEIGL, *Chemistry of Specific, Selective and Sensitive Reactions*, Academic Press, New York, 1949, p. 305.
3 H. WIELAND, *Ber.*, 46 (1913) 3296; 52 (1919) 886; G. W. MONIER-WILLIAMS, *Analyst*, 56 (1931) 397; I. M. KOLTHOFF and G. E. NEPONEN, *J. Am. Chem. Soc.*, 55 (1933) 1443.
4 H. RIEHM, *Z. Anal. Chem.*, 81 (1930) 439.
5 K. GREBLER and J. V. KARABINOS, *J. Research Natl. Bur. Standards*, 49 (1952) 463.
6 C. L. WHITMAN and M. I. FAUTH, *Anal. Chem.*, 30 (1958) 1672.

96. Esters and amides of nitric acid (—ONO$_2$ and $>$N—NO$_2$ groups)

Test through fusion with benzoin [1]

Nitrates of organic bases, esters of nitric acid, as well as amines of nitric acid (nitroamines) split off nitrous acid when fused at about 160° with benzoin (m.p. 137°). The nitrous acid can be detected in the gas phase by means of the Griess color reaction. The pertinent procedure is given on p. 297 in connection with the test for aliphatic nitro compounds, which likewise yield nitrous acid when fused with benzoin; the latter acts as H-donor in this pyrohydrogenolysis.

Limits of Identification are in the range of 10–50 γ.

1 Unpublished studies.

97. Thiocyanates (—SCN group)

(1) Test through reductive cleavage [1]

If hydrochloric acid–alcohol solutions of aliphatic or aromatic thiocyanates are warmed with Devarda or Raney nickel alloy, the nascent hydrogen or the more reactive hydrogen in nickel hydride (comp. p. 83) occasions the reductive cleavage:

$$(R,Ar)SCN + 4 H^0 \rightarrow (R,Ar)H + H_2S + HCN \tag{1}$$

The detection of the concurrent production of hydrogen sulfide (through blackening of lead acetate paper) and of hydrogen cyanide (through bluing of copper acetate–benzidine acetate, comp. p. 546) therefore makes possible a specific test for organic thiocyanates.

Procedure. A small amount of the solid test material or a drop of its solution is placed in a micro test tube and followed by several cg of Raney nickel alloy and two drops of 5 N hydrochloric acid. After the stormy evolution of hydrogen has subsided, a dry wad of cotton impregnated with lead acetate is placed in the upper portion of the tube and a piece of filter paper moistened with hydrogen cyanide reagent is laid over the mouth of the tube. On warming (water bath), the lower end of the cotton turns black or brown and the reagent paper becomes blue if thiocyanates are present.

The test revealed:

thiocyano-2,4-dinitrobenzene	A: 5 γ; B: 50 γ
ethylenedithiocyanate	A: 2.5 γ; B: 5 γ

A, through formation of H_2S; B, through formation of HCN

It is noteworthy that the reductive cleavage leading to hydrogen cyanide proceeds to a lesser extent with aromatic thiocyanates than with aliphatic thiocyanates. Probably besides (1) the redox reaction (2) takes place:

$$(R,Ar)SCN + 8 H^0 + H^+ \rightarrow (R,Ar)H + H_2S + CH_3\overset{+}{N}H_3 \qquad (2)$$

In alkaline medium, the redox reaction leading to methylamine occurs exclusively in the case of aromatic and aliphatic thiocyanates.

1 F. FEIGL and D. HAGUENAUER-CASTRO, *Mikrochim. Acta*, (1962) 701.

(2) Test through formation of Prussian (Turnbull's) blue[1]

If organic thiocyanates in water–alcohol solution are warmed with sodium sulfite, and iron(III) chloride is then added along with acid, a blue precipitate of Prussian or Turnbull's blue appears. The formation of these compounds requires that a reaction has taken place in advance between the thiocyanate group and sodium sulfite which leads to a specific cleavage of the cyanide radical.

Procedure. A drop of the test solution, 4 drops of n-butanol, 2 drops of 10% sodium sulfite, and 5 drops of 2% sodium hydroxide solution are mixed in a micro test tube. The mixture is evaporated almost to dryness, the residue is moistened with 2 additional drops of sulfite solution and again taken almost to dryness. One drop of 10% ferric chloride solution and three drops of 15% sulfuric acid are added to the cold residue. A blue precipitate or color indicates a positive response.

Limits of identification:

110 γ ethyl thiocyanate	230 γ 4-thiocyanoacetanilide
100 γ thiocyanobenzene	540 γ 4-(4-thiocyanophenyl)-thiosemi-
100 γ 2-thiocyanonaphthalene	carbazide
60 γ thiocyanoaniline	190 γ N-(2-methyl-4-thiocyanophenyl)-N-
300 γ 4-thiocyanatophenazone	-4-chlorophenylthiocarbamide

Thiocyano-2,4-dinitrobenzene gives a weak response with this test.

1 R. POHLOUDEK-FABINI, D. GÖCKERITZ and M. SCHÜSSLER, *Mikrochim. Acta*, (1963) 668.

98. Amides of thiocarboxylic acids (—CSNH₂ group)

Test with alkali plumbite[1]

Thioacetamide is widely used in inorganic analysis as a source of hydrogen sulfide.[2] It reacts with alkali plumbite (prepared by adding excess alkali hydroxide to solutions of lead salts) to give black lead sulfide:

$$CH_3CSNH_2 + Pb(OH)_2 \rightarrow PbS + CH_3CONH_2 + H_2O \qquad (1)$$

This reaction alone is not indicative for thioacetamide specifically since quite a few other sulfur-bearing compounds also react with plumbite to give lead sulfide; examples include mercapto fatty acids, thiocarboxylic acids, thiourea, and many of its derivatives. However, alkali plumbite solutions contain free caustic alkali, and consequently partial hydrolysis of acetamide occurs on warming to 70–80° subsequent to reaction (*1*) and ammonia is split off:

$$CH_3CONH_2 + NaOH \rightarrow CH_3COONa + NH_3 \qquad (2)$$

The amides of other thiocarboxylic acids behave in this same manner. Accordingly, the *simultaneous* formation of lead sulfide and ammonia on warming with alkali plumbite is indicative of the presence of amides of thiocarboxylic acids.

Procedure. A minimal amount of the solid test material or one drop of its solution is brought in a micro test tube together with one drop of a plumbite solution. The mixture is heated in a water bath whereby the mouth of the tube is covered with a piece of filter paper moistened with Nessler's reagent. A positive response is indicated by the formation of black lead sulfide and the appearance of a brown fleck on the reagent paper.

Limits of identification: 2 γ thioacetamide (through PbS formation)
                                         50 γ   ,,   (through NH₃ formation)

1 Unpublished studies.
2 Comp. H. FLASCHKA, *Chemist-Analyst*, 44 (1955) 2.

99. Dithiocarbamates (—SCS—N— group)

Test by conversion into cupric salts soluble in organic solvents[1]

The dithiocarbamide group is present in the stable water-soluble and water-insoluble salts of the unstable dithiocarbamic acid (I) and its N-substituted derivatives (II) and (III).

$$S\,C\!\!\begin{array}{l}{}^{\nearrow SH}\\{}_{\searrow NH_2}\end{array}\;(I) \qquad\qquad S\,C\!\!\begin{array}{l}{}^{\nearrow SH}\\{}_{\searrow NHR}\end{array}\;(II) \qquad\qquad S\,C\!\!\begin{array}{l}{}^{\nearrow SH}\\{}_{\searrow NR_1R_2}\end{array}\;(III)$$

In (II), R denotes an alkyl or aryl group; in (III) R_1 and R_2 are alkyl groups only, and R_1 may be the same as R_2.

Water-soluble salts of (II) and (III) with primary or secondary aliphatic amines are produced by addition reactions of carbon disulfide:

$$CS_2 + 2NH_2R \longrightarrow SC\underset{\diagdown NHR}{\overset{\diagup SH.NH_2R}{}} \tag{1}$$

$$CS_2 + 2NHR_1R_2 \longrightarrow SC\underset{\diagdown NR_1R_2}{\overset{\diagup SH.NHR_1R_2}{}} \tag{2}$$

If alkali hydroxide or ammonia is present, the corresponding water-soluble alkali or ammonium salts of (II) and (III) result. With primary aromatic amines, carbon disulfide does not undergo a reaction analogous to (1); instead, symmetrical diarylthioureas and hydrogen sulfide are formed (compare p. 249). If, however, a mixture of ammonia and primary aromatic amines is allowed to react with carbon disulfide, ammonium salts of monoarylated dithiocarbamic acid result:[2]

$$CS_2 + NH_2R + NH_3 \longrightarrow SC\underset{\diagdown NHR}{\overset{\diagup SH.NH_3}{}} \tag{3}$$

A characteristic feature of water-soluble dithiocarbamates and their N-substituted derivatives is their precipitation as brown cupric salts, which dissolve in water-immiscible organic liquids to produce red-brown solutions.* This property, which has been utilized for the colorimetric determination of copper,[4] can also be employed in the detection of water-soluble dithiocarbamates and their N-substituted derivatives.

Procedure. A drop of the neutral test solution is treated in a micro test tube with a drop of an acetic acid–copper chloride solution and then shaken out with two or three drops of chloroform. The presence of dithiocarbamates is shown by the reddish-brown assumed by the chloroform layer.

Reagent: 1% solution of $CuCl_2$ mixed with an equal volume of 1:1 acetic acid.

A positive response was given by:

1.2 γ ammonium dithiocarbamate
2.5 γ sodium diethyldithiocarbamate
2 γ ammonium phenyldithiocarbamate
2 γ ammonium *p*-aminophenyldithiocarbamate
1.8 γ piperidinium piperidyldithiocarbamate

---

* A notable exception is presented by the copper salt of diethanolaminedithiocarbamic acid, which is soluble only in oxygen-containing organic liquids, but not in chloroform, benzene, *etc.*[3]

1 Unpublished studies, with D. GOLDSTEIN.
2 S. H. LOSANITSCH, *Ber.*, 24 (1891) 3021, 40 (1907) 2970.
3 E. GEIGER and H. G. MUELLER, *Helv. Chim. Acta*, 26 (1943) 996.
4 Comp. F. J. WELCHER, *Organic Analytical Reagents*, Vol. IV, Van Nostrand, New York, 1948, Chap. 3.

100. Sulfamic acids (—NHSO₃H group)

Test through splitting out of sulfuric acid[1]

Organic sulfamic acids are N-substituted aliphatic or aromatic derivatives of the inorganic amidosulfonic acid usually known as sulfamic acid. Their behavior toward nitrous acid is the same as that shown by the parent compound. This means that the following reactions occur at room temperature:

$$NH_2SO_3H + HNO_2 \rightarrow H_2O + H_2SO_4 + N_2$$

$$(R,Ar)NHSO_3H + HNO_2 \rightarrow (R,Ar)OH + H_2SO_4 + N_2$$

Since the barium salts of the sulfamic acids are soluble in water, the occurrence of the above reaction in the presence of barium ions leads to the precipitation of barium sulfate. This fact may be made the basis of a reliable test for this class of compounds containing an NH group (*cf.* detection of cyclohexanesulfamic acid, p. 620).

A more general test for organic sulfamic acids is based on the fact that they undergo saponification when evaporated with caustic alkali:

$$\begin{matrix} X \diagdown \\ X \diagup \end{matrix} N-SO_3Na + NaOH \longrightarrow \begin{matrix} X \diagdown \\ X \diagup \end{matrix} NH + Na_2SO_4$$

(X=R,Ar,H)

The resulting alkali sulfate can be detected through the precipitation of violet-tinted barium sulfate (Wohlers effect) on the subsequent addition of a conc. solution of potassium permanganate containing barium chloride and hydroxylamine hydrochloride. For details of the Wohlers effect and details of the test see p. 84.

The test revealed: 40 γ phenylsulfamic acid (NH₄ salt)
25 γ N,N-dicyclohexylsulfamic acid (Na salt).

1 Unpublished studies.

101. Arylsulfohydroxamic acids

Identification through tests for the SO₂ and NHOH component

Consistent with the structural formula $ArSO_2NHOH$, these acids are derivatives of arylsulfonic acid and of hydroxylamine. Accordingly, they are desulfurized when warmed with Raney nickel alloy and caustic alkali with production of nickel sulfide, which in turn yields hydrogen sulfide on treatment with mineral acid. (Compare the detection of aromatic sulfonic acids, p. 234). The *Identification limit* of this partial test is 1 γ benzenesulfohydroxamic acid. The hydroxylamine component can be revealed by the procedure given on p. 286, which involves treatment of the acetic acid solution with iodine to yield nitrous acid which gives a red color with Griess reagent (see p. 90).

Positive responses to the tests for both SO_2 and NHOH groups constitute a reliable indication of the presence of benzenesulfohydroxamic acid (m.p. 127°) and furthermore these responses to both these tests serve to differentiate it from benzhydroxamic acid, $C_6H_5CONHOH$ (m.p. 124°).

102. N-Halosulfonamides (—SO_2NHHal group)

Test with phenothiazine[1]

A test for phenothiazine (thiodiphenylamine) is described on p. 521; it is based on the production of a red-violet thiazine dyestuff through reaction with chloramine T. This color reaction occurs with all N-halosulfonamides and can be employed for their detection.

Procedure. A drop of the aqueous test solution is placed on filter paper that has been impregnated with a 2% solution of phenothiazine in ether or benzene. A positive response is indicated by the appearance of a red-violet stain or ring on the paper within 1–2 min.

The test revealed: 1 γ chloramine T   15 γ dichloramine
                              1 γ chloramine B

1 F. FEIGL and D. HAGUENAUER-CASTRO, *Chemist-Analyst*, 51 (1962) 7.

F. ORGANIC COMPOUNDS of P, As, Sb, Bi, and Hg

103. Triphenylphosphine, -arsine, -stibine, or -bismuthine

Test though formation of polyiodides[1]

If a colorless solution of triphenylphosphine in chloroform, benzene or carbon disulfide is treated with a dilute violet solution of iodine (in the same solvent) added drop by drop, the color is discharged first and then a yellow solution follows as more of the reagent is dropped in. The initial product is probably triphenylphosphine diiodide which then yields the polyiodide which dissolves in chloroform, *etc.* to form a yellow solution:

$$(C_6H_5)_3P + 2\ I \rightarrow (C_6H_5)_3PI_2$$

$$(C_6H_5)_3PI_2 + x\ I_2 \rightarrow (C_6H_5)_3PI_2 \cdot x\ I_2$$

If a drop of the yellow solution that contains excess iodine is placed on filter paper, the solvent evaporates along with the free iodine and leaves a brown or yellow stain that becomes lighter or disappears on standing because of loss of iodine from the polyiodide. However, the color is restored by spotting the area with a drop of iodine solution. The polyiodide is likewise formed by the action of iodine vapor on solid triphenylphosphine.

Triphenylarsine, -stibine and -bismuthine behave like the phosphine.

Procedure   A drop of the solution of the material in ether, chloroform, benzene, *etc.* is placed on filter paper. The fleck is held in iodine vapor generated by warming dry iodine in a small beaker. A brown or yellow stain appears at once if triphenyl phosphine (-arsine *etc.*) is present.

Limits of identification:

2.5 γ triphenylphosphine	5 γ triphenylstibine
2.5 γ triphenylarsine	5 γ triphenylbismuthine

When conducting the test it should be noted that many organic bases yield brown polyiodides. It is well, when examining unknown mixtures, to suspend the test material in dilute sulfuric acid and then to extract the suspension with chloroform, benzene, *etc.* The bases will remain in the water layer as sulfates.

It is remarkable that the stable triphenylphosphonium compounds $[(C_6H_5)_3PR]Hal$ show the same behavior towards iodine vapors as triphenylphosphine.[2]

1 Unpublished studies, with D. GOLDSTEIN; see also *Mikrochim. Acta* (1965), in press.
2 Private communication from Prof. P. SENISE, São Paulo.

104. Arsonic acids (—AsO(OH)₂ group)

Test with phenylhydrazine[1]

Acidic solutions of phenylhydrazine undergo an oxidative cleavage with arsenic acid:[2]

$$C_6H_5NHNH_2 + 2\ H_3AsO_4 \rightarrow 2\ H_3AsO_3 + H_2O + N_2 + C_6H_5OH$$

This reaction occurs even at room temperature; if the reaction mixture is heated, the phenol volatilizes with the water vapor.

Aliphatic and aromatic arsonic acids behave similarly to the inorganic parent compound. Since the resulting phenols can be readily detected in the gas phase through the indophenol reaction with 2,6-dichloroquinone-4-chloroimine (see p. 185), the detection of arsonic acids based on these facts is made specific provided steam-volatile phenols are not present.

Procedure. The test is made in a micro test tube. A little of the solid or a drop or two of its aqueous solution is mixed with several cg of phenylhydrazine chloride, and a drop of water is added. The tube is placed in a boiling water bath and its mouth is covered with a disk of filter paper impregnated with a saturated benzene solution of 2,6-dichloroquinone-4-chloroimine. After several minutes heating, the paper is exposed to ammonia vapors. If the response is positive, a blue stain develops.

The following amounts were detected:

5 γ phenylarsonic acid   5 γ *p*-nitrophenylarsonic acid
5 γ *p*-hydroxyphenylarsonic acid   10 γ 4'-dimethylaminoazobenzene-4-arsonic
                                                                  acid

1 F. FEIGL and E. JUNGREIS, *Talanta*, 1 (1958) 367.
2 M. OECHSNER DE CONINCK, *Compt. Rend.*, 126 (1898) 1042.

105. Stibonic acids (—SbO(OH)₂ group)

Test through heating with phenylhydrazine chloride and cetyl alcohol[1]

The procedure described in Sect. 104 is not applicable to stibonic acids. Most of the compounds in this category are not soluble enough in water to give a reaction or they respond only faintly even when large amounts are taken for the test. On the other hand, stibonic acids exhibit a mode of reaction that is not shown by arsonic acids, namely the rapid production of a red material on warming with solid phenylhydrazine chloride and cetyl alcohol (m.p. 49.3°) The chemical basis of this color reaction is not known; probably it involves an oxidation of the phenylhydrazine and the cetyl alcohol acts as solvent for the reaction product. The latter is soluble in benzene giving a red solution.

Procedure. A micro test tube is used. A little finely ground test material and several cg of phenylhydrazine chloride and cetyl alcohol are warmed in a boiling water bath. A positive response is indicated by the development of a red color within 10 sec. The reaction mass is treated with several drops of water and 1 drop of benzene, and shaken. A red color appears in the benzene layer.

Because of the insolubility of the stibonic acids tested, no *limits of identification* could be established. However, very distinct positive results were found with amounts in the neighborhood of 1 mg of the following compounds: benzenestibonic acid, (*m,p*)-toluenestibonic acid, *p*-chlorobenzenestibonic acid, stibanilic acid, *p*-stibonobenzoic acid, ethyl *p*-stibonobenzoate.

Under the above conditions mg amounts of arsonic acids give a faintly red-orange color after 2 min. heating.

1 M. van der Plas-Kronenburg (Rio de Janeiro), unpublished studies.

106. Arylmercuric salts (Ar—HgX group)

Test with benzenesulfonamide[1]

Sulfonamides react in solution with arylmercuric salts; the NH_2 group is mercurated, for instance:[2]

$$C_6H_5SO_2NH_2 + C_6H_5HgAc \rightarrow C_6H_5SO_2NHHgC_6H_5 + HAc$$

This reaction occurs rapidly if a dry mixture of the amide (m.p. 150°) and of the phenylmercuric acetate (m.p. 149°) is heated in a boiling water bath. Vapors of acetic acid are formed and may be detected with acid–base indicator paper. Phenylmercuric nitrate (m.p. 178–184°) behaves analogously, *i.e.* nitric acid is evolved. In contrast, phenylmercuric chloride, bromide, and iodide, and also tolylmercuric chloride, all with melting points above 230°, do not react.

When mixed with sodium acetate and benzenesulfonamide and heated to around 150°, the non-reactive arylmercuric salts yield acetic acid. This result obviously is brought about through the formation of the reactive acetate by the transformation:

$$Ar—HgX + NaAc \rightarrow Ar—HgAc + NaX.$$

A test for arylmercuric salts based on these effects is described here.

Procedure. A drop of the benzene solution of the sample is mixed in a micro test tube with several mg of benzenesulfonamide and anhydrous sodium acetate and heated to 90° to remove the solvent. The tube is then immersed to half its length in a glycerol bath that has been preheated to 150°; a piece of Congo red paper is laid over the mouth of the test tube. (The aqueous alcoholic solution of Congo red used for preparing the indicator paper should be as concentrated as possible.) A positive response is indicated by the development of a blue stain on the indicator paper within 1–2 min.

Limits of identification:

20 γ phenylmercuric acetate	30 γ phenylmercuric bromide
30 γ phenylmercuric chloride	20 γ phenylmercuric nitrate
30 γ phenylmercuric chloroacetate	30 γ tolylmercuric chloride

1 F. Feigl, D. Haguenauer-Castro and E. Libergott, *Microchem. J.*, 5 (1961) 565.
2 C. M. Suter, *The Organic Chemistry of Sulfur*, New York, 1944, p. 596.

107. Mercury diaryls (Ar$_2$Hg group)

Test through conversion into arylmercuric acetate[1]

Mercury diaryls can be converted into arylmercuric acetate by refluxing with glacial acetic acid:[2]

$$Ar—Hg—Ar + CH_3COOH \rightarrow Ar—HgOOCCH_3 + ArH$$

It has been found that a single evaporation of mercury diaryls with glacial acetic acid produces sufficient arylmercuric acetate to give a positive response to the test for acetic acid vapor when pyrolyzed with benzenesulfonamide (comp. Sect. 106).

Procedure. A small amount of the sample or a drop of its benzene solution is placed in a micro test tube along with 1 drop of glacial acetic acid. The benzene is removed if need be by keeping the tube in boiling water for 10 min. The excess acetic acid is eliminated by immersing the test tube as deep as possible in a glycerol bath at 130° for some time and then keeping it in an oven at this temperature (130°) until the vapors no longer turn Congo red paper blue. It is well to run a parallel trial with a drop of glacial acetic acid to guarantee that the acid has been entirely removed. Several mg of benzenesulfonamide and a drop of benzene are added to the acetic acid-free residue and then the benzene is volatilized by placing the test tube in water at 90°. After most of the benzene has volatilized, the tube is placed in boiling water and the mouth of the test tube is covered with Congo red paper. If the response is positive the paper turns blue within 2 min. at most.

The rest revealed:

60 γ diphenylmercury	100 γ di-*p*-tolylmercury
	100 γ di-*o*- ,, ,,

1 F. Feigl, D. Haguenauer-Castro and E. Libergott, *Microchem. J.*, 5 (1961) 565.
2 E. Dreher and R. Otto, *Ann.*, 154 (1870) 117.

108. Alkyl esters of non-carboxylic acids

Test through pyrolysis with sodium thiosulfate[1]

When esters of this type are heated at 160–180° with sodium thiosulfate, sulfur dioxide is liberated which can be detected in the gas phase by appropriate color reactions.

The underlying pyrolytic reactions are the same as those outlined on p. 171, namely the formation and pyrolysis of Bunte salts, *i.e.*

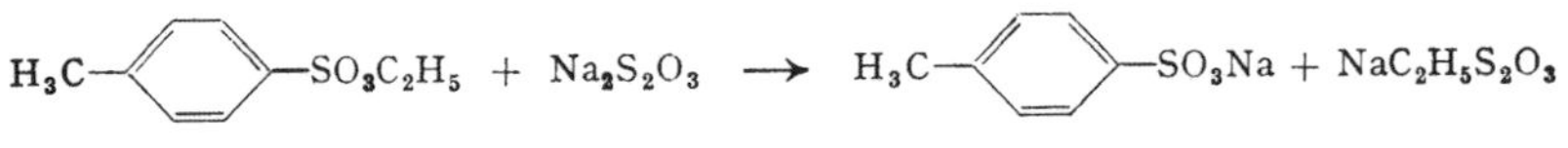

$$H_3C—C_6H_4—SO_3C_2H_5 + Na_2S_2O_3 \rightarrow H_3C—C_6H_4—SO_3Na + NaC_2H_5S_2O_3$$

$$2\ NaC_2H_5S_2O_3 \rightarrow Na_2SO_4 + C_2H_5 S S C_2H_5 + SO_2.$$

The procedure to be followed is described on p. 171.

The following amounts were detected:

5γ potassium ethyl sulfate 20γ ethyl p-toluenesulfate

A strong response was also given by alkyl esters of phosphoric and thiophosphoric acid*.

It may be expected that these acids can be differentiated, by esterification with diazomethane followed by the test for alkyl esters of non-carboxylic acids, from carboxylic acids.

---

* Up to the present only esters of primary alcohols have been examined. By analogy with the behavior of primary halogenoalkyls (see p. 171), it may be expected that esters of secondary and tertiary alcohols as well as phenyl esters of non-carboxylic acids will not react.

1 F. FEIGL and V. ANGER, *Chemist-Analyst*, 49 (1960) 13.

109. Methyl esters of non-carboxylic acids

Test by formation of nitromethane[1]

Dimethyl sulfate reacts with an aqueous solution of sodium nitrite to yield nitromethane:[2]

$$(CH_3)_2SO_4 + 2\ NaNO_2 \rightarrow 2\ CH_3NO_2 + Na_2SO_4 \tag{1}$$

Methyl iodide reacts in analogous manner:[3]

$$CH_3I + NaNO_2 \rightarrow CH_3NO_2 + NaI \tag{2}$$

Since nitromethane can be easily detected through the color reaction with 1,2-naphthoquinone-4-sulfonic acid in alkaline solution (see p. 558), the realization of (*1*) and (*2*) provides an indirect but selective test for dimethyl sulfate or methyl iodide. The tests are not very sensitive, probably because reactions (*1*) and (*2*) yield also the isomeric methyl nitrite (CH_3ONO), which does not react with napthoquinonesulfonic acid.

Procedure. A drop of a 0.5% solution of sodium 1,2-naphthoquinone-4-sulfonate and several mg of solid $NaNO_2$ are mixed in a micro test tube with a drop of the alcoholic test solution. The tube is kept at 60° in a water bath for about 5 min. Several cg of calcium oxide are then introduced. Up to about 300 γ of the methyl compounds gives a blue color; smaller amounts yield a violet shade.

Limits of identification: 80γ dimethyl sulfate 100γ methyl iodide

Methyl iodide·and dimethyl sulfate can be differentiated through the fact that alkaline hydrolysis of the latter yields alkali sulfate which can be detected readily through the Wohlers effect (see p. 84).

1 F. FEIGL and D. GOLDSTEIN, *Anal. Chem.*, 29 (1957) 1522.
2 W. STEINKOPF, *Ber.*, 41 (1908) 4457.
3 K. MEYER and O. STUEBER, *Ber.*, 5 (1872) 203.

Chapter 4

Detection of Structures and Certain Types of Organic Compounds

General remarks

The tests for functional groups included in the preceding chapter and those designed to identify individual compounds as described in the following chapter have clear-cut analytical objectives. In contrast, the goals of another subdivision of organic qualitative analysis are not always so definite; it deals with the detection of structures and types of organic compounds. What is comprised in this latter subdivision becomes somewhat clearer if it is remembered that the preponderant majority of all organic analytical procedures are based on reactions that involve particular functional groups and lead to products that can be recognized either directly or indirectly. The part of the molecule attached to the particular functional group frequently exerts a distinct influence on the speed and extent of the reaction, and likewise affects the color and the solubility of the reaction products. Very notable effects that can be of analytical use may arise in case the molecule contains in addition a functional group that is different in character and that in its turn can be detected. In such instances a better and more far-reaching statement regarding the nature of the compound under examination can be made from the responses to two group tests applied to the test material than is possible through the detection of a single functional group. As will be shown in Chap. 5, this system of multiple tests can sometimes accomplish the identification of individual organic compounds. The presence of two different functional groups in the molecule of an organic compound can however result in its reacting with appropriate partners in ways that are both different and of analytical value. *p*-Nitroso aromatic amines of the general formula $\diagdown$N$-\langle\bigcirc\rangle-$NO are a case in point. These compounds respond positively not only to the tests for the two N-bearing groups but they themselves produce a color with palladium chloride, that is given by neither of the two N-bearing groups.

Another instance involves the behavior of aniline- and phenolsulfonic acids. The selective tests described in Chap. 3 can be employed for the detection of NH_2, OH, and SO_3H groups in these compounds. Moreover, the ability of aniline and phenol to be brominated is retained (formation of tribromoaniline or tribromophenol). This fact is shown by the finding that in acid solution the sulfonic group is oxidatively removed as sulfuric acid by bromine water:

$$(HO,H_2N)-\hspace{-2pt}\bigcirc\hspace{-2pt}-SO_3H + Br_2 + H_2O \longrightarrow (HO,H_2N)-\hspace{-2pt}\bigcirc\hspace{-2pt}-Br + H_2SO_4 + HBr$$

The above exchange reaction is *not* shown by the parent benzenesulfonic acid; accordingly it is induced by the *para*-situated NH_2 or OH groups and the precipitation of barium sulfate makes possible the detection of aniline- or phenolsulfonic acids. The selectivity of the formation of sulfuric acid is demonstrated by the finding that treatment of the amide of sulfanilic acid with bromine water yields no sulfuric acid.

The influence of structure is clearly shown in the analytical behavior of *m*-dinitro compounds which give a color reaction with potassium cyanide. No other mono- or poly-nitro compound behaves similarly. Accordingly, there is justification in the cases mentioned above for speaking of typical structures that show characteristic reactions.

In this chapter examples are also given of the detection of certain types of organic compounds which are characterized by the presence of two or more different functional groups. Pertinent compounds can be detected by the application and the positive response of selective spot tests described in Chap. 3.

1. Vicinal polyalcohols $\left(\diagdown C(OH)-(HO)C\diagup \text{type}\right)$

(1) Test through increase in acidity[1]

A test for acidic compounds is described on p. 114; it is based on the liberation of nitrous acid from alkali nitrite followed by the Griess color reaction.

Boric acid is so weak that the liberation of nitrous acid becomes evident only after 10–15 min. However, an increase in acidity sufficient to be revealed immediately by indicators *etc.* can be brought about by adding 1,2-dihydroxy compounds. This well-known effect[2] is due to the formation of an inner-complex ester of orthoboric acid in which one OH group of the glycol becomes strongly acidic:

$$2 \begin{array}{c} -C-OH \\ | \\ -C-OH \end{array} + H_3BO_3 \longrightarrow \left[\begin{array}{c} -C-O\diagdown \diagup O-C- \\ \quad\quad B \\ -C-O\diagup \diagdown O-C- \end{array}\right]^- H^+ + 3\ H_2O$$

The increase in acidity combined with the liberation of nitrous acid from nitrites permits the detection of polyalcohols with vicinal OH groups. The test described here should not be applied directly in the presence of free acids. Under such circumstances, it is necessary to evaporate the test material with ammonia water to produce inactive ammonium salts. Attention must be given to the fact that o-diphenols as well as *peri*-dihydroxy-naphthalene and its derivatives likewise increase the acidity of boric acid.

Procedure.   Several drops of the Griess reagent solution are placed in a depression of a spot plate. A small amount of boric acid is added and stirred with a glass rod. Then one drop of the test solution or a pinch of the solid sample is added. An immediate orange color indicates the presence of a compound with two or more adjacent hydroxyl groups.

Limits of identification have not yet been determined.

Positive responses were obtained with: glycol, propyleneglycol, glycerol, pyrocatechol, gallic acid and 2,3-dihydroxybenzoic acid (both as ammonium salts), chromotropic acid (as disodium salt).

1 Y. NOMURA, *Bull. Chem. Soc. Japan*, 32 (1959) 893.
2 J. BOESEKEN, *Advan. Carbohydrate Chem.*, 4 (1949) 193–210.

(2) Tests by oxidation with periodic acid

Polyhydroxy alcohols such as glycol, glycerol, erythritol, mannitol, *etc.* with vicinal OH groups can be oxidatively broken down to formaldehyde and formic acid by excess periodic acid in the cold; the periodic acid is reduced to iodic acid.[1,2]

$$CH_2OH$$
$$[CHOH]_{n-1} + n\,HIO_4 \rightarrow 2\,CH_2O + (n-1)\,HCOOH + n\,HIO_3 + H_2O$$
$$CH_2OH$$

Sensitive tests are available for the organic products of this oxidation. Accordingly, the presence of polyhydroxy compounds can be established by treating the sample with periodate and then testing for (a) formaldehyde or (b) formic acid. The test described under (b) is more sensitive than (a), since, as the equation shows, 1 molecule of a polyvalent alcohol furnishes only 2 molecules of CH_2O as against (n—1) molecules of HCOOH.

The reaction is given by almost all glycols and by many carbohydrates that do not normally react with fuchsin–sulfurous acid; monosaccharides give a positive reaction. Saccharose reacts only after boiling, when the acid probably breaks up the disaccharide into the two monosaccharides, which in turn give the glycol reaction.*

---

* According to J. M. BAILEY (*J. Lab. Clin. Med.*, 54 (1954) 158) serine, ethanolamine and its alkyl derivatives split off CH_2O when treated with HIO_4.

The treatment with periodic acid also permits a differentiation between tartaric and citric acid or their salts. The former, which has two adjacent OH groups, is probably oxidized to glyoxylic acid, and reacts with fuchsin–sulfurous acid due to cleavage of formaldehyde; citric acid shows no change.

(a) Oxidation with periodate and detection of formaldehyde[3]

The test with fuchsin–sulfurous acid (see p. 195) can be used to detect the formaldehyde after destroying the excess periodate and iodate. Most polyhydroxy alcohols can then be detected, provided aldehydes are absent.

Procedure. A drop of the aqueous or alcoholic solution of the polyhydroxy alcohol is mixed in a micro crucible with a drop of 5% potassium periodate solution and a drop of 10% sulfuric acid and allowed to stand for 5 min. The excess periodic acid is then reduced with a few drops of saturated sulfurous acid, and the sample is treated with a drop of fuchsin–sulfurous acid (preparation see p. 195). After a short time, a red to blue color appears. When testing for polysaccharides, the procedure is the same, except that the contents of the covered crucible are heated to boiling and then likewise left for about 5 min.

The following amounts were detected:

5	γ glycol	25	γ arabinose
2.5	γ glycerol	25	γ saccharose (with boiling)
5	γ mannitol	12.5	γ dextrin (with boiling)
25	γ glucose	50	γ starch (with boiling)
12.5	γ fructose	100	γ tartaric acid
25	γ lactose	100	γ mucic acid (dissolved in dioxane)

The following gave a positive reaction: erythritol, cherry gum, gum arabic, whereas citric acid, inositol, pentaerythritol, pentaacetylglucose, acetylcellulose, gave no response.

1 L. MALAPRADE, *Compt. Rend.*, 185 (1927) 1132.
2 L. MALAPRADE, *Bull. Soc. Chim.*, [4], 43 (1928) 683.
3 Unpublished studies, with R. ZAPPERT.

(b) Oxidation with periodate and detection of formic acid[1]

The detection of the formic acid produced when polyhydric alcohols are oxidized with periodic acid is based on the redox reaction of formic acid with bromine:

$$HCOOH + Br_2 \rightarrow 2 HBr + CO_2$$

The carbon dioxide can be detected by baryta water (precipitation of $BaCO_3$). The formaldehyde, produced along with the formic acid is also oxidized by bromine to carbon dioxide, so that all of the carbon of the polyhydroxy compound is converted into carbon dioxide.

Procedure. The microdistillation apparatus (Fig. 29, *Inorganic Spot Tests*) is used. A little of the sample is mixed with 2 drops of 5% potassium periodate and 2 drops of 1 N sulfuric acid and warmed slightly. Bromine water is then added until a distinct yellow is obtained. A glass bead is introduced and the stopper put in place. The delivery tube is inserted into a small test tube containing baryta water. (The reagent is protected against the carbon dioxide of the air by a layer of paraffin oil.) The distillation apparatus is gently warmed. If the liquid in the receiver becomes turbid, polyhydroxy compounds are indicated.

The following were detected:

2.5	γ glycerol	5	γ arabinose	6	γ lactose
3	γ mannitol	2.5	γ mannose	6	γ dextrin
5	γ glucose	2.5	γ maltose	3	γ mucic acid
5	γ fructose	7	γ saccharose	20	γ starch

Malonic acid behaves like polyhydric alcohols under the conditions of this test[2].

1 O. FREHDEN and K. FUERST, *Mikrochem. ver. Mikrochim. Acta*, 26 (1930) 36.
2 M. P. FLEURY, *Chim. Anal.*, 35 (1953) 197.

2. 7-Dehydrosterols

Test with trichloroacetic acid and lead tetraacetate[1]

Compounds containing the characteristic grouping of 7-dehydrosterols (I) are converted into the corresponding triols (II) by oxidation with lead tetraacetate in glacial acetic acid. If this oxidation occurs in the presence of trichloroacetic acid, a red color develops[2]. The chemistry of this reaction is not known, but it probably rests on the production of a solvate of the triol with trichloroacetic acid. Esters of 7-dehydrosterols react in the same manner as the corresponding sterols.

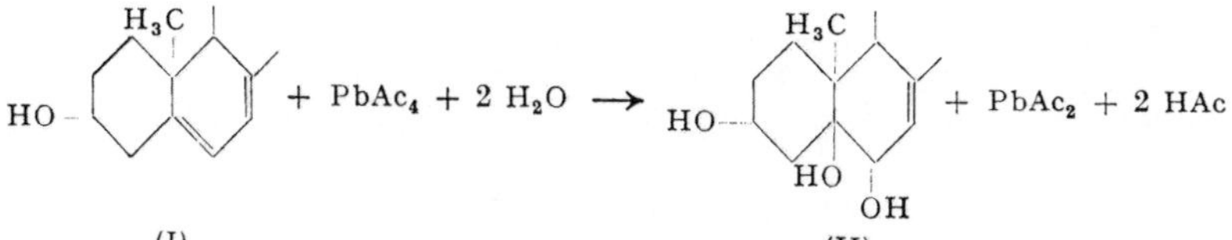

(I)   (II)

Procedure. The material to be tested is dissolved in several drops of chloroform, 1 ml of trichloroacetic acid added, and then treated with a drop of lead tetraacetate solution. If 7-dehydrosterols or their esters are present, a pink-violet color appears which fades quickly.

Reagents: 1)   90% trichloroacetic acid and 10% water.
2)   0.25% solution of lead tetraacetate in glacial acetic acid.
Limits of identification:

2 γ ergosterol	1 γ 7-dehydrocholesterol
1 γ ergosteryl acetate	1 γ ergosteryl butyrate

The following procedure is more sensitive: A grain of the sample or a drop of its chloroform solution is mixed with a drop of CHCl₃ and a drop of the reagent solution. A pink color appears if 7-dehydrosterols or their esters are present.

Reagent: 1 ml of 90% trichloroacetic acid is treated with a drop of 0.25% solution of lead tetraacetate in glacial acetic acid.

The reagent solution must be freshly prepared.

Limit of identification: 0.1 γ ergosterol

1 A. v. CHRISTIANI and V. ANGER, *Ber.*, 72 (1939) 1124.
2 O. ROSENHEIM, *Biochem. J.*, 23 (1928) 47.

3. Phenoxy compounds $\left(C_6H_5O\text{—type} \right)$

Test with hydrazoic acid and N,N'-diphenylbenzidine [1]

When phenol and sodium azide are added to a solution of small amounts of N,N'-diphenylbenzidine in concentrated sulfuric acid, a blue color appears. This oxidation of the diphenylbenzidine is probably achieved by the action of benzoquinoneimine formed through the following reactions:

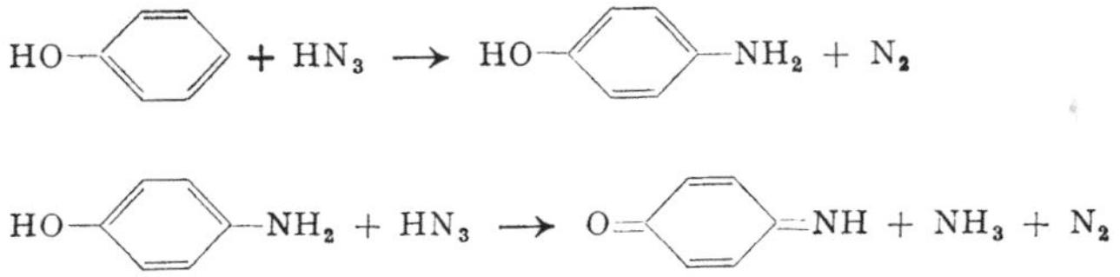

Phenol derivatives which yield phenol by the hydrolytic action of concentrated sulfuric acid behave like the parent compound. In this category belong aliphatic and aromatic phenol ethers, phenol esters of carboxylic acids and o- and p-hydroxybiphenyl. o-Cresol behaves similarly to phenol, but the p-isomer does not.

The following phenolic compounds are inactive: thymol, salicyl alcohol, salicylaldehyde, salicylic acid, sulfosalicylic acid, the three isomeric nitro- and amino-phenols, α- and β-naphthol. Accordingly the test described here seems to be almost specific within the class of phenolic compounds.

Vapors of phenol show the color reaction when they come into contact with a drop of a mixture of the sulfuric acid solution of N,N'-diphenyl-benzidine plus sodium azide, if the apparatus shown in Fig. 23, *Inorganic Spot Tests*, is employed.

C-Nitroso compounds (nitrosobenzene is an exception) and quinones do not respond to the procedure described below.

Procedure. A small amount of the undissolved sample or a drop of its solution is united on a spot plate with about 20 mg of sodium azide. Then 1–2 drops of the reagent are added and mixed by blowing with a pipet. According to the amount of phenoxy compounds, a blue color develops within half a minute, and deepens. A blank test without addition of the sample is recommended.

Reagent: Solution of 5 mg N,N′-diphenylbenzidine in 100 ml of concentrated sulfuric acid (freshly prepared).

The following amounts were detected:

0.25 γ phenol	2 γ diphenyl phthalate
0.5 γ anisole	2.5 γ phenyl acetylsalicylate
0.8 γ phenetole	2.5 γ 1,3-diphenoxy-2-propanol
5 γ phenyl benzoate	2 γ phenyl salicylate
2.5 γ diphenyl ether	20 γ phenoxyacetic acid

1 F. Feigl, V. Gentil and D. Haguenauer-Castro, *Microchem. J.*, 4 (1960) 445.

4. Polyhydroxybenzenes ($C_6H_{6-n}(OH)_n$ type)

Test by fusion with oxalic acid[1]

When phloroglucinol, pyrogallol, resorcinol, or pyrocatechol is mixed with oxalic acid and heated to 160°, colored melts result in a short time. The chemistry of these fusion reactions is not known. Perhaps the colored products are due to condensation reactions between the polyhydroxy-benzenes and vapors of formic acid (split off from oxalic acid) since the latter likewise shows this same behavior towards large amounts of polyhydroxy-benzenes.

Procedure. A minimal amount of the test material or a drop of its solution is mixed with several cg of crystallized oxalic acid in a micro test tube. The mixture is taken to dryness if need be. The test tube is then immersed in a glycerol bath, that has previously been brought to 160°, and kept there for 3–4 min. After cooling, the color of the mass is examined. Phloroglucinol gives a red, pyrogallol a grey, resorcinol a yellow-brown, and pyrocatechol a pink or rose color. Hydroquinone yields no colored reaction mass.

The procedure revealed:

2 γ phloroglucinol	25 γ pyrogallol
5 γ pyrocatechol	25 γ resorcinol

1 E. Jungreis (Jerusalem), unpublished studies.

5. *m*-Dihydroxybenzene derivatives (![OH / H / OH structure] type)

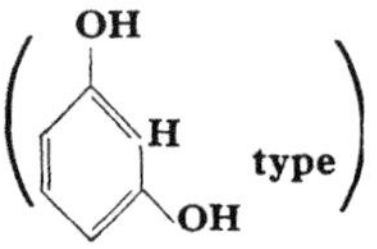

Test with furfural[1]

In hydrochloric acid solution, *m*-dihydroxybenzene derivatives condense with furfural with intermediate formation of an orange product (I) that after a short time turns green, and after dilution with water becomes blue. The blue product may be a hydroxypentamethine derivative (II).

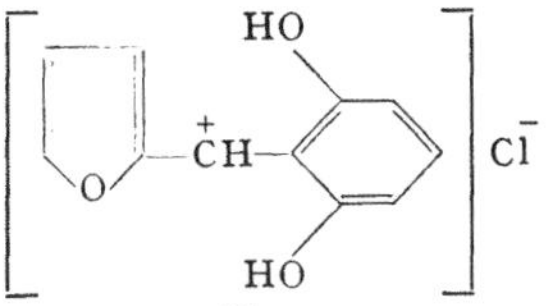

(I) 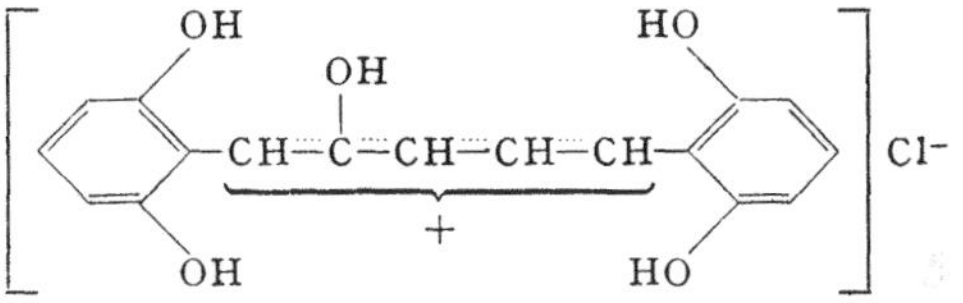   (II)

On treatment with ammonia the blue hydroxypentamethine compound is transformed into a pink quinoid compound.

The reaction is given by *m*-dihydroxybenzene compounds whose 2-position is not substituted. Naphthoresorcinol yields only the orange color (see p. 425) but not the green or blue products.

Procedure. A drop of the test solution is treated in a depression of a spot plate with a drop of a 0.1% solution of furfural in methanol and then with a drop of *concentrated* hydrochloric acid. A positive response is indicated by the appearance of a green to blue color which follows a transient orange color.

Limits of identification:

    1 γ resorcinol   0.5 γ orcinol   0.05 γ phloroglucinol

1 V. ANGER and S. OFRI, unpublished studies.

6. *m*-Dihydroxyaryl compounds (*m*-Ar(OH)$_2$ type)

Test with glutaconic aldehyde[1]

In strong hydrochloric acid solution, *m*-dihydroxyaryls condense with glutaconic aldehyde to yield red-violet compounds:

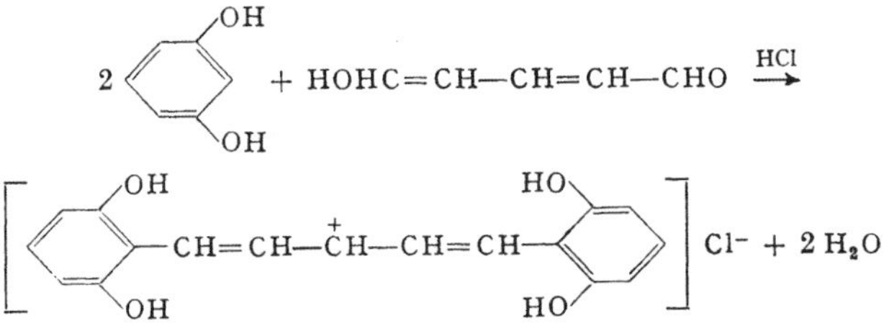

The glutaconic aldehyde is produced during the test through alkaline cleavage of pyridylpyridinium dichloride (comp. p. 246).

Procedure. A drop of the test solution is mixed in a depression of a spot plate with a drop of a 1% solution of pyridylpyridinium dichloride and a drop of 1 N sodium hydroxide, whereupon five drops of concentrated hydrochloric acid are added. A positive response is shown by the appearance of a red-brown to red-violet color.

Aromatic amines may not be present (see p. 245).

Limits of identification:

1 γ resorcinol	red-brown	0.1 γ phloroglucinol	violet-red
0.5 γ orcinol	,,		
0.5 γ naphthoresorcinol	,,		

1 V. ANGER and S. OFRI, *Mikrochim. Acta*, (1964) 987.

7. *vic*-Trihydroxyaryls (... type)

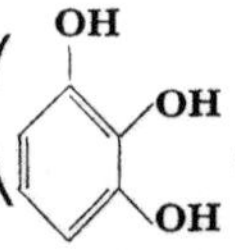

Test with rhodanine and ammonia[1]

An intense red-violet color ensues when rhodanine and ammonia are added to pyrogallol and its derivatives with free hydroxyl groups. The mechanism of this reaction is still unknown. Hydroquinone yields a green reaction product and if the latter is present along with the red-violet compound, then one drop of the reaction mixture should be placed on filter paper. Capillary separation occurs; the red-violet product remains at the site of the fleck while the green product migrates outward.

Procedure.   A drop of the test solution is treated in a micro test tube with a drop of a saturated aqueous solution of rhodanine and a drop of concentrated ammonia water. The mixture is gently shaken. If pyrogallol or its derivatives are present, a red-violet color appears.

Limits of identification:

0.1 γ pyrogallol   0.1 γ gallic acid   2 γ tannic acid

1 V. ANGER and S. OFRI, *Mikrochim. Acta*, (1964) 918.

8. *p*-Hydroxy- and *p*-alkoxy-benzene derivatives

Test with barbituric acid[1]

Pertinent compounds (I) as well as *p*-hydroxy- and *p*-alkoxycinnamic alcohols of the general formula (II):

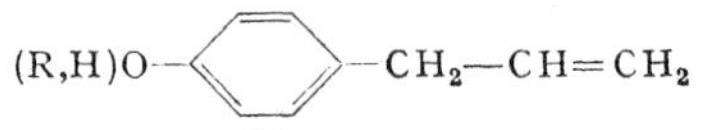

$$(R,H)O-\langle\bigcirc\rangle-CH_2-CH=CH_2 \qquad (R,H)O-\langle\bigcirc\rangle-CH=CH-CH_2OH$$

$$(I) \qquad\qquad (II)$$

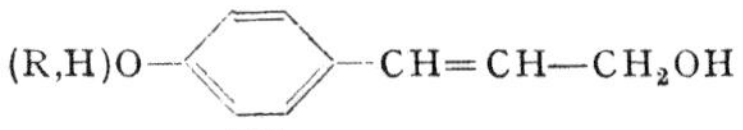

react with barbituric acid to give compounds that fluoresce green or yellow in ultraviolet light. See p. 169 regarding the chemistry of the reaction.[2]

Procedure. A drop of a saturated solution of barbituric acid in 2 N hydrochloric acid is placed on filter paper and then a drop of the alcoholic test solution is added. If any of the above compounds is present, a fluorescing spot appears (green in the case of p-hydroxy compounds; yellow with alkoxy compounds).

p-Hydroxy and p-alkoxyaldehydes react in the same fashion and consequently may not be present. A test for these compounds with naphthoresorcinol is described on p. 345.

Limits of identification:

1 γ eugenol	green fluorescence	10 γ estragole yellow fluorescence
1 γ coniferyl		10 γ eugenol methyl
alcohol ,, ,,		ether ,, ,,

1 V. ANGER and S. OFRI, private communication.
2 V. ANGER, *Anal. Chem. (Proceedings Feigl Anniversary Symposium, Birmingham)*, 1962, p. 58.

9. Methylene ethers of o-diphenols $\left(\bigcirc\begin{smallmatrix}-O\\\\-O\end{smallmatrix}\hspace{-4pt}{>}CH_2 \text{ type}\right)$

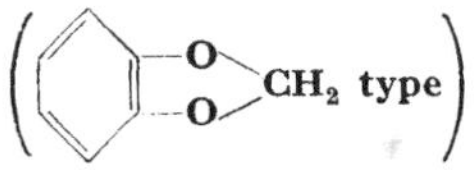

Test through hydrolytic splitting out of formaldehyde[1]

When cyclic methylene ethers of o-diphenols are heated alone, or along with concentrated sulfuric acid, to 170–180° formaldehyde results:

$$Ar\begin{smallmatrix}O\\\\O\end{smallmatrix}CH_2 + H_2O \longrightarrow Ar\begin{smallmatrix}OH\\\\OH\end{smallmatrix} + CH_2O$$

In the case of dry heating, this hydrolysis is brought about by the water split out of the test material. It acts as superheated steam at the site of its production and accomplishes hydrolyses which cannot be brought about by boiling water, and which are realized but slowly by warming with dilute acids or alkalis. Also when the material is heated with concentrated sulfuric acid, it is the water in the acid which as quasi superheated steam is responsible for the hydrolysis shown above. The resulting formaldehyde can be easily detected in the vapor phase through the color (violet) that appears in a mixture of chromotropic acid and concentrated sulfuric acid. The chemistry of this color reaction is discussed on p. 434.

A study of the detection of cyclic bound OCH_2O groups by means of the formaldehyde test showed that the same limits of identification are obtained when the material is heated dry or in contact with concentrated sulfuric acid. However, only the latter procedure is recommended because the alkaloids codeine and corydaline, which only contain one and two methoxy groups respectively, give a decided formaldehyde reaction when heated dry to 170°, but this is by no means the case when they are heated to this temperature with concentrated sulfuric acid.*

When carrying out the test it is best to start with an ethereal solution because interferences from water-soluble compounds which yield formaldehyde under the conditions of the test are averted. Such materials include: hexamine and its salts, glycolic acid, monochloroacetic acid, pyramidone, organic sulfoxyl compounds.

Procedure. The apparatus shown in Fig. 23, *Inorganic Spot Tests*, is used. A little of the solid or a drop of the test solution is placed in the bulb and taken to dryness if necessary. 1–2 drops of concentrated sulfuric acid are introduced. The knob of the stopper is charged with a drop of freshly prepared reagent solution and the stopper is put in place. The apparatus is immersed to a depth of 0.5 cm in a bath previously heated to 170°. At this temperature, after 1–10 min., depending on the amount of formaldehyde generated, a more or less intense violet color develops in the suspended drop. If the latter is wiped onto a spot plate, even slight colors are easily visible.

Reagent: A pinch of the purest available chromotropic acid or its sodium salt is placed in a centrifuge tube and well stirred with 2 or 3 ml of concentrated sulfuric acid. The suspension is then centrifuged. The supernatant liquid, which usually is turbid, serves for the test.

The test revealed:

0.1 γ piperine	0.1 γ berberine	0.4 γ safrole
0.2 γ narcotine	0.1 γ chelidonine	0.5 γ narceine
0.2 γ hydrastine		

A positive response was given also by heliotropin, piperonylic acid, isosafrole and apiole.

The procedure may also be used with the insoluble salts of the alkaloids listed above. This was established for the molybdates. The precipitation can be carried out in the bulb of the apparatus and the precipitate then collected on the bottom of the vessel by centrifuging. (The precipitate need not be washed.) After drying at 110°, two drops of concentrated sulfuric acid are introduced and the vessel is warmed as prescribed.

* The assumption that a splitting off of formaldehyde on dry heating rests on a disproportionation of methoxy groups, namely $\geq C{-}OCH_3 \rightarrow \geq CH + CH_2O$, cannot be valid or hold in general. It has been found that quinine and veratrole, containing respectively one and two methoxy groups, show no splitting out of formaldehyde, and papaverine with four methoxy groups exhibits no more than a slight effect of this kind.

1 F. FEIGL and L. HAINBERGER, *Mikrochim. Acta*, (1955) 806; comp. also O. R. HANSEN, *Acta Chem. Scand.*, 7 (1953) 1125.

10. 1,2-Dioxo compounds (—CO—CO— type)

(1) Test through formation of phenylosazones[1]

1,2- and o-dioxo compounds condense rapidly with phenylhydrazine in mineral acid solution to yield yellow or red water-insoluble phenylosazones:

$$\begin{array}{l} -C=O \\ | \\ -C=O \end{array} + 2\,NH_2NHC_6H_5 \longrightarrow \begin{array}{l} -C=N-NHC_6H_5 \\ | \\ -C=N-NHC_6H_5 \end{array} + 2\,H_2O$$

Sugars and α-hydroxy ketones do not react under the same conditions.

The test based on the formation of osazones is restricted to colorless or faintly colored samples. It cannot be applied in the presence of certain aromatic and α,β-unsaturated aldehydes which form yellow phenylhydrazones (*i.e.* cinnamic aldehyde, pyridinealdehydes, p-dimethylaminobenzaldehyde). The presence of such aromatic aldehydes can be detected through the tests with thiobarbituric acid, described on p. 205.

Procedure. A small amount of the solid sample or a drop of its aqueous alcoholic solution is mixed in a test tube with a drop of 2 N hydrochloric acid and a drop of 1% phenylhydrazine hydrochloride. The tube is immersed for 1–2 min. in a boiling water bath. A positive response is indicated by the appearance of a yellow or red color.

The following were detected:

0.5 γ glyoxal	yellow	5 γ 1,2-naphthoquinone-4-
5 γ diacetyl	,,	sulfonic acid red
5 γ benzil	,,	0.5 γ ninhydrin ,,
1 γ dihydroxytartaric acid	,,	0.5 γ phenanthraquinone ,,
5 γ rhodizonic acid	,,	5 γ 1,2-naphthoquinone ,,
5 γ alloxan	,,	

Dimethylglyoxime and diacetylmonoxime give a positive response (*Id. Lim.*: 5 γ) because they are hydrolyzed to the respective 1,2-dioxo compounds.

1 F. FEIGL, L. BEN-DOR and S. YARIV, *Israel J. Chem.*, 2 (1964) 138.

(2) Test by condensation with thiophene[1]

When dissolved in concentrated sulfuric acid, 1,2-diketones, such as benzil, isatin, ninhydrin (triketohydrindene), condense with thiophene (and thiophene derivatives with a free α-position) to yield colored compounds.[2] In these color reactions, which provide sensitive tests for thiophene (see p. 491), the sulfuric acid functions both as dehydrating agent and as oxidant. The condensation of the diketo group and thiophene can be represented by the net reaction[3]:

The term indophenine has become a generic expression and refers to the colored product (red, violet, blue) obtained by condensation of thiophene or thiophene compounds that contain two nuclear hydrogen atoms in the 2,5- or 2,3-position with 1,2-dicarbonyl compounds.

The test based on the formation of indophenine leads to the desired goal only with those diketones whose solutions in concentrated sulfuric acid are stable. For instance, diacetyl is excluded because it is carbonized by concentrated sulfuric acid. Attention must be given to the fact that cold concentrated sulfuric acid produces a variety of colors with many (even colorless) compounds, particularly those which are polycyclic or highly substituted.[4] To detect 1,2-diketones in the presence of such compounds, a comparison test with a solution of the sample in concentrated sulfuric acid should be conducted without addition of thiophene.

In concentrated sulfuric acid, mercaptans react with thiophene to yield a green coloration.[5] Since disulfides show no reaction, it is necessary when mercaptans are present to convert them into disulfides by evaporation with hydrogen peroxide. The presence of mercaptans is easily established by the test given on p. 219.

Procedure.   A drop of the solution to be tested (if possible in alcohol) is placed in a micro test tube and evaporated to dryness in a water bath. Alternatively, fractions of a milligram of the solid may be taken. The material or residue is dissolved or suspended in three drops of concentrated sulfuric acid and treated with two drops of thiophene solution. Depending on the quantity of 1,2-diketone present, a characteristic color appears at once or within fifteen minutes at most.

Reagent: 0.3% solution of thiophene in purest benzene.

The test revealed:

5 γ benzil	violet to pink	10 γ phenanthraquinone	blue to blue-green
1.5 γ ninhydrin	,, ,,	5 γ isatin	,, ,,
		4 γ rhodizonic acid	blue-green

Alloxan in amounts up to 25 γ gives a blue color; smaller quantities yield pink colorations. Color reactions are also given by esters of phenylglyoxylic and mesoxalic acid, which contain a 1,2-dicarbonyl group.

1 Unpublished studies, with D. GOLDSTEIN.
2 V. MEYER, *Ber.*, 16 (1883) 2073; comp. also W. STEINKOPF, *Die Chemie des Thiophens,* Steinkopf, Dresden, 1941, p. 125–133.
3 W. SCHLENK and O. BLUM, *Ann.*, 433 (1923) 85.
4 N. CAMPBELL, *Qualitative Organic Chemistry*, Macmillan, London, 1939, p. 44.
5 G. DENIGÈS, *Compt. Rend.*, 108 (1889) 350.

(3) Test with p-nitrophenylhydrazine[1]

When Schiff bases of *p*-nitrophenylhydrazine with carbonyl compounds of the type shown in (I) are alkalized, they undergo prototropic rearrangement into water-soluble akali salts of p-quinoidal nitronic acids (II):

$$\underset{\text{(I)}}{\overset{|}{\underset{|}{C}}=N-NH-\overset{}{\bigcirc}-NO_2} + KOH \longrightarrow \underset{\text{(II)}}{\overset{|}{\underset{|}{C}}=N-N=\overset{}{\bigcirc}=NO_2K} + H_2O$$

The alkali salts are intensely colored (blue, red, violet) if the chain of conjugated double bonds contained in (II) is lengthened by additional conjugated bonds. This occurs when p-nitrophenylhydrazones of aromatic and α,β-unsaturated carbonyl compounds are alkalized. This finding is the basis of an exceptionally sensitive test for compounds of this kind (comp. p. 193).

When Schiff bases of p-nitrophenylhydrazine with 1,2-diketones are alkalized, double prototropic rearrangement ensues with formation of alkali salts of quinoid nitronic acids in which two NO_2K groups are the terminal members of a system of continuous conjugated double bonds:

$$\overset{|}{\underset{|}{C}}=N-NH-\bigcirc-NO_2$$
$$\overset{|}{\underset{|}{C}}=N-NH-\bigcirc-NO_2 \quad + 2\ KOH \longrightarrow$$
$$\overset{|}{\underset{|}{C}}=N-N=\bigcirc=NO_2K$$
$$\overset{|}{\underset{|}{C}}=N-N=\bigcirc=NO_2K \quad + 2\ H_2O$$

As expected, these dialkali salts are intensely colored and a very sensitive test for 1,2-diketones has been based on their production. The absence of aryl- and α,β-unsaturated carbonyl compounds is necessarily assumed.

Hydrazones and oximes of 1,2-diketones behave like the parent compound with respect to p-nitrophenylhydrazine; obviously the NOH or N–NH$_2$ groups are eliminated in these instances.

Procedure.   As outlined on p. 194.

The following were revealed:

0.0001 γ isatin	0.05 γ diacetyl	0.05 γ ninhydrin
0.0001 γ 5,7-dibromoisatin	0.01 γ benzil	5 γ glucosazone
0.0001 γ triquinol	0.05 γ diacetylmonoxime	
0.001 γ glyoxal	0.5 γ dimethylglyoxime	

1 F. FEIGL, V. ANGER and G. FISCHER, *Mikrochim. Acta*, (1962) 878.

(4) Test through formation of imidazoles[1]

1,2-Diketones, in contrast to o-quinones, react with aldehydes in the presence of ammonia to yield imidazoles:[2]

$$\underset{R'-CO}{\overset{R-CO}{|}} + 2\ NH_3 + R''-CHO \longrightarrow R'-\underset{\underset{H}{\diagdown N \diagup}}{\overset{R-C-N}{\underset{\|}{C}}\ \overset{\|}{\underset{}{C}}-R''} + 3\ H_2O$$

The imidazoles formed show the test for tertiary amines described on p. 251. This behavior permits a selective test for 1,2-diketones.

Procedure.   In a micro test tube a few grains of paraformaldehyde are mixed with one drop of the test solution and one drop of conc. ammonia. After waiting 10–15 min., the mixture is evaporated to complete dryness. To the cooled residue some crystals of citric acid and 2 drops of acetic anhydride are added. The bottom of the tube is immersed to the depth of a few mm in a water bath heated to 80–85°. A red-violet color, appearing within 10 min., indicates a positive response. A blank test is recommended. Alkali salts must be absent.

Limits of identification:

0.5 γ glyoxal	5 γ dihydroxytartaric acid
2 γ diacetyl	3 γ isatin
5 γ dimethylglyoxime	5 γ furylglyoxal
10 γ benzil	3 γ nitrofurylglyoxal

1 F. FEIGL, L. BEN-DOR and S. YARIV, *Israel J. Chem.*, 2 (1964) 140.
2 I. L. FINAR, *Organic Chemistry*, Vol. II, 3rd Ed., Longmans Green, London, 1964, p. 447.

11. Aliphatic 1,2-dioxo compounds (R—CO—CO—R type)

Test by conversion into colored inner-complex nickel dioxime salts[1]

Aliphatic and monocyclic hydroaromatic *o*-dioxo compounds react with hydroxylamine or with hydroxylamine hydrochloride plus alkali to produce 1,2-dioximes which form water-insoluble red or yellow inner-complex nickel salts.[2] These salts are also formed directly from the components:[3]

$$\begin{array}{c} -C=O \\ | \\ -C=O \end{array} + 2\,NH_2OH + {}^{1}/_{2}\,Ni^{2+} + OH^- \longrightarrow \begin{array}{c} O \\ -C=N \\ | \diagdown Ni/2 \\ -C=N \diagup \\ | \\ OH \end{array} + 3\,H_2O$$

This reaction permits the differentiation of aliphatic (or monocyclic hydro-aromatic) dioxo compounds from aromatic dioxo compounds, such as benzil, phenanthraquinone, camphorquinone, *etc.* The latter are not dioximated by hydroxylamine under the conditions described here.

Procedure.   A drop of the test solution is treated in a centrifuge tube with a drop of hydroxylamine hydrochloride solution and warmed briefly on the water bath. A drop of the clear solution is placed on filter paper and spotted with a drop of 5% nickel acetate solution. A more or less intense yellow or red appears, either at once or after fuming over ammonia.

Reagent: 1 g hydroxylamine hydrochloride plus 1 g sodium acetate in 2 ml water.

Limits of identification:

0.5 γ diacetyl	red	10 γ indane-1,2-dione	orange
5 γ cyclohexane-1,2-dione	,,		

1 M. ISHIDATE, *Mikrochim. Acta,* 3 (1938) 283.
2 F. J. WELCHER, *Organic Analytical Reagents,* Vol. 3, Chap. VI, Van Nostrand, New York, 1946.
3 W. N. HIRSCHEL and J. A. VERHOEFF, *Chem. Weekblad,* 20 (1923) 319.

12. Aromatic 1,2-dioxo compounds and o-quinones
(Ar—CO—CO—Ar type)

Test by conversion into oxazine dyes[1]

The action of 2-amino-5-dimethylaminophenol (I) on aromatic o-diketones such as phenanthraquinone (II), involves first a mutual reduction and oxidation to the system of compounds (III) and (IV). The interaction continues as a condensation between (III) and (IV), and the deep blue oxazine dyestuff with the quinoidal cation (V) results:

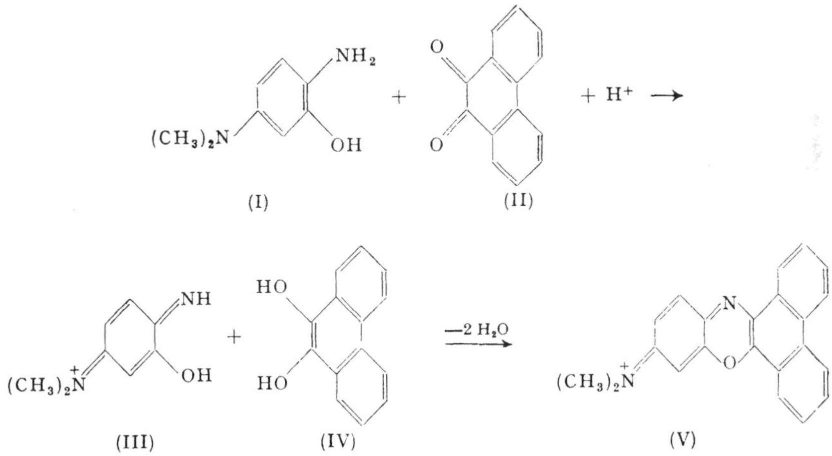

The formation of this dye occurs only with aromatic dioxo compounds which are not oximated when subjected to the prodecure described in Sect. 11. Accordingly, aliphatic 1,2- and aromatic o-dioxo compounds can be distinguished from each other by the dioxime formation on the one hand, and through the oxazine formation on the other.

Care must be taken that the sample contains no oxidizing agent, since the latter will react with the reagent and form colored quinoidal compounds.

Procedure. One drop of the test solution is treated in a micro test tube with 2 drops of freshly prepared reagent solution. A more or less intense blue appears either immediately or after gentle warming, depending on the quantity of dioxo compound present. When dealing with small quantities it is well to run a blank and to view the color against a white background.

Reagent: 0.05 g of 2-nitroso-5-dimethylaminophenol is suspended in 5 ml glacial acetic acid and shaken with zinc dust (with cooling) until decolorized. The filtrate is diluted to 10 ml with glacial acetic acid. The reagent must be freshly prepared; it turns light blue on standing in contact with the air.

The test revealed:

2 γ benzil   1 γ camphorquinone   0.3 γ phenanthraquinone

1 M. ISHIDATE, *Mikrochim. Acta*, 3 (1938) 283.

13. 1,2-Dioxo compounds and quinones

Test by catalytic acceleration of the formaldehyde–o-dinitrobenzene reaction[1]

Formaldehyde is among the organic compounds which function as hydrogen donors in alkaline solution and therefore reduce o-dinitrobenzene to the violet alkali salt of the *aci*-form of o-nitrophenylhydroxylamine (see p. 133). However, the reaction

$$\text{(ring)}\begin{matrix}-NO_2\\-NO_2\end{matrix} + 2\ CH_2O + 4\ OH^- \rightarrow \text{(ring)}\begin{matrix}=NO^-\\=NO_2{}^-\end{matrix} + 2\ HCOO^- + 3\ H_2O \qquad (1)$$

proceeds very slowly in alkali carbonate solution even though considerable amounts of aldehyde are present. The reaction is hastened by 1,2-dioxo compounds to such an extent that a sensitive test for these diketones becomes possible.

The catalytic acceleration of (*1*) by α-diketones may rest on the fact that in alkaline solution they are reduced by formaldehyde to hydroxyketones as shown in (*2*), and the latter in turn reduce the o-dinitrobenzene to the violet o-quinoid alkali salt as shown in (*3*). The diketone is thus regenerated and again enters into (*2*). The redox reactions (*2*) and (*3*) which occur repeatedly proceed faster than the non-catalzyed reaction (*1*) which leads to the same endproduct. Accordingly, the sum of (*2*) and (*3*) is a net reaction identical with (*1*) which does not show the catalyzing and constantly regenerated diketone as a participant:

$$2\ \begin{matrix}-CO\\-CO\end{matrix} + 2\ CH_2O + 2\ OH^- \rightarrow 2\ \begin{matrix}-CHOH\\-CO\end{matrix} + 2\ HCOO^- \qquad (2)$$

$$2 \; \begin{array}{c} -CHOH \\ | \\ -CO \end{array} \; + \; \underset{-NO_2}{\overset{-NO_2}{\bigcirc}} \; + \; 2\,OH^- \; \longrightarrow \; \underset{-NO_2^-}{\overset{-NO^-}{\bigcirc}} \; + \; 2 \; \begin{array}{c} -CO \\ | \\ -CO \end{array} \; + \; 3\,H_2O \qquad (3)$$

$$\underset{-NO_2}{\overset{-NO_2}{\bigcirc}} \; + \; 2\,CH_2O \; + \; 4\,OH^- \; \longrightarrow \; \underset{-NO_2^-}{\overset{-NO^-}{\bigcirc}} \; + \; 2\,HCOO^- \; + \; 3\,H_2O \qquad (2{+}3)$$

When the catalytic acceleration is employed in analysis, the conditions prescribed in the procedure must be carefully maintained so that the un-catalyzed reaction proceeds so slowly that there is no difficulty in observing the catalytic effect produced by even microgram amounts of diketones.

The catalysis reaction is not specific for α-diketones because p- and o-quinones likewise act as catalysts. This effect is probably the result of redox reactions analogous to (2), namely

$$O{=}\bigcirc{=}O \; + \; 2\,CH_2O \; + \; 2\,OH^- \; \longrightarrow \; HO{-}\bigcirc{-}OH \; + \; 2\,HCOO^-$$

$$\underset{O}{\overset{}{\bigcirc}}{=}O \; + \; 2\,CH_2O \; + \; 2\,OH^- \; \longrightarrow \; \underset{OH}{\overset{-OH}{\bigcirc}} \; + \; 2\,HCOO^-$$

in which these quinones lead to hydroxy compounds, which function as hydrogen donors toward o-dinitrobenzene with re-formation of the quinones, which then act again with formaldehyde.

The test described here is subject to interference by those organic compounds which function *per se* as hydrogen donors to o-dinitrobenzene and thus lead to the color reaction. Consequently, a test must first be made of the behavior of the sample as prescribed in the procedure but without the addition of formaldehyde. If no or only a pale violet color results, the absence of interfering compounds is assured and the test can be conducted with the addition of formaldehyde. If the result is now a violet or a more intense shade, the presence of catalytically effective α-diketones or quinones is indicated.

A simple method for detecting o-diketones in the presence of large quantities of organic reducing materials, which give the color reaction in the absence of formaldehyde, consists in oxidizing the sample with alkali hypobromite and then carrying out the catalysis reaction. (Compare the test for inositol in the presence of reducing sugars, p. 419.) When using this procedure, it should be remembered that it does not avert the interference due to quinones and polyphenols (which are oxidized to quinones).

Procedure. A drop of the aqueous or benzene test solution is mixed in a micro test tube with one drop each of 25% sodium carbonate solution, 4% formaldehyde, and 5% solution of o-dinitrobenzene in benzene. The mixture is warmed in boiling water and shaken from time to time. Depending on the amount of catalyzing material present, a more or less intense violet color appears within 1–4 min.

A red color is produced if p-dinitrobenzene is used. The limits of identification are higher or lower than with the isomer, depending on the nature of the sample. It is likely that 3,4-dinitrobenzoic acid could be used here with success, since it has been employed for the chromatographic detection of reducing sugars[2].

The procedure revealed:

0.05 γ diacetyl	0.002 γ phenanthraquinone
2.5 γ furil	0.01 γ 2-methyl-1,4-naphthoquinone (vitamin K_3)
2 γ benzil	0.5 γ sodium 1,2-naphthoquinone-4-sulfonate
30 γ isatin	0.5 γ sodium rhodizonate
0.5 γ ninhydrin	0.002 γ 3-nitrophenanthraquinone
0.05 γ anthraquinone	0.5 γ sodium anthraquinone-2-sulfonate

A strong response was observed with dehydroascorbic acid, p-benzoquinone and chloranil. The fact that vitamin K_3 catalyzes the reaction suggests that the same will be true of K_1 and K_2, since they too contain a naphthoquinone nucleus.

1 F. FEIGL and CL. COSTA NETO, *Anal. Chem.*, 28 (1956) 397.
2 F. WEYGAND and H. HOFMANN, *Ber.*, 83 (1950) 405.

14. Quinones

(1) Test with rhodanine and ammonia [1]

Numerous quinones give colored condensation products with rhodanine; the color is green or blue with p-quinones, and violet to red with o-quinones. The green condensation products are probably identical with the compounds formed with the hydroquinones (see p. 428), while the blue and violet compounds perhaps are condensation products formed from quinones and rhodanine.

Dibenzoquinones, such as anthraquinone and phenanthraquinone, and dihydroxybenzoquinone do not react. Hydroquinone behaves similarly to quinone. It is notable that "Entobex" (4,7-phenanthroline-5,6-quinone) reacts.

Procedure. A drop of the test solution is treated in a micro test tube with a drop of a saturated aqueous solution of rhodanine and 1 drop of concentrated ammonia water; the mixture is shaken gently. A positive response is indicated by the appearance of a color.

Limits of identification:

0.2 γ	p-benzoquinone	green	2	γ 2,5-xylo-p-quinone	blue
1	γ p-toluquinone	,,	0.2	γ p-naphthoquinone	blue-green
2	γ chloranil	,,	0.2	γ o- ,, ,,	violet
1	γ 2,3-dichloro-5,6-dicyano-		1	γ o-naphthoquinone-4-	
	quinone	,,		sulfonic acid	,,
0.2 γ	2-methylnaphthoquinone	,,	2	γ "Entobex"	,,
5	γ 3-tolu-o-quinone diacetate	,,			

The following give no color reaction: chloranilic acid, tetrahydroxyquinone, phylloquinone, triquinone, rhodizonic acid, alloxan, acenaphthoquinone, anthraquinone, phenanthraquinone, duroquinone.

1 V. ANGER and S. OFRI, *Mikrochim. Acta*, (1963) 915.

(2) Test with 1-phenyl-3-methylpyrazol-5-one and ammonia [1]

Quinones react with the pyrazolone whereby colored products are formed. The following condensation seems probable:

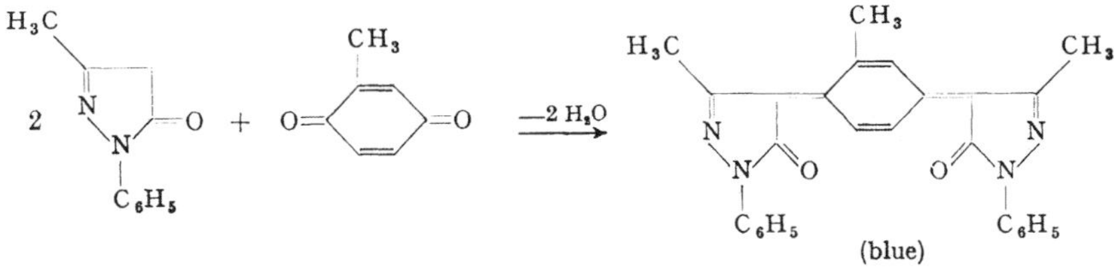

p-Benzoquinone yields a violet product, similarly hydroquinone, while pyrocatechol yields a blue product; pyrogallol and its derivatives do not react. It is notable that triquinol, whose quinone character is only slightly evident because of hydration, does react.

Procedure. A drop of the test solution is mixed in a micro test tube with a drop of the reagent solution and 1 drop of concentrated ammonia water. If compounds that respond positively are present, a green, blue, or violet color gradually develops (in 5–10 min. if very slight amounts are involved). It is advisable to run two blanks: the first without the test solution, the second without reagent solution.

Reagent: Saturated solution of 1-phenyl-3-methylpyrazole-5-one in methanol–water (1 : 1).

Limits of identification:

1 γ	p-toluquinone	blue	0.5 γ	p-benzoquinone	violet
2 γ	2,5-xylo-p-quinone	,,	2	γ 2-methylnaphthoquinone	,,
1 γ	p-naphthoquinone	,,	20	γ triquinol	,,
2 γ	o-naphthoquinone-4-sulfonic		1	γ o-naphthoquinone	green
	acid	,,	2	γ "Entobex"	,,
1 γ	3-tolu-o-quinone diacetate	,,	5	γ chloranil	,,
1 γ	2,3-dichloro-5,6-dicyanoquinone	,,			

The following give no reaction: chloranilic acid, tetrahydroxyquinone, phyllo-quinone, rhodizonic acid, alloxan, acenaphthoquinone, anthraquinone, phenan-thraquinone, duroquinone.

1 V. ANGER and S. OFRI, *Mikrochim. Acta*, (1963) 917.

15. *o*-Quinones $\left(\text{type}\right)$

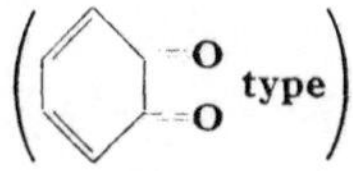

Test through sintering with guanidine carbonate[1]

There is rapid formation of a blue-violet product when a mixture of orange phenanthraquinone and colorless guanidine carbonate is heated to 180–200°. The chemistry of this reaction has not yet been experimentally cleared up, but there are several bases for constructing a plausible explanation. In the first place, the reaction is limited to *ortho*-quinones and their homologs since anthraquinone does not react nor does benzil or 1,2-naphthoquinone-4-sulfonate. The guanidine carbonate can be replaced by the base or its acetate, but not by salts of guanidine with strong acids nor by derivatives of guanidine. The assumption that on heating to 160° the guanidine carbonate loses ammonia to form melamine which then reacts with the phenanthra-quinone does not fit the facts, since melamine is inactive in this respect. Accordingly, the following condensation seems probable:

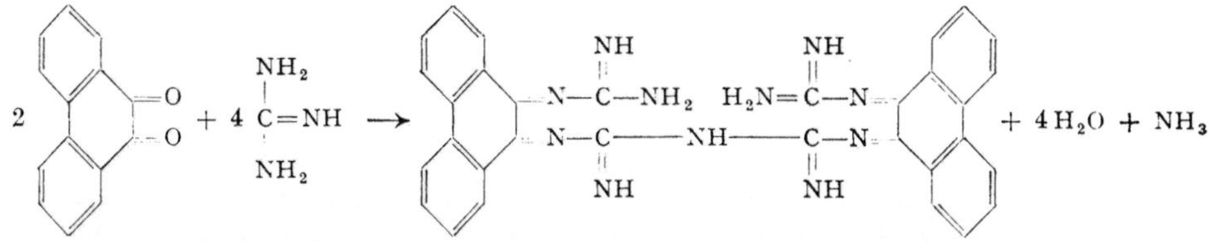

In accord with the structure as phenanthraguanidine assumed here, the product contains a continuous system of conjugated double bonds, which is in conformity with the intense color quality. The blue-violet product dis-solves in ether and other O-containing organic liquids to give a red solution; it is basic in character, as indicated by the discharge of the color when the ethereal solution is treated with a mineral acid and formation of a blue precip-itate.

Procedure. A drop of the benzene or ethereal test solution is mixed in a micro test tube with several cg of guanidine carbonate and the solvent evaporated. The test tube is placed in a bath preheated to 160° and the temperature raised to 180–190°. A positive response is indicated by the appearance of a colored product

Limit of identification: 1.5 γ phenanthraquinone (blue-violet)
                                 5   γ "Entobex" (red)

A positive response was given by *o*-naphthoquinone (violet) and acenaphthoquinone (dark).

1 Unpublished studies, with D. GOLDSTEIN.

16. Naphthoquinones

Test with 2,4-dihydroxythiazole[1]

Colored reaction products are formed from naphthoquinones, 2,4-dihy-
droxythiazole and sodium carbonate.

The mechanism of this color reaction is still unknown. It appears to be
specific for naphthoquinones. Naphthoquinones that are alkylated on the
quinone ring do not show this reaction.

Procedure. A drop of the solution to be tested is united in a micro test
tube with a drop of a saturated solution of the thiazole in water and a drop of a
5% sodium carbonate solution. The mixture is kept in boiling water for 2–3 min.
A green or violet color indicates a positive response.

Limits of identification and colors:

0.5 γ *o*-naphthoquinone   violet   0.5 γ *p*-naphthoquinone green
1   γ 1,2-naphthoquinone-4-sulfonic acid   ,,

2-Methylnaphthoquinone and phylloquinone do not react.

1 V. ANGER and S. OFRI, *Mikrochim. Acta*, (1963) 917.

17. *p*-Naphthoquinones $\left(\vcenter{\hbox{(naphthoquinone structure)}} \text{ type} \right)$

Test with 2-aminothiophenol[1]

Certain derivatives of 1,4-naphthoquinone yield a red product when they
react with 2-aminothiophenol in dilute hydrochloric acid. The color changes
to blue on treatment with concentrated hydrochloric acid. The mechanism
of the color reaction is unknown as yet. Other quinones as well as *p*-naphtho-
quinones with electron-donor groups do not give this reaction.

Procedure. A drop of the reagent solution is placed on filter paper and
spotted with a drop of the methanolic test solution. A red color that turns blue
when spotted with a drop of concentrated hydrochloric acid indicates a positive response.

Reagent: 0.5 ml of 2-aminothiophenol is mixed with 4 ml of concentrated hydrochloric acid and the mixture then diluted to 50 ml with water.

Limits of identification: 0.5 γ of each of the following compounds: *p*-naphthoquinone, 2-methylnaphthoquinone, 2-chloronaphthoquinone, 2,3-dichloronaphthoquinone, 2,3-dibromonaphthoquinone.

1 E. SAWICKI and W. C. ELBERT, *Anal. Chim. Acta*, 23 (1960) 205.

18. Anthraquinones $\left(\begin{array}{c} \text{O} \\ \text{type} \end{array} \right)$

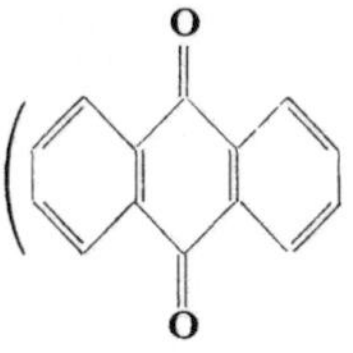

Test through reduction to anthrahydroquinone [1]

If a solution or suspension of anthraquinone is warmed with an appropriate hydrogen donor, the yellow anthrahydroquinone is formed: [2]

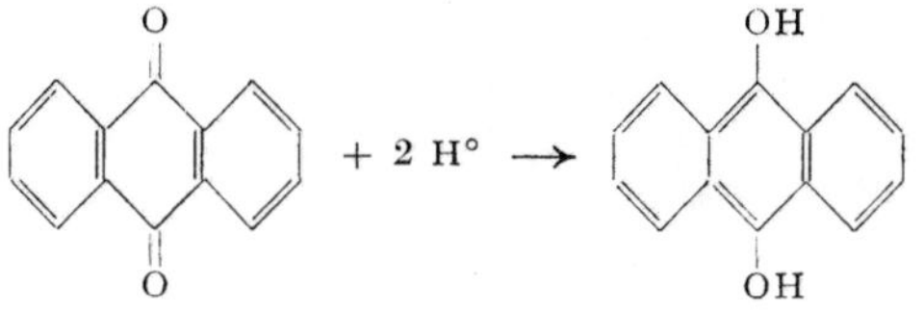

$+ 2 \text{ H}° \longrightarrow$

The aqueous solution of the alkali salts of anthrahydroquinone is faintly red in the cold and deep red when heated; [3] they are easily reoxidized to anthraquinone in contact with air. However, when an alkaline solution containing an excess of sodium hydrosulfite ($Na_2S_2O_4$) is used as reductant, the autoxidation is inhibited. Hence a sensitive test for anthraquinone has been developed on the basis of heating it with a strongly alkaline solution of sodium hydrosulfite. Derivatives of anthraquinone behave like the parent compound, a fact that is readily observed in the case of colorless or faintly colored derivatives (see anthraquinonesulfonic acids, p. 358).*

Although anthraquinone is but slightly soluble in organic liquids, it is readily sublimed, and the test given here may be applied to the sublimate. Consequently, the sublimable anthraquinone may be separated and differentiated from the non-subliming anthraquinonesulfonic acids.

Procedure.   A little of the solid or the evaporation residue of a drop of the test solution is treated in a micro test tube with 1 drop of 5 *N* sodium hydroxide and several cg of sodium hydrosulfite. The mixture is heated over a micro flame. A more or less red color appears if the response is positive.

* The violet alkaline solutions of polyhydroxyanthraquinones (alizarin, purpurin, quinalizarin, *etc.*) also undergo reduction to derivatives of anthrahydroquinone. The color change to red is readily observed even when only small amounts of the compound are involved. The same is true of dyestuffs with anthraquinone structure (see p. 672).

Limits of identification:

     0.5 γ anthraquinone   1 γ 1-aminoanthraquinone

     2   γ 2-chloro-3-methylanthraquinone

1 Unpublished studies, with D. GOLDSTEIN.

2 E. GRANDMOUGIN, *Ber.*, 39 (1906) 3563.

3 K. H. MEYER, *Ann.*, 379 (1911) 43.

19. Carbohydrates

Test through transformation to furfurals

Carbohydrates (di- and polysaccharides) are hydrolyzed by heating with strong mineral acids or with oxalic acid. Monosaccharides are formed:

$$(C_6H_{10}O_5)_x + x\ H_2O \rightarrow x\ C_6H_{12}O_6$$

On further heating, the monosaccharides (pentoses), are partly dehydrated to furfural or similar aldehydes, such as hydroxymethylfurfural in the case of hexoses:

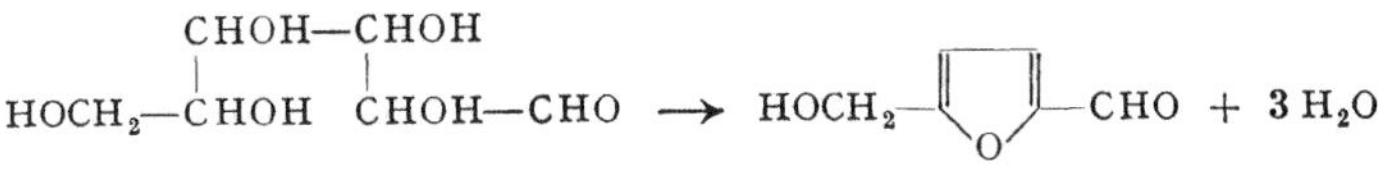

$$\begin{array}{c} CHOH{-}CHOH \\ | \qquad\qquad | \\ HOCH_2{-}CHOH \quad CHOH{-}CHO \end{array} \rightarrow HOCH_2{-}\!\!\bigcirc\!\!{-}CHO + 3\ H_2O$$

These aldehydes are volatile with steam and react with aniline to give violet Schiff bases (see p. 444). The hydrolysis and the dehydration are best accomplished with syrupy phosphoric acid.

Procedure.[1] A pinch of the material being studied is placed in a micro crucible or a drop of the test solution is taken to dryness there. A drop of syrupy phosphoric acid is added and a disk of filter paper moistened with a drop of a 10% solution of aniline in 10% acetic acid is placed over the mouth of the crucible, and weighted down with a watch glass. The bottom of the crucible is cautiously heated for 30–60 sec. with a micro flame (spattering must be avoided). A pink to red stain appears on the reagent paper.

The rest revealed:

2.5 γ glucose	5 γ agar-agar	2.5 γ sorbose
2.5 γ fructose	25 γ saccharose	2.5 γ maltose
3.0 γ starch	2.5 γ lactose	2.5 γ arabinose
2.5 γ galactose		

A positive reaction was given by 5 γ of ethyl- and methyl cellulose, acetyl cellulose, and gum tragacanth.

1 F. FEIGL, J. E. R. MARINS and Cl. COSTA NETO, unpublished studies; see also O. FREHDEN and L. GOLDSCHMIDT, *Mikrochim. Acta*, 2 (1937) 184.

20. Reducing sugars

(1) Test with triphenyltetrazolium chloride

If the colorless alkaline solution of triphenyltetrazolium chloride is heated with a reducing sugar, red triphenylformazan precipitates:[1]

$$C_6H_5-C \begin{array}{c} {}^{+} \\ N=N-C_6H_5 \\ | \\ N-N-C_6H_5 \end{array} + H_2O + 2\,e \longrightarrow C_6H_5-C \begin{array}{c} N-NH-C_6H_5 \\ N=N-C_6H_5 \end{array} + OH^-$$

A sensitive test for reducing sugars has been based on this reaction.[2] In contrast to most other tests for reducing sugars, aldehydes do not interfere. Furthermore, other reductants, such as hydrazine, hydroxylamine, sulfites, tartaric and citric acid, are completely inactive toward this reagent. Ascorbic acid, which on the basis of its chemical constitution can be regarded as a reducing sugar, gives the same reaction.

Procedure.[2] A drop of the test solution is mixed, in a micro test tube or micro crucible, with two drops of 0.5% aqueous triphenyltetrazolium chloride solution and one drop of 0.5 N sodium hydroxide. The mixture is boiled for 1–2 min. A red color or precipitate indicates a positive response.

Limit of identification: 0.2 γ glucose, fructose, lactose, mannose, arabinose, ascorbic acid.

1 R. KUHN and D. JERCHEL, *Ber.*, 74 (1941) 949.
2 A. BONDI (Rehovoth), unpublished studies; see also A. N. MATTSON and C. O. TENSEN, *Anal. Chem.*, 22 (1950) 183; *Science*, 106 (1947) 294.

(2) Other tests

Almost all the macro tests for reducing sugars may be applied as spot tests. Since they all depend on reducing properties (*m*-phenylenediamine is an exception), other reducing substances must be absent.

The *limits of identification* (drop size=0.04 ml) for a number of tests for reducing sugars are given in Table 14.[1]

o-Dinitrobenzene (see p. 134) and the 3,4- and 3,5-isomers of dinitrophthalic acid[2] give a sensitive color reaction (violet) when heated with reducing sugars in aqueous Na_2CO_3. The *Idn. Limits* are 5 γ glucose with either of the *o*-dinitro compounds

The detection of reducing sugars through reduction of silver oxide may also be carried out as a spot reaction on paper.[3] Strips of filter paper are soaked in 0.2 N AgNO$_3$ solution and dried. Single drops of the alkaline test solution and of a sodium hydroxide solution of approximately the same strength are placed next to each other on the paper. Two brown flecks develop. After about 1 min., the paper is placed in ammonia water.

The unchanged silver oxide dissolves completely, while a black fleck of finely divided silver is left in the stain produced by the sample containing reducing sugar.

Limit of identification: 0.1 γ glucose.

TABLE 14

REDUCING SUGARS

Reagent	Color change	Identification limit
Nylander's solution (alkaline bismuth solution containing tartrate)	Blackening	10 γ
Ammoniacal AgNO$_3$ solution	Blackening	0.1 γ
Magnesium hypoiodite [Mg(OH)$_2$ and KIO]	Decolorization	5 γ
Alkaline 0.001% methylene biue solution	Decolorization	1 γ
Alkaline dinitroacetanilide (1,3,4) solution	Violet	2 γ
m-Phenylenediamine hydrochloride (evaporated with 1 drop test solution)	Yellow-green fluorescence	0.5 γ

The test cannot be applied in the presence of organic compounds containing CS- or SH groups. Under the prescribed conditions, they form black Ag$_2$S, which likewise is not soluble in ammonia.

1 Unpublished studies, with G. FRANK.
2 T. MOMOSE, A. INABA, K. INOUE, K. MIYAHARA and T. MORI, *Chem. Pharm. Bull. Tokyo*, 12 (1964) 14.
3 F. FEIGL, *Chem. Ind. (London)*, 57 (1938) 1161.

21. Hexoses

Test with 5-hydroxy-1-tetralone [1]

If an aqueous solution of a hexose is heated with an alcoholic solution of 5-hydroxy-1-tetralone in the presence of concentrated sulfuric acid, a faint brown color appears that shows fluorescence. When the mixture is diluted with much water, the brown fades away and the green fluorescence is then distinctly visible in daylight. Examination under ultraviolet light renders the fluorescence still more evident.

The fluorescent compound produced in the reaction is probably benzo-naphthenedione (V). On reaction with glucose, 5-hydroxy-1-tetralone (I) may combine initially with the aldehyde group of the sugar by virtue of the activated CH$_2$ group shown in (Ib), followed by dehydration to produce an active CH$_2$ group in the 3-position of the bound glucose (III), and this may then combine with the CO group of the tetralone. Three carbons (R) of glucose may be eliminated before or after the ring closure, and the resulting dihydrobenzonaphthenedione (IV) may be oxidized to benzonaphthene-dione (V).

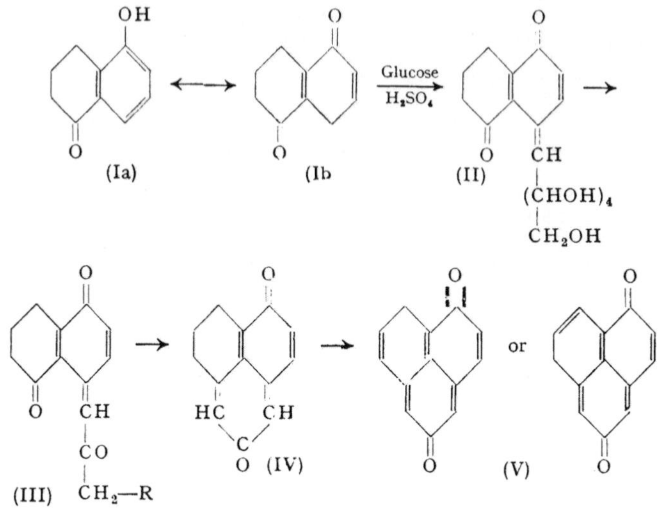

Procedure. A drop of the sugar solution is placed in a micro test tube and mixed with a drop of 0.1% alcoholic solution of 5-hydroxy-1-tetralone (for preparation, see reference 1) and 5 drops of concentrated sulfuric acid. The mixture is kept in a boiling water bath for 30 min., cooled, and then diluted with 1 ml water. The resulting green fluorescence is viewed in ultraviolet light that has been filtered from visible light. If the sugar is in excess with respect to the reagent, the mixture exhibits a muddy brown color. In this case, it is necessary to repeat the test with a more dilute test solution.

The procedure gave a positive result with the following amounts of sugars. If the fluorescence was observed in daylight, ten times these quantities were required to give reliable results:

0.2 γ glucose	0.3 γ maltose	0.2 γ dextrin
0.25 γ mannose	0.2 γ lactose	0.2 γ inulin
0.3 γ galactose	0.25 γ saccharose	0.35 γ agar-agar
0.2 γ fructose	0.2 γ starch (soluble)	0.3 γ cellulose

Only a few compounds were found to interfere. Glycerol gives a positive result when present in amounts exceeding 1 mg per drop, and a few micrograms of glyceraldehyde also show a like fluorescence.

Negative results were obtained with specimens known to contain pentoses, ascorbic acid, glucuronic acid, galacturonic acid, polyhydric alcohols, amino acids, aldehydes, ketones, organic acids, phenols, ethers, proteins.

1 T. Momose and Y. Ohkura, *Chem. & Pharm. Bull.* (*Tokyo*), 7 (1959) 31; 4 (1956) 209; 6 (1958) 412.

22. Ketohexoses

Test with stannous chloride, sulfuric acid and urea[1]

A blue color results when ketohexoses are heated with a sulfuric acid solution of urea containing stannous chloride. Aldohexoses give a red color with this reagent but only after prolonged heating. The mechanism of these color reactions is still unkown.

Procedure.[2]   A drop of the sugar solution is treated in a micro crucible with 6–20 drops of the reagent solution and then heated over a micro flame. After cooling, a blue color appears if ketohexoses are present. The red color due to aldohexoses develops only after longer boiling and is less intense.

Reagent solution: 4 g urea and 0.2 g $SnCl_2$ are dissolved by heating with 10 ml of 40% sulfuric acid.

The test revealed: 8 γ fructose   15 γ saccharose
                   10 γ inulin   8 γ sorbose

1 J. H. FOULGER, *J. Biol. Chem.*, 99 (1932) 207; *Compt. Rend.*, 196 (1933) 2984.
2 Unpublished studies, with R. ZAPPERT.

23. *o*-Hydroxyaryl carbonyl compounds (⟩–OH ⟩–CO type)

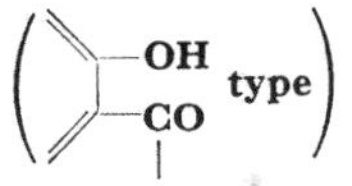

Test by formation of fluorescing aldazines and ketazines[1]

In neutral or acetic acid solution, hydrazine condenses at both of its NH_2 groups with *o*-hydroxyaldehydes and *o*-hydroxyketones to produce crystalline, light yellow insoluble Schiff bases:

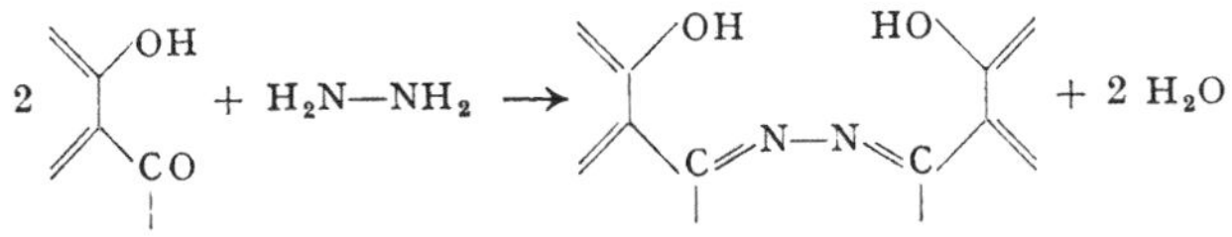

Without exception, the solid but not the dissolved aldazines and ketazines of *o*-hydroxyaldehydes and ketones exhibit an intense yellow-orange fluorescence in ultraviolet light. The phenolic OH group situated in the *ortho* position is essential because the Schiff bases of the *meta*- and *para*-isomers do not fluoresce. Furthermore, the symmetric structure with respect to the =N—N= group and this group itself appear to be essential, since the phenylhydrazone of salicylaldehyde which lacks these features does not fluoresce.

The condensation of *o*-hydroxyaldehydes with hydrazine to produce fluorescing aldazines occurs almost instantaneously, and there seems to be no steric hindrance even when the aldehyde has a complicated structure. In contrast, all studies indicate that the condensations of o-hydroxyketones with hydrazine to produce the fluorescing ketazines proceed very slowly.* A fundamental advantage with respect to the analytical application of this fluorescence reaction, which is sensitive and highly selective for o-hydroxy-aldehydes and ketones, is provided by the resistance of the aldazines and ketazines (once they have been formed) to dilute acids.

Procedure.   A drop of the solution of the test material in alcohol, acetone, dioxane *etc.*, is placed on filter paper and spotted with a drop of hydrazine solution. If preferred, filter paper impregnated with hydrazine salt may be used and the moist reagent paper is spotted with the test solution. According to the quantity of o-hydroxyaldehyde present, a more or less intense yellow-orange fluorescing fleck appears at once or within a minute or two. The fluorescing fleck persists even when the paper is bathed in acetic or dilute (1 N) mineral acid.

If the highest sensitivity is not needed, the solid test material may be spotted directly with the reagent solution.

Reagent: 5 g hydrazine sulfate and 10 g sodium acetate in 100 ml water.

Positive results were obtained with:

1 γ salicylaldehyde	1.1 γ 4-methylhematommic acid
3.5 γ psoromic acid	6.5 γ β-orcinol aldehyde
2.5 γ thamnol	2.2 γ protocetraric acid
1.8 γ hematommic acid methyl ester monobenzyl ether	
0.8 γ 6-methyl-4,5-dihydroxysalicylaldehyde	
0.1 γ 2,4-dihydroxyacetophenone (after evaporation with hydrazine solution)	

A positive response was also given by norstictic acid, salazinic acid[2] and pyridox-al.

* No conclusive trials have been made as yet to learn whether this finding may be used as the basis for differentiating o-hydroxyaldehydes and ketones.

1 Unpublished studies, with Y. HASHIMOTO (Kyoto); J. MUELLER (Vienna), personal communication.
2 See Y. ASAHINA, *Acta Phytochim.* (*Japan*), 8 (1934) **33**, regarding the constitution of these depsides and depsidones; comp. W. B. MORS, *Rev. Brasil. Biol.*, 12 (1952) 389.

24. o-Hydroxyarylaldehydes $\left(\begin{array}{c} \text{OH} \\ \\ \text{CHO} \end{array} \text{type}\right)$

(1) Test by conversion into yellow, fluorescing alkali phenolate [1]

Salicylaldehyde is sparingly soluble in water but dissolves readily in organic liquids. It reacts with alkali hydroxides to produce water-soluble phenolates, whose yellow color in daylight, and blue-green fluorescence in ultraviolet light, may be related to the chelate nature of salicylaldehyde and its alkali salts:

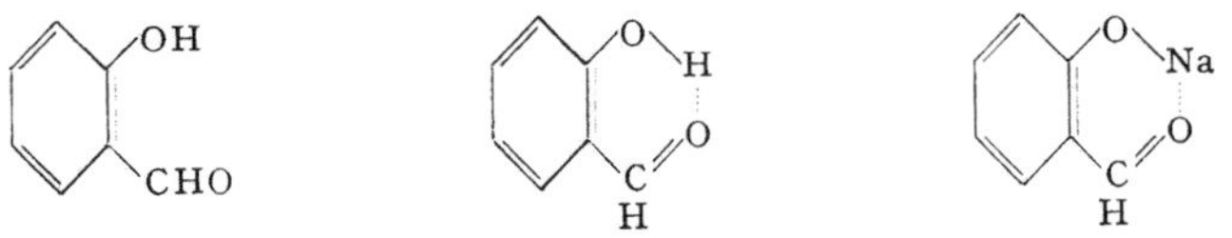

The formation of alkali salts of salicylaldehyde occurs instantly if a drop of its solution is placed on filter paper impregnated with caustic alkali. A yellow fleck results (*Idn. Limit*: 6 γ aldehyde); it displays an intense blue-green fluorescence in ultraviolet light. If the test is carried out with an ethereal solution of the aldehyde which is so dilute that no visible yellow fleck remains, the blue-green fluorescence may be detected down to a dilution of 1:1,000,000. Its isomers, namely *m*- and *p*-hydroxybenzaldehydes, are likewise soluble in alkali hydroxide (the *m*-compound with honey-yellow color) but the solutions show no fluorescence in ultraviolet light.

The test for salicylaldehyde through formation of the fluorescing alkali salt takes advantage of the fact that this aldehyde (b.p. 197°) has a distinct vapor pressure even at room temperature, and it increases distinctly when the material is moderately warmed. Accordingly, the fluorescing phenolate is formed immediately when the vaporizing aldehyde comes into contact with alkali hydroxide. This test in the gas phase, though less sensitive than the test with dissolved aldehyde, is almost specific because it is not impaired by the presence of other compounds whose alkaline solutions are colored or fluorescent. Salicylic acid is an exception (see p. 478).

Procedure. One drop of the solution in ether or another readily volatilizable solvent is placed in a micro crucible, which is then covered with a disk of filter paper impregnated with 1 N potassium hydroxide. A watch glass is placed on the filter paper, and the crucible is allowed to stand at 25–50° for 5–10 min. If salicylaldehyde is present, a circle, which fluoresces blue-green under the ultraviolet lamp, is formed.

Limit of identification:

     0.05 γ salicylaldehyde*   0.05 γ orthovanillin

Test (*1*) and also the following Test (*2*) are given by compounds which split off salicylaldehyde when hydrolyzed by acids or when dry heated. Examples are: salicylaldoxime and its salts, helicin (glucoside of salicylaldehyde), salicyl alcohol, and salicin (glucoside of salicyl alcohol). In the latter two instances, the salicyl alcohol is converted to salicylaldehyde by autoxidation.

* This identification limit, and also that for Test (*2*), was determined through saponification of helicin (see p. 655), because commercial salicylaldehyde always contains impurities.

1 Unpublished studies, with W. A. MANNHEIMER.

(2) Test through conversion into fluorescing aldazines [1]

Salicylaldehyde dissolved in acetic acid, alcohol, acetone, *etc.* quantitatively reacts with a solution of hydrazine or a hydrazine salt to yield the whitish-yellow water-insoluble salicylaldazine, which shows a strong orange fluorescence in ultraviolet light (comp. p. 341).

$$H_2NNH_2 + 2\ C_6H_4(OH)CHO \rightarrow HOC_6H_4CH{=}N{-}N{=}CHC_6H_4OH + 2\ H_2O$$

The precipitation occurs also from mineral acid solutions of hydrazine sulfate; it is complete after buffering with alkali acetate. The precipitating ability of salicylaldehyde stands in sharp contrast to the lack of reactivity shown under the same conditions by the isomeric *m-* and *p-*hydroxybenzaldehydes. This divergence is probably due, as shown in the partial structural formulas, to the fact that only the condensation product of salicylaldehyde with hydrazine leads to a 6-membered ring (chelate ring) as a result of the coordination of the H atom of the phenolic OH group to the tervalent nitrogen atom ligated with the carbon atom.

A further indication of the singularity of the chelation is the fact that only the salicylaldazine exhibits a luminous greenish-yellow fluorescence, whereby traces of this compound may be detected in ultraviolet light.

Solid salicylaldazine is fluorescent but surprisingly its solutions in ether, chloroform, *etc.* show no fluorescence. If a drop of the non-fluorescing solution is placed on filter paper and the solvent allowed to evaporate, a yellow-green fluorescing spot is left.

The precipitation of salicylaldazine provides a sensitive and specific test for hydrazine (*Inorganic Spot Tests*, Chap. 3), but also a means of detecting salicylaldehyde. If one drop of hydrazine solution containing alkali acetate is treated on paper with a drop of an ethereal solution of salicylaldehyde, a yellow fluorescent fleck results. (When small amounts of salicylaldehyde are involved, the given order of addition must be observed.) It is better to take advantage of the vapor tension of salicylaldehyde, as in Test (*1*), and to accomplish the aldazine formation from the gas phase of this aldehyde.

Procedure.   One drop of the ethereal test solution is placed in a microcrucible, which is then covered with a disk of filter paper that has been moistened with hydrazine solution. A watch glass is placed on the paper and the assembly allowed to stand for 5 min. at 25–50°. If salicylaldehyde is present, a circle, which fluoresces yellow in ultraviolet light, forms on the paper.

Reagent: Cold saturated solution of hydrazine sulfate or chloride, to which several grams of sodium acetate have been added.

Limit of identification: 0.05 γ salicylaldehyde   0.05 γ orthovanillin.

A positive response is also given by 2-hydroxynaphthaldehyde.

1 Unpublished studies, with W. A. MANNHEIMER.

25. *p*-Hydroxyarylaldehydes $\left(\text{HO}\!-\!\!\left\langle\!\!\!\bigcirc\!\!\!\right\rangle\!\!-\!\text{CHO type}\right)$

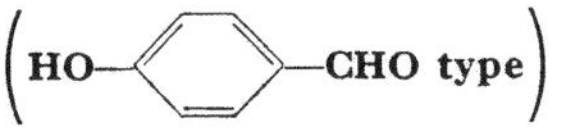

Test with thiobarbituric acid and hydrochloric acid[1]

The condensation of barbituric acid with aromatic aldehydes (see Sect. 26) is paralleled by the behavior of thiobarbituric acid. The condensation products with *p*-hydroxyarylaldehydes are yellow in acidic solution and fluoresce green. When made alkaline with ammonia, the color turns to red and the fluorescence becomes yellow. Derivatives of *p*-hydroxycinnamaldehyde react analogously, however in acid solution the color is orange-yellow and the fluorescence red; on alkalization the color becomes violet and the fluorescence turns blue. *p*-Hydroxyarylallyl compounds show a similar behavior to that of the corresponding *p*-hydroxycinnamaldehydes, and the same is true of *p*-hydroxycinnamic alcohols (see p. 323).

Procedure. A drop of a saturated solution of thiobarbituric acid in 2 *N* hydrochloric acid is placed on filter paper and spotted with a drop of the alcoholic test solution. If reactive compounds are present, a yellow or yellow-orange color appears that fluoresces green or red in u.v. light. After a drop of concentrated ammonia water is added, the color changes to red or violet and the fluorescence becomes yellow or blue.

Limit of identification:

0.1 γ vanillin	5 γ eugenol
0.1 γ coniferylaldehyde	5 γ coniferyl alcohol

1 V. ANGER and S. OFRI, *Z. Anal. Chem.*, 203 (1964) 426.

26. *p*-Hydroxy- and *p*-alkoxyaldehydes $\left(\text{(H,R) O}\!-\!\!\left\langle\!\!\!\bigcirc\!\!\!\right\rangle\!\!-\!\text{CHO type}\right)$

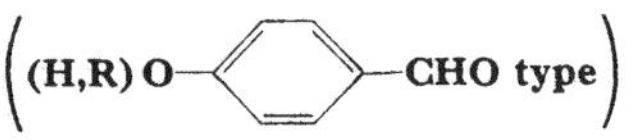

Test with barbituric acid[1]

Barbituric acid condenses with aromatic aldehydes to yield difficultly soluble compounds that usually are faintly colored. It has been found that aromatic *p*-hydroxyaldehydes react with barbituric acid to give products that fluoresce intensely green only in acid solution. p-Alkoxyaldehydes produce intensely yellow compounds that fluoresce yellow in ultraviolet light and this fluorescence remains on alkalization with ammonia. p-Dimethylaminobenz-aldehyde condenses with barbituric acid to give a product that fluoresces red. A selective test for these aldehydes is based on these behaviors.

Procedure. A drop of a saturated solution of barbituric acid in 2 N hy-drochloric acid is placed on filter paper and followed by a drop of the alcoholic test solution. If p-hydroxyaldehydes are present, a green fluorescence appears that disappears when concentrated ammonia water is added. A yellow fluores-cence that is resistant to ammonia indicates the presence of p-alkoxyaldehydes.

Limit of identification and color of fluorescence:

0.01 γ vanillin	green	0.5 γ p-methoxycinnamaldehyde	yellow
0.2 γ piperonal	,,	0.5 γ p-dimethylaminobenzaldehyde	orange-red
0.01 γ coniferylaldehyde	,,		

No fluorescence was shown by: benzaldehyde, o-nitrobenzaldehyde, furfural, crotonaldehyde, cinnamaldehyde, salicylaldehyde, orthovanillin, o-phthalalde-hyde, p-dimethylaminocinnamaldehyde.

1 V. ANGER and S. OFRI, *Z. Anal. Chem.*, 203 (1964) 427.

27. Arylvinyl- and p-hydroxy-(alkoxy-)aryl-aldehydes

$$\left(\text{Ar—CH=CH—CHO and (R,H)O} \left\langle \bigcirc \right\rangle \text{—CHO type} \right)$$

Test with orcinol–hydrochloric acid [1]

In contrast to phloroglucinol and naphthoresorcinol, only orcinol reacts in hydrochloric acid with compounds of the above types. Hereby orange or violet condensation products are formed. These different colors permit a distinction between the two types of aldehydes. The chemistry of the reac-tion is analogous to that of the test with phloroglucinol described on p. 204.

Procedure. A drop of the test solution is treated in a depression of a spot plate with a drop of a 0.1% alcoholic solution of orcinol and several drops of concentrated hydrochloric acid. An orange color signals the presence of cinnamal-dehyde and its derivatives or p-hydroxybenzaldehydes.

Limits of identification:

1 γ vanillin	violet	0.1 γ cinnamaldehyde	orange
5 γ piperonal	,,		
5 γ anisaldehyde	,,		

Furfural gives a reaction color that turns green.

1 V. ANGER and S. OFRI, *Z. Anal. Chem.*, 203 (1964) 428.

28. Polyhydroxyanthraquinones $\left(\left\{\begin{array}{c}O\\ \\ \\ O\end{array}\right\}(OH)_n \text{ type}\right)$

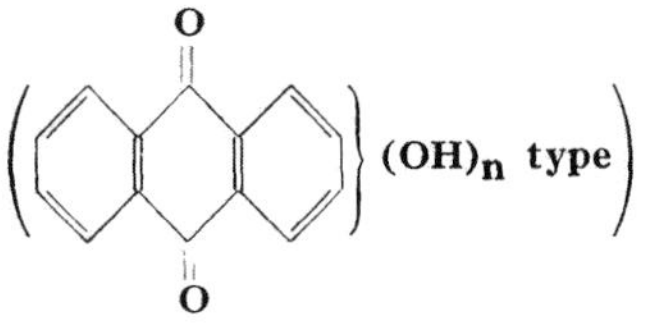

Test by formation of zirconium color lakes[1]

If alizarin is added to acid solutions of zirconium salts, red-violet precipitates or colorations result. This fact is the basis of sensitive tests for zirconium.[2] In conformity with the principle of group action of organic reagents,[3] the color reaction with zirconium is not restricted to alizarin. All polyhydroxyanthraquinones with two hydroxyl groups in the *ortho* position behave similarly to alizarin, but those lacking this structure do not react. When alkaline solutions of polyhydroxyanthraquinones are mixed with solutions of zirconium salts, colored precipitates are obtained in every instance, and these are not altered by even prolonged washing with hot water. If treated with dilute hydrochloric acid, they leave colored products only if polyhydroxyanthraquinones with OH groups in the *ortho* position were present. Consequently, the divergent reaction picture presented by polyhydroxyanthraquinones toward zirconium salts in acid and basic surroundings makes possible a general test for these compounds and also provides information regarding the relative position of the OH groups in the molecule.

The colored products of the reaction between zirconium salts and polyhydroxyanthraquinones in acid medium are stoichiometrically definable phenolates only in extreme cases[4] and at high dilutions. Otherwise, color lakes are formed by the action of polyhydroxyanthraquinones on hydrolysis products (hydrosols) of zirconium salts or alkaline gels of $Zr(OH)_4$. The lakes are not definite "daltonian" compounds but are adsorption products of the polyhydroxyanthraquinones with the sol or gel particles in which the surface zirconium atoms take over the salt-like bonding without leaving the phase association of the sol or gel.[5] The bonding scheme (I) may be considered for the acid-resistant lakes and (II) for the acid-labile lakes:

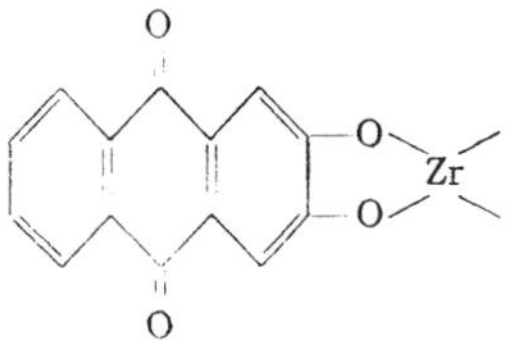

(I) 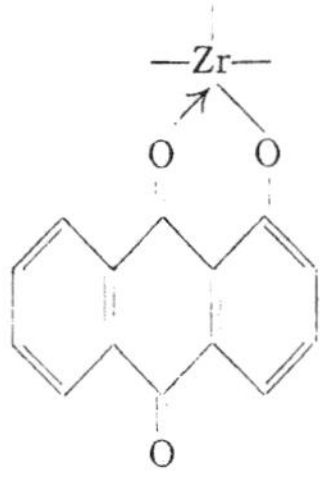   (II)

In the general test for polyhydroxyanthraquinones, the alkaline solution should be treated with dilute zirconium salt solution, the mixture warmed gently, the precipitate isolated by centrifuging or filtering, and washed with water. If a violet or red residue is left, the presence of polyhydroxyanthraquinones is indicated. Should dilute hydrochloric acid turn the residue yellow or discharge its color, then the polyhydroxyanthraquinone did not have OH groups *ortho* to each other, but if the color is not altered polyhydroxyanthraquinones with OH groups *ortho* to each other are present. The following procedure is recommended for the direct detection of polyhydroxyanthraquinones with OH groups in the *ortho* position.

Procedure.   A drop of the test solution (in acetone) and a drop of a 4% solution of zirconium nitrate in dilute hydrochloric acid are mixed on a spot plate. A colored precipitate or a characteristic color change indicates a positive response.

The following compilation presents the results obtained with a variety of polyhydroxyanthraquinones and the respective *limits of identification*.

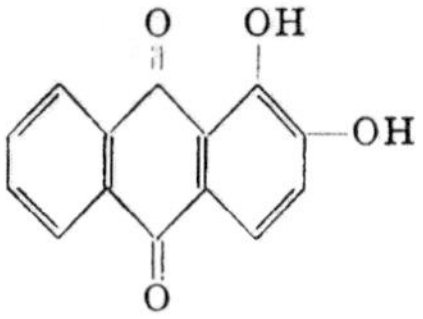

Alizarin (1,2-dihydroxyanthraquinone)	Yellow-purple 0.7 γ	
Anthragallol (1,2,3-trihydroxyanthraquinone)	Orange-brown 2.5 γ	
Purpurin (1,2,4-trihydroxyanthraquinone)	Orange-reddish 0.1 γ	
Hystazarin (2,3-dihydroxyanthraquinone)	Orange-red 2.0 γ	
Rufigallic acid (1,2,3,5,6,7-hexahydroxyanthraquinone)	Yellow-violet 1.0 γ	
Quinalizarin (1,2,5,8-tetrahydroxyanthraquinone)	Red-violet 0.8 γ	

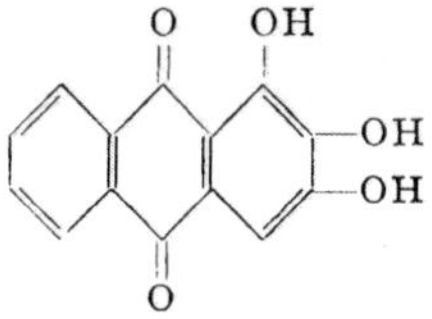

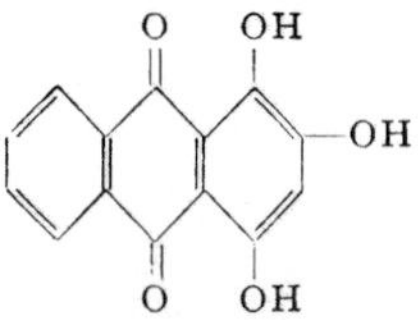

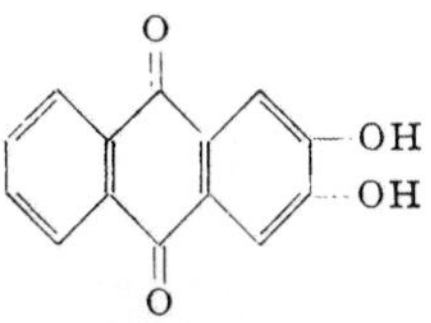

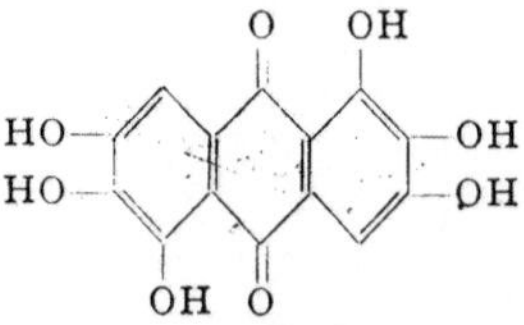

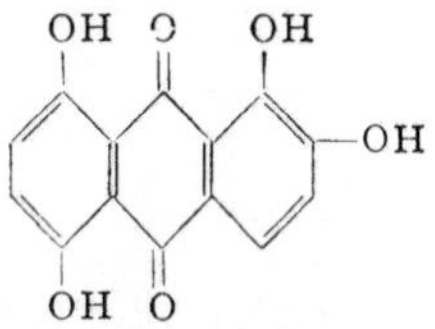

The following polyhydroxyanthraquinones gave no color change in acid solution, but yielded orange to red-violet precipitates in alkaline solution: 1,8-

dihydroxy-3-methylanthraquinone; 1,3-dihydroxy-2-methylanthraquinone; 1,8-dihydroxy-3-methyl-6-methoxyanthraquinone; 1-hydroxy-2-methylanthraquinone; 1,8-dihydroxy-3-hydroxymethyl-anthraquinone; 1,8-dihydroxy-2-methylanthraquinone; 1,8-dihydroxyanthraquinone-3-carboxylic acid; 1,4-dihydroxy-2-methylanthraquinone; 1,6,8-trihydroxy-3-methylanthraquinone; 1,3,6,8-tetrahydroxyanthraquinone.

1 W. B. Mors and B. Zaltman, *Bol. Inst. Quim. Agr. (Rio de Janeiro)*, **34** (1954) 7.
2 F. Feigl, *Spot Tests in Inorganic Analysis*, Elsevier, Amsterdam, 1958, p. 200.
3 F. Feigl, *Specific, Selective and Sensitive Reactions*, Academic Press, New York, 1949, Chapter VI.
4 H. A. Liebhaffsky and E. H. Winslow, *J. Am. Chem. Soc.*, **69** (1947) 1130; **71** (1949) 3630.
5 F. Feigl, ref. 3, p. 547.

29. Acyloins (Ar—CHOH—CO—Ar type)

(1) Test through pyrohydrogenolysis with Lindane[1]

A test for aliphatic halogen compounds is based on the finding that hydrogen halides are evolved on fusion with benzoin (see p. 147). The converse of this pyrohydrogenolysis employing the aliphatic hexachlorocyclohexane (Lindane) makes possible the detection of acyloins.

The test revealed: 2.5 γ benzoin, 15 γ furoin, 5 γ anisoin.

1 F. Feigl and D. Haguenauer-Castro, *Microchem. J.*, **6** (1962) 171.

(2) Test through pyrohydrogenolysis with benzoic acid hydrazide[1]

A test for acid hydrazides is described on p. 279; it is based on the production of ammonia during the fusion with benzoin. The pyrohydrogenolysis with benzoic acid hydrazide permits the detection of acyloins.

The test revealed: 20 γ benzoin, 50 γ furoin, 30 γ anisoin.

1 F. Feigl and D. Haguenauer-Castro, *Microchem. J.*, **6** (1962) 171.

(3) Test through pyrolysis with phenylmercuric nitrate[1]

Nitrous acid is formed if phenylmercuric nitrate is heated to 140–150° with benzoin (m.p. 137°). The reaction is based on the fact that benzoin in its acidic reductone form initially releases nitric acid and then reduces the latter to nitrous acid:

$$2 \ C_6H_5HgNO_3 + C_6H_5C(OH) = C(OH)C_6H_5 \rightarrow$$
$$(C_6H_5Hg)_2[C_6H_5CO = COC_6H_5] + 2 \ HNO_3 \tag{1}$$

$$HNO_3 + C_6H_5C(OH) = C(OH)C_6H_5 \rightarrow C_6H_5COCOC_6H_5 + H_2O + HNO_2 \tag{2}$$

The production of nitrous acid through reactions (*1*) and (*2*) and hence the presence of benzoin can be readily detected by means of the Griess color reaction.

Procedure. A tiny portion of the solid or a drop of its solution in alcohol, benzene, *etc.* is placed in a micro test tube along with several mg of phenylmercuric nitrate; the solvent is evaporated if need be. A disk of filter paper moistened with Griess reagent is laid over the mouth of the tube which is then immersed in a glycerol bath preheated to 120°. The temperature is raised to 150–160°. A positive response is indicated by the appearance of a red spot on the reagent paper.

Limit of identification: 5 γ benzoin or furoin.

1 Unpublished studies.

30. 5-Hydroxyflavonols

Tests by formation of fluorescent metal compounds[1]

Derivatives of flavone (I) and flavonol (II) with OH and OCH$_3$ groups in various positions of the two rings are important yellow plant pigments (usually as glucosides).

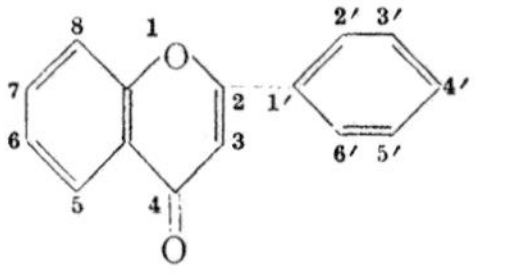

(I) 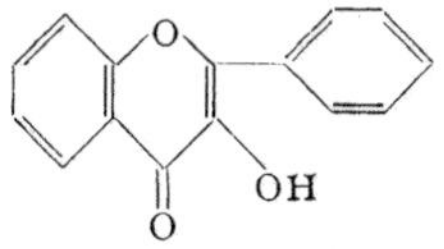   (II)

Among the flavone pigments, some of which are still used as mordant dyes, are morin (5,7,2',4'-tetrahydroxyflavonol) and the isomer quercetin (5,7,3',4'-tetrahydroxyflavonol) whose structural formulas are (III) and (IV).

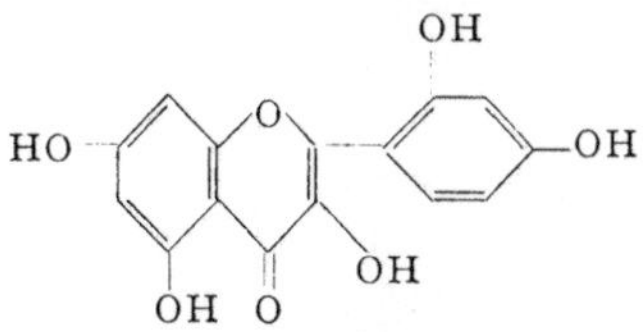

(III) 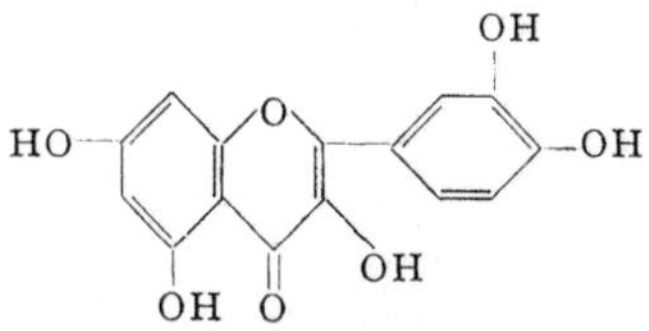   (IV)

The ability of hydroxy derivatives of flavone and flavonol to act as mordant dyes, *i.e.* to be fixed by alumina and other metal oxyhydrates, is related to the chelating position of OH groups to CO groups, as can be seen in (III) and (IV). Probably (and analogous to the case of the hydroxyanthraquinones, see Sect. 28) the OH group in the 5-position is particularly active.

When the hydrogen is replaced by certain metal atoms, chelate bonding occurs with production of inner complex salts or adsorption compounds, which display an intense yellow-green or blue-green fluorescence in ultra-violet light. This is the basis of sensitive tests for these metals,[2] since solutions of morin in alcohol, alkali hydroxide, *etc.* do not fluoresce. Conversely, morin can be detected, with high sensitivity, through the formation of fluorescent metal compounds. The best procedure is to allow the test solution to react with an alkaline alkali beryllate solution or an acid solution of zirconium chloride. The former yields a yellow-green and the latter a blue-green fluorescence (Procedures I and II). When morin reacts with alkali beryllate,[3] the fluorescent product is a soluble compound, in which the beryllium is a constituent of an inner-complex anion, as shown in (V) and (Va). The zirconium–morin reaction[4] in hydrochloric acid solution may involve a chemical adsorption of morin on the surface of the colloidal dispersed hydrolysis products of the zirconium chloride; (VI) presents a schematic picture of the hydrosol particles of the resulting adsorption compound.

When Quercetin behaves like morin. It is very likely that other hydroxyflavonols with an OH group in the 5-position will also yield fluorescing compounds on treatment with alkaline beryllate or acid solutions of zirconium salts.

Procedure I. Single drops of acid 0.01% zirconium chloride solution are placed in adjacent depressions of a spot plate. One is treated with a drop of the test solution, the other with a drop of water. Both are then viewed in ultraviolet light. If morin is present, a yellow-green fluorescence is seen, whose hue depends on the amount present. Very small amounts of morin give a blue-green fluorescence.

Limit of identification: 0.01 γ morin.

Procedure II. Single drops of alkaline beryllate solution are placed in adjoining depressions of a spot plate. One is treated with a drop of the neutral or alkaline test solution, the other with a drop of water or of the solvent used for the morin. A more or less intense yellow-green fluorescence appears in u.v. light.

Reagent: 0.01% beryllium sulfate solution is treated with drops of 0.5 N alkali hydroxide until the initial precipitate has disappeared.

Limit of identification: 0.005 γ morin.

1 Unpublished studies.
2 Comp. F. J. WELCHER, *Organic Analytical Reagents*, Vol. IV, Van Nostrand, New York, 1948, p. 370.
3 H. L. ZERMATTEN, *Proc. Acad. Sci. Amsterdam*, 36 (1933) 899.
4 G. CHARLOT, *Anal. Chim. Acta*, 1 (1947) 233.

31. 1,2-Dicarboxylic acids $\left(\begin{array}{l}\text{C—COOH}\\ \,| \\ \text{C—COOH}\end{array}\ \text{type}\right)$

Test by melting with resorcinol[1]

Acids of this type* (one COOH group may be substituted by a SO_3H group) or their derivatives, such as esters, anhydrides, or imides, form dyes of the fluorescein type (sacchareins) when melted with resorcinol, or heated with resorcinol and concentrated sulfuric acid. The dyes show a vivid green-yellow fluorescence in alkaline solution. The reaction with succinic acid is:

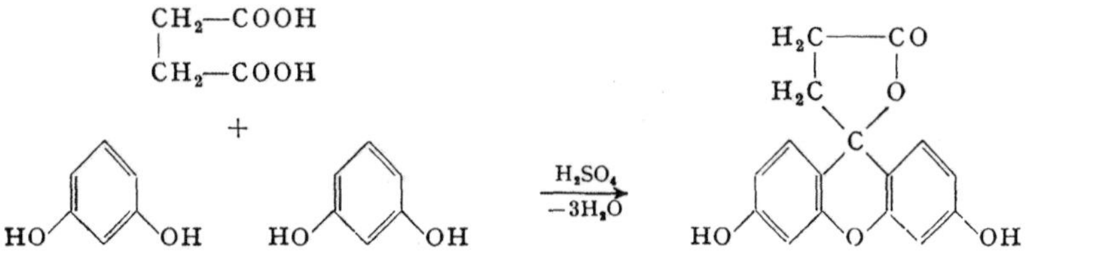

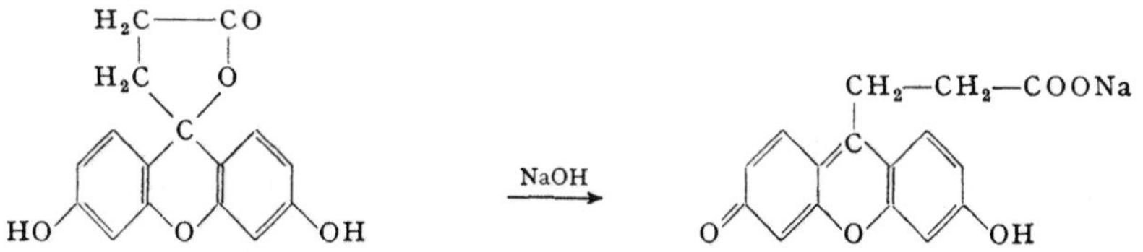

1,2-Dicarboxylic acids with a free hydroxyl adjacent to a COOH group:

$$\begin{array}{l}>\text{C—COOH}\\ \ \ \ | \\ \text{—C(OH)—COOH}\end{array}$$

react with resorcinol in a different manner. They lose formic acid, *i.e.* CO and H_2O, through the action of hot concentrated sulfuric acid, to form semialdehydes of malonic acid or its homologs (*1*), and these in their tautomeric enolic form then condense with resorcinol to produce umbelliferone or its homologs (*2*):

$$\begin{array}{l}\text{COOH}\\ |\\ \text{CH}_2\\ |\\ \text{CHOH}\\ |\\ \text{COOH}\end{array} \xrightarrow[-\text{HCOOH}]{H_2SO_4} \begin{array}{l}\text{CHO}\\ |\\ \text{CH}_2\\ |\\ \text{COOH}\end{array} \rightleftharpoons \begin{array}{l}\text{CHOH}\\ \|\\ \text{CH}\\ |\\ \text{COOH}\end{array} \qquad (1)$$

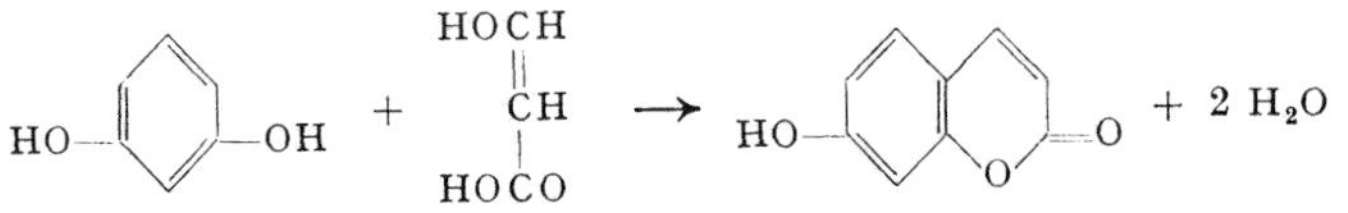

$$\text{HO}\!-\!\langle\bigcirc\rangle\!-\!\text{OH} \;+\; \begin{array}{c}\text{HOCH}\\ \|\\ \text{CH}\\ |\\ \text{HOCO}\end{array} \;\longrightarrow\; \text{HO}\!-\!\langle\bigcirc\bigcirc\rangle\!-\!\text{O} \;+\; 2\,\text{H}_2\text{O} \qquad (2)$$

The resulting umbelliferones are almost colorless, but have an appreciable fluorescence even in daylight. The ultraviolet fluorescence is a brillant blue in alkaline solution.

Procedure. A few mg of the sample is placed in a micro crucible, or a drop of a solution is evaporated there to dryness. A little sublimed resorcinol and a few drops of pure concentrated sulfuric acid are added and the mixture kept at 130° for 5 min., either on an asbestos plate or, better still, on an aluminum block (*Inorganic Spot Tests*, Fig. 6). The crucible plus contents is then dropped into water (50 ml beaker). The solution is made alkaline with sodium hydroxide. In the presence of substances containing one of the pertinent groups, a fluorescence occurs, which is especially bright in ultraviolet light.

A blank test should always be made because if the temperature has exceeded 130°, even a blank shows a fluorescence, which is green by daylight, and green-blue in ultraviolet light. Apparently the resorcinol partially decomposes at the higher temperature, and gives rise to dicarboxylic acids.

Limits of identification and colors and fluorescences are given in Table 15.

* Carboxyl groups in the *peri* position behave like those in the *ortho* position (see Table 15)

1 F. FEIGL, V. ANGER and O. FREHDEN, *Mikrochemie*, 17 (1935) 29.

TABLE 15

DICARBOXYLIC ACIDS

Name	Limits of identification by	
	color	fluorescence in u.v. light
Succinic acid	5 γ yellow	5 γ emerald-green
Succinic anhydride (or imide)	5 γ ,,	5 γ ,,
Tricarballylic acid	5 γ yellow-rose	5 γ grass-green
Phthalic acid	5 γ yellow-brown	5 γ light green
Trimellitic acid trimethyl ester	2.5 γ yellow	2.5 γ ,, ,,
Malic acid		1 γ light blue
Citric acid	15 γ ,,	5 γ grass-green
Tartaric acid		25 γ dark blue-green
Dihydroxymaleic acid	40 γ red	15 γ blue-gray
Asparagine	5 γ wine-red	10 γ dark green
Saccharin	10 γ yellow	5 γ greenish yellow
Naphthalic acid	5 γ yellow-brown	5 γ dark green

32. α-Halogenofatty acids (—CHHalCOOH type)

Test through conversion into α-aminocarboxylic acid [1]

The action of ammonia on α-halogenofatty acids or their alkali salts brings about the exchange of NH_2 for the halogen with consequent production of the corresponding ammonium or alkali salt of the α-aminocarboxylic acid:

$$RCHHalCOOH + 3 NH_3 \rightarrow RCH(NH_2)COONH_4 + NH_4Hal$$

$$RCHHalCOONa + 2 NH_3 \rightarrow RCH(NH_2)COONa + NH_4Hal$$

The occurrence of these reactions and hence the presence of α-halogenofatty acids can be revealed through the positive response to tests for α-aminocarboxylic acids. The color reaction outlined on p. 370 can be recommended; it is based on the formation of colored complex cobalt salts from Schiff bases derived from α-aminocarboxylic acids and pyridine-2-aldehyde. The amination of α-halogenofatty acids is readily accomplished by fuming with ammonium carbonate.

Procedure. A small quantity of the solid sample or a drop of its alcoholic solution is mixed in a micro test tube with several cg of ammonium carbonate and taken to dryness if necessary. The test tube is then immersed as far as possible in a glycerol bath that has been preheated to 120°. The heating should be continued until no residue remains in a second tube that originally contained an approximately equal amount of ammonium carbonate. The rest of the procedure is that described on p. 370.

The following *limits of identification* were found:

10 γ monochloroacetic acid
10 γ monoiodoacetic acid
10 γ α-monochloropropionic acid
10 γ α-monobromoisovaleric acid
15 γ α-monobromobutyric acid
20 γ α-monobromocaproic acid
25 γ α-monobromocaprylic acid
30 γ α-monobromocapric acid

1 F. FEIGL and S. YARIV, *Anal. Chim. Acta*, 29 (1963) 581.

33. Uronic acids

Test with o-aminophenol [1]

Uronic acids are aldehydecarboxylic acids of the sugar series. The most important are: D-glucuronic acid (I), D-galacturonic acid (II), and D-mannuronic acid (III):

(I)   (II)   (III)

The uronic acids show the color reactions of the carbohydrates. They can be differentiated through the behavior on fusion with *o*-aminophenol and oxalic acid with addition of pimelic acid as diluent. Colored products result that are soluble in the melt, in water, and in amyl alcohol. The mechanism of the color reaction is not known.

Procedure.   A drop of the test solution is placed in a small porcelain crucible along with a drop of a 2.5% solution of *o*-aminophenol in alcohol–acetone (1 : 1) followed by enough oxalic acid–pimelic acid mixture (2 : 1) to produce a thick sludge. The suspension is kept at 120° for about 10 min. in a drying oven. After cooling, the fusion cake is dissolved in 3 ml of water. A blue-violet color, that remains unchanged on standing, is a positive response. If the solution is shaken with several drops of amyl alcohol, the alcohol layer turns blue.

Limit of identification:

1 γ galacturonic acid,   glucuronic acid,   mannuronic acid

1 R. WAGNER, *Mikrochim. Acta*, (1963) 289.

34. Salicylic acid derivatives

Test through hydrolysis[1]

Esters of salicylic acid derived from alcohols and phenols invariably have a pleasing odor proving that they have a distinct vapor pressure even at room temperature. They can be saponified by bringing their vapors into contact with caustic hydroxides or magnesium oxide; the resulting salicylate exhibits fluorescence in ultraviolet light.

If several mg of a salicylic ester are placed in a micro crucible that is then covered with a disk of filter paper moistened with caustic alkali and held in place with a watch glass, a circle that fluoresces blue-violet will develop within 10–15 min. at room temperature.

A more sensitive test for salicylic esters is to place a drop of the ethereal solution in the apparatus (see *Inorganic Spot Tests*, p. 54) and then drive off the ether by brief immersion in boiling water. The knob charged with a drop of caustic alkali or magnesia suspension is fitted into place, and the residual ester is kept at 130° for about 10 min. The vapors are saponified in the hanging drop with production of fluorescent salicylate.

The method revealed:

1 γ methyl salicylate (oil of wintergreen)   1.5 γ phenyl salicylate (salol)

Nonvolatile derivatives of salicylic acid that can be cleaved by hydrolysis, such as the amide, anilide, and hydrazide, when warmed to around 120–140° with syrupy phosphoric acid yield vapors of salicylic acid which can be detected as outlined above. The same is true of phenyl salicylate (salol). If the response to this test is positive, the following compounds can be detected by supplementary tests:

Salicylamide: production of ammonia on warming with caustic alkali (comp. p. 256);

Salicylanilide: production of aniline on heating with benzoin (see p. 147);

Salicylic acid hydrazide: release of ammonia on heating with benzoin (see p. 279);

Phenyl salicylate: release of phenol by pyrohydrolysis (comp. p. 216).

From the foregoing, it may be deduced that the particular derivatives of salicylic acid can be identified even though they constitute components of a mixture.

1 Unpublished studies, with H. BLOHM

35. Phenoxyacetic acids $\left(\bigotimes\!\!-O\!-\!CH_2\!-\!COOH \text{ type}\right)$

Test with concentrated sulfuric acid and chromotropic acid[1]

Brief heating of phenoxyacetic acid (or its halogen-substituted derivatives) with concentrated sulfuric acid and chromotropic acid to 150° yields the violet color characteristic of formaldehyde. The chemistry of the color reaction is analogous to that of monochloroacetic acid discussed on p. 461. Through its water content, concentrated sulfuric acid brings about the hydrolysis of phenoxyacetic acid to phenol and glycolic acid:

$$C_6H_5OCH_2COOH + H_2O \rightarrow CH_2OHCOOH + C_6H_5OH$$

The glycolic acid is cleaved to yield formaldehyde, which reacts with the chromotropic acid present.

When the test is applied directly, there is interference by compounds which split out formaldehyde when heated with concentrated sulfuric acid. However, the test becomes quite selective for phenoxyacetic acid and halogenated phenoxyacetic acids if use is made of their solubility in benzene.

The interfering compounds can thus be taken out of the reaction theatre, including those which caramelize when heated with concentrated sulfuric acid and so impair the recognition of the color reaction.

Procedure. A micro test tube is used. A drop of the test solution (benzene) is taken to dryness, 2 ml concentrated sulfuric acid added and several mg of solid chromotropic acid or its sodium salt mixed in. The mixture is kept at 250° for about 2 min. A more or less intense violet color indicates the presence of phenoxyacetic acid or its halogen derivatives.
Limit of identification: 0.05 γ 2,4-dichlorophenoxyacetic acid.

The following acids react in analogous fashion: phenoxyacetic, *o*- and *p*-chlorophenoxyacetic, 2,4-dibromophenoxyacetic acid.

Phenoxyacetic acid and its halogen derivatives can be detected by the procedure given for monochloroacetic acid, in which the formaldehyde split out is detected in the vapor phase by contact with chromotropic acid and concentrated sulfuric acid. However, the sensitivity is then only about one tenth as great.

1 V. H. FREED, *Science*, 107 (1948) 98.

36. Phenol- and α-naphthol-sulfonic acids

Test through pyrohydrolytic release of a volatile phenol[1]

Succinic acid loses water at 200–240° to yield the anhydride. If this dehydration is carried out in the presence of phenol- or α-naphtholsulfonic acids, the water which is set free as quasi superheated steam pyrohydrolyzes the sulfonic acid:

$$C_6H_4(OH)SO_3H + H_2O \rightarrow C_6H_5OH + H_2SO_4$$

The resulting phenols are volatile with the steam which arises from the dehydration of succinic acid and can be detected by the indophenol reaction with 2,6-dichloroquinone-4-chloroimine (see p. 185).

Volatile phenols and phenol esters must be absent. The latter are pyrohydrolyzed to phenols. It is possible to separate the sulfonic acids from phenols by extracting the acidic test solution with ether in which the sulfonic acids do not dissolve. The test is suitable for benzene- and α-naphtholsulfonic acids which on pyrolysis yield volatile phenols with a free *para* position.

Procedure. The test is conducted as outlined on p. 216, but succinic acid is used, and the temperature is 220°.
Limit of identification: 0.5 γ sodium *p*-phenolsulfonate.

When heated at 250° 1-naphthol-4- and 1-naphthol-5-sulfonic acid release α-naphthol.

1 F. FEIGL and V. ANGER, *Mikrochim. Acta*, (1960) 410.

37. Naphthalenesulfonic acids (⟮⟯ SO₃H type)

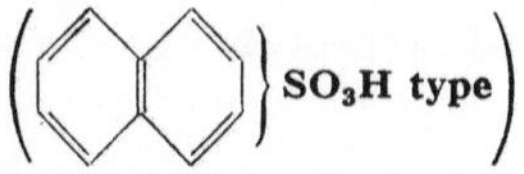

Test through reductive release of naphthalene[1]

It was pointed out on p. 233 that aromatic sulfonic acids are reductively cleaved when warmed with Raney nickel alloy and caustic alkali; nickel sulfide results and there is replacement of a sulfonic acid group by hydrogen. Accordingly, α- and β-naphthalenesulfonic acids thus yield naphthalene which is volatile with steam. As little as 150 γ of this hydrocarbon can be detected directly by its characteristic odor. When dealing with smaller amounts, it is advisable to extract the naphthalene produced by reductive cleavage and then to test the ethereal extract for naphthalene by the Le Rosen procedure given on p. 137.

Procedure.   A drop of the aqueous test solution is placed in a micro test tube followed by 2 drops of 5 N sodium hydroxide and several cg of Raney nickel alloy. The mixture is kept at around 60° for about 5 min. After cooling, the system is shaken with 1 ml of ether; the ether extract is transferred by means of a pipette into the apparatus shown (Fig. 23, p. 54, *Inorganic Spot Tests*) and evaporated there. The rest of the directions are given on p. 234.

Limit of identification: 10 γ naphthalenesulfonic acid (α and β).

1 Unpublished studies.

38. Anthraquinonesulfonic acids (⟮⟯ SO₃H type)

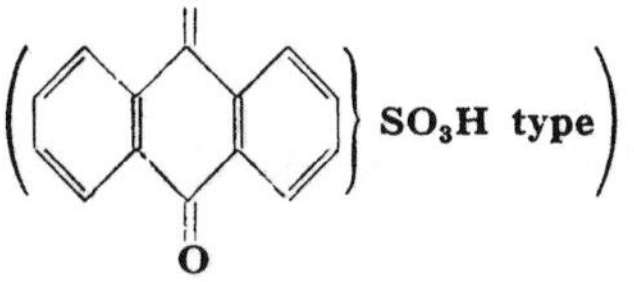

(1) *Test by conversion into dihydroxyanthraquinones*[1]

The classic method for preparing alizarin is to fuse anthraquinone-2-sulfonic acid with caustic alkali in contact with air. Alizarin is formed more quickly if alkali nitrate or chlorate is used as oxygen donor:

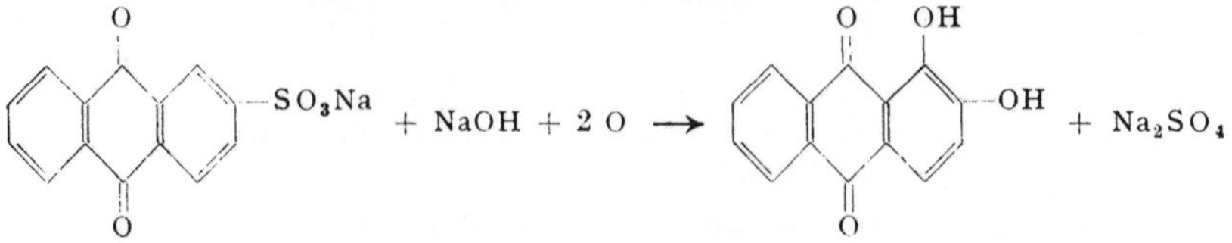

The corresponding dihydroxyanthraquinones are formed in an analogous manner from anthraquinonedi- and trisulfonic acids. Sintering at temperatures below 200° is sufficient and if excess alkali hydroxide is used, violet alkali salts of the dihydroxyanthraquinones result. Since anthraquinonesulfonic acids and their alkali salts are practically colorless, the production of the colored phenolates in the sintering reaction is easily discerned.

Procedure. A drop of the test solution or a little of the solid is placed in a depression of a spot plate. A drop or two of sodium nitrate solution (1 g in 10 ml of 20% sodium hydroxide) is added. The mixture is taken to dryness on a hot plate and kept at 160–180° for several minutes. The residue turns more or less violet according to the amount of anthraquinone present.

Limit of identification: 2 γ anthraquinone-2-sulfonic acid.

1 Unpublished studies, with V. GENTIL.

(2) Detection through catalysis of the glucose–alkali hydroxide reaction [1]

An intense red-brown color appears if a strongly alkaline sugar solution is warmed. The chemistry of this reaction is not known. The change is slow with dilute caustic or carbonate solution, and does not occur at all with ammonia or ammonium carbonate. In these latter cases, the reaction is speeded up by anthraquinonemono- and disulfonic acid. This catalytic effect is so marked that it serves as the basis of a test for these acids.

Procedure. One drop of the aqueous solution of the sulfonic acid or its alkali salt is treated in a micro test tube with one drop of a 10% solution of glucose in 1 : 1 ammonia. A yellow or red-brown color appears, the shade depending on the amount of sulfonic acid present.

The *limits of identification* are:

0.4 γ anthraquinone-2-sulfonic acid   30 γ anthraquinone-1,5-disulfonic acid

1 Unpublished studies, with Cl. COSTA NETO.

(3) Test through conversion into anthrahydroquinonesulfonic acids [1]

Anthraquinone is rapidly converted into a red water-soluble alkali salt of anthrahydroquinone by warming with a strong alkaline solution of sodium hydrosulfite (see Sect. 18). Mono- and disulfonic acids of anthraquinone as well as its nitro derivatives behave like the parent compound in this respect.

Procedure. As given on p. 336.

The test revealed:

1 γ anthraquinone-1-sulfonic acid   2 γ anthraquinone-1,5-disulfonic acid
1 γ ,,   -2-   ,,   2 γ ,,   -2,6-   ,,
1 γ 1-nitroanthraquinone-5-sulfonic acid

When anthraquinonesulfonic acids are to be detected in the presence of anthraquinone the test here described is not sufficient. A test for SO_3H groups in aromatic compounds is needed in addition. The procedure is described on p. 234.

1 Unpublished studies, with D. GOLDSTEIN.

39. N-Alkylanilines

Test through formylation[1]

An aldehyde group can be introduced into the *para*-position of mono- and dialkylanilines by heating with a dry mixture of oxalic acid dihydrate and hexamine. This formylation involves two partial reactions. The hexamine condenses in its trialkyl-form:

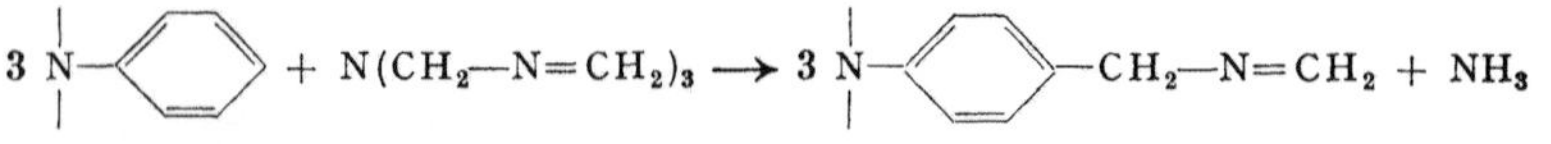

$$3 \; N\!\!-\!\!\langle\ \rangle + N(CH_2\!-\!N\!=\!CH_2)_3 \rightarrow 3 \; N\!\!-\!\!\langle\ \rangle\!-\!CH_2\!-\!N\!=\!CH_2 + NH_3 \qquad (1)$$

Thereupon, the condensation product, in its isomeric form as a Schiff base of monomethylamine, is hydrolyzed by the water of crystallization of the oxalic acid to give the aldehyde:

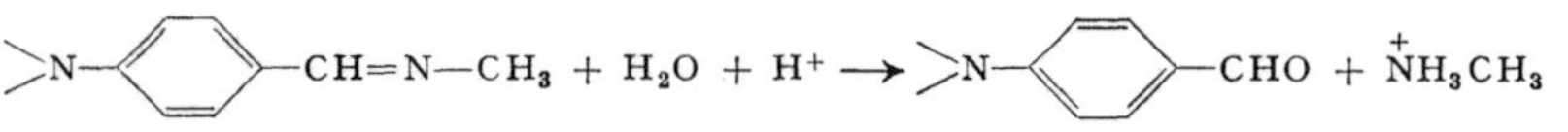

$$>\!N\!\!-\!\!\langle\ \rangle\!-\!CH\!=\!N\!-\!CH_3 + H_2O + H^+ \rightarrow \; >\!N\!\!-\!\!\langle\ \rangle\!-\!CHO + \overset{+}{N}H_3CH_3 \quad (2)$$

The chemistry of this formylation is analogous to the formylation of phenols discussed on p. 181.

The realization of (*1*) and (*2*) and hence the presence of mono- and dialkylaniline can be recognized through the orange Schiff base produced by the condensation of the resulting aromatic aldehyde with *o*-dianisidine.

Procedure.   A drop of the ethereal or alcoholic test solution is added to a micro test tube containing several cg of a 1 : 1 oxalic acid–hexamine mixture, and the solvent is driven off. Solid salts of alkylanilines may also be tested. The tube is placed in a glycerol bath, previously heated to 150°, and the temperature is raised to 160°. After about 2 min., the tube is removed from the bath and the cooled residue treated with 2 drops of water. The resulting solution or suspension is placed on a filter paper moistened with an ethereal solution of benzidine. A more or less intense orange stain indicates a positive response.

The test revealed:

2 γ dimethyl(ethyl)aniline   2 γ monomethylaniline   3 γ monoethylaniline

The test is not directly applicable in the presence of aldehydes that give colored Schiff bases with benzidine, or of phenols which are formylated by hexamine. In such cases, it is necessary to separate the bases beforehand, which is easily accomplished by precipitation with phosphomolybdic acid from acid solution. An ethereal solution of the bases is readily obtained. by adding alkali to the phosphomolybdates and shaking with ether.

When mono- and dialkylanilines are to be detected in the presence of aniline, the latter is converted to phenol by warming with nitrous acid ($NaNO_2 + HCl$). If the solution is then made basic and extracted with ether, the test for the alkylated anilines can be conducted on the ethereal solution

1 F. FEIGL and E. JUNGREIS, *Analyst*, **83** (1958) 669.

40. N,N-Dialkylanilines

(1) *Test through conversion into p-nitrosodialkylaniline*[1]

Nitrous acid readily nitrosates dimethyl- and diethylaniline in the *para*-position. For example:

$$\langle\!\!\!\bigcirc\!\!\!\rangle\!\!-N(CH_3)_2 + HNO_2 \longrightarrow ON\!\!-\!\!\langle\!\!\!\bigcirc\!\!\!\rangle\!\!-N(CH_3)_2 + H_2O$$

Sensitive tests for dimethyl- or diethylaniline can be based on this fact in combination with the production of dimethyl- or diethylamine from the corresponding *p*-nitroso compounds by treatment with alkali (compare p. 558), whereby monomethyl(ethyl)amine is formed.

Procedure. The gas absorption apparatus (Fig. 23, *Inorganic Spot Tests*) is charged with a drop of the acidified test solution. A little solid potassium nitrite is added and if need be a drop of hydrochloric acid. The nitrous vapors are driven off by vigorous warming. The remainder of the procedure is as given on p. 558.

Limit of identification: 3 γ dimethyl- or diethylaniline.

1 Unpublished studies, with L. HAINBERGER.

(2) *Test with sodium chlorite*[1]

A violet color appears if sodium chlorite is added to a weakly acidic alcohol–water solution of a dialkylaniline. The chemistry of the color reaction in which the chlorite functions as oxidant is not known as yet. Probably a quinoidal compound is produced; for example:

$$2\,\langle\!\!\!\bigcirc\!\!\!\rangle\!\!-N(CH_3)_2 + 2\,O \longrightarrow (CH_3)_2\overset{+}{N}\!\!=\!\!\langle\!\!\!\bigcirc\!\!\!\rangle\!\!-\!\!\langle\!\!\!\bigcirc\!\!\!\rangle\!\!=\!\!\overset{+}{N}(CH_3)_2 + 2\,\overset{-}{O}H$$

Aniline and monoalkylanilines do not show this reaction in dilute solution.

Procedure. A drop of the water–alcohol test solution is treated in a depression of a spot plate with 1 drop of sodium chlorite (5% solution of the commercial 80% salt) and 1 drop of 10% hydrochloric acid.

The test revealed: 1 γ dimethyl- or diethylaniline.

1 J. POPA, E. GRIGORE and F. M. ALBERT, Z. Anal. Chem., 193 (1963) 324.

41. Diphenylamine and derivatives

(1) Test through fusion with oxalic acid[1]

When diphenylamine is fused with oxalic acid at 140–180° the dyestuff aniline blue (diphenylamine blue) is formed (compare the test for oxalic acid, p. 457). Carbazole and many other derivatives of diphenylamine with substituents in the benzene nucleus or in the NH group behave like the parent compound. Exceptions hitherto found are nitro derivatives; no information about acridan and thiodiphenylamine is available.

Procedure. A small quantity of the solid sample or a drop of its solution is mixed in a micro test tube with several cg of oxalic acid dihydrate and taken to dryness if need be. The tube is placed in a glycerol bath preheated to 190°. A positive response is indicated by the appearance of a blue or blue-green product within a few minutes.

Limits of identification:

0.5 γ	diphenylamine	2.5 γ	N,N-diphenylformamide
0.5 γ	N-methyldiphenylamine	5 γ	N,N'-diphenylbenzidine
2.5 γ	N-acetyldiphenylamine	5 γ	tropaeolin 00
15 γ	phenyl-1-naphthylamine	0.5 γ	carbazole
3 γ	*asym*-diphenylurea	1 γ	N-ethylcarbazole
3 γ	4,4-diphenylsemicarbazide	1 γ	N-ethyl-4-diethylaminocarbazole
5 γ	ethyl N,N-diphenylcarbamate	1.5 γ	N-*p*-toluenesulfonylcarbazole
5 γ	N-phenylanthranilic acid		
5 γ	*asym*-N,N-diphenylhydrazine chloride		
2 γ	barium diphenylamine-4-sulfonate		

1 F. FEIGL and D. GOLDSTEIN, Anal. Chem., 32 (1960) 861.

(2) Test with p-nitrosophenol and hydrochloric acid[1]

It is well known that diphenylamine and its derivatives are oxidized in strongly acidic surroundings to give blue meriquinoidal compounds (comp. p. 363). The same oxidation action is shown by p-nitrosophenol in concentrated hydrochloric acid but not in concentrated sulfuric acid. Accordingly, it must be assumed that the oxidant is generated from p-nitrosophenol and concentrated hydrochloric acid. A possible mechanism is the formation of quinonechloroimine:

$$HO-\langle\!\!\!\!\bigcirc\!\!\!\!\rangle-NO \rightleftharpoons O=\langle\!\!\!\!\bigcirc\!\!\!\!\rangle=NOH + HCl \rightleftharpoons O=\langle\!\!\!\!\bigcirc\!\!\!\!\rangle=N-Cl + H_2O$$

Procedure. A drop of the test solution is treated in a depression of a spot plate with a drop of freshly prepared 0.05% solution of *p*-nitrosophenol in concentrated hydrochloric acid. A blue color appears if diphenylamine or analogous compounds are present (thiodiphenylamine is an exception).

Limits of identification:

       0.5 γ thiodiphenylamine (red)   0.1 γ diphenylamine
       0.1 γ N,N'-diphenylbenzidine

1 V. ANGER and S. OFRI, unpublished studies.

(3) Test through oxidation to quinoneimines [1]

The familiar and very sensitive color reaction for nitrite and nitrate with diphenylamine and concentrated sulfuric acid may be reversed to detect diphenylamine, and also its derivatives which have no substituent in the *para* position. This test is quite selective if the solubility characteristics of diphenylamine and its derivatives are taken into consideration. The pertinent compounds are very weak bases which, in contrast to primary aromatic and mixed aliphatic–aromatic secondary amines, are not soluble in dilute mineral acids. Accordingly, they may be separated from other organic bases by digesting with dilute mineral acids. On the other hand, because of the weak basic character of the NH group, the introduction of sulfonic and carboxylic groups into one of the aromatic rings results in acidic compounds which are soluble in dilute alkalies. Such derivatives of aromatic secondary amines can therefore be separated from water-insoluble aromatic bases by digestion with dilute alkalis.

Procedure. A pinch of the test material is treated in a micro crucible with a drop of concentrated nitric acid (d = 1.4). If solutions in organic solvents or alkali hydroxides are being tested (see above), a drop of the liquid is brought to dryness in the crucible and a drop of concentrated nitric acid then added. The color obtained is more or less blue or blue-green.

The test revealed:

0.25 γ diphenylamine   0.3 γ diphenylamine-4-sulfonic acid
0.17 γ N,N'-diphenylbenzidine   0.4 γ diphenylamine-2,2'-dicarboxylic acid
0.3 γ phenyl-1-naphthylamine

A direct bluing effect should not be expected with N-substituted derivatives of diphenylamine when treated with concentrated nitric acid. (4,4-Diphenylsemicarbazide is an exception. Its limit of identification is 5 γ.) It appears that these compounds are saponified when warmed with concentrated sulfuric acid to 150–160° with splitting out of diphenylamine:

$$OC \overset{N(C_6H_5)_2}{\underset{NH_2}{\big<}} + H_2O \longrightarrow CO_2 + NH_3 + NH(C_6H_5)_2$$

If a little alkali nitrate is added to the cold reaction mass, a blue color results. This behavior was observed not only in the case of *asym*-diphenylurea (10 γ) but also with N-methyldiphenylamine (2.5 γ) and N-acetyldiphenyl-amine (5 γ). Only a yellow color results when ethyl N,N-diphenylcarbamate or N,N-diphenylanthranilic acid is subjected to this treatment.

It is notable that the following compounds give a blue color even at room temperature when treated with concentrated sulfuric acid with addition of alkali nitrite:

$$OC\begin{array}{c} \diagup N(C_6H_5)_2 \\ \diagdown NH_2 \end{array}$$
asym-Diphenylurea
(110 γ)

$$HN{=}C\begin{array}{c} \diagup N(C_6H_5)_2 \\ \diagdown NH_2 \end{array}$$
asym-Diphenylguanidine
(100 γ)

$$OC\begin{array}{c} \diagup N(C_6H_5)_2 \\ \diagdown OC_2H_5 \end{array}$$
Ethyl N,N-diphenylcarbamate
(15 γ)

$$OC\begin{array}{c} \diagup N(C_6H_5)_2 \\ \diagdown NHNH_2 \end{array}$$
4,4-Diphenylsemicarbazide
(2.5 γ)

$$C_6H_4\begin{array}{c} \diagup N(C_6H_5)_2 \\ \diagdown COOH \end{array}$$
N,N-Diphenylanthranilic acid
(4 γ)

It is likely, in the cases just noted, that the —N(C$_6$H$_5$)$_2$ group is split off as N-nitrosodiphenylamine through the action of nitrous acid:

$$OC\begin{array}{c} \diagup N(C_6H_5)_2 \\ \diagdown NH_2 \end{array} + 2\,HNO_2 \longrightarrow CO_2 + ON{-}N(C_6H_5)_2 + N_2 + 2\,H_2O$$

Accordingly, saponification of the N-nitrosodiphenylamine yields diphenyl-amine and nitrous acid, and hence the expected blue color.

So far as is now known, N-substituted diphenylamines can therefore be detected by warming with concentrated sulfuric acid to 150–160° and subsequent addition of alkali nitrite, as well as by their behavior toward concentrated sulfuric acid and alkali nitrite. (See also p. 127).

1 Unpublished studies, with D. GOLDSTEIN.

42. Halogenated anilines

Test through conversion into aniline[1]

If an ethereal or alcoholic solution of an halogenated aniline or the aqueous solution of its salt is warmed with caustic alkali and Devarda's alloy, aniline results through the action of nascent hydrogen. For example:

$$C_6H_2Br_3NH_2 + 6\,H^0 + 3\,NaOH \rightarrow C_6H_5NH_2 + 3\,NaBr + 3\,H_2O$$

The resulting aniline is volatile with steam and can be detected in the gas phase by the color reactions with *p*-dimethylaminobenzaldehyde or sodium 1,2-naphthoquinone-4-sulfonate (comp. Chaps. 3 and 2).

Procedure. A tiny portion of the solid or a drop of its solution in organic liquids or acids is treated in a micro test tube with 1 or 2 drops of 5% caustic alkali solution and several mg of Devarda's alloy. Gentle warming is used at first until the vigorous evolution of hydrogen has subsided, and then the mouth of the test tube is covered with a piece of filter paper moistened with 0.5% aqueous solution of sodium 1,2-naphthoquinone-4-sulfonate solution. The test tube is placed in boiling water. A red-violet stain appears on the yellow paper if the response is positive.

Limit of identification: 10 γ tribromoaniline.

1 F. FEIGL, *Anal. Chem.*, 33 (1961) 1118.

43. β-Hydroxyethylamines ($>$N—CH$_2$—CH$_2$—OH type)

Test through dry heating with sodium chloroacetate[1]

Quaternized β-hydroxyethylamines (ethanolamines) yield acetaldehyde when dry heated:

$$[R_3\overset{+}{N}CH_2CH_2OH]X^- \rightarrow [R_3\overset{+}{N}H]X^- + CH_3CHO \quad (R = H, CH_3, CH_2COOH, etc.)$$

The quaternization of β-hydroxylamines and the subsequent cleavage can be brought about by heating the amines with sodium chloroacetate to 240–250°. The initial product is a betaine compound which in turn decomposes with delivery of acetaldehyde:

$$R_2NCH_2CH_2OH + ClCH_2COONa \rightarrow R_2\overset{+}{N}CH_2CH_2OH + NaCl \qquad (1)$$
$$\underset{CH_2COO^-}{|}$$

$$\underset{\underset{CH_2COO^-}{|}}{R_2\overset{+}{N}CH_2CH_2OH} \rightarrow \underset{\underset{CH_2COO^-}{|}}{R_2\overset{+}{N}H} + CH_3CHO \qquad (2)$$

The occurrence of reactions (*1*) and (*2*) and hence the presence of ethanolamines can be established by the detection of the resulting acetaldehyde in the gas phase. The color reaction (blue color) with a solution of sodium nitroprusside and morpholine (see Chap. 3) is recommended.

It should be noted that the test does not require the isolation of the ethanolamines per se; actually reactions (*1*) and (*2*) which are essential to the test, can be accomplished also with their salts with inorganic acids.

Procedure.[2] A little sodium chloroacetate is placed in a micro test tube along with a drop of the alcoholic test solution. The solvent is removed on the water bath. The test tube is then placed in a glycerol bath previously heated to 245°. The mouth of the tube is covered with a disk of filter paper impregnated with the reagent solution (for preparation see p. 438). A blue stain on the reagent paper indicates a positive response.

Sodium chloroacetate is prepared by adding a concentrated alcoholic solution of chloroacetic acid to a concentrated solution of sodium hydroxide. The precipitate is filtered off, washed with ethanol, and dried.

The following *identification limits* were obtained:

30 γ triethanolamine	75 γ diethylethanolamine
50 γ diethanolamine	30 γ hydroxyethyl-propylenediamine
50 γ N-α-methylbenzyldiethanolamine	120 γ hydroxyethyl-2-heptadecenyl-
25 γ N-hydroxyethyl-morpholine	glyoxalidine

Another test for the $>$N—CH$_2$CH$_2$OH group is described on p. 166.

1 M. J. ROSEN, *Anal. Chem.*, 27 (1955) 114.
2 R. A. ROSELL (Rio de Janeiro), unpublished studies.

44. Amino acids

Test by reaction with fused potassium thiocyanate [1]

A test for amines through the detection of hydrogen sulfide produced by fusion with dried potassium thiocyanate was described on p. 241. However, the procedure succeeds only if salts of the respective amines are used. These salts, which are substituted salts of ammonium, initially yield substituted ammonium salts of thiocyanic acid and on heating give substituted thioureas. The latter, in the isothiourea form, are decomposed into hydrogen sulfide and substituted cyanamides. According to this conception, aliphatic and aromatic aminocarboxylic acids, as well as aromatic aminosulfo acids, behave like salts of amines. The same is true of compounds in which an acidic hydrogen (NH, OH, COOH, AsO$_3$H$_2$ group) is present and also (in another position) a basic nitrogen atom, so that the compounds may be regarded as amino acids, using the term in the widest sense.

The following test for amino acids can be carried out in the presence of free amines. Salts of amines and ammonium salts, as well as solid carboxylic acids must be absent. The same is true of organic compounds which split off water at the temperature of the fusion, since superheated water reacts with potassium thiocyanate with evolution of hydrogen sulfide.

Procedure. One drop of the test solution is evaporated to dryness in a micro test tube, and held for a short time at 110°. An excess of well dried potassium thiocyanate is added and the mixture heated in a bath to about 200–250°. A filter paper moistened with a 10% solution of lead acetate is placed over the mouth of the test tube. In the presence of amino acids a black stain can be observed on the paper. If alkaline solutions of amino acids are to be tested (in case of previous extraction of the bases with ether or chloroform) a drop of the solution is evaporated together with one drop of 1 : 1 hydrochloric acid, and the dried (110°) residue treated as described.

The following amounts were detected:

15 γ glycine (aminoacetic acid)	25 γ anthranilic acid
50 γ acetylglycine (aceturic acid)	50 γ sulfanilic acid.

The following amino acids also react: phenylalanine (α-amino-β-phenylpropionic acid); tyrosine (β-(p-hydroxyphenyl)-alanine); aspartic acid (aminosuccinic acid); methionine (2-amino-4-methylthiobutanoic acid); leucine (α-aminoisocaproic acid); tyramine (4-hydroxyphenethylamine); 4-aminosalicylic acid, 1-amino-2-naphthol-4-sulfonic acid; methyl red (p-dimethylaminoazobenzene-o-carboxylic acid) and H-acid (1-amino-8-naphthol-3,6-disulfonic acid).

In accordance with their amino acid nature, sulfanilamide (p-aminobenzenesulfonamide), sulfadiazine (2-sulfanilamidopyrimidine) and other "sulfa drugs" give a positive response. Barbituric acid and its derivatives, with the exception of 5-nitrobarbituric acid, also respond positively.

The behavior of alkali salts of ethylenediaminetetraacetic acid (see p. 505) is interesting. Whereas the acidic disodium salt yields hydrogen sulfide immediately with molten potassium thiocyanate, the tetrasodium salt is unreactive. Accordingly, these alkali salts can readily be distinguished.

1 H. E. FEIGL, unpublished studies.

45. Aliphatic aminocarboxylic and -sulfonic acids

(1) Test through pyrolysis with guanidine carbonate[1]

If amino acids of this kind are heated to 160° with guanidine carbonate, this strongly basic compound brings about a salification of the COOH- or SO₃H group. The condensation reaction that is characteristic for primary aliphatic amines can then ensue:

$$2 \text{ R—NH}_2 \rightarrow \text{R—NH—R} + \text{NH}_3$$

and the resulting ammonia can be readily detected by Nessler's reagent.

The test described here is not impaired by aromatic amino acids. On the other hand, ammonium salts and salts of primary aliphatic amines must not be present, since they too yield ammonia when heated with guanidine carbonate.

Procedure. A small amount of the solid is intimately mixed in a micro test tube with several cg of guanidine carbonate and a disk of filter paper moistened with Nessler reagent is placed over the mouth of the tube. The latter is then immersed in a glycerol bath preheated to 160°. A positive response is indicated by the appearance of a brown or yellow fleck on the reagent paper.

The test revealed 5–10 γ of acids in this category.

1 Unpublished studies.

(2) Test through demasking the acidic groups[1]

If amino acids of this type are heated to 130–140° with paraformaldehyde, the resulting formaldehyde methenylates the NH_2 group:

$$-NH_2 + CH_2O \rightarrow -N=CH_2 + H_2O$$

In contrast to the NH_2 group, the NCH_2 group has no basic character, and thus the methenylation is accompanied by the freeing (demasking) of the COOH- and SO_3H groups that are more or less salified in the amino acid. The acids that are left behind after the fuming with formaldehyde yield hydrogen cyanide when dry-heated with mercuric cyanide (see p. 147) and can be sensitively detected in this gas phase through the color reaction with copper–benzidine acetate (cf. Chap. 5).

A prerequisite for the application of the test described here is the absence of compounds that react directly with mercuric cyanide with evolution of hydrogen cyanide. Such materials are listed on p. 148.

Procedure. A small amount of the solid or a drop of the solution is placed in a micro test tube and taken to dryness if need be. Several cg of paraformaldehyde are introduced and the test tube is then kept for 5 min. in a drying oven at 120°. After an additional heating for 10 min. at 140° and cooling several cg of mercuric cyanide and a drop of acetone are added. The mixture is evaporated to dryness, and then the tube is immersed about 0.5 cm in a glycerol bath that has been preheated to 160°. The mouth of the tube is covered with a piece of paper moistened with hydrogen cyanide reagent. A blue stain appears quickly if the response is positive.

The test revealed 5–10 γ of acids in this category.

1 Unpublished studies.

46. Aliphatic aminocarboxylic acids

Test through oxidation with chloramine[1]

Aliphatic aminocarboxylic acids are oxidatively decarboxylated and deaminated by the action of alkali hypochlorite (compare p. 369). Accordingly, these acids can be detected by adding small amounts of sodium hypochlorite and noting its disappearance. Chloramine T gives better results. It functions as a hypochlorite donor because of the hydrolysis:

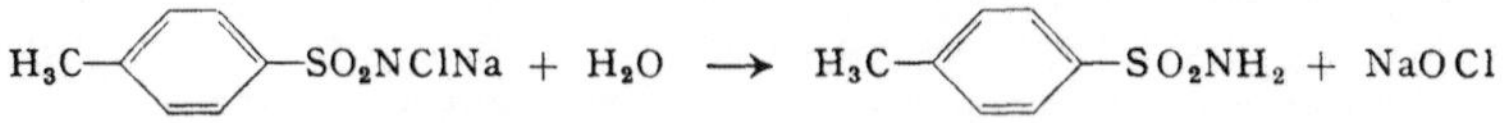

$$H_3C-\!\!\left\langle\;\right\rangle\!\!-SO_2NClNa + H_2O \rightarrow H_3C-\!\!\left\langle\;\right\rangle\!\!-SO_2NH_2 + NaOCl$$

and can be readily detected through its oxidation of thio-Michler's ketone to give a blue quinoidal disulfide (see p. 65).

The procedure given here can be applied only in the absence of compounds that are oxidized by chloramine T. Pertinent examples of such interfering substances are: thiol compounds, *p*-aminophenol, nitroso compounds; inorganic reductants must likewise be absent.

Procedure. A drop of water and a drop of the test solution are placed in adjoining depressions of a spot plate. Each is treated with a drop of 0.01% water solution of chloramine T and kept for 3–4 min. in an oven previously heated to 120°. The plate is allowed to cool, and the two specimens are then treated with a drop of 0.1% alcoholic solution of thio-Michler's ketone. A positive response is shown by the development of a yellow, green, blue or faintly violet color. The blank will show a deep violet. Water-insoluble amino acids should be dissolved in 0.2 N hydrochloric acid, and the resulting solution neutralized with a slight amount of calcium carbonate before proceeding with the test.

The test revealed:

17 γ glycine	0.5 γ asparagine	0.5 γ tyrosine
5 γ aspartic acid	5 γ β-alanine	25 γ arginine chloride

1 Unpublished studies, with R. A. ROSELL.

47. α-Aminocarboxylic acids $\left(-\overset{\diagup NH_2}{\underset{\diagdown COOH}{CH}}\ \text{type}\right)$

(1) Test by conversion to aldehyde[1]

α-Aminocarboxylic acids are both deaminated and decarboxylated on treatment with alkali hypochlorite or hypobromite. An aldehyde with one C atom less than in the original amino acid results:

$$RCH(NH_2)COOH + NaClO \rightarrow RCHO + NH_3 + CO_2 + NaCl$$

The aldehyde may then be detected with fuchsin–sulfurous acid (see p. 195). When an alkaline hypohalogenite solution is used, a partial oxidation of the aldehyde cannot be avoided. Therefore, the test is not very sensitive.

Procedure. A little of the test substance is united with a few drops of hypochlorite solution in a micro crucible and warmed gently. When the reaction is complete, an excess of fuchsin–sulfurous acid is added drop by drop. A red color appears if amino acids are present.

Reagents: 1) Sodium hypochlorite (saturated solution).
2) Fuchsin–sulfurous acid (for preparation see p. 195).

The rest revealed:

60 γ glycine	100 γ L-aspartic acid	30 γ diiodotyrosine
100 γ L-asparagine	100 γ tyrosine	60 γ D-arginine
100 γ α-alanine		

1 O. FREHDEN and L. GOLDSCHMIDT, *Mikrochim. Acta*, 2 (1937) 186.

(2) Detection through condensation with pyridine-2-aldehyde[1]

If about 0.5 mg of an α-amino acid and some pyridine are added to an aqueous solution of pyridine-2-aldehyde containing cobalt chloride, an intensely blue or violet color appears immediately. (Only 2-aminoisobutyric acid yields a yellow color.) The color reaction results from the formation of the Schiff base (I) formed by the condensation of amino acids and pyridine-2-aldehyde. (It is essential that the NH_2 group be set free through salification of the COOH group by pyridine). Condensed chelate rings are formed (as shown in II) by replacement of the H atom of the carboxyl group in (I) by cobalt and its coordinative bonding to two nitrogen atoms.

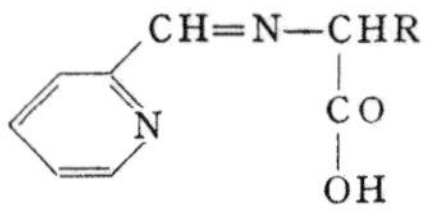

(I) 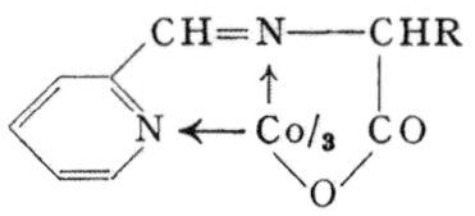   (II)

The intense color of the cobalt salt is obviously a result of the chelate bonding. This reaction is analogous to the reaction described on p. 376 in connection with the detection of *o*-aminophenol.

Procedure. A drop of the test solution is brought together with a drop of the reagent and a drop of pyridine in a micro test tube. The mixture is heated for 1–3 min. in a boiling water bath and then allowed to cool. If the amount of amino acid present exceeds approximately 80 γ, a blue or violet color develops. Smaller amounts, from around 10 γ downward, yield pink, orange, or yellow shades, especially in the vicinity of the identification limit.

Reagent: Five drops of a 0.5% aqueous solution of pyridine-2-aldehyde are mixed with 1 drop of 0.1 M cobalt nitrate. The reagent keeps for about a week.

The test revealed:

3 γ cystine	2 γ lysine	1 γ L-leucine
3 γ cysteine	1 γ glycine	5 γ aspartic acid
2.5 γ methionine	3 γ tyrosine	5 γ asparagine
5 γ α-alanine		

A positive response was also given by monoiodo-L-tyrosine, valine, citrulline, arginine, phenylalanine, tryptophan, 2-amino-*n*-octanoic acid.

1 F. FEIGL and S. YARIV, *Anal. Chim. Acta*, 29 (1963) 581.

48. Proteins

(1) Test with tetrabromophenolphthalein ethyl ester[1]

Solid or dissolved tetrabromophenolphthalein ethyl ester:

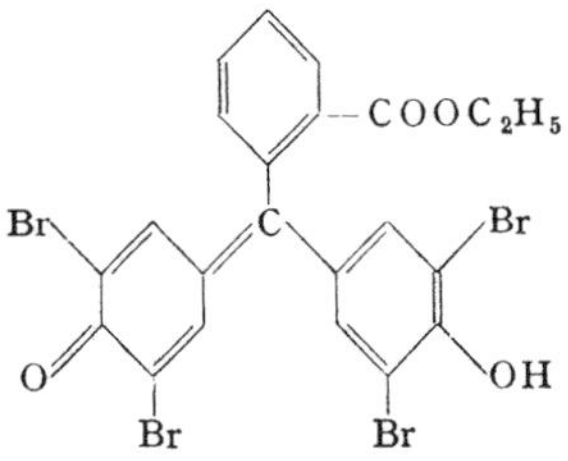

is yellow but its water-soluble alkali salts are blue.[2] The latter are decomposed by dilute acetic acid with regeneration of the phenol. If the ester is brought into contact with proteins, which are generally in the colloid state, a blue color appears. Apparently this is due to the formation of a salt-like adsorption compound which, however, unlike the alkali salts, is not decomposed by dilute acetic acid. The phenomenon known as "protein error" of indicators (displacement of the transformation range of pH indicators) appears to be involved. This effect is shown very strongly by tetrabromophenolphthalein ester, particularly by its stable potassium salt, whose blue solution is turned yellow (color of the free ester) by dilute acetic acid, but the blue persists in the presence of proteins. It then changes to yellow only on the addition of more concentrated acetic acid or mineral acids.

The reaction seems to be specific for *native proteins*. Protein fission products, such as amino acids, di- and tripeptides, or peptones, do not react.

Alkaloids, when present in large amounts, show a behavior similar to that of native proteins. Probably this is due to the formation of adsorption compounds of colloidally dispersed alkaloid with tetrabromophenolphthalein ester, which are resistant to dilute acetic acid.[3]

Procedure. A drop of the test solution is mixed on a spot plate with a drop of the blue reagent solution and then acidified with a drop of 0.2 N acetic acid. A blank turns yellow but the blue or greenish color persists if the sample contains protein.

Reagent: 0.1% solution of the potassium salt of tetrabromophenolphthalein ethyl ester in alcohol.*

The test revealed:

0.5 γ egg albumin	0.5 γ casein	0.5 γ salmine
0.5 γ hemoglobin	5 γ edestin	1 γ gliadin
0.35 γ serum albumin	0.5 γ clupeine	

A pathologically increased protein content may be detected in a drop of urine by this reaction.

* A better reagent is the isopropyl ester whose yellow solution changes to blue on the addition of proteins. This ester is marketed by Loba Chemie, Vienna.

1 F. Feigl and V. Anger, *Mikrochim. Acta*, 2 (1937) 107.
2 R. Nietzki and E. Burckhardt, *Ber.*, 30 (1897) 175.
3 Comp. F. Feigl, *Chemistry of Specific, Selective and Sensitive Reactions*, Academic Press, New York, 1949, p. 484.

(2) *Test after hydrolytic or thermal fission to amino and imino compounds*

Proteins are hydrolyzed by strong mineral acids to yield fission products such as polypeptides, amino acids, *etc.* Since tryptophan is a component of nearly all proteins, it is broken down to indole and indole derivatives. These amino and imino compounds may be detected either by melting with fluorescein chloride (see p. 238) or better by condensation with *p*-dimethylaminobenzaldehyde to yield colored Schiff bases. The condensation of indole bases, which have been formed by the acid splitting of proteins, apparently plays the chief role in the following test with *p*-dimethylaminobenzaldehyde (see test for pyrrole, p. 381).

Procedure.[1] The sample to be tested for proteins (solid or in solution) is mixed in a micro crucible with several drops of a saturated glacial acetic acid solution of *p*-dimethylaminobenzaldehyde and one drop of fuming hydrochloric acid. A violet color indicates the presence of proteins.

The following *identification limits* were obtained:

1 γ dried peptone	10 γ edestin	5 γ casein
20 γ blood albumin	30 γ egg albumin	60 γ pancreatin
100 γ hide powder		

Pepsin is very insensitive.

Thyroglobin gives a positive albumin reaction, whereas the reaction with pure thyroxine results negatively. Hypophysis preparations give a positive response. As little as 1% albumin (*e.g.* edestin) can be distinctly detected in mixtures with carbohydrates.

Dry heating also splits proteins. The resulting pyrrole and pyrrole derivatives volatilize with the combustion products. If they are brought into contact with filter paper that has been impregnated with a 5% solution of *p*-dimethylaminobenzaldehyde in concentrated hydrochloric acid (or trichloroacetic acid) a violet color appears.[2] The pyrolysis can be conducted in a micro crucible or a micro test tube. (Compare differentiation of animal and vegetable fibers, p. 686). The *identification limit* is around 100 γ protein.

1 O. Frehden and L. Goldschmidt, *Mikrochim. Acta*, 1 (1937) 351.
2 Unpublished studies, with E. Silva (Pernambuco).

49. Aniline- and phenolsulfonic acids

Test through release of sulfuric acid[1]

When aqueous solutions of sulfanilic acid (or the isomeric metanilic and orthanilic acid) are treated with barium chloride there is no apparent change because the alkaline-earth salts of arylsulfonic acids are soluble in water. If an acidified aqueous solution of the anilinesulfonic acids is treated with an excess of bromine water, light yellow tribromoaniline precipitates and the solution then contains free sulfuric acid:[2]

$$H_2N-\langle\rangle-SO_3H + 3\ Br_2 + H_2O \longrightarrow H_2N-\langle\rangle-Br + 3\ HBr + H_2SO_4$$

Since tribromoaniline is readily soluble in ether, extraction of the suspension with this solvent (which also takes up the excess bromine) yields a clear solution in which the resulting SO_4^{2-} ions can be detected by precipitation of barium sulfate. Phenolsulfonic acids behave similarly to the anilinesulfonic acids; 2,4,6-tribromophenol and sulfuric acid result.

Sulfanilamide and sulfonamides react with bromine water to give insoluble brominated substitution products, which, for the most part, are soluble in ether, but there is no liberation of sulfuric acid.

There is no exchange of SO_3H groups by bromine atoms in benzene- and naphthalenesulfonic acids or their nitro derivatives.[3] Naphthol- and naphthylaminesulfonic acids likewise remain unchanged. Sulfosalicylic acid in amounts above 0.5 mg shows distinct formation of barium sulfate.

Procedure. A drop of the test solution is placed in a micro test tube and acidified with hydrochloric acid. Saturated bromine water (or a saturated solution of bromine in 5% potassium bromide) is added drop by drop; a precipitate or turbidity is produced. The suspension is shaken with a few drops of ether and drops of 3% barium chloride solution are allowed to flow through the ether layer. An immediate precipitate or a turbidity which develops within a few minutes indicates the presence of sulfanilic acid.

Bromine water containing barium chloride may also be used. In this case a positive result is signalled by a precipitate or turbidity in the water layer after extraction with ether.

Limit of identification: 1 γ sulfanilic acid.

Since the phenolsulfonic acids likewise exchange the SO_3H group for bromine on treatment with bromine water, the identification or detection of anilinesulfonic acids requires the supplementary detection of NH_2 groups.

This can be accomplished through the formation of orange Schiff bases by the action of acetic acid solutions of p-dimethylaminobenzaldehyde (see p. 243). If the test is conducted on a porcelain plate, as little as 0.05 γ sulfanilic acid can be detected.

1 Unpublished studies, with L. HAINBERGER.
2 A. SCHMITT, Ann., 120 (1859) 136.
3 R. L. DATTA and J. C. BHOUMIK, J. Am. Chem. Soc., 43 (1921) 303.

50. Aniline derivatives containing strongly acidic groups in the para-position

Detection by oxidation with potassium bromate and nitric acid[1]

If compounds of this kind, with the general formula 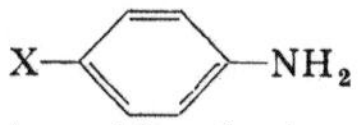 X—⟨ ⟩—NH₂ (X=COOH, SO₃H, SO₂H, AsO₃H₂) are heated with a nitric acid solution of potassium bromate, a blue color appears that quickly turns brown and then yellow. The mechanism of the color reaction, which undoubtedly is based on an oxidation, has not been clarified as yet. A plausible explanation could be that the hydrazo compound (I) is formed initially and gives rise in turn to the azo compound (II). A blue quinhydrone-like molecular compound may possibly be formed as an unstable intermediate product. The following scheme would apply:

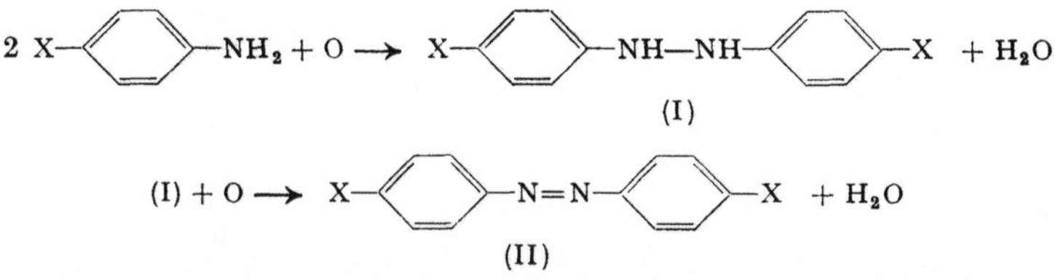

$$2\ X\!-\!\langle\ \rangle\!-\!NH_2 + O \rightarrow X\!-\!\langle\ \rangle\!-\!NH\!-\!NH\!-\!\langle\ \rangle\!-\!X\ + H_2O$$

(I)

$$(I) + O \rightarrow X\!-\!\langle\ \rangle\!-\!N\!=\!N\!-\!\langle\ \rangle\!-\!X\ + H_2O$$

(II)

Procedure.   A drop of the test solution is mixed in a micro test tube with a drop of 0.1 M potassium bromate and a drop of 6 M nitric acid; the mixture is then heated in a boiling water bath. A positive response is indicated by the appearance of a blue color which changes to brown and eventually to yellow.

The test revealed:

| 50 γ p-aminobenzoic acid | 10 γ arsanilic acid | 0.5 γ sulfanilic acid |

1 E. JUNGREIS (Jerusalem), private communication.

51. Anilinearsonic and anilinestibonic acids

(1) Test through reductive release of aniline[1]

If the alkaline test solution is warmed with Devarda alloy, the action of the nascent hydrogen causes the splitting off of aniline:

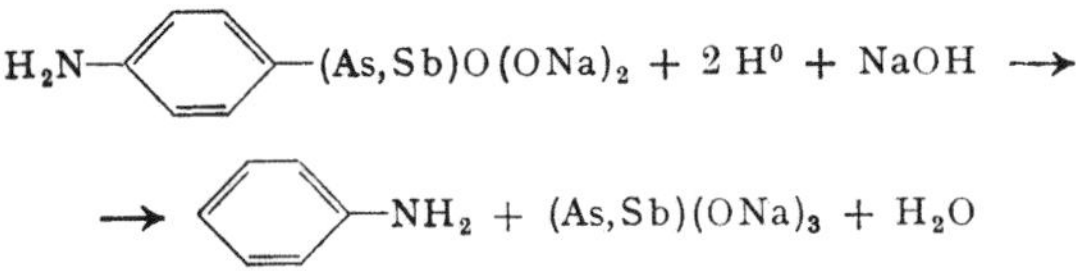

$$H_2N-\langle\bigcirc\rangle-(As,Sb)O(ONa)_2 + 2 H^0 + NaOH \longrightarrow$$

$$\longrightarrow \langle\bigcirc\rangle-NH_2 + (As,Sb)(ONa)_3 + H_2O$$

The aniline produced by the above redox reaction is volatile with steam and can be readily detected in the gas phase by means of color reactions with p-dimethylaminobenzaldehyde or sodium 1,2-naphthoquinone-4-sulfonate as described on p. 153.

Sulfanilic acid is not affected when warmed with Devarda alloy and caustic hydroxide. Therefore, the redox reaction permits the detection of aniline-arsonic and stibonic acids in the presence of anilinesulfonic acids.

Nitrobenzenearsonic and stibonic acids show the same behavior as the arsanilic and stibanilic acids since they are converted into the latter by reduction with Devarda alloy.

Procedure. A drop of the alkaline test solution and several mg of Devarda alloy are placed in a micro test tube. The mouth of the tube is covered with a disk of filter paper moistened with 1 drop of an acetic acid solution of p-dimethyl-aminobenzaldehyde or of a freshly prepared 2.5% aqueous solution of 1,2-naphthoquinone-4-sulfonate. After the vigorous evolution of hydrogen has subsided, the test tube is placed in boiling water. Depending on the amount of aniline split off, a yellow or red-violet color appears on the reagent paper within 2–3 min.

The test revealed:

6 γ arsanilic acid	5 γ p-nitrobenzenestibonic acid
6 γ p-aminobenzenestibonic acid	6 γ p-nitrobenzenearsonic acid

1 F. FEIGL, *Anal. Chem.*, **33** (1961) 1119.

(2) Test through pyrohydrolytic release of aniline[1]

If hydrated sodium thiosulfate is heated it melts in its water of crystallization and then becomes anhydrous at about 160–180°. If alkali salts of anilinearsonic or stibonic acids are present during this dehydration, the superheated water that is being driven off brings about the following pyrohydrolysis:

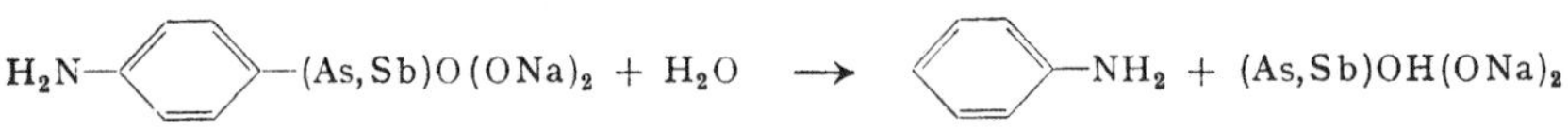

$$H_2N-\langle\bigcirc\rangle-(As,Sb)O(ONa)_2 + H_2O \longrightarrow \langle\bigcirc\rangle-NH_2 + (As,Sb)OH(ONa)_2$$

The aniline formed can be detected in the gas phase by the color reactions described in (1).

Procedure. A small amount of the solid or a drop of its aqueous solution is placed in a micro test tube along with several cg of $Na_2S_2O_3.5\ H_2O$ and taken to dryness if need be. A disk of filter paper moistened with one of the reagents prescribed in (1) is laid across the mouth of the tube and the latter is then placed in a glycerol bath that has been preheated to 120°. The temperature is gradually raised to 160°. A yellow or red-violet stain on the reagent paper signals a positive response.

The test revealed:

$5\ \gamma$ arsanilic acid $5\ \gamma$ p-aminobenzenestibonic acid

1 Unpublished studies.

52. o- and peri-Hydroxyarylamines

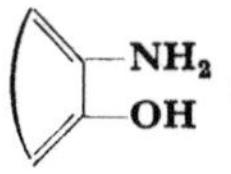

Test through condensation with pyridine-2-aldehyde[1]

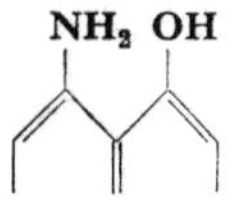

When solutions of o-aminophenol, pyridine-2-aldehyde and cobalt chloride are united, the initial result is a yellow color that goes through orange to red. The following explanation seems probable for the color reaction. The first stage is condensation of the organic reactants to give the (yellow) Schiff base (I) which contains an acid OH group in chelate position with respect to two nitrogen atoms capable of coordination. Salification by cobalt leads to the formation of two chelate rings held together by the metal atom as shown in (II). Consequences of the formation of two condensed rings are the intense color and the acid-resistance of the cobalt compound in contrast to the unsalified anil (I) which is yellow and easily destroyed by dilute acids.

Only the mode of linkage of the cobalt is shown in (II) but not its valence; information about the formula of the cobalt salt is given by the fact that the red color develops gradually which points to autoxidation, and furthermore red or violet precipitates are obtained from the red solution on addition of voluminous anions such as $[CdI_4]^{2-}$, $[Hg(CNS)_4]^{2-}$, and $[TlBr_4]^{-}$. This finding indicates that trivalent cobalt, along with the chelate rings, is a constituent of a voluminous complex cation of the formula (III).

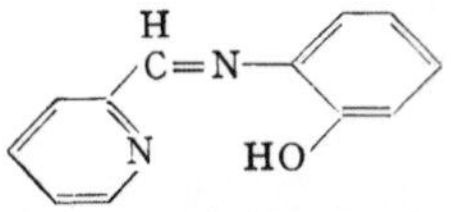

(I) $(C_{12}H_{10}N_2O)$ 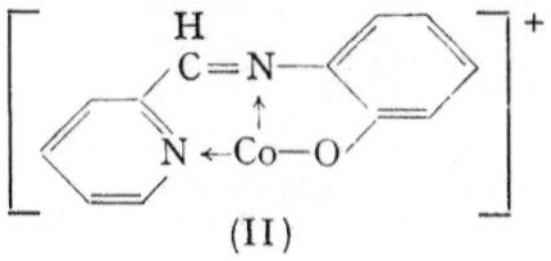   (II) $[Co(C_{12}H_9N_2O)_2]^+$ (III)

The formation of (III) provides a selective test for *o*- and *peri*-hydroxy-arylamines and their differentiation from their isomers. It should be noted that the production of (III) serves as the basis of spot tests for pyridine-2-aldehyde.

Procedure. A drop of the test solution and a drop of the reagent solution are brought together in a micro test tube and heated. A positive response is indicated by the appearance of a red or pink color that is resistant to mineral acids.

    Reagent: 5 ml of 0.5 *M* pyridine-2-aldehyde is mixed with 10 mg of cobalt sulfate.

The procedure revealed:

    0.5 γ *o*-aminophenol   0.1 γ 2,4-diaminophenol   1 γ H-acid   1 γ K-acid

The red colorations produced with H- and K-acids are discharged by the addition of mineral acids, whereas those obtained with aminophenols and 1,2-aminonaphthols are acid-resistant. Accordingly, it seems likely that it is possible to differentiate between *ortho*- and *peri*-hydroxyarylamines.

1 F. Feigl, E. Jungreis and S. Yariv, *Z. Anal. Chem.*, 200 (1964) 36.

53. Aromatic tertiary N-ring bases

(1) Test with methyl iodide and sodium thiosulfate [1]

A test for the CH_2Hal group is described on p. 171; it is based on the fact that compounds of this kind react with sodium thiosulfate to yield the so-called Bunte salts, which in turn split off sulfur dioxide when they are dry-heated. The methiodides of aromatic tertiary ring bases show analogous behavior.

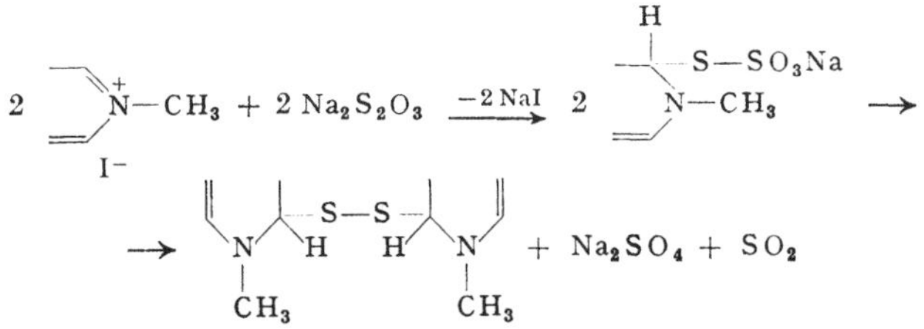

The pertinent tertiary bases can therefore be detected by quaternizing them with methyl iodide and then heating the resulting methiodide with sodium thiosulfate to 160–180°. A positive response is indicated by the evolution of SO_2, which can be detected by the bluing of Congo paper moistened with hydrogen peroxide or by other tests for sulfur dioxide.

The quaternization with methyl iodide followed by the pyrolysis with sodium thiosulfate is easily realizable with pyridine, quinoline, isoquinoline and thiazole bases which have a distinct aromatic nature. Methiodides of bases without heterocyclic bonded nitrogen do not react. Primary and secondary bases weaker than ammonia should not be present because with methyl iodide some of them form hydriodides of weak secondary (tertiary) bases, which react with sodium thiosulfate at 150° (see p. 111). It is obvious that when tertiary ring bases of the mentioned type that are already quaternized are pyrolysed with sodium thiosulfate formation of SO_2 occurs.

Procedure.   The test is conducted in a micro test tube. A drop of the ethereal solution of the base is treated with several drops of methyl iodide and the quaternization is accomplished as described in (*3*). Several cg of sodium thiosulfate are added and the test is completed as given on p. 171.

The *limits of identification* of tertiary ring bases which are readily converted into reactive quaternary compounds are in the region of 10 γ. This was found to be the case with the following cyclic bases: quinine hydrochloride, α,α'-dipyridyl, *o*-phenanthroline, isonicotinic acid hydrazide, sulfathiazole.

1 F. FEIGL, V. ANGER and D. GOLDSTEIN, *Helv. Chim. Acta*, **43** (1960) 2139.

(2) *Test by hydrogenation to secondary ring bases*[1]

Tertiary six-membered ring bases are converted into secondary ring bases by treatment with zinc and hydrochloric acid (compare p. 525):

$$\text{(structure)} + 2\,H^0 \longrightarrow \text{(structure)} \tag{1}$$

The hydrogenation (reduction) is so extensive within a few minutes in the case of many tertiary ring bases that the resulting secondary bases can be detected. A suitable test involves their inclusion in benzene-soluble copper salts of dithiocarbamic acids (see p. 250). This salt formation probably proceeds from the components:

$$CS_2 + HN\langle \text{ } + \tfrac{1}{2}\,Cu^{2+} + NH_3 \longrightarrow S C \langle \text{ } + NH_4^+ \tag{2}$$

Many tertiary ring bases can be detected in small amounts through (*1*) and (*2*) followed by extraction of the copper dithiocarbamate with benzene.

Certain pyridine and quinoline derivatives can not be detected by this procedure. Examples are: pyridinedicarboxylic acids (2,3 and 3,4), 6-hydroxymethyl-picolinic acid, acridine, sedulone, 6-nitroquinoline, vitamin B_6 (pyridoxine). The reduced solution of the first of these exceptions shows no change on the addition of carbon disulfide and ammoniacal copper salt solution. Reduced 6-nitroquinoline and vitamin B_6 yield brown precipitates, which are not extractable with benzene. Probably the hydrophilic groups in these ring bases prevent the solubility of the copper dithiocarbamates in benzene. Moreover, OH and COOH groups in *para* position to the cyclic N atom seem to impair the reduction.

The procedure given here may not be applied in the presence of secondary aliphatic amines since they behave like secondary ring bases. Consequently, the test solution should first be acidified and treated with carbon disulfide and strongly ammoniacal copper sulfate solution. The mixture is shaken with benzene. If the benzene layer does not become colored, the positive response after reduction with zinc indicates the presence of tertiary ring bases.

Procedure.   One drop of the strongly acid test solution is treated in a micro test tube with several grains of zinc (10 mesh) and warmed for about 5 min. in a water bath. After the metal has dissolved, the solution is cooled and one drop of 5% copper sulfate solution is added and then ammonia (drop by drop) until the blue color appears. The solution is shaken with two drops of a mixture of 1 volume of carbon disulfide and 3 volumes of benzene. A positive response is indicated by a yellow or brown color in the benzene layer.

Limits of identification:

30 γ pyridine	5 γ α,α'-dipyridyl	100 γ nicotinic acid
2 γ quinoline	50 γ quinine chloride	40 γ β-pyridylcarbinol tartrate
3 γ isoquinoline	20 γ cinchonine	20 γ papaverine chloride
50 γ quinaldine		

A positive response was obtained with (α,β,γ)-picoline, picolinic acid, coramine (N,N-diethylnicotinamide), β-naphthoquinoline, nicotine, α-nicotyrine, 3-bromoquinoline nitrate.

As shown by the schematic representation of the hydrogenation of tertiary ring bases, the hydrogen is taken up by the cyclic N atom and a C atom *ortho* to this nitrogen atom. These hydrogenated products are hydrogen donors in alkaline surroundings, and consequently they can be detected with certainty by the color reaction with *o*-dinitrobenzene (see p. 133).

1 Unpublished studies, with C. STARK-MAYER.

54. Aromatic heterocyclic ring bases

Test with methyl iodide and sodium 1,2-naphthoquinone-4-sulfonate [1]

Quaternized aromatic ring bases act toward sodium 1,2-naphthoquinonesulfonate in the same way as compounds containing active CH_2 groups and in alkaline solution give deeply colored water-soluble quinoid compounds (see p. 153). The mechanism of the color reaction in the case of alkylpyridinium salts and corresponding compounds may be:

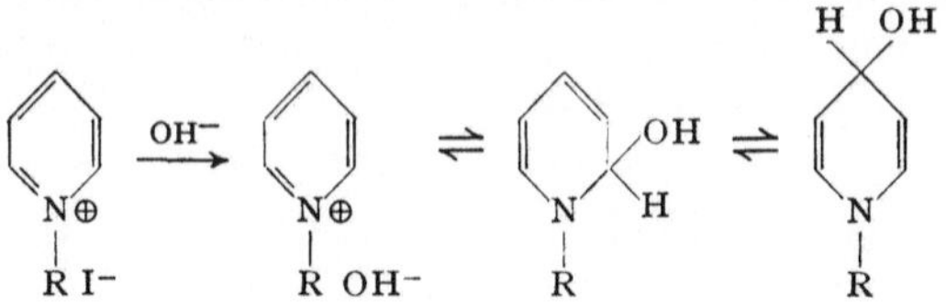

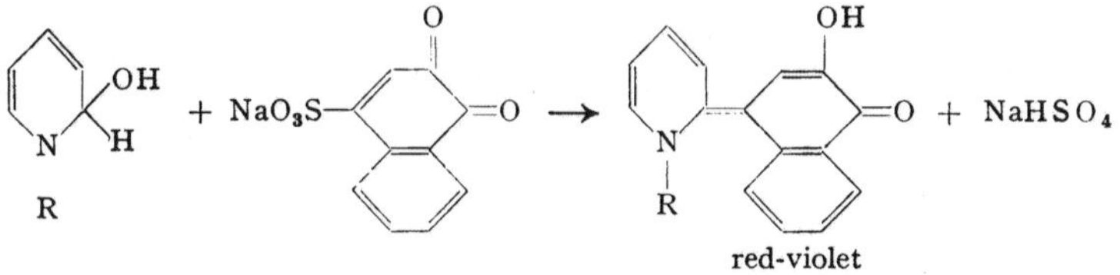

red-violet

The color reaction in the case of compounds that correspond to α- and γ-alkylpicolinium salts may possibly be represented by the following formulation:

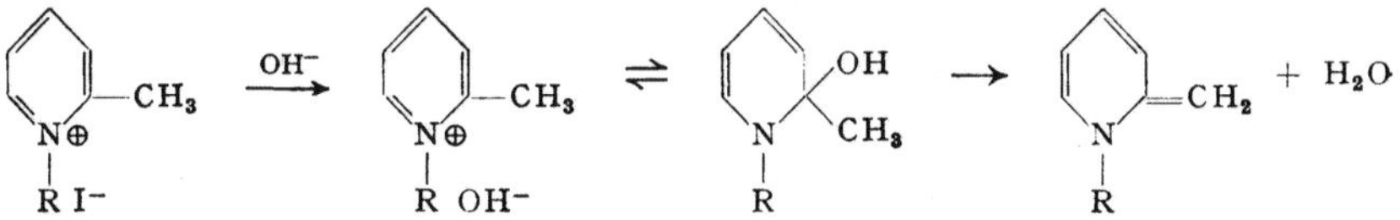

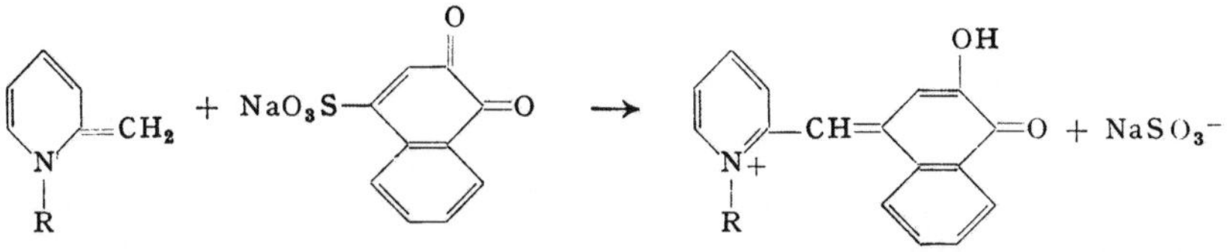

blue-violet

Since aromatic ring bases are readily converted into the corresponding quaternary ammonium salts by alkylating agents, they can be detected in this way.

The quaternization may be accomplished also with dimethyl sulfate, ethyl bromide or iodide, ethyl p-toluenesulfonate, and other suitable agents.

Pyrone derivatives react analogously under formation of oxonium salts.

Procedure. A little of the test substance (either the solid or a drop of solution) is mixed in a tall micro crucible with 5 or 6 drops of methyl iodide or dimethyl sulfate, and heated to gentle boiling on an asbestos plate. Substances hard to convert to quaternary compounds should be kept at 100° for some hours with methyl iodide in a closed capillary tube in a water bath. The quaternary compound is treated with 2 or 3 drops of a saturated alcoholic solution of sodium 1,2-naphthoquinone-4-sulfonate, and the mixture made alkaline with $0.5N$ sodium hydroxide. The appearance of a color or a change of color indicates that the test is positive. On acidifying with $1 N$ acetic acid there is always a change of color.

Limits of identification are given in Table 16.

TABLE 16

TERTIARY RING BASES

Name	Methylating agent	Color	Limit of identi- fication
Pyridine	CH_3I, $(CH_3)_2SO_4$	red to red-violet	12 γ
α-Picoline	CH_3I, $(CH_3)_2SO_4$	dark blue-violet	12 γ
Quinoline	CH_3I, $(CH_3)_2SO_4$, C_2H_5Br	dark brown-green to black-green	25 γ
Quinaldine	CH_3I, $(CH_3)_2SO_4$	dark blue-violet to blue-green	12 γ
2,6-Dihydroxypyridine-4-carboxylic acid	CH_3I	black-green	25 γ
2,6-Dimethylpyrone	CH_3I, $(CH_3)_2SO_4$, C_2H_5Br	brown-violet, then green-black	25 γ
Chelidonic acid	CH_3I	dark olive green to green-black	25 γ
Cinchonine	CH_3I	cherry red	100 γ

1 F. FEIGL and O. FREHDEN, *Mikrochemie*, 16 (1934) 84.

55. Pyrrole derivatives

Test with p-dimethylaminobenzaldehyde [1]

When a mixture of pyrrole (I) and a weakly acid alcoholic solution of p-dimethylaminobenzaldehyde (II) is warmed, a red-violet color develops. This reaction is based on the fact that pyrrole can react in its tautomeric forms (Ia) and (Ib) known as pyrrolenine.[2]

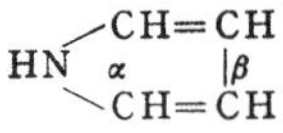

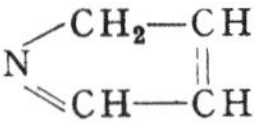

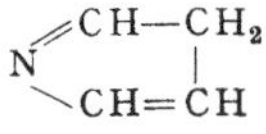

The CH_2 groups in (Ia) or (Ib) permit condensation with the aldehyde. For example, (Ib) reacts with loss of water:

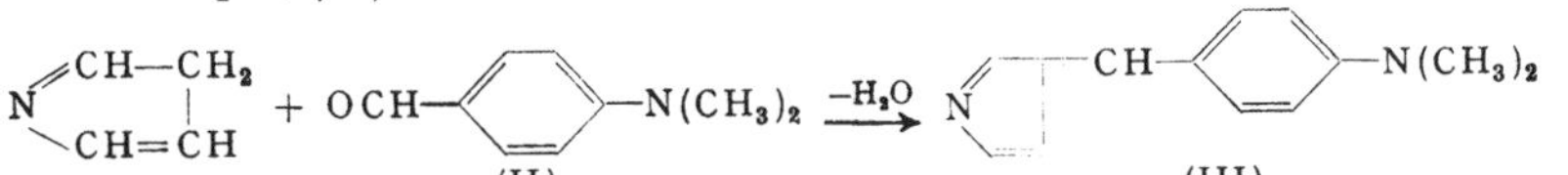

The compound (III) produced in (1) combines with H⁺ ions and rearranges into the quinoidal red-violet compound (IV):

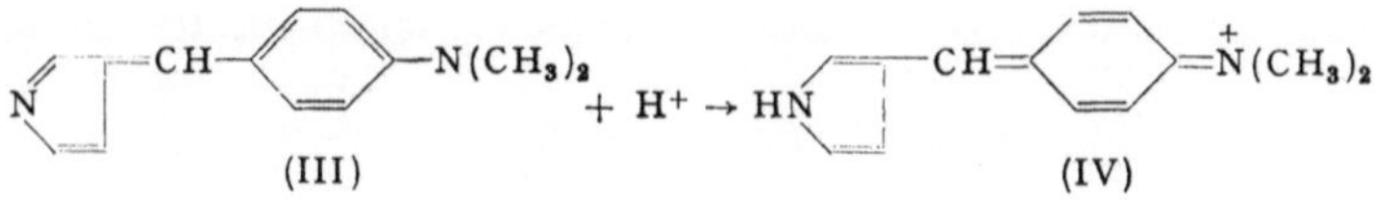

$$\text{(III)} \quad + \text{ H}^+ \to \quad \text{(IV)} \qquad (2)$$

The procedure described here takes advantage of the fact that reactions (1) and (2) occur almost immediately and without warming if a solution of p-dimethylaminobenzaldehyde in conc. hydrochloric acid is used as reagent.

The derivatives of pyrrole show the same ability to undergo this condensation provided they have an intact CH group in the α- or β-position relative to the cyclic NH group. Accordingly, indole (benzopyrrole), skatole (β-methylindole), and tryptophan (β-indolylalanine) react analogously to pyrrole, whereas carbazole (dibenzopyrrole) does not react.

It should be noted that primary aromatic and aliphatic amines also condense with p-dimethylaminobenzaldehyde to yield colored Schiff bases. These products however are never violet, but yellow, orange-red, sometimes brown. Furthermore, the sensitivity of the color reaction based on the formation of Schiff bases, particularly in strong acid solution, is considerably lower[3] than that of the color reaction of pyrrole and pyrrole derivatives, which rests on another kind of condensation.

Procedure.[4]   One drop of the aqueous, ethereal, or alcoholic test solution is mixed with one or two drops of the reagent solution on a spot plate. If pyrrole or reaction-capable pyrrole derivatives are present, a violet color will appear at once or after a short time.

Reagent: 5% solution of p-dimethylaminobenzaldehyde in conc. HCl.

The procedure revealed: 0.04 γ pyrrole; 0.06 γ indole; 0.1 γ tryptophan.

1 H. FISCHER and F. MEYER-BETZ, *Z. Physiol. Chem.*, 75 (1911) 232; see also E. SAL-KOWSKY, *Biochem. Z.*, 103 (1920) 185.
2 H. FISCHER and H. ORTH, *Die Chemie des Pyrrols*, Vol. I, Springer, Leipzig, 1934, p. 66; see also N. V. SIDGWICK and R. W. J. TAYLOR, *Organic Chemistry of Nitrogen*, Oxford, 1937, p. 484.
3 Comp. O. FREHDEN and L. GOLDSCHMIDT, *Mikrochim. Acta*, 1 (1937) 338.
4 Unpublished studies, with Cl. COSTA NETO.

56. Indole derivatives

Test with glutaconic aldehyde[1]

Indole derivatives with free α- or β-position in the pyrrole ring condense in acid solution with glutaconic aldehyde (formed through hydrolysis of pyridylpyridinium chloride, see p. 246) to red-violet polymethine dyestuffs:

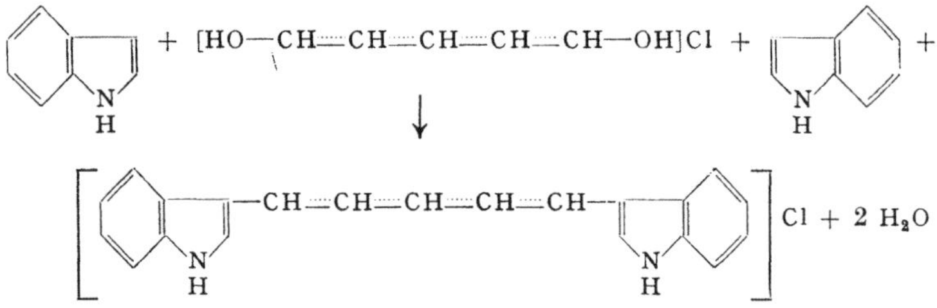

Indole derivatives with nitrogen-bearing side chains, such as tryptophan and gramine, do not give this reaction. Therefore the test is especially useful for the detection of indoles in the presence of tryptophan. Phloroglucinol and aromatic amines interfere (see pp. 245 and 321).

Procedure. A drop of the test solution is treated in a micro test tube with a drop of a 1% solution of pyridylpyridinium dichloride in water, a drop of 1 N sodium hydroxide, and then with 1 drop of concentrated hydrochloric acid. If indoles are present a red-violet to red (skatole) color appears.

Limits of identification: 0.1 γ of indole, 2-methylindole, 7-methylindole and skatole.

1 V. ANGER and S. OFRI, *Mikrochim. Acta*, (1964) 772.

57. Carbazole derivatives

Test with sulfuric acid and alkali nitrate[1]

Carbazole is closely related to diphenylamine from which it is obtained by vapor phase heating. Like diphenylamine, carbazole and its derivatives react with concentrated sulfuric acid plus alkali nitrate or nitrite to yield colored products, which are never blue, in contrast to the oxidation products of diphenylamines. The mechanism of this reaction is presumably analogous to that of the diphenylamine–nitrate (nitrite) reaction (see p. 301).

Procedure. A drop of the test solution is evaporated on a spot plate and a drop of concentrated sulfuric acid and several mg of alkali nitrate are added.

The test revealed:

1 γ carbazole	green	0.5 γ N-ethyl-3-diethyl-	
0.5 γ N-ethylcarbazole	,,	aminocarbazole	violet
10 γ 3-nitrocarbazole	,,	2.5 γ N-ethyl-4-amino-	
2.5 γ N-p-toluenesulfonylcarbazole	,,	carbazole	pale pink

These findings indicate that carbazole and its derivatives may be characterized by color reactions, and differentiated from diphenylamine and its derivatives. However, when these classes of materials are presented as mixtures it is not possible to distinguish them from each other in this way.

It is rather characteristic that carbazole and many carbazole derivatives produce colored solvates in concentrated sulfuric acid; this is not true for diphenylamine and its derivatives and use can be made of this fact for differentiation purposes.

1 D. GOLDSTEIN (Rio de Janeiro), unpublished studies.

58. Pyridine derivatives

(1) Test by conversion into polymethine dyes

The action of cyanogen bromide and primary aromatic amines on pyridine results in the formation of colored Schiff bases of glutaconic aldehyde (polymethine dyes).[1] The reaction requires the hydrolytic opening of the pyridine ring with formation of glutaconic aldehyde; this takes place instantly after the intermediate addition of cyanogen bromide to pyridine:

$$\text{(pyridine-Br}^-,\text{CN)} + 2\text{ H}_2\text{O} \longrightarrow \underset{\text{OCH}}{\overset{\text{CH}}{\text{HC}}}\underset{\text{CHOH}}{\overset{\text{CH}}{\text{CH}}} + \text{NH}_2\text{CN} + \text{HBr}$$

If primary aromatic amines are present at this hydrolysis, they condense with the resulting glutaconic aldehyde to yield colored Schiff bases:[2]

$$\underset{\text{OCH}}{\overset{\text{CH}}{\text{HC}}}\underset{\text{CHOH}}{\overset{\text{CH}}{\text{CH}}} + 2\text{ NH}_2\text{R} \longrightarrow \underset{\text{RN=CH}}{\overset{\text{CH}}{\text{HC}}}\underset{\text{CHNHR}}{\overset{\text{CH}}{\text{CH}}} + 2\text{ H}_2\text{O}$$

Derivatives of pyridine with free α,α'-positions behave analogously to the parent compound; they give substituted glutaconic aldehydes and their colored Schiff bases. The formation of polymethine dyes constitutes the basis of numerous tests and colorimetric methods for determining pyridine and its derivatives.[3] The condensation with o-tolidine is quite serviceable.

Procedure.[4] A spot plate is used. One drop of the test solution is mixed with one drop of saturated bromine water and one drop of 2% potassium cyanide solution (production of cyanogen bromide). A drop of an aqueous suspension of o-tolidine is added. Depending on the amount of pyridine or reactive pyridine derivatives, a red to pink color appears at once or within a few minutes.

Reagent: 1 g sodium acetate is added to 100 ml of saturated aqueous solution of o-tolidine dichloride to precipitate the base in finely divided condition. A suitable suspension is obtained by shaking.

The procedure revealed:

0.2 γ pyridine	6 γ coramine
2 γ β-picoline	1 γ nicotinamide
1 γ pyridine-4-aldehyde	0.5 γ β-pyridylcarbinol tartrate
0.1 γ β-methylaminopyridine	2 γ isonicotinic acid hydrazide

A procedure for the formation of polymethine dyes based on the action of volatile BrCN on filter paper spotted with the test solution and ethereal benzidine solution has been described recently.[5].

1 W. KOENIG, *J. Prakt. Chem.*, [2], 69 (1904) 105; comp. also P. KARRER, *Organic Chemistry*, 4th Engl. ed., Elsevier, Amsterdam, 1950, p. 807.
2 Comp. H. A. WAISMAN and C. A. ELVEHJEM, *Ind. Eng. Chem., Anal. Ed.*, 13 (1941) 221.
3 Comp. M. PESEZ and P. POIRIER, *Méthodes et Réactions de l'Analyse Organique*, Masson et Cie., Paris, 1954, p. 126; see also I. V. KULIKOV and T. N. KRESTOVOS-DVIGENSKAJA, *Z. Anal. Chem.*, 79 (1930) 454.
4 Unpublished studies, with D. GOLDSTEIN.
5 T. BROUWER, *Chem. Weekblad*, 58 (1962) 530.

(2) Test with cyanogen bromide and barbituric acid or phloroglucinol

As outlined in (*1*), pyridine and likewise its substituted derivatives containing organic radicals in the β- or γ-position are cleaved by the action of cyanogen bromide to yield glutaconic aldehyde or its corresponding derivatives. Glutaconic aldehyde condenses with barbituric acid[1] or phloroglucinol (see p. 321) to give polymethine dyestuffs. This series of reactions permits the detection of these kinds of pyridine derivatives.

Procedure.[2]   A drop of the neutral test solution (treated with sodium bicarbonate if need be) is mixed in a depression of a spot plate with a drop of cyanogen bromide solution. After 5 min. a drop of the barbituric or phloroglucinol reagent is added. If positive-reacting pyridine derivatives are present a color appears.

Reagents: 1) Cyanogen bromide solution. Saturated bromine water is added to a 10% solution of potassium cyanide until a permanent yellow color is obtained. The solution is then decolorized by dropwise addition of potassium cyanide solution.

2) Barbituric acid solution. A 3% solution of primary potassium phosphate is saturated with barbituric acid.

3) Phloroglucinol solution. Concentrated hydrochloric acid is saturated with phloroglucinol.

Limits of identification:

A) with barbituric acid reagent:

0.5 γ γ-picoline	orange-violet	0.5 γ pyridine-3-aldehyde	orange-pink
2 γ β- ,,	orange	0.5 γ ,, -4- ,,	,,
0.1 γ nicotinic acid	,,	0.1 γ isonicotinic acid	pink
0.01 γ pyridine	red-violet	0.1 γ nicotinamide	,,
0.1 γ isonicotinamide	pink-violet		

B) with phloroglucinol reagent:

0.01 γ pyridine	blue	0.5 γ β-picoline	green-violet
0.5 γ pyridine-3-aldehyde	,,	2 γ γ- ,,	green-grey
0.5 γ ,, -4- ,,	,,	0.1 γ nicotinamide	brown
0.1 γ nicotinic acid	violet		

Isonicotinic acid and isonicotinamide give no color reaction with the phloro-glucinol reagent solution.

No reaction with either of the reagents was observed with: α-picoline, 4-hydroxypyridine, 2-aminopyridine, 4-aminopyridine, 3,4-diaminopyridine, 2-amino-3-nitropyridine, pyridine-3,5-dicarboxylic acid, pyridoxine, α,α'-dipyridyl, quinoline, isoquinoline, isonicotinic acid hydrazide.

1 E. ASMUS and H. GARSCHAGEN, Z. Anal. Chem., 139 (1937) 81.
2 V. ANGER and S. OFRI, private communication.

59. Pyridine-aldehydes

Test with phenylhydrazine[1]

Addition of a phenylhydrazine salt (chloride, sulfate, acetate, *etc.*) to an aqueous solution containing any of the three isomeric pyridine-aldehydes results in an intense yellow color due to the formation of the respective phenylhydrazone:

$$C_5H_5NCHO + NH_2NHC_6H_5 \rightarrow C_5H_5NCH=N-NHC_6H_5 + H_2O$$

Limits of identification: 0.5 γ pyridine-(2, 3 or 4)-aldehyde (test on a spot plate).

1 F. FEIGL and L. BEN-DOR, *Talanta*, 10 (1963) 1111.

60. α-Carboxylic acids of aromatic heterocyclic bases

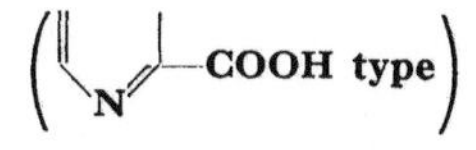

Test through formation of colored iron(II) salts[1]

Of the three isomeric pyridinemonocarboxylic acids, only the α-compound, picolinic acid (I), reacts in aqueous or alcoholic solution with ferrous ions to give a light yellow color, which is stable to acetic acid or dilute mineral acids.[2] The color reaction may be due to the production of the inner-complex cation (II):

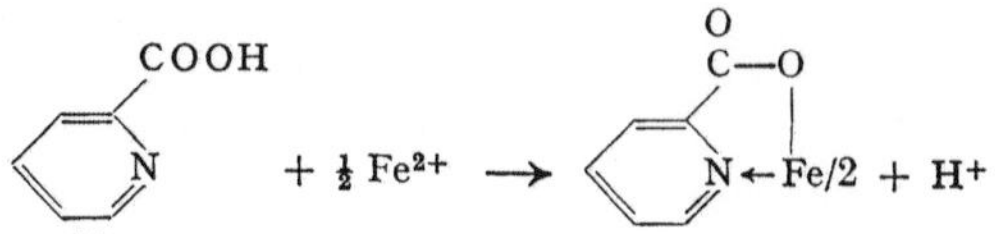

Analogous color reactions with ferrous ions are given by derivatives of mono- and polycarboxylic acids of pyridine, quinoline, pyridazine, pyrimidine, pyrazine, quinoxaline and benzothiazole, that have a COOH group adjacent to the cyclic nitrogen atom. It is noteworthy that the color of the complex salt deepens to violet when more carboxylic groups are introduced into the ring. Accordingly, α-carboxylic acids of aromatic heterocyclic bases can be detected through the color reaction with ferrous ions. The ferric ions that are almost always present in ferrous salts disturb the reaction, and consequently should be masked by adding alkali fluoride.

Procedure.   One drop of the test solution is mixed in a micro test tube with a drop of a solution of 1 g of $FeSO_4.(NH_4)_2SO_4.6 H_2O$ and 0.5 g of KF in 100 ml of 0.1 N acetic acid. A positive response is indicated by the appearance of a yellow-orange to red-violet color.

The test revealed:

5 γ picolinic acid	10 γ pyridine-2,3-dicarboxylic acid
7 γ 6-hydroxymethyl-picolinic acid	2 γ quinoxaline-2,3-dicarboxylic acid

The following acids also give color reactions with ferrous salt:

Yellow-orange:

4-methylpyridine-2-carboxylic-	4-methyl-5-ethylpyridine-2,3-dicarb-
4,6-dimethylpyridine-2-carboxylic-	oxylic-
quinoline-2-carboxylic-	quinoline-2,3-dicarboxylic-
4-methylquinoline-2-carboxylic-	pyridazine-3-carboxylic-
quinolinic acid 3-methyl ester	pyrazinecarboxylic-
pyridine-2,4-dicarboxylic-	5-methylpyrimidine-4-carboxylic-
,, -2,5- ,,	4,6-dimethylpyrimidine-2-carboxylic-
,, -2,6- ,,	2,3-dimethylpyrazine-5-carboxylic-
pyridine-2-carboxylic-5-acetic-	4-methylpyridine-2,3-dicarboxylic-
	6- ,, -2,4- ,,

Red-violet:

6-phenylpyridine-2-carboxylic	benzothiazole-2-carboxylic-
pyridine-2,3,5,-tricarboxylic-	pyridine-2,3,5,6-tetracarboxylic-
,, -2,3,6- ,,	,, -2,3,4,6- ,,
,, -2,4,5- ,,	pyridinepentacarboxylic-
,, -2,4,6- ,,	pyrazine-2,3-dicarboxylic-
4-methylpyridinetetracarboxylic-	,, -2,5- ,,
pyrimidine-4-carboxylic-	,, -2,6- ,,
5,6-dimethylpyrazine-2,3-dicarboxylic-	pyrazinetetracarboxylic-
pyrazinetricarboxylic-	pyridazinecarboxylic-

1 F. FEIGL and V. ANGER, unpublished studies.
2 H. WEIDEL, *Ber.*, 12 (1879) 1994.
3 Z. SKRAUP, *Monatsh.*, (1886) 212; H. LEY, C. SCHWARTE and O. MÜNNICH, *Ber.*, 57 (1924) 349; *cf.* also BEILSTEIN, *Handb. org. Chem.*, 4th Ed., Berlin, 1931.

61. 8-Hydroxyquinoline and derivatives (⟨structure⟩ type)

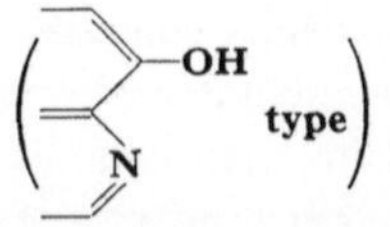

Test through fluorescence by chemical adsorption on magnesium hydroxide[1]

The group in question is present in 8-hydroxyquinoline (I) and its derivatives (II–VIII). As shown by the acidic phenolic OH group and the basic tertiary N atom in the quinoline ring system, 8-hydroxyquinoline ("oxine") and its derivatives are ampholytes, which can be brought into solution by either acids or bases. (The acidic character is strengthened in the halogen substituted derivatives as well as in the sulfonic acids.)

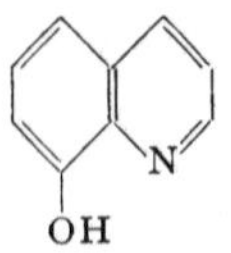

(I) 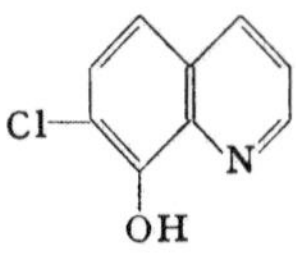   (II) 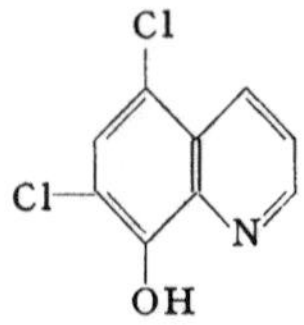   (III) 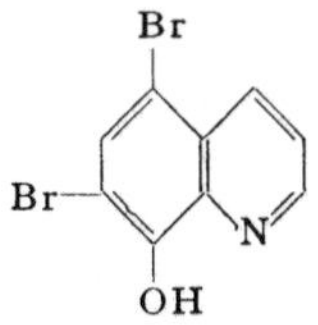   (IV) 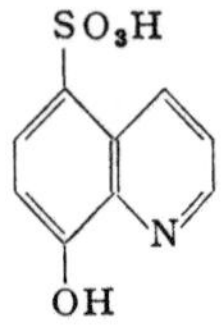   (V)

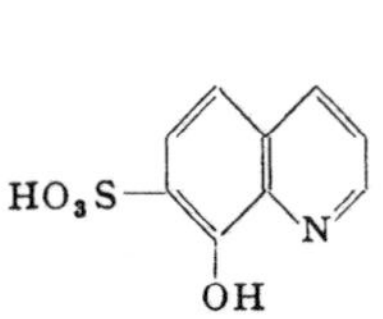

(VI) 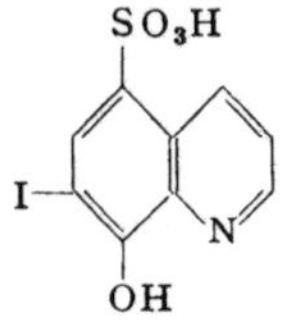   (VII) 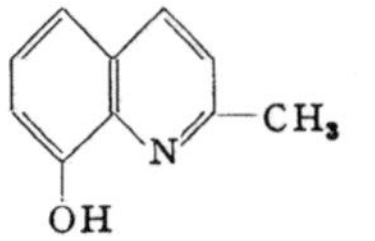   (VIII)

Alcoholic solutions of (I)–(IV) quantitatively precipitate numerous metal ions from neutral, ammoniacal-tartrated, and acetate-buffered solutions.[2] The precipitates consist of inner complex salts in which, as shown by (IX) and (X), the hydrogen of the phenol group is replaced by one equivalent of metal with coordinative bonding on the cyclic N atom:

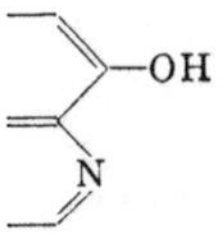

(IX) 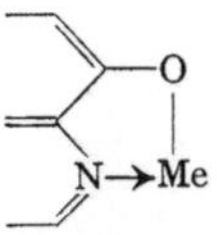   (X)

The yellow precipitates produced by (I)–(IV) with Al^{3+}, Zn^{2+}, and Mg^{2+} ions fluoresce intensely yellow-green or bluish white if the solids or their solutions in organic liquids (immiscible with water) are viewed in ultraviolet light. The derivatives (V)–(VIII) are not precipitants because of the solubilizing influence of the SO_3H group. 8-Hydroxyquinaldine (VIII) has no precipitating power toward Al^{3+} ions, presumably because of steric hindrance by the CH_3 group.[3] However, a binding of the reactive group (IX) to metal atoms as pictured in (X) is established not only in the production of stoichiometrically defined salts but also when oxine and its derivatives are adsorbed on the surface of water-insoluble oxides or hydroxides of the metals just mentioned.* Proof is provided by the fact that products with the same fluorescence color result in both cases, *i.e.*, when there is salt-formation as an independent phase and also when chemical adsorption occurs in which no new phase is produced. The latter is the case with (V)–(VIII), which do not enter into precipitation reactions with the parent metal ions of the oxides. The fluorescence resulting when there is salt-formation or chemical adsorption on contact with metal oxides can therefore be utilized as a general test to detect the group (IX) in oxine and its derivatives.

Procedure.   About 0.2 g of magnesium oxide is placed on a watch glass or in a depression of a spot plate. A drop of the test solution (in water, alcohol, ether, *etc.*) is added. The suspension is placed under a quartz lamp. Depending on the quantity of active material present, a light yellow or whitish blue fluorescence appears at once or in a short while. A blank with magnesium oxide and the particular solvent is advisable when small amounts of the hydroxyquinoline are suspected.

An aqueous suspension of freshly prepared $Mg(OH)_2$ may be used in place of MgO.

Limits of identification:

0.5 γ 8-hydroxyquinoline (I) (aqueous alcohol solution)
0.6 γ 7-chloro-8-hydroxyquinoline (II) (acetone)
0.8 γ 5,7-dichloro-8-hydroxyquinoline (III) (acetone)
1 γ 5,7-dibromo-8-hydroxyquinoline (IV) (acetone)
0.5 γ 8-hydroxyquinoline-5-sulfonic acid (V) (water)
0.5 γ 8-   ,,   -7-   ,,   (VI) (water)
1 γ 7-iodo-8-hydroxyquinoline-5-sulfonic acid (VII) (water)
0.4 γ 2-methyl-8-hydroxyquinoline (8-hydroxyquinaldine) (VIII) (alcohol)

The above procedure can also be applied to acidic solutions of 8-hydroxyquinoline or its derivatives; the MgO then acts as neutralizer as well as Mg^{2+}-donor and adsorbent.

It should be noted that 4,8-dihydroxyquinoline-2-carboxylic acid (xanthurenic acid), whose aqueous solution shows a yellow fluorescence, is not adsorbed by magnesium oxide to yield a fluorescing product.

---

* A schematic representation of the surface reaction for the chemical adsorption of oxine on alumina is discussed on p. 527.

1 F. FEIGL, *Mikrochem. Ver. Mikrochim. Acta*, 39 (1952) 404.
2 Comp. R. BERG, *Die analytische Verwendung des Oxychinolins (Oxin) und seiner Derivate*, 2nd. ed., Enke, Stuttgart,1938. See also F. J. WELCHER, *Organic Analytical Reagents*, Vol. I, Van Nostrand, New York, 1947, Chap. XIII.
3 L. L. MERRITT and I. K. WALKER, *Ind. Eng. Chem., Anal. Ed.*, 16 (1944) 387.

62. Urea and derivatives $\left(OC \begin{smallmatrix} \diagup NH- \\ \diagdown NH- \end{smallmatrix} type \right)$

Test by conversion to diphenylcarbohydrazide[1]

When urea is heated to 150–200° with an excess of phenylhydrazine, diphenylcarbohydrazide (diphenylcarbazide) is formed.[2]

$$OC \begin{smallmatrix} \diagup NH_2 \\ \diagdown NH_2 \end{smallmatrix} + 2\ NH_2NHC_6H_5 \longrightarrow OC \begin{smallmatrix} \diagup NH-NHC_6H_5 \\ \diagdown NH-NHC_6H_5 \end{smallmatrix} + 2\ NH_3 \qquad (1)$$

Since the reaction temperature is above the melting point (131°) of urea, some biuret is produced but it too forms diphenylcarbazide:

$$NH(CONH_2)_2 + 4\ NH_2NHC_6H_5 \longrightarrow 2\ OC \begin{smallmatrix} \diagup NH-NHC_6H_5 \\ \diagdown NH-NHC_6H_5 \end{smallmatrix} + 3\ NH_3 \qquad (2)$$

Diphenylcarbazide on reaction with numerous metal ions yields colored inner-complex salts, many of them readily soluble in organic liquids.[3] In particular, an ammoniacal nickel solution is recommended; it gives a violet inner-complex salt, soluble in chloroform.

Procedure.   One drop of the aqueous test solution is evaporated to dryness in a micro test tube. A drop of phenylhydrazine is added and the mixture is kept at 195° for 5 min. After cooling, five drops of 1 : 1 ammonia and of 10% nickel sulfate solution are added, and the mixture is shaken with 10 drops of chloroform. A red-violet color indicates the presence of urea.

Limit of identification: 10 γ urea.

The mono- and disubstituted derivatives of urea also form diphenylcarbazide when heated with phenylhydrazine. The pertinent reaction schemes are analogous to (1), amines being split out in the case of methylurea, acetylmethylurea, *asym*-diphenylurea, *m*-tolylurea, *sym*-di-*m*-tolylurea, allylurea, *tert*-amylurea, *tert*-butylurea.

Idn. Limits of 100–300 γ were obtained.

Likewise, urethans, *i.e.* esters of carbamic acid or esters of carbonic acid, produce diphenylcarbazide when treated with phenylhydrazine. Again, a scheme analogous to (1) applies, in that ammonia or amine along with ethanol is split off. *Idn. Limits* of 100–300 γ were obtained with ethylurethan, phenylurethan, *n*-butyl carbamate.

It seems that N,N'-dialkyl- and N,N'-diaryl-carbamates, as well as cyclic derivatives of urea, are resistant to phenylhydrazine. This lack of reactivity was shown by N,N'-diphenyl carbamate, N,N'-di-*n*-butyl carbamate, 6,8-dichlorobenzoyleneurea.

When thiourea and its derivatives are heated with phenylhydrazine, the sulfur analog of diphenylcarbazide is formed, namely diphenylthiocarbazide which likewise forms a violet inner-complex nickel salt. However, the detection of thiourea through the condensation with phenylhydrazine is much less sensitive (*Idn. Limit*: 800 γ).

Thiourea and its derivatives can be detected by heating the solid sample to about 200°. Hydrogen sulfide is evolved and is easily detected by its action on lead acetate paper (*Idn. Limit:* 1 γ thiourea).[4]

1 F. FEIGL, V. GENTIL and D. GOLDSTEIN, unpublished studies.
2 S. SKINNER and S. RUHEMANN, *J. Chem. Soc.*, (1888) 53; see also K. H. SLOTTA and K. R. JACOBY, *Z. Anal. Chem.*, 77 (1929) 344.
3 Compare F. J. WELCHER, *Organic Analytical Reagents*, Vol. III, Van Nostrand, New York, 1947, p. 430.
4 H. E. FEIGL, unpublished studies.

63. Thiourea and derivatives $\left(\text{SC} \begin{smallmatrix} \nearrow \text{NH-} \\ \searrow \text{NH-} \end{smallmatrix} \text{type} \right)$

Test by conversion into rhodanine[1]

A test has been described on p. 463 for monochloroacetic acid based on the reaction with ammonium thiocyanate:

$$\text{CH}_2\text{ClCOOH} + 2\ \text{NH}_4\text{CNS} + \text{H}_2\text{O} \longrightarrow \underset{\text{S}}{\text{SC}} \overset{\text{HN}-\text{CO}}{\underset{\diagup}{\mid \qquad \mid}} \text{CH}_2 + \text{NH}_4\text{Cl} + 2\ \text{NH}_3 + \text{CO}_2 \quad (1)$$

The resulting rhodanine (rhodanic acid) gives a blue-violet color with sodium 1,2-naphthoquinone-4-sulfonate (Ehrlich–Herter reagent) in alkaline solution (comp. p. 153).

If thiourea, the isomer of ammonium thiocyanate, is warmed with mono-chloroacetic acid, rhodanic acid is not produced but instead the reaction:

$$\text{CH}_2\text{ClCOOH} + \text{SC(NH}_2)_2 \rightarrow \text{SCNCH}_2\text{COOH} + \text{NH}_4\text{Cl} \qquad (2)$$

yields isothiocyanoacetic acid[2] which in alkaline medium gives an orange-red color with the Ehrlich–Herter reagent. However, rhodanic acid can be formed likewise in the thiourea–monochloroacetic acid system if the heating is done at 180°. At this temperature, the tautomer equilibrium:

$$\text{SC(NH}_2)_2 \rightleftharpoons \text{NH}_4\text{CNS}$$

lies so far toward the right that reaction (*1*) occurs rather than (*2*). For the same reason, N-alkylated (arylated) thioureas, which are isomeric with the corresponding substituted ammonium thiocyanates, give rhodanic acid when fused with monochloroacetic acid.

The procedure given here is reliable provided compounds which give colored condensation products with the reagent because of their active NH_2 or CH_2 groups are absent.

Procedure. A small amount of the solid or the evaporation residue from its solution in water, alcohol, ether, *etc.* is mixed in a micro test tube with a drop of a 10% alcoholic solution of monochloroacetic acid and brought to dryness if need be. The test tube is then kept for 1–2 min. in a glycerol bath at 180°. (If considerable amounts of thiourea are present, the melt is yellow.) The cooled mass is treated with 1 drop of 0.5% solution of sodium 1,2-naphthoquinone-4-sulfonate and made basic with 1 drop of 2 N NaOH. A more or less intense blue-violet color indicates a positive response.

Limit of identification: 2.5 γ thiourea.

1 Unpublished studies, with V. Gentil.
2 J. Volhard, *J. Prakt. Chem.*, 29 (1874) 9.

64. Quinonechloroimines $\left(O=\!\!\left\langle\begin{array}{c}\end{array}\right\rangle\!\!=\!N\!-\!Cl\ \text{type}\right)$

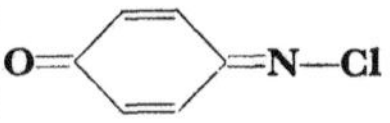

(1) Test with diphenylamine–hydrochloric acid[1]

Quinonechloroimines yield an intensely blue coloration with diphenylamine in the presence of hydrochloric acid. Probably meriquinoid benzidine blue is formed by oxidation (see p. 301). A similar reaction is given by *p*-nitrosophenol (comp. p. 362).

Procedure. A drop of the alcoholic test solution is treated in a depression of a spot plate with 5 drops of a saturated solution of diphenylamine in concentrated hydrochloric acid. A positive response is indicated by the appearance of a blue color.

Limits of identification:

　　0.1 γ 2,6-dichloroquinone-4-chloroimine
　　0.5 γ 2,6-dibromoquinone-4-chloroimine

Chloramine T does not respond.

1 V. Anger and S. Ofri, *Mikrochim. Acta*, (1964) 112.

(2) Test with phenol and ammonia[1]

Quinonechloroimines yield blue indophenols with phenols and ammonia (see p. 185). This reaction serves as the basis for a test for quinonechloroimines.

Procedure. A drop of the test solution is placed on filter paper and fumed over phenol and then over ammonia. A positive response is shown by the development of a blue fleck.

Limit of identification: 0.5 γ 2,6-di(bromo, chloro)quinone-4-chloroimine.

1 Unpublished studies, with L. Ben-Dor.

65. Aliphatic 1,2-dioximes $\left(\begin{array}{c}\text{HON} \quad \text{NOH} \\ || \quad || \\ \text{R—C—C—R}\end{array}\text{ type}\right)$

Test through pyrolytic reduction[1]

If aliphatic dioximes are mixed with zinc and dry-heated, pyrrole results. It can be detected in the gas phase by the color reaction with p-dimethylaminobenzaldehyde (see p. 381).

The following partial reactions underlie the formation of pyrrole:

(1) During the pyrolysis of aliphatic dioximes, water is produced as quasi superheated steam, which reacts with zinc dust to give hydrogen.

(2) The nascent hydrogen produced as per (1) reduces the aliphatic dioximes to the corresponding diamines:

$$\begin{array}{c}\text{R—C=NOH} \\ | \\ \text{R—C=NOH}\end{array} + 8\,\text{H}° \rightarrow \begin{array}{c}\text{R—CH—NH}_2 \\ | \\ \text{R—CH—NH}_2\end{array} + 2\,\text{H}_2\text{O}$$

(3) The diamines formed as per (2) are substituted ethylenediamines which, when subjected to dry heating, exhibit the same behavior as ethylenediamine, i.e. the initial product is a piperazine, which pyrolyzes to give pyrrole (see detection of ethylenediaminetetraacetate, p. 507).

Through realization of (1) to (3), there results a reliable means of distinguishing aliphatic from aromatic dioximes, since the latter do not form volatile pyrrole derivatives under the conditions of the test.

Procedure. A small amount of the solid or the residue from 1 drop of its solution is mixed in a micro test tube with about 1 cg of zinc dust. The tube is fastened in an asbestos support and its mouth is covered with a disk of filter paper impregnated with a 5% solution of p-dimethylaminobenzaldehyde in 20% trichloroacetic acid–benzene solution. The bottom of the tube is heated in a free flame for 1–2 min. A violet stain on the agent paper indicates a positive response.

The procedure revealed:

5 γ methylethylglyoxime	25 γ 1,2-cyclohexanedione dioxime
5 γ dimethylglyoxime	25 γ 1,2-cycloheptanedione dioxime

It is remarkable that diacetyl monoxime behaves like dimethylglyoxime with respect to this procedure.

1 Unpublished studies, with V. GENTIL.

66. o-Nitrosophenols $\left(\begin{array}{c}\text{—OH} \\ \text{—NO}\end{array}\text{ type}\right)$

Test by formation of cobaltic chelate compounds[1]

Dissolved o-nitrosophenols (I) give an equilibrium with their tautomeric o-quinone oxime forms (II):

$$\text{(I)} \quad \text{(II)} \quad \text{(III)}$$

The quinone oxime group combines with a number of metal ions to produce water-insoluble inner-complex salts or water-soluble complex compounds yielding inner-complex metal-containing anions, provided the molecule also contains hydrophilic groups such as CO_2H, SO_3H, *etc.* The most familiar of such salts are those of trivalent cobalt, which contain the chelate ring (III).

The water-insoluble inner-complex salts of trivalent cobalt are brown-red. They dissolve in chloroform, carbon tetrachloride, *etc.* to give brown-red solutions, a fact first noted with the classic instance of α-nitroso-β-naphthol.[2] The water-soluble salts containing inner-complex bound trivalent cobalt have the same color; they cannot be taken up in chloroform. Organic compounds containing the groups (I) and (II) are widely used as reagents in inorganic qualitative and quantitative analysis.[3]

The ability to form chelate compounds of trivalent cobalt can be utilized for the selective detection of *o*-nitrosophenols. Information concerning the presence or absence of hydrophilic groups in the nitrosophenol molecule is also furnished by the behavior of the cobalt salt formed toward organic solvents that are not miscible with water.

Procedure.   A drop of the test solution is placed on filter paper along with a drop of 0.1% cobalt nitrate solution and the paper is then held over ammonia water. A positive response is signalled by a brown-red to yellow stain, which is resistant to 2 N sulfuric acid.

The test revealed:

1 γ α-nitroso-β-naphthol	5 γ nitroso-R-salt
1 γ β- ,, -α- ,,	0.25 γ dinitrosoresorcinol

1 F. FEIGL, *Anal. Chem.*, 27 (1955) 1316.
2 M. ILLINSKY and G. V. KNORRE, *Ber.*, 18 (1885) 699.
3 Comp. F. J. WELCHER, *Organic Analytical Reagents*, Vol. III, Van Nostrand, New York, 1947, p. 299 ff.

67. *p*-Nitrosophenols $\left(\text{ON}\!-\!\langle\ \rangle\!-\!\text{OH type} \right)$

Test with thymol–sulfuric acid[1]

p-Nitrosophenols and -naphthols condense in concentrated sulfuric acid with phenols to give indophenols. The latter are blue in concentrated sulfuric acid but the color changes to red if such solutions are diluted with water. Alkalization with ammonia again yields a blue form of the indophenol (see p. 186). It is better to use thymol than phenol as the phenol reactant, since the former is more stable and is less affected by concentrated sulfuric acid and the oxygen of the air.

p-Nitrosonaphthol differs from *p*-nitrosophenol in that addition of ammonia to its red solution gives a yellow and not a blue color.

Procedure. A small amount of the solid material or the evaporation residue from a drop of the test solution is mixed in a depression of a spot plate with 2 drops of a 0.1% solution of thymol in concentrated sulfuric acid; a blue-violet to green color appears if the response is positive. The color changes to red on the addition of 2 drops of water. Addition of concentrated ammonia changes the red to blue in the case of *p*-nitrosophenol and its homologs.

Limits of identification:

0.2 γ *p*-nitrosophenol (blue-violet to red to blue)

1   γ *p*-nitrosonaphthol (green to red-brown to yellow).

1 V. ANGER and S. OFRI, *Mikrochim. Acta*, (1964) 111.

68. *o*-Nitrosonaphthol derivatives $\left(\begin{array}{c} \text{NO} \\ \text{OH} \quad \text{type} \end{array} \right)$

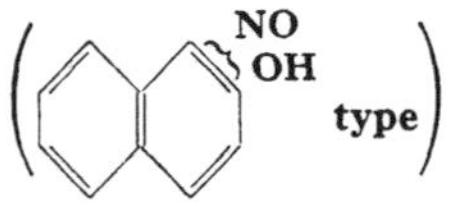

Test with β-naphthol[1]

o-Nitrosonaphthols react with β-naphthol and nitric acid to form red-violet products. See p. 421 and p. 557 regarding the mechanism. *p*-Nitrosonaphthol and nitrosophenols do not give this reaction.

Procedure. A drop of the alcoholic test solution is united in a micro test tube with a drop of a 1% solution of β-naphthol and the mixture is warmed briefly in a water bath. Then 5–8 drops of nitric acid (s.g. 1.4) are introduced. A red color appears if *o*-nitrosonaphthols are present.

Limits of identification:

0.5 γ α-nitroso-β-naphthol	0.5 γ nitroso-R-salt
0.5 γ β- ,, -α- ,,	

1 V. ANGER and S. OFRI, *Mikrochim. Acta*, (1964) 111.

69. *p*-Nitrosoarylamines $\left(>\text{N}-\left<\text{=}\right>-\text{NO type} \right)$

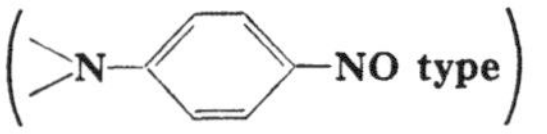

(1) Test with paladium chloride

A selective and sensitive test for palladium is based on the formation of a colored precipitate when an alcoholic solution of a *p*-nitroso aromatic amine is added to weakly acidic solutions of palladium salts[1]. The product is an addition compound of the two components. The possible coordination centers[2] are indicated in the following formulas:

$$\text{>N-}\langle\;\rangle\text{-N=O} \qquad \text{>N-}\langle\;\rangle\text{-N=O} \qquad \text{>N-}\langle\;\rangle\text{-N=O}$$

$$\frac{PdCl_2}{2} \qquad\qquad \frac{PdCl_2}{2} \qquad\qquad \frac{PdCl_2}{2}$$

The formation of the colored addition products can also be used as the basis for a sensitive and highly selective test for p-nitroso aromatic amines[3].

The following facts should be kept in mind when carrying out this test. Many amines and aminophenols form addition compounds with palladium chloride. Though nearly all of them are yellow and cannot be confused with the red nitrosopalladium compounds, they may consume all of the reagent and hence small amounts of nitroso amines may escape detection.

1,2- and 2,1-Nitrosonaphthols and their derivatives react with $PdCl_2$, forming insoluble, brown-violet inner-complex palladium salts. Such phenols can be removed if the sample is treated with aqueous caustic alkali and extracted with ether. The nitrosoamines are taken up by the ether, whereas the phenols remain in the aqueous layer as alkali phenolates.

Procedure.   One drop of the test solution or a small amount of the solid and one drop of the palladium chloride solution are mixed in a depression of a spot plate. The immediate production of a colored precipitate or a color is taken as a positive response.

Reagent: 0.1 g $PdCl_2$ and 0.2 g NaCl are dissolved in several ml of water and diluted to 100 ml.

This procedure revealed 0.005 γ of:

p-nitrosodimethylaniline	bright red	p-nitrosoaniline	brown
p-nitrosodiethylaniline	,,		
p-nitrosodiphenylamine	,,		

1 J. H. YOE and L. G. OVERHOLSER, *J. Am. Chem. Soc.*, 61 (1939) 2058; 63 (1941) 3224.
2 F. FEIGL, *Chemistry of Specific, Selective and Sensitive Reactions*, Academic Press, New York, 1949, p. 326.
3 Unpublished studies, with J. E. R. MARINS.

(2) Test with ascorbic acid and p-dimethylaminobenzaldehyde[1]

p-Nitroso-aromatic amines are reduced by ascorbic acid to the corresponding p-amino compounds, which condense with p-dimethylaminobenzaldehyde to a Schiff base or, more correctly, its quinoidal cation:

$$RR_1N-\langle\;\rangle\text{-NO + ascorbic acid} \rightarrow RR_1N-\langle\;\rangle\text{-NH}_2 + \text{dehydroascorbic acid}$$

$$(1)$$

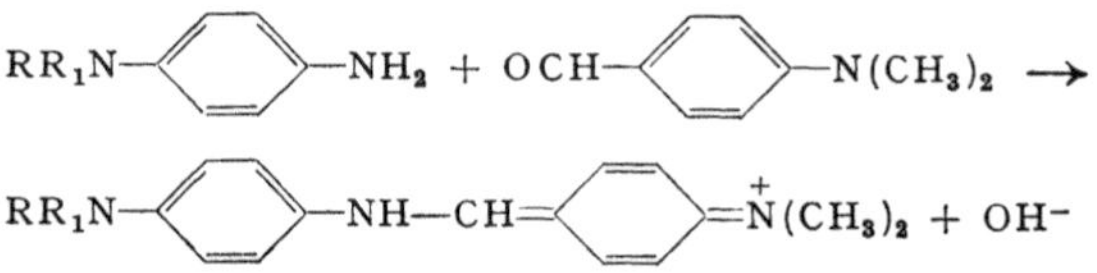

$$RR_1N\langle\!\!\!-\!\!\!\rangle\!-\!NH_2 + OCH\!-\!\langle\!\!\!-\!\!\!\rangle\!-\!N(CH_3)_2 \longrightarrow$$

$$(2)$$

$$RR_1N\langle\!\!\!-\!\!\!\rangle\!-\!NH\!-\!CH\!=\!\langle\!\!\!-\!\!\!\rangle\!=\!\overset{+}{N}(CH_3)_2 + OH^-$$

The realization of (*1*) and (*2*), which do not occur with other C-nitroso compounds, may be made the basis of a test for N-substituted *p*-nitroso aromatic amines, provided compounds which condense with *p*-dimethyl-aminobenzaldehyde are absent. Primary aromatic amines and compounds with reactive CH_2 groups are in this category. Consequently, a preliminary study must be made of the behavior of the sample toward an alcoholic solution of the aldehyde. If no color results, the test is applicable.

Procedure. A drop of the alcoholic test solution is treated in a micro test tube with a drop of a 1% solution of *p*-dimethylaminobenzaldehyde in concentrated acetic acid. A few mg of ascorbic acid are added and the mixture warmed in a boiling water bath. Depending on the amount of *p*-nitroso aromatic amine present, a violet or pink color appears at once or within 1–2 min.

The following were detected:

0.1 γ *p*-nitrosodimethylaniline   0.3 γ *p*-nitrosodiphenylamine
0.1 γ *p*-nitrosodiethylaniline

1 F. FEIGL and D. GOLDSTEIN, *Microchem. J.*, 1 (1957) 176.

70. *m*-Dinitro compounds $\left(\begin{array}{c}\text{NO}_2\\ \text{type}\\ \text{NO}_2\end{array}\right)$

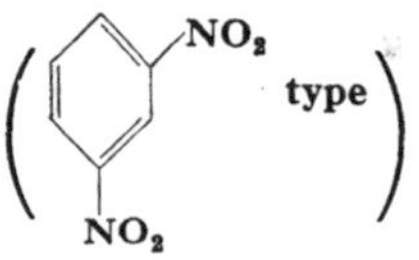

Test by reaction with potassium cyanide[1]

If *m*-dinitro compounds are warmed with a solution of alkali cyanide, a red-brown to violet color or precipitate is formed. The color is permanent in the presence of dilute acids. This behavior is in contrast to that of nitrophenols, whose color change is reversed when acids are added. The constitution of these products of the cyanide reaction is only partly known. Presumably they are substituted phenylhydroxylamines.[2,3] For example, 2,4-dinitrophenol reacts with potassium cyanide to produce the potassium salt of 4-nitro-2-hydroxylamino-3-cyanophenol (meta-purpuric acid):

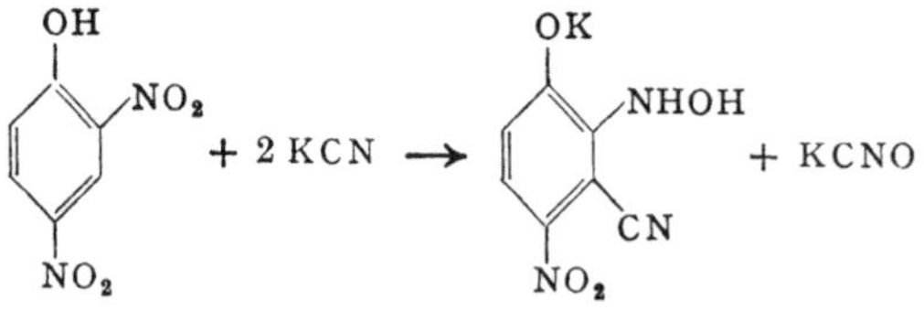

$$\begin{array}{ccc}
\text{OH} & & \text{OK}\\
\langle\text{NO}_2\rangle + 2\,KCN \longrightarrow & \langle\text{NHOH, CN}\rangle + KCNO\\
\text{NO}_2 & & \text{NO}_2
\end{array}$$

1,8-Dinitronaphthalene differs from the other *peri*-naphthalene derivatives and behaves like a *m*-dinitronaphthalene with potassium cyanide. Mononitrobenzene and (*o,p*)-dinitrobenzene derivatives and also 1,5-dinitronaphthalene do not react; therefore *m*-dinitro compounds may be detected even in the presence of other aromatic nitro compounds.

Procedure.   A drop of a solution or a few grains of the solid is mixed with a drop of 10% potassium cyanide solution in a micro crucible and heated gently over a micro burner. A violet or red color which remains unchanged even on the addition of a few drops of 2 *N* hydrochloric acid is a positive response.

Limits of identification:

2 γ picryl chloride	red-brown	10 γ dinitrobenzene	violet
2 γ picric acid	,,	1 γ chlorodinitro-	
2 γ picramine	,,	benzene	red
2 γ dinitrochlorobenzoic acid	,,	1 γ dinitrophenol	,,
1 γ dinitroaminobenzoic acid	,,	2 γ dinitronaphthol	dark brown

1 V. ANGER, *Mikrochim. Acta*, 2 (1937) 6.
2 L. PFAUNDLER and A. OPPENHEIM, *Z. Chem.*, [2] 1 (1865) 470.
3 W. BORSCHE and E. BÖCKER, *Ber.*, 37 (1904) 1844.

71. Benzaldehydesulfonic acids

Detection through tests for the aldehyde and sulfonic groups[1]

The detection of benzaldehydesulfonic acids and their substituted derivatives can be based on the positive response to the tests for aromatic aldehydes and aromatic sulfonic acids described in Chap. 3. The former involves the condensation with thiobarbituric acid to yield colored products; the latter employs the formation of nickel sulfide through the action of Raney nickel alloy in caustic alkaline surroundings. The respective procedures are outlined on p. 205 and 234.

Limits of identification:

5 γ methoxybenzaldehydesulfonic acid through the aldehyde test
10 γ   ,,   ,, through the test for aromatic sulfonic
                                     acids

1 Unpublished studies.

72. Aminophenols

Detection through tests for the amino and phenol components[1]

The three isomeric aminophenols are both primary amines and phenols that have a free position *ortho* to the phenol group. Reliable methods for the detection of these functional groups are given in Chap. 3. The tests are based on color reactions of the NH_2 group with sodium 1,2-naphthoquinone-4-sulfonate or with p-dimethylaminobenzaldehyde, and on the formation of cobalt(III) chelate compounds of nitrosophenol through the reaction with sodium cobaltinitrite. Accordingly, the presence of aminophenols can be established by positive responses to these tests used in sequence.

Procedure. As described on p. 243 and p. 184.

1.5 γ (o-, m- or p-) aminophenol was detected by the color reaction with naphthoquinonesulfonate.

5 γ o-, 0.5 γ m- and 10 γ p-aminophenol were detected by the reaction with sodium cobaltinitrite.

1 Unpublished studies.

73. Phenylhydrazinesulfonic acids

Detection through tests for the arylhydrazine and sulfonic groups[1]

The presence of compounds of this category can be established by the positive responses to the tests for the arylhydrazine component and for the aromatically bound sulfonic group. The conversion into diazonium salts through oxidation with selenious acid is characteristic for the former (see p. 272). The procedure described on p. 234 can be used for the detection of the aromatically bound SO_3H group. It is based on the formation of nickel sulfide by the action of Raney nickel alloy and caustic alkali.

Limit of identification:
0.06 γ phenylhydrazinesulfonic acid (through oxidation with SeO_2)
10   γ ,,   ,, (through the formation of NiS)

1 Unpublished studies.

74. Nitro- and aminobenzaldehydes

Detection through tests for CHO, and NO_2 or NH_2 groups[1]

Compounds of these categories give a positive response to tests for aromatic aldehydes through condensation with thiobarbituric acid (see p. 205). If tests for NO_2 or NH_2 groups are also conducted and likewise receive positive responses, it is very likely that nitro- or aminobenzaldehydes are present. The test for NO_2 groups is based on the color reaction on fusion with diphenylamine or tetrabase (see p. 295). The condensation reaction with sodium 1,2-naphthoquinone-4-sulfonate (violet color) will serve for the detection of aromatically bound NH_2 groups. For this procedure see p. 153.

The following were revealed:

5 γ nitrobenzaldehyde	through the aldehyde reaction
1 γ ,,	through the color reaction for NO_2 groups
5 γ aminobenzaldehyde	through the aldehyde reaction
2 γ ,,	through the color reaction for NH_2 groups

1 Unpublished studies.

Chapter 5

Identification of Individual Organic Compounds

General remarks

The detection of functional groups and of characteristic compound types, the topics of the preceding chapters, was formerly known as the *structural-analytical* method of organic analysis, an exceedingly apt designation.[1] By its side was placed the *ultimate-analytical* method, *i.e.*, the qualitative and quantitative determination of the elements composing the compound. In addition there is the final identification of individual organic compounds by *molecular-analytical* methods. The latter are based on the measurement of physical properties, which are related to the architecture and size of the molecule of organic compounds. Included are the determination of the melting- and boiling points and density of compounds, and also optical examinations. The sample itself serves for the melting- and boiling point determinations, or it may be mixed with known materials to obtain characteristic points such as eutectic temperatures. This latter method has recently been used to characterize organic materials and mixtures in micro quantities, an advance whose usefulness is steadily becoming more apparent.[2] Solutions in various solvents or melts serve for determinations of the molecular weight through measurement of the elevation of the boiling point or lowering of the freezing point. The determination can be carried out on derivatives of the compound in question since in some cases they offer more favorable properties. The optical methods include the measurement of the refractive index, optical activity, ultraviolet and infrared absorption, Raman spectrum, optical characteristics of the crystals, *etc.*

Physical laboratory methods of identification are very reliable. Their employment, however, assumes, as a rule, the previous isolation of organic compounds in a state of high purity, an operation that may be costly with regard to time and material. Consequently, the importance of chemical identifications, which succeed rapidly and consume but little amounts of the mixtures to which they may often be applied directly, should not be underestimated. Rather they should be held in high esteem, and especially

because they have the added advantage of not requiring special and sometimes very expensive apparatus. Comparatively few cases have been known
hitherto in which it is possible, through a single characteristic reaction, to
make a direct specific test for individual organic compounds. The chief
reason is that the majority of organic compounds with typical groups have
homologous relatives, and when the particular group can be identified by a
suitable reaction, the same type of reaction yielding similar reaction products
may be given by the homologous compounds as was shown in Chap. 3. Consequently, at first sight, there seems to be little prospect of success for searches
directed toward the discovery of specific or selective tests for individual
organic compounds. This situation is responsible for the fact that for a long
time the detection of functional groups often served particularly to disclose
the possibility of arriving by preparative methods at derivatives which, after
isolation, can be characterized by physical methods and thus allow reliable
conclusions to be drawn regarding the starting material. However, recent
findings show that it is possible to develop tests for individual compounds
through appropriate modification and combination of well established group
reactions which can be accomplished quickly. Efforts along these lines,
together with studies to discover new analytically useful reactions of organic
compounds, are aided by experiences and considerations which in part are
often overlooked and consequently not used.

First of all, it should be noted that, in the great majority of cases where
recourse is taken to qualitative organic analysis, the analyst is not faced with
an inextricable artificial mixture of compounds. As a rule, the available
information about the origin, method of preparation, action, and use of the
material in question can provide valuable clues as to the lines along which
the tests should be made. In fact, the analytical goal is often well defined by
the question as to whether particular compounds are present or absent. In
such instances, tests for functional groups if need be in combination with
preliminary tests (both conducted within the technique of spot test analysis),
may be entirely adequate, and the intensity of the response to a characteristic test may even permit conclusions as to how much of a particular organic
compound is present.

Another point, often given too little consideration, is that the reactivity of
certain groups in organic compounds is sometimes strongly affected by the
remainder of the molecule or by the groups it contains. This influence may
show itself through widely differing reaction rates, sometimes through complete loss of reactivity, and through alterations in the solubility and in the
basic or acidic character, as well as in the color or fluorescence of the reaction
products. Obviously such peculiarities in the response to tests for functional
groups can sometimes serve for the detection or identification and differentiation of individual compounds. In the search for specific and selective

organic reagents for inorganic analysis, special consideration has been given to the activity of particular salt-forming groups and to the influence exerted on this action by the remainder of the molecule, as well as by the reaction milieu. The findings[3] in this area of study will also be useful to the chemical methods of organic analysis dealing with the detection of functional groups and individual compounds. This point is discussed in Chap. 1 and is confirmed by numerous examples to be found throughout the text.

Of greatest importance for the sure detection of individual compounds in the presence of compounds that react analogously is the possibility of removing the former from a common reaction theater. This possibility is always provided when only the lowest member of an homologous series is volatile or sublimable at room temperature or on slight warming, or if cleavage through heat or chemical action yields a volatile or sublimable product which in turn can be identified in the vapor phase by a suitable test. In such cases, a group reaction, which is only selective in solution, can become completely specific for an individual compound if the test is carried out on the vapor phase.

The most extensive use of a separation of reaction theaters occurs in chromatography. The solution components are collected in definite, separate zones through adsorption on a powdered adsorber (column chromatography) or on porous paper (paper chromatography). Specific tests for minute amounts of individual compounds can then be accomplished in these zones, either directly or after elution, by means of tests which must be sensitive but not necessarily specific or selective.* Furthermore, the products of selective group reactions can be adsorptively separated from aqueous solution or from solution in organic solvents, and in this way individual compounds can be detected in certain zones by suitable tests.

The following sections contain descriptions of spot tests that in part are characteristic for individual compounds and in part respond to those representatives of a particular class of compounds that have practical importance and are most likely to occur in the samples presented for analysis. In addition, examples are given of the identification of compounds containing several functional groups that can be detected by selective tests. In such cases the application and positive response to tests for several different functional groups often lead to unequivocal identifications. When individual compounds are to be identified, it is recommended that note be taken of the response of the sample to a variety of tests, if available. Greater certainty in judging the nature of the sample is secured by such multiplicity of findings

---

* The great importance of spot tests for paper chromatography was not adequately appreciated for a long time. NEU[4] has dealt with this topic properly. An increased use of spot tests may be expected in gas chromatography[5] and especially in thin-layer chromatography developed by STAHL[6].

and the truth of the verdict will also be assisted by conducting "pattern trials" in which the typical reaction picture as shown by the pure compound in various dilutions is available for comparison.

There are not strict boundaries between tests for functional groups, tests for characteristic types of compounds and tests for establishing the presence of individual compounds. In every case, the essential feature is the occurrence of definite chemical reactions involving certain atom groups. The distinction employed in this text is justified in that the tests used for identifying individual compounds are often much more selective than the tests given in Chaps. 3 and 4 for functional groups and for structures and types of compounds which in their turn are more selective than the preliminary tests outlined in Chap. 2.

1 Concerning this and the following designations see H. STAUDINGER, *Anleitung zur organischen qualitativen Analyse*, 2nd ed., Springer, Berlin, 1929.
2 L. KOFLER and A. KOFLER, *Thermo-Mikro-Methoden zur Kennzeichnung organischer Stoffe und Stoffgemische*, Verlag Chemie, Weinheim, 1954.
3 See F. FEIGL, *Chemistry of Specific, Selective and Sensitive Reactions*, Academic Press, New York, 1949, Chap. VI.
4 R. NEU, *Mikrochim. Acta*, (1957) 196; (1958) 267.
5 Comp. E. BAYER, *Gas Chromatography*, Elsevier, Amsterdam, 1960.
6 E. STAHL, *Pharmazie*, 11 (1956) 633; *Chem. Ztg.*, 82 (1958) 323; comp. also G. MACHATA, *Mikrochim. Acta*, (1960) 79.

1. Acetylene

(1) Test by formation of cuprous acetylide [1]

Acetylene, either gaseous or dissolved in water or organic solvents, reacts with colorless ammoniacal solutions of cuprous salts to produce cuprous acetylide (carbide or more correctly cuproacetylenic carbide):

$$Cu_2^{2+} + C_2H_2 + 2\ NH_3 \rightarrow C_2Cu_2 + 2\ NH_4^+$$

This salt, a monohydrate,[2] appears as an amorphous red-brown to red-violet precipitate, depending on the degree of dispersion. It is soluble in dilute mineral acids and in solutions of compounds which yield stable cuprous complexes with regeneration of acetylene.[3] An analogous behavior toward cuprous salts is shown by homologs of acetylene (acetylenic hydrocarbons), which in conformity with the general formula $C_nH_{2n+1}C\equiv CH$ contain a terminal triply bound carbon linked to an acidic hydrogen atom, or, more correctly, the H atom becomes acidic through the triply bound C atom.

The detection of acetylene through precipitation of Cu_2C_2 from ammoniacal cuprous solutions is impaired by S^{2-} ions because of the formation of black cuprous sulfide. Mercapto compounds similarly precipitate yellow cuprous salts of the particular mercaptans (sometimes mixed with copper sulfide). The hydrides of phosphorus and arsenic, which always are present in commercial acetylene, and which impart to it the unpleasant odor usually associated with this gas, do not interfere with the test.

Procedure. A drop of the ammoniacal cuprous solution is treated in a depression of a spot plate with a drop of the acetylene solution. According to the quantity of acetylene present, a red-brown precipitate or a brown-violet color appears in the practically colorless solution.

Limit of identification: 1 γ acetylene.

If filter paper moistened with the reagent is used, 0.3 γ C$_2$H$_2$ can be detected.

The test can also be made by allowing gaseous acetylene to react with a hanging drop of the reagent solution in a gas-evolution apparatus.

Reagents: 1) Solution of 1.5 g cupric chloride and 3 g ammonium chloride in 20 ml concentrated ammonia, diluted to 50 ml with water.

2) 5 g hydroxylamine hydrochloride in 50 ml water.

1 ml of (1) and 2 ml of (2) are mixed before use.

The *identification limit* is about 5 γ acetylene.

The sensitivity of the test was established by means of a saturated water solution prepared from purified acetylene. In the dilutions, the solubility (at 18°) was taken as 0.118 g per 100 ml water.[4]

1 L. Ilosvay von Nagy Ilosva, *Ber.*, 32 (1899) 2697; see also E. Pietsch and A. Kotowski, *Z. Angew. Chem.*, 44 (1931) 309.

2 J. Schreiber and H. Reckleben, *Ber.*, 41 (1908) 3816; 44 (1911) 220.

3 Comp. J. A. Nieuwland and R. R. Vogt, *Chemistry of Acetylene*, New York, 1925.

4 *Handbook of Chemistry and Physics*, Cleveland, 1951.

(2) Test by protective layer effect on silver chromate[1]

When acetylene (gaseous, or dissolved in water, acetone, *etc.*) reacts with a nitric acid or ammoniacal solution of silver chromate, an orange or yellow flocculent precipitate is formed. The product* is an addition compound of silver acetylide and silver chromate:

$$4\ Ag^+ + CrO_4^{2-} + C_2H_2 \rightarrow Ag_2C_2 \cdot Ag_2CrO_4 + 2\ H^+$$

If conducted as a spot test, this reaction will reveal 2.5 γ acetylene in aqueous solution. A more sensitive and reliable test is provided by the reaction on a suspension of red-brown silver chromate in water or dilute acetic acid. The addition product is then formed through a topochemical reaction on, or very close to, the surface of the silver chromate or via its

---

* A compound, C$_2$H$_2$·Ag$_2$O·Ag$_2$CrO$_4$, which probably should be viewed as Ag$_2$C$_2$. Ag$_2$CrO$_4$·H$_2$O, has been prepared from silver chromate and acetylene.[']

dissolved portions. A result of this envelopment of the acid-soluble Ag_2CrO_4 by the acid-insoluble $Ag_2C_2.Ag_2CrO_4$ is that the unused silver chromate is markedly protected against rapid and complete solution in dilute nitric or sulfuric acid. This protective layer effect [2] can be applied in the detection of acetylene.

If hydrogen sulfide, mercapto compounds and hydrogen halides are absent, this test can be carried out as a spot reaction in a porcelain crucible (Procedure I) or in a larger volume of liquid (Procedure II), with employment of "analytical flotation". The latter operation makes use of the fact that small amounts of precipitate, suspended in water, are often gathered into the interface and thus made more visible when the suspension is shaken with an inert organic liquid that is immiscible with water. [3]

Procedure I. Single drops of the reagent are placed in two small porcelain crucibles, and the red-brown Ag_2CrO_4 is precipitated by adding one drop of 6 N acetic acid to each. One drop of the solution being tested for acetylene is added to one of the suspensions, and a drop of water to the other, which serves as a blank. The mixtures are stirred from time to time over a period of 1–10 min. depending on the quantity of acetylene expected. The blank is then treated with drops of 1 : 10 nitric acid until the Ag_2CrO_4 disappears and a clear yellow solution is obtained. The same number of drops of dilute nitric acid is then added to the actual test; for the sake of certainty, a drop or two extra may be added. If acetylene is present, a residue of Ag_2CrO_4 remains; the amount will be in accord with the quantity of acetylene present.

Reagent: Ammoniacal solution of silver chromate. Well washed Ag_2CrO_4 (prepared by mixing K_2CrO_4 and $AgNO_3$ solutions), is added to (1 : 5) ammonia water, vigorously shaken, and filtered. The yellow solution is stored in a closed brown bottle. If a turbidity develops on standing, more ammonia should be added, or the suspension can be filtered.

Limit of identification: 1 γ acetylene.

Procedure II. Two test tubes (capacity about 3 ml) fitted with glass stoppers are used. One drop of amoniacal silver chromate solution is placed in each tube and the red-brown silver chromate is precipitated by adding one drop of 6 N acetic acid. One drop of the aqueous test solution is added to one tube, and a drop of water to the other. Each mixture is then diluted with a drop or two of water, the tubes are stoppered and shaken for about one minute. Both suspensions are then covered with 1 ml of amyl alcohol. The blank is then treated with successive drops of 1 : 10 nitric acid (vigorous shaking) until the Ag_2CrO_4 disappears and no particles of precipitate remain in the water–amyl alcohol interface. The suspension being tested for acetylene is then shaken with the same number of drops of nitric acid plus one extra for the sake of certainty. If acetylene was present, the water–amyl alcohol interface will contain a red-brown film after separation of the two liquids.

Reagent: As in Procedure I.

Limit of identification: 1 γ acetylene.

1 Unpublished studies, with D. GOLDSTEIN.
2 F. FEIGL, *Chemistry of Specific, Selective and Sensitive Reactions*, Academic Press, New York, 1949, p. 663.
3 F. FEIGL, *op. cit.* p. 442.
4 E. EDWARDS and L. HODGKINSON, *Chem. News*, 90 (1904) 140.

2. Naphthalene

Test with chloranil[1]

The colorless aromatic hydrocarbons with condensed benzene rings react with light yellow chloranil to give colored addition compounds in the molecular ratio 1:1 or 1:2. These products can be formed in concentrated solutions of the components in benzene, chloroform, *etc.*, in melts of the hydrocarbons, and also through sintering conducted below the melting points of the components.[2]

The brick-red addition compound of chloranil and naphthalene is formed even when the hydrocarbon vapor comes into contact with solid chloranil (m.p. 290°). Naphthalene sublimes even at 50–70° and the vapors bring about the color reaction with chloranil. If conducted under these conditions, the color reaction is quite selective for naphthalene since, with the exception of methyl- and ethylnaphthalene, none of the hydrocarbons referred to above react. If the procedure described here is used, it must be remembered that sublimable phenols and amines likewise give colored products with chloranil.

Procedure. The gas absorption apparatus (Fig. 23, *Inorganic Spot Tests*) is used. A little of the solid or a drop of its ethereal solution is placed in the bulb and the solvent allowed to evaporate. A drop of a saturated solution of chloranil in petroleum ether or benzene is placed on the knob of the stopper and the solvent removed by evaporation. The apparatus is closed and placed in water at around 40°. The temperature is raised to 50–60°. If naphthalene is present, the light yellow film on the knob turns more or less brick-red, the shade being determined by the amount of naphthalene present.

Limit of identification: 25 γ naphthalene.

1 F. FEIGL, V. GENTIL and C. STARK-MAYER, *Mikrochim. Acta*, (1957) 350.
2 H. HAAKH, *Ber.*, 42 (1909) 4594.

3. Anthracene and phenanthrene

Test by conversion into anthraquinone and phenanthraquinone[1]

In contrast to other aromatic hydrocarbons, anthracene is not nitrated by concentrated nitric acid but is converted to anthraquinone.[2] Its isomer, phenanthrene, is nitrated, but when only small amounts are involved phenanthraquinone is also formed. Even a single evaporation with concentrated nitric acid yields enough of the respective quinone to permit an indirect test for these hydrocarbons to be based on this result.

A sensitive test for anthraquinone and phenanthraquinone is based on the catalytic action of these compounds, as well as other 1,2-diketones, on the color reaction between formaldehyde and 1,2-dinitrobenzene as outlined on p. 330.

Procedure.   The test is conducted in a micro test tube. One drop of the test solution (benzene, ether) is evaporated to dryness and a drop or two of concentrated nitric acid (s.g. 1.4) is added to the residue. The evaporation is repeated. One drop of 25% sodium carbonate solution is added, followed by one drop of 4% formaldehyde solution, and one drop of a 5% solution of 1,2-dinitrobenzene in benzene. The mixture is placed in boiling water and shaken occasionally. After 1–4 min., a violet color appears, the shade depending on the amount of anthracene or phenanthrene present.

The *limits of identification* are: 2 γ anthracene, 3 γ phenanthrene.

This procedure will reveal anthracene even though large amounts of naphthalene are present. The latter is only nitrated by concentrated nitric acid. It should be noted that benzene-soluble benzoins react directly with 1,2-dinitrobenzene without the addition of formaldehyde. A separate test should consequently be run with this fact in mind.

1 F. FEIGL and C. COSTA NETO, *Anal. Chem.*, 28 (1956) 397.
2 E. SCHMIDT, *Ann.*, 9 (1832) 241.

4. Anthracene

Test through conversion into anthraquinone[1]

As mentioned in Sect. 3, evaporation with concentrated nitric acid converts anthracene into anthraquinone. The anthraquinone produced by this oxidation can be detected by reductive conversion into the red water-soluble sodium salt of anthrahydroquinone as described on p. 336.

Limit of identification: 3 γ anthracene.

1 Unpublished studies, with V. GENTIL.

5. Phenanthrene

Test through oxidation to phenanthraquinone[1]

Wet oxidation of phenanthrene yields phenanthraquinone which can be specifically detected by heating with guanidine carbonate; the resulting blue-violet compound gives a red solution in ether (see p. 334). When phenanthrene is oxidized it yields also diphenic acid and fluorenone (diphenyl ketone). Oxidation with dilute potassium permanganate in glacial acetic acid is satisfactory for qualitative purpose even though only a part of the phenanthrene is converted to phenanthraquinone by this treatment.

Procedure. The test is made in a micro test tube. A drop of the test solution in glacial acetic acid is united with a drop of 0.05 N potassium permanganate and the mixture is kept in a boiling water bath for 5 min. After cooling, several drops of benzene are added, shaken, and the benzene solution is then transferred to another test tube by means of a pipette. Several cg of guanidine carbonate are added and the benzene evaporated. The remainder of the procedure is given on p. 334.

Limit of identification: 20 γ phenanthrene.

Anthracene does not interfere with this test.

1 Unpublished studies, with V. GENTIL.

6. Chloroform and bromoform

(1) Test through formylation of phenol[1]

A test is described for phenols on p. 181. It is based on the action of chloroform with alkali phenolate to give salicylaldehyde (Reimer–Tiemann formylation), which is then condensed with hydrazine to yield salicylaldazine. The latter is soluble in various organic liquids and exhibits a strong yellow-green fluorescence in the solid state. The formation of salicylaldazine can be applied to detect chloroform and also bromoform which reacts analogously.

The sensitivity of this test is relatively low in comparison with the detection of phenol. The reason probably is that it is impossible to avoid a loss of chloroform (bromoform) through the saponification:

$$CHHal_3 + 4\ NaOH \rightarrow HCOONa + 3\ NaHal + 2\ H_2O$$

Procedure. A drop of the alcoholic test solution is united in a micro test tube with a drop of 20% caustic alkali and a drop of 5% phenol solution. The mixture is kept in a boiling water bath for 2 min. After cooling, a drop of a 5% solution of hydrazine chloride is added, the mixture acidified with concentrated acetic acid and about 5 drops of ether added. After shaking, the ethereal solution is transferred dropwise to a filter paper by means of a pipette or glass capillary. A positive response is indicated by the development of a yellow-greenish stain on the paper after evaporation of the ether. The stain is strongly fluorescent, especially in ultraviolet light.

Limit of identification: 25 γ chloroform and 50 γ bromoform.

1 Unpublished studies, with V. GENTIL.

(2) Test through pyroammonolytic formation of hydrogen cyanide[1]

When guanidine carbonate is heated to 180–190°, ammonia is released[2]. If this decomposition occurs in the presence of vapors of chloroform (b.p. 61.3°) or bromoform (b.p. 149.6°), the following ammonolysis occurs:

$$CH(Hal)_3 + NH_3 \rightarrow HCN + 3 HHal$$

Tests for chloroform and bromoform can be based on the detection of the resulting hydrogen cyanide by the test described on p. 546.

Procedure.   A micro test tube is used. A drop of chloroform or bromoform is added to several mg of guanidine carbonate, and the mouth of the test tube is covered with a disk of filter paper that has been moistened with a freshly prepared solution of equal parts of saturated copper acetate and benzidine acetate solutions. The test tube is placed in a glycerol bath which has been preheated to 180–190°. A positive response is shown by the development of a blue stain on the reagent paper within 1–2 min.

This procedure revealed 500 γ CHBr$_3$ or 1 mg CHCl$_3$ in benzene solution.

The pyro-ammonolytic formation of hydrogen cyanide may likewise be employed for the detection of guanidine carbonate. The identification limit is 30 γ when a 10% solution of bromoform in benzene is used.

1 F. FEIGL, D. GOLDSTEIN and D. HAGUENAUER-CASTRO, *Rec. Trav. Chim.*, 19 (1960) 531.
2 Compare BEILSTEIN, *Handbuch der Organischen Chemie*, Vol. 3 (1927), p. 83.

7. Bromoform

(1) Test through oxidative cleavage with benzoyl peroxide[1]

If bromoform is warmed with benzoyl peroxide (m.p. 103°) bromine is released, probably by the reaction:

$$CHBr_3 + 2 (C_6H_5CO)_2O_2 \rightarrow CO_2 + HBr + 2 (C_6H_5CO)_2O + Br_2$$

This reaction occurs even at water bath temperature and more quickly at 110–120°. The bromine can be detected in the vapor phase by its reaction on fluorescein paper (p. 70).

Procedure. A drop of the test solution is treated in a micro test tube with a drop of 10% solution of benzoyl peroxide in benzene. The mouth of the tube is covered with a piece of moistened fluorescein paper. The tube is placed in boiling water. If bromoform was present, a red stain appears on the paper, the depth of the color depending on the amount of bromoform.

Limit of identification: 50 γ bromoform.

1 Unpublished studies, with E. SILVA.

(2) Test by conversion to acetylene [1]

If bromoform is warmed with zinc activated by copper, acetylene results:[2]

$$2 \text{ CHBr}_3 + 3 \text{ Zn}^0 \rightarrow 3 \text{ ZnBr}_2 + \text{CH}\equiv\text{CH}$$

The acetylene may be detected in the gas phase through the formation of red cuprous acetylide (see Sect. 1). Tetrachloro- and tetrabromoethane also react in this manner but the sensitivities are lower.

Procedure. Several cg of powdered zinc, previously washed with dilute hydrochloric acid to remove the film of zinc oxide, are placed in a micro test tube together with a drop of 1% alcoholic copper(II) chloride and a drop of the alcoholic test solution. The tube is placed in a heated water bath, and a disk of filter paper moistened with freshly prepared Ilosvay's reagent (for preparation see p. 405) is placed over the mouth of the test tube. A positive response is indicated by the appearance of a red stain on the paper.

Limit of identification: 80 γ bromoform.

This procedure also revealed 200 γ tetrachloro- or tetrabromoethane.

1 Unpublished studies, with C. STARK-MAYER.
2 P. CAZENEUVE, *Compt. Rend.*, 113 (1891) 1054; *Bull. Soc. Chim. France*, [3] 7 (1892) 69.

8. Carbon tetrachloride

(1) Detection by pyrohydrolysis [1]

Manganous sulfate monohydrate releases its water of crystallization in the region 150–200° [2]. If this anhydrization occurs in the presence of carbon tetrachloride vapors (b.p. 76°), the steam brings about the hydrolysis which cannot be effectuated in the wet way:

$$\text{CCl}_4 + 2 \text{ H}_2\text{O} \rightarrow \text{CO}_2 + 4 \text{ HCl}$$

This pyrohydrolysis can also be accomplished with a mixture of CCl_4- and C_6H_6 vapors. The resulting hydrogen chloride is readily detected by acid-base indicator paper.

The vapors of chloroform, *sym*-dichloro(bromo)ethane and *sym*-tetrachloro(bromo)ethane also undergo pyrohydrolysis in the same way, but much larger amounts of these compounds are needed.

Procedure. The test is conducted in a micro test tube (75 × 7 mm). About 10 mg of $MnSO_4.4 H_2O$ and a drop of the benzene test solution are united. The tube is immersed to a depth of about 1 cm in a glycerol bath preheated to 240°. A disk of Congo red paper is placed over the mouth of the test tube. If $CHCl_3$ is present, a blue stain appears on the red indicator paper within 1–3 min.

Limit of identification: 5 γ carbon tetrachloride.

1 F. FEIGL, D. GOLDSTEIN and D. HAGUENAUER-CASTRO, *Rec. Trav. Chim.*, 79 (1960) 531.
2 C. DUVAL, *Inorganic Thermogravimetric Analysis*, Elsevier, Amsterdam, 1963, p. 316.

(2) *Detection through topochemical-pyrolytic release of chlorine*[1]

If a drop of carbon tetrachloride is volatilized in a micro test tube and the remaining vapors are then heated by placing the test tube in a glycerol bath at 250°, no change is discernible. However, if this heating is conducted in the presence of several cg of pulverized quartz sand, chlorine is split off within a short time, and it can be detected by the bluing of Congo red paper that has been moistened with hydrogen peroxide, according to $Cl_2 + H_2O_2 \rightarrow 2 HCl + O_2$. It is likely that the result is due to disproportionations which occur on the surface of the quartz:

$$2\ CCl_4 \rightarrow \begin{matrix} CCl_3 \\ | \\ CCl_3 \end{matrix} + 2\ Cl^0 \quad (1) \qquad \begin{matrix} CCl_3 \\ | \\ CCl_3 \end{matrix} \rightarrow \begin{matrix} CCl_2 \\ || \\ CCl_2 \end{matrix} + 2\ Cl^0 \quad (2)$$

The same effect is obtained with finely powdered SnO_2, ThO_2, TiO_2, Ta_2O_5, Nb_2O_5, carbon and $BaSO_4$. The disproportionations (*1*) and (*2*) were observed when vapors of carbon tetrachloride passed through a strongly heated tube.[2] Therefore, it may be assumed that when the carbon tetrachloride vapors come into contact with the surface of the above-mentioned solids, a catalytic acceleration of the disproportionations occurs.

Procedure. Several cg of quartz dust and 1 drop of the solution under examination are placed in a micro test tube (75 × 7 mm). The test tube is immersed to a depth of about 1 cm in a glycerol bath that has been preheated to 150°. A disk of Congo red paper moistened with 5% hydrogen peroxide is placed over the mouth of the test tube. A blue stain on the red paper will appear within 2–6 min. if the response is positive.

This test revealed carbon tetrachloride in 1 drop of a 0.5% solution in chloroform. However, it failed when applied to a 1:1 solution in benzene.

The temperature must not rise above 250°, since chloroform likewise decomposes with release of chlorine when exposed to higher temperatures. However, *sym*-dichloro(bromo)ethane and *sym*-tetrachloro(bromo)ethane show the same behavior as carbon tetrachloride.

1 F. Feigl. D. Goldstein and D. Haguenauer-Castro, *Rec. Trav. Chim.*, 29 (1960) 531.
2 K. A. Hofmann and E. Seiler, *Ber.*, 38 (1905) 3058.

9. Methanol

Test by conversion to formaldehyde [1]

The specific test for formaldehyde with chromotropic acid (see p. 434) can be used for the detection of methanol since in acid solution the latter is readily oxidized to formaldehyde by permanganate:

$$5\ CH_3OH + 2\ MnO_4^- + 6\ H^+ \rightarrow 5\ CH_2O + 2\ Mn^{2+} + 8\ H_2O$$

If the experimental conditions prescribed here are maintained, *i.e.*, slight acidity and brief reaction time, ethanol is oxidized by permanganate solely to acetaldehyde without production of noticeable quantities of formaldehyde, as is the case when less mild oxidizing conditions prevail. The excess permanganate and the manganese dioxide produced by the reaction:

$$3\ Mn^{2+} + 2\ MnO_4^- + 2\ H_2O \rightarrow 5\ MnO_2 + 4\ H^+$$

can be removed, prior to the color test, by adding sodium sulfite.

Procedure. A drop of the test solution is mixed with a drop of 5% phosphoric acid and a drop of 5% potassium permanganate solution in a test tube for 1 min. A little solid sodium bisulfite is then added, with shaking, until the mixture is decolorized. If any brown precipitate of the higher oxides of manganese remains undissolved, a further drop of phosphoric acid should be added and a very little more sodium bisulfite. After the solution has become colorless, 4 ml of 12 N sulfuric acid and a little finely powdered chromotropic acid are added; the mixture is well shaken, and then heated to 60° for 10 min. A violet color, that deepens on cooling, indicates the presence of methanol.

Limit of identification: 3.5 γ methanol.

No reaction was given by: ethanol, propanol, isopropanol, *n*-butanol, *tert.*-butanol, *n*-amyl alcohol, isoamyl alcohol, amylene hydrate, glycol, propylene glycol, erythritol, adonitol, mannitol, dulcitol, acetaldehyde, butyraldehyde, isobutyraldehyde, isovaleraldehyde, methylglyoxal, acetone, oxalic acid, lactic acid, tartaric acid, citric acid, glucose.

Color reaction were given by: glycerol (yellow, and green fluorescence), furfural (brownish), arabinose, fructose, lactose, saccharose (yellow).

1 E. Eegriwe, *Mikrochim. Acta*, 2 (1937) 329.

10. Ethanol

Test through oxidation to acetaldehyde

Ethanol is oxidized by permanganate in the presence of sulfuric acid to give acetaldehyde. The latter can be detected in the vapor phase through the blue color it produces when it comes in contact with a solution of sodium nitroprusside containing morpholine or piperidine. (Compare the detection of secondary aliphatic amines, p. 251). The oxidation of ethanol always produces some acetic acid along with the acetaldehyde and therefore the sensitivity of the test based on the formation of acetaldehyde is relatively low.

Methanol is oxidized to formaldehyde, which does not react with nitroprusside. It is thus possible to distinguish between ethanol and methanol. Because of the difficulty of regulating the amount of oxidant required, the detection of small quantities of ethanol in methanol is uncertain. The same holds for the detection of ethanol in the presence of other materials oxidizable by permanganate. Because of its alcohol content, ordinary ether responds to this test. Since this result is obtained even with the pure ether intended for anesthesia, it appears that the oxidation.

$$C_2H_5OC_2H_5 + O \rightarrow C_2H_5OH + CH_3CHO$$

is possible. Therefore, the procedure given here is of limited applicability.

Procedure.[1]   One drop of the test solution is treated in a micro test tube with a drop of permanganate and the mixture is shaken. The mouth of the tube is covered with a disk of filter paper moistened with a drop of sodium nitroprusside–morpholine solution (see p. 438). If ethanol is present, a blue stain appears on the paper at once or within a few minutes. The depth of the color and the time required for it to develop depend on the quantity of ethanol involved.

Reagent: Equal volumes of 1 N permanganate and 1 : 1 sulfuric acid.

Limit of identification: 3 γ ethanol.

1 F. FEIGL and C. STARK, *Chemist-Analyst*, 45 (1956) 39.

11. Cyclohexanol and cyclohexanone

(1) Test through oxidation to adipic acid[1]

If cyclohexanol (I) or cyclohexanone (II) are evaporated with nitric acid, the oxidation produces adipic acid (III).[2]

$$
\text{(I) } \underset{\text{OH}}{\bigcirc} \quad \text{or} \quad \text{(II) } \underset{\text{O}}{\bigcirc} \xrightarrow{\text{HNO}_3} \quad \begin{array}{c} \text{H}_2\text{C---CH}_2 \\ | \qquad | \\ \text{H}_2\text{C} \quad \text{CH}_2 \\ | \qquad | \\ \text{HOOC} \quad \text{COOH} \end{array} \text{ (III)}
$$

The formation of adipic acid and hence indirectly the presence of (I) or (II) can be detected by heating the residue with mercuric cyanide; hydrogen cyanide results and may be identified through its color reaction with copper–benzidine acetate (see p. 546).

The test described here is characteristic for cyclohexanol and cyclohexanone in the absence of compounds that split off hydrogen cyanide when pyrolyzed with mercuric cyanide (comp. Chap. 2, sect. 39).

Procedure. The test is conducted in a micro test tube. A drop of the alcoholic or ethereal solution of the sample is taken to dryness, the evaporation residue is treated with a drop or two of nitric acid (1 : 2) and the evaporation is repeated to remove the residual nitric acid completely. Several cg of mercuric cyanide are added to the evaporation residue, and the mouth of the test tube is covered with a piece of filter paper that has been moistened with the cyanide reagent. The tube is then placed in a glycerol bath and heated to 140°. A blue stain on the paper indicates a positive response.

Limit of identification: 6 γ cyclohexanol or cyclohexanone.

It was stated in Chap. 2 that compounds which split off water at temperatures up to 190° can be detected by conducting the heating in the presence of thiobarbituric acid. Hydrogen sulfide is evolved and blackens lead acetate paper. Adipic acid is among such water-losing compounds because it is transformed into its anhydride. Compare p. 150 regarding the procedure for this test (*Identification:* 2.5 γ adipic acid).

1 Unpublished studies, with J. R. AMARAL.
2 Compare P. KARRER, *Lehrbuch der organischen Chemie*, 13th ed., Stuttgart, 1959, p. 303.

(2) Differentiation through dehydrogenation or dehydration [1]

If cyclohexanone (b.p. 156°) is heated with elementary sulfur, CH_2–CH_2 groups are rapidly dehydrogenated with evolution of hydrogen sulfide as shown by blackening of lead acetate paper. Cyclohexanol (b.p. 160°) is not attacked by sulfur at this temperature even though its structure is very similar to that of cyclohexanone. Accordingly these two cyclohexane derivatives can be differentiated and cyclohexanone can be detected in the presence of cyclohexanol. The dehydrogenation can be conducted in a micro test tube covered with a piece of lead acetate paper and immersed in a glycerol bath heated to 160°. The detection of cyclohexanone is not sensitive; nevertheless it can be detected in 1 drop of a 10% solution in ether or benzene. Cyclopentanone (b.p. 120°) behaves in the same manner as cyclohexanone but the dehydrogenation is less because of the greater volatility.

Cyclohexanol can be differentiated from cyclohexanone and detected in the presence of the latter by heating to 180° with thiobarbituric acid; hydrogen sulfide is evolved. This result is due to inner-molecular loss of water by cyclohexanol to yield cyclohexene[2], the released superheated water pyrohydrolyzing the thiobarbituric acid with production of hydrogen sulfide:

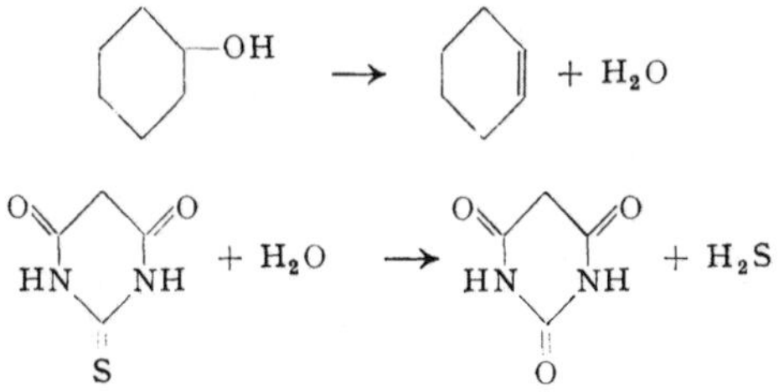

The dehydrogenation in the presence of thiobarbituric acid is conducted analogously to the dehydrogenation in the presence of sulfur as described above. The test succeeds with 1 drop of a 3% ethereal or benzene solution of cyclohexanol.

1 F. FEIGL, V. ANGER and E. LIBERGOTT, unpublished studies.
2 Compare BEILSTEIN, *Handbuch der organischen Chemie*, 4th ed., Vol. 5 (1922), p. 63.

12. Glycerol

(1) Test by conversion to acrolein[1]

When heated with a dehydrating agent, such as potassium bisulfate, glycerol produces acrolein, an unsaturated aldehyde:[2]

$$CH_2(OH)CH(OH)CH_2(OH) \rightarrow H_2C{=}CH{-}CHO + 2\ H_2O$$

If acrolein is treated with an aqueous solution of sodium nitroprusside containing piperidine or morpholine, a blue color appears which turns violet-red[3] with alkali. The chemistry of this reaction is unknown.

The acrolein formed by dehydration of glycerol may alternatively be detected by reaction with dianisidine to give a colored Schiff base (see p. 197).

The procedure given here makes use of the volatility of acrolein and its color reactions. It cannot be employed in the presence of ethylene glycol or lactic acid, since they decompose under the prescribed conditions to yield acetaldehyde which shows the same color reactions.

Procedure. A small amount of the test material or a drop of the test solution is placed in the hard glass tube described in *Inorganic Spot Tests*, Fig. 28, and mixed with finely powdered potassium bisulfate. A piece of filter paper moistened with reagent (1) is placed over the open end of the tube and covered with the glass cap. The acrolein, produced on heating, colors the paper a deep gentian blue. On treating with 2 N sodium hydroxide, the blue area changes to a peach blossom color.

When the alternative reagent (2) is used, a brown-red to yellow stain is formed on the paper.

Reagents: 1) Freshly prepared mixture of one drop of 5% sodium nitro-prusside and one drop of 20% piperidine or morpholine.
2) Saturated solution of *o*-dianisidine in glacial acetic acid.

Limit of identification: 5 γ glycerol.

1 F. FEIGL and O. FREHDEN, *Mikrochim. Acta*, 1 (1937) 137.
2 C. A. KOHN, cf. W. FRESENIUS, *Z. Anal. Chem.*, 30 (1891) 619.
3 L. SIMON, *Compt. Rend.*, 125 (1897) 1105; L. LEWIN, *Ber.*, 32 (1899) 3388; E. RIMINI, *Gazz. Chim. Ital.*, 30, I (1900) 279.

(2) Test by formation of 8-hydroxyquinoline[1]

A mixture of aniline, glycerol, concentrated sulfuric acid and a weak oxidant, *e.g.* As_2O_5, yields quinoline when heated. If derivatives of aniline with a free *ortho* position are used in this Skraup[2] synthesis, corresponding derivatives of quinoline can be prepared. For example, 8-hydroxyquinoline results if *o*-aminophenol is used. In this synthesis,[3] the glycerol is dehydrated to acrolein by the concentrated sulfuric acid (*1*). The acrolein condenses, with ring-closure,[4] with *o*-aminophenol to produce 8-hydroxy-1,2-dihydro-quinoline (*2*). The latter is then oxidized to 8-hydroxyquinoline (*3*).

$$CH_2OH-CHOH-CH_2OH \quad \longrightarrow \quad H_2C=CH-CHO + 2\,H_2O \qquad (1)$$

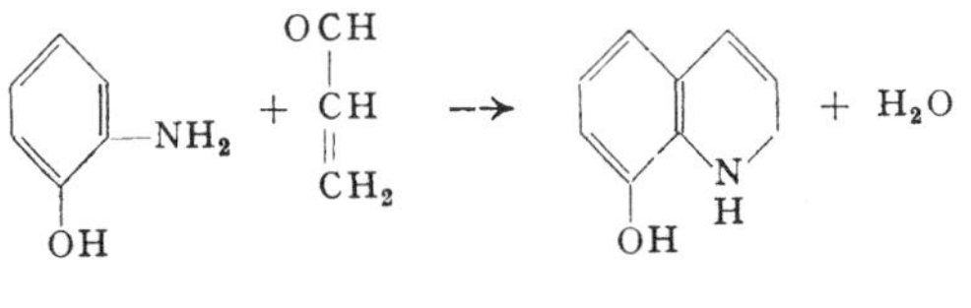

$$(2)$$

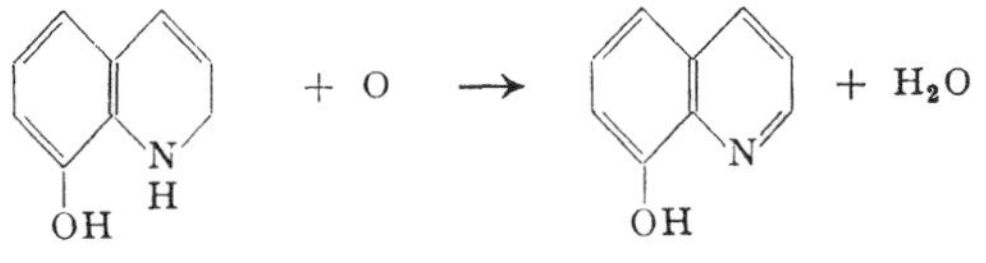

$$(3)$$

8-Hydroxyquinoline (oxine) reacts, under suitable conditions, with many metal ions to form water-insoluble inner-complex salts.[5] The oxinates of colorless metal ions, either as solids or dissolved in organic liquids, exhibit an intense fluorescence, which is yellow-green in most instances (see p. 388).

Mg oxinate precipitates even from very dilute solutions:

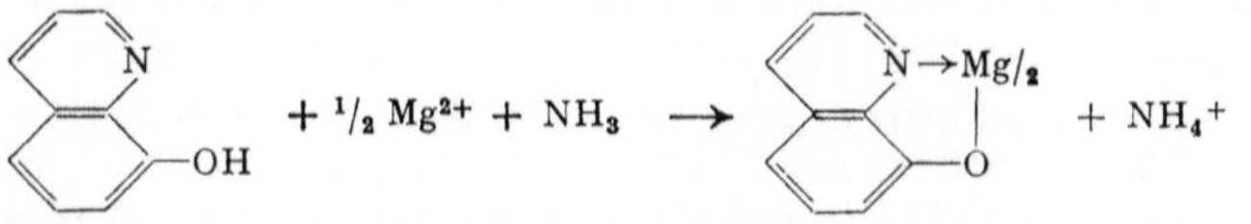

$$+ \,^1/_2\, Mg^{2+} + NH_3 \quad \longrightarrow \qquad\qquad + \; NH_4^+ \qquad (4)$$

The Skraup synthesis, as represented in (1)–(3), can be accomplished with one drop of a very dilute aqueous or alcoholic solution of glycerol, and the detection of the resulting Mg oxinate through its fluorescence reaction (4) thus provides a test for glycerol. Only crotonaldehyde interferes; it yields 2-methyl-8-hydroxyquinoline, which behaves like 8-hydroxyquinoline.

Procedure. Two drops of a 2% alcoholic solution of o-aminophenol are evaporated at 110° in a micro test tube. One drop of the test solution is added, followed by four drops of a 1% solution of arsenic acid in concentrated sulfuric acid. The tube is kept at 140° for 15 min. and then cooled to room temperature. Five drops of conc. sodium hydroxide, one drop of 2 N magnesium sulfate solution and three drops of concentrated ammonia are added, with agitation and cooling in cold water. A bluish-green fluorescence in ultraviolet light indicates glycerol. If a turbidity develops, centrifuge.

Limit of identification: 0.5 γ glycerol.

1 Unpublished studies, with W. A. MANNHEIMER.
2 Z. H. SKRAUP, *Monatsh.*, 3 (1882) 536.
3 E. C. WAGNER and J. K. SIMONS, *J. Chem. Educ.*, 13 (1936) 265.
4 See P. KARRER, *Organic Chemistry*, 4th Engl. ed., Elsevier, Amsterdam, 1950, p. 881.
5 Comp. F. J. WELCHER, *Organic Analytical Reagents*, Vol. 1, Van Nostrand, New York, 1947, p. 264.

(3) Other tests for glycerol

The catalysis by glycerol of the decomposition of oxalic acid at 100° to 110°, to form carbon dioxide and formic acid, may be applied as a test.[1] The acids formed may be detected by the decolorization of paper impregnated with sodium carbonate–phenolphthalein solution (*Idn. Limit*: 5 γ glycerol). Alternatively, the monoformin (monoformic ester of glycerol) which is an intermediate compound, may be detected by the test for esters given on p. 214 (*Idn. Limit*: 40 γ glycerol).

1 O. FREHDEN and C. H. HUANG, *Mikrochim. Acta*, 2 (1937) 20.

13. Inositol

Detection through oxidation with nitric acid

When inositol (I) is warmed with concentrated nitric acid it is converted to leuconic acid (II). Omitting the intermediate stages, the reaction may be represented:[1]

$$\text{(I)} \quad + 7\,O \quad \longrightarrow \quad \text{(II)} \quad + CO_2 + 6\,H_2O$$

The initial product is hexahydroxybenzene and subsequently the alicyclic polyketones tetrahydroxyquinone, rhodizonic acid, triquinoyl, and croconic acid appear. Depending on the conditions there probably results a mixture of leuconic acid with varying quantities of the other ketonic compounds.

The keto compounds remaining after inositol has been fumed with nitric acid have the following characteristic properties: (1) they or their hydrates are acids which form red water-insoluble alkaline-earth salts,[2] (2) they hasten the color reaction between formaldehyde and 1,2-dinitrobenzene (see test for 1,2-diketones, p. 330). These behaviors constitute the basis of the spot tests described here.

Procedure I.[3]   The test is conducted on a spot plate. A drop of the aqueous test solution is taken to dryness along with a drop of concentrated nitric acid. A drop of 5% calcium chloride solution and a drop of concentrated ammonia are added to the residue. After evaporation, the spot plate is kept at 180°. According to the quantity of inositol present, a brown-red to light red residue is obtained almost at once or after 4–10 min. at the most.

Limit of identification: 2.5 γ inositol.

This test is specific for inositol which otherwise shows the furfural reaction of carbohydrates (p. 337). However, considerable amounts of reducing sugars interfere because of caramelization during heating and the resulting yellow makes it difficult to discern the pink color due to inositol. Even so, it is possible to detect 10 γ inositol in the presence of 1000 γ glucose.

Procedure II.[4]   The evaporation with concentrated nitric acid is carried out in a micro test tube. The residue is treated with one drop of 4% formaldehyde and one drop of 25% sodium carbonate, and then with one drop of a 5% solution of 1,2-dinitrobenzene in benzene. The test tube is placed in boiling water and shaken. Depending on the amount of inositol, a more or less intense violet color develops within 1–4 min.

Limit of identification: 5 γ inositol.

This procedure cannot be applied directly in the presence of reducing sugars, since they give a violet color with alkaline 1,2-dinitrobenzene (compare p. 133). The interference by reducing sugars can be averted by oxidation with sodium hypobromite. The procedure is as follows: a drop of the aqueous solution is treated with one drop of strong bromine water and one drop of 0.5 N sodium hydroxide. The mixture is warmed in the water bath for 1–2 min. The excess hypobromite is then decomposed by a drop of 10% hydrogen peroxide and the liquid is taken to dryness. The residue is carried through Procedure II. It is possible to detect 10 γ inositol in the presence of 1 mg glucose in this manner.

1 N. SEIDEL, *Chemiker-Ztg.*, (1887) 316, 676; comp. P. FLEURY and P. BALATRE, *Les Inositols*, Masson et Cie., Paris, 1947, Chap. IV.
2 Comp. G. DENIGÈS, L. CHELLE and A. LABAT, *Précis de Chimie Analytique*, Vol. I, 7e ed., Maloine, Paris, 1930, p. 208.
3 F. FEIGL and V. GENTIL, *Mikrochim. Acta*, (1955) 1004.
4 F. FEIGL and Cl. COSTA NETO, *Anal. Chem.*, 28 (1956) 397.

14. p-Cresol

Test with p-nitrophenyldiazonium chloride and magnesium oxide[1]

The test is based on the formation of a blue color lake when p-cresol is coupled with diazotized p-nitroaniline and the product then brought into contact with magnesium oxide in alkaline surroundings:

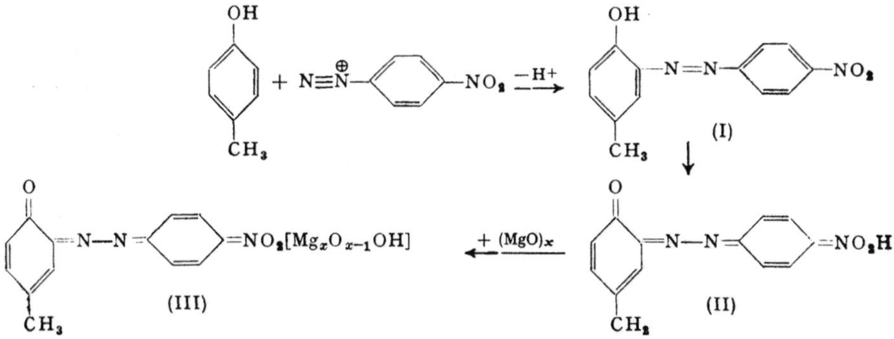

It seems that the formation of the blue magnesium lake requires the rearrangement of (I) into the isomeric nitronic acid (II) which then combines with the magnesium oxide to yield a salt that is adsorbed on the surface of the MgO-particles (III).

Procedure. Several cg of magnesium oxide are treated in a micro test tube in succession with 1 drop of the test solution, 1 drop of the nitrobenzene diazonium salt solution, and 1 drop of 5% potassium hydroxide solution. If p-cresol is present, a blue color appears almost immediately. Even minute amounts of the color lake can be readily seen if the colored mixture is placed on filter paper.

Reagent: 0.1 g of *p*-nitroaniline antidiazotate is dissolved in 100 ml of water. Then 10 drops of concentrated HCl are added.

Limit of identification: 0.2 γ *p*-cresol.

The following compounds behave like *p*-cresol: resorcinol, α-naphthol, chromotropic acid, H-acid, malonamide, pyrrole.

1 F. FEIGL and V. ANGER, *Anal. Chem.*, 33 (1961) 89.

15. α-Naphthol

Test with potassium p-nitrophenylantidiazotate[1]

In contrast to the other phenols, α-naphthol reacts with potassium *p*-nitrophenylantidiazotate when warmed in 2% sodium carbonate solution to yield the corresponding azo dyestuff:

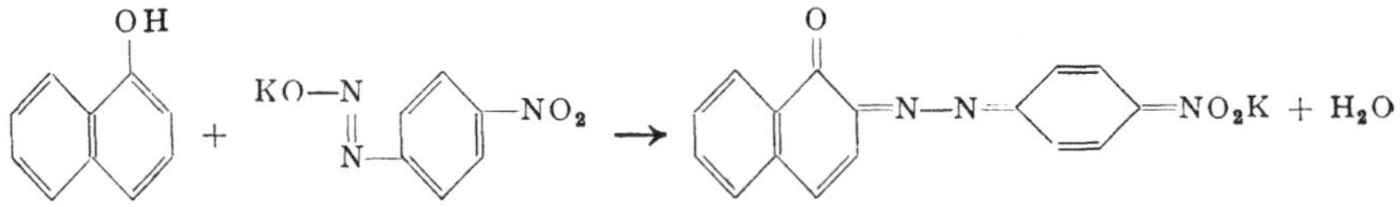

whose alkali salt is soluble in alcohols to give a blue color. This salt can be extracted with isoamyl alcohol, a finding that is the basis of a specific test.

Procedure. A micro test tube is used. One drop of the test solution is treated with 3 drops of potassium *p*-nitrophenylantidiazotate (30 mg in 10 ml ethanol), 2 drops of 2% sodium carbonate solution, and 3 drops of isoamyl alcohol. The mixture is shaken. A blue color in the isoamyl alcohol layer shows the presence of α-naphthol.

Limit of identification: 0.5 γ α-naphthol.

1 F. FEIGL and V. ANGER, *Z. Anal. Chem.*, 193 (1963) 274.

16. β-Naphthol

Test with 2-nitroso-1-naphthol[1]

β-Naphthol condenses with 2-nitroso-1-naphthol in its keto-form to give an almost colorless product that is oxidized to an intensely red-violet compound by nitric acid. The reaction may perhaps be represented:[2]

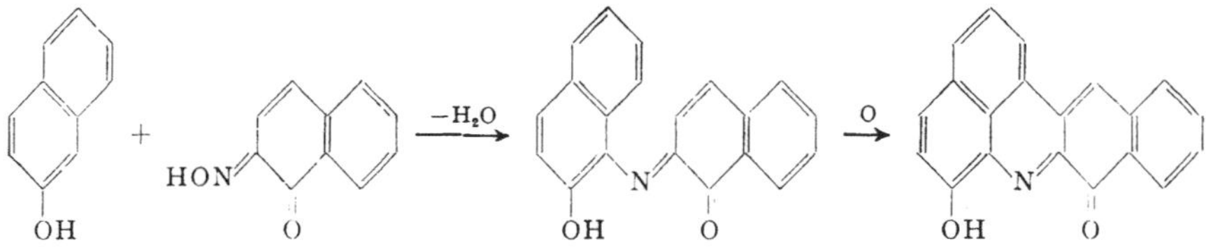

This reaction is selective for β-naphthol and is not given even by its sulfonic acids. Accordingly, an excellent test for β-naphthol can be based on the reaction.

Procedure. A drop of the test solution is treated in a micro test tube with a drop of a 0.05% ethanolic solution of 2-nitroso-1-naphthol and the test tube is then kept for 2–3 min. in a water bath at 60–80°. A drop of concentrated nitric acid is then introduced; the development of a red-violet color indicates the presence of β-naphthol.

Limit of identification: 0.1 γ β-naphthol.

1 V. ANGER, unpublished studies.
2 V. ANGER, *Oesterr. Chem. Ztg.*, 62 (1961) 355.

17. Pyrocatechol

(1) Test with metaldehyde[1]

Pyrocatechol gives a red-violet to pink color on treatment with metaldehyde in approximately 12 N sulfuric acid. The chemistry of this color reaction is not known. The concentrated sulfuric acid may produce a phenol–aldehyde condensation which is followed by oxidation to a quinoid compound. Compare the Le Rosen test, p. 137.

The following information is available concerning the behavior (color reaction) of other aromatic hydroxy compounds toward metaldehyde and sulfuric acid:

gradual greenish: phenol
yellow to reddish: orcinol
yellow: hydroquinone, hydroxyhydroquinone, α- and β-naphthol, naphthoresorcinol, protocatechualdehyde, 2-hydroxy-3-methoxybenzaldehyde, vanillin, veratraldehyde, piperonal
red: guaiacol, veratrole
orange-red: 2,7-dihydroxynaphthalene
yellow to reddish-yellow: 2,3,4-trihydroxybenzaldehyde
no reaction: 1,8- and 2,3-dihydroxynaphthalene, salicylic acid, protocatechuic acid, gallic acid, pyrogallic acid, quinic acid, 3,4-dihydroxycinnamic acid

Procedure. A drop of the aqueous test solution is treated on a spot plate with a little solid metaldehyde and one ml of approximately 12 N sulfuric acid. The mixture is stirred. Depending on the quantity of pyrocatechol, a violet or red color appears immediately or after several minutes.

Limit of identification: 4 γ pyrocatechol.

1 E. EEGRIWE, *Z. Anal. Chem.*, 125 (1943) 241.

(2) Test with phloroglucinol[1]

A green to blue-green color, which is stable for some time, appears if phloroglucinol and caustic alkali are added to an aqueous solution of pyrocatechol. The chemical basis of this color reaction is not known.

The behavior of other aromatic hydroxy compounds toward alkaline phloroglucinol has been reported as:

no reaction: phenol, resorcinol, orcinol, guaiacol, veratrole, α- and β-naphthol, 1,8-, 2,3- and 2,7-dihydroxynaphthalene, vanillin, veratraldehyde, piperonal, salicylic acid, quinic acid

yellow-red to orange: hydroquinone

brownish to brown-red: pyrogallol

reddish-yellow: naphthoresorcinol

light orange-violet: 2,6-dihydroxynaphthalene

yellow: 2-hydroxy-3-methoxybenzaldehyde, guaiacolaldehyde

yellow, gradually turning olive-brown: protocatechualdehyde

olive-green, turning red-brown: 2,3,4-trihydroxybenzaldehyde

yellow-green, later red-brown: protocatechuic acid

green-yellow, later red-yellow: 3,4-dihydroxycinnamaldehyde

transient greenish-yellow, then yellow: 3,4-dihydroxyphenylalanine

Procedure. One drop of the aqueous test solution, in a small test tube, is united with a little solid phloroglucinol, washed down with several drops of water and the mixture then rendered alkaline by shaking with a drop of 1 N sodium hydroxide. The total volume should be about 0.5 ml. Depending on the quantity of pyrocatechol present, a green or blue-green color appears immediately or within a minute or two.

Limit of identification: 0.5 γ pyrocatechol.

1 E. EEGRIWE, *Z. Anal. Chem.*, 125 (1943) 241.

(3) Test with 1-phenyl-3-methylpyrazol-5-one and ammonia[1]

Pyrocatechol yields a blue color when treated with phenylmethylpyrazolone and ammonia; the chemistry of this reaction has not been clarified. Several quinones react in this same manner under these conditions (see p. 333).

Procedure. The test is conducted in a micro test tube. A drop of the test solution is treated with 1 drop of the reagent and with a drop of concentrated ammonia. The gradual appearance of a blue color signals a positive response.

Reagent: Saturated solution of 1-phenyl-3-methylpyrazol-5-one in 1 : 1 water–methanol

Limit of identification: 1 γ pyrocatechol.

1 V. ANGER and S. OFRI, *Mikrochim. Acta*, (1963) 916.

18. Resorcinol

Test with pyrocatechol and alkali hydroxide[1]

If a dilute aqueous solution of resorcinol is treated with solid pyrocatechol and dilute sodium hydroxide, a transient blue-green color appears and then gradually a pink to violet-red color develops on the surface of the solution, slowly progressing downward and increasing in intensity. The production of the red color is characteristic for resorcinol. The chemistry of this color reaction is not known.

Orcinol gives a transient olive-green, going to brownish on standing; pyrogallol is colored olive-greenish briefly, then olive-yellow, and brownish on standing; phloroglucinol, permanently olive-green to blue-green; hydroxyhydroquinone orange-reddish, then orange-yellowish to light brownish on standing; phenol, *o,o'-* and *p,p'*-dihydroxydiphenyl, α- and β-naphthol, α,α'- and β,β'-dinaphthol, 1,8- and 2,7-dihydroxynaphthalene, no reaction; naphthoresorcinol, olive-greenish; 4-(*n*-hexyl)-resorcinol, transient blue-green, on the surface an olive-green ring, which is unchanged even after 5 minutes; 2,4-dihydroxybenzaldehyde, transient weak green, becoming weakly orange-yellowish on standing; 2,3,4-trihydroxybenzaldehyde, greenish-yellow, olive-yellow; 2,4,6-trihydroxybenzaldehyde, yellow, green, blue-green, olive-green, and an olive-green ring; benzoic acid, salicylic acid, *m-* and *p*-hydroxybenzoic acid, protocatechuic acid, *o*-phthalic acid, 1-hydroxy-2-naphthoic acid, no reaction. The following additional color changes were observed: α-resorcylic acid gradually green, on standing, a greenish-yellow ring; β-resorcylic acid, transient olive-greenish, then the upper half of the liquid slowly turns pale orange-pink; 2,4,6-trihydroxybenzoic acid, olive-greenish, green, blue-green, and a blue-green ring.

Procedure. One drop of the aqueous test solution is treated, in a small test tube, with solid pyrocatechol and, with shaking, made up to 2 ml with water. One drop of 0.33 *N* sodium hydroxide is then added, the mixture swirled once, and then allowed to stand quietly. The appearance of a pink to violet-red color in the upper half of the liquid indicates the presence of resorcinol.

Limit of identification: 1 γ resorcinol.

1 E. EEGRIWE, *Z. Anal. Chem.*, 125 (1943) 241.

19. Orcinol

Test with chloroform and sodium hydroxide[1]

A green fluorescence results if chloroform is added to a hot alkaline solution of orcinol.[2] Since resorcinol does not react under these conditions it may be assumed that the CH_3 group in orcinol (I) is involved in the fluorescence reaction. This group is transformed into a methylene group in the tautomeric form (II) of orcinol. Reaction of one molecule each of (I) and (II) with chloroform and alkali yields a polymethine dyestuff:

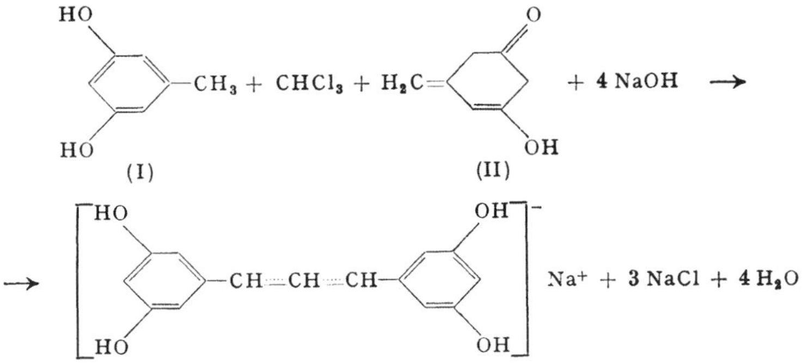

This assumption is supported by the finding that *m*-cresol reacts likewise though with much less sensitivity (100 γ).

Procedure.   A drop of the sample is united in a micro test tube with a drop each of N sodium hydroxide and of chloroform; the mixture is kept in a boiling water bath for 2–3 min. The presence of orcinol is indicated by a green fluorescence in daylight or in ultraviolet light.

Limit of identification: 1 γ orcinol

1 V. ANGER and S. OFRI, unpublished studies.
2 H. SCHWARZ, *Ber.*, 13 (1880) 543.

20. Naphthoresorcinol

Test with furfural[1]

Furfural–hydrochloric acid reacts with *m*-dihydroxybenzene derivatives (see p. 321) to yield initially orange condensation products that rapidly go over into green to blue polymethines. The change is accelerated by adding water. Only naphthoresorcinol gives a stable orange coloration. This finding can be employed in a selective detection of naphthoresorcinol.

Procedure.   A drop of the test solution is treated in a depression of a spot plate with a drop of a 0.1% solution of furfural in methanol and 1 drop of concentrated hydrochloric acid. The presence of naphthoresorcinol is indicated by the development of an orange color that does not change its hue on standing.

Limit of identification: 0.1 γ naphthoresorcinol.

1 V. ANGER and S. OFRI, *Z. Anal. Chem.*, 203 (1964) 429.

21. Hydroquinone

(1) Test with o-phthalaldehyde[1]

o-Phthalaldehyde, dissolved in concentrated sulfuric acid, gives with hydroquinone a blue to blue-violet color at once or within a few minutes, depending on the amount of hydroquinone present. Compounds such as quinic acid and the glucoside arbutin, which are split by concentrated sulfuric acid to yield hydroquinone, react in the same way. The chemistry of this color reaction has not been clarified.

The results with other aromatic hydroxy compounds are:

guaiacol	yellow	menthol	orange-reddish
eugenol	,,	thymol	red
m-hydroxybenzoic acid	,,	1,4-dihydroxy-naphthalene	,,
hydroxyquinone-carboxylic acid	,,	1,5-dihydroxy-naphthalene	violet-red
β-resorcylic acid	,,	1,8-dihydroxy-	
cresol (o,m)	,,	naphthalene	golden yellow
p,p'-dihydroxy-diphenyl	,,	2,6-dihydroxy-naphthalene	brownish red
o,o'-dihydroxy-diphenyl	olive greenish	2,7-dihydroxy-naphthalene	red-brown to violet-brown
pyrocatechol	dark brown	α-naphthol	brown-red
pinacol hydrate	yellowish	β- ,,	yellow-red
pyrogallol	red-brown	1-hydroxy-2-naphthoic acid	orange
p-cresol	brownish yellow		
naphthoresorcinol	,, ,,	3-hydroxy-2-naphthoic acid	reddish brownish
gallic acid	intense yellow		
terpinol hydrate	brownish red	quinhydrone	violet
resorcinol	yellow, reddish yellow		
orcinol	yellow, red-yellow		
hydroxyhydroquinone	red, violet-red		
salicylic acid	yellowish, orange-pink		
β,β'-dinaphthol	reddish, brownish red		
o-hydroxydiphenyl	yellow with green fluorescence		
m- ,,	orange-red, brownish red		
phenol	reddish, yellow-red, brown-red		
α-resorcylic acid	greenish yellowish		
quinone	violet-brownish to violet		

No reaction was shown by: phloroglucinol, 2.5-dihydroxybenzaldehyde, p-hydroxybenzoic acid, protocatechuic acid, pyrogallolcarboxylic acid.

Procedure. A drop of the aqueous test solution is treated in a small test tube with some solid o-phthalaldehyde and 1.5 ml of concentrated sulfuric acid.

A violet color, which appears at once or within a few minutes, indicates the presence of hydroquinone.

Limit of identification: 7 γ hydroquinone.

1 E. EEGRIWE, *Z. Anal. Chem.*, 125 (1943) 241.

(2) Test with phloroglucinol[1]

Small quantities of hydroquinone give an orange-pink color with phloroglucinol in alkaline solution. Larger quantities yield a more yellow-red to orange-red color. The chemistry of the color reaction is not known.

The following statements refer to the behavior of other aromatic hydroxy compounds: phenol, resorcinol, orcinol, arbutin, no reaction; hydroxyhydroquinone, a brownish color; pyrogallol, light violet-orange to reddish-brown; pyrocatechol, green to blue-green; 2,5-dihydroxybenzaldehyde, yellow; 1,4-dihydroxynaphthalene, gradual orange-yellow; quinone and quinhydrone, same reaction as hydroquinone.

Procedure. One drop of the aqueous test solution is treated in a small test tube with a little solid phloroglucinol, rinsed down with several drops of water until the total volume is about 1 ml. Then a drop of 0.5 N sodium hydroxide solution is added with shaking. According to the quantity of hydroquinone, an orange-red to orange-pink color appears.

Limit of identification: 0.5 γ hydroquinone.

1 E. EEGRIWE, *Z. Anal. Chem.*, 125 (1943) 241.

(3) Test with 1-phenyl-3-methylpyrazol-5-one and ammonia[1]

If a solution of hydroquinone is shaken in the presence of air with a solution of phenylmethylpyrazolone and ammonia, a violet color gradually appears. With respect to the color reaction it may be supposed that at first a condensation occurs with the reactive portion of the pyrazolone as represented in (*1*):

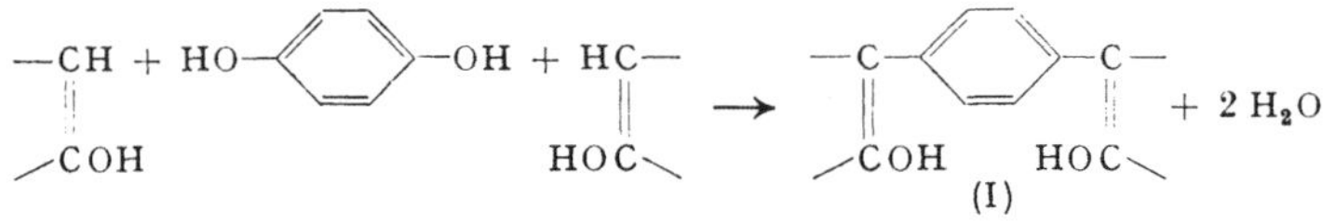

$$-CH + HO-\underset{}{\bigcirc}-OH + HC- \quad \longrightarrow \quad -C-\underset{}{\bigcirc}-C- \;+\; 2\,H_2O \qquad (1)$$
$$\underset{\diagup COH}{} \qquad\qquad HOC\diagdown \qquad\qquad \underset{\diagup COH}{}\quad HOC\diagdown$$
$$(I)$$

The product (I) is converted to product (II) by the oxygen of the air:

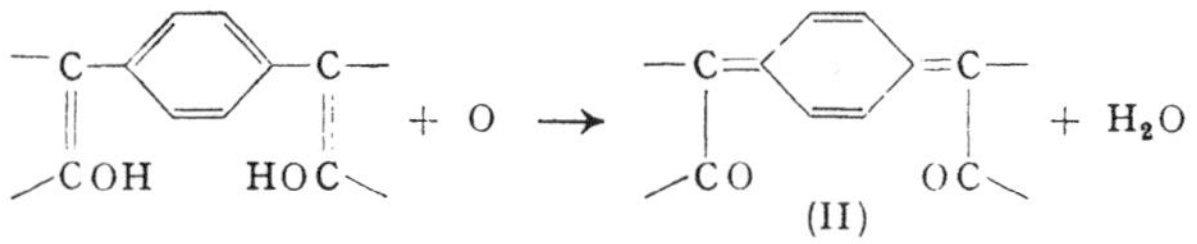

$$-C-\underset{}{\bigcirc}-C- \;+\; O \quad \longrightarrow \quad -C=\underset{}{\bigcirc}=C- \;+\; H_2O \qquad (2)$$
$$\underset{\diagup COH}{}\quad HOC\diagdown \qquad\qquad \underset{\diagup CO}{}\qquad OC\diagdown$$
$$(II)$$

The products (I) and (II) formed by (*1*) and (*2*) yield a violet ammonium salt of a meriquinoid compound:

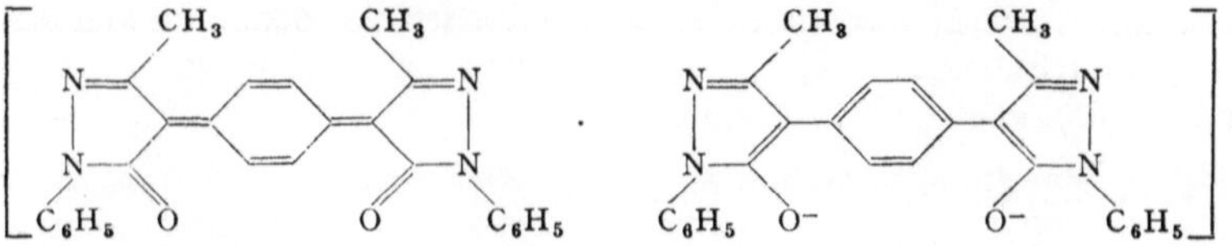

$$\left[\begin{array}{c} \text{CH}_3 \quad \text{CH}_3 \\ N=\!\!\!<\!\!\!\!\!\!\diagdown\!\!\!\!=N \\ | \qquad \qquad | \\ N\!-\!\!\!\!\!\diagup \diagdown\!\!\!\!-N \\ C_6H_5 \; O \qquad O \; C_6H_5 \end{array} \cdot \begin{array}{c} \text{CH}_3 \quad \text{CH}_3 \\ N=\!\!\!<\!\!\!\!\!\!\diagdown\!\!\!\!=N \\ | \qquad \qquad | \\ N\!-\!\!\!\!\!\diagup \diagdown\!\!\!\!-N \\ C_6H_5 \; O^- \qquad O^- \; C_6H_5 \end{array} \right] (NH_4^+)_2$$

A violet product is likewise given by *p*-benzoquinone; other quinones and pyrocatechol react with formation of green to blue compounds. Pyrogallol does not react.

Procedure. A drop of the test solution is shaken in a micro test tube with 1 drop of the reagent and 1 drop of ammonia. A positive response is signaled by the development of a violet color. If only small amounts of hydroquinone are involved the color reaction occurs after 5–10 min.

Reagent: Saturated solution of 1-phenyl-3-methylpyrazol-5-one in water–methanol (1 : 1).

Limit of identification: 0.2 γ hydroquinone.

1 V. ANGER and S. OFRI, *Mikrochim. Acta*, (1963) 916.

(4) Test with rhodanine and ammonia [1]

An intense green color results if a solution of hydroquinone is shaken (free access of air) with rhodanine:

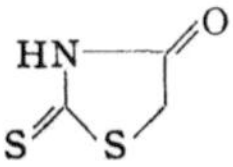

$$\begin{array}{c} HN\!-\!\!\!\!\!\diagdown \quad \diagup\!\!O \\ \diagup \qquad \diagdown \\ S \qquad S \end{array}$$

The chemistry of this color reaction is perhaps analogous to that given in Test (*3*). The corresponding green meriquinoid compound would then have the following structure:

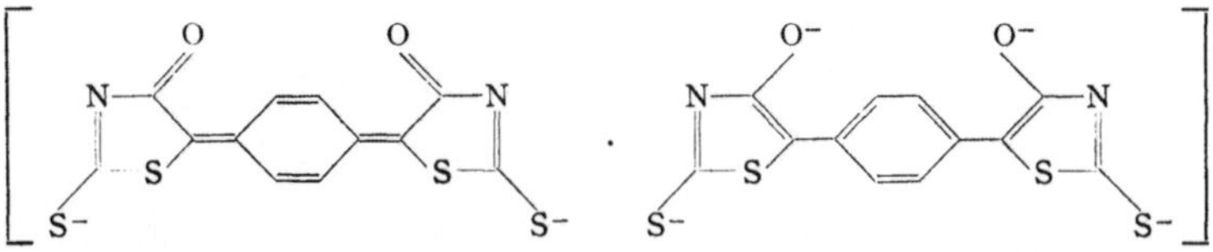

$$\left[\begin{array}{c} O \qquad O \\ N\!-\!\!\!\!\!\diagdown \qquad \diagdown\!\!\!\!-N \\ \diagup \; S \qquad S \; \diagup \\ S^- \qquad S^- \end{array} \cdot \begin{array}{c} O^- \qquad O^- \\ N\!-\!\!\!\!\!\diagdown \qquad \diagdown\!\!\!\!-N \\ \diagup \; S \qquad S \; \diagup \\ S^- \qquad S^- \end{array} \right] (NH_4^+)_6$$

Quinones react in the same way with rhodanine and ammonia, but in some cases, for instance benzoquinone, they can be masked by adding hydroxylamine hydrochloride (oxime formation). Pyrogallol and its derivatives give violet reaction products. Pyrocatechol and phloroglucinol do not interfere.

Procedure. A drop of the test solution is gently shaken in a micro test tube along with one drop each of a saturated aqueous solution of rhodanine and ammonia. A green color gradually appears if hydroquinone is present.

Limit of identification: 0.2 γ hydroquinone.

1 V. ANGER and S. OFRI, *Mikrochim. Acta*, (1963) 915.

22. Pyrogallol

Test with phloroglucinol[1]

As mentioned in Sect. 17 and 21, a caustic alkaline solution of phloroglucinol reacts with pyrocatechol and hydroquinone to give characteristic blue-green or yellow-red colors. The same colors are produced in ammoniacal solutions. In addition, phloroglucinol reacts in ammoniacal solution with pyrogallol to give a violet-orange to violet-reddish color. The chemistry of the color reaction is not known.

Procedure. A drop of the aqueous test solution is treated on a spot plate with a drop of phloroglucinol solution and a drop of concentrated ammonia. Depending on the quantity of pyrogallol present, a violet or a pink tending toward violet appears after about 30 sec.

Reagent: 0.05 g phloroglucinol in 25 ml water (freshly prepared).

Limit of identification: 1 γ pyrogallol.

The following information is available regarding the behavior of ammoniacal phloroglucinol solutions toward aromatic hydroxy compounds:

pyrocatechol	blue-green	hydroquinone	yellow
resorcinol	yellow, olive yellow	α-naphthol	bluish-violet
naphthoresorcinol	red-yellow	β- ,,	weak violet
1,8-dihydroxynaphthalene	grey		
2,6- ,,	yellow to orange-yellowish		
orcinol	light orange-yellowish		
hydroxyhydroquinone	grey-brown, olive brown		
protocatechuic acid	yellow, olive green, green		
pyrogallolcarboxylic acid	olive green, brown-olive		

No reaction was shown by: phenol, 2,3- and 2,7-dihydroxynaphthalene, vanillin, veratraldehyde, salicylic acid, quinic acid.

Gallic acid gives the same orange-reddish color as is produced by ammonia alone.

1 E. EEGRIWE, *Z. Anal. Chem.*, 126 (1943) 134.

23. Hydroxyhydroquinone

Test with p-phthalaldehyde[1]

A violet color forms if hydroxyhydroquinone is treated with *p*-phthalaldehyde (terephthalaldehyde) and concentrated sulfuric acid and, after dilution with water, the mixture is made weakly basic. Probably a condensation product is formed; its composition and constitution are unknown.

The behavior of other phenols and phenolcarboxylic acids is:

thymol	faint yellow	pyrocatechol	olive brownish
resorcinol	orange	3,4-hydroxycinnamic	
hydroquinone	brownish	acid	,, ,,
quinic acid	yellow	pyrogallol	brownish yellow
protocatechuic acid	intense yellow	tannic acid	,, ,,
pyrogallolcarboxylic acid	,, ,,	gallic acid	olive green
phloroglucinol	orange-yellowish turbidity		
2,4,6-trihydroxybenzoic			
acid	orange turbidity		

No reaction was shown by: phenol, orcinol, p-hydroxybenzoic acid, α- and β-resorcylic acid.

Procedure.   A drop of the aqueous test solution is treated in a small test tube with solid p-phthalaldehyde, and 0.5 ml concentrated sulfuric acid added.

The mixture is gently warmed, cooled, diluted with water, and (with cooling) then made alkaline with 8 N sodium hydroxide. The final volume should be about 7 ml. Depending on the quantity of hydroxyhydroquinone present, a violet to pink-violet color appears.

Limit of identification: 5 γ hydroxyhydroquinone.

1 E. EEGRIWE, *Z. Anal. Chem.*, 126 (1943) 134.

24. Phloroglucinol

(1) Test with salicylaldehyde[1]

Phloroglucinol in concentrated hydrochloric acid reacts with salicylaldehyde to yield orange-red condensation products. See p. 204 regarding the chemistry of the reaction. This behavior permits a selective detection of phloroglucinol; only naphthoresorcinol shows the same color reaction. (*Idn. Limit: 1 γ*).

Procedure.   A drop of the test solution is treated in a depression of a spot plate with 1 drop of a 10% methanolic solution of salicylaldehyde and several drops of concentrated hydrochloric acid. An orange color appears if phloroglucinol is present.

Limit of identification: 0.5 γ phloroglucinol.

1 V. ANGER and S. OFRI, *Z. Anal. Chem.*, 203 (1964) 429.

(2) Test through reaction with allyl compounds[1]

A test for allyl compounds is described on p. 169; it is based on the finding that they give a red color with a hydrochloric acid solution of phloroglucinol. The chemistry of this color reaction is discussed on p. 169; the reaction may be employed likewise for the detection of phloroglucinol and to differentiate it from its isomers.

Procedure. A micro test tube is used. A drop of the alcoholic test solution is treated in a boiling water bath with a drop of concentrated hydrochloric acid and a drop of alcoholic eugenol solution. A purple or pink color indicates the presence of phloroglucinol.

Limit of identification: 0.1 γ phloroglucinol.

1 Unpublished studies, with E. LIBERGOTT.

25. Pentachlorophenol

Test by conversion into chloranil

Colorless pentachlorophenol is easily converted to chloranil by brief warming with concentrated nitric acid:[1]

This reaction, whose occurrence is revealed by the yellow color of the chloranil, is the basis of a test for nitric acid[2] and also of a colorimetric method for determining chloranil.[3] However, this procedure is neither sensitive (*Idn. Limit*: 12 γ chloranil) nor unequivocal, because many phenols are easily converted to yellow nitro compounds by nitric acid. It is better to detect the resulting chloranil by means of a citric acid solution of tetrabase buffered with sodium acetate. A blue oxidation product of tetrabase results (Comp. p. 448). If small amounts of pentachlorophenol are suspected, the nitrous acid or nitrogen oxides must be destroyed by adding urea to prevent oxidation of the tetrabase by nitrogen oxides.

Procedure.[4] One drop of the test solution is evaporated in a micro test tube. The residue is treated with a drop of concentrated nitric acid and the mixture is briefly warmed over a flame or kept for 2 min. in the boiling water bath. After cooling, several mg of urea are added, followed by a drop of a solution of tetrabase in citric acid[5] and a pinch (tip of knife blade) of solid sodium acetate. The mixture is then reheated in boiling water. A blue color indicates the formation of chloranil and hence the presence of pentachlorophenol in the sample.

Reagent: 2.5 g tetrabase and 10 g citric acid are dissolved in 10 ml water and diluted to 500 ml.

Limit of identification: 2.5 γ pentachlorophenol.

1 H. BILTZ and W. GIESE, *Ber.*, 37 (1904) 4018.
2 A. BARETTO, *Rev. Quim. Ind. (Brazil)*, 10 (1941) No. 115, 12.
3 W. DEICHMANN and L. J. SCHAFER, *Ind. Eng. Chem., Anal. Ed.*, 14 (1942) 310.
4 Unpublished studies, with J. E. R. MARINS.
5 Comp. R. J. CARNEY, *J. Am. Chem. Soc.*, 34 (1912) 325.

26. Salicyl alcohol (saligenin)

Test by pyrolytic conversion into salicylaldehyde[1]

When saligenin is warmed with a sulfuric acid solution of alkali chromate, the CH_2OH group is oxidized to the CHO group.

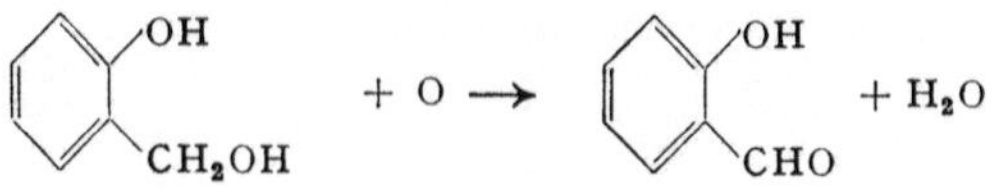

The salicylaldehyde may react in the vapor state with hydrazine to give the fluorescing salicylaldazine (compare p. 341).

The oxidation of salicyl alcohol to salicylaldehyde can be accomplished in simple fashion and without significant further oxidation to salicylic acid by heating the salicyl alcohol (m.p. 100°) with manganese dioxide to 240°. This is an interesting topochemical redox reaction of the molten or vaporizing salicyl alcohol with solid MnO_2.*

Procedure.   A drop of the alcoholic solution is taken to dryness in a micro test tube along with several cg of manganese dioxide paste. The mouth of the tube is covered with a disk of filter paper moistened with hydrazine solution and the mixture is heated for 5 min. in a bath at 240–250°. If salicyl alcohol is present, a stain is formed on the paper; it fluoresces yellow in ultraviolet light.

Reagents: 1) Manganese dioxide paste. A warm manganous solution is treated with excess alkali hypobromite; the precipitate is washed
and centrifuged.

2) Hydrazine solution. 10 g hydrazine sulfate and 10 g sodium
acetate per 100 ml water.

Limit of identification: 2 γ salicyl alcohol.

* There is marked oxidation to salicylaldehyde if vapors of salicyl alcohol come into contact with air at about 150° (unpublished observation).

1 F. Feigl and C. Stark, *Mikrochim. Acta*, (1955) 996.

27. Carbon monoxide

Test with phosphomolybdic acid and palladium chloride[1]

When carbon monoxide is passed through a solution of phosphomolybdic acid there is no discernible effect. However, if a small amount of palladium chloride is added, there is immediate formation of molybdenum blue:

$$2 MoO_3 + CO \rightarrow Mo_2O_5 + CO_2 \tag{1}$$

The catalytic hastening is brought about by the reaction:

$$Pd^{2+} + CO + H_2O \rightarrow Pd^0 + CO_2 + 2\ H^+ \qquad (2)*$$

The metallic palladium produced (even in minimal quantities) absorbs and at the same time activates carbon monoxide, and it is this activated[2] CO which enters into reaction (1) and produces the molybdenum blue. This effect, which is an example of heterogeneous catalysis, is the basis of an extremely sensitive and selective test for palladium (compare *Inorganic Spot Tests*, Chap. 3) and for the test for carbon monoxide given here.

A test for carbon monoxide is of value because certain types of organic acids yield CO when decomposed in the wet way. This is accomplished by concentrated sulfuric acid[3] or syrupy phosphoric acid. The most familiar examples are formic and oxalic acid:

$$HCO_2H \rightarrow H_2O + CO$$
$$H_2C_2O_4 \rightarrow H_2O + CO_2 + CO$$

α-Hydroxy- and α-ketocarboxylic acids are broken down in various ways by warming with concentrated sulfuric or phosphoric acid; carbon monoxide may be produced directly or via formic and oxalic acid:

$$RCH(OH)COOH \rightarrow RCHO + HCO_2H$$
$$R(CO)COOH \rightarrow RCOOH + CO$$
$$2\ R(CO)COOH \rightarrow R\!-\!R + H_2C_2O_4 + 2\ CO$$

At room temperature, tertiary carboxylic acids are broken down quantitatively with production of the respective carbinols (tertiary alcohols):

$$\begin{array}{c} R \\ \diagdown \\ R\!-\!C\!-\!COOH \\ \diagup \\ R \end{array} \rightarrow \begin{array}{c} R \\ \diagdown \\ R\!-\!C\!-\!OH + CO \\ \diagup \\ R \end{array} \qquad (R = \text{alkyl or aryl})$$

In general, the lower members of the α-hydroxy acids undergo considerable decomposition even at 80–100°, while the higher members require temperatures around 140–160°.

Syrupy phosphoric acid must be used in the test for carbon monoxide split out of organic compounds given here, because concentrated sulfuric acid when heated with organic compounds yields sulfur dioxide which reduces phosphomolybdic acid to molybdenum blue. Although it rarely happens, there is danger of splitting off volatile reducing agents when organic materials undergo pyrolysis in the presence of syrupy phosphoric acid. This possibility can be checked by means of filter paper moistened with phosphomolybdic acid. If the paper does not turn blue, the positive response obtained by the

---

* This reaction is the basis of a well-known macro test[4] for CO conducted on filter paper impregnated with a 1% solution of $PdCl_2$.

procedure given here is certain proof of the splitting off of carbon monoxide.

Procedure.[5]   A drop of the test solution is taken to dryness in the bulb of the apparatus described in *Inorganic Spot Tests*, Fig. 23. Alternatively, a little of the solid sample may be placed in the apparatus. A drop of syrupy phosphoric acid that has been previously heated to 250–300° and then cooled, is introduced. The knob of the stopper is dipped into the reagent solution, and put into place. The closed apparatus is heated with a micro flame until the mass solidifies. Care must be taken that the drop on the knob does not go to dryness. After cooling, the drop is wiped onto filter paper and moistened with a drop of water. If carbon monoxide was produced, a blue to blue-green color appears. A blank test is advisable.

Reagent: 1) 0.02 g $PdCl_2$ is dissolved in 2 drops of concentrated hydrochloric acid and brought to 10 ml with water.

2) Cold saturated solution of phosphomolybdic acid in water.
The reagent consists of a mixture of 2 ml (1) and 8 ml (2).

Limits of identification: 5 γ oxalic acid   10 γ tartaric acid
10 γ citric acid   25 γ mandelic acid

1 Unpublished studies, with D. GOLDSTEIN.
2 C. ZENGHELIS, *Z. Anal. Chem.*, 49 (1910) 429.
3 A. BISTRZYCKI and B. V. SIEMIRADSKI, *Ber.*, 39 (1906) 51; 41 (1908) 1665.
4 For literature see J. SCHMIDT, *Das Kohlenoxyd*, Akad. Verlag, Leipzig, 1935, p. 186.
5 F. FEIGL, CL. COSTA NETO and J. E. R. MARINS, unpublished studies.

28. Formaldehyde

(1) Test with chromotropic acid[1]

When formaldehyde is warmed with chromotropic acid (1,8-dihydroxy-naphthalene-3,6-disulfonic acid) in strong sulfuric acid solution, a violet-pink color develops. The chemistry of this color reaction is not known with certainty. Since aromatic hydroxy compounds condense with formaldehyde[2] to yield colorless hydroxydiphenylmethanes, it is probable that the initial step consists of a condensation of the phenolic chromotropic acid with form-aldehyde as shown in (1) followed by oxidation to a *p*-quinoidal compound as shown in (2). (See the Le Rosen test described on p. 137).

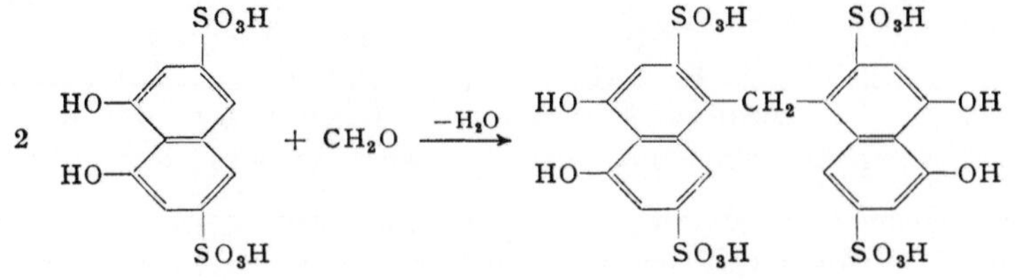

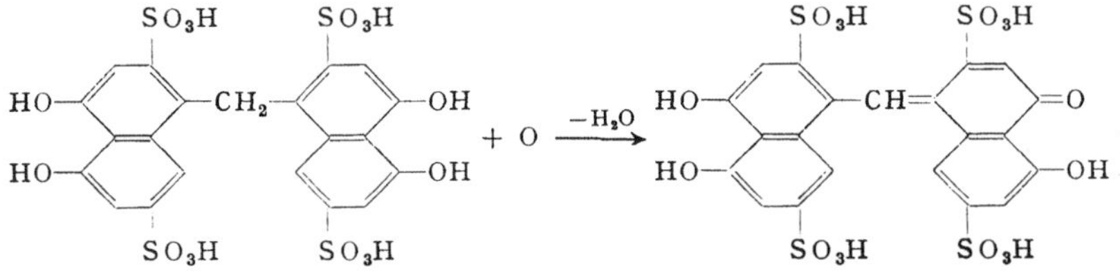

$$ (2) $$

Concentrated sulfuric acid participates in both (*1*) and (*2*). In the former it functions as a dehydrant to bring about the condensation; in (*2*) it is an oxidant and is reduced to sulfurous acid.

Acet-, propion-, butyr-, isobutyr-, and isovaleraldehydes, oenanthal, crotonaldehyde, chloral hydrate, glyoxal, and aromatic aldehydes give no reaction with a sulfuric acid solution of chromotropic acid. Glyceraldehyde, furfural, arabinose, fructose, and saccharose give a yellow coloration. Other sugars, acetone, and carboxylic acids do not react. Large amounts of furfural give a reddish color.

Procedure. A drop of the test solution is mixed with 2 ml 12 N sulfuric acid in a test tube, a little solid chromotropic acid or its sodium salt is added, and the tube heated for 10 min. in a water bath at 60°. A bright violet color appears in the presence of formaldehyde.

Limit of identification: 0.14 γ formaldehyde.

A violet to pink color revealed 0.5 γ formaldehyde in the presence of:

0.047	mg fructose	*i.e.*	94	times the quantity
0.99	mg saccharose		198	,, ,, ,,
0.19	mg furfural		380	,, ,, ,,
2.16	mg arabinose		4326	,, ,, ,,
3.48	mg lactose		6956	,, ,, ,,
5.05	mg glucose		10100	,, ,, ,,

The detection of formaldehyde with chromotropic acid can be applied also to compounds which split off formaldehyde on treatment with acids. Instances are: hexamine, formaldoxime, trimethylene-α-mannitol, etc.[3]

It is in harmony with the chemistry of the formaldehyde–chromotropic acid test that the greatest sensitivity is attained in concentrated sulfuric acid.[4] Therefore, smaller amounts of solid hexamine can be detected compared with corresponding amounts of formaldehyde. For this reason, when minimal amounts of formaldehyde are to be detected, it is recommended to evaporate the sample to dryness together with ammonia in order to obtain hexamine. If conducted within the technique of spot test analysis, these procedures have *identification limits* of 0.003 γ formaldehyde.

The fact that formaldehyde gives the chromotropic acid tests in the vapor phase, makes possible a convenient method of detecting compounds which split out formaldehyde when hydrolyzed or pyrolyzed. It is interesting to note that naphthalenesulfonic acid–formaldehyde condensates, which are important tanning agents, do not belong in this category.

1 E. EEGRIWE, Z. Anal. Chem., 110 (1937) 22.
2 Regarding the condensation of aromatic hydroxy compounds with formaldehyde see W. WOLFF, Ber., 26 (1893) 83; J. BRESLAUER and A. PICTET, Ber., 40 (1907) 3786; A. CASTIGLIONI, Z. Anal. Chem., 113 (1938) 428.
3 Regarding the hydrolytic cleavage with liberation of formaldehyde comp. C. L. HOFFPAUIR, G. W. BUCKALOO and J. D. GUTHRIE, Ind. Eng. Chem., Anal. Ed., 15 (1943) 605.
4 F. FEIGL and L. HAINBERGER, Mikrochim. Acta, (1955) 110.

(2) Detection through demasking effects

Formaldehyde reacts with cyanide ions or ethylenediamine to yield water-soluble addition or condensation products:

$$CN^- + CH_2O \longrightarrow H_2C {\overset{\displaystyle O^-}{\underset{\displaystyle CN}{\big<}}}$$

$$\begin{matrix} H_2C-NH_2 \\ | \\ H_2C-NH_2 \end{matrix} + 2\,CH_2O \longrightarrow \begin{matrix} H_2C-N=CH_2 \\ | \\ H_2C-N=CH_2 \end{matrix} + 2\,H_2O$$

These reactions also occur when formaldehyde acts on certain complex salts, whose anion or cation contain CN groups or ethylenediamine molecules. This is shown by the immediate precipitation of the red nickel salt if formaldehyde is added to a solution of $Na_2[Ni(CN)_4]$ containing dimethylglyoxime, in which the concentration of Ni^{2+} ions is too low to react with dimethylglyoxime.[1] Similarly, the addition of formaldehyde to a solution of silver ethylenediamine chromate yields a red-brown precipitate of Ag_2CrO_4. Therefore, formaldehyde releases the metal ions from the equilibria:

$$[Ni(CN)_4]^{2-} \rightleftharpoons Ni^{2+} + 4\,CN^- \quad \text{and} \quad [Ag\ en]^+ \rightleftharpoons Ag^+ + en$$

and thus enables these ions to give their normal reactions with dimethylglyoxime and chromate ions, respectively. Even small amounts of formaldehyde are sufficient to realize these effects.

The detection of formaldehyde through its demasking action cannot be applied to acidic solutions since H^+ ions likewise set the metals free by reacting with cyanide ions or with ethylenediamine. Acid solutions can be neutralized with calcium carbonate, or the formaldehyde can be expelled by heating the test solution and then detected in the vapor phase by means of the complex salts solutions. However, the sensitivity is less.

When formaldehyde reacts with silver ethylenediamine chromate solution, the initial product is red-brown silver chromate. If considerable amounts of formaldehyde are present, the precipitate quickly turns black because of the secondary reactions:

$$Ag_2CrO_4 + CH_2O + 3\ OH^- \rightarrow 2\ Ag^0 + HCOO^- + CrO_4^{2-} + 2\ H_2O$$
$$Ag_2CrO_4 + 2\ OH^- \rightarrow Ag_2O + CrO_4^{2-} + H_2O$$

The OH⁻ ions participating in these reations are furnished by the alkalinity of the ethylenediamine–chromate reagent solution.

Procedure I   (demasking of nickel ions).[2] A drop of the neutral test solution is placed on a strip of the reagent paper. The resulting stain is pink or red according to the amount of formaldehyde present. When amounts below 2 γ are suspected, a blank should be run with a drop of water on a separate strip of reagent paper and with rapid drying of the fleck.

Reagent paper: About 0.5 g of freshly precipitated nickel dimethylglyoxime is suspended in 100 ml water, 0.1 g KCN is added, shaken vigorously, and filtered after 24 h. The equilibrium solution should be stored in borosilicate bottles. Filter paper is moistened with the solution and dried in an oven or under an infrared lamp. It should be stored in a closed container.

Limit of identification: 0.5 γ formaldehyde

Among the aliphatic and aromatic aldehydes, only acetaldehyde likewise gives a positive reaction. Acidic solutions can be made basic before the test by adding ammonia.

Procedure II   (demasking of silver).[3] One drop of the neutral test solution is treated on a spot plate with a drop of the reagent solution. Formaldehyde yields an immediate red precipitate, which turns black.

Reagent: 3 g freshly prepared silver chromate, washed with water, is suspended in 20 ml water and 1 g ethylenediamine is added. The suspension is boiled for 5 min. and filtered. The filtrate is made up to 100 ml with water, allowed to stand for one hour, and filtered if necessary.

Limit of identification: 3.5 γ formaldehyde.

Procedure II is less sensitive than I but has the advantage of not giving a response with acetaldehyde.

1 F. Feigl and H. J. Kapulitzas, *Z. Anal. Chem.*, 82 (1930) 417.
2 P. W. West and B. Sen, *Anal. Chem.*, 27 (1955) 1460.
3 Cl. Costa Neto, comp. *Chem. Abstr.*, 54 (1960) 1160.

(3) Test with J-acid[1]

When formaldehyde reacts with J-acid (I) dissolved in concentrated sulfuric acid, condensation occurs followed by dehydration and oxidation. These partial reactions thus produce a yellow xanthylium dyestuff that fluoresces yellow in ultraviolet light. Its cation has the structure (II).

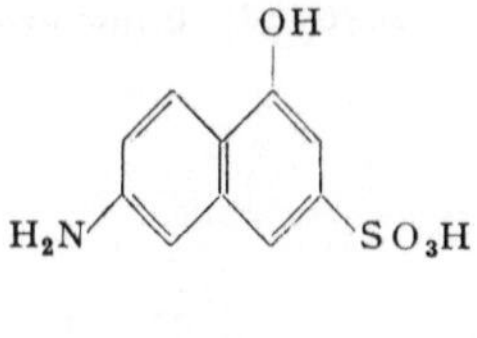

OH

H₂N   SO₃H

(I)

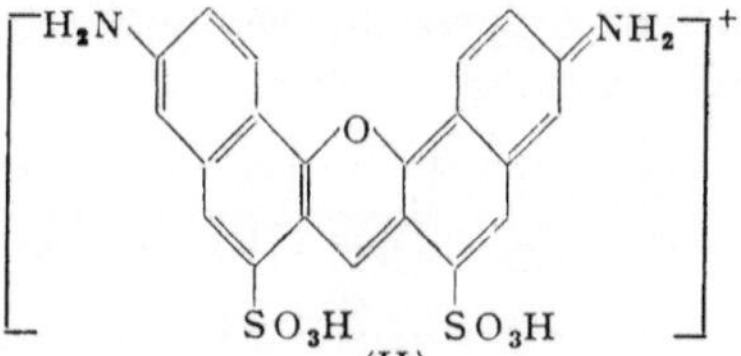

[H₂N ... NH₂]⁺

O

SO₃H   SO₃H

(II)

The yellow solution of the dyestuff in concentrated sulfuric acid turns blue on the addition of water. Since the reaction of formaldehyde with J-acid occurs even at room temperature, there is no interference by compounds that do not yield formaldehyde until they are heated. Acrolein however does react with J-acid and the compounds described here; the product fluoresces yellow-green (*Idn. Limit* 0.05 γ).

Procedure. A drop of a 0.1% solution of J-acid in concentrated sulfuric acid is placed on glass filter paper and 1 μl of the aqueous test solution is added. If the response is positive, a yellow fluorescence appears within 15 sec. Confirmation is obtained if the yellow solution turns blue after the addition of water.

Limits of identification: 0.01 γ formaldehyde (through fluorescence).

The color change to blue is distinctly visible when the quantity of formaldehyde exceeds 2 γ. In contrast, the color obtained with acrolein shows no change. Therefore, it is possible in this way to detect formaldehyde in the presence of acrolein.

1 E. SAWICKI, T. W. STANLEY and J. PFAFF, *Chemist-Analyst*, 51 (1962) 9.

29. Acetaldehyde

Detection with piperidine or morpholine, and sodium nitroprusside[1]

When acetaldehyde is added to a solution of sodium nitroprusside containing piperidine, or better morpholine, a blue color appears. This color reaction occurs likewise if the reagent contains another secondary aliphatic amine instead of piperidine or morpholine. A test for such amines is based on this finding. Compare p. 251.

Even small amounts of acrolein (see p. 416), crotonaldehyde, tiglic aldehyde, and cinnamaldehyde react similarly to acetaldehyde.[2]

Procedure. One drop of the solution to be tested is placed on a spot plate or filter paper. One drop of the reagent solution is added. A more or less intense blue color appears, the shade depending on the quantity of acetaldehyde involved.

Reagent: Freshly prepared mixture of equal volumes of 20% aqueous solution
           of morpholine and 5% aqueous solution of sodium nitroprusside.

Limit of identification: 1 γ acetaldehyde.

The color reaction is obtained likewise if acetaldehyde vapor comes into contact with filter paper moistened with the reagent. This is the basis of indirect tests for compounds which yield acetaldehyde when oxidatively cleaved or by another way. Pertinent examples are to be found in this book.

If it is desired to detect acetaldehyde vapor by means of the nitroprusside reagent, the absence of ethanol vapors must be assured, since they are distinctly oxidized to acetaldehyde, when heated with access of air.

1 L. LEWIN, *Ber.*, 32 (1899) 3388.
2 J. DOEUVRE, *Bull. Soc. Chim. France*, [4] 39 (1926) 1102.

30. Chloral and bromal

(1) Test through formylation of phenol[1]

The water-soluble chloral (trichloroacetaldehyde) is saponified by caustic alkali whereby chloroform is formed:

$$CCl_3CHO + NaOH \rightarrow CHCl_3 + HCOONa$$

When the saponification is carried out in presence of phenol, salicylaldehyde is produced, which condenses with hydrazine to salicylaldazine, easily detectable by its yellow-green fluorescence (compare p. 181). Bromal (tribromoacetaldehyde) behaves like chloral.

Procedure. To a drop of the aqueous test solution in a micro test tube add a drop each of 5% alcoholic phenol and 20% sodium hydroxide. For further treatment see test for chloroform (p. 409).
Limit of identification: 50 γ chloral.

The relatively low sensitivity is due to saponification of the resulting chloroform to alkali formate (comp. p. 409).

1 Unpublished studies, with V. GENTIL.

(2) Test through conversion into acetaldehyde[1]

A macro test for chloral hydrate consists in subjecting it to the action of nascent hydrogen $(Zn+H_2SO_4)$ to yield acetaldehyde:

$$CCl_3CHO + 6 H^0 \rightarrow CH_3CHO + 3 HCl \tag{1}$$

If the faintly warmed vapors of acetaldehyde are brought into contact with a solution of sodium nitroprusside containing piperidine or morpholine, a blue color results (compare Sect. 29).

The conversion of chloral into acetaldehyde can likewise be accomplished by warming with formaldehyde–sodium sulfoxylate (Rongalite),[2] which here functions as hydrogen donor for the realization of (1):

$$CH_2OH—SO_2Na + H_2O \rightarrow CH_2OH—SO_3Na + 2 H^0 \qquad (2)*$$

The test carried out in the technique of spot test analysis cannot be applied in the presence of acetaldehyde or acetaldehyde–bisulfite, nor of compounds which give off hydrogen sulfide or sulfur dioxide when treated with zinc and dilute acid, since these products enter into color reactions with sodium nitroprusside (compare *Inorganic Spot Tests*, Chap. 4).

Procedure.   About 0.5 g of Rongalite is treated in a micro test tube with a drop of the test solution and the mouth of the tube is covered with a disk of filter paper moistened with a drop of the freshly prepared reagent solution (see Sect. 29). The same quantity of granulated zinc (40 mesh) and 1–2 drops of dilute sulfuric acid may be used in place of the Rongalite. The test tube is kept in boiling water for about three minutes. If chloral was present, a blue stain appears on the yellow reagent paper.

   Limit of identification: 10 γ chloral.

Bromal hydrate reacts analogously with respect to its conversion into acetalde-hyde and hence the procedure for the detection of chloral will also apply.

---

   * This equation does not mean formation of hydrogen by pure hydrolysis; this kind of hydrolysis occurs only in the presence of an appropriate hydrogen acceptor.

1 T. JONA, *Z. Anal. Chem.*, 52 (1913) 230.
2 F. FEIGL and C. STARK, *Mikrochim. Acta*, (1955) 996.

31. Citral

Test by alkaline cleavage[1]

The unsaturated aldehyde citral (I) occurs in many ethereal oils. When gently warmed with aqueous caustic alkali, it splits to yield acetaldehyde and methylheptenone (II):[2]

$$(CH_3)_2C=CH(CH_2)_2C=CHCHO + H_2O \rightarrow (CH_3)_2C=CH(CH_2)_2CO + CH_3CHO$$

$$\quad (I) \qquad CH_3 \qquad\qquad\qquad (II) \qquad CH_3$$

Consequently, an indirect test for citral can be obtained through detection of the resulting acetaldehyde by the color reaction with sodium nitroprusside–morpholine mixture (Sect. 29). The sensitivity is low because the treatment of citral with caustic alkali also causes polymerization of the unsaturated aldehyde as shown by browning of the reaction mixture. Compensating for this disadvantage is the specificity, since no other instance is known in which acetaldehyde is produced by the alkaline hydrolysis of an organic compound.

Procedure.   The test is conducted in a micro test tube. A drop of the oil or a drop of its solution in ether, *etc.* is treated with a drop or two of 0.5 *N* alkali. The mouth of the tube is covered with a disk of filter paper that has been moistened with the acetaldehyde reagent. The test tube is placed in boiling water. If citral is present, the paper turns blue.

Limit of identification: 130 γ citral.

1 Unpublished studies, with E. SILVA.
2 Comp. P. KARRER, *Organic Chemistry*, 4th English ed., Elsevier, Amsterdam, 1950, p. 171.

32. Glyoxal

(1) Test with o-aminophenol and calcium oxide[1]

The colorless Schiff base obtained through condensation of *o*-aminophenol and glyoxal, *i.e.* glyoxal-bis(2-hydroxyanil) is a selective and sensitive reagent for calcium ions, with which it gives a water-soluble red inner-complex salt.[2] If *o*-aminophenol is brought together with glyoxal and calcium oxide, the calcium salt is formed even in the cold:

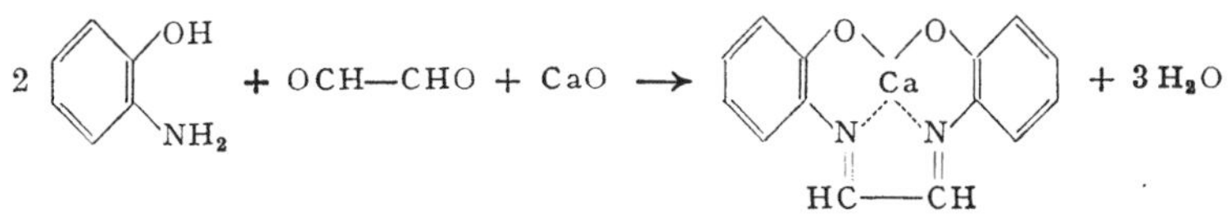

This reaction provides the basis of a specific test for glyoxal since other 1,2-diketones, such as diacetyl and benzil, do not react. No information is available regarding the behavior of methylglyoxal.

The condensation occurs only if an excess of *o*-aminophenol is present; otherwise the calcium salt does not form. Probably in this case there is condensation with only one of the CHO groups of glyoxal. There is no danger that the test here described will fail when applied to a dilute solution of glyoxal.

Procedure.   A drop of the test solution is united in a micro test tube with a little freshly sublimed *o*-aminophenol and ignited calcium oxide. Depending on the amount of glyoxal present, a more or less intense red color appears at once or within a few minutes.

Limit of identification: 3 γ glyoxal.

This limit can be lowered to 1 γ glyoxal if the mixture of *o*-aminophenol and a drop of the test solution is kept in a boiling water bath for 1 min. and then treated with calcium oxide.

1 F. FEIGL and D. GOLDSTEIN, *Z. Anal. Chem.*, 163 (1958) 30.
2 D. GOLDSTEIN and C. STARK-MAYER, *Anal. Chim. Acta*, 19 (1958) 437.

(2) Test with 2-hydrazinobenzothiazole[1]

When glyoxal is warmed with 2-hydrazinobenzothiazole, which is soluble in inorganic acids, a yellow fluorescing compound results. The mechanism of the reaction probably consists in the formation of a dihydrazone which undergoes oxidation:

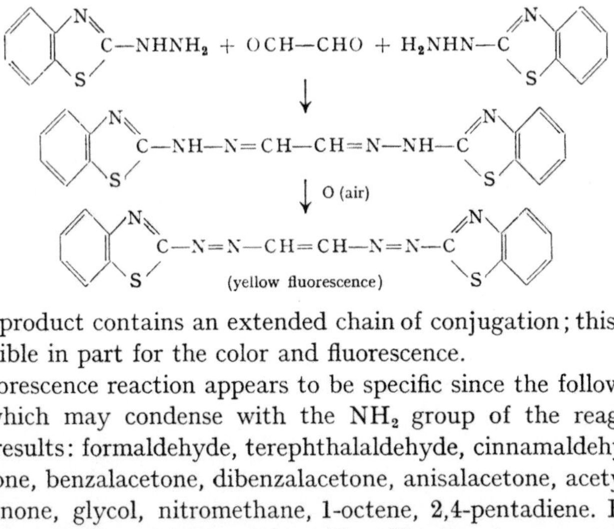

The final product contains an extended chain of conjugation; this probably is responsible in part for the color and fluorescence.

The fluorescence reaction appears to be specific since the following compounds which may condense with the NH₂ group of the reagent gave negative results: formaldehyde, terephthalaldehyde, cinnamaldehyde, acrolein, acetone, benzalacetone, dibenzalacetone, anisalacetone, acetylacetone, anthraquinone, glycol, nitromethane, 1-octene, 2,4-pentadiene. Especially noteworthy were the negative results with α-dicarbonyl compounds such as methylglyoxal, diacetyl, phenanthraquinone, acenaphthenequinone.

Procedure. One drop (0.3 ml) of the reagent solution is placed on the filter paper. A drop of the aqueous test solution is placed in the center of the spot. A positive response is indicated by a stain that shows yellow fluorescence in u.v. light.

Limit of identification: 0.3 γ glyoxal.

1 E. SAWICKI and W. EBERT, *Talanta*, 5 (1960) 63.

(3) Test by transformation into glyoxaline[1]

Glyoxal combines with ammonia to give the heterocyclic compound glyoxaline (imidazole). The mechanism of the reaction is uncertain; one suggestion is[2] that one molecule of glyoxal breaks down into formic acid and formaldehyde, and that the latter reacts as follows:

$$
\begin{array}{c}
\text{NH}_3 \\
\text{HCO} \\
|\quad + \text{H}_2\text{CO} \\
\text{HCO} \\
\text{NH}_3
\end{array}
\longrightarrow
\begin{array}{c}
\text{HC}\!-\!\!-\!\text{N} \\
|\qquad | \\
\text{HC}\quad\text{CH} \\
\diagdown\text{N}\diagup \\
\text{H}
\end{array}
+ 3\,\text{H}_2\text{O}
$$

Glyoxaline being a tertiary amine gives a positive response to the Ohkuma test described on p. 251. The above transformation therefore permits the detection of glyoxal provided that there are in the mixture no tertiary amines or compounds which form products containing tertiary nitrogen atoms when heated with ammonia. The latter is true for pyruvic acid which forms uvitonic acid.[3] A good test for pyruvic acid is described on p. 485.

Procedure. In a micro test tube one drop of the test solution is mixed with one drop of conc. ammonia and the mixture is evaporated to dryness. For further treatment see p. 252. A positive response is indicated by the appearance of a violet color extractable with ether.

Limit of identification: 1 γ glyoxal.

If pyruvic acid is present or if the residue after evaporation with ammonia is colored, it is best to extract the residue with ether, which gives an ethereal solution of glyoxaline. The test is then carried out with the residue which remains after evaporation of the ether.

1 F. FEIGL and S. YARIV, *Talanta*, 12 (1965) 159.
2 K. HOFMANN, *Imidazole and its Derivatives*, Interscience, New York, 1953.
3 A. W. K. DE JONG, *Rec. Trav. Chim. Pays Bas*, 23 (1904) 137.

33. *o*-Phthalaldehyde

Test with barbituric acid and ammonia[1]

o-Phthalaldehyde reacts with barbituric acid in acid solution to give a non-fluorescing condensation product, which turns pink when made alkaline with ammonia and then shows an intense yellow fluorescence. Since no other aromatic aldehyde reacts analogously, this behavior makes possible a selective test for *o*-phthalaldehyde.

Procedure. A drop of saturated solution of barbituric acid in 2 N hydrochloric acid is placed on filter paper and then spotted with a drop of the alcoholic test solution. The spot must not show any fluorescence in u.v. light. A positive response is indicated if the addition of a drop of concentrated ammonia produces a pink stain that fluoresces yellow in u.v. light.

Limit of identification: 0.5 γ *o*-phthalaldehyde

1 V. ANGER and S. OFRI, *Z. Anal. Chem.*, 203 (1964) 427.

34. Furfural

(1) Test through condensation with aniline[1]

Solutions of furfural in organic liquids, or in water (in which it dissolves to the extent of 8%), react with aniline or aniline salts (the acetate is best) to produce a red precipitate or color. In this reaction, the first stage consists of the formation of a light yellow Schiff base[2] through the condensation reaction (1). This then reacts as shown in (2) with a second aniline molecule, with cleavage of the furfural ring and formation of a dianil of hydroxy-glutaconic dialdehyde:[3]

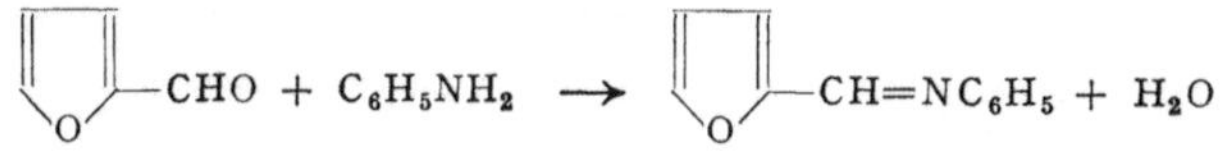

$$\text{[furan]}-CHO + C_6H_5NH_2 \rightarrow \text{[furan]}-CH=NC_6H_5 + H_2O \qquad (1)$$

$$\text{[furan]}-CH=NC_6H_5 + C_6H_5NH_2 \rightarrow C_6H_5NH-CH \quad HC=NC_6H_5 \qquad (2)$$

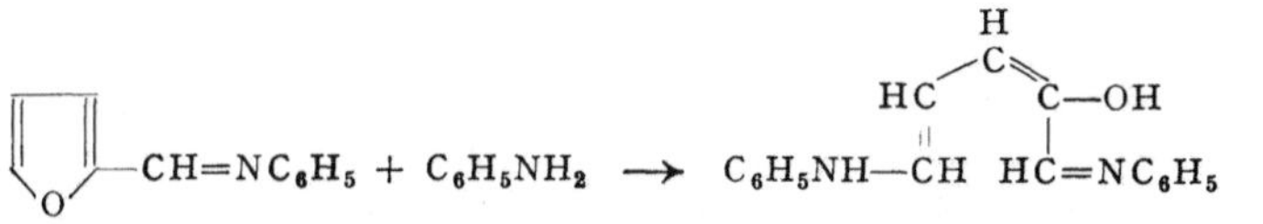

The dianil belongs to the class of polymethine dyes discussed on pp. 245 and 384 in connection with the tests for pyridine and its derivatives.

Although furfural boils at 160°, its vapor pressure at room temperature is quite marked; this aldehyde is also readily volatilized with steam. Therefore, it can be sensitively detected in the gas phase by the production of the red dianil when the vapors come into contact with aniline acetate. The test is not impaired by other volatile aliphatic or aromatic aldehydes.

Procedure.[4] One drop of the test solution (in water, alcohol, ether, *etc.*) is placed in a micro crucible, which is then covered with a disk of filter paper moistened with a 10% solution of aniline in 10% acetic acid. A watch glass is placed on the paper and the crucible is heated to 40°. A positive response is indicated by the appearance of a pink or red stain on the reagent paper.

Limit of identification: 0.05 γ furfural.

1 H. Schiff, *Ber.*, 20 (1887) 540.
2 G. de Chalmot, *Ann.*, 271 (1892) 12.
3 Comp. T. Zincke and G. Muehlhausen, *Ber.*, 38 (1905) 3824; W. Dieckmann and L. Beck, *Ber.*, 38 (1905) 4122.
4 Unpublished studies, with J. E. R. Marins.

(2) Test with phloroglucinol[1]

Like other aromatic aldehydes, furfural reacts with phloroglucinol–hydrochloric acid to give an orange condensation product. However, the latter, in contrast to all the other orange to red-violet reaction products formed with aldehydes and phloroglucinol, rapidly changes to green or blue. See p. 204 regarding the mechanism of the reaction. This behavior is characterististic for furfural and can be employed for its selective identification.

Procedure. A drop of the test solution is mixed in a depression of a spot plate with a drop of saturated phloroglucinol in conc. hydrochloric acid. If furfural is present, a green color appears that is either unchanged or turned to blue by the addition of several drops of water.

Limit of identification: 0.01 γ furfural.

1 V. Anger and S. Ofri, *Z. Anal. Chem.*, 203 (1964) 428.

35. Acetone

Test with salicylaldehyde and sodium hydroxide[1]

In strong alkaline solution, acetone condenses with two molecules of salicylaldehyde to give an intense red product:

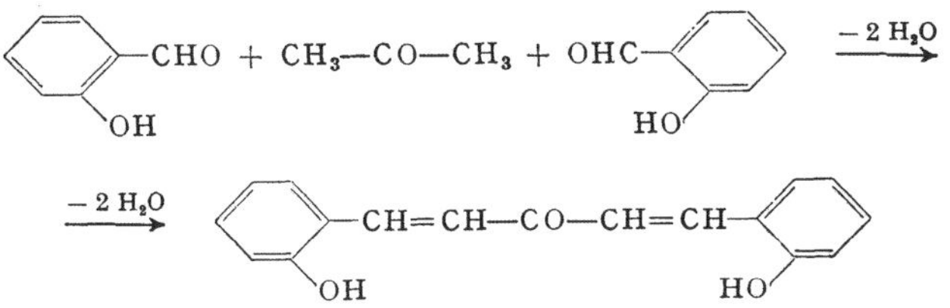

This reaction can be used in the form of a drop reaction as a specific test for acetone.

Procedure.[2] A drop of a 10% alcoholic solution of salicylaldehyde and a drop of 40% sodium hydroxide solution (in water) are placed in a micro test tube. A drop of the test solution is added. The tube is kept for 2–3 min. in water at 70–80°. Depending on the amount of acetone present, a deep red to orange color appears.

Limit of identification: 0.2 γ acetone.

1 R. Fabinyi, *Chem. Zentr.*, 1900 II, 302.
2 V. Anger and S. Ofri, *Z. Anal. Chem.*, 206 (1964) 186.

36. Diacetyl

Test with dicyanodiamide[1]

A test for the guanidyl group described on p. 263 is based on the production of an orange color when such compounds react with diacetyl and caustic alkali or lime.[2] The converse of this test with dicyanodiamide as reagent permits the reliable detection of diacetyl.

The chemistry of the color reaction is unknown, but certain things can be said about its occurrence. If diacetyl is warmed with alkali hydroxide, water is lost and the initial product is an aldol, and the latter, again with loss of water, yields 2,5-dimethylbenzoquinone (xyloquinone):[2]

$$
\begin{array}{ccc}
& \overset{\textstyle CO}{H_3C\diagup\diagdown CO-CH_3} & \\
H_3C-CO\diagdown\diagup CH_3 & & \\
& \underset{\textstyle CO}{} &
\end{array}
\longrightarrow
\begin{array}{c}
\overset{\textstyle CO}{H_2C\diagup\diagdown CO-CH_3} \\
H_3C-\underset{\textstyle\underset{HO}{|}}{C}\diagdown\underset{CO}{\diagup} CH_3
\end{array}
$$

Since neither neutral solutions of diacetyl nor the quinone produced as end-product by alkali hydroxide give a color reaction with guanidyl compounds, it may be assumed that the aldol functions as a participant here. Most probable is a condensation with both NH_2 groups of the guanidyl compound.

Procedure. A micro test tube is used. A drop of the test solution is treated in a micro test tube with several mg of dicyanodiamide and calcium oxide, and the mixture kept in a boiling water bath for 1–3 min. If the response is positive, the calcium oxide is tinted orange. The solution itself is colored when considerable amounts of diacetyl are involved.

Limit of identification: 2 γ diacetyl.

1 Unpublished studies, with C. STARK-MAYER.
2 A. HARDEN and D. NORRIS, *J. Physiol.*, 42 (1911) 332.

37. Acetylacetone

Test through conversion into dimethyldiacetyldihydropyridine[1]

When warmed with formaldehyde in the presence of ammonium acetate, acetylacetone condenses to yield 2,6-dimethyl-3,5-diacetyl-1,4-dihydropyridine:[2]

$$2\ CH_3COCH_2COCH_3 + CH_2O + NH_3 \longrightarrow
\begin{array}{c}
CH_3CO{-}\!\!\diagup\diagdown\!\!{-}COCH_3 \\
CH_3{-}\diagdown\underset{\underset{H}{N}}{\diagup}{-}CH_3
\end{array}
+ 3\,H_2O$$

The compound resulting from this condensation is deep yellow and shows a green fluorescence and accordingly provides the basis for a selective test for acetylacetone. Benzoylacetone yields only a yellow coloration but no fluorescence. (*Idn. lim.* 20 γ.) Acetoacetic ester gives no color but the condensation product fluoresces blue. The reaction described here was previously reported for the detection and determination of formaldehyde.[2]

Procedure. A drop of the alcoholic test solution is mixed with a drop of the reagent solution in a micro test tube. The test tube is kept for 3 min. in water at

60–80°. A yellow color and a green fluorescence in u.v. light indicates the presence of acetylacetone.

> *Reagent:* 25 g of ammonium acetate are dissolved in 100 ml water; 3 ml glacial acetic acid and 0.5 ml 1% formaldehyde are added.

Limit of identification: 2 γ acetylacetone.

1 V. ANGER and S. OFRI, *Z. Anal. Chem.*, 206 (1964) 187.
2 T. NASH, *Biochem. J.*, 55 (1953) 416.

38. *p*-Benzoquinone

Test through conversion into quinonedioxime[1]

Quinonedioxime (I) can react as the tautomeric *p*-nitrosophenylhydroxyl-

$$HON=\langle\!\!\!\!\bigcirc\!\!\!\!\rangle=NOH \quad \rightleftharpoons \quad ON-\langle\!\!\!\!\bigcirc\!\!\!\!\rangle-NHOH$$

$$(I) \qquad\qquad\qquad (II)$$

amine (II) and as such produces blue-violet intercalation complexes with prussic salts (see Chap. 3, Sect. 84). Hydroxylamine readily converts *p*-benzoquinone into the dioxime. Through combination of the two reactions conducted under the conditions described here (the use of the heat-stable zinc salt of pentacyanoammineferroic acid is imperative), it is possible to arrive at a rapid and selective test for benzoquinone.

Procedure.   A drop of the test solution is treated in a micro test tube with a drop of the reagent solution and kept in boiling water for 1 min. If the response is positive, a blue precipitate appears. The reaction is more discernible if the content of the tube is transferred to a filter paper; a sharply outlined blue stain is formed.

> *Reagent:* To a mixture of equal parts of 1% sodium pentacyanoammineferroate and 2% zinc sulfate add 5% of solid hydroxylamine hydrochloride and shake until it is dissolved. The reagent must be thoroughly shaken before each test.

Limit of identification: 2 γ *p*-benzoquinone.

It is remarkable that in the specified time, 100 γ *p*-toluquinone give the same reaction as 2 γ benzoquinone. Other quinones do not react.

1 V. ANGER, *Mikrochim Acta*, (1962) 95.

39. Chloranil

(*1*) *Test by oxidation of tetrabase*[1]

Chloranil (I) acts as an oxidant under suitable conditions. It is reduced to tetrachlorohydroquinone (II) or to the corresponding anion (IIa):

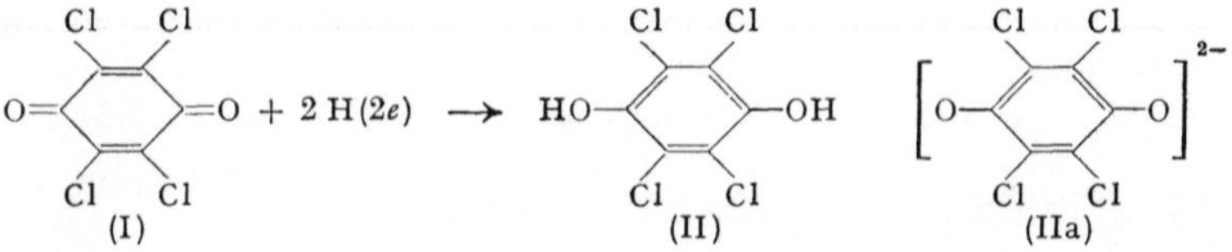

$$O=\underset{\underset{Cl}{\underset{Cl}{\bigotimes}}}{\overset{Cl\quad Cl}{\bigotimes}}=O + 2\,H(2e) \longrightarrow HO-\underset{\underset{Cl}{\underset{Cl}{\bigotimes}}}{\overset{Cl\quad Cl}{\bigotimes}}-OH \qquad \left[O-\underset{\underset{Cl}{\underset{Cl}{\bigotimes}}}{\overset{Cl\quad Cl}{\bigotimes}}-O\right]^{2-}$$

(I)   (II)   (IIa)

For example, the addition of alcoholic chloranil to an acidified alkali iodide or bromide solution liberates iodine or bromine, or, if introduced into an acetic acid solution of tetrabase (III), there is immediate oxidation to a blue basic diphenylmethane dyestuff which contains the quinoidal cation (IV). Consequently, tetrabase may be regarded as the air-stable leuco-form of the dye (IV).

When a dilute ethereal solution of chloranil is treated with a dilute ethereal solution of tetrabase, there is no noticeable change. If, however, a drop of the light yellow mixture is placed on filter paper or in a depression of a spot plate, evaporation of the ether leaves a deep blue residue which is identical with the blue oxidation product (IV) of the tetrabase, which can be produced by other means.[2] The same effect may be observed if alcohol, chloroform, benzene, *etc.* is used as the mutual solvent. Accordingly, the non-reactivity is due to the formation of stable solvates, and the color reaction occurs only after the evaporation of the solvent. The blue product may also be formed by a solid–solid reaction, namely by grinding dry chloranil with dry tetrabase:

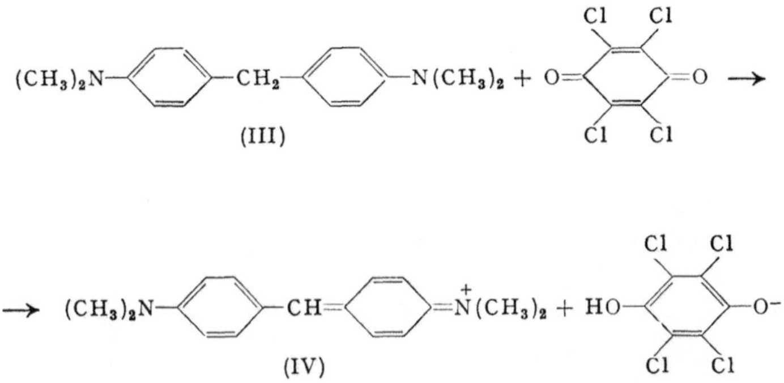

$$(CH_3)_2N-\bigcirc-CH_2-\bigcirc-N(CH_3)_2 + O=\underset{\underset{Cl}{\underset{Cl}{\bigotimes}}}{\overset{Cl\quad Cl}{\bigotimes}}=O \longrightarrow$$

(III)

$$\longrightarrow (CH_3)_2N-\bigcirc-CH=\bigcirc=\overset{+}{N}(CH_3)_2 + HO-\underset{\underset{Cl}{\underset{Cl}{\bigotimes}}}{\overset{Cl\quad Cl}{\bigotimes}}-O^-$$

(IV)

The redox reaction between chloranil and tetrabase in the absence of solvents can be made the basis of a sensitive test for chloranil, even though bromanil and iodanil react analogously. Only benzoyl peroxide (and probably all organic peroxide compounds) oxidize tetrabase to the blue diphenylmethane dyestuff. Accordingly, the test is quite specific.

Procedure. A minimum quantity of the solid or a drop of its solution in ether, benzene, *etc.*, is placed in a depression of a spot plate. One drop of an ethereal solution of tetrabase is added and stirred. A blue residue is left as the solvent evaporates. Alternatively, a drop of the ethereal test solution can be placed on filter paper, which has previously been spotted with a drop or two of the reagent solution. According to the quantity of chloranil present, a deep blue to light blue fleck is left.

Limit of identification: 0.25 γ chloranil.

1 F. FEIGL, V. GENTIL and J. E. R. MARINS, *Anal. Chim. Acta*, **13** (1955) 210.
2 R. MOEHLAU and M. HEINZE, *Ber.*, **35** (1902) 358; comp. J. B. COHEN, *Practical Organic Chemistry*, London, 1949, p. 268.

(2) *Differentiation of chloranil and bromanil*[1]

Test (*1*) does not distinguish between chloranil (m.p. 290°) and bromanil (m.p. 300°). This problem can be solved through the behavior of these compounds toward pyrrole. In ethereal solution, union of the components yields orange colors, a fact previously known about chloranil.[2] These non-sensitive color reactions occur also when benzene, toluene, petroleum ether, *etc.* is used as solvent. If a drop of the colored solution is placed on filter paper and the solvent allowed to evaporate, a blue stain is left if chloranil is present, whereas in the case of bromanil the intial red fleck becomes colorless. Chloranil solutions which are so dilute that they give no orange coloration when pyrrole is added, nevertheless leave a blue stain on filter paper. Chloranil can thus be sensitively detected, even in the presence of bromanil.

The chemistry of the color reaction in solution and after evaporation of the solvent is not known. The following seems plausible: The orange colorations are due to the formation of addition compounds in ether, benzene, *etc.* The bromanil compound is unstable after the solvent has volatilized and decomposes into its components, while a blue condensation product of chloranil and pyrrole (in its pyrrolenine-form, p. 381) is produced after the solvent has evaporated. This view is supported by the observation that the blue stain remaining after evaporation of the chloranil–pyrrole mixture is not washed away with an orange color when spotted with ether *etc.*

Procedure. An area is outlined on filter paper with a pencil, and a drop of the ethereal or benzene test solution is placed within its bounds. After the solvent has evaporated, the spot is treated with a drop of a freshly prepared benzene or ethereal solution of pyrrole. A more or less intense blue stain remains, surrounded by a blue ring. The shade depends on the amount of chloranil present.

Limit of identification: 2 γ chloranil.

Alternatively, a drop of the benzene or ethereal test solution may be placed on a disk of filter paper. After the solvent has disappeared, the fleck is fumed with pyrrole vapors by placing the disk on a micro crucible in which 1–2 drops of pyrrole are warmed with a micro flame. This procedure is about 6–8 times less sensitive than direct spotting, but it has the advantage of permitting the use of even impure pyrrole, whose ethereal or benzene solution is brown.

1 F. Feigl, V. Gentil and C. Stark-Mayer, *Mikrochim. Acta*, (1957) 348.
2 R. Ciusa, *Gazz. Chim. Ital.*, 41 (1907) 666.

40. Phenanthraquinone

Test through oxidation to diphenic acid[1]

If phenanthraquinone is taken to dryness along with dilute hydrogen peroxide, the resulting diphenic acid, like other acidic compounds, yields hydrogen cyanide when dry-heated with mercuric cyanide (see Chap. 2, Sect. 39). The realization of the following series of reactions

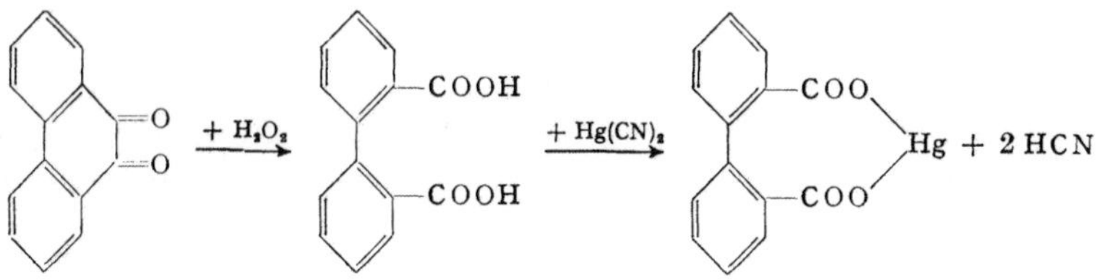

and hence the presence of phenanthraquinone can be revealed by the color reaction of hydrogen cyanide with copper–benzidine acetate (see p. 546) provided no acidic compounds are present in the sample. The test is not impaired by anthraquinone, naphthalene, phenanthrene, or anthracene.

It is essential to use dilute hydrogen peroxide for the following reasons: a) concentrated hydrogen peroxide acts on phenanthrene also producing diphenic acid via the quinone; b) concentrated hydrogen peroxide contains stabilizers (urea, acetamide, *etc.*) which remain after the evaporation and react with mercuric cyanide to yield hydrogen cyanide (comp. Chap. 7, Sect. 15). When dilute hydrogen peroxide is used, there is no danger of an oxidation of phenanthrene and the amount of evaporation residue is so slight that no interference need be feared.

Procedure. A drop of the benzene test solution is taken to dryness in a micro test tube, a drop of 3% hydrogen peroxide is added and the evaporation is repeated. Several cg of mercuric cyanide and a drop of acetone are introduced and after evaporation the open end of the test tube is covered with a piece of filter paper moistened with the hydrogen cyanide reagent solution. The test tube is immersed to a depth of around ½ cm in a glycerol bath that has been heated to about 130°. The temperature is then raised to 160°. A positive response is indicated by the development of a blue stain on the paper.

Limit of identification: 3 γ phenanthraquinone.

1 Unpublished studies, with V. GENTIL.

41. Rhodizonic acid

Test by formation of barium rhodizonate[1]

The dark brown sodium salt of rhodizonic acid (formula see p. 497) is stable when dry. Its yellow solution is an excellent reagent for Ba, Ca, and Pb ions with which it gives red and red-violet precipitates, respectively (see *Inorganic Spot Tests*, Chap. 3). The formation of the barium salt may be employed for the detection of rhodizonic acid.

Analogous reactions are given by tetrahydroxyquinone and croconic acid, whose structures are similar to that of rhodizonic acid (compare p. 419).

Procedure. If the sample is a neutral or alkaline solution of alkali salts of rhodizonic acid, a drop of the solution is treated with a drop of 2% barium chloride solution on a spot plate. Depending on the quantity of rhodizonate, a red precipitate or a pink coloration results. In case acid solutions are to be tested, a drop is treated on a spot plate with as small an amount as possible of barium carbonate which then becomes coated with a red or pink layer.

Limit of identification: 0.1 γ rhodizonic acid.

1 Unpublished studies, with V. GENTIL.

42. Formic acid

(1) Test by conversion to formaldehyde[1]

Formic acid is readily reduced to formaldehyde by nascent hydrogen:

$$HCO_2H + 2 H^0 \rightarrow CH_2O + H_2O$$

The formaldehyde can then be identified by the chromotropic acid test (see p. 434).

Procedure. A drop of the test solution is mixed in a test tube with a drop of 2 N hydrochloric acid; magnesium powder is added until no more gas is liberated. Three milliliters of 12 N sulfuric acid and a little chromotropic acid are then added; the tube is kept for 10 min. at 60° in a water bath. A violet-pink appears if formic acid is present.

Limit of identification: 1.4 γ formic acid.

Solutions of the following compounds were tested with chromotropic acid after treatment with magnesium and hydrochloric acid:

No reaction: glycolic-, glyoxylic-, oxalic-, malic-, citric-, malonic-, salicylic-, uric-, protocatechuic acids; alloxan, arabinose, galactose.

Yellow color: glyceric acid, pyruvic acid.

Yellow to orange color: fructose, saccharose, raffinose.

Yellow to green color: rhamnose.

Glucose interferes with the test because of partial breakdown to formic acid. Consequently, small amounts of formic acid cannot be reliably detected in the presence of much grape sugar.

1 E. EEGRIWE, *Z. Anal. Chem.*, 110 (1937) 22.

(2) *Test by reaction with mercuric chloride*[1]

When formic acid or an alkali formate is warmed with mercuric chloride in acetic acid–acetate buffered solution, white, crystalline mercurous chloride (calomel) precipitates:

$$2\ HgCl_2 + HCO_2^- \rightarrow Hg_2Cl_2 + CO_2 + 2\ Cl^- + H^+$$

Small amounts of mercurous chloride can be detected by the reaction with ammonia (blackening due to finely divided mercury):

$$Hg_2Cl_2 + 2\ NH_3 \rightarrow HgNH_2Cl + NH_4Cl + Hg^0$$

If the following conditions are provided, no reaction is given even by large quantities of acetic, glycolic, lactic, oxalic, tartaric, citric and malic acids. Therefore, the test is recommended for detecting formic acid (formate) in mixtures with carboxylic and sulfonic acids (or their alkali salts).

Procedure.[2] A drop of the acid, neutral, or weakly basic diluted test solution is placed in a micro crucible and one drop of 10% mercuric chloride solution and one drop of buffer solution are added. The mixture is brought to dryness in the oven at 100° (exclude light). The evaporation residue is taken up in a drop of water and a drop of 0.1 N ammonia added. According to the formate content, a more or less intense black to grey color appears. The buffer solution contains 1 ml glacial acetic acid and 1 g sodium acetate per 100 ml water.

Limit of identification: 5 γ formic acid.

This method revealed 5 γ formate in one drop of saturated sodium oxalate solution. This corresponds to a ratio 1 : 370.

1 J. W. HOPTON, *Anal. Chim. Acta*, 8 (1953) 429.
2 Unpublished studies, with D. GOLDSTEIN.

(3) *Test through pyrolytic decomposition with mercuric cyanide*[1]

If metal formates are heated in mixture with mercuric cyanide to 160–180°, hydrogen cyanide is evolved:

$$2 \text{ HCOOMe} + \text{Hg(CN)}_2 \rightarrow 2 \text{ HCN} + \text{Hg}^0 + (\text{COOMe})_2$$

$$\text{Me} = \text{Na, K, } \tfrac{1}{2} \text{ Ca, } \tfrac{1}{2} \text{ Pb, } etc.$$

The hydrogen cyanide released as per the above equation can be sensitively detected in the gas phase through the color reaction described in Sect. 126. It is noteworthy that the reaction occurs as a solid body reaction at the contact sites of the two solid reactants. Perhaps a small amount of the mercuric cyanide sublimes and vapors of this compound react with solid formate. The alkali formates are the most reactive.

Procedure. A drop of the aqueous test solution is taken to dryness along with several cg of mercuric cyanide in a micro test tube. An intimate mixture of solid formate and mercuric cyanide may be tested directly if desired. The test tube is immersed to a depth of about 0.5 cm in a glycerol bath that has been heated to 140° and the mouth of the tube is covered with a piece of filter paper that has been moistened with the hydrogen cyanide reagent. The temperature of the bath is then raised to 180°. A blue stain appears if formate is present.

Limit of identification: 2.5 γ sodium formate.

1 F. FEIGL and D. HAGUENAUER-CASTRO, *Chemist-Analyst*, 50 (1961) 102.

(4) *Test with N-methylquinaldinium toluene-p-sulfonate*[1]

p-Alkylquinaldinium salts condense with formic acid to give a cyanine dyestuff:

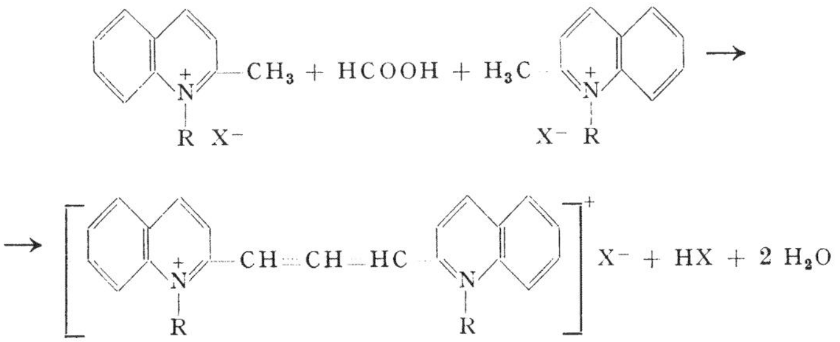

This reaction is selective for formic acid under certain conditions. Among the various compounds that are suitable as reagents in this case, the most appropriate has been found to be N-methylquinaldinium toluene-*p*-sulfonate.

Procedure. One drop of the aqueous test solution is treated in a micro test tube with 10 drops of reagent solution, one drop of pyridine, and one drop of acetic anhydride. After gentle mixing, the tube is kept in boiling water for

1 min. A positive response is shown by a blue to green color. The blank remains yellow.

Reagent: 5% solution of N-methylquinaldinium toluene-*p*-sulfonate in dimethylsulfoxide.

Limit of identification: 0.04 γ formic acid.

Negative results were given by the following compounds: formaldehyde, acetaldehyde, propionaldehyde, glyoxal, glutaraldehyde, butyraldehyde, benzaldehyde, glyoxylic acid, dihydroxyacetone, diacetyl, acetic, oxalic, malonic, butyric, lactic, malic, citric, benzoic, salicylic, piperonylic acids, ethyl formate, allyl alcohol, anisyl alcohol.

1 E. Sawicki, T. W. Stanley, J. Pfaff and J. Ferguson, *Analytical Chemistry, 1962* (Proceedings Feigl Anniversary Symposium, Birmingham, England), p. 62.

43. Acetic acid

(1) Test with lanthanum nitrate and iodine[1]

When lanthanum salts are mixed under suitable conditions with iodine and ammonia in the presence of acetic acid or alkali acetates, a dark blue precipitate or solution results.[2] This probably is due to the adsorption of iodine on basic lanthanum acetate; however, this effect occurs only on a suitable variety of this substrate.

Nitrates, chlorides, bromides, and iodides do not interfere with the detection of small amounts of acetate, even when present in 30 to 40 times excess, but they do weaken the intensity of the blue color. Sulfates, however, interfere in relatively small amounts, and similarly all anions that form insoluble salts with lanthanum (*e.g.*, phosphates), and all cations that give precipitates with ammonia. Sulfates and phosphates may be removed by adding barium nitrate and the test carried out on the filtrate.

Propionates react similarly to acetates. (*Idn. Limit* 50 γ).

Procedure. A drop of the test solution is mixed on a spot plate with a drop of a 5% solution of lanthanum nitrate and a drop of 0.01 N iodine solution. A drop of 1 N ammonia is added, and in a few minutes (in the presence of acetates) a blue to blue-brown ring develops around the drop of ammonia.[3]

Limit of identification: 50 γ acetic acid.

1 D. Krueger and E. Tschirch, *Ber.*, 62 (1929) 2776; 63 (1930) 826.
2 A. Damour, *Compt. Rend.*, 43 (1857) 976.
3 D. Krueger and E. Tschirch, *Mikrochemie*, 8 (1930) 218.

(2) Test by formation of indigo[1]

Acetone is formed by the dry distillation of calcium acetate:

$$\begin{array}{c} H_3C-COO \\ \\ H_3C-COO \end{array}\hspace{-1em}>Ca \;\longrightarrow\; \begin{array}{c} H_3C \\ \\ H_3C \end{array}\hspace{-1em}>CO + CaCO_3$$

When acetone, in alkaline solution, is allowed to react with o-nitrobenzaldehyde, indigo is formed (see test for methyl ketones, p. 209). Starting with small amounts of calcium acetate, indigo can be formed by the action of acetone vapor on a reagent paper impregnated with an alkaline solution of o-nitrobenzaldehyde. In this way, acetic acid can be decisively detected in the presence of formic acid. It is interesting to note that acetone vapor as obtained hy heating acetone or acetone–water mixtures at 100°, reacts slowly and incompletely. The rapid formation of indigo in this test is probably due to the fact that the acetone vapor formed in the distillation of calcium acetate reacts at a higher temperature.

Acetates of other alkaline-earth metals and likewise zinc, magnesium, lead and alkali metal acetates react in the same way as calcium acetate when subjected to dry distillation, i.e., they yield acetone.[2]

Propionic acid and other fatty acids cannot be detected by the indigo reaction; when their calcium salts are subjected to dry distillation, higher ketones (without the CH_3CO group) are formed, but no acetone. This test is less sensitive in the presence of other fatty acids because the acetone is mixed with higher ketones as well as mixed ketones, which are the chief products in the distillation of the calcium salts.

It must be noted that the acetic acid test is unsuccessful in the presence of large amounts of copper, silver, and mercury salts. Chromates and manganese dioxide are, however, without effect.

Procedure. The solid sample is mixed with calcium carbonate, or a drop of the acid solution is evaporated to dryness with excess $CaCO_3$ in the ignition tube (Fig. 28, *Inorganic Spot Tests*). The open end of the tube is covered with a strip of filter paper moistened with freshly prepared saturated solution of o-nitrobenzaldehyde in 2 N sodium hydroxide. The tube is then hung through the asbestos plate and gradually heated with a micro flame. The acetone vaporizes and colors the yellow paper blue or blue-green. When very small amounts of acetate are involved, it is advisable to spot the stain with a drop of 1:10 hydrochloric acid. The original yellow color of the paper is discharged and the blue of the indigo becomes more visible.

Limit of identification: 60 γ acetic acid.

1 F. FEIGL, J. V. SANCHEZ and R. ZAPPERT, *Mikrochemie*, 17 (1935) 165.
2 L. ROSENTHALER, *Pharm. Acta Helv.*, 25 (1950) 366.

(3) Other tests for acetic acid

(a) The acetaldehyde formed by the dry distillation of a mixture of acetate with $Ca(OH)_2$ and calcium formate may be identified by the color reaction with sodium nitroprusside and piperidine (see p. 438).[1] The procedure for this test is the same as for Test (2) (*Idn. Limit*: 10 to 15 γ acetic acid).

(*b*) A drop of acetate solution, that is exactly neutralized, may be identified by the formation of the red-brown complex ferric acetate on addition of $FeCl_3$ (*Idn. Limit*: 10 γ acetic acid).[2] It is best to neutralize acid solutions by heating with $CaCO_3$; alkaline solutions by heating with excess $Zn(NO_3)_2$, and then filtering or centrifuging.

1 J. V. Sanchez, *Chem. Abstr.*, 30 (1936) 4432.
2 Unpublished studies, with R. Seboth.

44. Cinnamic acid

Test by oxidative formation of benzaldehyde[1]

If cinnamic acid is warmed with alkali permanganate solution, benzaldehyde is split off oxidatively,[2] and its vapors give a yellow Schiff base on contact with *o*-dianisidine:

$$C_6H_5CH=CHCOOH + O_2 \rightarrow C_6H_5CHO + OCHCOOH \tag{1}$$

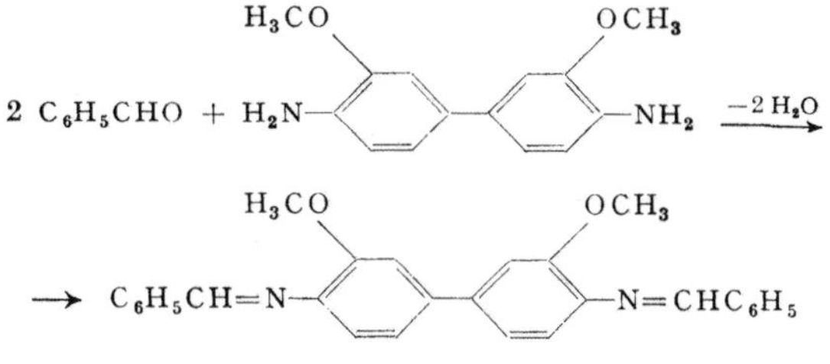

$$(2)$$

Reactions (*1*) and (*2*) are best realized by warming with neutral dilute permanganate solution, although further oxidation of the aldehyde to benzoic acid cannot be completely prevented.

Procedure. A tiny amount of the solid or a drop of its benzene solution is mixed with a drop of neutral 0.5 N potassium permanganate in a micro test tube, whose mouth is then covered with a disk of filter paper impregnated with a freshly prepared saturated ethereal solution of pure *o*-dianisidine. The development of a yellow stain when the reaction mixture is heated in a boiling water bath indicates a positive response.

Limit of identification: 12 γ cinnamic acid.

The test likewise revealed the following derivatives of cinnamic acid:
15 γ cinnamic acid N-cyanomethylamide and 30 γ cinnamaldazine.

1 E. Jungreis (Jerusalem), unpublished studies.
2 A. Michael and W. W. Garner, *Am. Chem. J.*, 35 (1906) 265.

45. Oxalic acid

(1) Test through conversion into oxamide[1]

If oxalic acid is evaporated with excess ammonium hydroxide the residue consists of ammonium oxalate, which on dry heating (150–280°) yields water and oxamide. A mixture of the latter and thiobarbituric acid (m.p. 235°) gives a brick-red condensation product on heating to 140–150° (see p. 536). The formation of oxamide can be conducted in the presence of thiobarbituric acid; a specific test for oxalic acid results.

Procedure. A drop of the solution being tested for oxalic acid is evaporated in a micro test tube along with 1 or 2 drops of concentrated ammonia. Several mg of thiobarbituric acid are added and the test tube is placed in a bath that has been preheated to 130°. The temperature is then increased to 140–160°. If oxalic acid was present, a red product, soluble in alcohol, is rapidly formed.

Limit of identification: 1.6 γ oxalic acid.

1 Unpublished studies, with D. HAGUENAUER-CASTRO.

(2) Test by formation of diphenylamine (aniline) blue[1]

When oxalic acid is heated to 240–250° with diphenylamine (m.p. 54°; b.p. 302°), diphenylamine blue (also known as aniline blue) results. It is assumed[2] that oxalic acid is broken down at the reaction temperature to give formic acid (1), which reacts with diphenylamine to form leucoaniline blue (2), which undergoes autoxidation to aniline blue (3).

$$(COOH)_2 \rightarrow HCOOH + CO_2 \qquad (1)$$

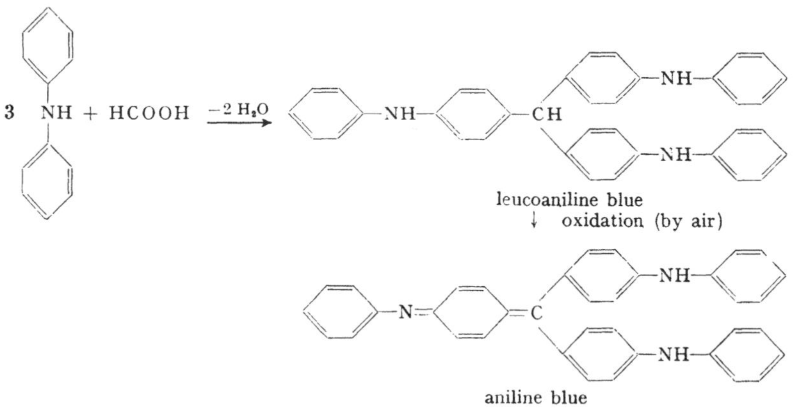

leucoaniline blue
↓ oxidation (by air)

aniline blue

Considerable quantities of malic acid interfere because it yields oxalic acid when heated to 250°. Formic, acetic, propionic, tartaric, citric, succinic, dihydroxymaleic, benzoic, phthalic, tricarballylic, glycolic, and glyoxylic acids do not react under the conditions of the test, so that the formation of the dye in the reaction with diphenylamine is very selective for oxalic acid.

Oxidizing materials, both organic and inorganic, must be absent if the fusion test with diphenylamine is used because they transform the latter into a blue dye (see p. 301).

Procedure. A tiny crystal of the sample (a solution must be evaporated to dryness) is melted with a little diphenylamine in a micro test tube over a free flame. After cooling, the melt is taken up in a drop of alcohol; a blue color indicates the presence of oxalic acid.

Limit of identification: 5 γ oxalic acid.

If oxalate ions are to be detected in a mixture containing other anions precipitated by Ca^{2+} ions, it is advisable to proceed as follows: The acetic acid solution is treated with $CaCl_2$ solution; the precipitate is collected by centrifuging, and freed from water, either by drying or by washing with alcohol and ether. A little of the precipitate is mixed with diphenylamine in a test tube, then a drop of syrupy phosphoric acid is added, and the mixture heated over a free flame. Calcium phosphate is formed and oxalic acid is liberated which can react with the diphenylamine. The liquid turns blue, but the color fades on cooling. If the melt is taken up in alcohol, a brilliant blue solution is obtained. If taken up in water, the excess diphenylamine is precipitated and made light blue by adsorption of the dye. The dye can be extracted from the aqueous solution with ether, which increases the sensitivity of the test.

The oxalates of thorium and other rare earths which can be precipitated from mineral acid solution behave similarly to calcium oxalate.

1 F. Feigl and O. Frehden, *Mikrochemie*, 18 (1935) 272.
2 Comp. P. Karrer, *Organic Chemistry*, 4th Engl. ed., Elsevier, Amsterdam, 1950, p. 614.

46. Esters of oxalic acid

Test through conversion into oxamide [1]

The solid esters of oxalic acid respond to the test for oxalic acid described in Sect. 45. A supplementary method can be based on the transformation of the esters into oxamide.

If a mixture of an ester of oxalic acid, urea, and thiobarbituric acid is heated to 120°, the red (water- and ether-soluble) product characteristic for oxamide is formed (see Sect. 116). Probably oxamide results through the reaction of the ester with urea below the decomposition temperature of the latter (135°). The following reaction may be involved:

$$\begin{matrix} COOR \\ | \\ COOR \end{matrix} + CO(NH_2)_2 \longrightarrow \begin{matrix} CONH_2 \\ | \\ CONH_2 \end{matrix} + OC\begin{matrix} \diagup OR \\ \diagdown OR \end{matrix}$$

The test described here permits the detection of esters of oxalic acid in the presence of esters of other dicarboxylic acids. The mono- and dihydrazide of oxalic must be absent since these compounds likewise yield oxamide when dry-heated with urea (see p. 556).

Alkyl oxamates behave like oxamide due to the disproportionation:

$$2 \text{ ROOC—CONH}_2 \rightarrow (\text{COOR})_2 + (\text{CONH}_2)_2$$

Procedure. A small quantity of the ester or a drop of its alcoholic or ethereal solution is mixed in a micro test tube with a few mg of urea and a few cg of thiobarbituric acid. This mixture is heated in a glycerol bath to 120°. A positive reponse is indicated by the appearance of a red product.

Limits of identification:

2.5 γ cyclohexyl oxalate	3.5 γ diphenyl oxalate
5 γ n-butyl oxamate	2.5 γ dimethyl oxalate

1 Unpublished studies, with D. HAGUENAUER-CASTRO.

47. Malonic acid

Test through conversion into barbituric acid[1]

If malonic acid is fused with urea, barbituric acid results though in poor yield:

$$
\begin{array}{ccccc}
\text{COOH} & \text{NH}_2 & & \text{CO—NH} & \\
| & | & & | \quad | & \\
\text{CH}_2 & + \text{ CO} & \rightarrow & \text{H}_2\text{C} \quad \text{CO} & + 2 \text{ H}_2\text{O} \\
| & | & & | \quad | & \\
\text{COOH} & \text{NH}_2 & & \text{CO—NH} &
\end{array}
$$

Since barbituric acid is detected selectively and sensitively through reaction with pyridylpyridinium dichloride (see p. 542), combination of the two reactions provides the basis for a specific test for malonic acid, since esters and amides of malonic acid do not react.

Procedure. A little of the solid or the evaporation residue of a drop of its solution is taken for the test and dissolved in a micro test tube in a saturated methanolic solution of urea. The alcohol is driven off and the residual mixture is heated for several minutes in a glycerol bath at 120°. A drop of a 1% solution of pyridylpyridinium dichloride in dimethyl formamide is added, and the mixture then heated at 120° for several minutes. If malonic acid is present, a reddish-blue color develops on cooling; it fluoresces red under u.v. light.

Limit of identification: 50 γ malonic acid.

1 V. ANGER and S. OFRI, *Talanta*, 10 (1963) 1302.

48. Succinic acid

Test through formation of pyrrole[1]

Pyrrole is formed when succinimide is subjected to dry heating with metallic zinc (zinc dust distillation).[2] Ammonium succinate behaves similarly, since it is tranformed into succinimide when dry heated:

$$
\begin{array}{c}
CH_2-CO_2NH_4 \\
|\\
CH_2-CO_2NH_4
\end{array}
\longrightarrow
\begin{array}{c}
CH_2-CO \\
|\quad\quad\rangle NH + NH_3 + 2\ H_2O \\
CH_2-CO
\end{array}
\qquad (1)
$$

$$
\begin{array}{c}
CH_2-CO \\
|\quad\quad\rangle NH + 2\ Zn^\circ \\
CH_2-CO
\end{array}
\longrightarrow
\begin{array}{c}
CH=CH \\
|\quad\quad\rangle NH + 2\ ZnO \\
CH=CH
\end{array}
\qquad (2)
$$

The pyrrole resulting from this succession of reactions (*1*) and (*2*) can be detected in the gas phase by the blue color produced with an acid solution of *p*-dimethylaminobenzaldehyde (see p. 381). Accordingly, a test for succinic acid can be based on this color reaction.

To arrive at the maximum sensitivity, the zinc dust distillation should be conducted in the presence of ammonium chloride, and a trichloroacetic acid rather than a hydrochloric acid solution of *p*-dimethylaminobenzaldehyde should be used. The ammonium chloride, through the reaction:

$$ZnO + 2\ NH_4Cl \rightarrow ZnCl_2 + H_2O + 2\ NH_3$$

removes the film of zinc oxide from the surface of the metallic zinc and so facilitates reaction (*2*). The use of trichloroacetic acid (b.p. 197°) instead of aqueous hydrochloric acid (b.p. 110°) assures a heat-stable acid milieu for the pyrrole reaction, which is not the case with a hydrochloric acid solution, because of the high temperature of the zinc dust distillation.

It is remarkable that succinic acid yields pyrrole if dry heated with ammonium chloride and zinc; accordingly, it behaves in the same manner as ammonium succinate of succinimide. Obviously the latter is formed via ammonium succinate in the sintering reactions.

$$
\begin{array}{c}
CH_2-CO_2H \\
|\\
CH_2-CO_2H
\end{array}
+ 2\ NH_4Cl \xrightarrow{-2\ HCl}
\begin{array}{c}
CH_2-CO_2NH_4 \\
|\\
CH_2-CO_2NH_4
\end{array}
\xrightarrow[-2H_2O]{-NH_3}
\begin{array}{c}
CH_2-CO \\
|\quad\quad\rangle NH \\
CH_2-CO
\end{array}
$$

Consequently, it is possible to detect both succinimide (see Sect. 118) and succinic acid through the formation of pyrrole by zinc dust distillation in the presence of ammonium chloride. It is to be expected that derivatives of succinic acids will react analogously.

Procedure. A drop of the test solution or a small amount of the solid is taken to dryness in a micro test tube along with a drop of a 0.5% solution of ammonium chloride and several mg zinc dust. The mouth of the tube is covered with a disk of filter paper moistened with a 5% solution of p-dimethylaminobenzaldehyde in a 20% benzene solution of trichloroacetic acid. The bottom of the test tube is heated vigorously with a micro flame for about 1 min. Depending on the amount of succinic acid or succinimide, a blue-violet or pink stain appears on the reagent paper.

Limit of identification: 5 γ succinic acid.

The test described above is not applicable in the presence of malic, mucic, or oxalic acid. The first yields hydroxypyrrole under the prescribed conditions and this behaves like the parent compound toward the aldehyde. If oxalic acid is dry-heated along with NH_4Cl, ammonium oxalate is formed initially and then oxamide. The latter is reduced by nascent hydrogen to ethylenediamine which yields pyrrole when pyrolyzed (see Sect. 84). The possible presence of malic acid can be detected by the procedure given in Sect. 55 and of mucic acid as outlined in Sect. 57.

1 F. FEIGL, V. GENTIL and C. STARK-MAYER, *Mikrochim. Acta*, (1957) 344.
2 C. A. BELL, *Ber.*, 13 (1880) 817.

49. Monochloroacetic acid

(1) Test with concentrated sulfuric acid and chromotropic acid[1]

If monochloroacetic acid (b.p. 189°) is warmed with concentrated sulfuric acid to 150–180°, considerable amounts of formaldehyde result. This effect can be ascribed to two partial reactions in which concentrated sulfuric acid participates without however appearing in the corresponding equations. The first step is hydrolysis of chloroacetic acid to glycolic acid:

$$CH_2ClCOOH + H_2O \rightarrow CH_2OHCOOH + HCl \qquad (1)$$

This hydrolysis is accomplished by the water contained in the concentrated sulfuric acid. Under the conditions of the test, the water is in a state comparable to superheated steam and accordingly is particularly reactive. The glycolic acid produced in (1) is — like other α-hydroxycarboxylic acids — cleaved by the concentrated sulfuric acid, functioning as a dehydrant,[2] with formation of formaldehyde:

$$CH_2OHCOOH \rightarrow H_2O + CO + CH_2O \qquad (2)$$

Formaldehyde reacts with chromotropic acid and concentrated sulfuric acid to give a red-violet quinoid compound (comp. p. 434), the concentrated sulfuric acid serving as condensing and oxidizing agent. Therefore, if chloroacetic acid is heated with concentrated sulfuric acid in the presence of chromotropic acid, the color characteristic of formaldehyde appears. In the test described here, the concentrated sulfuric acid accordingly acts as dehydrating agent, oxidant, and water donor. This is supported by the finding that compounds which lose water when heated above 150° act in the same fashion as concentrated sulfuric acid. They include cellulose and hydrated sulfates of zinc or manganese.

Procedure. The test is conducted in a micro test tube. A tiny crystal of the sample, or a drop of its aqueous solution is kept at 105° for 30 min, and then treated with 1–2 drops of concentrated sulfuric acid. One or two tiny particles of chromotropic acid are added and the mixture heated over a micro flame or in a glycerol bath at 170°. A more or less intense violet color appears in accordance with the amount of monochloroacetic acid present.

The heating may also be carried on in the gas absorption apparatus (Fig. 23, *Inorganic Spot Tests*). The formaldehyde in the vapor phase is allowed to come into contact with a drop of concentrated sulfuric acid containing some chromotropic acid.

Limit of identification: 5 γ monochloroacetic acid.

It is characteristic of the test described here that the color reaction does not occur at temperatures below 150°. Free formaldehyde and the majority of compounds yielding formaldehyde show this color reaction even at room temperature or after brief warming in the water bath. It is therefore advisable, if no color appears or only a pale coloration, to continue the heating in a glycerol bath previously brought to 170°. If a violet color results or if the shade deepens, the presence of monochloroacetic acid is assured.

1 F. FEIGL and R. MOSCOVICI, *Analyst*, 80 (1955) 803.
2 A. BISTRZYCKI and B. V. SIEMIRADSKI, *Ber.*, 39 (1906) 52.

(2) Test through formation of nitromethane[1]

The classic method or preparing nitromethane is based on the successive reactions:[2]

$$CH_2ClCOOH + NaNO_2 \rightarrow NaCl + CH_2NO_2COOH \rightarrow CH_3NO_2 + CO_2$$

Even though the yield of nitromethane, which is removed from the reaction theater by steam distillation, is only around 33% of the theory,[3] the extremely sensitive test for this compound with sodium 1,2-naphthoquinone-4-sulfonate (see p. 558) permits an indirect test for monochloroacetic acid. No separation of the nitromethane is needed; it is sufficient to add the reagent to the mixture of monochloroacetic acid and alkali nitrite containing an excess of calcium oxide. The test given here is specific in the absence of compounds that contain active CH_2- and NH_2 groups since they form colored products directly with the reagent (see p. 153).

Procedure.   A drop of the aqueous test solution is mixed in a micro test tube with several mg of calcium oxide and sodium nitrite. After warming for 2–3 min. in the water bath, 1 or 2 drops of a 0.5% solution of sodium 1,2-naphtho-quinone-4-sulfonate is added. A more or less intense blue or violet color appears.

Limit of identification: 5 γ monochloroacetic acid.

1 F. FEIGL and D. GOLDSTEIN, *Anal. Chem.*, 29 (1957) 1522.
2 W. STEINKOPF, *Ber.*, 41 (1908) 4457; W. STEINKOPF and G. KIRCHHOFF, *Ber.*, 42 (1909) 3439.
3 H. LIEB and W. SCHOENIGER, *Anleitung zur Darstellung organischer Präparate mit kleinen Substanzmengen*, Springer, Vienna, 1956, p. 67.

(3) Test through formation of rhodanine [1]

Rhodanine is formed by heating monochloroacetic acid with ammonium thiocyanate [2]:

$$CH_2ClCOOH + 2\ NH_4CNS + H_2O \longrightarrow \underset{\displaystyle \overset{\textstyle SC \quad CH_2}{\diagdown S \diagup}}{\overset{\textstyle HN\!-\!\!-CO}{|\qquad |}} + NH_4Cl + CO_2 + 2\ NH_3$$

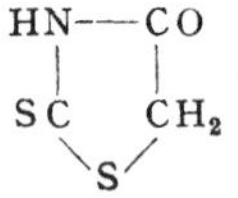

Rhodanine is among those compounds which, by virtue of reactive CH_2 groups, form *para*-quinoidal colored condensation products with sodium 1,2-naphthoquinone-4-sulfonate (see Chap. 2, Sect. 44). The compound resulting from rhodanine is blue-violet. It is formed by conducting the above reaction and then treating the resulting reaction mass with the reagent and alkali hydroxide. This constitutes the basis of the present test for monochloroacetic acid (and also monobromoacetic acid), which is reliable provided amines and compounds with reactive CH_2 groups are absent.

Procedure.   The test is made in a micro test tube. A drop of the aqueous, alcoholic, *etc.* test solution is mixed with several mg of ammonium thiocyanate and warmed in a boiling water bath for 2 min. The cooled mixture is treated with 1–2 drops of 0.5% solution of sodium 1,2-naphthoquinone-4-sulfonate and then made alkaline with 1% NaOH solution. If monochloro(bromo)acetic acid was present, a more or less blue color appears.

Limit of identification: 5 γ monochloroacetic acid.

1 Unpublished studies, with V. GENTIL.
2 M. NENCKI, *J. Prakt. Chem.*, [2] 16 (1876) 2.

50. Dichloroacetic acid

Test by formation of glyoxylic acid [1]

If dichloroacetic acid is mixed with Nessler reagent (alkaline solution of alkali iodomercurate), there is immediate deposition of finely divided mercury. This effect is based on the formation of alkali glyoxylate:

$$CHCl_2COO^- + 2\ OH^- \rightarrow OCHCOO^- + 2\ Cl^- + H_2O \qquad (1)$$

which because of its aldehydic nature reduces iodomercurate:

$$OCHCOO^- + [HgI_4]^{2-} + 3\ OH^- \rightarrow Hg^0 + 4\ I^- + 2\ H_2O + (COO)_2^{2-} \qquad (2)$$

The present test applies the fact that dichloroacetic acid (b.p. 193°) is somewhat volatile with steam, and consequently reactions (1) and (2) occur when its vapors come into contact with Nessler reagent. The test is not very sensitive because of the incomplete volatilization of dichloroacetic acid with steam but this disadvantage is counterbalanced by the selectivity when the goal is the detection of dichloroacetic acid in a mixture with other acids.

Procedure.   The aqueous or acetone test solution is mixed in a micro test tube with 1 cg of oxalic acid dihydrate and taken to dryness. The test tube is then placed in a glycerol bath previously heated to 160°, and the mouth of the tube is covered with a disk of qualitative filter paper moistened with Nessler reagent. The development of a dark stain on the paper within a few minutes indicates a positive response.

Limit of identification: 50 γ dichloroacetic acid.

1 Unpublished studies, with E. JUNGREIS.

51. Trichloroacetic acid

Test through formylation of phenol [1]

A test for chloroform is based on the Reimer–Tiemann formylation of phenol in aqueous alkali solution, followed by the formation of yellow-green fluorescent salicylaldazine through hydrazine (see p. 181). These processes can be employed for the detection of compounds which split off chloroform when treated with caustic alkali. Pertinent examples are chloral (see p. 439) and also trichloroacetic acid:

$$CCl_3COOH + 2\ NaOH \rightarrow CHCl_3 + Na_2CO_3 + H_2O$$

The above reaction occurs rapidly; however, losses of $CHCl_3$ are inevitable in view of the subsequent saponification:

$$CHCl_3 + 4\ NaOH \rightarrow HCOONa + 3\ NaCl + 2\ H_2O$$

Therefore, the test described here is not very sensitive. It cannot be applied directly in the presence of chloral.

Procedure. As described in Sect. 30 for the detection of chloral and bromal. *Limit of identification:* 50 γ trichloroacetic acid.

1 Unpublished studies, with V. GENTIL.

52. Glycolic acid

(1) Test with 2,7-dihydroxynaphthalene and sulfuric acid[1]

When glycolic acid ($CH_2OHCOOH$) is heated with a solution of 2,7-dihydroxynaphthalene (I) in concentrated sulfuric acid, a violet to violet-red color gradually develops. This color reaction probably depends on the condensation of the formaldehyde (split off from the glycolic acid by the action of the concentrated sulfuric acid[2]) with the 2,7-dihydroxynaphthalene, in the position *ortho* to one of the OH groups:

$$HOCH_2COOH \longrightarrow CH_2O + CO + H_2O$$

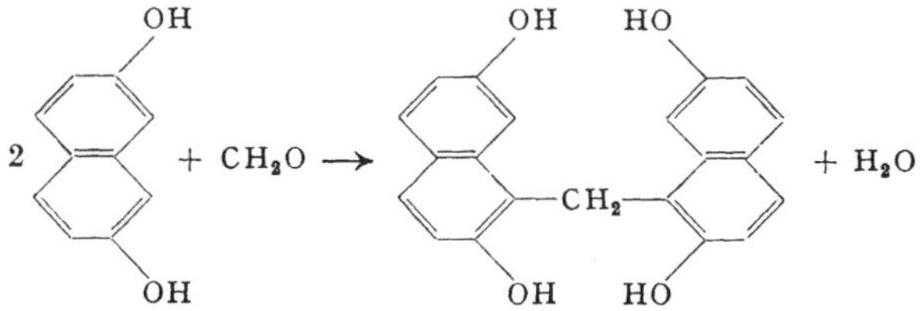

The colorless product, 2,2',7,7'-tetrahydroxy-1,1'-dinaphthylmethane (II), dissolved in sulfuric acid, is gradually oxidized to a deep red-violet dyestuff, whose constitution is not known.[3] It is probable that (II) is oxidized to a quinoidal compound by the concentrated sulfuric acid (compare p. 435).

Neither formic, acetic, oxalic, succinic, citric, benzoic, nor salicylic acid interfere with the test. Both lactic acid[4] and malic acid give a yellow color and green fluorescence with the reagent, whereas tartaric acid gives an olive to dark green color. This reaction permits the detection of lactic acid[4]. Aldehydes such as salicylaldehyde, anisaldehyde or acetaldehyde react with 2,7-dihydroxynaphthalene similarly to formaldehyde with formation of oxidizable condensation products. Under the conditions of the test, certain glycols give yellow colors.[5]

Procedure. A drop of the test solution is mixed in a micro test tube with 2 ml of the reagent solution and heated for 10–15 min. in a water bath. A red to violet-red appears, according to the amount of glycolic acid present.

Reagent: 1 mg 2,7-dihydroxynaphthalene dissolved in 10 ml concentrated sulfuric acid.

Limit of identification: 0.2 γ glycolic acid.

A freshly prepared reagent is yellow, with a green fluorescence, but both color and fluorescence disappear on heating for a short time, or after standing overnight in a stoppered bottle. The reagent is gradually colored violet by the action of a number of inorganic oxidizing agents.

The test is also applicable in the presence of citric acid. The glycolic acid can be detected by the appearance of an orange to orange-red color and a green fluorescence. In this way, 1 γ glycolic acid may be detected in the presence of 20,000 times the quantity of citric acid.

1 E. Eegriwe, *Z. Anal. Chem.*, 89 (1932) 123.
2 G. Denigès, *Bull. Trav. Soc. Pharm. Bordeaux*, 49 (1909) 193.
3 Comp. W. Wolff, *Ber.*, 26 (1893) 83.
4 A. Bondi (Rehovoth), private communication. See also Ch. weizmann, E. Bergmann and Y. Hirshberg, *J. Am. Chem. Soc.*, 58 (1936) 1675.
5 R. S. Pereira, *Mikrochemie Ver. Mikrochim. Acta*, 35/36 (1951) 398.

(2) Detection with chromotropic acid and sulfuric acid [1]

As stated in Test *1* and also with respect to the detection of aminoacetic acid (Sect. 79), glycolic acid yields formaldehyde when warmed with concentrated sulfuric acid. Since formaldehyde is readily detected by the violet color it produces with chromotropic acid (see Sect. 28), a mixture of chromotropic acid and sulfuric acid can serve as a reagent for glycolic acid. The absence of formaldehyde and compounds which split off formaldehyde is assumed.

Procedure.   A drop of the test solution is taken to dryness at 105–110° in a micro test tube. The residue is treated with 2 or 3 drops of the reagent (preparation see p. 435) and the mixture is kept for 3 min. in a boiling water bath. A violet color develops if glycolic acid was present.

Limit of identification: 0.2 γ glycolic acid.

1 Unpublished studies, with C. Stark-Mayer.

53. Lactic acid

(1) Test with p-hydroxydiphenyl and sulfuric acid [1]

When lactic acid is gently warmed with concentrated sulfuric acid, it decomposes into acetaldehyde and formic acid; the latter immediately breaks down:

$$CH_3CH(OH)COOH \rightarrow CH_3CHO + HCOOH (\rightarrow H_2O + CO)$$

The acetaldehyde reacts with *p*-hydroxydiphenyl (I), probably by condensation at the position *ortho* to the OH group, and forms di-(*p*-hydroxydiphenyl)-ethane (II):

$$2\ HO-\hspace{-4pt}\bigcirc\hspace{-6pt}\bigcirc\hspace{-4pt}\text{ + CH}_3\text{CHO} \longrightarrow \text{H}_2\text{O} + \text{H}_3\text{C}-\text{CH}\begin{array}{c} HO-\bigcirc\hspace{-6pt}\bigcirc \\[6pt] HO-\bigcirc\hspace{-6pt}\bigcirc \end{array}$$

(I)   (II)

In sulfuric acid solution, (II) is slowly oxidized to a violet product of unknown constitution. Therefore, analogous to the detection of glycolic acid by means of 2,7-dihydroxynaphthalene (Sect. 52), the present test involves an aldehyde–phenol reaction in which concentrated sulfuric acid functions as condensing and oxidizing agent. In fact, metaldehyde, paraldehyde, aldol, and propionaldehyde react similarly to acetaldehyde with (I) and sulfuric acid to give deep violet products. With formaldehyde the color is blue-green, with butyraldehyde red, and with oenanthal orange. Accordingly, α-hydroxybutyric acid and pyruvic acid behave like lactic acid.

Procedure. A drop of the test solution and 1 ml concentrated sulfuric acid are heated for 2 min. in a dry test tube in a water bath at 85°. After cooling under the tap to 28°, a pinch of solid p-hydroxydiphenyl is added, the mixture is swirled several times, and left for 10–30 min. The violet color appears gradually and deepens after some time.

Limit of identification: 1.5 γ lactic acid.

1 E. EEGRIWE, *Z. Anal. Chem.*, 95 (1933) 323 ff.

(2) *Test with o-hydroxydiphenyl and sulfuric acid*[1]

The acetaldehyde formed on heating lactic acid with concentrated sulfuric acid can also be detected by the blue fluorescence with o-hydroxydiphenyl. Formaldehyde, metaldehyde, paraldehyde, and the next higher homologs of the aldehyde series behave analogously to acetaldehyde. Pyruvic acid does not interfere. The chemical basis of the fluorescing reaction is not known.

Procedure. A drop of the test solution is mixed with a crystal of o-hydroxydiphenyl and 0.5–1 ml concentrated sulfuric acid in a test tube and heated for 2 min. at 85° in a water bath. It is then examined for a blue fluorescence while holding the test tube against black paper.

Limit of identification: 1 γ lactic acid.

1 E. EEGRIWE, *Z. Anal. Chem.*, 95 (1933) 323 ff.

(3) *Test through transformation into pyruvic acid*[1]

Lactic acid is easily oxidized to pyruvic acid:

$$\text{CH}_3\text{CHOHCOOH} + [\text{O}] \rightarrow \text{CH}_3\text{COCOOH} + \text{H}_2\text{O}$$

When the oxidation is carried out with lead dioxide, followed by the specific condensation of pyruvic acid with β-naphthylamine (outlined in Sect. 64), lactic acid can be detected.

Procedure. In a micro centrifuge tube a drop of the test solution is mixed with some mg of lead dioxide, and heated on the water bath for 15 min. After addition of 3 drops of water and centrifugation the clear liquid is transferred to a micro test tube. The remainder of the procedure is the same as described on p. 485.

Limit of identification: 10 γ lactic acid.

1 F. FEIGL and S. YARIV, *Talanta*, 12 (1965) 159.

54. Glyceric acid

Test with naphthoresorcinol and sulfuric acid [1]

When an aqueous solution of glyceric acid is heated with concentrated sulfuric acid containing a little dissolved naphthoresorcinol, an intense blue appears.

The chemistry of this color reaction, which occurs also with esters of the acid, probably can be regarded from the same standpoint as the aldehyde–phenol reactions for glycolic and lactic acid.

Glycolic acid and glycine give a brown color; lactic, α- and β-hydroxybutyric, erythronic, gluconic, glyoxylic, pyruvic, levulinic, malic, saccharic, citric, mesoxalic, and dihydroxytartaric acid do not react under the conditions of the test. Tartronic and quinic acids give a green color, glucuronic acid a yellow color with a greenish fluorescence, tartaric acid a green to blue-green color, and malic acid a yellow color and a blue fluorescence. [2]

Procedure. A drop of the test solution is mixed with 0.75 ml of the reagent solution and heated for 30–50 min. in a water bath at 90°. A light to deep blue color appears when glyceric acid was present.

Reagent: 0.01 g naphthoresorcinol dissolved in 100 ml 96% sulfuric acid.

Limit of identification: 10 γ glyceric acid.

1 E. EEGRIWE, *Z. Anal. Chem.*, 95 (1933) 323 ff.
2 Comp. C. NEUBERG and H. LUSTIG, *Exptl. Med. Surg.*, 1 (1934) 14.

55. Malic acid

Test with β-naphthol and sulfuric acid [1]

A bluish fluorescence is produced when malic acid is heated with concentrated sulfuric acid containing a little β-naphthol. The chemistry of this reaction is not known. Probably the malic acid is split by the concentrated sulfuric acid (phosphoric acid behaves analogously) to produce an aldehyde, which then condenses with the β-naphthol.

The following acids do not interfere: oxalic, tartaric, citric, succinic, cinnamic, benzoic, salicylic, acetic, formic. A few hydroxy acids such as glycolic and tartaric cause a more or less intense green fluorescence.

Procedure. To detect malate ion in a drop of a solution or in a mixture with calcium salts of citric and other organic acids, the test material is treated with 1 ml of the naphthol reagent and heated briefly on the water bath. If malic acid was present, a yellowish color with a blue fluorescence appears.

When dilute solutions of malic acid or its salts are involved, it is best to add a little oxalate and then calcium acetate. The precipitate of calcium oxalate functions as collector for the slight quantities of calcium malate. The fluorescence produced by the action of β-naphthol is visible, even in the presence of calcium sulfate, provided the quantity of malic acid is not exceedingly small.

Reagent: 0.0025 g β-naphthol dissolved in 100 ml of 96% sulfuric acid.

Limit of identification: 10 γ malic acid.

1 E. EEGRIWE, *Z. Anal. Chem.*, 89 (1932) 121.

56. Tartaric acid

(1) Test with sulfuric acid and gallic acid[1]

If calcium tartrate is heated with concentrated sulfuric acid, containing a little gallic acid, a blue color results. Very small amounts of tartaric acid give a blue-green to yellow-green color. The concentrated sulfuric acid splits tartaric acid (I) and produces glycolaldehyde (II):

$$HOOC(CHOH)_2COOH \longrightarrow H_2O + CO_2 + CO + OHCCH_2OH$$
$$(I) \qquad\qquad\qquad\qquad\qquad\qquad (II)$$

The aldehyde then forms a colored condensation product with gallic acid, $C_6H_2(OH)_3COOH$. The concentrated sulfuric acid probably functions as dehydrant and oxidant. (Compare the Le Rosen test, p. 137.)

Colored products are likewise produced by glycolic, tartronic, glyceric, and glyoxylic acids, and also formaldehyde, and carbohydrates. No color is formed by oxalic, citric, succinic, lactic, malic, cinnamic, and salicylic acid and the fatty acids.

Procedure. A very small portion of the calcium precipitate is used for this test. It should be filtered, washed, and dried on a hardened filter paper (to avoid interference from fibers of cellulose which also react), transferred to a test tube, treated with 1 ml of a 0.01% solution of gallic acid in concentrated sulfuric acid, and heated to 120–150° over a free flame. When only an extremely small amount of the calcium precipitate is available, it is best to dissolve it on the filter in 2 N sulfuric acid and to use a drop of the filtrate for the test.

The following colors are obtained with different amounts of tartaric acid:

100 γ blue   10 γ blue-green   5 γ bluish green   2 γ yellowish green

1 E. EEGRIWE, *Z. Anal. Chem.*, 89 (1932) 121.

(2) Test with β,β′-dinaphthol and sulfuric acid[1]

A green fluorescence appears if tartaric acid is heated with concentrated sulfuric acid containing $β,β′$-dinaphthol:

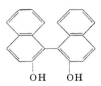

The following compounds give no color reaction with this reagent: lactic, α-hydroxybutyric, β-hydroxybutyric, erythronic, pyruvic, levulinic, oxalic, saccharic, citric, salicylic, m-hydroxybenzoic, p-hydroxybenzoic, protocatechuic, gallic, α, β- and γ-resorcylic, pyrogallolcarboxylic, quinic, and cinnamic acids; formaldehyde and carbohydrates.

Various colors are given by: glycolic and glyoxylic acids (red-brown to brown), glyceric and mesoxalic acids (gray), gluconic, glucuronic, dihydroxy-tartaric, tartronic, and malic acids (green).

Procedure. The solid or a drop of its solution is treated with a little solid $β,β′$-dinaphthol or a few ml of its 0.05% solution in concentrated sulfuric acid, and kept for half an hour in a water bath at 85°. When tartaric acid is present, a more or less strong luminous green fluorescence gradually appears during the heating, and deepens on cooling. The violet fluorescence of the reagent itself disappears gradually.

As little as 10 γ of tartaric acid can be detected if a drop of the test solution is heated for 20–30 min. with 1–2 ml of the reagent solution; 0.05 g tartaric acid can be detected in the presence of any amount of oxalic or succinic acid, and in the presence of 1000 times the amount of citric acid, or 150 times the amount of cinnamic acid, or 10 times the amount of malic acid.

Limit of identification: 10 γ tartaric acid.

1 E. EEGRIWE, *Z. Anal. Chem.*, 95 (1933) 326.

57. Mucic acid

Detection through pyrolytic formation of pyrrole[1]

Dry distillation of ammonium mucate yields pyrrole, pyrrole-α-carboxylic acid, carbon dioxide, ammonia and water.[2] The production of pyrrole can be represented by the net reaction:

$$
\begin{array}{l}
\text{COONH}_4 \\
\quad | \\
\text{CHOH—CHOH} \\
\quad | \qquad\qquad \longrightarrow \\
\text{CHOH—CHOH} \\
\quad | \\
\text{COONH}_4
\end{array}
\qquad
\begin{array}{l}
\text{CH} = \text{CH} \\
\quad | \qquad \text{>NH} \; + \; 2\,\text{CO}_2 \; + \; 4\,\text{H}_2\text{O} \; + \; \text{NH}_3 \\
\text{CH} = \text{CH}
\end{array}
$$

The formation of ammonium mucate and its decomposition as given above can be realized in simple fashion by melting or sintering mucic acid along with urea or better biuret at 170–180°. Both compounds function as NH_3 donor and local overheatings and other decompositions, which occur in the dry distillation, are avoided by this treatment. The pyrrole, which results in the pyrolysis, can be sensitively detected in the gas phase by means of the color reaction with p-dimethylaminobenzaldehyde (see p. 381).

Procedure. A little of the solid or a drop of the solution is mixed in a micro test tube with several mg of biuret and taken to dryness if necessary. A disk of filter paper moistened with a benzene solution of p-dimethylaminobenzaldehyde and trichloroacetic acid is placed over the mouth of the test tube that is placed in a glycerol bath preheated to 130° and the temperature raised to 180°. If the response is positive, a violet stain appears on the reagent paper.

Limit of identification: 1 γ mucic acid.

The test described here can be used in the indirect detection of various sugars (dulcitol, galactose, quercitol, *etc.*) and also vegetable products which yield mucic acid on oxidation with nitric acid or on acid hydrolysis.

Oxalic, malic and succinic acid must be absent because they yield pyrrole or pyrrole derivatives under the conditions of the test.

1 Unpublished studies, with D. GOLDSTEIN.
2 C. A. BELL, *Ber.*, 10 (1877) 1866.

58. Citric acid

(*1*) *Test through conversion into pentabromoacetone*[1]

Alkali permanganate converts citric acid into acetonedicarboxylic acid:

$$
\begin{array}{l}
\text{CH}_2\text{COOH} \\
\quad | \\
\text{HOOC—C—OH} \quad + \,[\text{O}] \; \longrightarrow \\
\quad | \\
\text{CH}_2\text{COOH}
\end{array}
\qquad
\begin{array}{l}
\text{CH}_2\text{COOH} \\
\quad | \\
\text{CO} \qquad\qquad + \; \text{CO}_2 \; + \; \text{H}_2\text{O} \\
\quad | \\
\text{CH}_2\text{COOH}
\end{array}
$$

If this oxidation is conducted in the presence of bromine, colorless penta-bromoacetone precipitates:

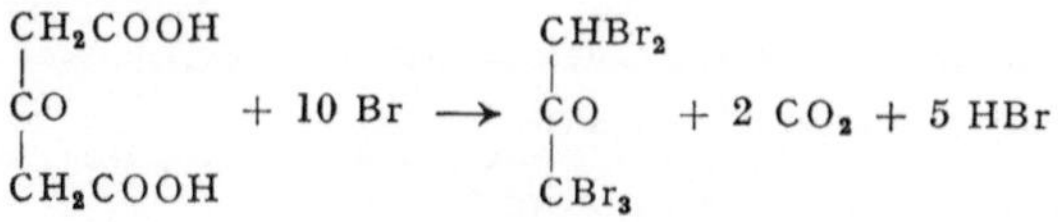

$$
\begin{array}{ccc}
CH_2COOH & & CHBr_2 \\
| & & | \\
CO & + 10\ Br \rightarrow & CO \quad + 2\ CO_2 + 5\ HBr \\
| & & | \\
CH_2COOH & & CBr_3
\end{array}
$$

The test described here may be applied also to soluble citrates. Oxalic, tartaric, malic, and succinic acids and their salts do not interfere. Organic compounds that form precipitates through reaction with bromine must be absent, or the spurious precipitate must be removed beforehand by centrifuging.

Procedure.[2]   A drop of the test solution is mixed in a micro test tube with one drop each of 0.01 N $KMnO_4$ and saturated bromine water and faintly warmed. After cooling, the excess bromine and the MnO_2 formed eventually are destroyed by the addition of some solid sulfosalicylic acid. A positive response is indicated by the appearance of a white precipitate or turbidity.

Limit of identification: 6 γ citric acid.

1 A. WÖHLK, *Z. Anal. Chem.*, 41 (1902) 91.
2 V. GENTIL (Rio de Janeiro), unpublished studies.

(2) *Test by conversion into ammonium citrazinate by fusion with urea*[1]

Citrazinic acid (2,6-dihydroxypyridine-4-carboxylic acid) can be prepared from citric acid by various methods.[2] Its ammonium salt, either solid or dissolved, has an intense blue fluorescence, which can be applied to reveal citrazinic acid produced from citric acid.[3] This salt is formed in amounts adequate for analytical purposes by fusing citric acid (m.p. 153°) with urea (m.p. 130°) at 150°. It is essential that the ammonia, produced in the conversion of urea into biuret at this temperature:

$$2\ CO(NH_2)_2 \rightarrow H_2NOC—NH—CONH_2 + NH_3$$

reaches the citric acid as quasi superheated ammonia. The net reaction (in which the triamide of citric acid is perhaps an intermediate) is:

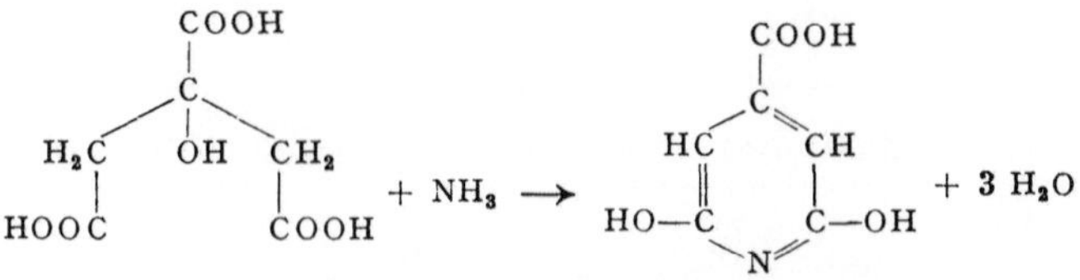

When this reaction is used analytically, it should be determined whether the sample has a self-fluorescence or becomes fluorescent when heated to 150°

with no addition of urea. Tartaric acid in amounts below 500 γ shows no fluorescence. No reports regarding the behavior of other hydroxy acids are available. The quenching of the fluorescence of ammonium citrazinate by the addition of dilute mineral acid provides a check, since the blue fluorescence is revived by adding ammonia.

Procedure.[4]   One drop of the solution to be tested is evaporated to dryness in a micro test tube or a little of the solid may be used. About 100–200 mg of urea are added, and the tube is kept for 2 min. in a glycerol bath heated to 150°. After cooling, the solidified melt is taken up in 2 or 3 drops of water and examined in ultraviolet light for fluorescence. A blank without urea, and examination of the behavior of the fluorescing solution after addition of 1 : 5 hydrochloric acid (see above) are advisable.

Limit of identification: 2 γ citric acid.

It is very remarkable that the solid, non-melting alkali and alkaline-earth citrates likewise yield blue fluorescent products when subjected to this procedure.[5] Probably the respective salts of citrazinic acid are produced through solid body reactions. Consequently, the fusion reaction with urea can be employed also for the detection of salts of citric acid.

Aconitic acid likewise responds to the test described here.

1 A. STEIGMANN, *J. Soc. Chem. Ind.*, 62 (1943) 176.
2 A. BEHRMAN and A. W. HOFMANN, *Ber.*, 17 (1884) 2688; see also: T. H. EASTERFIELD and W. J. SELL, *J. Chem. Soc.*, 65 (1894) 29; E. E. LEINIGER, *Anal. Chem.*, 24 (1952) 1967.
3 F. FEIGL and V. ANGER, *Mikrochemie*, 17 (1935) 35.
4 Unpublished studies, with V. GENTIL.
5 H. E. FEIGL, unpublished studies.

59. Ascorbic acid (vitamin C)

Ascorbic acid (I), a lactone of an unsaturated sugar acid in which two enol groups produce the acidity, is converted, by even weak oxidizing agents, into dehydroasorbic acid (II) which actually is not an acid but a neutral lactone. It may be reconverted to ascorbic acid by reduction:

$$
\begin{array}{c}
\overset{\displaystyle\mathrm{O}}{\overbrace{\mathrm{O\!=\!C\!-\!COH\!=\!COH\!-\!CH\!-\!CHOH\!-\!CH_2OH}}} \quad (I) \\[2pt]
\text{oxidation} \updownarrow \text{reduction} \\[2pt]
\overset{\displaystyle\mathrm{O}}{\overbrace{\mathrm{O\!=\!C\!-\!CO\!-\!CO\!-\!CH\!-\!CHOH\!-\!CH_2OH}}} \quad (II)
\end{array}
$$

Stronger oxidizing agents seriously oxidize (I) or (II). Ascorbic acid is therefore an active reductant, which can be detected by color tests based on redox reactions with soluble or insoluble inorganic and organic participants.

(1) Test by reduction of manganese dioxide[1]

Ascorbic acid reduces manganese dioxide to a manganous salt. Paper impregnated with finely divided MnO_2 is prepared by treating filter paper with potassium permanganate solution whereby part of the cellulose is oxidized, and finely divided MnO_2 is precipitated in the paper. The color ranges from brown to almost white according to the MnO_2 content. Larger quantities of ascorbic acid are revealed directly by the white fleck that results when a drop of the test solution is placed on the brown paper. Practically colorless reagent paper should be used for minute quantities of ascorbic acid. The removal of MnO_2, in this case, is made visible by bathing the spotted paper in benzidine solution which forms "benzidine blue" with traces of MnO_2 (compare *Inorganic Spot Tests*, Chap. 3). Consequently, the whole surface will turn blue, except those spots reduced by ascorbic acid.

Procedure.   A drop of the acetic acid, neutral, or alkaline test solution is placed on the reagent paper. After 1–2 min. the paper is bathed in benzidine solution. A white fleck on the blue paper indicates the presence of ascorbic acid.

Reagents: 1) Manganese dioxide paper. 1 ml 0.2 N $KMnO_4$ is diluted to 1000 ml. Filter paper is soaked in this solution for 15 min. After draining, it is dried in a stream of heated air.

2) Saturated solution of *o*-tolidine chloride diluted just before use with an equal volume of water.

Limit of identification: 0.03 γ ascorbic acid (in 0.004 ml).

Other reducing compounds in weakly acid solution exert the same effect on finely divided manganese dioxide, *i.e.*, they reduce it to a manganous salt. Citric acid is an example. To detect ascorbic acid in the presence of the latter (in fruit juices, for instance) the test solution is shaken beforehand with an excess of calcium carbonate. A drop of the suspension is then applied to the reagent paper. This preliminary treatment produces insoluble unreactive calcium citrate while the action of ascorbic acid is affected but slightly.

1 F. FEIGL and H. T. CARDOSO, *Rev. Brasil. Biol.*, 2 (1942) 117.

(2) Test by reduction of ammonium phosphomolybdate[1]

The complexly bound MoO_3 molecules in the yellow water-soluble phosphomolybdic acid, $H_3PO_4.12\,MoO_3$.aq., are far more easily reduced than the hexavalent molybdenum of normal molybdate ions. Accordingly, certain reducing agents, which in acid solution are without effect toward MoO_4^{2-} ions react with the complex phosphomolybdate as shown by the formation of blue colloidally dispersed lower molybdenum oxide ("molybdenum blue"). Since even traces of the latter are easily discernible, phosphomolybdic acid is a good reagent for revealing reducing compounds (see p. 130). The enhanced reducibility of the complex'y bound molybdenum is retained in the insoluble yellow ammonium phosphomolybdate, but this is true only with respect to strong reductants. Ascorbic acid is in this category and can be detected by spotting it on paper impregnated with $(NH_4)_3PO_4.12 MoO_3$. Of course, ascorbic acid likewise reduces the water-soluble phosphomolybdic acid.[2]

Procedure. A drop of the acid, neutral, or alkaline test solution is placed on the reagent paper. According to the amount of ascorbic acid, a blue or green fleck appears on the yellow paper immediately, or after several minutes.

> *Reagent:* Ammonium phosphomolybdate paper. Filter paper is immersed in a saturated alcoholic solution of phosphomolybdic acid, drained, and dried in a current of cold air. It is then bathed in concentrated ammonium nitrate solution containing several drops of nitric acid. After washing with water, the paper is dried in a blast of heated air. It will keep for several days in the dark. On long standing and exposure to daylight, the yellow reagent paper turns blue because of reduction by the cellulose.

Limit of identification: 0.1 γ ascorbic acid (in 0.01 ml).

Since citric acid gives no reaction under the conditions just prescribed, is is possible to detect ascorbic acid in the presence of as much as 1000 parts of citric acid.

Uric acid and urates react readily with ammonium phosphomolybdate, but mineral acids prevent this reduction. Consequently, ascorbic acid can be detected in urine, provided the sample is acidified beforehand.

1 F. Feigl and H. T. Cardoso, *Rev. Brasil. Biol.*, 2 (1942) 117.
2 N. Bagssonoff, *Compt. Rend.*, 173 (1921) 466; *Nature*, 139 (1937) 468.

(3) Test through precipitation of cuprous ferrocyanide[1]

Ascorbic acid reduces ammoniacal cupric solutions; the blue color is discharged. However, Cu_2^{2+} ions are detected with greater sensitivity through appropriate precipitation reactions from ammoniacal solutions. In contrast to the corresponding cupric salts, both cuprous ferrocyanide and cuprous ferricyanide are insoluble in ammonia. Accordingly, if ammoniacal cupric solutions are reduced by ascorbic acid in the presence of alkali ferri- or ferrocyanide, white $Cu_3[Fe(CN)_6]$ or $Cu_4[Fe(CN)_6]$ precipitates immediately.

Hydrazine and hydroxylamine salts as well as thiol compounds must be absent because they show reducing actions similar to that of ascorbic acid.

Procedure. A watch glass on black paper, or a black spot plate is used. A drop of the ammoniacal, neutral, or weakly acid test solution is mixed with a drop of reagent solution. A white precipitate or turbidity appears, according to the amount of ascorbic acid present.

Reagent: A slight excess of 0.2 N potassium ferricyanide is added to 0.5 *N* copper sulfate and the resulting light brown precipitate is dissolved in ammonia. The emerald green solution keeps for several days.

Limit of identification: 0.5 γ ascorbic acid.

1 F. FEIGL and M. STEINHAUSER, *Mikrochemie Ver. Mikrochim. Acta,* 35 (1950) 553.

(4) Test with chloranil [1]

Ascorbic acid is readily oxidized by chloranil. In alcohol or water–alcohol solution, the redox reaction is discernible through the discharge of the yellow color of the chloranil solution, provided rather large quantities of the reactants are present. The redox reaction occurs also in dilute solutions, and permits the detection of the ascorbic acid through the consumption of chloranil. Since very dilute chloranil solutions are almost colorless, it is necessary to use the highly sensitive test for chloranil with tetrabase (compare Sect. 39), which is based on the oxidation of the latter to a blue diphenylmethane dyestuff. Accordingly, if the acetic acid test solution is treated with a little chloranil, warmed, and tetrabase then added, a negative response to the chloranil test shows that the latter has been consumed and consequently the presence of ascorbic acid is indicated.

As in the other tests for ascorbic acid based on redox reactions, the absence of other reducing agents is necessary. Citric, tartaric, oxalic and formic acid do not react with chloranil.

Procedure. A drop of the aqueous neutral or acetic acid test solution is treated in a micro test tube with a drop of 0.001% alcoholic chloranil solution and the mixture is warmed for 5 min. in the water bath. One drop of tetrabase solution is then added, and the heating continued for 2–3 min. A comparison test with a drop of water or dilute acetic acid should be made. The presence of ascorbic acid is indicated if, in contrast to the blank, no blue color appears or if a distinctly weaker blue results.

Reagent: 0.2 g tetrabase is dissolved in 1 : 1 acetic acid, and the volume made up to 10 ml with saturated sodium acetate solution.

Limit of identification: 0.5 γ ascorbic acid.

1 H. E. FEIGL and J. E. R. MARINS, unpublished studies.

(5) Other tests for ascorbic acid

a) The brown solution of ferric ferricyanide (equal parts of 1% $Fe_2(SO_4)_3$, 0.4% $K_3Fe(CN)_6$ and 8% acetic acid) gives a blue precipitate or color with ascorbic acid (*Idn. Limit:* 3 γ ascorbic acid).[1] Considerable quantities of cysteine, pyrogallol and glutathione react similarly to ascorbic acid.

Alkaline solutions of ascorbic acid are not stable in air. However, this autoxidation proceeds slowly enough that tests based on redox reactions with ammoniacal solutions of ascorbic acid are possible. This is the basis of the reactions *b–d*.[2]

b) A red precipitate of cuprous *p*-dimethylaminobenzylidenerhodanine appears when an ammoniacal solution of ascorbic acid is added to a mixture of 0.5 *N* copper sulfate plus an equal volume of 1 *N* sodium pyrophosphate plus several drops of alcoholic *p*-dimethylaminobenzylidenerhodanine. The reagent solution keeps for several hours. A drop is placed on filter paper and spotted with one drop of the ammoniacal test solution (*Idn. Limit*: 0.05 *γ*).

c) A drop of $Na_2CuP_2O_7$ solution (preparation, see *b*), to which has been added several crystals of KCNS, is mixed on a watch glass or black spot plate with a drop of the ammoniacal test solution. Cuprous thiocyanate is formed as a white precipitate or turbidity (*Idn. Limit*: 2 *γ*).

Hydrazine or hydroxylamine interfere with tests *a–c*.

d) A white fleck results when paper impregnated with a little PbO_2 or Tl_2O_3 is spotted with the ammoniacal test solution. The reduction of these brown higher oxides is made more evident by bathing the dried paper in a solution of *o*-tolidine hydrochloride.

e) The ninhydrin test for amino acids can also serve for the detection of ascorbic acid; a reddish-brown fleck or ring is obtained (*Idn. Limit*: 10 *γ*). Large amounts of glucose interfere, but citric acid does not. Amino acids and primary aliphatic amines must be destroyed by adding a drop of 0.5% $NaNO_2$ to the acid test solution and warming.

f) The test for reducing sugars by means of triphenyltetrazolium chloride (p. 338) can be employed for the detection of ascorbic acid. In view of the instability of alkaline solutions of ascorbic acid, the test solution must be added to a boiling mixture of two drops of triphenyltetrazolium chloride solution and one drop of 0.5% NaOH (*Idn. Limit*: 0.2 *γ*).[3]

g) Ascorbic acid is one of the few organic compounds which reduce selenious acid at room temperature to red elementary selenium[4] (*Idn. Limit* 1 *γ*). Thiourea and many of its derivatives (see p. 136) as well as organic derivatives of hydrazine show the same behavior. Strong inorganic reductants must be absent because they form selenium with SeO_2.

h) The selective test for reducing compounds with *p*-nitrosodimethylaniline plus *p*-dimethylaminobenzaldehyde (see p. 134) can serve to detect ascorbic acid (*Idn. Limit* 1 *γ*).[5]

1 H. TAUBER, *Mikrochemie*, 17 (1935) 111.
2 F. FEIGL and M. STEINHAUSER, *Mikrochemie Ver. Mikrochim. Acta*, **35** (1950) 553.
3 A. BONDI (Rehovoth), unpublished studies.
4 V. E. LEVINE, *Proc. Soc. Exptl. Biol. Med.*, 35 (1936) 231.
5 F. FEIGL and D. GOLDSTEIN, *Microchem. J.*, 1 (1957) 177.

60. Salicylic acid

Test by conversion into fluorescent alkali salicylates[1]

Both solid and dissolved alkali and alkaline-earth salts of salicylic acid exhibit a strong blue-violet fluorescence in ultraviolet light. This effect can be observed even with minute quantities, provided the salicylate is present alone. The fluorescence test is not reliable if applied directly, say by adding alkali to solutions containing salicylic acid, since the alkali salts of many carboxylic acids and phenols, as well as numerous other organic compounds, also fluoresce. However, salicylic acid can be separated from accompanying materials by converting it into methyl- or ethyl salicylate. These esters are formed by heating solutions of salicylic acid in methanol or ethanol with concentrated sulfuric acid, which acts as dehydrant in the reaction:

$$C_6H_4(OH)COOH + ROH \rightarrow C_6H_4(OH)COOR + H_2O$$

Although these esters boil at 223° and 232°, respectively, they are sufficiently volatile at room temperature or on gentle heating to permit their saponification in the gas phase on contact with caustic alkali or magnesia, with consequent formation of the fluorescing salt.

Salicylic acid can also be adequately separated from organic admixed materials by merely heating with concentrated sulfuric acid. The loss of water from one or two molecules leads to the salicylic anhydride (I) or (II):

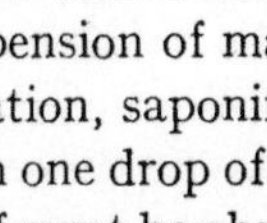

(I) 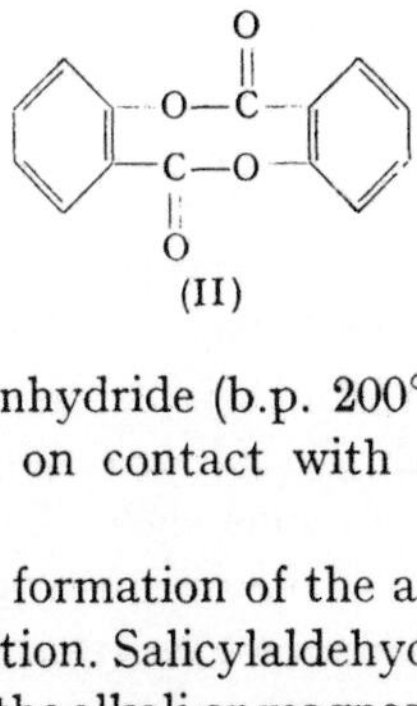   (II)

The vapor pressure of this sublimable anhydride (b.p. 200°) is sufficient at 130° to produce fluorescent salicylates on contact with a caustic alkali solution or suspension of magnesia.

The esterification, saponification, and formation of the anhydride can be carried out with one drop of the test solution. Salicylaldehyde or compounds which split it off must be absent because the alkali or magnesium salts of this phenolic aldehyde exhibit a blue-green fluorescence in ultraviolet light.

Procedure.   The apparatus (Fig. 23, *Inorganic Spot Tests*) is used. A drop of the test solution is evaporated to dryness in the bulb (it is not essential to evaporate alcoholic solutions). A drop of concentrated sulfuric acid is then added. The knob of the stopper is dipped into 3% potassium hydroxide solution or an aqueous suspension of magnesium oxide or hydroxide, and put into position. The closed apparatus is placed in an oven at 130° for 5 min. If salicylic acid was present, a violet fluorescence will be observed on the knob.

Limit of identification: 5 γ salicylic acid.

This esterification procedure will reveal acetylsalicylic acid (aspirin) $C_6H_4(OCOCH_3)COOH$, with an *identification limit* of 7 γ, following removal of the CH_3CO group by evaporation with dilute caustic alkali. The procedure can also be applied to alkali salts of salicylic acid.

1 Unpublished studies, with H. BLOHM.

61. Phenoxyacetic acid

Test through pyroammonolytic release of phenol [1]

If guanidine carbonate is heated above 200°, considerable amounts of biguanide are formed and ammonia, water, and carbon dioxide are split off. If phenoxyacetic acid is present during the heating, the hot ammonia vapors bring about the splitting:

$$C_6H_5OCH_2COOH + NH_3 \rightarrow CH_2(NH_2)COOH + C_6H_5OH$$

The occurrence of this reaction, which cannot be accomplished by the wet method, can be recognized through detection of the evolved phenol vapors by means of the indophenol reaction (see p. 185). It should be noted that phenol esters behave in the same manner as phenoxyacetic acid when pyrolyzed with guanidine carbonate.

Procedure. A small amount of the solid test material or a drop of its solution in benzene is placed in a micro test tube along with several cg of guanidine carbonate and taken to dryness if necessary. A piece of filter paper impregnated with a saturated benzene solution of 2,6-dichloroquinone-4-chloroimine is placed over the mouth of the tube. The latter is immersed in a glycerol bath whose temperature is brought to 240°. A positive response is indicated by the appearance of a blue fleck on the reagent paper within 1–2 min.

Limit of identification: 50 γ phenoxyacetic acid.

1 Unpublished studies, with D. HAGUENAUER-CASTRO.

62. Coumarin

(1) Test through conversion into fluorescing alkali salt of o-hydroxycinnamic acid [1]

Coumarin dissolves in caustic alkali with cleavage of the pyrone ring and formation of alkali salts of *o*-hydroxycinnamic acid or their anion:

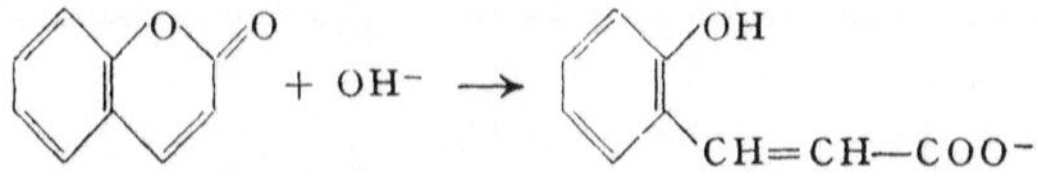

No fluorescence in ultraviolet light is exhibited by the freshly prepared alkaline solution or by solutions stored in the dark for months. However, when irradiated with ultraviolet light, these solutions give a yellow-green fluorescence within a few minutes, and the intensity increases to a maximum. (Prolonged exposure to daylight also produces the fluorescence.) This photo-effect may be demonstrated easily if a fresh alkaline solution of coumarin is brought on filter paper, and half of the spot is covered with black paper, and irradiated with ultraviolet light. The exposed portion begins to fluoresce within one minute, and the intensity increases. If the rest of the spot is uncovered after 6–8 min., the strongly fluorescing segment stands out sharply from the other portion. If the irradiation is then resumed, the portion of the spot originally shielded also begins to fluoresce, and after sufficient exposure the entire area shows a uniform fluorescence.

It is logical to assume that when coumarin is dissolved in alkali, the initial product is the non-fluorescing *cis*-form (*a*) of *o*-hydroxycinnamic acid, which is converted by irradiation into an isomeric fluorescing *trans*-form (*b*):

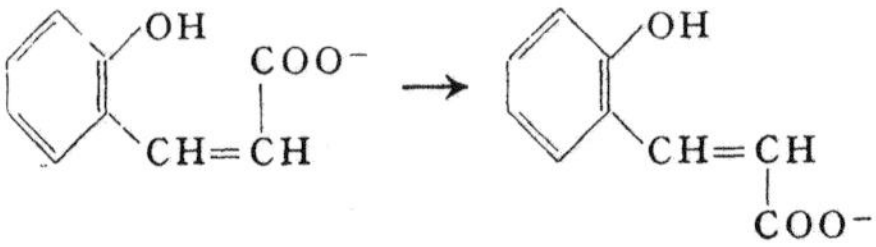

(*a*)   (*b*)

Comparison of (*a*) and (*b*), *i.e.* of the anions of coumarinic and coumaric acids, shows that only in (*b*) is the H atom of the phenol group so placed that its chelation (6-membered ring) to an unsaturated C atom is not hindered by the COOH group. It is probable that the chelated bonding is causally related to the fluorescence.

The tests described here demonstrate that the photo-effect also appears in the alkaline solutions of such coumarin derivatives which have no free OH groups in the benzene ring. Substituents seem to have an influence on the rate at which the fluorescence appears. For instance, if a drop of an alkaline solution of bergapten is placed on filter paper and exposed to ultraviolet rays, the start of the fluorescence can be discerned only after about 10 min. and the increase in the intensity of the fluorescence is much slower as compared with an equimolar coumarin solution.

Coumarin derivatives, which have free OH groups in the benzene ring (umbelliferone, esculetin, daphnetin, *etc.*) are known to fluoresce deep blue in the solid state and in alkaline solution. No photo-effect occurs in alkaline solutions of these compounds.

The fluorescence following ultraviolet irradiation of alkaline solutions of coumarin can be discerned with as little as 0.005 γ coumarin if a drop of the solution is exposed on filter paper. The corresponding dilution limit is 1 : 10,000,000. This type of test is not specific for coumarin since its derivatives containing no hydroxyl groups show the same behavior, and since hydroxycoumarins make it difficult to establish any fluorescence produced by irradiation because they possess a self-fluorescence.

A specific test for coumarin can be attained if use is made of the fact that this compound has a distinct vapor pressure at room temperature or when gently warmed. Therefore, the formation of (a) by contact of the coumarin vapor with caustic alkali and its conversion to the fluorescing product (b) on irradiation with ultraviolet light can be accomplished.

Procedure. A drop of the ethereal solution is evaporated in a micro test tube or a tiny portion of the solid is taken. The mouth of the test tube is covered with a disk of filter paper moistened with 1 N caustic alkali and the bottom of the tube is placed in hot water. After several minutes the paper is held under the quartz lamp. A yellow-green fluorescing circle appears in a short time.

Limit of identification: 0.5 γ coumarin.

1 F. FEIGL, H. E. FEIGL and D. GOLDSTEIN, *J. Am. Chem. Soc.*, 77 (1955) 4162.

(2) Test with 2,6-dichloroquinone-4-chloroimine[1]

As outlined in test (1) coumarin forms salts of o-hydroxycinnamic acid (coumarinic acid) with caustic alkali. If ammonia is used, this cleavage is achieved only by heating. The o-hydroxycinnamic acid, a phenol having a free *para*-position, reacts with 2,6-dichloroquinone-4-chloroimine in alkaline medium to form a blue indophenol dye (see p. 185). Coumarin (a lactone) does not give this reaction. The indophenol dye formed is more acidic than the analogous dyes from phenols, and their blue color is therefore maintained at pH 5.

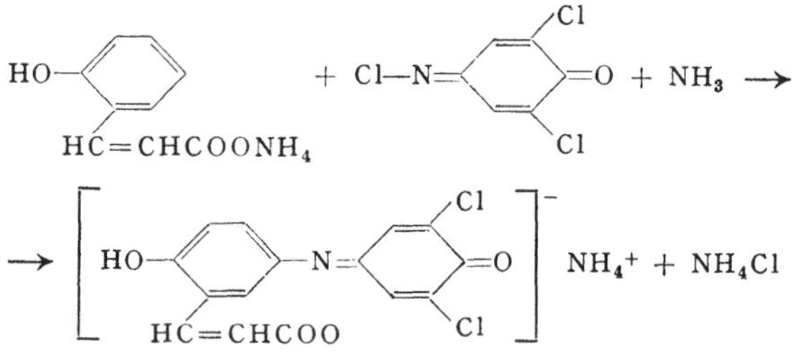

Procedure. A drop of the sample is mixed in a micro test tube with a drop of conc. ammonia. The tube is placed in a glycerol bath pre-heated to 105° and warmed until about a third of the liquid remains in the tube. This liquid is placed on a filter paper which is impregnated with a solution of 0.1% 2,6-dichloroquinone-4-chloroimine in chloroform and subsequently dried. A blue stain rapidly appears if coumarin is present. The stain remains blue when spotted with a buffer solution (pH 5).

Limit of identification: 1 γ coumarin.

1 V. ANGER, *Mikrochim. Acta*, (1964) 1126.

63. Glyoxalic (glyoxylic) acid

(1) Test with pyrogallolcarboxylic acid and sulfuric acid[1]

When glyoxalic acid (I) and pyrogallolcarboxylic acid (II) are brought together in the presence of excess concentrated sulfuric acid, a blue color develops. The chemistry of this reaction, which is specific for glyoxalic acid, is not known. Probably, in analogy with the tests for glycolic, lactic, and glyceric acid, an aldehyde–phenol condensation occurs through direct partici-pation of the aldehyde group of the glyoxalic acid.

$$OCH\!-\!COOH.H_2O \quad or \quad \substack{HO \\ HO}\!>\!CH\!-\!COOH \qquad HO\!-\!\langle \rangle\!-\!COOH$$

$$(I) \qquad\qquad\qquad (II)$$

Since the condensation of aldehydes and phenols always occurs *ortho* to a phenolic OH group, the case of glyoxalic acid may possibly also involve the anhydrization of this OH group with the COOH group of the glyoxalic acid with production of a phenol ester. The assumption that the formation of a phenol ester plays a role in the intense color effect is supported by the fact that propionaldehyde, butyraldehyde, isobutyraldehyde, isovaler-aldehyde, and aromatic aldehydes react to give only a yellow color. Form-aldehyde and acrolein yield an orange coloration.

Mesoxalic acid [$HO_2C\!-\!CO\!-\!CO_2H.H_2O$ or $HO_2C\!-\!C(OH)_2\!-\!CO_2H$] be-haves similarly to glyoxalic acid, due to decomposition by concentrated sulfuric acid into carbonic acid and glyoxalic acid:

$$HO_2C\!-\!C(OH)_2\!-\!CO_2H \rightarrow H_2O + CO_2 + OCH\!-\!COOH$$

Dihydroxytartaric acid also reacts for the same reason.

Procedure. A drop of the test material is treated in a micro test tube with a little solid pyrogallolcarboxylic acid and a drop or two of concentrated sulfuric acid. The mixture is cooled by plunging the test tube into water and 0.2–0.7 ml additional sulfuric acid is added. The mixture is then kept at 40° in warm water for about 30 min. A dark to light blue color indicates a positive response.

Limit of identification: 1 γ glyoxalic acid.

Solutions of the following compounds gave:

No reaction: arabinose, glucose, saccharose, urea, hippuric acid, alloxan, alloxantin, allantoin, uracil, uric acid, diphenylamine, glycolic, lactic, α- and β-hydroxybutyric, isobutyric, glyceric, gluconic, pyruvic, levulinic, oxalic, malonic, tartronic, malic, tartaric, citric, *o-*, *m-*, *p-*hydroxybenzoic and protocatechuic acid.

Yellow color: glucuronic acid.
Olive green: fructose.

1 E. EEGRIWE, *Z. Anal. Chem.*, 100 (1935) 34.

(2) *Test by conversion into glycolic acid* [1]

Nascent hydrogen reduces glyoxalic acid to glycolic acid:

$$OCHCOOH + 2\ H^0 \rightarrow CH_2(OH)COOH$$

which can be detected by the delicate test with 2,7-dihydroxynaphthalene (Sect. 52). Before the reduction essential to this test is carried out, it is necessary to be sure that no glycolic acid, is present. Oxalic acid is likewise reduced to glycolic acid by nascent hydrogen. To test for and, if necessary, to remove oxalic acid the test solution should be treated with several drops of saturated $CaSO_4$ solution to precipitate calcium oxalate.* The reduction to glycolic acid is then carried out in the filtrate or centrifugate.

Procedure. A drop of the test solution is treated in a micro test tube with a drop of 1 N sulfuric acid and a little magnesium powder. After the metal has dissolved, two ml of the reagent are added and the test tube kept in boiling water for 14–20 min. A red to violet color indicates a positive response.

Reagent: For preparation see p. 465.

Limit of identification: 0.5 γ glyoxalic acid.

* Brief heating with acetic anhydride completely decomposes (dehydrates) oxalic and formic acid.[2] No study has been made as yet of this behavior for the elimination of these acids.

1 V. P. CALKINS, *Ind. Eng. Chem. Anal. Ed.*, 15 (1943) 762.
2 H. KRAUSE, *Ber.*, 52 (1919) 426.

(3) *Test with phenylhydrazine and oxidizing agents* [1]

If glyoxalic acid is warmed with phenylhydrazine in the presence of mineral acid, and an oxidizing agent (potassium ferricyanide, alkali persulfate, hydrogen peroxide, *etc.*) is then added to the cold mixture, a red color appears. The mechanism of this reaction is unknown. It seems logical to assume that two reactions of phenylhydrazine play a role, namely: the formation of glyoxalic acid phenylhydrazone (I) as shown in (1); and the conversion of phenylhydrazine into a diazonium salt (II) through the action of the particular oxidant as shown in (2):

$$C_6H_5NHNH_2 + CHOCOOH \rightarrow C_6H_5NHN=CHCOOH + H_2O \qquad (1)$$
$$(I)$$

$$C_6H_5NHNH_2.HX + 2O \rightarrow C_6H_5—\overset{+}{N}\equiv N \; \overset{-}{X} + 2H_2O \qquad (2)$$
$$(II)$$

The red water-soluble compound is the product of a coupling of (I) and (II) with simultaneous decarboxylation[2] leading to 1,5-diphenylformazan:

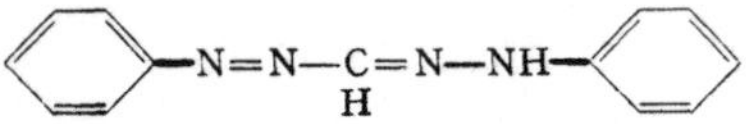

The color reaction with phenylhydrazine and oxidants appears to be specific for glyoxalic acid. Glyceric, oxalic, tartaric, citric, malonic, and mandelic acid produce no change in a mixture of phenylhydrazine and hydrogen peroxide (or other oxidants). In contrast, under the conditions of the test, glycolic acid probably reacts in the same manner since an oxidation to glyoxalic acid can occur:

$$CH_2OHCOOH + [O] \rightarrow CHOCOOH + H_2O$$

This assumption is strengthened by the fact that the glyoxalic acid test is positive when applied to a solution of oxalic acid which has been reduced with zinc or magnesium and which assuredly contains glycolic acid.

Procedure.[3]   A drop of the mineral acid test solution and a drop of freshly prepared 1% phenylhydrazine chloride solution are placed in a depression of a spot plate. The plate is kept at 110° for 5 min. Then one drop of concentrated hydrochloric acid and one drop of 3% hydrogen peroxide are added to the cold reaction mixture. A red to pink tint appears at once, the color depending on the glyoxalic acid content of the sample.

Limit of identification: 1 γ glyoxalic acid.

1 M. PAGET and R. BERGER, *Chem. Abstr.*, 32 (1938) 4908.
2 M. MATSUI, M. OKADA and M. ISHIDATE, *Anal. Biochem.*, 12 (1965) 143.
3 Unpublished studies, with D. GOLDSTEIN.

64. Pyruvic acid

(1) Test with o-hydroxydiphenyl and sulfuric acid after reduction to lactic acid[1]

Pyruvic acid is reduced to lactic acid by nascent hydrogen (magnesium plus acid):

$$CH_3COCOOH + 2 H^0 \rightarrow CH_3CHOHCOOH$$

The lactic acid can be detected by conversion to acetaldehyde and the subsequent color reaction with hydroxydiphenyl (see Sect. 53). The higher homologous α-ketonic acids behave similarly to pyruvic acid.

Procedure.   A drop of the test solution is placed in a dry test tube together with a little magnesium powder and a small drop of concentrated sulfuric acid is allowed to flow down the wall of the tube. After the magnesium has dissolved, a crystal of o-hydroxydiphenyl and 0.5 ml of conc. sulfuric acid are added. The mixture is kept at 85° for 2 min. in a water bath and examined for any color reaction.

Limit of identification: 3 γ pyruvic acid.

1 E. Eegriwe, *Z. Anal. Chem.*, 95 (1934) 325.

(2) Test through condensation with β-naphthylamine[1]

Pyruvic acid condenses with β-naphthylamine as follows:[2]

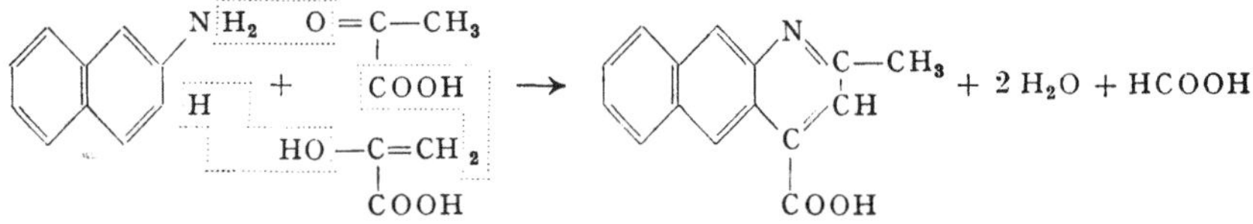

The α-methyl-β-naphthocinchoninic acid formed contains a tertiary nitrogen atom and thus shows the color reaction with a mixture of acetic anhydride and citric acid as outlined on p. 251. The above condensation seems to be specific for pyruvic acid and permits its sensitive detection.

Procedure.   In a micro test tube one drop of the test solution is mixed with one drop of 2.5% alcoholic solution of β-naphthylamine. After 10–15 min. the mixture is evaporated to dryness and the residue is heated for 5 min. at 120°. The further treatment with acetic anhydride and citric acid is the same as described on p. 251.

Limit of identification: 0.2 γ pyruvic acid.

1 F. Feigl and S. Yariv, *Talanta*, 12 (1965) 159.
2 L. F. Fieser and M. Fieser, *Organic Chemistry*, 5th Ed., 1956, p. 480.

65. Acetoacetic ester

Test through formation of 1,4-dihydro-2,6-lutidinedicarboxylic ester[1]

Acetoacetic ethyl ester condenses with aldehydes and ammonia to give 1,4-dihydropyridinedicarboxylic esters:

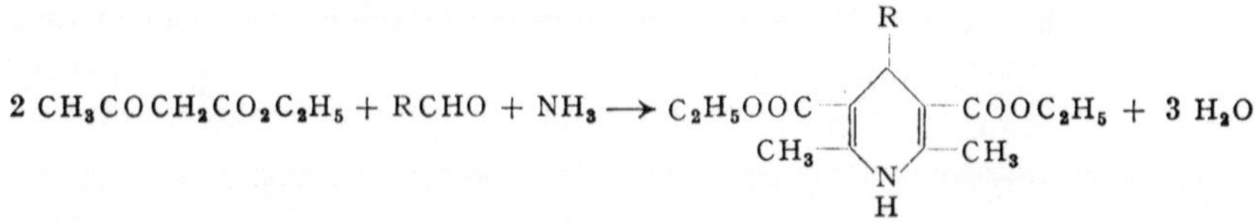

$$2\ CH_3COCH_2CO_2C_2H_5 + RCHO + NH_3 \longrightarrow C_2H_5OOC \overset{R}{\underset{\underset{H}{N}}{\diagdown}} COOC_2H_5 + 3\ H_2O$$

Compounds of this type, though practically colorless, nevertheless fluoresce intensely blue in u.v. light. Their formation makes possible a selective test for acetoacetic ester.

Procedure.   A drop of the alcoholic test solution is mixed in a micro test tube with a drop of the reagent, and heated in a water bath at 60–80° for 3 min. The presence of acetoacetic ester is indicated by a blue fluorescence in u.v. light.

Reagent: 25 g of ammonium acetate are dissolved in 100 ml water and 3 ml glacial acetic acid and 0.5 ml of 1% formaldehyde solution are added.

Limit of identification: 10 γ acetoacetic ethyl ester.

Ethylacetoacetic ethyl ester does not react nor does malonic ester.

1 V. ANGER and S. OFRI, *Z. Anal. Chem.*, 206 (1964) 287.

66. Phosgene

Test by conversion to diphenylcarbohydrazide[1]

Due to its two mobile chlorine atoms, phosgene reacts quickly and quantitatively with an excess of alkyl- and arylhydrazines to form symmetric carbohydrazides.[2] For example, diphenylcarbohydrazide (diphenylcarbazide) is produced from phosgene and phenylhydrazine (in an inert solvent):

$$\begin{matrix} C_6H_5-NH-NH_2 \\ C_6H_5-NH-NH_2 \end{matrix} + \begin{matrix} Cl \\ Cl \end{matrix} CO \longrightarrow \begin{matrix} C_6H_5-NH-NH \\ C_6H_5-NH-NH \end{matrix} CO + 2\ HCl \qquad (1)$$

Diphenylcarbazide, in neutral, ammoniacal, or weakly acid solution, reacts with copper ions to form water-insoluble, deep violet inner-complex copper diphenylcarbazide.[3] This product can be extracted with water-immiscible organic liquids such as ether, chloroform, *etc.* The reaction is:

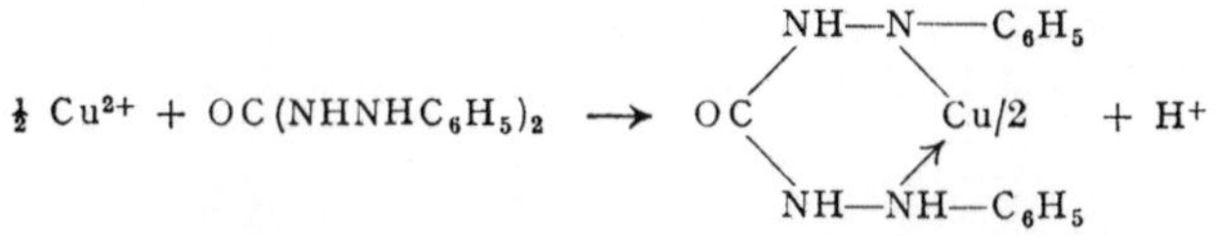

$$\tfrac{1}{2}\ Cu^{2+} + OC(NHNHC_6H_5)_2 \longrightarrow OC \overset{NH-N-C_6H_5}{\underset{NH-NH-C_6H_5}{\diagdown}} Cu/2 + H^+ \qquad (2)$$

Reaction (2) permits the detection not only of small quantities of copper, but also of small amounts of diphenylcarbazide. Consequently, the realization of (1) and (2) can serve for the selective detection of phosgene. Since free phenylhydrazine does not keep well, it is advisable to use its stable cinnamate for the reaction with phosgene.

Procedure. A drop of the test solution is mixed in a micro crucible with a granule of phenylhydrazine cinnamate, and after 5 min. a drop of 1% copper sulfate solution is added. A red-violet to pink color appears, according to the amount of phosgene present.

Alternatively, filter paper is impregnated with copper sulfate solution and dried. Just before the test, a little solid phenylhydrazine cinnamate is rubbed on it. A drop of the test solution is added, followed, when the solvent has evaporated, by a drop of water. In the presence of phosgene, a red-violet stain forms.

Limit of identification: 0.5 γ phosgene.

The test may be applied to the detection of phosgene in commercial $CHCl_3$ and CCl_4. Chloroform for anesthesia, and pure carbon tetrachloride contain only traces of $COCl_2$ in comparison with the technical products.

1 V. ANGER and S. WANG, *Mikrochim. Acta*, 3 (1938) 24.
2 G. HELLER, *Ann.*, 263 (1891) 277.
3 Comp. F. FEIGL and F. L. LEDERER, *Monatsh.*, 45 (1925) 115; see also P. CAZENEUVE, *J. Pharm. Chim.*, 12 (1900) 150.

67. Benzoyl peroxide

The characteristic redox reaction of inorganic and organic derivatives of hydrogen peroxide, in which iodine is liberated from acidified alkali iodide solution, is not applicable to the very stable benzoyl peroxide because of its slight solubility in water and alcohol. However, this peroxide can react in characteristic fashion in nonaqueous solution, or as solid or melt, and deliver oxygen to appropriate reaction partners.

(1) Test through oxidation of tetrabase[1]

Benzoyl peroxide reacts with tetrabase (see p. 448) in benzene, ether, *etc.* solution to yield a blue quinoidal diphenylmethane dye:

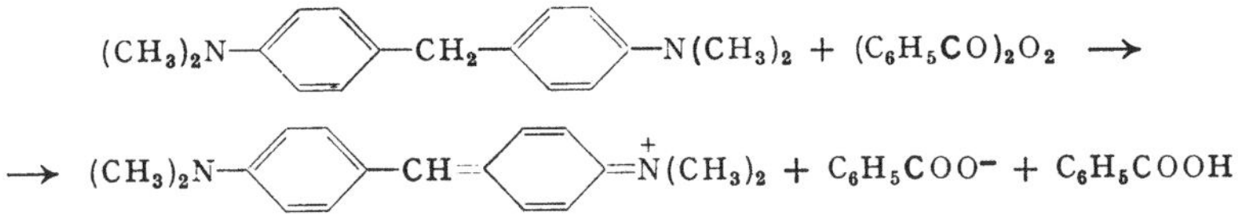

It is noteworthy that this oxidation proceeds rapidly only after evaporating the mutual solvent.

The test based on this effect cannot be used in the presence of considerable amounts of nitro compounds or quinones. They react with tetrabase to give orange-yellow to orange-red molecular compounds (comp. Chap. 3,

Sect. 91) and so interfere with the detection of small quantities of benzoyl peroxide.

Procedure. A drop of a 5% solution of tetrabase in benzene is placed on filter paper and then a drop of the benzene or ethereal solution of the sample is added. After the solvent has evaporated, a blue stain is left, the color depending on the quantities of benzoyl peroxide present.

Limit of identification: 0.5 γ benzoyl peroxide.

1 Unpublished studies, with CL. COSTA NETO.

(2) Test with diethylaniline[1]

If benzoyl peroxide is warmed with an excess of diethylaniline, acetaldehyde results (comp. Chap. 3, Sect. 58) because of the redox reaction:

$$C_6H_5N(C_2H_5)_2 + 2 (C_6H_5CO)_2O_2 \rightarrow 2 CH_3CHO + C_6H_5NH_2 + 2 (C_6H_5CO)_2O$$

The acetaldehyde can be detected by the blue color which appears when it comes into contact with a solution of sodium nitroprusside containing morpholine (see Sect. 29). This indirect test is specific for benzoyl peroxide.

Procedure. The test is made in a micro centrifuge tube (Emich tube). One drop of the benzene or ethereal test solution and one drop of diethylaniline are mixed and the open end of the tube is covered with a disk of filter paper that has been moistened with the reagent solution (p. 439). The tube is placed in a water bath at 90° and after the benzene has volatilized the water is brought to boiling. A deep to light blue stain appears on the colorless or light yellow paper.

Limit of identification: 3 γ benzoyl peroxide.

1 Unpublished studies, with CL. COSTA NETO.

(3) Detection by heating with hexamine[1]

The redox reaction between benzoyl peroxide and hexamine (see p. 499) can likewise be employed for the detection of benzoyl peroxide. Procedure I is used and it is applied to a mixture of several cg of hexamine and a drop of the test solution (benzene as solvent).

Limit of identification: 10 γ benzoyl peroxide.

1 F. FEIGL, D. GOLDSTEIN and D. HAGUENAUER-CASTRO, *Z. Anal. Chem.*, 178 (1961) 422.

68. Carbon disulfide

(1) Test with sodium azide and iodine[1]

An aqueous solution containing sodium azide (NaN_3) and iodine (as KI_3) shows no change even on long standing. However, the addition of carbon disulfide brings about an immediate reaction, with vigorous evolution of nitrogen and consumption of iodine:

$$2 \text{ NaN}_3 + \text{I}_2 \rightarrow 2 \text{ NaI} + 3 \text{ N}_2 \tag{1}$$

The carbon disulfide seemingly takes no part in this reaction but acts as a catalyst. Superficially this catalysis is similar to that described in *Inorganic Spot Tests*, Chap. 4, in the tests for inorganic thio compounds, and in a test for organic compounds containing the C=S and C—SH groups (p. 219).

The action of carbon disulfide on the iodine–azide is due to a typical intermediate reaction catalysis. Sodium azide and carbon disulfide form the sodium salt of azidodithiocarbonic acid,[2] as shown in (2). It is followed by the redox reaction (3) through which CS_2 is regenerated and again enters (2). Summation of the rapid intermediate reactions (2) and (3) gives (1), in which the catalytically active CS_2 no longer appears:

$$2 \text{ NaN}_3 + 2 \text{ CS}_2 \rightleftarrows 2 \text{ CS(SNa)N}_3 \tag{2}$$

$$2 \text{ CS(SNa)N}_3 + \text{I}_2 \rightarrow 2 \text{ CS}_2 + 2 \text{ NaI} + 3 \text{ N}_2 \tag{3}$$

Carbon disulfide can accordingly be detected by the evolution of nitrogen or through the loss of color of the iodine–sodium azide mixture. Mercaptans and thioketones must be absent because they behave analogously.

Procedure. The test for carbon disulfide in an organic liquid, *e.g.* alcohol, can be carried out by mixing one drop of the test solution with one drop of the reagent on a watch glass and noting the evolution of bubbles of nitrogen. When materials which consume iodine are present (hydrogen sulfide, *etc.*), the test solution should be treated beforehand with drops of iodine dissolved in potassium iodide or alcohol, until the iodine color is no longer discharged. The sodium azide–iodine mixture can then be added.

Reagent: 3 g sodium azide dissolved in 100 ml 0.1 N iodine solution.

Limit of identification: 0.14 γ carbon disulfide.

1 F. FEIGL and E. CHARGAFF, *Z. Anal. Chem.*, 74 (1928) 376.
2 Comp. F. SOMMER, *Ber.*, 48 (1915) 1833.

(2) Test with formaldehyde and plumbite solution [1]

Carbon disulfide is converted into trithiocarbonate on treatment with alkali hydroxide:

$$3 \text{ CS}_2 + 6 \text{ KOH} \rightarrow 2 \text{ K}_2\text{CS}_3 + \text{K}_2\text{CO}_3 + 3 \text{ H}_2\text{O}$$

The trithiocarbonate reacts with heavy metals to give insoluble salts, which can be readily decomposed with formation of sulfide.

Appreciable amounts of carbon disulfide can be rapidly converted into lead sulfide by treatment with concentrated caustic alkali and a plumbite solution. However, this does not apply to small amounts of carbon disulfide. The formation of trithiocarbonate and hence of PbS can be accelerated by formaldehyde for a reason that is not understood. Acetaldehyde, benzaldehyde, arabinose, glucose, and lactose show the same effect.[2] This action affords a means of detecting small amounts of carbon disulfide.

Procedure.   A drop of the test solution is placed on a spot plate, 2 or 3 drops of 40% formaldehyde solution added and then a drop of strongly alkaline plumbite solution is stirred in. A black precipitate or a brown to black color is formed, depending on the amount of carbon disulfide present.

Limit of identification: 3.5 γ carbon disulfide.

The test must be modified when hydrogen sulfide is present, because the formation of black lead sulfide is then no longer characteristic for carbon disulfide. The hydrogen sulfide can be destroyed by adding bromine water: $H_2S + 4\,Br_2 + 4\,H_2O \rightarrow H_2SO_4 + 8\,HBr$. The excess bromine is removed by adding sulfosalicylic acid (see p. 72).

Procedure.   A drop of the test solution is treated on a spot plate with concentrated bromine water, added drop by drop, until the yellow color is permanent. The solution is decolorized by adding solid sulfosalicylic acid. Then 2 or 3 drops of a 40% solution of formaldehyde are added, followed by a drop of a strongly alkaline solution of plumbite, and the mixture stirred. A black precipitate, or a black to brown color, appears if carbon disulfide was present.

Limit of identification: 10 γ carbon disulfide.

1 F. FEIGL and K. WEISSELBERG, *Z. Anal. Chem.*, 83 (1931) 101.
2 L. ROSENTHALER, *Pharm. Zentralhalle*, 74 (1933) 288.

(3) Test through formation of copper dimethyldithiocarbamate [1]

Carbon disulfide reacts with secondary aliphatic amines to yield water-soluble dithiocarbamates of the respective amines (comp. p. 250). These products give water-insoluble yellow-brown copper salts that yield solutions of the same color in benzene, chloroform or ether. The colored solutions can be obtained directly from the components. For example, if water saturated with carbon disulfide (0.2% CS_2) is added to an ammoniacal copper solution containing dimethylamine, extraction with benzene yields a yellow benzene layer at once. The same effect is secured if the blue copper solution is shaken with a solution of CS_2 in benzene.

The test based on this reaction is characteristic for carbon disulfide and compounds which split off carbon disulfide on treatment with acids (xanthates, trithiocarbonates).

Procedure. A drop of the liquid (benzene, chloroform, *etc.*) to be tested is treated in a micro test tube with 1 or 2 drops of the blue reagent solution and shaken, if need be, with 2 drops of benzene. If carbon disulfide is present, a yellow-brown to light yellow color appears in the benzene layer.

 Reagent: 9 ml of 0.2% copper sulfate solution is treated with 1 ml concentrated ammonia and 0.5 g piperidine or dimethylamine chloride. The solution keeps.

Limit of identification: 3 γ carbon disulfide.

1 Unpublished studies, with CL. COSTA NETO.

69. Thio-Michler's ketone

Test with chloramine T [1]

Tetrabase and thio-Michler's ketone behave similarly toward manganese dioxide; in both cases blue-violet oxidation products are formed. However, they differ with respect to chloramine T which oxidizes thio-Michler's ketone but not tetrabase. A blue quinoidal dyestuff is formed (comp. p. 65).

Procedure. A small amount of solid chloramine T is placed in a depression of a spot plate and a drop of the benzene solution of the test material is added. A violet coloration appears initially; a blue product remains after evaporation of the benzene.

Limit of identification: 0.2 γ thio-Michler's ketone.

This test revealed 0.2 γ thio-Michler's ketone in the presence of 2 mg tetrabase.

1 Unpublished studies, with E. LIBERGOTT.

70. Thiophene

It was pointed out (in Chap. 4) that 1,2-diketones in concentrated sulfuric acid solution condense with thiophene to yield colored quinoidal compounds that are known as indophenines. The tests for 1,2-diketones based on this behavior are the converse of the tests given here for thiophene.

(1) Test by condensation with isatin [1]

Isatin (I), dissolved in concentrated sulfuric acid, condenses with thiophene (II), a reaction accompanied by a union of two thiophene molecules:[2]

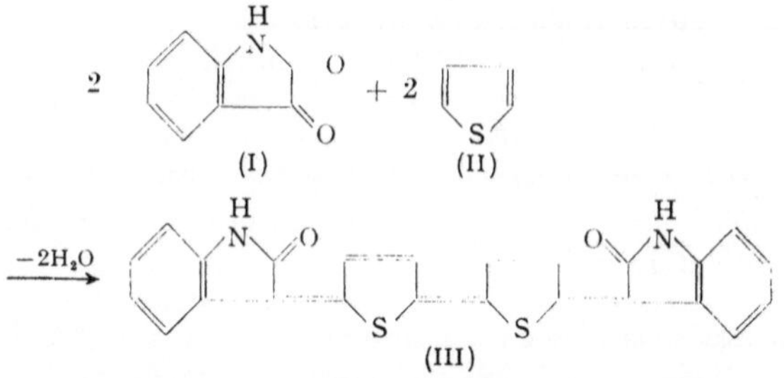

The sulfuric acid does not function solely as dehydrating agent. This fact is supported by the finding that no formation of indophenine occurs in syrupy phosphoric acid, which can be substituted for concentrated sulfuric acid in many condensations. The production of the dyestuff probably requires that the thiophene be sulfonated as well as oxidized[3] to α,α'-dithienyl (VI) by the concentrated sulfuric acid. The addition of a trace of nitric acid to the isatin–sulfuric acid is often recommended, but this device is not necessary, particularly if the reaction is conducted with gentle heating.

β-Methylthiophene (IV), β,β'-dimethylthiophene (V), α,α'-dithienyl (VI) and thiophthene (VII) analogously condense with isatin to yield blue dyestuffs. Since α,β'-dimethyl- and α,α'-dimethylthiophene do not react[4], it is likely that the condensation with isatin occurs in the α,α' or α,β-position of thiophene and thiophene homologs, as shown in the structure of (III).

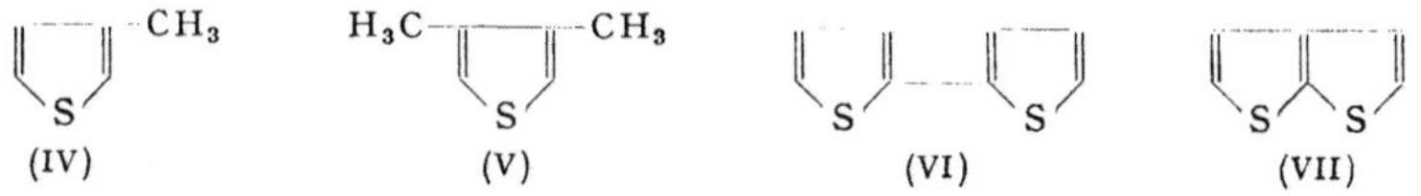

The indophenine reaction can be carried out as a drop reaction on a spot plate, in a porcelain micro crucible, or in a micro test tube. It should be noted that it is difficult to see a faint blue coloration in colored solutions, and that many colorless organic compounds give colored solutions in concentrated sulfuric acid. These interferences are avoided if use is made of the facts that thiophene boils at 84° and that the indophenine reaction occurs very quickly when thiophene vapor comes in contact with the isatin solution.

Procedure.[5] The apparatus shown in Fig. 23, *Inorganic Spot Tests*, is used. A drop of the test solution (in thiophene-free benzene, *etc.*) is placed in the bulb, and the knob of the stopper is dipped into the yellow reagent. The apparatus is stoppered and placed in an oven at 80–90°. If thiophene is present the suspended drop turns blue in 1–2 min.; the intensity depends on the quantity of thiophene.

Reagent: 0.2% solution of isatin in concentrated sulfuric acid.

Limit of identification: 1.5 γ thiophene.

If the indophenine reaction is carried out with a 0.2% solution of benzil in concentrated sulfuric acid,[6] no heating is necessary. It is possible to detect 5 γ thiophene in this way.

1 V. MEYER, *Ber.*, 15 (1882) 3813; 16 (1883) 1465.
2 W. SCHLENK and O. BLUM, *Ann.*, 433 (1923) 85.
3 A. TÖHL, *Ber.*, 27 (1894) 665; R. AUWERS and T. V. BREDT, *Ber.*, 27 (1894) 1746.
4 W. STEINKOPF and H. HEMPEL, *Ann.*, 495 (1932) 144.
5 Unpublished studies, with D. GOLDSTEIN.
6 V. MEYER, *Ber.*, 16 (1883) 2973.

(2) *Test by condensation with ninhydrin* [1]

The color reactions for thiophene given in Test (*1*) involve a condensation reaction with 1,2-diketones, dissolved in concentrated sulfuric acid, in which the latter functions both as dehydrant and oxidant. Polyketones which contain two adjacent CO groups in open or closed carbon chains likewise appear capable of forming quinoidal dyestuffs with thiophene. The behavior of ninhydrin (triketohydrindene hydrate) strengthens this supposition. Even very dilute solutions of this triketone in concentrated sulfuric acid give a red color almost immediately following the addition of a trace of thiophene. It is probable that the red dyestuff, which is deep violet in greater concentrations, is formed by an analogous succession of reactions and probably has a constitution analogous to those of the dyestuffs formed by the reaction of thiophene with isatin or benzil.

The color reaction with ninhydrin is so sensitive that purified benzene, supposedly free of thiophene, often gives a pale but nevertheless discernible pink on the addition of ninhydrin dissolved in sulfuric acid.

If the hanging drop method given in Test (*1*) is used, ninhydrin solution will reveal 3 γ thiophene at room temperature. The reagent loses its activity at 80–90°. Good results are obtained by adhering to the following directions.

Procedure. A spot plate or a micro test tube is used. One drop of a freshly prepared 0.01% solution of ninhydrin in concentrated sulfuric acid is covered with a drop of the solution being tested. According to the amount of thiophene, a deep violet to pink color appears.

Limit of identification: 0.2 γ thiophene.

1 Unpublished studies, with D. GOLDSTEIN.

71. Dixanthogen

Test by conversion into alkali xanthate [1]

When an alcoholic, nearly colorless, solution of dixanthogen [diethyl dithio-bis(thionoformate)] is treated with zinc and alkali hydroxide, the nascent hydrogen formed reduces the disulfide:

$$S=\overset{\overset{\displaystyle O\,C_2H_5}{|}}{C}-S-S-\overset{\overset{\displaystyle O\,C_2H_5}{|}}{C}=S \quad + 2\,H^{\circ} + 2\,NaOH \quad \longrightarrow \quad 2\,S=C\overset{\displaystyle \diagup S\,Na}{\diagdown O\,C_2H_5} \quad + 2\,H_2O$$

This transformation can be detected by the rapid appearance of the yellow alkali xanthate or by the color reaction of xanthates with molybdic acid (see p. 226). Instead of zinc, ascorbic acid may be used as a reductant.

Procedure.   In a micro test tube a drop of the alcoholic test solution is mixed with a drop of 1% aqueous ascorbic acid solution and 2 drops of an alcoholic sodium hydroxide solution. The mixture is heated in a water bath, whereupon a yellow color appears. After cooling, 2 drops of chloroform are added followed by a drop of acidified ammonium molybdate solution (2% molybdate with the same volume of 2 N HCl). After shaking, a violet color appears in the chloroform layer. The color fades after a time.

Limit of identification: 20 γ dixanthogen (yellow color)
                                            5 γ ,,   (violet color)

1 F. FEIGL and L. BEN-DOR, *Chemist-Analyst*, 54 (1965) 7.

72. Benzenesulfonic acid

Test through dry heating with sodium amide [1]

When alkali salts of benzenesulfonic acid are dry heated with sodium amide, aniline is formed: [2]

$$C_6H_5SO_3Na + NaNH_2 \rightarrow C_6H_5NH_2 + Na_2SO_3$$

The aniline vapor can be detected by allowing it to react with p-dimethylaminobenzaldehyde, whereby a bright yellow Schiff base is formed (comp. p. 243).

Procedure.   A drop of the aqueous neutral or alkaline test solution is evaporated to dryness in a micro test tube. After addition of several cg of pure sodium amide the open end of the tube is covered with a disk of filter paper impregnated with benzene containing 2% p-dimethylaminobenzaldehyde and 10% trichloroacetic acid. The tube is immersed in a glycerol bath preheated to 110° and the temperature is then raised to 160°. A positive response is indicated by the appearance of a yellow stain on the reagent paper.

Limit of identification: 5 γ benzenesulfonic acid.

1 F. FEIGL and V. ANGER, unpublished studies.
2 C. L. JACKSON and J. F. WING, *Ber.*, 19 (1886) 902.

73. α-Naphthalenesulfonic acid

Test through reductive cleavage [1]

When α-naphthalenesulfonic acid is treated with sodium amalgam in neutral or weakly acid solution naphthalene and sulfurous acid are formed:[2]

$$C_{10}H_7SO_3H + 2\ H^0 \rightarrow C_{10}H_8 + H_2SO_3$$

Both cleavage products are easily detectable. The former with chloranil (see Sect. 2), the latter in the gas phase with appropriate reagent papers.

Derivatives of α-naphthalenesulfonic acid behave like the parent compound with respect to the formation of sulfur dioxide. β-Naphthalenesulfonic acid remains unaltered when treated with sodium amalgam.

Procedure. A drop of the weakly acid or neutral test solution is placed in a micro test tube and several cg of pulverized sodium amalgam are added. When the evolution of hydrogen stops, several drops of dilute hydrochloric acid are added and the mouth of the test tube is covered with a piece of Congo paper moistened with 5% hydrogen peroxide. The tube is then immersed in boiling water. The appearance of a blue stain on the indicator paper signals the formation of sulfur dioxide.

Limit of identification: 20 γ α-naphthalenesulfonic acid.

Alternatively, after reduction the solution is shaken with several drops of ether and the ether extract tested for naphthalene as described in Sect. 2.

Limit of identification: 60 γ α-naphthalenesulfonic acid.

1 F. FEIGL and V. ANGER, unpublished studies.
2 P. FRIEDLÄNDER and P. LUCHT, *Ber.*, 26 (1893) 3030.

74. Sulfosalicylic acid

(1) Test through dry heating with alkali nitrate[1]

If non-volatile nitrogen-free organic compounds are ashed with alkali nitrate, nitrous acid results.[2] The reason is that organic materials or their carbonaceous pyrolytic cleavage products reduce nitrate to nitrite, and the latter yields nitrous acid through:

$$KNO_2 + H_2O \rightarrow KOH + HNO_2$$

The water essential to this hydrolysis is formed, as superheated steam, during the pyrolysis of organic compounds containing hydrogen and oxygen. Surprisingly, a mixture of alkali nitrate and sulfosalicylic acid (m.p. 158°) was found to yield nitrous acid even at 100°. As little as 200 γ sulfosalicylic acid can be detected in this way, if the Griess color reaction for nitrite is applied.

The production of nitrous acid from alkali nitrate by means of sulfosalicylic acid is due to a topochemical solid-body reaction. The initial step is the formation of nitric acid at the point of contact of the alkali nitrate and the strong sulfosalicylic acid, followed by oxidation of the latter by the nitric acid with production of nitrous acid. Since nitric acid volatilizes, its reduction to nitrous acid is incomplete because of insufficient contact with the sulfosalicylic acid. This assumption is in accord with the observation that the highest sensitivity for the test for sulfosalicylic acid was obtained when the nitric acid, produced by dry heating a mixture of alkali nitrate and sulfosalicylic acid, was detected along with the nitrous acid. A reagent suitable for this double or reinforced effect is the colorless solution of N,N'-diphenylbenzidine in concentrated sulfuric acid, which reacts with both oxidizing acids to give a blue color.[3]

The procedure permits the detection of sulfosalicylic acid and its differentiation from salicylic acid.

Procedure. The test is conducted in the gas absorption apparatus of Fig. 23, *Inorganic Spot Tests*. Several cg of dry pulverized potassium nitrate are placed in the bulb along with a drop of the alcoholic or ethereal test solution and taken nearly to dryness. The knob of the stopper is moistened with a freshly prepared solution of 10 mg of N,N'-diphenylbenzidine in 50 ml of concentrated sulfuric acid, and the closed apparatus is kept in boiling water for 2–3 min. If sulfosalicylic acid is present, the drop suspended on the knob turns blue.

Limit of identification: 3.5 γ sulfosalicyclic acid.

1 F. FEIGL, J. R. AMARAL and V. GENTIL, *Mikrochim. Acta*, (1957) 730.
2 F. FEIGL and C. STARK-MAYER, *J. Chem. Educ.*, **34** (1957) 457.
3 H. RIEHM, *Z. Anal. Chem.*, **81** (1930) 439.

(2) *Test by conversion to salicylic acid*[1]

If sulfosalicylic acid is heated above its melting point (120°), phenol and salicylic acid result.[2] These compounds can be detected in the gas phase through the Gibbs' color reaction and the fluorescence reaction described on p. 185 and 478. The limits of identification obtained by heating at 180° are in the region of 1.5 γ sulfosalicylic acid.

1 Unpublished studies, with CL. COSTA NETO.
2 L. MENDIUS, *Ann.*, **103** (1853) 45, 50.

75. 1,2-Naphthoquinone-4-sulfonic acid

Test through condensation with aniline[1]

It was pointed out in Chap. 2, Sect. 44), that the sodium salt of 1,2-naphthoquinone-4-sulfonic acid is a sensitive reagent for compounds containing reactive CH_2 and NH_2 groups since it condenses with them to yield colored products. This behavior can be employed likewise as a sensitive test for 1,2-naphthoquinone-4-sulfonic acid.

Procedure. A drop of the aqueous test solution is brought together with a

drop of a saturated aqueous solution of aniline on filter paper or in a depression o
a spot plate. A positive response is signalled by the formation of a red-violet
stain or a violet solution.

Limit of identification: 1 γ 1,2-naphthoquinone-4-sulfonic acid.

1 Unpublished studies, with V. GENTIL.

76. Phenylsulfuric acid

Test through liberation of volatile phenol[1]

Phenylsulfuric acid, which is stable against caustic alkali, may be saponified by heating with dilute mineral acids:[2]

$$C_6H_5OSO_3H + H_2O \rightarrow C_6H_5OH + H_2SO_4$$

The phenol formed can be detected in the gas phase by the indophenol
reaction with 2,6-dichloroquinone-4-chloroimine described on p. 185.
This permits a specific test for phenylsulfuric acid, provided volatile phenols
are absent. The latter can be expelled if necessary by prolonged heating
with sodium bicarbonate solution.

Procedure. A drop of the test solution is warmed in a micro test tube with
a drop of 2 N sulfuric acid. The open end of the test tube is covered with a piece
of the reagent paper (see p. 185). If the response is positive, the paper turns
blue when exposed to ammonia vapors.

Limit of identification: 0.3 γ phenylsulfuric acid.

1 F. FEIGL and V. ANGER, *Mikrochim. Acta*, (1960) 409.
2 E. BAUMANN, *Ber.*, 11 (1878) 1908.

77. Ethylenediamine and propylenediamine

Test by precipitation with sodium rhodizonate[1]

If the yellow-brown aqueous solution of sodium rhodizonate (I) is added
to a neutral aqueous solution of an ethylenediamine salt (chloride, sulfate,
etc.), a dark violet precipitate of ethylenediamine rhodizonate (II) is formed:

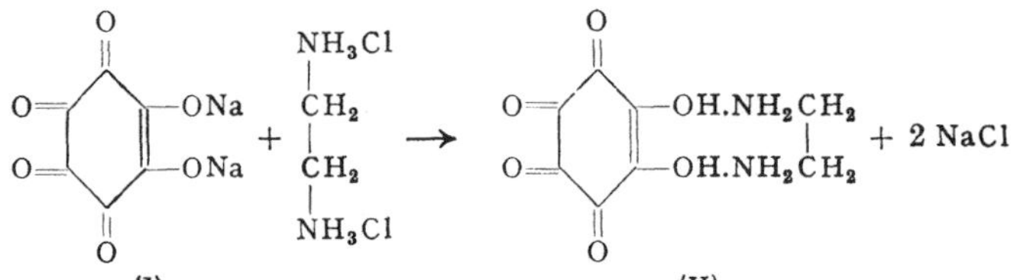

It has not been established whether this salt is the sole product. There appears to be a side reaction in which a cyclic Schiff base is produced through a condensation of ethylenediamine with two adjacent CO groups of the rhodizonic acid. Propylenediamine salts behave similarly to ethylenediamine. Ammonium-, di- and triethanolamine salts, hydrazine and hydroxylamine salts, and also the chlorides of o-phenylenediamine and 1,8-naphthylenediamine do not react under the conditions prescribed here. The chlorides of primary aromatic monoamines and of pyridine react only in very concentrated solution. On the other hand, a violet crystalline precipitate is given by benzidine chloride even at high dilution. After the crystalline ethylenediamine rhodizonate has once been formed, it is rather stable against dilute acids but not against ammonia. The latter destroys this violet salt as well as the violet benzidine rhodizonate.

Pure insoluble benzidine sulfate reacts with sodium rhodizonate, but an excess of sulfate ions prevents the reaction. In contrast, the ethylenediamine rhodizonate reaction is not altered in the presence of excess alkali sulfate.

Sodium rhodizonate apparently reacts likewise with salts of N-substituted derivatives of ethylenediamine to give colored salts. This is shown by the behavior of N,N'-dibenzylethylenediamine (see Chap. 7).

Procedure.   A drop of the neutral test solution and a drop of sodium rhodizonate solution are brought together on a spot plate. According to the quantity of ethylenediamine present, a violet precipitate appears at once or after brief standing. If the yellow-brown reagent solution is decolorized by adding one drop of the buffer solution, small amounts of the reaction product are more easily seen. It is better to carry out the reaction on paper. One drop of the reagent is placed on the paper, then a drop of the test solution, and after several minutes the color of the reagent is discharged by adding a drop of the buffer solution.

Reagents: 1) 1% aqueous sodium rhodizonate solution (freshly prepared).
2) Buffer solution (pH = 2.79). 1.9 g sodium bitartrate and 1.5 g tartaric acid per 100 ml.

Limit of identification: 0.3 γ ethylenediamine or propylenediamine.

1 Unpublished studies, with H. E. FEIGL.

78. Hexamethylenetetramine (urotropine, hexamine)

(1) Test by reaction with β-naphthylamine[1]

A test for the NH_2 group in organic compounds is described in Chap. 2. It is based on dry heating with hexamine, whereby ammonia is formed:

$$6 \,(R,Ar)NH_2 + (CH_2)_6N_4 \rightarrow 6 \,(R,Ar)N{=}CH_2 + 4 \,NH_3$$

This methenylation using β-naphthylamine permits a specific and sensitive test for hexamine and its salts. Solid salts of hexamine also react with fused β-naphthylamine.

Procedure. A small amount of the solid sample or a drop of its solution in chloroform is united in a micro test tube with several cg of β-naphthylamine. Its open end is covered with a disk of filter paper moistened with Nessler reagent. The mixture is heated in a glycerol bath up to 130°. The rapid appearance of a brown stain on the paper is a positive response.

Limit of identification: 2.5 γ hexamine.

1 Unpublished studies.

(2) Test through pyrohydrolysis [1]

When conducted in the wet way, the hydrolysis:

$$(CH_2)_6N_4 + 6 H_2O \rightarrow 6 CH_2O + 4 NH_3$$

occurs only in the presence of acids. In the dry way, this hydrolysis may be accomplished without acids if the hexamine is heated along with hydrates which lose the water of crystallization at elevated temperatures. Appropriate water donors include hydrated manganous sulfate and hydrated oxalic acid, which lose their water in the temperature range 110–200°. Both formaldehyde and ammonia result when manganous sulfate is used. If oxalic acid hydrate is employed, the ammonia is fixed as oxalate and the vapor contains only formaldehyde, which can be detected by the black stain of elementary mercury it produces on filter paper moistened with Nessler reagent.

Procedure. A small quantity of the sample or the evaporation residue from a drop of its solution is mixed with about 1 cg of oxalic acid dihydrate in a micro test tube. The tube is placed in a glycerol bath that has been preheated to 120°. The mouth of the tube is covered with filter paper moistened with Nessler re-agent. If hexamine is present, a grey or black stain appears after a few minutes.

Limit of identification: 5 γ hexamine.

1 F. FEIGL, D. HAGUENAUER-CASTRO and E. JUNGREIS, unpublished studies.

(3) Test through oxidative cleavage [1]

In the absence of water, hexamine can be oxidatively cleaved according to the following reactions:

$$(CH_2)_6N_4 + 6 O \rightarrow 6 CH_2O + 2 N_2 \tag{1}$$

$$(CH_2)_6N_4 + 12 O \rightarrow 6 CO_2 + 4 NH_3 \tag{2}$$

The redox reaction (*1*) is brought about by warming with benzoyl peroxide and the resulting formaldehyde is revealed by the blackening of filter paper moistened with Nessler solution (Procedure I). Redox reaction (*2*) occurs if the material is heated with manganese dioxide or lead dioxide; the resulting ammonia is revealed by means of acid–base or Nessler indicator paper (Procedure II).

Procedure I.   A little of the sample or a drop of its solution in chloroform is placed in a micro test tube; 3 drops of a 5% solution of benzoyl peroxide in benzene are added, and the mouth of the tube is covered with a disk of qualitative filter paper moistened with Nessler solution. The test tube is immersed to ¾ of its length in a glycerol bath which has been preheated to 100°, the temperature of which may be raised to 130° without danger of an explosion. The oxidation (reaction *1*) begins even during the volatilization of the solvent. A positive response is indicated by the development of a grey or black stain on the reagent paper.

Limit of identification: 20 γ hexamine.

Procedure II.   A little of the solid or a drop of its solution in chloroform is placed in a micro test tube along with several cg of lead peroxide, and a disk of acid–base indicator is placed over the mouth of the tube. The test tube is then placed (½–1 cm deep) in a glycerol bath which has been preheated to 170°. A change in color of the indicator paper demonstrates the positive outcome of the test; the color change occurs within 1–3 min. The ammonia may also be revealed by a disk of filter paper moistened with Nessler solution. A brown or yellow stain results.

Limit of identification: 20 γ hexamine.

Procedures I and II may be applied to the detection of hexamine salts with organic acids (camphorate, mandelate, salicylate) which are used in medicine.

1 F. FEIGL, D. GOLDSTEIN and D. HAGUENAUER-CASTRO, *Z. Anal. Chem.*, 178 (1961) 421.

79. Aminoacetic acid (glycine)

(*1*) *Test by conversion into glycolic acid* [1]

NH_2 groups in organic compounds can be exchanged for OH groups by warming with nitrous acid. Accordingly, hydroxyacetic acid (glycolic acid) can be formed from glycine:

$$CH_2NH_2COOH + HNO_2 \rightarrow CH_2OHCOOH + H_2O + N_2 \qquad (1)$$

If gently warmed with concentrated sulfuric acid, glycolic acid is split with production of formaldehyde:

$$CH_2OHCOOH \rightarrow CH_2O + CO + H_2O \qquad (2)$$

As shown by (*1*) and (*2*) formaldehyde is produced, which can be detected by the color reaction with chromotropic acid described in Sect. 28. Other amino acids do not interfere because only glycine is convertible into glycolic acid.

The test described here may not be applied in the presence of compounds which yield formaldehyde, such as hexamethylenetetramine and its salts, or sulfoxylate compounds.

Procedure. The test is conducted in a micro test tube. One drop of the test solution is mixed with one drop of a 1% solution of sodium nitrite and one drop of concentrated hydrochloric acid. The mixture is heated briefly to boiling. After evaporation at 105–110°, the residue is cooled and treated with 3 drops of the freshly prepared reagent. The mixture is kept in a boiling water bath for 5–10 min. A violet color indicates the presence of glycine.

Reagent: A suspension of chromotropic acid in concentrated sulfuric acid is gently warmed and then centrifuged. The supernatant liquid is used.

Limit of identification: 10 γ glycine.

1 Unpublished studies, with J. R. AMARAL; see also F. FEIGL, V. GENTIL and C. STARK-MAYER, *Mikrochim. Acta*, (1957) 348.

(2) Test through degradation to formaldehyde[1]

When α-amino acids are treated with alkali hypohalogenite, they are decarboxylated and deaminated, whereby aldehydes are formed with one C atom less than the original amino acid (see p. 369). In the case of glycine*, the degradation which leads to formaldehyde is easily achieved with chloramine T. The use of this derivative of hypochlorite has the advantage that the oxidation of formaldehyde (which may be detected through the chromotropic acid tests) is diminished.

The test described below is specific for glycine within the class of amino acids. Compounds which split off CH_2O when treated with chromotropic–sulfuric acid must be absent or destroyed beforehand.

Procedure. One drop of a 0.5% aqueous solution of chloramine T (freshly prepared) is added to one drop of the test solution in a micro test tube and the mixture warmed for 3–5 min. in a hot water bath. After cooling, 5 drops of concentrated sulfuric acid and a very small quantity of solid chromotropic acid are added. A positive response is indicated by the appearance of a violet color, either immediately or after heating for a short time in a water bath.

Limit of identification: 3 γ glycine.

* α-Alanine and tyrosine are also degraded by means of chloramine T to aldehydes with one C atom less than the original compound.

1 Unpublished studies with J. R. AMARAL; see also F. FEIGL, V. GENTIL and C. STARK-MAYER, *Mikrochim. Acta*, (1957) 348.

(3) Test through pyrolytic formation of carbon[1]

If glycine is heated to 220–250° for several minutes, water and ammonia are split out and a black residue remains that looks like amorphous carbon.

Probably there is initial production of light yellow glycine anhydride $(C_2H_3ON)_x$ and its colorless isomer dioxopiperazine.[2] It is known that glycine anhydride carbonizes above 250° without melting; there may be thermal decomposition with loss of ammonia:

$$(C_2H_3ON)_3 \rightarrow 3\ H_2O + 6\ C + NH_3 + N_2$$

Thermogravimetric studies[3] showed that the black residue does not consist of pure carbon. This finding indicates that only the glycine anhydride carbonizes while the dioxopiperazine produced remains unchanged. Since assuredly there are not more than a few organic compounds that carbonize when heated briefly at 250°, this behavior can serve as the basis of a test for glycine. Though this test is not sensitive it is nevertheless quite selective, and may even be specific within the amino acids.

Procedure. A little of the solid test material or a drop of the aqueous solution is placed in a micro test tube or crucible and taken to dryness if necessary. The vessel is then placed in a glycerol bath that has been preheated to 250° and the heating is continued for about 5 min. A black residue indicates a positive response.

Limit of identification: **40** γ glycine.

1 F. Feigl, D. Goldstein and A. C. M. Perissé, unpublished studies.
2 Comp. Beilstein, *Handb. Org. Chem.*, Vol. 4, (1922) pp. **336, 340**.
3 Conducted in the laboratory of Prof. Cl. Duval, Paris.

80. α-Aminopropionic acid (α-alanine)

Test by degradation to acetaldehyde[1]

The isomeric α- and β-aminopropionic acids can be distinguished from each other by warming with alkali hypobromite or better chloramine T. Only α-alanine is decarboxylated and deaminated by this treatment whereby acetaldehyde results:*

$$2\ CH_3CH(NH_2)COOH + 5\ NaBrO \rightarrow 5\ NaBr + 2\ CO_2 + N_2 + 3\ H_2O + 2\ CH_3CHO$$

Acetaldehyde vapors can be detected by the blue color that appears on contact with sodium nitroprusside and morpholine (see p. 438).

The test given here is specific for α-alanine among amino acids, assuming the absence of other compounds which split off acetaldehyde.

Procedure. A drop of the aqueous test solution is mixed with a drop of a freshly prepared 0.5% solution of chloramine T in a micro test tube, whose mouth is then covered with a disk of filter paper moistened with freshly prepared ni-

* Analogous to other β-amino acids, β-alanine is broken down by the hypohalogenite via the β-keto acids to the corresponding ketones. See L. Birkhofer and R. Bruns, *Ber.*, 90 (1957) 2536.

troprusside–morpholine reagent. The tube is placed in a boiling water bath. A more or less intense blue color appears within 1–3 min., depending on the amount of α-alanine present.

Limit of identification: 4 γ α-aminopropionic acid.

1 Unpublished studies, with J. R. AMARAL; see also F. FEIGL, V. GENTIL and C. STARK-MAYER, *Mikrochim. Acta*, (1957) 348.

81. Hippuric acid

Test through pyrohydrolytic formation of aminoacetic acid[1]

When hippuric acid (benzoylglycine) is heated together with manganese sulfate monohydrate at 150–180°, the following pyrohydrolysis occurs:

$$C_6H_5CONHCH_2COOH + H_2O \rightarrow C_6H_5COOH + H_2NCH_2COOH$$

The aminoacetic acid (glycine) obtained can be detected by the formation of formaldehyde when treated with chloramine T. (Comp. Sect. 79.)

Procedure. In a micro test tube some manganese sulfate monohydrate is added to one drop of the aqueous test solution and taken to dryness. The tube is immersed in a heated glycerol bath and kept there for about 5 min. increasing the temperature to 160°. After cooling, the rest of the procedure is the same as described on p. 501.

Limit of identification: 12 γ hippuric acid.

Esters of hippuric acid and the amide behave like the parent compound. (See also the detection of hippuran, Chap. 7.)

1 Unpublished studies, with D. GOLDSTEIN.

82. Tyrosine

(1) Test with α-nitroso-β-naphthol[1]

Tyrosine, $HOC_6H_4CH_2CH(NH_2)COOH$, as well as other phenols, gives a deep purple color on the addition of α-nitroso-β-naphthol and nitric acid (p. 421). The rather complicated mechanism is discussed elsewhere.[2] The test is characteristic for tyrosine among the α-aminocarboxylic acids.

Procedure. A drop of the test solution is heated in a micro crucible with a drop of 0.2% alcoholic solution of α-nitroso-β-naphthol. A drop of concentrated nitric acid is added to the hot solution. A purple color appears in the presence of tyrosine. The color fades within a few minutes.*

Limit of identification: 0.05 γ tyrosine.

* It has been reported[4] that the color can be stabilized by adding ferric ammonium sulfate.

Proteins, which almost without exception contain tyrosine as an integral constituent, can be readily detected by the color reaction here described.

Tyrosine may be detected in pathological urine or serum as follows:[3] 0.5 ml of sample is diluted with 1 ml water and 1 ml 20% trichloroacetic acid is added. The mixture is well stirred and filtered or centrifuged to remove the precipitated albumin. As little as 0.05 γ tyrosine can be detected in a drop of the filtrate or clear centrifugate.

1 O. GERNGROSS, K. VOSS and H. HERFELD, *Ber.*, **66** (1933) 435.
2 V. ANGER, *Oesterr. Chemiker-Ztg.*, **62** (1961) 354.
3 K. NOSAKA, *Mikrochim. Acta*, 1 (1937) 79.
4 A. MACIAG and R. SCHOENTAL, *Mikrochemie*, 24 (1938) 250.

(2) Detection through tests for NH$_2$-, COOH-, and phenolic OH groups [1]

As shown by its structure, tyrosine is a phenol with a free *o*-position and also an α-aminocarboxylic acid. Accordingly, tyrosine responds to the phenol test with sodium cobaltinitrite (see p. 183), and to the the test for acidic compounds through pyrolysis with mercuric cyanide (splitting off of hydrogen cyanide, see p. 147), and it also yields ammonia when pyrolyzed with hexamine (see p. 119). Therefore, tyrosine can be identified by positive responses to all three of these tests.

Limits of identification: 5 γ tyrosine through the phenol test

5 γ ,, ,, pyrolysis with mercuric cyanide

12 γ ,, ,, pyrolysis with hexamine

1 Unpublished studies, with E. LIBERGOTT.

83. Tryptophan

(1) Test with bromine [1]

When tryptophan in aqueous solution is treated with bromine water (hypobromite) a red product is formed which is soluble in organic liquids immiscible with water. Since tryptophan is an α-aminocarboxylic acid it is deaminated and decarboxylated by this treatment (see p. 369), β-indolyl-acetaldehyde being formed:

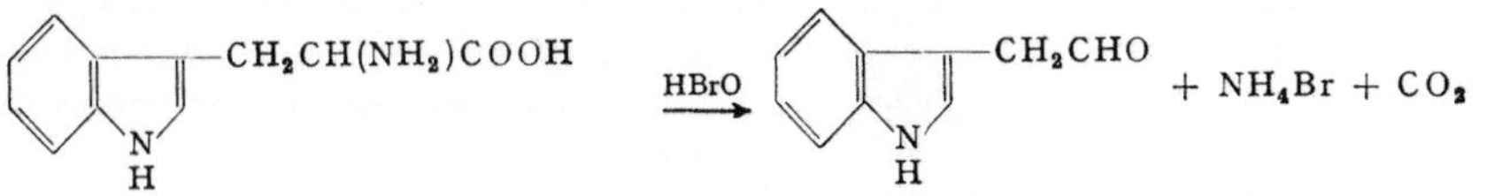

It is probable that three molecules of this aldehyde condense, with participation of oxidizing bromine water, to a dyestuff.

Procedure.[2]   A micro test tube is used. To a drop of the test solution a drop of highly diluted bromine water is added and heated for half a minute in boiling water. When not too small amounts of tryptophan are present, a pink coloration is formed. In the case of very small amounts of tryptophan, extraction with ether is recommended. The ether layer then shows a pinkish color.

Limit of identification: 5 γ tryptophan.

1 F. G. HOPKINS and S. W. COLE, *J. Physiol.*, 27 (1901) 427.
2 V. ANGER and G. FISCHER, unpublished studies.

(2) Test through conversion into β-indolyllactic acid[1]

Indole derivatives such as gramine (β-dimethylaminomethylindole) and tryptophan (β-indolylalanine) that have amino nitrogen in the side chain do not show the color reaction characteristic of indoles with glutaconic aldehyde (see p. 382). However, the amino group of tryptophan is readily replaced by hydroxyl through the action of nitrous acid:

$$C_8H_6N-CH_2CH(NH_2)CO_2H + HNO_2 \rightarrow C_8H_6N-CH_2CH(OH)CO_2H + H_2O + N_2$$

The resulting β-indolyllactic acid reacts with glutaconic aldehyde to yield an orange color. Gramine cannot be deaminated with nitrous acid.

Procedure.   The slightly acidic aqueous test solution is treated with a little solid sodium nitrite, and a little solid sulfamic acid is introduced after several minutes in order to destroy nitrous acid. Then a drop of 1% pyridyl pyridinium dichloride solution is added, followed by a drop of 1 N NaOH, and then a drop of concentrated HCl. An orange color indicates the presence of tryptophan.

Limit of identification: 3 γ tryptophan.

1 V. ANGER and S. OFRI, *Mikrochim. Acta*, (1964) 732.

84. Ethylenediaminetetraacetic acid

(1) Test by masking the formation of nickel dimethylglyoxime*

The alkali salts of ethylenediaminetetraacetic acid (I) have great technical importance as water softeners, for cleaning iron surfaces, for preventing the precipitation of water-insoluble soaps, *etc.*[1] These actions are due to the fact that in accord with its structure the diamino acid is a powerful complexing agent. Many metal ions react with the di- or tetraalkali salts** of this acid

---

* The test is based on a method for the determination of ethylenediaminetetraacetic acid.[4]

** Compare p. 367 regarding the differentiation of these alkali salts.

to produce water-soluble alkali salts in which the respective metal atom is included in an inner-complex anion. Anions of the coordination structure (II) are formed with bivalent metals:

(I)   (II)

The concentration of free Me^{2+} ions is so small in solutions of (II) that hydroxide-, phosphate-, etc. precipitations do not occur. The same is true in solutions of the salts of this diamino acid with tervalent and quadrivalent metals. Consequently, ethylenediaminetetraacetic acid or its anions act as masking agents in many analytical procedures.[2] Since the precipitation of the red nickel salt of dimethylglyoxime is prevented when ethylenediaminetetraacetic acid or its alkali salts are present, a convenient test for this acid can be based on this effect. Although the anions of the diamino acid prevent the precipitation of nickel dimethylglyoxime, once the latter has been formed the precipitate is quite resistant to ammoniacal or acetic acid solutions of the alkali salts of ethylenediaminetetraacetic acid.

Procedure.[3] Single drops of 0.008% solution of hydrated nickel sulfate are placed in adjacent depressions of a spot plate, and a drop of the solution to be tested is added to one, and a drop of water to the other. Then one drop of dilute ammonia and one drop of alcoholic 1% dimethylglyoxime solution are added to each. A distinct precipitate of red nickel dimethylglyoxime will appear in the blank, whereas the other mixture will remain unaltered if ethylenediaminetetraacetic acid is present.

Limit of identification: 1.7 γ ethylenediaminetetraacetic acid.

1 See *The Versenes*, Bersworth Chemical Company, Framingham, Mass., 1951; *Technical Bulletin*, May and October 1951, The Alrose Chemical Company, Providence, Rhode Island.
2 G. SCHWARZENBACH et al., *Helv. Chim. Acta*, 28 (1945) 828, 1132; 29 (1946) 364, 811, 1338; 30 (1947) 1798; 31 (1948) 331, 456, 495, 678; *Chimia*, 2 (1948) 56; comp. also R. PŘIBIL, *Collection Trav. Chim. Tchécoslov.*, 14 (1949) 320.
3 Unpublished studies, with D. GOLDSTEIN.
4 A. DARBEY, *Anal. Chem.*, 24 (1952) 373.

(2) Test by masking the formation of zinc 8-hydroxyquinoline[1]

As noted in Test (*1*), ethylenediaminetetraacetate ions prevent precipitation of Ni-dimethylglyoxime, because this metal is included in the anionic portion of a water-soluble chelated compound of this masking agent. A similar sequestering action also affects zinc ions with respect to their precipitability with 8-hydroxyquinoline (oxine), with which Zn^{2+} ions normally give a light yellow precipitate which fluoresces yellow in ultraviolet light. If this zinc oxinate is dissolved in dilute mineral acid, the non-fluorescing solution yields the fluorescing salt on the addition of excess ammonia. However, if an excess of ethylenediaminetetraacetic acid is present, no precipitation occurs and consequently there is no fluorescence. This behavior permits the detection of ethylenediaminetetraacetic acid.

Procedure. A drop of the test solution is placed in a depression of a spot plate, and a drop of water in an adjoining depression. One drop of the solution of zinc oxinate in dilute hydrochloric acid is stirred into each. A drop of 1 : 1 ammonia is then introduced. An intense fluorescence, clearly visible in ultraviolet light, appears at once in the blank test.

Reagent: 0.025 g zinc oxinate* dissolved in 100 ml dilute hydrochloric acid.

Limit of identification: 8 γ ethylenediaminetetraacetic acid.

* Preparation: The warm ammoniacal solution of a zinc salt is treated with an alcoholic solution of oxine. The light yellow precipitate is washed with water, then with alcohol, and dried.

1 Unpublished studies, with D. GOLDSTEIN.

(3) *Test by pyrolytic formation of pyrrole* [1]

As mentioned on p. 166, pyrrole is formed when substituted ethylenediamines are dry-heated (with intermediate production of piperazine). The resulting mixture of vapors yields a red-violet condensation product with *p*-dimethylaminobenzaldehyde. Ethylenediaminetetraacetic acid and its soluble alkali salts show the same behavior when dry heated as the parent compound and other derivatives of ethylenediamine.

Procedure. As given on p. 393.

Limit of identification: 10 γ ethylenediaminetetraacetic acid.

1 Unpublished studies, with V. GENTIL.

85. Taurine (2-aminoethanesulfonic acid)

Test through demasking of the SO_3H groups [1]

A test was described in Chap. 3, Sect. 42, for aliphatic sulfonic acids; it is based on the splitting out of sulfurous acid on fusion with benzoin. Taurine does not show this behavior since in line with the isomerism equilibrium

$$\begin{array}{ccc} CH_2-NH_2 & & CH_2-\overset{+}{N}H_3 \\ | & \rightleftharpoons & | \\ CH_2-SO_3H & & CH_2-SO_3^- \\ (I) & & (II) \end{array}$$

it is present almost exclusively in the ammonium salt form (II). For the same reason it yields no hydrogen cyanide when dry-heated with mercuric cyanide, and hence does not respond to this characteristic test for acidic compounds (comp. p. 148). This masking of the SO_3H group can be overcome by fuming with formalin or paraformaldehyde; the sulfo form (I) is withdrawn from the isomerism equilibrium by the condensation reaction:

$$\begin{matrix} CH_2-NH_2 \\ | \\ CH_2-SO_3H \end{matrix} \; + \; CH_2O \; \longrightarrow \; \begin{matrix} CH_2-N=CH_2 \\ | \\ CH_2-SO_3H \end{matrix} \; + \; H_2O$$

The SO_3H group is unsalified in the Schiff base of taurine and accordingly it can enter into the pyrolysis reactions with mercuric cyanide or benzoin. Combination of the demasking with these pyrolysis reactions permits a selective test for taurine.

When using the pyrolytic production of hydrogen cyanide subsequent to demasking with formaldehyde as outlined in Procedure I, it should be noted that aminocarboxylic acids may not be present since they behave similarly to taurine; likewise the test material must contain no acidic compounds that react directly with mercuric cyanide to yield hydrogen cyanide on dry-heating. A pertinent preliminary test is therefore necessary. Procedure II which makes use of the pyrolytic release of sulfur dioxide after demasking with formaldehyde is less equivocal. In this case, interference is occasioned only by aliphatic sulfonic acids or by aliphatic compounds containing —SH or —S—S— groups, whose absence can be confirmed by a preliminary test through fusion with benzoin (reductive production of H_2S).

Procedure I. A drop of the test solution is taken to dryness in a micro test tube along with a drop of formalin, or the evaporation residue is mixed with several cg paraformaldehyde and the mixture is kept for 20 min. at 140° in an oven in order to expel the excess of paraformaldehyde. Several cg mercuric cyanide and a drop of acetone are then introduced and taken to dryness. The mouth of the test tube is covered with a disk of filter paper moistened with cyanide reagent (see p. 546) and the test tube is immersed about ½ cm deep in a glycerol bath that has been preheated to 160°. A positive response is indicated by the development of a blue stain on the paper.

Limit of identification: 10 γ taurine.

Procedure II. The demasking with formalin or paraformaldehyde is accomplished as described in Procedure I. Several cg of benzoin are then added and the test tube is placed about ½ cm deep in a glycerol bath that has been preheated to 130°. The mouth of the test tube is covered with a disk of filter paper moistened with ferri ferricyanide solution (for preparation see p. 86) and the temperature raised to around 180°. The appearance of a blue stain on the paper indicates a positive response.

Limit of identification: 10 γ taurine.

1 F. Feigl, *Anal. Chim. Acta*, 26 (1962) 424.

86. Benzylamine

Test with p-nitrobenzaldehyde[1]

If a colorless dilute alcoholic solution of benzylamine and *p*-nitrobenzal-dehyde is treated with alkali hydroxide and warmed, an intense red-brown color appears. The color reaction is obviously due to the formation of the Schiff base (I):

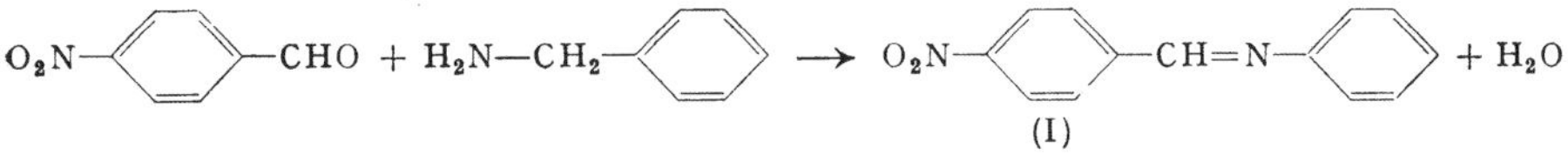

$$O_2N-\langle\rangle-CHO + H_2N-CH_2-\langle\rangle \longrightarrow O_2N-\langle\rangle-CH=N-\langle\rangle + H_2O$$

(I)

This product undergoes prototropic rearrangement in the presence of alkali hydroxide to yield a *p*-quinoid nitronic acid salt (II):

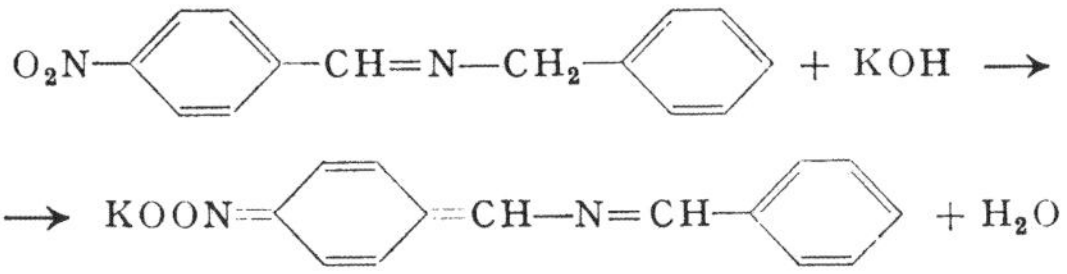

$$O_2N-\langle\rangle-CH=N-CH_2-\langle\rangle + KOH \longrightarrow$$

$$\longrightarrow KOON=\langle\rangle=CH-N=CH-\langle\rangle + H_2O$$

(II)

A selective test for benzylamine can be based on these reactions.

Procedure. A drop of the test solution is treated in a micro test tube with a drop of a saturated alcoholic solution of *p*-nitrobenzaldehyde and kept at 80° in a preheated water bath. After 5 min, a drop of 2% alcoholic sodium or potassium hydroxide is added. A red-brown color indicates a positive response.

Limit of identification: 10 γ benzylamine.

1 F. FEIGL and V. ANGER, *Anal. Chim. Acta*, 24 (1961) 494.

87. α-Naphthylamine

Detection through condensation with nitrosoantipyrine[1]

α-Naphthylamine, as well as its chloride or sulfate, condenses in acetic acid solution with nitrosoantipyrine to yield a violet azo dye:

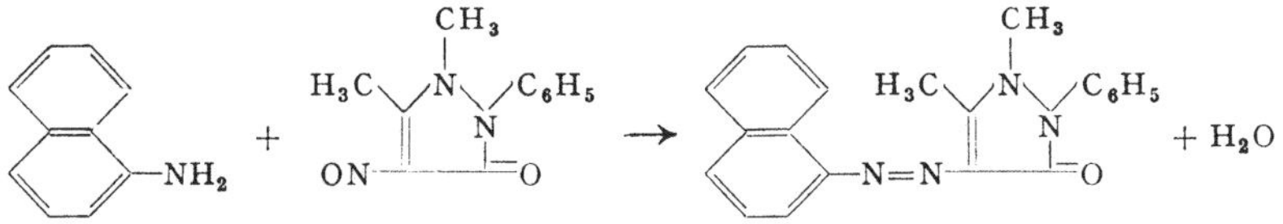

Aniline chloride (sulfate) does not react with nitrosoantipyrine but the acetate reacts similarly to give a violet color. β-Naphthylamine does not react in either acetic acid solution or in the form of its salts. Therefore, α-naphthylamine can be detected in the presence of any quantity of the β-isomer.

Procedure.   One drop of the acetic acid test solution (base or hydrochloride) is treated in a micro test tube with a drop of the reagent solution. The mixture is heated in the water bath for 1–2 min. Depending on the quantity of α-naphthylamine, a more or less intense violet color appears.

Reagent: 1 g antipyrine is dissolved in 20 ml 1:1 acetic acid and 0.6 g sodium nitrite added. After standing 10 min., 0.5 g sodium azide is added, and the volume brought up to 150 ml with 1:1 acetic acid. The solution will keep for several days.

Limit of identification: 0.5 γ α-naphthylamine.

α-Naphthylamine-sulfonic acids likewise react with nitrosoantipyrine. However, the sensitivity is much lower than with α-naphthylamine and its salts. Brown condensation products result, as shown in the case of the 1,3- and 1,4-naphthylamine-sulfonic acids. The limit of identification is 75 γ α-naphthylamine-sulfonic acid. Brown-red condensation products are given by H-acid (8-amino-1-naphthol-3,6-disulfonic acid) and by 1,8-naphthylenediamine. The violet color is apparently characteristic for α-naphthylamine and a shift of the color toward brown occurs when acidic or basic groups are introduced into the aromatic ring.

1 F. FEIGL, Cl. COSTA NETO and E. SILVA, *Anal. Chem.*, 27 (1955) 1319.

88. *o*-Phenylenediamine

(1) Detection through quinoxaline condensation [1]

o-Phenylenediamine condenses with 1,2-diketones to yield quinoxalines[2]. If the sodium salt of dihydroxytartaric acid is used in this condensation, the sodium salt of quinoxaline-2,3-dicarboxylic acid results:

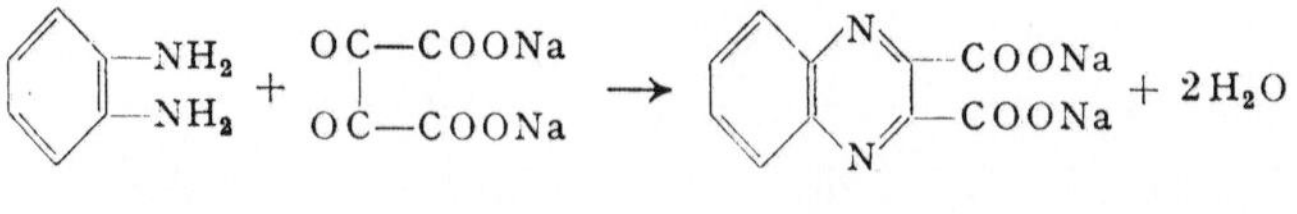

This acid, which contains the group requisite for the formation of colored inner-complex Fe(II) salts (see p. 386), has been recommended as a reagent for a delicate test for iron[3].

Since the quinoxaline condensation and the formation of the violet iron salt can be accomplished in a single operation, this succession of reactions constitutes the basis of a characteristic test for *o*-phenylenediamine.

Procedure.   The test is conducted in a micro test tube. One drop of the test solution and several cg of dihydroxytartaric acid are treated with 1 drop of 0.5% ferrous ammonium sulfate in 0.1 N acetic acid containing a little sodium fluoride, and the mixture is heated in a boiling water bath. According to the quantity of o-phenylenediamine present, a violet color appears almost at once.

Limit of identification: 5 γ o-phenylenediamine.

1 V. ANGER, unpublished studies.
2 Compare P. KARRER, *Lehrbuch der Organischen Chemie*, 13th ed., Stuttgart, 1959, p. 495.
3 H. LEY, C. SCHWARTE and O. MÜNNICH, *Ber.*, 57 (1924) 354.

(2) Detection through formation of 2-methylbenzimidazole[1]

o-Phenylenediamine condenses with anhydrous acetic acid on warming:[2]

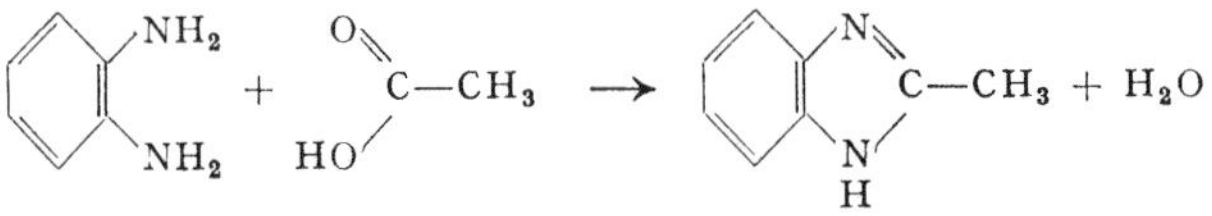

The resulting 2-methylbenzimidazole is a tertiary amine and accordingly responds to the color reaction described on p. 251 on treatment with a mixture of acetic anhydride and citric acid. Therefore if o-phenylenediamine or a salt of this base is added to this mixture and warmed, the formation of 2-methylbenzimidazole is followed by the almost simultaneous color reaction.

The procedure given here permits the detection of o-phenylenediamine and its differentiation from its two isomers. It is assumed that other aromatic o- and 1,2-aliphatic diamines will behave in the same fashion.

Procedure. A small quantity of the solid (base or salt) or a drop of its solution in acetone or water is taken to dryness in a micro test tube. Several crystals of citric acid and 3 drops of acetic anydride are added and the mixture is heated in boiling water. A positive response is signaled by the development of a red or pink color within 5 min. It is advisable to run a comparison blank.

Limit of identification: 2 γ o-phenylenediamine.

1 F. FEIGL and A. DEL'ACQUA, *Z. Anal. Chem.*, 204 (1964) 421.
2 A. LADENBURG, *Ber.*, 8 (1875) 677.

(3) Test with salicylaldehyde and cobaltous salts[1]

When an alcoholic–aqueous solution of salicylaldehyde containing cobaltous salts is added to o-phenylenediamine a yellow color appears which turns brownish-red, and then orange. By making the solution alkaline an intense red color is formed immediately. The color reaction is based on the fact that first the Schiff base (I), and then the inner-complex cobalt salt (II) is formed.

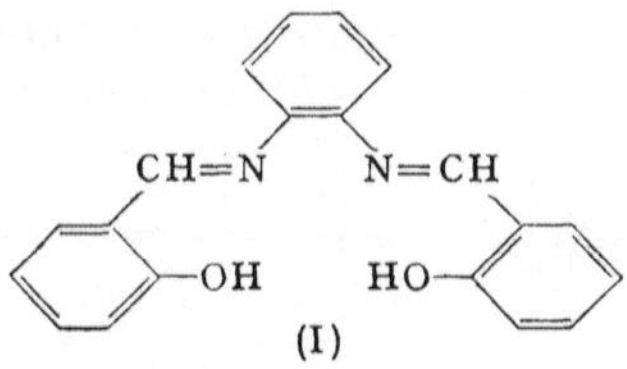

(I) 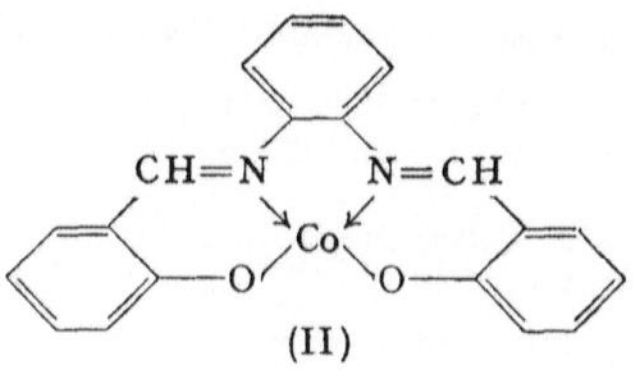   (II)

m- and *p*-Phenylenediamine do not show the color reaction, because the respective Schiff bases cannot form cobalt chelates. Derivatives of *o*-phenylenediamines as well as *o*-diaminonaphthalenes behave like the parent compound. 1,8-Diaminonaphthalene gives a brown cobalt salt under the test conditions.

The test described here permits the detection of *o*-nitroaniline and *o*-dinitrobenzene after reduction to the respective *o*-diamines by means of Raney nickel alloy and hydrochloric acid.

Procedure.   In a micro test tube a drop of the test solution is mixed with a drop of the reagent. If the test solution is acidic, a drop of pyridine should be added. In the presence of reactive diamines a brown coloration appears within 5 min. After acidifying the solution with dilute acetic acid, the color changes to orange-red.

Reagent: 3 drops of salicylaldehyde are mixed with 10 drops of 0.5 M Co(NO$_3$)$_2$ and diluted with 5 ml ethanol.

The following could be detected: 5 γ 1,2-diaminonaphthalene
    1 γ *o*-phenylenediamine   1 γ 2,3-   ,,   ,,
    1 γ 4-nitro-*o*-phenylenediamine   20 γ 1,8-   ,,   ,,

1 S. YARIV, *Israel J. Chem.*, 2 (1964) 29.

89. *p*-Phenylenediamine

Test by conversion into phenylene blue[1]

When *p*-phenylenediamine is mixed with alkali persulfate in the presence of aniline, in slightly acid solution, the indamine dye, phenylene blue, is formed immediately:

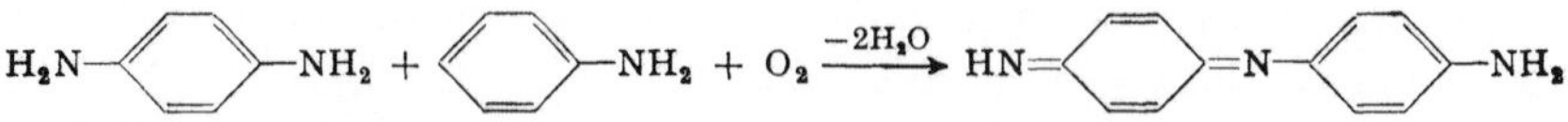

Salts of the basic dye result of course in the acid solution.

As the equation indicates, essential features are the production of benzoquinonediimide and its oxidative condensation with an aromatic amine, which is not substituted in the position *para* to the nitrogen. Accordingly, N-mono- or N-dialkylated *p*-diamines with a free NH_2 group behave similarly to *p*-phenylenediamine; the aniline can be replaced by other aromatic monoamines with a free position *para* to the NH_2 group. Bandrowski's base is an exception. Despite its *p*-phenylenediamine structure (see p. 674) it does not react. However, it is easily split by reduction to yield *p*-phenylenediamine and can be detected in this manner (see p. 675). Analogous oxidation and condensation reactions with phenols, which are not substituted in the *para* position, lead to indophenol dyes.

The formation of phenylene blue in accord with the preceding equation has been employed on the macro scale for the characterization of aniline.[2] It was shown in this connection that the dye base, produced by alkalization, can be extracted with carbon tetrachloride to yield a red solution.

Procedure. A drop of the acetic acid test solution is mixed with a drop of aniline water (one drop $C_6H_5NH_2$ in 50 ml H_2O) and several crystals of potassium persulfate are added. A dark to light blue color appears at once, according to the amount of *p*-phenylenediamine present.

Limit of identification: 0.5 γ *p*-phenylenediamine.

1 O. HEIM, *Ind. Eng. Chem., Anal. Ed.*, 7 (1935) 146.
2 K. W. MERZ and A. KAMMERER, *Arch. Pharm.*, 86 (1953) 198.

90. *p*-Aminodimethylaniline

(1) Test through conversion into Wurster's red[1]

p-Aminodimethylaniline may easily be oxidized to the so-called Wurster's red. The latter is an addition compound (meriquinoidal compound) of the base with its quinoidal oxidation product formed according to:

$$H_2N\!-\!\!\langle\bigcirc\rangle\!-\!N(CH_3)_2 + [O] + H^+ \longrightarrow HN\!=\!\langle\bigcirc\rangle\!=\!\overset{+}{N}(CH_3)_2 + H_2O$$

For analytical purposes the oxidation with dilute ferric chloride solution is recommended.

Procedure. A drop of the alcoholic neutral or slightly acid test solution is mixed in a depression of a spot plate with one drop of a 0.01% ferric chloride solution. (When large amounts of the base are expected, a 0.5% solution of the oxidant is recommended.) A positive response is indicated by the appearance of a red-violet color.

Limit of identification: 2 γ *p*-amino-N,N-dimethylaniline.

1 E. JUNGREIS (Jerusalem), unpublished studies.

91. Tetrabase (4,4'-tetramethyldiaminodiphenylmethane)

Tests with manganese dioxide[1]

There is no specific test for tetrabase but this colorless compound can be reliably identified through positive responses to two selective tests that are based on its behavior toward manganese dioxide. In acetic acid medium there is oxidation to a blue *p*-quinoidal compound as shown in (*1*) while dry heating with MnO_2 at 110–120° occasions oxidative release of CH_3 groups accompanied by formation of formaldehyde as shown in (*2*):

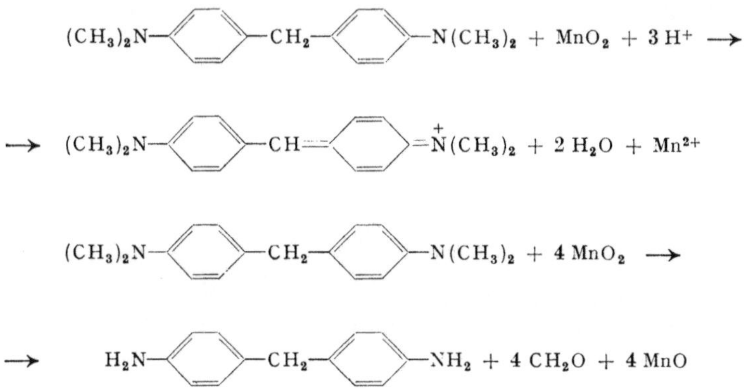

$(CH_3)_2N$⟨⟩$-CH_2-$⟨⟩$-N(CH_3)_2 + MnO_2 + 3 H^+ \longrightarrow$

$\longrightarrow (CH_3)_2N$⟨⟩$-CH=$⟨⟩$=\overset{+}{N}(CH_3)_2 + 2 H_2O + Mn^{2+}$ (*1*)

$(CH_3)_2N$⟨⟩$-CH_2-$⟨⟩$-N(CH_3)_2 + 4 MnO_2 \longrightarrow$

$\longrightarrow H_2N$⟨⟩$-CH_2-$⟨⟩$-NH_2 + 4 CH_2O + 4 MnO$ (*2*)

The blue oxidation product resulting from reaction (*1*) can be detected by Procedure I, and the formaldehyde formed in reaction (*2*) can be detected by Procedure II.

The only material showing the same behavior as tetrabase is thio-Michler's ketone which is yellow-brown both in the solid state and in solution. Michler's ketone itself gives a positive response only to Procedure II. Benzidine, diphenylbenzidine, and diphenylamine respond positively only to Procedure I by being oxidized to blue quinoidal compounds.

Procedure I. A drop of the ethereal or benzene test solution is placed on filter paper impregnated with manganese dioxide (for preparation see p. 474) and the spot is exposed to vapors of acetic acid. A blue stain indicates a positive response.
Limit of identification: 0.5 γ tetrabase.

Procedure II. A drop of the benzene test solution is placed in a micro test tube and several cg of manganese dioxide are added. The mouth of the tube is covered with a disk of filter paper moistened with Nessler reagent and the tube is immersed in boiling water. If formaldehyde is produced, its vapors will give a black or grey stain on the paper.
Limits of identification: 2.5 γ tetrabase.

1 Unpublished studies, with D. HAGUENAUER-CASTRO.

92. Benzidine sulfate

Identification through tests for sulfuric acid and benzidine[1]

It was pointed out in Chap. 2, Sect. 12, in connection with the detection of thionic compounds, that barium sulfate precipitated from a saturated solution of permanganate has a violet color that is not discharged on the addition of reductants. This Wohlers effect is likewise observed if small amounts of water-insoluble benzidine sulfate are subjected to the procedure described on p. 84.

A supplementary test for benzidine (as sulfate) utilizes the formation of a violet product through the action of sodium rhodizonate (see Sect. 41). Tiny amounts of the test material are placed on filter paper and spotted with the reagent solution.

1 Unpublished studies.

93. o-Aminophenol

Test with glyoxal and calcium oxide[1]

The colorless Schiff base, namely glyoxal bis-2-hydroxyanil, produced by the reaction of o-aminophenol and glyoxal, affords a sensitive and specific test for calcium ions, with which it yields a red inner-complex salt.[2] If an appreciable amount of o-aminophenol is brought together with glyoxal and calcium oxide, the calcium salt is formed even at room temperature:

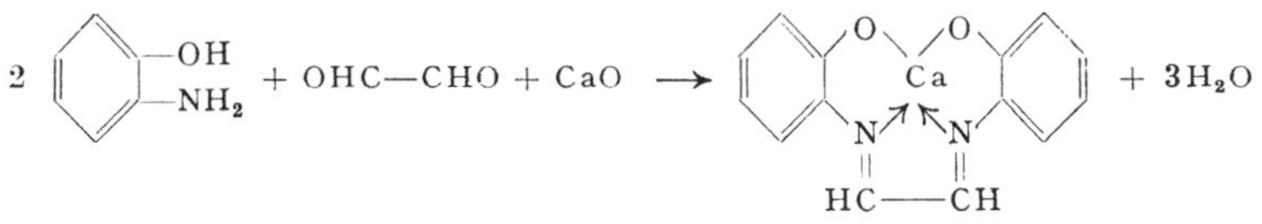

This reaction is the basis of a test for either glyoxal (comp. Sect. 32) or o-aminophenol. It must be noted that an excess of amine is essential to the condensation of both carbonyl groups. If an excess is not present, the mono-Schiff base results which is not a color reagent for calcium. When only small amounts of o-aminophenol are to be detected, even the use of a dilute solution of glyoxal does not prevent the formation of the mono-Schiff base, and consequently the test has a relatively low sensitivity.

Procedure. A drop of the ethereal or alcoholic test solution is treated in a micro test tube with a drop of 0.02% aqueous solution of glyoxal. The tube is kept in boiling water for about 2 min. and then cooled. Several cg of freshly ignited lime are added. A red or pink color develops.

Limit of identification: 25 γ o-aminophenol

o-Aminonaphthols do not respond to this test.

1 F. FEIGL and D. GOLDSTEIN, *Z. Anal. Chem.*, 163 (1958) 30.
2 D. GOLDSTEIN and C. STARK-MAYER, *Anal. Chim. Acta*, 19 (1958) 437.

94. *m*-Aminophenol

(1) Test through conversion into 7-hydroxyquinoline[1]

Application of the Skraup quinoline synthesis (heating with glycerol, sulfuric acid plus oxidant) to *o*-aminophenol leads to 8-hydroxyquinoline (see Sect. 12). Analogously, *m*- and *p*-aminophenol lead to 7- or 6-hydroxy-quinoline respectively. Among the three isomeric hydroxyquinolines, only 7-hydroxyquinoline shows a yellow-green fluorescence and the latter is especially visible in u.v. light. It has been found that this compound is even produced in amounts that can be detected through its fluorescence if *m*-aminophenol or its salts are heated with glycerol merely to around 120°. This finding constitutes the basis of the test described here for *m*-amino-phenol and its differentiation from the two isomers.

Procedure. A drop or two of glycerol and a small amount of the test material are placed in a micro test tube and heated to 120° in a preheated glycerol bath. A drop of alcohol is added to the cooled mixture and the contents of the test tube are viewed under a quartz lamp. Yellow-green fluorescence signals a positive response.

Limit of identification: 2 γ *m*-aminophenol.

1 E. JUNGREIS and V. LIPETZ, *Mikrochim. Acta*, (1963) 886.

(2) Test through dry-heating with mercuric cyanide[1]

It was stated in Chap. 2, Sect. 25, that dry-heating of acidic compounds of all kinds with Hg(CN)$_2$ produces HCN that can be readily detected in the gas phase. The three aminophenols react as phenols in this way at temperatures above 125°; even at 115° the evolution of HCN is so vigorous in the case of *m*-aminophenol that it can be distinguished from its isomers in this way. This exceptional behavior may be attributed perhaps to a more marked acidic behavior by the *meta*-isomer as compared with the others. It may also be taken into account here that the melting point (122°) of this isomer is quite close to the temperature (115°) just mentioned as suitable for the reaction, whereas the *ortho*- and the *para*-isomers melt at 170° and 186° respectively. According to the Hedvall rule[2], the highest reactivity of compounds is exhibited in the vicinity of their transformation points.

Procedure. An intimate mixture of mercuric cyanide and the test material is placed in a micro test tube, or a drop of the benzene solution is evaporated to dryness in the presence of the cyanide. The mouth of the tube is covered with a disk of filter paper moistened with the reagent for cyanide (see p. 546) and the tube is immersed in a glycerol bath preheated to 110°. The temperature is raised to 115°. A positive response is shown by the development of a blue stain on the reagent paper.

Limit of identification: 15 γ *m*-aminophenol.

m-Aminophenol can be detected by this procedure even in the presence of its isomers.

1 Unpublished studies.
2 A. J. HEDVALL, *Reaktionsfähigkeit fester Stoffe*, Leipzig, 1939.

95. *p*-Aminophenol

(1) Test through oxidation to p-quinone-4-chloroimine[1]

Excellent yields of *p*-quinone-4-chloroimine[2] are obtained by the action, under cooling, of chloride of lime or alkali hypochlorite on hydrochloric acid solutions of *p*-aminophenol. The agent is the chlorine liberated from the hypochlorite and hydrochloric acid:

$$HO\text{—}\langle\rangle\text{—}NH_2 + 4\ Cl^0 \rightarrow O=\langle\rangle=NCl + 3\ HCl \tag{1}$$

p-Quinone-4-chloroimine condenses with phenol in the presence of ammonia to give the blue ammonium salt of an indophenol dye (see p. 185):

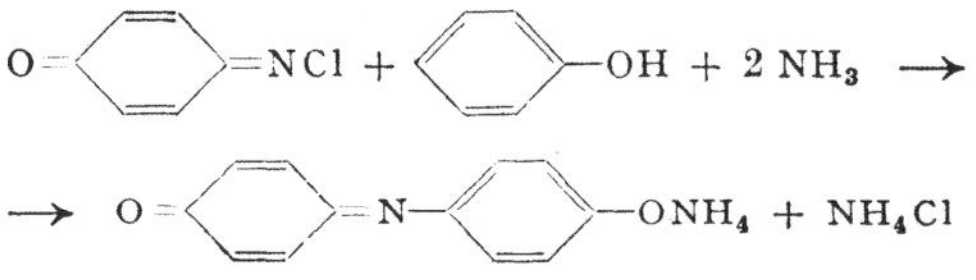

$$O=\langle\rangle=NCl + \langle\rangle\text{—}OH + 2\ NH_3 \rightarrow$$

$$\rightarrow O=\langle\rangle=N\text{—}\langle\rangle\text{—}ONH_4 + NH_4Cl \tag{2}$$

Reactions (1) and (2) can be realized in the technique of spot tests analysis if chloramine T is employed as chlorine donor, and accordingly there results a selective test for *p*-aminophenol and also a method for distinguishing it from the isomeric *o*-and *m*-aminophenol. The methyl- and ethyl ethers of *p*-aminophenol, known respectively as anisidine and phenetidine, behave like the parent compound because they are dealkylated under the conditions of the test. Phenacetin may likewise be converted to *p*-quinone-4-chloro-imine by chloramine T if mg quantities are present. Smaller quantities require preliminary saponification to aminophenols (see p. 634).

1-Amino-4-naphthol behaves in the same manner as *p*-aminophenol. In contrast, the oxidation does not appear to proceed with monosubstituted aminophenols, a finding confirmed by trials with 4-amino-3-methylphenol and 2,4-diaminophenol.

Procedure. A drop of the hydrochloric acid test solution is united in a micro test tube with a little solid chloramine T, and after a few minutes the mixture is extracted with several drops of ether. Successive drops of the ethereal solution are brought on filter paper that has previously been moistened with an ethereal solution of phenol. A blue stain appears when the fleck is exposed to ammonia vapors if p-aminophenol or its reactive derivatives were present.

Limit of identification: 1 γ p-aminophenol.

The above procedure allows the detection of N-alkylated p-aminophenols (or their ethers) after dealkylation. This can be achieved when the test material is twice taken to dryness with hydriodic acid (d = 1.8). 5 γ p-methylaminophenol sulfate could be detected in this way.

1 E. JUNGREIS (Jerusalem), unpublished studies.
2 R. SCHMITT, *J. Prakt. Chem.*, [2] 19 (1939) 315.

(2) Test with potassium iodate[1]

If an aqueous solution of the chloride or sulfate of p-aminophenol is brought into contact with potassium iodate a stable intense blue color results. This probably is due to the formation of an indamine dye by the following partial reactions:

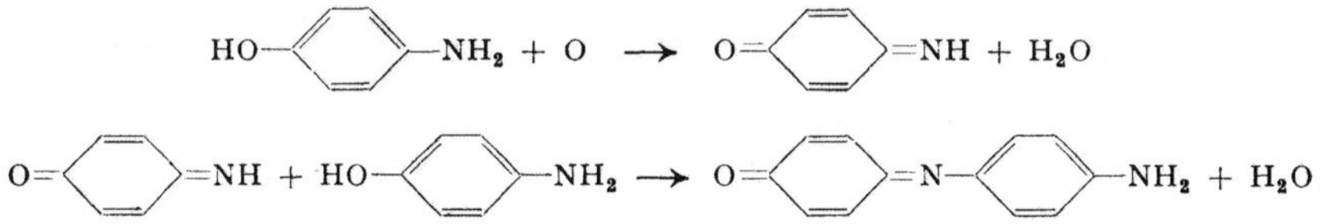

Since o- and m-aminophenol are not affected by potassium iodate, this oxidation of p-aminophenol permits its detection and also differentiation from the isomers.

Procedure.   A drop of the test solution and several mg KIO_3 are placed in a micro test tube. A blue color indicates a positive response.

Limit of identification: 0.5 γ p-aminophenol.

1 F. FEIGL, E. JUNGREIS and S. YARIV, *Z. Anal. Chem.*, 200 (1964) 38.

96. N,N-Diphenylanthranilic acid

Detection through tests for the $N(C_6H_5)_2$- *and the COOH group*

In line with its formula, $C_6H_4(COOH)N(C_6H_5)_2$, N,N-diphenylanthranilic acid yields hydrogen cyanide when dry-heated with mercuric cyanide (see p. 117). In addition, the acid shows the behavior characteristic of diphenylamine and its derivatives, namely oxidation to blue quinoneimine when treated with concentrated sulfuric acid and alkali nitrite (see p. 363). Therefore N,N-diphenylanthranilic acid can be identified through positive responses to both of these tests. It should be noted that diphenylamine must be absent, because it gives a positive response to the partial tests here applied. Previous separation based on different solubilities is easily accomplished.

Limits of identification:

4 γ N,N-diphenylanthranilic acid through oxidation to quinoneimine.

6 γ ,,   ,,   ,,   ,,   pyrolysis with mercuric cyanide.

1 Unpublished studies, with E. LIBERGOTT.

97. Piperidine

Test with sodium 1,2-naphthoquinone-4-sulfonate[1]

A test has been given in Chap. 2, Sect. 44 for reactive NH_2- and CH_2 groups that is based on the reaction of sodium 1,2-naphthoquinone-4-sulfonate with compounds containing such groups to yield intensely colored p-quinoid condensation products. Piperidine also shows this behavior.

This reaction, which leads to a brick-red compound, can be accomplished by warming acidic aqueous solutions of a piperidine salt with sodium carbonate. Piperidine (b.p. 106°) is then evolved with water vapor. The procedure described here permits an unequivocal differentiation from piperazine (b.p. 146°) which reacts in the same way as piperidine in solution, and in the gas phase, but does not volatilize under the conditions prescribed for the test. The non-volatilization of piperazine is due not only to the higher boiling point of this base but also to the formation of piperazine carbonate.

The test described here assumes the absence of ammonium salts and salts of aromatic primary amines that are volatile with steam, since ammonia as well as these amines yield color reactions with sodium 1,2-naphthoquinone-4-sulfonate. The test for ammonia can be made by means of Nessler reagent while the possible presence of aryl amines can be determined through the formation of yellow Schiff bases on treatment with p-dimethylaminobenzaldehyde (see Chap. 3, Sect. 50).

Procedure. A drop of the acidic test solution and several cg of sodium carbonate are brought together in a micro test tube. The latter is immersed to within about 1 cm below its mouth in water that has been heated to 90°. The mouth of the tube is covered with a disk of filter paper moistened with a drop of 2.5% aqueous solution of sodium 1,2-naphthoquinone-4-sulfonate. A red stain on the yellow reagent paper indicates a positive response.

The following were detected: 500 γ piperidine, deep red stain in 1 min.

50	γ	,,	,, ,, ,, ,,	3 ,,
25	γ	,,	,, ,, ,, ,,	6 ,,
12.5	γ	,,	,, pink ,, ,,	8 ,,

The test clearly revealed 20 γ piperidine in mixture with 10 mg piperazine.

1 F. FEIGL, *Chemist-Analyst*, 50 (1961) 15.

98. Piperazine

Test through conversion to carbonate with subsequent thermal liberation[1]

It was pointed out in Sect. 97 that if acidic solutions of piperidine are warmed with sodium carbonate the base is volatilized, whereas this same treatment applied to acidic solutions of piperazine does not result in the liberation of this base, which instead is left as its carbonate. The latter yields free piperidine only on dry-heating and it can be detected in the gas phase through the color reaction with sodium 1,2-naphthoquinone-4-sulfonate. This difference in behavior makes possible the detection of piperazine in the presence of piperidine since the latter is completely removed when the acidic solution is taken to dryness with excess sodium carbonate, while the piperazine remains as carbonate. The test described here is not impaired by ammonium salts and primary amines since they behave in the same manner as salts of piperidine.

Procedure.   A drop of the acidic test solution is taken to dryness along with a drop of a 5% solution of sodium carbonate in a micro test tube; the residue is kept at 105° for 5 min. in an oven. The test tube is then immersed in a glycerol bath preheated to 150° and the mouth of the tube is covered with a disk of filter paper moistened with a 2.5% solution of the reagent. The temperature is then raised to 220°. A positive response is indicated by the development of a pink or red-violet stain, the shade depending on the quantity of piperazine present.

Limit of identification: 20 γ piperazine in the presence of 10 mg piperidine.

1 F. FEIGL, *Chemist-Analyst*, 50 (1961) 18.

99. Carbazole

Test with p-nitrosophenol and sulfuric acid

Carbazole reacts with p-nitrosophenol in 'concentrated sulfuric acid to give a deep blue indophenol-like compound[1]:

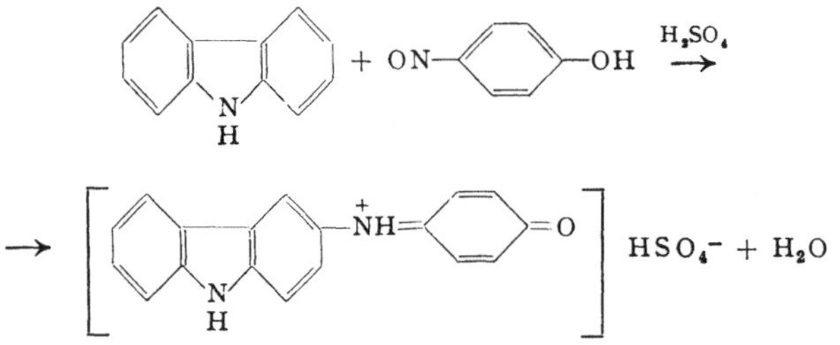

Since this reaction proceeds in the cold and since diphenylamine shows but slight reaction, this interaction provides the basis of a selective test for carbazole.

Procedure.[2] A slight amount of the solid test material or the evaporation residue from a drop of the test solution is treated in a depression of a spot plate with a drop of a freshly prepared 0.05% solution of p-nitrosophenol in concentrated sulfuric acid. The appearance of a blue color, which does not disappear when a drop of water is added, indicates the presence of carbazole.

Limit of identification: 0.2 γ carbazole.

1 L. HAAS, DRP 227 323.
2 V. ANGER and S. OFRI, private communication.

100. Thiodiphenylamine (phenothiazine)

Test with chloramine T[1]

If solid chloramine T (sodium N-chloro-p-toluenesulfonamide) is added to an alcoholic solution of thiodiphenylamine, a violet product appears immediately. After dilution with water, this colored material can be extracted into ether, benzene, chloroform, or carbon disulfide. Likewise, if a benzene solution of thiodiphenylamine is added to a concentrated solution of chloramine T and shaken, the benzene layer becomes violet and the water layer colorless. If solid thiodiphenylamine is moistened with aqueous chloramine T solution, the solid turns violet. The chemistry of these color reactions cannot be stated with certainty as yet because the colored product has not been isolated in pure form. However, it is likely that the two compounds produced by the hydrolysis of the chloramine T:

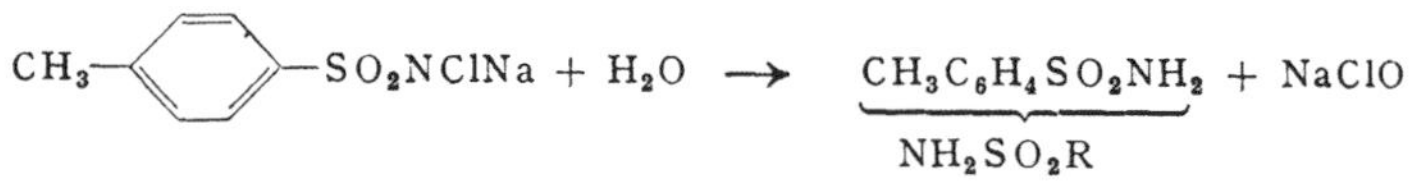

$$CH_3\text{—}\langle\ \rangle\text{—}SO_2NClNa + H_2O \rightarrow \underbrace{CH_3C_6H_4SO_2NH_2}_{NH_2SO_2R} + NaClO$$

react with thiodiphenylamine. The following redox- and oxidative condensation reactions seem plausible for the formation of thiazine dyes:

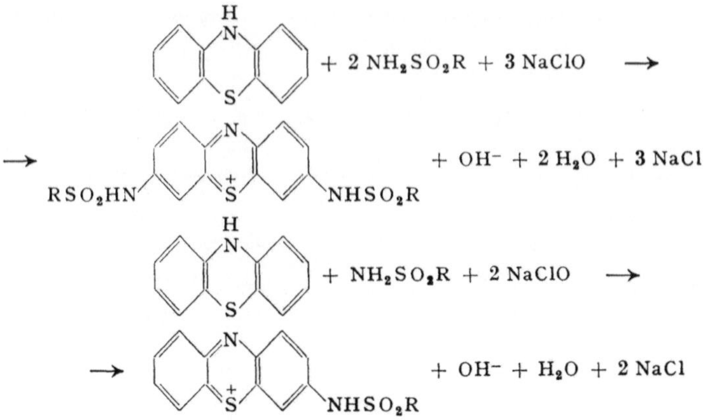

(thiodiphenylamine structure) $+ 2\,NH_2SO_2R + 3\,NaClO \longrightarrow$

$\longrightarrow$ (oxidized thiazine structure: RSO_2HN—, —$NHSO_2R$) $+ OH^- + 2\,H_2O + 3\,NaCl$

(thiodiphenylamine structure) $+ NH_2SO_2R + 2\,NaClO \longrightarrow$

$\longrightarrow$ (thiazine structure: —$NHSO_2R$) $+ OH^- + H_2O + 2\,NaCl$

This assumption is in agreement with the finding that a color reaction (red-violet) also occurs in acid solution and that the resulting colored products, like other thiazine dyes, can be extracted from aqueous solution or suspension by ether or benzene.

The detection with chloramine T can be accomplished with alcoholic solutions of thiodiphenylamine by Procedure I or with the benzene solution by Procedures II or III.

Procedure I. A drop of the alcoholic test solution is treated in a micro test tube with several mg of chloramine T. A blue-violet color appears at once, the intensity increasing with the quantity of thiodiphenylamine present.
Limit of identification: 0.6 γ thiodiphenylamine.

Procedure II. Several drops of a freshly prepared 10% aqueous solution of chloramine T are placed in a micro test tube and 1 drop of the benzene solution of the test material is added. The test tube is then placed in a water bath pre-heated to 60–70°. The benzene layer becomes violet if the response is positive. The color deepens on extraction.
Limit of identification: 0.5 γ thiodiphenylamine.

Procedure III. A drop of the ethereal or benzene solution is placed on filter paper and as soon as the solvent has evaporated the area is spotted with a drop of freshly prepared 10% aqueous solution of chloramine T. A violet stain or ring appears within 1–2 min., depending on the amount of thiodiphenylamine present.
Limit of identification: 0.03 γ thiodiphenylamine.

1 F. FEIGL and D. HAGUENAUER-ÇASTRO, *Anal. Chem.*, **33** (1961) 1412.

101. Pyridine

(1) Test through formation of polymethine dyes from the vapor phase[1]

A test for pyridine and its derivatives with free α-positions is described in Chap. 4, Sect. 58. It is based on the action of bromocyanogen and *o*-tolidine with solutions of the compounds in question to yield a red polymethine dye. Since pyridine and water form an azeotropic mixture (b.p. 95°) the color reaction may also be accomplished by bringing the vapor into contact with a solution of the reagents. The result is a selective test for pyridine and also for the reactive pyridine bases (β- and γ-picoline) which likewise are volatile with steam. The corresponding volatility of β-pyridinecarbinol is too small for it to show analogous behavior.

Procedure. The apparatus shown in Fig. 23, *Inorganic Spot Tests*, is used. A drop of the test solution is placed in the bulb; if the solution is acid add a drop or two of dilute alkali. The knob of the stopper is charged with a drop of the reagent and put in place. The closed apparatus is kept in boiling water for 3–5 min. If pyridine is present, the hanging drop turns red or pink. Even slight colorations can be seen if the drop is wiped onto filter paper.

Reagent: A freshly prepared mixture of one drop each of 2% KCN, saturated bromine water, and suspension of *o*-tolidine base (compare p. 384).

Limit of identification: 2.5 γ pyridine.

1 Unpublished studies, with D. GOLDSTEIN.

(2) Test through conversion into piperidine [1]

Studies of the behavior of caustic alkaline pyridine solutions when heated with Devarda's alloy or Raney nickel alloy showed that only in the second case were vapors evolved which produced a violet-red stain when they came into contact with filter paper moistened with a solution of sodium 1,2-naphthoquinone-4-sulfonate, thus indicating the formation of piperidine (comp. Sect. 97). Consequently, the reaction with the alloy does not involve nascent hydrogen but rather the nickel hydride present in the Raney nickel:

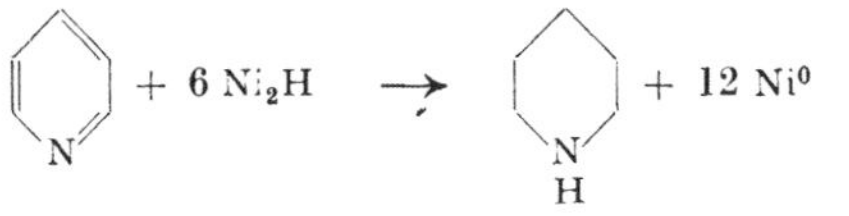

The realization of the above reactions permits the detection of pyridine, provided volatile arylamines, which likewise show a color reaction with the naphthoquinonesulfonate, are absent.

Procedure. A drop of the alkaline test solution is treated in a micro test tube with a few mg of Raney nickel alloy and the mouth of the tube is covered with a piece of filter paper moistened with a freshly prepared solution of sodium 1,2-naphthoquinone-4-sulfonate. After the stormy evolution of hydrogen has abated, the tube is placed in boiling water. A positive response is shown by the development of a red-violet stain on the reagent paper within 1–3 min.

Limit of identification: 2.5 γ pyridine.

1 F. FEIGL, *Anal. Chem.*, **33** (1961) 1121.

102. Pyridine-2-aldehyde

Test through conversion into pyridine-2-aldoxime[1]

A sensitive test (also a colorimetric determination) for ferrous ions is based on their behavior toward pyridine-2-aldoxime (I). The violet complex cation (II) is produced in neutral, faintly acid, or alkaline media.[2]

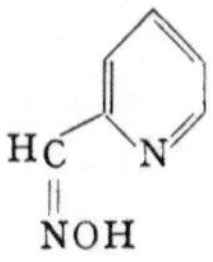

The color reaction just mentioned permits the detection of 5 γ of the aldoxime.[1]

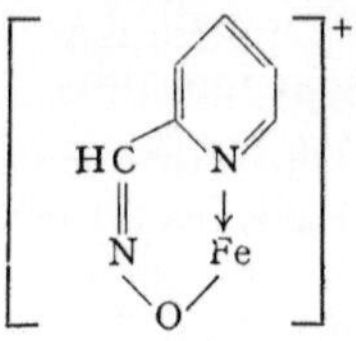

The violet complex ion (II) is formed likewise if pyridine-2-aldehyde is oximated in the presence of ferrous ions; this reaction can be carried out by alkalization with pyridine of a mixture of the aldehyde, ferrous salts, and hydroxylamine salts. Pyridine-2-aldehyde may be detected in this way, and it may likewise be distinguished from its isomers on the same basis. Because of the distance of the NOH group from the heterocyclic N atom in the aldoximes of the isomers, they are incapable of chelating the iron ions.

Procedure. A few mg of hydroxylamine hydrochloride, a drop of pyridine, and a drop of a 1% solution of ferrous ammonium sulfate are added to a drop of the test solution in a depression of a spot plate. A violet color signals a positive response.

Limit of identification: **4** γ pyridine-2-aldehyde.

1 F. FEIGL and L. BEN-DOR, *Talanta*, 10 (1963) 1111.
2 H. HARTKAMP, *Z. Anal. Chem.*, 170 (1959) 399. See also, Y. ISRAELI and E. JUNGREISS, *Bull. Res. Counc. Israel*, 11 A2 (1962) 121.

103. 3-Pyridinesulfonic acid

Test through reduction and desulfonation with Raney nickel alloy[1]

When warmed with Raney nickel alloy and caustic alkali, pyridine is converted into piperidine which can be detected in the gas phase through the color reaction with sodium 1,2-naphthoquinone-4-sulfonate (see Sect. 101). Pyridinesulfonic acid behaves like aromatic sulfonic acids when treated with Raney nickel alloy and caustic alkali, *i.e.* the SO_3H group is split off with formation of nickel sulfide (comp. p. 230). Accordingly, two readily recognizable products result when pyridinesulfonic acid is warmed with Raney nickel alloy and caustic alkali.

Procedure. A drop of the strongly alkaline test solution is brought together with several cg of Raney nickel alloy in a micro test tube. After the turbulent evolution of hydrogen has subsided a disk of filter paper that has been moistened with a drop of freshly prepared aqueous solution of sodium 1,2-naphthoquinone-4-sulfonate is placed over the mouth of the tube. The latter is immersed in boiling water. If the release of piperidine is established through the production of a violet stain on the reagent paper, the reaction mixture is cooled and 1 or 2 drops of concentrated hydrochloric acid are introduced. A disk of lead acetate paper is laid over the mouth of the test tube which is then heated in boiling water. The formation of hydrogen sulfide is made evident by the development of a black or brown stain on the reagent paper.

The test revealed:

30 γ 3-pyridinesulfonic acid through conversion into piperidine
10 γ ,,   ,, through formation of nickel sulfide.

1 Unpublished studies.

104. Quinoline

Test through reduction and subsequent oxidation or bromination [1]

If quinoline (b.p. 239°) or its isomer isoquinoline (b.p. 240°) is treated with zinc and mineral acid, the reduction results in cyclic secondary ring bases. This is readily demonstrated through reactions that are characteristic of secondary amines (see p. 250). The reduction leads to dihydroquinoline[2] and dihydroisoquinoline, respectively. The hydrogen is taken up by the cyclic N atom and a C atom in the position *ortho* to this nitrogen atom.*

The reduced acid solution of quinoline, when treated with alkali persulfate in the presence of copper salts, yields a brown precipitate or color. The reactions involved (reduction and catalytic oxidation) may be represented by the scheme:

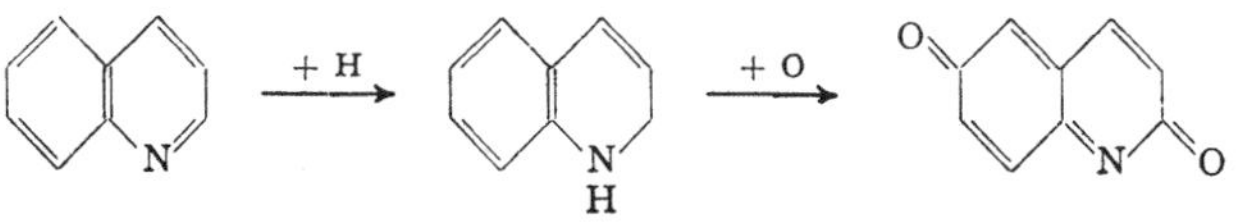

This permits a test for quinoline. Since isoquinoline does not show this behavior, a differentiation of the isomers is possible.

Procedure I. A drop of the test solution containing considerable amounts of hydrochloric acid is treated in a depression of a spot plate with 4 or 5 granules of 10 mesh zinc. After 1–2 min. any unused zinc is taken out with a glass rod. A drop of 1% copper sulfate is added along with about 20 mg of potassium

* Private communication from Prof. P. KARRER, Zurich.

persulfate. The mixture is agitated by blowing onto its surface. Depending on the quantity of quinoline involved, a brown to yellow precipitate or color appears within 1–2 min. The color attains its maximum after 3–4 min.

Limit of identification: 20 γ quinoline.

If isoquinoline solutions are carried through this procedure, no visible change occurs, no matter what their concentration.

When the reduced solution of quinoline is treated with bromine a red compound, probably a polybromide of dihydroquinoline, is formed. This behavior permits a test for quinoline.

Procedure II. The reduction (hydrogenation) is conducted as in Procedure I. A drop of the reduced solution is placed on filter paper and the spot is held over strong bromine water. According to the quantity of quinoline present, the moist spot turns pink or red. The color fades in an oven (110°) but does not disappear, and is restored by renewed exposure to bromine vapor.

Limit of identification: 2.5 γ quinoline.

Procedures I and II are primarily suited for distinguishing between quinoline and isoquinoline, provided only one of them is present. The redox reaction with persulfate can be used to detect quinoline in the presence of isoquinoline. For example, Procedure I revealed 50 γ quinoline in the presence of 2000 γ isoquinoline when applied to one drop of the test solution.

The behavior of 10% solutions of the following quinoline derivatives was examined by Procedures I and II: quinoline iodomethylate, 2-methyl-quinoline (quinaldine), β-naphthoquinoline, 6-nitroquinoline, 3-bromo-quinoline nitrate, cinchophen, 4-hydroxy-7-chloroquinoline, 4-hydroxy-7-chloroquinoline-3-carboxylic acid, 3-cyanoquinoline and acridine. Only quinaldine and quinoline iodomethylate behave analogously to quinoline. However, the response to the reaction with persulfate or bromine was definitely weaker in the case of quinaldine.

1 F. Feigl and V. Gentil, *Anal. Chem.*, 26 (1956) 1309.
2 W. Koenig, *Ber.*, 14 (1881) 98.

105. Isoquinoline

Test with 2,4-dinitrochlorobenzene[1]

Aliphatic and aromatic primary, secondary and tertiary amines condense with 2,4-dinitrochlorobenzene to give quinoidal compounds (see p. 240). Quinoline and isoquinoline likewise react; a yellow addition product is probably formed. These two bases show a distinct difference in this respect: isoquinoline is far more reactive than its isomer. Isoquinoline reacts in equimolar dilutions, whereas quinoline does not.

Procedure. A drop of a 1% ethereal solution of each base is placed on filter paper and spotted with a drop of a 1% alcoholic solution of 2,4-dinitrochloroben-zene. The papers are kept at 100° for 3 min. in an oven, and are then exposed to strong ammonia vapors. Isoquinoline shows a much more intense yellow stain than quinoline, which yields no more than a faint coloration.

Limit of identification: 2 γ isoquinoline.

The limit for quinoline is about 10 times as great.

1 J. R. AMARAL (Rio de Janeiro), unpublished studies.

106. 8-Hydroxyquinoline

Test by chemical adsorption from the vapor phase on alumina or magnesia[1]

8-Hydroxyquinoline (I), usually called "oxine", is widely used in analysis as precipitant for many metal ions from acetate-buffered or tartrated ammoniacal solutions.[2] The resulting oxinates, in the case of Al, Mg and Zn, are yellow inner-complex salts of the coordination structure (II):

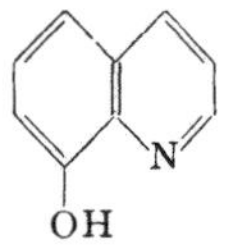

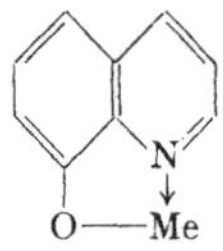

$Me = \frac{1}{2} Mg, \frac{1}{2} Zn, \frac{1}{3} Al$

The above oxinates give a strong yellow to blue-green fluorescence in ultra-violet light. This effect is probably related to the formation of salts of the amphoteric oxine since not only metal oxinates but practically all salts of oxine with inorganic and organic acids fluoresce powerfully in the solid form (but not in aqueous solution)[3], whereas oxine itself exhibits no fluorescence either as solid or in solution. However, there is a salt-like binding of oxine to metal atom not only when formula-pure oxinates precipitate but also when small quantities of oxine are chemically adsorbed on the surface of metal oxides. In other words, metal oxides (oxyhydrates, hydroxides) on contact with oxine dissolved in water or organic liquids are brought to fluoresce with the same hue as is displayed by the corresponding metal oxinate.[4] The chemical adsorption of oxine (HOx) on solid aluminium hydroxide (and analogously on Al_2O_3, MgO, ZnO, *etc.*) can be schematically represented as a surface reaction:

$$[Al(OH)_3]_x + HOx \rightarrow [Al(OH)_3]_{x-1}.Al(OH)_2Ox + H_2O \qquad (1)$$
yellow fluorescence

If Al_2O_3, MgO, ZnO, *etc.* and a little solid oxine are placed side by side under a watch glass, the oxide begins to fluoresce vigorously within a few minutes, because oxine (m.p. 79°) has sufficient vapor pressure even at room temperature to bring oxine vapor into contact with the oxide. The surface reaction just represented schematically occurs under these conditions and fluorescence ensues. This vapor phase adsorption serves as the basis for the following test, which is strictly specific for oxine. Its halogen derivatives, as well as sulfonated 8-hydroxyquinoline, which have the same reactive group as the parent compound, do not exert enough vapor pressure, even when heated to 150°, to give the vapor phase test.

The optimum conditions for the adsorption of oxine from the gas phase are observed in the following procedure in which directions are also included for the liberation of oxine from its salts with mineral acids.

Procedure. The test is conducted in the apparatus shown in *Inorganic Spot Tests*, Fig. 23. The glass knob is coated with Al_2O_3 or MgO by dipping it into a thick paste of the oxide in water and then drying in an oven. A drop of the ethereal or chloroform solution to be tested is transferred to the bulb of the apparatus and the solvent is evaporated by brief warming in boiling water. The apparatus is then closed with the stopper carrying the adsorbent and kept at 110° for 5 min. in an oven. The stopper is then examined under a quartz lamp. The presence of oxine is indicated by a yellow-green fluorescence exhibited by the adsorbent in ultraviolet light.

When dealing with aqueous solutions of oxine or its salts a drop is placed in the apparatus and about 0.5 g of anhydrous sodium sulfate is added to bind the water. If it is necessary to liberate the oxine from its salts, solid ammonium carbonate should be added before the sodium sulfate.

Limit of identification: 0.25 γ 8-hydroxyquinoline.

1 F. FEIGL, *Mikrochemie Ver. Mikrochim. Acta*, 39 (1952) 404.
2 R. BERG, *Die analytische Verwendung des o-Oxychinolins (Oxin) und seiner Derivate*, Enke, Stuttgart, 1938; F. J. WELCHER, *Organic Analytical Reagents*, Vol. I, Van Nostrand, New York, 1947, p. 264.
3 Comp. F. FEIGL, H. ZOCHER and C. TÖRÖK, *Monatsh.*, 81 (1950) 214.
4 See F. FEIGL, *Chemistry of Specific, Selective and Sensitive Reactions*, Academic Press. New York, 1949, p. 676.

107. 8-Hydroxyquinolinesulfonic acids

Test through formation of fluorescing adsorption compounds and through. reductive release of the SO_3H group[1]

A fairly reliable identification can be attained if two sensitive and selective tests are made for the characteristic functional groups contained in the sulfonic acids and positive results are obtained. One of these tests is based on the production of fluorescing adsorption compounds of 8-hydroxyquinoline and its derivatives with magnesium hydroxide. The other test involves the reductive release of aromatically bound SO_3H groups through warming with Raney nickel alloy and caustic alkali to yield nickel sulfide that is decomposed by acids to give hydrogen sulfide. The appropriate procedures are given on pp. 389 and 230.

Limits of identification:
8-hydroxyquinoline-5-sulfonic acid:

0.5 γ by the fluorescence reaction	1 γ by the formation of NiS

7-iodo-8-hydroxyquinoline-5-sulfonic acid:

1 γ by the fluorescence reaction	4 γ by the formation of NiS.

Through these tests it is possible to distinguish between 8-hydroxyquinolinesulfonic acids and 8-hydroxyquinoline sulfate (quinosol) since the sulfate does not react with Raney nickel plus caustic alkali.

1 Unpublished studies.

108. Urea

(1) Test through the Hofmann degradation [1]

If amides of carboxylic acids are warmed with alkali hypochlorite, they undergo the so-called Hofmann degradation, and yield an amine with one carbon atom less: [2]

$$R—CONH_2 + NaClO \rightarrow NaCl + CO_2 + R—NH_2$$

Accordingly, urea (the diamide of carbonic acid) should yield hydrazine:

$$CO(NH_2)_2 + NaClO \rightarrow NaCl + CO_2 + NH_2NH_2 \tag{1}$$

However, this product cannot be detected directly because (1) is followed by the further oxidation:

$$NH_2NH_2 + 2\,NaClO \rightarrow 2\,NaCl + 2\,H_2O + N_2 \tag{2}$$

Summation of (1) and (2) gives the well-known redox reaction:

$$CO(NH_2)_2 + 3\,NaClO \rightarrow 3\,NaCl + CO_2 + 2\,H_2O + N_2 \tag{3}$$

which is the basis of the gas volumetric method of determining urea.

The validity of (1) as an intermediate reaction of (3) can be demonstrated [3] by allowing hypochlorite to act on urea in the presence of salicylaldehyde. The hydrazine is then immediately converted to the insoluble salicylaldazine, which is stable against hypochlorite, and consequently reaction (2) is prevented. Since salicylaldazine even in small amounts is readily revealed by its yellow-green fluorescence in ultraviolet light (see p. 341), the Hofmann degradation can be made the basis of a selective test for urea.

Procedure. A drop of the test solution is treated in a micro test tube with one drop each of sodium hypochlorite solution, salicylaldehyde solution and 20% sodium hydroxide. After 3 min. in a boiling water bath, the cooled mixture is acidified with glacial acetic acid. A precipitate or turbidity, which exhibits a strong yellow-green fluorescence in ultraviolet light, indicates the presence of urea. It is advisable to place a drop of the acetic acid mixture on filter paper and to view the spot under ultraviolet light.

Reagents: 1) 0.5 N sodium hypochlorite.*
2) 5 drops of salicylaldehyde in 100 ml ethanol.

Limit of identification: 0.5 γ urea.

Among the derivatives of urea, methylurea and acetylmethylurea, after treatment with hypochlorite, *etc.*, likewise exhibited an aldazine fluorescence. No aldazine reaction was found with *asym*-diphenylurea, *m*-tolylurea, semicarbazide chloride, *sym*-di-*m*-tolylurea, phenylurea, allylurea.

* For preparation, see *Organic Reactions*, 3 (1949) 281.

1 F. FEIGL, J. R. AMARAL and V. GENTIL, *Mikrochim. Acta*, (1957) 726.
2 Comp. E. S. WALLIS and J. F. LANE, in *Organic Reactions*, 3 (1949) 267.
3 P. SCHESTAKOW. *Chem. Zentr.*, (1905) I, 1227.

(2) Test by enzymatic hydrolysis to ammonia

The hydrolysis of urea by acids and alkalis:

$$OC(NH_2)_2 + H_2O \rightarrow CO_2 + 2\,NH_3$$

is not hastened essentially by warming. In contrast, urease, an enzyme which occurs in soya beans, brings about complete hydrolysis of urea rapidly even at room temperature. A neutral or alkaline medium is required. The evolution of ammonia from urea by urease provides not only the basis of a quantitative micro determination of urea[1] but, with the aid of the Nessler reaction, it may also be used for the detection of urea.[2] This reaction yields a red-brown precipitate or turbidity, or a yellow color, due to $HgNH_2I.HgI_2$,[3] which results when a basic solution of alkali mercuriiodide is added to solutions of ammonium salts or comes in contact with ammonia vapor:

$$[HgI_4]^{2-} \rightleftarrows HgI_2 + 2\,I^-$$
$$HgI_2 + NH_3 + KOH \rightarrow KI + HgNH_2I + H_2O$$
$$HgNH_2I + HgI_2 \rightarrow HgNH_2I.HgI_2$$

The following test for urea is reliable in the absence of ammonium salts and salts of volatile aliphatic amines, and guanidine, which give precipitates

with Nessler solution. These interferences can be avoided (see below). Urethanes and ureides do not interfere.

Procedure. The test is conducted in a depression of a spot plate or in a
micro crucible. A drop of the neutral or alkaline test solution is introduced,
several mg of urease added, and stirred in. After 2–5 min., a drop of Nessler solution is introduced. Depending on the amount of urea, a brown precipitate or turbidity, or a yellow color appears. In case ammonium salts are present, a drop of
the test solution should be taken to dryness along with a drop of dilute alkali.
The ammonium salts are thus completely destroyed with release of ammonia,
and without significant hydrolysis of urea. The evaporation residue is taken up
in a drop of water, and the test continued as described. Stable solid urease products can be purchased.

Limit of identification: 1 γ urea.

1 J. B. Gibbs and P. L. Kirk, *Mikrochemie*, 16 (1934) 25.
2 Unpublished studies, with D. Goldstein; see also G. L. Baker and L. H. Johnson,
 Anal. Chem., 24 (1952) 1625.
3 M. L. Nichols and C. Willits, *J. Am. Chem. Soc.*, 56 (1934) 69.

109. Guanidine

Test through dry heating with phenanthraquinone [1]

A test was described on p. 334 for phenanthraquinone based on the fact
that on heating with guanidine carbonate to 180–190° it yields a blue-violet
product which dissolves in ether to give a red color. This reaction can be
applied for the direct detection of guanidine, or its carbonate or acetate.
If other salts (chloride, nitrate, oxalate, *etc.*) are present, the test material
should be ground with sodium carbonate before the heating with anthraquinone, to give the reactive guanidine carbonate *.

Among the derivatives of guanidine, only dicyanodiamidine (see Sect. 131)
behaves, after reaction with sodium carbonate, analogously to guanidine
carbonate. Consequently, the test described here is highly selective for
guanidine and its salts.

Procedure. The test is made in a micro test tube. A drop of the test solution
(if need be after addition of a little sodium carbonate) is taken to dryness, and a
drop of a 0.1% benzene solution of phenanthraquinone is added. After volatilizing the solvent, the test tube is placed in a bath that has been preheated to 160°
and the temperature raised to 180°. After cooling, a little ether is added. If the
solution has a pink to red color, the reaction is positive.

Limit of identification: 5 γ guanidine carbonate.

* This method does not succeed with guanidine thiocyanate, because heating of
alkali thiocyanate with guanidine results in the amine reaction described on p. 241
and the guanidine is consumed.

1 Unpublished studies, with D. Goldstein.

110. Biguanide

Test through splitting into guanidine and urea [1]

If the tests described in this chapter show the absence of guanidine, dicyanodiamidine (guanylurea) and dicyanodiamide (cyanoguanidine), the presence of biguanide is still not excluded. The detection of the latter is possible through the splitting into guanidine and urea which occurs when the compound is heated with lime [2] or sodium carbonate:

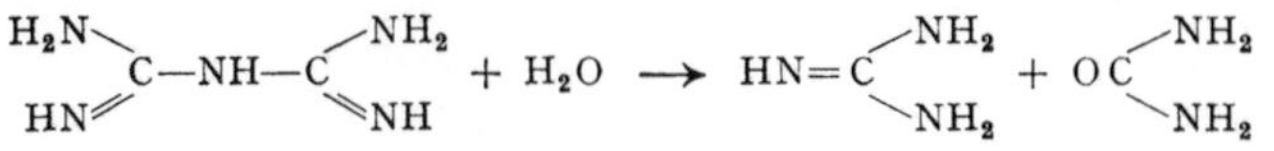

$$\begin{array}{c} H_2N \\ HN \end{array} \!\! C\!-\!NH\!-\!C \!\! \begin{array}{c} NH_2 \\ NH \end{array} + H_2O \;\longrightarrow\; HN\!=\!C \!\! \begin{array}{c} NH_2 \\ NH_2 \end{array} + OC \!\! \begin{array}{c} NH_2 \\ NH_2 \end{array}$$

After this treatment and evaporation, the resulting guanidine responds to the color reaction with phenanthraquinone on heating to 190° (see p. 334).

It should be noted that the procedure given here permits the detection of salts of biguanide in the presence of salts of dicyanodiamide, whereas salts of guanidine and dicyanodiamidine must be absent.

Procedure. The test is conducted in a micro test tube. Several cg of sodium carbonate are added to a drop of the acid or neutral test solution, and the mixture is taken to dryness. After cooling, a drop of freshly prepared 1% solution of phenanthraquinone in benzene is added. The remainder of the procedure is as given in Sect. 109.

Limit of identification: 12.5 γ biguanide.

1 Unpublished studies, with D. GOLDSTEIN.
2 F. EMICH, *Monatsh.*, 12 (1891) 11.

111. Melamine

Detection through tests for basicity, behavior toward caustic alkali and resin formation [1]

Melamine (I) is soluble in water to the extent of about 0.02% at room temperature; because of the nitrogen atoms of the triazine ring it is a base, a fact that can be verified by binding of acids. The precipitation of red nickel dimethylglyoxime from Ni-dimethylglyoxime equilibrium solution can be employed for this test. See Chap. 2, Sect. 24 regarding the mechanism and the procedure.

Limit of identification: 7.5 γ melamine.

As triamine of cyanuric acid, melamine is saponified when warmed with caustic alkali and yields alkali cyanurate (II) and ammonia:

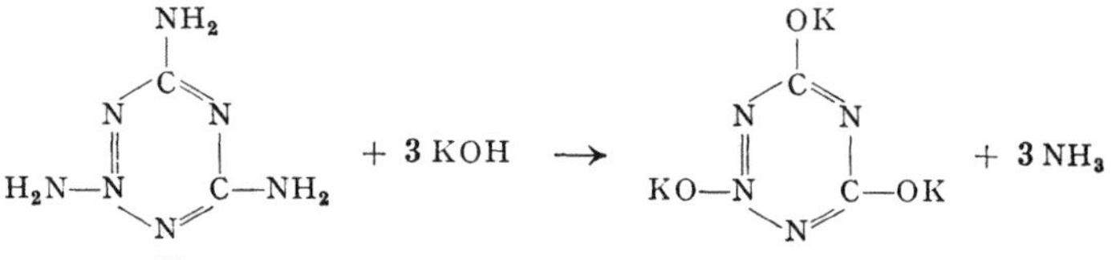

$$
\begin{array}{c}
\text{(I)} \\
\end{array}
\quad + \text{ 3 KOH } \longrightarrow \quad
\begin{array}{c}
\text{(II)} \\
\end{array}
\quad + \text{ 3 NH}_3
$$

The detection of the resulting ammonia (Nessler reaction) permits the detection of 7.5 γ melamine.

The positive response to the test for basicity combined with the saponifiability constitute a strong indication of the presence of melamine. An additional though less sensitive test can be based on the fact that, because of the free NH_2 groups, melamine condenses at about 140° with formaldehyde to yield water-insoluble melamine resin. A selective test for melamine can be based on this resin formation. The test is of particular interest since it is the first instance of the analytical employment of the formation of a plastic.

Procedure. A drop of the saturated solution is diluted with an equal volume of water and taken almost to dryness in a micro test tube. Several cg of paraformaldehyde are added and the mixture is gradually brought to 140°. After cooling, a drop or two of water is added. A positive response is indicated by the production of an amorphous mass that adheres to the walls of the test tube.

Limit of identification: 30 γ melamine.

1 Unpublished studies, with E. LIBERGOTT.

112. Formamide

Test by dehydration [1]

When amides of carboxylic acids are heated with phosphorus pentoxide they are dehydrated and the corresponding nitriles are formed.[2] In the case of formamide (b.p. 210°) dehydration therefore leads to hydrogen cyanide:

$$ \text{HCONH}_2 \xrightarrow{-\text{H}_2\text{O}} \text{HCN} $$

The detection of hydrogen cyanide through the color reaction with copper acetate–benzidine acetate (comp. p. 546) permits the recognition of the above dehydration, and therefore of formamide. Under the conditions described here, only the following compounds yield hydrogen cyanide: ammonium formate (m.p. 116°), formoxime (b.p. 84°) and glyoxime (m.p. 178°). The latter yields dicyanogen (see Sect. 116).

Procedure. A few cg of phosphorus pentoxide are placed in a micro test tube and after addition of a drop of the test solution the tube is immersed in a glycerine bath preheated to 140–150°. A piece of filter paper moistened with the reagent for cyanide (p. 546) is placed on the open end of the tube. A positive response is indicated by the almost immediate appearance of a blue fleck.

Limit of identification: 10 γ formamide.

1 L. BEN-DOR and R. REISFELD, *Chemist-Analyst*, 54 (1965) 7.
2 R. B. WAGNER and H. D. ZOOK, *Synthetic Organic Chemistry*, Wiley, New York, 1953, p. 596.

113. N,N-Dimethylformamide

Identification through basic and acidic saponification [1]

When a drop of the test solution together with a drop of alkali hydroxide is evaporated to dryness in a micro crucible, dimethylamine is volatilized and alkali formate remains. The latter can be identified through the redox reaction with mercuric chloride described in Sect. 42.

Limit of identification: 2 γ N,N-dimethylformamide.

Complete acidic saponification of N,N-dimethylformamide leading to dimethylamine sulfate is achieved by the following procedure: One drop of the test solution together with one drop of diluted sulfuric acid is heated in a micro test tube in a boiling water bath for 10 min. After cooling, the copper dithiocarbamate test for secondary aliphatic amines, as described on p. 250, is applied.

Limit of identification: 3 γ N,N-dimethylformamide.

1 F. FEIGL and R. REISFELD, *Chemist-Analyst*, 54 (1965) 63.

114. Formanilide

Test by pyrohydrogenolysis [1]

On heating a mixture of formanilide and benzoin (m.p. 137°) to 170°, aniline vapors are evolved due to pyrohydrogenolysis:

$$HCONHC_6H_5 + C_6H_5COCH(OH)C_6H_5 \rightarrow C_6H_5NH_2 + CH_2O + C_6H_5COCOC_6H_5$$

The aniline, and therefore the occurrence of the above redox reaction, can be detected by the color reaction with sodium 1,2-naphthoquinone-4-sulfonate. The test is specific for formanilide in the absence of volatile compounds containing reactive CH_2 and NH_2 groups, which show the color reaction with the sulfonate. A preliminary heating of the sample without benzoin is recommended to prove the presence or absence of pertinent volatile compounds.

Procedure. A drop of the alcoholic test solution is evaporated to dryness in a test tube and a minimum amount of solid benzoin is added. The tube is immersed in a glycerine bath preheated to 160–170°, and is then covered with a piece of filter paper moistened with 1% soln. of the reagent (see p. 153). The appearance of a violet spot is an indication of a positive result.

Limit of identification: 50 γ formanilide.

1 F. FEIGL and L. BEN-DOR, *Anal. Chim. Acta*, 23 (1965) 190.

115. Benzamide

Test through Hofmann degradation [1]

Acid amides are converted into amines containing one carbon atom less when warmed with alkali hypobromite (Hofmann degradation)[2]. The net reaction is:

$$(R,Ar)CONH_2 + NaBrO \rightarrow (R,Ar)NH_2 + NaBr + CO_2$$

Accordingly, benzamide yields aniline when subjected to this treatment; this base is volatile with steam and can be detected through the color reaction with sodium 1,2-naphthoquinone-4-sulfonate (see p. 153).

Procedure. A tiny portion of the solid test material or a drop of its solution is treated in a micro test tube with a drop of sodium hypobromite solution. The latter is prepared by the dropwise addition of dilute sodium hydroxide to dilute bromine water until the color is just discharged. The mouth of the tube is covered with a disk of filter paper moistened with a freshly prepared 2% solution of sodium 1,2-naphthoquinone-4-sulfonate. The test tube is placed in boiling water. A positive response is indicated by the development of a red or pink stain on the yellow reagent paper.

Limit of identification: 20 γ benzamide.

1 Unpublished studies, with E. LIBERGOTT.
2 Comp. R. B. WAGNER and H. D. ZOOK, *Synthetic Organic Chemistry*, Wiley, New York, 1953, p. 674.

116. Oxamide

(1) Test through sintering with thiobarbituric acid [1]

If oxamide (m.p. 419° with decomposition) is heated to 140–160° with barbituric acid (m.p. 248°) or thiobarbituric acid (m.p. 235°), a yellow or red material forms rapidly. These products give the respective colors when dissolved in ethanol. In view of the high melting points of oxamide and the two acids, this reaction is undoubtedly a solid–solid reaction, at least at the start.

The fact that the CH_2 group in these acids as well as intact NH_2 groups in oxamide are essential leads to the conclusion that a condensation between these groups occurs:

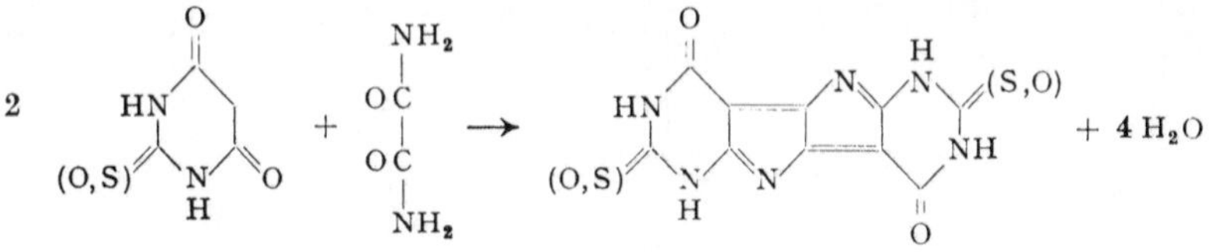

The condensation product, written in the tautomeric form, represents a system of 8 conjugated double bonds, which accounts for the intense color.

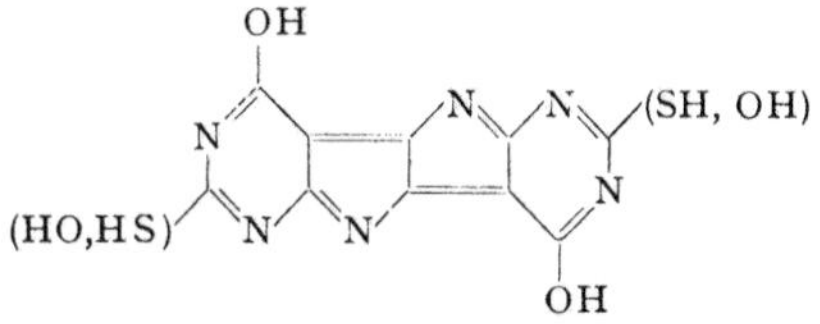

The sintering color reaction described here succeeds with minimal quantities of oxamide and permits its specific detection; other acid amides yield sometimes yellow, but never red, products with thiobarbituric acid. It should be stressed that water-soluble hexamine behaves like oxamide.

Procedure.[2]　A little of the solid is mixed in a micro test tube with several mg of thiobarbituric acid and 1–2 drops of glycerol. The tube is placed in a bath preheated to 120° and the temperature is raised to 140–160°. If oxamide is present, a deep red to light pink product appears, the shade depending on the amount of oxamide involved. The product gives a solution of the same color in ethanol.

Limit of identification: 2.5 γ oxamide.

When barbituric acid is used in place of the thiobarbituric acid, the product is yellow-orange to light yellow. The *limit of identification* is 6 γ oxamide.

The limits of identification were determined by starting with standard solutions of ammonium oxalate. Known volumes were evaporated and the residues then dry-heated at 180°. The conversion into oxamide is quantitative.*

---

* Alkyl oxamates behave like oxamide propably due to the disproportionation

$$2 \begin{array}{c} COOR \\ | \\ CONH_2 \end{array} \rightarrow (COOR)_2 + (CONH_2)_2$$

---

1 Unpublished studies, with D. HAGUENAUER-CASTRO.
2 Ch. WEISSMANN (Zürich), private communication.

(2) Test through dry heating with phloroglucinol[1]

A black or brown product results if a dry mixture of oxamide and excess phloroglucinol is heated at 160–200°. The chemistry of the reaction is perhaps analogous to the reaction of oxamide with thiobarbituric acid as outlined on p. 536. Oxanilide, as well as other amides, does not react; hexamine and alkyl oxamates behave similarly to oxamide.

Procedure. A small quantity of the sample is mixed in a micro test tube with about 1 cg of phloroglucinol. The tube is kept for some minutes in a glycerol bath that has previously been heated to 180°. If oxamide was present, the mixture changes from its original light color to black or brown, according to the amount of oxamide present. The reaction can be observed more easily in the presence of some cg cetyl alcohol.

Limit of identification: 2 γ oxamide.*

This limit was determined by evaporating 1 drop of a uniform alcoholic suspension of oxamide.

* See footnote to test (1).

1 Unpublished studies, with E. JUNGREIS.

(3) Test through dehydration to dicyanogen[1]

If oxamide is heated to 120–160° with P_2O_5, dehydration ensues:[2]

$$(CONH_2)_2 \rightarrow (CN)_2 + 2\ H_2O \tag{1}$$

The gaseous dicyanogen which results can be detected by the red color it yields on contact with a concentrated potassium cyanide solution containing oxine (see Sect. 127). Simple pyrolysis cannot take the place of the heating with P_2O_5 because the water vapor hydrolyzes the dicyanogen:

$$(CN)_2 + H_2O \rightarrow HCN + HCNO \tag{2}$$

whereas the P_2O_5 binds the water through formation of phosphoric acid. The test given here, which is based on the dehydration, is almost specific. Only glyoxime, $(-CH=NOH)_2$, which is isomeric with oxamide, likewise yields dicyanogen when gently heated with P_2O_5. Since glyoxime is readily soluble in ether while oxamide is not, a separation of these isomers before the test is readily accomplished, if necessary.

It may be noted that in contrast to oxamide, glyoxime is dehydrated by acetic anhydride;[3] the $(CN)_2$ formed permits the detection of glyoxime.

Procedure. A little of the solid is mixed in a micro test tube with an excess of phosphorus pentoxide. The mouth of the tube is covered with a disk of oxine paper (see p. 549) that has been moistened with 25% KCN solution. The test tube is placed in a glycerol bath that has been heated to 160°. If dicyanogen is released, a circular pink or red stain develops on the yellow paper.

It is not possible to determine experimentally the limit of identification of this test because solutions of oxamide cannot be prepared. However, on the basis of the *identification limit* for the dicyanogen test, it appears that less than 1γ oxamide can be revealed.

If desired, use can be made instead of the hydrolysis of dicyanogen to hydrogen cyanide as shown in equation (2). This hydrolysis occurs when the $(CN)_2$ released by P_2O_5 comes in contact with an aqueous solution of copper–benzidine acetate. A blue color results (see p. 546).

1 F. Feigl and Cl. Costa Neto, *Mikrochim. Acta*, (1955) 969.
2 T. Wallis, *Ann.*, 345 (1906) 362; E. Ott, *Ber.*, 52 (1919) 663.
3 B. Lack, *Ber.*, 17 (1884) 1571.

117. Oxanilide and oxanilic acid

Test through pyroammonolytic formation of oxamide[1]

It was shown on p. 260 that anilides of carboxylic acids yield aniline when heated to 220–250° in the presence of guanidine carbonate. Oxanilide conforms to this behavior. The basis of this effect is that the ammonia produced by the pyrolysis of the guanidine brings about an ammonolysis of the anilide; oxamide is formed thus from oxanilide:

$$\begin{array}{c} CONHC_6H_5 \\ | \\ CONHC_6H_5 \end{array} + 2\ NH_3 \longrightarrow \begin{array}{c} CONH_2 \\ | \\ CONH_2 \end{array} + 2\ C_6H_5NH_2$$

An analogous transformation occurs with oxanilic acid $(HO_2C–CONHC_6H_5)$.

Since the ammonolytic production of aniline occurs with all acid anilides, the subsequent detection of aniline is not adequate for characterizing oxanilide or oxanilic acid. This goal can be reached through the detection of the oxamide produced by applying the test involving condensation with thiobarbituric acid as described in Sect. 116. Since this condensation proceeds as a solid–solid reaction, good contact is imperative. Accordingly, at least part of the melamine, which results from the heating with guanidine carbonate, must be removed by digestion with dilute hydrochloric acid in order that the surface of the oxamide may be laid bare.

Procedure. A micro test tube is used. A little of the solid (or its suspension) is mixed with several cg of guanidine carbonate and the mixture is brought to dryness if need be. The tube is placed in a glycerol bath that has previously been brought to 200° and the temperature is raised to 220°. After 2–3 min., the reaction mass is allowed to cool and several drops of dilute hydrochloric acid are added. The mixture is warmed in a water bath and then centrifuged. The acid layer is poured off and replaced with alcohol and again centrifuged. The remainder of the procedure, *i.e.* treatment with thiobarbituric acid, is described on p. 536.

Limits of identification: 4 γ oxanilide and 2 γ oxanilic acid.

The limits of identification were determined on suspensions of oxanilide in warm benzene and with alcoholic solutions of oxanilic acid.

1 Unpublished studies, with D. HAGUENAUER-CASTRO.

118. Succinimide

Test through conversion to pyrrole[1]

If succinimide is distilled with zinc dust it yields pyrrole, which can be detected in the gas phase through contact with *p*-dimethylaminobenz-aldehyde, with which it gives a blue-violet color. The chemistry of the pyrolysis, which likewise occurs with ammonium succinate, is given on p. 460, where the Procedure is also outlined.

Limit of identification: 2.5 γ succinimide.

When subjected to zinc dust distillation, N-bromo-, chloro-, or iodo-succinimide behave in the same manner as succinimide.

It should also be noted that succinimide responds to the test for acid amides described on p. 257 and likewise to the test for dicarboxylic acids given on p. 352.

1 F. FEIGL, V. GENTIL and C. STARK-MAYER, *Mikrochim. Acta*, (1957) 344.

119. N-Bromo(chloro-)succinimide

Test with fluorescein[1]

The N-halogenated succinimides can be detected by the redox reaction with alkaline thallous salt solution (see p. 127). Their behavior toward fluorescein or fluorescein containing potassium bromide can also be employed. If filter paper impregnated with fluorescein is spotted with 1 drop of the aqueous or alcoholic solution of N-bromosuccinimide, a red stain or ring of eosin appears on the yellow paper.

Limit of identification: 3 γ N-bromosuccinimide.

N-Chlorosuccinimide does not give this eosin reaction, which is characteristic for free bromine or HBrO.[2] However, if filter paper impregnated with potassum bromide and fluorescein is used, N-chlorosuccinimide reacts in the same manner because of the intervention[3] of the redox reaction:

$$Cl^0 + Br^- \rightarrow Br^0 + Cl^-$$

The reagent paper is prepared by bathing filter paper in an alcoholic solution of fluorescein that is saturated with potassium bromide at 50°.

1 F. FEIGL and R. A. ROSELL, Z. Anal. Chem., 159 (1958) 337.
2 D. GANASSINI, Chem. Zbl., (1904) I, 1172.
3 O. FREHDEN and C. H. HUANG, Mikrochem. Ver. Mikrochim. Acta, 26 (1939) 41.

120. Isatin

Test with 4-nitrophenylhydrazine and magnesium hydroxide[1]

When isatin (I, Ia) is treated with 4-nitrophenylhydrazine (II, IIa) the

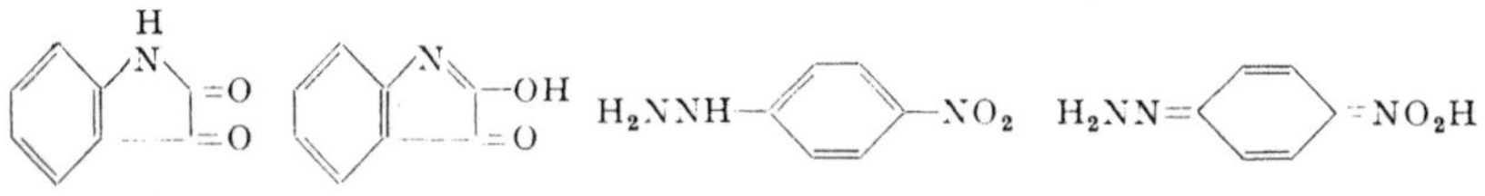

isatin β-4-nitrophenylhydrazone is formed. The violet alkaline solution of this hydrazone reacts with Mg^{2+} ions to yield a blue precipitate.

It is probable that the essential part of the blue product is the chelate compound

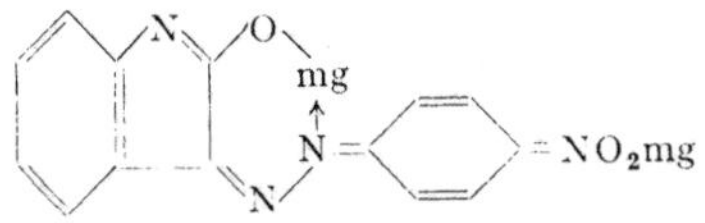

$$mg = \tfrac{1}{2}Mg$$

in which the enolized CO group in isatin (Ia) and the enolized NO_2 group in *p*-nitrophenylhydrazine (IIa) are salified by magnesium. A chelate binding of this kind may also be established through chemical adsorption on the surface of $Mg(OH)_2$ leading to a color lake. It is in line with this assumption that $Mg(OH)_2$ or MgO (ignited) turn blue in contact with the alcoholic solution of the hydrazone. The specific test for isatin described here is based on the formation of the blue product.

Procedure. A small amount of the solid test material or a drop of its solution is heated in a micro test tube with a drop of an alcoholic solution of 4-nitrophenylhydrazine for 5 min. After cooling, a drop of 10% sodium hydroxide and a drop of 1% magnesium nitrate are added. A positive response is indicated by the appearance of a blue precipitate. When the amount of isatin is less than 20 γ, extraction with ether to dissolve the hydrazine is recommended.

Limit of identification: 0.5 γ isatin.

1 F. FEIGL and D. GOLDSTEIN, Talanta, 4 (1960) 209.

121. Barbituric acid

Test through transformation to violuric acid [1]

Barbituric acid reacts almost instantaneously with nitrous acid to give violuric acid (5-isonitrosobarbituric acid or alloxan 5-oxime):

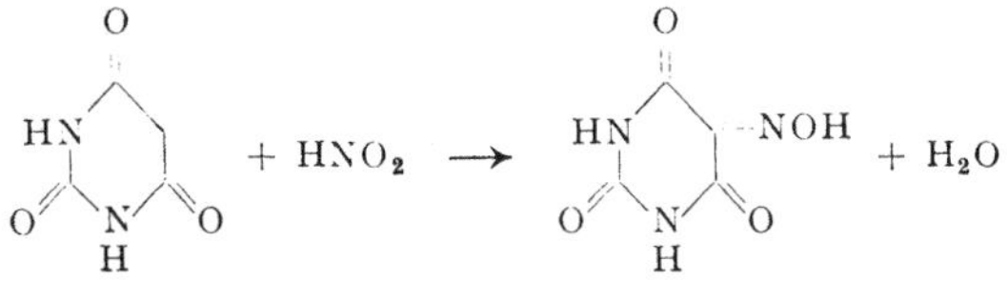

Since violuric acid yields red-violet alkali salts, the isonitrosation of barbituric acid can be used to detect the latter.

Procedure. A little of the solid or a drop of its solution is mixed in a depression of a spot plate with 1 drop each of 2 N acetic acid and saturated sodium nitrite solution. Depending on the amount of barbituric acid that was present, a more or less intense red-violet color appears.

Limit of identification: 10 γ barbituric acid.

1 Unpublished studies, with E. JUNGREIS.

122. Barbituric and thiobarbituric acid

(1) Test through sintering with oxamide or ethyl oxamate [1]

A test for oxamide is given in Sect. 116 which is based on the formation of a colored product when oxamide is dry-heated (120–160°) with barbituric acid or thiobarbituric acid. These products give yellow or red solutions, respectively, in ethanol. This sintering reaction can be applied for the detection (and differentiation) of barbituric and thiobarbituric acid. The oxamide may be replaced by the fusible ethyl oxamate which yields the amide besides ethyl oxalate through disproportionation.

Procedure. A drop of the test solution or a small quantity of the solid is placed in a micro test tube along with several mg of oxamate. After taking the mixture to dryness, the tube is placed in a bath that has been preheated to 140–160°. A yellow or red product is formed rapidly if the response is positive.

Limits of identification: 10 γ barbituric acid,   6 γ thiobarbituric acid.

Derivatives of barbituric acid which have no free CH_2 group in the 5-position (veronal for instance) do not give the above color reaction.

1 Unpublished studies, with D. HAGUENAUER-CASTRO.

(2) Test with sulfadiazine [1]

Sulfadiazine reacts in aqueous solution with barbituric or thiobarbituric acid to give yellow or red polymethine dyes, which fluoresce in characteristic manner in ultraviolet light. These color- or fluorescence reactions make possible the detection and differentiation of barbituric and thiobarbituric acid. No color reaction is given under the same conditions by the derivatives of these acids which do not have a free CH_2 group in the 5-position.

The formation of a polymethine dye from barbituric (thiobarbituric) acid can be represented:

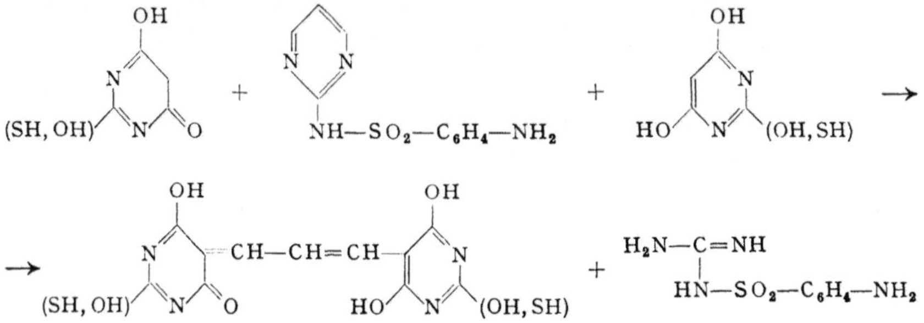

Procedure. A tiny portion of the material under test or a drop of its solution is treated in a micro test tube with several mg of sulfadiazine (along with a drop of water if need be) and the mixture is warmed in boiling water. A yellow or red color indicates a positive response. If a drop of the solution is then placed on filter paper, a colored stain appears and the latter under u.v. light fluoresces blue-green (barbituric acid) or orange to yellow (thiobarbituric acid).

Limits of identification: 3 γ barbituric acid   5 γ thiobarbituric acid

1 F. FEIGL, V. ANGER and V. GENTIL, *Clin. Chim. Acta*, 5 (1960) 153.

(3) Test with pyridylpyridinium dichloride[1]

When warmed with a solution of pyridylpyridinium dichloride in dimethylformamide, barbituric and thiobarbituric acid form the respective pentamethine dyes directly, and without the necessity of a preceding splitting off of glutaconic aldehyde as is essential in other instances. In this respect these two acids differ from all other compounds that form pentamethine dyestuffs with glutaconic aldehyde (resulting from the alkaline cleavage of pyridylpyridinium dichloride, see p. 246). The reaction can be represented:

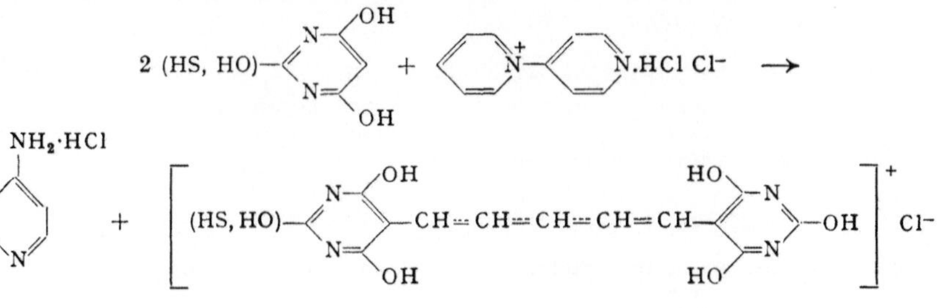

The pentamethine dye obtained from barbituric acid is blue; its fluorescence in daylight is so intensely red that it can show all colors between reddish blue and red depending on the illumination and the layer thickness. It fluoresces red under a quartz lamp. The pentamethine dye formed from thiobarbituric acid is pure blue; it does not fluoresce. It is notable that the color intensity of the solutions of these pentamethine dyes in dimethylformamide is temperature-dependent; the intensity is greatest in the cold.

Procedure. A little of the solid or the evaporation residue from a drop of its solution is dissolved in 1–2 drops of dimethylformamide in a micro test tube. A drop of a 1% solution of pyridylpyridinium dichloride in dimethylformamide is added, and the mixture is heated for 2–3 min. at 120° in a glycerol bath. A red or blue color that develops on cooling indicates a positive response.

Limits of identification:
0.5 γ barbituric acid (reddish blue to red)   0.5 γ thiobarbituric acid (blue).

1 V. Anger and S. Ofri, *Talanta*, 10 (1963) 1302.

123. Alloxan

(1) Test through pyroammonolytic formation of murexide[1]

Murexide is the water-soluble violet ammonium salt of purpuric acid. It is formed by the action of ammonia on colorless alloxantin:[2]

$$\begin{array}{l} \text{HN—CO} \quad \text{OC—NH} \\ | \quad | \qquad | \quad | \\ \text{OC} \quad \text{CH—O—C} \quad \text{CO} \quad + 2\,\text{NH}_3 \\ | \quad | \quad \text{HO} \quad | \\ \text{HN—CO} \quad \text{OC—NH} \end{array} \rightarrow \begin{array}{l} \text{HN—CO} \qquad \text{OC—NH} \\ | \quad | \qquad\qquad | \quad | \\ \text{OC} \quad \text{C} = \text{N} - \text{C} \quad \text{CO} \quad + 2\,\text{H}_2\text{O} \\ | \quad | \qquad\qquad || \quad | \\ \text{HN—CO} \quad \text{NH}_4\text{O—C—NH} \end{array}$$

In aqueous solution, the formation of murexide occurs only with comparatively large amounts of alloxan or alloxantin, which in the former case is preceded by the disproportionation of the alloxan. In contrast, the murexide reaction is very sensitive if conducted by dry-heating of alloxan or alloxantin, along with ammonium carbonate. The latter serves as ammonia donor through its thermal decomposition.

Procedure. A little of the test material or a drop of its solution is mixed in a micro test tube with several cg of ammonium carbonate and taken to dryness. The tube is then kept for several minutes in a glycerol bath previously heated to 110°. Over-heating must be avoided. A red-violet color indicates a positive response. When dealing with very small amounts, the faint color can be intensified by adding a drop of tap water to the residue.

Limit of identification: 2.5 γ alloxan.

1 Unpublished studies, with E. Jungreis.
2 M. Pesez and P. Poirier, *Méthodes et Réactions de l'Analyse Organique*, Vol. Masson et Cie., Paris, 1954, p. 11.

(2) Test by pyrohydrolytic fission [1]

When alloxan is heated with oxalic acid dihydrate at 150–160°, the latter releases its crystal water as superheated steam. This brings about the pyrohydrolytic fission:

$$
\begin{array}{ccc}
\text{HN—CO} & & \text{COOH} \\
| \quad | & & | \\
\text{OC} \quad \text{CO} + 2\,H_2O & \rightarrow & CO(NH_2)_2 + \text{CO} \\
| \quad | & & | \\
\text{HN—CO} & & \text{COOH}
\end{array}
\qquad (1)
$$

The resulting mesoxalic acid loses carbon dioxide to yield glyoxalic acid:

$$
HOOC—CO—COOH \rightarrow HOOC—CHO + CO_2 \qquad (2)
$$

The latter is volatile with steam and can be detected in the gas phase through its redox reaction with Nessler reagent; finely divided mercury is produced:

$$
HOOC—CHO + [HgI_4]^{2-} + 4\,OH^- \rightarrow (COO)_2^{2-} + Hg^0 + 4\,I^- + 3\,H_2O \qquad (3)
$$

The present test is based on the realization of (1)–(3).

Procedure. A little of the solid or a drop of its solution is mixed in a micro test tube with several cg of hydrated oxalic acid and taken to dryness if need be. The mouth of the tube is covered with a disk of filter paper moistened with Nessler reagent. The tube is placed in a glycerol bath previously heated to 150°. A black or grey stain develops within several minutes if alloxan is present.

Limit of identification: 20 γ alloxan.

1 Unpublished studies, with C. STARK-MAYER.

(3) Test through conversion to violuric acid [1]

Hydroxylamine transforms alloxan almost immediately into violuric acid (alloxan 5-oxime or isonitrosobarbituric acid):

$$
\begin{array}{ccc}
\text{HN—CO} & & \text{HN—CO} \\
| \quad | & & | \quad | \\
\text{OC} \quad \text{CO} + NH_2OH & \rightarrow & \text{OC} \quad \text{C}=NOH + H_2O \\
| \quad | & & | \quad | \\
\text{HN—CO} & & \text{HN—CO}
\end{array}
$$

As pointed out in Sect. 121, violuric acid is one of the few colorless organic acidic compounds which yield colored salts with ammonium or alkali metal ions. Therefore, the present test is almost specific for alloxan, and also for alloxantin, because of its alloxan component.

Procedure. A little of the solid or a drop of its solution is mixed in a depression of a spot plate with several cg of hydroxylamine hydrochloride and about the same amount of sodium carbonate. A positive response is indicated by the development of a violet color, whose intensity depends on the amount of alloxan present.

Limit of identification: 20 γ alloxan.

1 Unpublished studies, with E. JUNGREIS.

124. Parabanic acid (oxalylurea)

Detection through sintering with thiobarbituric acid[1]

If the colorless mixture of parabanic acid (m.p. 243°) and thiobarbituric acid (m.p. 235°) moistened with ethanol or water is heated to 105–120°, there results a red-brown product which dissolves in ether or alcohol to give a yellow solution. The chemistry of this solid-body reaction is not known. In view of the fact that the reaction picture is the same as in the reaction between thiobarbituric acid and oxamide (see p. 536) it is probable that the latter is formed under the conditions of the test by the hydrolysis of parabanic acid:

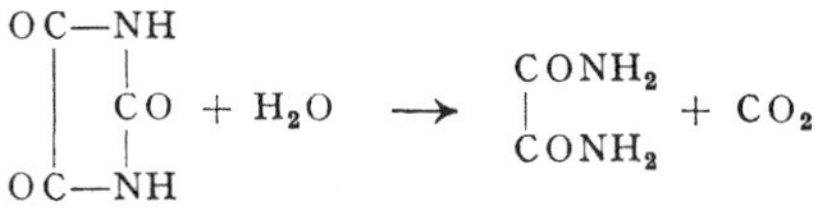

$$\begin{array}{c} OC-NH \\ |\qquad | \\ |\quad CO \\ |\qquad | \\ OC-NH \end{array} + H_2O \quad \longrightarrow \quad \begin{array}{c} CONH_2 \\ | \\ CONH_2 \end{array} + CO_2$$

In accordance with this assumption no reaction occurs between parabanic and thiobarbituric acid in the absence of water or alcohol containing water.

The test given here cannot be employed in the presence of oxamide. The latter can be removed by digesting the sample with alcohol in which only the parabanic acid dissolves.

Procedure. A drop of the alcoholic aqueous test solution is placed in a micro test tube and several cg of thiobarbituric acid are added. The tube is immersed in a glycerol bath preheated to 105°; then the temperature is gradually raised to 120°. If the response is positive, the residue will be brown-red to yellow within a few minutes. It gives a yellow solution in 1 drop of alcohol or a red solution in a 1 : 1 water–pyridine mixture.

Limit of identification: 2.5 γ parabanic acid.

1 Unpublished studies, with D. HAGUENAUER-CASTRO.

125. Hydantoin and allantoin

Test through conversion into allanturic acid[1]

Hydantoin (I) (glycolylurea) and allantoin (II) (5-ureidohydantoin) are both soluble in hot water. They differ in that II, as a urea derivative with a free NH_2 group, gives the oxamide reaction when heated with dimethyl oxalate and thiobarbituric acid (see p. 150).

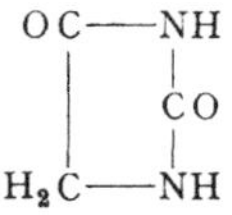

$$\begin{array}{c} OC-NH \\ |\qquad | \\ |\quad CO \\ |\qquad | \\ H_2C-NH \end{array} \qquad \begin{array}{c} OC-NH \\ |\qquad | \\ |\quad CO \\ |\qquad | \\ H_2NCONH-CH-NH \end{array} \qquad OC\begin{array}{c} \diagup NH_2 \\ \diagdown N=CHCOOH \end{array}$$

I (m.p. 220°) 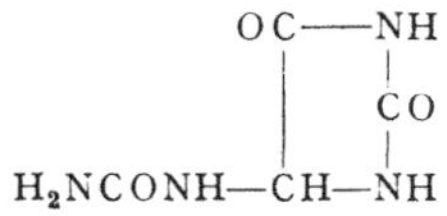   II (m.p. 235°) 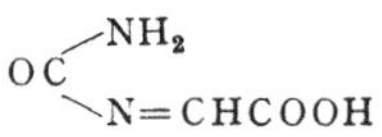   III

The same is true of allanturic acid (III) (glyoxalylurea), which is produced from I or II by oxidation with bromine and glacial acetic acid. This oxidation with subsequent application of the oxamide reaction is therefore characteristic for hydantoin, provided allantoin and other aliphatic compounds with free NH_2 groups are absent.

Uric acid likewise is oxidized to allantoin and allanturic acid by bromine and glacial acetic acid. This interference can be avoided by testing alcoholic solutions. Only hydantoin is soluble in ethanol.

Procedure. A drop of the alcoholic test solution is mixed in a micro test tube with 1 drop of glacial acetic acid and a microdrop of bromine. The mixture is warmed in a water bath for 10 min. and then brought to dryness. Several mg of dimethyl oxalate and thiobarbituric acid are added and the mixture is kept at 150° in a bath for 5 min. A positive response is indicated by the formation of a red product.

Limit of identification: 20 γ hydantoin or allantoin.

1 V. GENTIL (Rio de Janeiro), unpublished studies.

126. Hydrocyanic acid (Prussic acid)

(1) Test with copper acetate–benzidine acetate [1]

Benzidine (I) is transformed, in acetic acid solution, by numerous oxidants into "benzidine blue", which, as shown in (II), is an addition compound of 1 molecule of imine and 1 molecule of amine and an equivalent of acid:

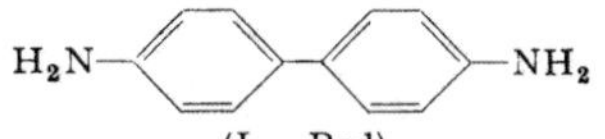

$$(I = Bzd) \qquad (II)$$

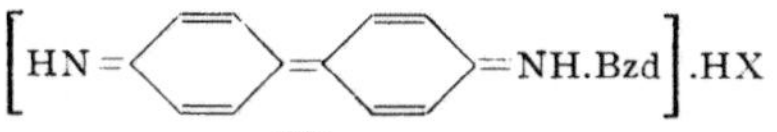

The oxidation potential of cupric ions is not adequate to convert (I) to (II), in other words, the concentration of benzidine blue in the equilibrium:

$$benzidine + Cu^{2+} \rightleftharpoons benzidine\ blue + Cu_2^{2+} \tag{1}$$

is so slight that the reaction is not discernible. However, if hydrogen cyanide is added to the system (1), cuprous ions are withdrawn from the equilibrium because of the irreversible reaction:

$$Cu_2^{2+} + 2\ CN^- \rightarrow Cu_2(CN)_2 \tag{2}$$

To re-establish the equilibrium cupric ions must react with benzidine according to (1). This continuous disturbance and restoration of the equilibrium, signifies that cyanide ions in reaction (2) raise the oxidation potential sufficiently so that reaction (1) proceeds to a considerable extent.

The test given here is characteristic for the volatile hydrogen cyanide liberated from soluble or insoluble inorganic cyanides by acids (in the latter case with addition of metallic zinc). It may also be employed to detect hydrogen cyanide in the pyrolytic cleavage products of nitrogenous organic compounds (see p. 49). No materials may be present which yield hydrogen sulfide when decomposed by acids or pyrolysis, since it reacts with the reagent to form black copper sulfide.

Procedure.   A drop of the test solution or a little of the solid is treated in a micro test tube with 1 or 2 drops of dilute sulfuric acid. In the case of acid-resistant cyanides, it is necessary to add some granulated zinc and more acid. The mouth of the test tube is covered with a disk of filter paper moistened with a drop of the reagent. Depending on the amount of hydrogen cyanide produced, a more or less intense blue appears on the reagent paper. Gentle warming in a water bath is advisable when small quantities of cyanide are suspected.

Reagent: Copper acetate–benzidine acetate solution.

  (a) 2.86 g copper acetate in 1 liter of water;
  (b) 675 ml of a solution of benzidine acetate saturated at room temperature plus 525 ml of water.

  Equal volumes of (a) and (b) are mixed just prior to use.

Limit of identification: 0.25 γ cyanide.

When testing for hydrogen cyanide during the pyrolysis of organic nitrogenous compounds, it is sufficient to heat a tiny quantity of the sample over a micro flame. If a blue stain appears on a disk of the moistened reagent paper placed over the mouth of the test tube, a positive response is indicated. It should be noted that this test will also reveal dicyanogen because on contact with the moist reagent paper it is hydrolyzed to HCN and HCNO (see p. 537).

In view of the fact that benzidine is a carcinogen[3] it is recommended to use as reagent a solution of 5% Cu–ethylacetoacetate* and 5% tetrabase in chloroform.[2] The sensitivity obtained is higher than with benzidine and the blue fleck formed with HCN is more stable.

 * This green inner-complex salt is available from Lobachemie, Vienna XIX.

1 For pertinent literature see F. Feigl, *Spot Tests in Inorganic Analysis*, 5th ed., Elsevier, Amsterdam, 1958, p. 276.
2 F. FEIGL and V. ANGER, *Analyst*, in press.
3 *Proc. Soc. Anal. Chem.*, 2, No. 4 (1965) 69.

(2) *Test by demasking of alkali palladium dimethylglyoxime*[1]

Bright yellow inner-complex palladium dimethylglyoxime (I) is not soluble in acids but it readily dissolves in caustic alkali to give a yellow solution of an alkali palladium dimethylglyoxime (II) in which the palladium is a constituent of an inner-complex anion.

$$\begin{array}{c} \text{CH}_3\text{—C=N} \\ | \quad\quad \searrow\text{Pd/2} \\ \text{CH}_3\text{—C=N} \nearrow \\ \text{OH} \end{array} \qquad \left[\begin{array}{c} \text{CH}_3\text{—C=N} \\ | \quad\quad \searrow\text{Pd/2} \\ \text{CH}_3\text{—C=N} \nearrow \\ \text{O}^- \end{array} \right] \text{M}^+$$

(I) Pd(DH)$_2$ (II) M$_2$ [PdD$_2$]

DH$_2$ = dimethylglyoxime   M = K, Na
DH(D) = univalent(bivalent) radical of DH$_2$

The inner-complexly bound dimethylglyoxime is masked in solutions of (II), *i.e.* no red Ni-dimethylglyoxime precipitate appears when Ni^{2+} ions are added to the solution. Likewise, the palladium is masked against practically all reagents that normally are characteristic for Pd^{2+} ions. Cyanide ions present an exception. If added in excess to the yellow solutions of (II), they cause immediate discharge of the color, and addition of nickel salt solutions (containing NH$_4$Cl) brings down red Ni-dimethylglyoxime.*

The demasking of the dimethylglyoxime is due to the reaction:

$$[\text{PdD}_2]^{2-} + 4\,\text{CN}^- \rightarrow [\text{Pd(CN)}_4]^{2-} + 2\,\text{D}^{2-}$$

Since the Ni-dimethylglyoxime reaction is higly sensitive, cyanide ions can be detected through their demasking effect. This test has the advantage that it may be conducted in alkaline solution.

Considerable quantities of ammonia gradually decolorize the yellow alkali palladium cyanide because of the demasking:

$$[\text{PdD}_2]^{2-} + 2\,\text{NH}_3 \rightarrow [\text{Pd(NH}_3)_2]^{2+} + 2\,\text{D}^{2-}$$

Accordingly, the test given here is not recommended for the detection of small amounts of cyanide in solutions that are strongly ammoniacal or that contain large amounts of ammonium salts.

Procedure. A drop of the alkaline test solution and a drop of alkali palladium dimethylglyoxime are brought together on a spot plate. A drop of the nickel–ammonium chloride solution is added. A red or pink precipitate appears.

Reagents: 1) Alkali palladium dimethylglyoxime solution. Pd-dimethylglyoxime (prepared by treating an acid solution of PdCl$_2$ with dimethylgloxime and thoroughly washing the precipitate) is shaken with 3 N alkali hydroxide. The undissolved portion is removed by filtering or centrifuging.

2) Nickel–ammonium chloride solution. 0.5 N nickel chloride saturated with ammonium chloride.

Limit of identification: 0.25 γ cyanide.

* The presence of ammonium salts prevents the precipitation of green Ni(OH)$_2$.

1 F. FEIGL and H. E. FEIGL, *Anal. Chim. Acta*, 3 (1949) 300.

127. Dicyanogen

Detection with potassium cyanide and 8-hydroxyquinoline (oxine)[1]

Dicyanogen, a gas, is a pyrolysis cleavage product of some nitrogenous organic compounds. It can be specifically detected through its action on a concentrated solution of potassium cyanide containing oxine. A deep red soluble product is formed immediately.[2] Its composition is not yet known.

The color reaction occurs only in the presence of much alkali cyanide, which leads to the supposition that—in analogy to polyhalides—polycyanide ions are formed from cyanide ions and dicyanogen:

$$CN^- + (CN)_2 \rightarrow [CN...(CN)_2]^-$$

It may be assumed that the irreversible saponification of dicyanogen:

$$(CN)_2 + H_2O \rightarrow HCN + HCNO$$

is prevented or at least lessened through the production of polycyanides, and that the coordinated dicyanogen oxidizes the oxine to a quinoidal product, perhaps according to:

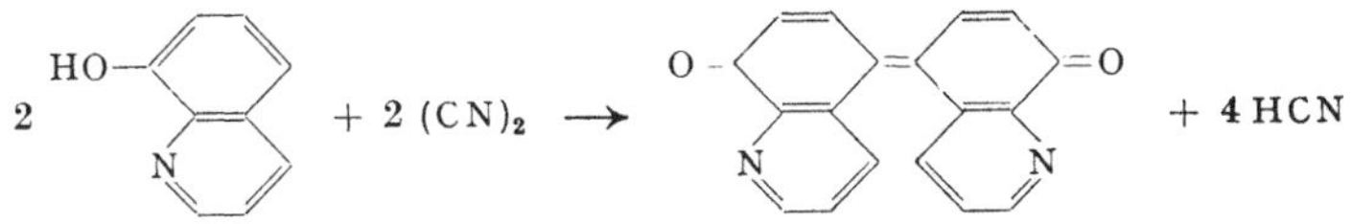

The assumption that coordinated dicyanogen may act as an oxidant is supported by the following observations:[3] (1) the colorless solution of phenolphthalin (produced by reduction of a solution of phenolphthalein in concentrated cyanide solution by zinc) is reddened (oxidized) on contact with dicyanogen; (2) a solution of α-naphthol in concentrated potassium cyanide gives a violet product on contact with dicyanogen. The color reaction is specific for dicyanogen and is not impaired by hydrogen cyanide.

Procedure.　A tiny particle of the dry sample is placed in a micro test tube and the mouth of the tube is covered with a disk of oxine paper that has been moistened with a drop of 25% potassium cyanide solution. When the tube is heated over a micro flame, a more or less intense red circular stain appears on the yellow reagent paper.

Reagent: Oxine paper is prepared by bathing filter paper in a 10% solution of oxine in ether and drying in the air. The reagent paper keeps.

Limit of identification: 1 γ dicyanogen.

This value was determined by dry heating a mixture of $Hg(CN)_2$ and $HgCl_2$, and assuming that the yield of dicyanogen is 50%, when the reaction:

$$Hg(CN)_2 + HgCl_2 \rightarrow Hg_2Cl_2 + (CN)_2$$

(which is recommended for the preparation of pure dicyanogen) is run under the conditions prescribed here.

With the aid of this test it was possible to show that the pyrolysis of uric acid and purine derivatives, guanidine and cyclic guanidine derivatives, and likewise pterins, dimethylglyoxime, furildioxime, yields dicyanogen.

According to the observations available the pyrolytic release of dicyanogen is fairly selective. However, we do not know the structural characteristics responsible for this effect. In this respect it is interesting to note that cyanides (nitriles) and aromatic thiocyanates which contain the CN group do not release $(CN)_2$ but only HCN when pyrolyzed. Possibly the cleavage of dicyanogen is always the first step when nitrogen-containing organic compounds are pyrolyzed, but when water is formed simultaneously then a hydrolysis occurs whereby hydrocyanic acid is formed.

1 F. FEIGL and L. HAINBERGER, *Analyst*, 80 (1955) 807; comp. also F. FEIGL and Cl. COSTA NETO, *Mikrochim. Acta*, (1955) 969.
2 A. S. KOMAROWSKY and N. S. POLUEKTOW, *Z. Anal. Chem.*, 96 (1934) 23.
3 Unpublished studies, with D. GOLDSTEIN.

128. Cyanogen halides

The test described here has been tried only with water-soluble cyanogen bromide (m.p. 52°; b.p. 61°). In all likelihood, the test will be applicable also for cyanogen chloride and iodide.

Test through formation of a polymethine dyestuff[1]

A test for pyridine and its derivatives with free *ortho*-positions is described on p. 384; it rests on the addition of cyanogen bromide to the cyclic N atom followed by hydrolysis to yield glutaconic aldehyde which condenses with *o*-tolidine to give a red to pink polymethine dyestuff. This reaction permits a sensitive means of detecting cyanogen bromide.

Procedure. A drop of the test solution is mixed in a micro test tube or a depression of a spot plate with a drop of a 10% solution of *o*-tolidine in pyridine. According to the amount of cyanogen bromide, a violet or red color appears at once or within several minutes.

Limit of identification: 0.3 γ cyanogen bromide.

1 Comp. W. N. ALDRIDGE, *Analyst*, 69 (1944) 262; 70 (1945) 474.

129. Cyanoacetic acid

Test through reaction with bromine[1]

When bromine acts on cyanoacetic acid in ethereal or aqueous solution dibromoacetamide results along with a compound (m.p. 86°) of unknown composition.[2] It has been found that the reaction product, which likewise is formed from the esters, amide, and anilide of cyanoacetic acid, behaves as a strong oxidant. It liberates iodine from acidified solutions of alkali iodides and yields a blue oxidation product when it reacts with yellow-brown thio-Michler's ketone (see detection of free halogens, p. 65). The production of this oxidizing compound serves as the basis of a selective test for cyanoacetic acid and its derivatives. Iodine-bearing organic materials which yield iodic acid when warmed with bromine must be absent (see p. 71). Compounds that set iodine free from acidified alkali iodide solutions must also be absent. Therefore, a preliminary test along these lines is essential. If the response is positive, they can usually be rendered harmless by evaporation of the sample with sulfurous acid.

Procedure. A drop of the test solution and a drop of bromine are brought together in a micro test tube and kept in boiling water until the excess bromine has been driven off. The complete removal is ensured by adding several cg of sulfosalicylic acid. A drop of starch solution containing alkali iodide is then introduced. A blue color indicates a positive response.

Limit of identification:

0.1 γ cyanoacetic acid amide   0.2 γ cyanoacetic acid anilide
0.2 γ cyanoacetic acid N-methylamide.

1 Unpublished studies with A. Del'Acqua.
2 Comp. W. Steinkopf, *Ber.*, 38 (1905) 2695.

130. Cyanamide

Test by conversion into guanidine[1]

Cyanamide adds ammonia to yield guanidine:[2]

$$N{\equiv}C-NH_2 + NH_3 \longrightarrow HN{=}C\begin{array}{c}{\nearrow}NH_2\\{\searrow}NH_2\end{array}$$

The resulting guanidine can be detected through its color reaction with sodium 1,2-naphthoquinone-4-sulfonate; a red *p*-quinoidal compound is formed.[3] Dicyanodiamide, which is the dimer of cyanamide, does not react, either when alone or after being treated with ammonia.

The test described here is specific for cyanamide, provided that compounds containing reactive CH_2- and NH_2 groups are absent, since such compounds likewise give color reactions with this reagent (see p. 153).

Procedure. A drop of the test solution is placed on filter paper, held over strong ammonia for 2–3 min. and then left in the air. After the excess ammonia has disappeared, the spotted area is treated with a drop of a freshly prepared 5% aqueous solution of sodium 1,2-naphthoquinone-4-sulfonate. Amounts of cyanamide up to 25 γ give a red stain at once; smaller amounts can be detected by warming the paper in an oven.

Limit of identification: 5 γ cyanamide.

1 F. FEIGL, D. GOLDSTEIN and E. LIBERGOTT, *Chemist-Analyst*, 53 (1964) 37.
2 Comp. P. KARRER, *Lehrbuch der Organischen Chemie*, 13th ed. (1959) p. 235.
3 M. X. SULLIVAN and W. C. HESS, *J. Am. Chem. Soc.*, 58 (1936) 47.

131. Dicyanodiamide (cyanoguanidine)

(1) Test through hydrolytic transformation into guanidine[1]

As stated in Sect. 109, only salts of dicyanodiamidine (guanylurea) and guanidine give a color reaction with phenanthraquinone when heated together with sodium carbonate at 180°. If this procedure fails and the presence of dicyanodiamide is suspected, this latter compound can be detected by transforming it into dicyanodiamidine through hydrolysis:

$$HN=C\begin{smallmatrix}\nearrow \overset{+}{N}H_3 \\ \searrow NHCN\end{smallmatrix} + H_2O \longrightarrow HN=C\begin{smallmatrix}\nearrow \overset{+}{N}H_3 \\ \searrow NHCONH_2\end{smallmatrix}$$

Dicyanodiamidine salts are hydrolyzed to guanidine in either the wet or the dry way or by heating with sodium carbonate:

$$HN=C\begin{smallmatrix}\nearrow NH_2 \\ \searrow NHCONH_2\end{smallmatrix} \longrightarrow HN=C\begin{smallmatrix}\nearrow NH_2 \\ \searrow NH_2\end{smallmatrix} + CO_2 + NH_3$$

When the dry heating is conducted in the presence of phenanthraquinone, the guanidine color reaction occurs.

Procedure. A little of the solid or a drop of its solution is united with a drop of concentrated hydrochloric acid in a micro test tube. The mixture is taken to dryness. The residue is intimately mixed with several mg of sodium carbonate and 1 drop of 0.1% benzene solution of phenanthraquinone. After evaporation of the solvent, the test tube is placed in a bath that has been preheated to 150°. The temperature is then increased to 190°. A positive result is indicated by the formation of a blue-violet product which is soluble in ether with a red color.

Limit of identification: 25 γ dicyanodiamide.

1 Unpublished studies, with D. GOLDSTEIN.

(2) Test through reductive conversion into guanidine [1]

Treatment of dicyanodiamide with zinc and hydrochloric acid yields guanidine along with hydrogen cyanide; the reduction proceeds further and the hydrogen cyanide is converted to methylamine[2]. The total reaction is therefore:

$$HN=C\begin{array}{c} {}^{\nearrow NH_2} \\ {}_{\searrow NHCN} \end{array} + 6\ H° \longrightarrow HN=C\begin{array}{c} {}^{\nearrow NH_2} \\ {}_{\searrow NH_2} \end{array} + CH_3NH_2$$

The resulting guanidine can be detected through the color reaction with sodium 1,2-naphthoquinone-4-sulfonate.

The above reactions provide the basis of the test given here for dicyanodiamide. The presence of cyanamide does not interfere because the latter is eliminated by reduction to methylamine and ammonia:

$$N\equiv C-NH_2 + 6\ H° \rightarrow CH_3NH_2 + NH_3$$

Procedure. A drop of the test solution is treated in a micro test tube with a drop of hydrochloric acid (1 : 1) and granular zinc. The mixture is kept in the water bath for 1–2 min. and the liquid is then poured off. The clear solution is treated with 20% sodium hydroxide solution added drop by drop until the initial precipitate of zinc hydroxide has been redissolved as zincate. A drop of a 2% freshly prepared solution of the sulfonate is then added and the mixture is kept in a water bath (90°) for 2–3 min. Then a drop of concentrated acetic acid is added. A red-violet color indicates a positive response. Because of the reddish color of the reagent solution, it is well to run a parallel blank test.

Limit of identification: 200 γ dicyanodiamide.

1 F. FEIGL, D. GOLDSTEIN and E. LIBERGOTT, *Chemist-Analyst*, 53 (1964) 37.
2 E. BAMBERGER and S. SEEBERGER, *Ber.*, 26 (1893) 1583.

132. Benzonitrile (phenyl cyanide)

Test through reduction to benzylamine [1]

If benzonitrile is warmed in alkaline medium with Devarda alloy, the resulting nascent hydrogen quickly reduces the nitrile to benzylamine:

$$C_6H_5CN + 4\ H° \rightarrow C_6H_5CH_2NH_2$$

Benzylamine is volatile with steam and therefore in the gas phase shows the color reaction typical of reactive CH_2- and NH_2 groups with sodium 1,2-naphthoquinone-4-sulfonate (see p. 153).

The application of this test assumes the absence of volatile primary aromatic bases such as aniline, tolidine, naphthylamine, nitro-, chloro-, bromo-aniline, and also piperidine. If the possible presence of such materials must be taken into account, a preliminary separation from the benzonitrile is imperative, for example by acidifying the test solution with dilute sulfuric acid and shaking with ether. The interfering bases remain in the water layer as sulfates, and the ether layer can then be tested for benzonitrile.

Procedure. A drop of the alcoholic or ethereal solution is mixed in a micro test tube with a drop of 5% caustic alkali solution and several mg of Devarda alloy. The suspension is cautiously warmed until the vigorous evolution of hydrogen has abated. The mouth of the test tube is then covered with a disk of filter paper moistened with a freshly prepared 0.5% solution of sodium 1,2-naphthoquinone-4-sulfonate. The test tube is then placed in boiling water. A positive response is indicated by the development of a brown-violet stain on the reagent paper within a few minutes.

Limit of identification: 3 γ benzonitrile.

1 F. FEIGL, *Anal. Chem.*, 33 (1961) 1113.

133. Phenylhydrazine

(1) *Test by oxidation to phenol* [1]

If phenylhydrazine is warmed with arsenic acid, the oxidation

$$C_6H_5NHNH_2 + 2\ H_3AsO_4 \rightarrow 2\ H_3AsO_3 + H_2O + N_2 + C_6H_5OH$$

yields phenol [2] which is volatile with steam and can be detected in the gas phase by the indophenol reaction with 2,6-dichloroquinone-4-chloroimine (see p. 185). A simple and rapid test for phenylhydrazine is based on these effects.

Procedure. A drop of the test solution is mixed in a micro test tube with a drop of a concentrated aqueous solution of arsenic acid. The mouth of the test tube is covered with a disk of filter paper impregnated with a saturated benzene solution of 2,6-dichloroquinone-4-chloroimine, and the tube is placed in boiling water for 3–4 minutes. If phenylhydrazine is present, a pale brown stain develops which turns blue when exposed to ammonia vapors. The blue color fades after a few minutes, but reappears on renewed exposure to ammonia.

Limit of identification: 5 γ phenylhydrazine.

This procedure can be applied also for the detection of phenylhydrazones and osazones which under the conditions of the test yield phenylhydrazine.

A preliminary separation is essential if phenols are present. The alkaline test solution is shaken with an organic liquid immiscible with water (ether, benzene, *etc.*). The alkali phenolate remains in the water layer.

1 F. FEIGL and E. JUNGREIS, *Talanta*, 1 (1958) 367.
2 M. OECHSNER DE CONINCK, *Compt. Rend.*, 126 (1898) 1042.

(2) Test by condensation with pyridine-2-aldehyde [1]

The three isomeric pyridine aldehydes can be detected through condensation with phenylhydrazine to produce yellow phenylhydrazones (see p. 386). Phenylhydrazine and its salts may likewise be detected by means of this condensation. The procedure may be applied to aqueous solutions of the salts or to the base itself dissolved in excess acetic acid.

Procedure. A drop of a 1% solution of pyridine-2-aldehyde is added to the test solution in a depression of a spot plate. A positive response is indicated by the almost immediate appearance of a yellow color.

Limit of identification: 0.5 γ phenylhydrazine.

1 F. FEIGL and L. BEN-DOR, *Talanta*, 10 (1963) 1111.

(3) Test through reduction with Devarda alloy and caustic alkali [1]

If phenylhydrazine or its salts are warmed with caustic alkali and Devarda alloy, the nascent hydrogen brings about reductive cleavage:

$$C_6H_5NHNH_2 + 2\ H^0 \rightarrow C_6H_5NH_2 + NH_3$$

The simultaneous formation of ammonia and steam-volatile aniline can be detected by means of Nessler indicator paper and the color reaction with sodium 1,2-naphthoquinone-4-sulfonate, respectively (see p. 153).

Naphthylhydrazine is cleaved in analogous fashion; however the resulting naphthylamine is much less volatile than aniline with water vapor; in fact the difference is so marked at 80° that a distinction is feasible.

Procedure. Two micro tests tubes are each charged with a drop of the test solution, 1 drop of 5 N caustic alkali, and several cg of Devarda alloy. The mouth of one tube is covered with a disk of filter paper moistened with Nessler reagent, the mouth of the other with filter paper moistened with freshly prepared solution of sodium naphthoquinone-sulfonate. The test tubes are immersed in a water bath preheated to 80°. The formation of a brown and a violet stain on the respective reagent papers within 3 minutes constitutes a positive response.

Limits of identification: 8 γ phenylhydrazine (aniline test)
                                 5 γ ,,   (Nessler test)

1 Unpublished studies, with E. LIBERGOTT.

(4) Test with vanillin and p-nitrosophenol [1]

The color reaction for aromatic aldehydes employing phenylhydrazine chloride and *p*-nitrosophenol (see p. 202) can be used in converse fashion for the detection of phenylhydrazine. Nitrophenylhydrazines do not give this reaction.

Procedure. A drop of the test solution is treated in a depression of a spot plate with a drop of a 0.5% alcoholic solution of vanillin and a drop of a freshly prepared 0.05% solution of p-nitrosophenol in concentrated hydrochloric acid. If phenylhydrazine is present, a blue-black color appears that goes over into brown.

Limit of identification: 0.1 γ phenylhydrazine chloride.

1 V. ANGER and S. OFRI, *Z. Anal Chem.*, 203 (1964) 428.

134. Hydrazobenzene

Test by rearrangement to benzidine [1]

On treatment with mineral acids, hydrazobenzene undergoes the well-known rearrangement to benzidine. The latter can readily be detected through oxidation to "benzidine blue" which is a meriquinoid compound of one molecule each of benzidine and its p-quinone imide. Manganese dioxide may be employed as the oxidant. This test for the benzidine formed is the converse of a highly sensitive means of detecting manganese (see *Inorganic Spot Tests*, Chapter 3).

Procedure. A drop of the alcoholic test solution together with a drop of 1:1 hydrochloric acid are placed in a micro test tube and kept in boiling water for several minutes. After cooling, several drops of 10% caustic alkali are added, and the reaction mass is extracted with ether. The ether extract is spotted on filter paper impregnated with manganese dioxide (for preparation see p. 474) and the paper is then held in strong acetic acid vapors. A blue stain indicates a positive response.

Limit of identification: 15 γ hydrazobenzene.

1 J. R. AMARAL (Rio de Janeiro), unpublished studies.

135. Oxalic acid hydrazides

Test through pyroammonolytic formation of oxamide [1]

The mono- and dihydrazides of oxalic acid exhibit the tests for acid hydrazides described on p. 274. A supplementary identification or detection is made possible through the ready transformation into oxamide. This change occurs if the material is heated to 135–140° along with urea. Ammonia is split off because of the formation of biuret (*1*) and the ammonia brings about the ammonolysis resulting in the formation of oxamide (*2*):

$$4\ CO(NH_2)_2 \quad \longrightarrow \quad 2\ H_2NOC\text{—}NH\text{—}CONH_2 + 2\ NH_3 \qquad (1)$$

$$\begin{array}{l} CONHNH_2 \\ | \qquad\qquad\quad + 2\ NH_3 \longrightarrow \\ CONHNH_2 \end{array} \quad \begin{array}{l} CONH_2 \\ | \qquad\qquad + 2\ NH_2\text{—}NH_2 \\ CONH_2 \end{array} \qquad (2)$$

The oxamide resulting from (1) and (2) can be detected through the red compound it produces when heated with thiobarbituric acid (see p. 535). Accordingly, the oxamide color reaction occurs when a mixture of oxalic acid hydrazide, urea and thiobarbituric acid is dry-heated. The temperature must not exceed 135–140° since otherwise the ammonia derived from the urea slowly decomposes the red product.

It is noteworthy that the ammonolysis (2) and likewise the formation of ammonium oxalate by the wet method through evaporation with ammonia water are not successful. The efficacy of the fusion with urea is therefore an impressive example of pyroammonolysis.

Procedure. A drop of the test solution or a small quantity of the solid is mixed in a micro test tube with several cg of urea and also of thiobarbituric acid. The test tube is placed in a bath that has been preheated to 110° and the temperature raised to 130–140°. A red color appears if the response is positive.

Limit of identification: 5 γ oxalic acid dihydrazide.

It must be remembered that the formation of oxamide through heating with urea occurs also with esters and alkylated amides of oxalic acid.

1 Unpublished studies, with D. HAGUENAUER-CASTRO.

136. p-Nitrosophenol

Test with benzalphenylhydrazone[1]

In concentrated hydrochloric or sulfuric acid, p-nitrosophenol reacts with arylaldehydephenylhydrazones to give intensively colored condensation products (see p. 281). Only p-nitrosophenol reacts in hydrochloric acid solution, but not p-nitrosonaphthol.

Procedure. A slight quantity of the test solid or the evaporation residue from a drop of the test solution is treated in a depression of a spot plate with a drop of 0.1% solution of benzalphenylhydrazone in concentrated hydrochloric acid. A deep red color indicates a positive response.

Limit of identification: 0.1 γ nitrosophenol.

1 V. ANGER and S. OFRI, *Mikrochim. Acta*, (1964) 111.

137. α-Nitroso-β-naphthol

Test with p-cresol[1]

The Gerngross test for tyrosine, which yields a red color (see p. 503), is given also by p-cresol and with still higher sensitivity. It has been found that *p*-cresol is a selective and sensitive reagent for α-nitroso-β-naphthol. The isomeric β-nitroso-α-naphthol does not react with this reagent.

Procedure. A drop of the test solution is treated in a micro test tube with a drop of approx. 0.1% solution of *p*-cresol in alcohol and warmed briefly in the water bath. Then 5 drops of nitric acid (s.g. 1.4) are added. A red color appears if α-nitroso-β-naphthol is present.

Limit of identification: 0.1 γ α-nitroso-β-naphthol.

1 V. ANGER and S. OFRI, *Mikrochim. Acta*, (1964) 112.

138. *p*-Nitrosodimethyl(ethyl)aniline

Test through release of dimethyl(ethyl)amine [1]

When dialkylated *p*-nitrosoanilines are heated with alkali hydroxide, the products are *p*-nitrosophenolate and dialkylamine:

$$ON-\!\!\!\left\langle\!\!\!\bigcirc\!\!\!\right\rangle\!\!\!-NRR + KOH \longrightarrow ON-\!\!\!\left\langle\!\!\!\bigcirc\!\!\!\right\rangle\!\!\!-OK + NHRR$$

Accordingly, *p*-nitrosodimethyl(ethyl)aniline yields dimethyl(ethyl)amine, which is volatile with steam. These amines condense with 2,4-dinitrochlorobenzene to yield yellow compounds which are soluble in chloroform. This constitutes the basis of a sensitive test for these amines and indirectly for compounds which furnish them as product of suitable reactions. See p. 240 regarding the chemistry of this color reaction.

Procedure. A drop of the test solution is placed in the bulb of the apparatus shown in Fig. 23, *Inorganic Spot Tests*, and one drop of 20% caustic alkali is added. A drop of a saturated solution of 2,4-dinitrochlorobenzene in alcohol is suspended on the knob of the stopper. The closed apparatus is placed in boiling water for several minutes. The hanging drop is then rinsed into a micro test tube by means of 1:1 acetic acid. Several drops of chloroform are added and the mixture shaken. A yellow color appears if *p*-nitrosodimethyl(ethyl)aniline was present.

Limit of identification: 2.5 γ *p*-nitrosodimethyl(ethyl)aniline.

1 Unpublished studies, with L. HAINBERGER.

139. Nitromethane

(1) Test with sodium 1,2-naphthoquinone-4-sulfonate [1]

Nitromethane reacts with alcoholic caustic alkali to yield water-soluble alkali salts of the *aci*-form of the nitro compound:

$$CH_3NO_2 + NaOH \rightarrow CH_2=NO_2Na + H_2O$$

The CH_2 group contained in the *aci*-form can react with sodium 1,2-naphtho-quinone-4-sulfonate,[2] which is a general reagent for reactive NH_2- and CH_2 groups (compare p. 153). The reaction with alkaline solutions of nitromethane can be written:

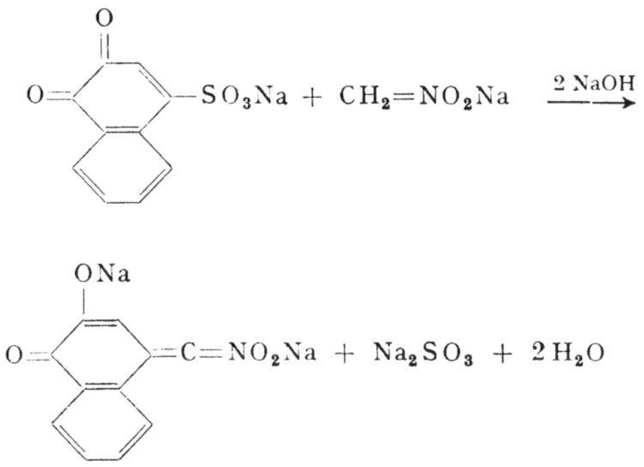

The *p*-quinoidal reaction product is blue-violet. Calcium oxide may be substituted for the caustic alkali in this reaction.

Nitroethane and other aliphatic and aromatic mononitro compounds like-wise yield *aci*-nitro compounds in alkaline milieu, but they contain no active CH_2 groups and accordingly give no color reaction with 1,2-naphthoquinone-4-sulfonic acid. Nevertheless, considerable amounts of nitroethane, nitro-propane, *etc.* prevent the reaction with nitromethane if only small amounts of the latter are present. This disadvantage can be avoided by using the fact that nitromethane forms an azeotrope with methanol that contains 12.5% by weight of nitromethane, and boils at 64.6°, whereas higher nitroparaffins form no azeotropes with methanol.[3]

Procedure.[4] A micro test tube is used. A drop of the alcoholic test solution is mixed with a drop of a 5% aqueous solution of sodium 1,2-naphthoquinone-4-sulfonate and then shaken with several mg of calcium oxide. A more or less intense blue or violet color appears, the shade depending on the amount of nitromethane present.

Limit of identification: 0.6 γ nitromethane.

1 F. TURBA, R. HAUL and G. UHLEN, *Z. Angew. Chem.*, 61 (1949) 74.
2 F. SACHS and M. CRAVERI, *Ber.*, 38 (1905) 3685.
3 Comp. L. R. JONES and J. A. RIDDICK, *Anal. Chem.*, 28 (1956) 1193.
4 F. FEIGL and D. GOLDSTEIN, *Anal. Chem.*, 29 (1957) 1522.

(2) *Test through conversion into formaldehyde*[1]

The acid-resistant water-insoluble primary nitroparaffins give water-soluble alkali salts of the *aci*-forms from which the nitro compounds are regenerated by the addition of dilute acids:

$$RCH_2NO_2 + OH^- \rightarrow RCH=NO_2^- + H_2O \qquad (1)$$

$$RCH=NO_2^- + H^+ \rightarrow RCH_2NO_2 \qquad (2)$$

The above ionic reactions, (1) and (2), should occur instantaneously, but this is not the case, because the regeneration of nitroparaffins as shown in (2) is merely the net result obtained from two partial reactions. The first of these is the rapid reaction (3) which yields the *aci*-form of the primary nitroparaffin. The second is the sluggish isomerization (4) whose speed determines the velocity of the entire process:[2]

$$RCH=NO_2^- + H^+ \rightarrow RCH=NO_2H \qquad (3)$$

$$RCH=NO_2H \rightarrow RCH_2NO_2 \qquad (4)$$

The intermediate formation of the $RCH=NO_2H$, which does not immediately isomerize, is the reason why the reaction series (3)–(4) can be realized only with stoichiometrically calculated quantities of acid, and even then it is not quantitative. If the alkali salts are treated with excess strong mineral acids, the nitroparaffin is not regenerated, but instead there is decomposition with production of aldehydes, nitrous acid, hydroxylamine, and carbon dioxide.[3] Decomposition reaction (5) may hold for the formation of the aldehyde; it involves the *aci*-form of the nitroparaffin:

$$2\ RCH=NO_2H \rightarrow 2\ RCHO + H_2O + N_2O \qquad (5)$$

This reaction proceeds to an extent that is adequate to permit its being used as the basis for the production of aldehydes from primary nitroparaffins.[4]

In the case of nitromethane, reaction (5) leads to formaldehyde, which can be detected by the color reaction with concentrated sulfuric acid and chromotropic acid (see p. 434). Since no other nitroparaffin yields formaldehyde, the following test is specific for nitromethane.

Procedure. A drop of the alcoholic test solution is mixed in a micro test tube with a drop of a 2% alcoholic solution of sodium hydroxide. After several minutes, 3–4 drops of a freshly prepared suspension of chromotropic acid in concentrated sulfuric acid are added and the mixture is warmed in a water bath. Depending on the amount of nitromethane present, a more or less intense violet color appears.

Limit of identification: 2.5 γ nitromethane.

1 F. Feigl and D. Goldstein, *Anal. Chem.*, 29 (1957) 1521.
2 A. Hantzsch and A. Veith, *Ber.*, 32 (1899) 607; E. Bamberger and E. Rust, *Ber.*, 35 (1902) 46.
3 J. U. Nef, *Ann.*, 280 (1894) 263.
4 K. Johnson and E. F. Degering, *J. Org. Chem.*, 8 (1943) 10.

140. Nitroethane

Test through production of acetaldehyde [1]

It was pointed out in Sect. 139 that decomposition of alkali salts of primary nitroparaffins by excess strong mineral acids leads to the respective aldehydes. Therefore nitroethane leads to acetaldehyde:

$$2\ CH_3CH=NO_2Na + 2\ H_2SO_4 \rightarrow 2\ CH_3CHO + N_2O + 2\ NaHSO_4 + H_2O$$

Since the vapors of acetaldehyde yield a blue color with sodium nitroprusside and piperidine (morpholine) as discussed on p. 438, the above reaction carried out at elevated temperature can be made the basis of a specific test for nitroethane.

Procedure. A drop of the alkaline methanolic test solution is mixed in a micro test tube with 2 drops of 1 : 1 sulfuric acid. Care must be taken that the upper portion of the test tube is not wetted by the liquids. A disk of filter paper moistened with the reagent (see p. 439) is placed over the mouth of the test tube that is then placed in a boiling water bath. Depending on the amount of nitroethane present, a more or less intense blue stain appears on the yellow paper at once or within 1–2 min.

Limit of identification: 7 γ nitroethane.

1 F. FEIGL and D. GOLDSTEIN, *Anal. Chem.*, 29 (1957) 1521.

141. *m*-Nitrophenol

Test through formation of 7-hydroxyquinoline [1]

It was pointed out on p. 516 that warming *m*-aminophenol or its salts with glycerol to 120° produces 7-hydroxyquinoline which fluoresces yellow-green. Since *m*-nitrophenol is readily reduced to the chloride of *m*-aminophenol by means of zinc dust or Devarda alloy and hydrochloric acid, this reduction in combination with subsequent heating with glycerol provides the basis of a test for *m*-nitrophenol and its differentiation from the isomeric *o*- and *p*-nitrophenols.

Procedure. A drop of the test solution is placed in a micro test tube and several mg of Devarda alloy are added. Several drops of hydrochloric acid (1 : 1) are introduced and warmed carefully. The contents of the tube are brought to dryness by immersing the tube as far as possible in a glycerol bath that has been heated to 110°. A drop or two of glycerol is added and the temperature kept at 120° for about 3 minutes. A positive response is indicated if, after adding a drop of alcohol, the system shows a yellow-green fluorescence in ultraviolet light.

Limit of identification: 3 γ *m*-nitrophenol.

1 E. JUNGREIS and V. LIPETZ, *Mikrochim. Acta*, (1963) 886.

142. p-Nitrobenzaldehyde

Test with benzylamine [1]

In alcoholic solution, p-nitrobenzaldehyde and benzylamine condense to yield a colorless Schiff base; the latter reacts with caustic alkali and forms an intensely colored nitronium salt through prototropic rearrangement. See Sect. 86 regarding the mechanism.

This color reaction is not shown by o- or m-nitrobenzaldehyde, nor by 2,4-dinitrobenzaldehyde.

Procedure. A drop of the test solution is treated in a micro test tube with a drop of benzylamine and the mixture is warmed in boiling water for 5 min. A drop of 2% alcoholic potassium hydroxide is added. A red or pink color indicates a positive response.

Limit of identification: 1 γ p-nitrobenzaldehyde.

1 F. FEIGL and V. ANGER, *Anal. Chim. Acta*, 24 (1961) 494.

143. m-Nitroaniline

Test through formation of dinitroacetanilide [1]

When m-nitroacetanilide is nitrated, it goes over almost exclusively into 2,3- or 3,4-dinitroacetanilide.[2] These compounds behave similarly to o-dinitrobenzene with respect to reducing agents (see p. 133) and yield intensely colored alkali salts of the tautomeric nitrohydroxylamino-benzene derivatives. Since m-nitroaniline is readily acetylated into m-nitroacetanilide by means of acetic anhydride this subsequent acetylation and nitration can be made the basis of a selective test for m-nitroaniline that is not given by the isomers.

Procedure. A slight amount of the test material or the evaporation residue from a drop of the test solution is treated in a micro test tube with 1 drop of acetic anhydride and warmed briefly. A drop of nitric acid (D = 1.5) is then added and the excess carefully removed by heating in a glycerol bath (120°). It is well to draw off the vapors from the micro test tube through a capillary connected to a water jet suction pump. The residue is treated with a drop of 2% solution of phenylhydrazine chloride and a drop of 2 N sodium hydroxide. If the response is positive, a violet color appears.

Limit of identification: 5 γ m-nitroaniline.

1 F. FEIGL and V. ANGER, *Z. Anal. Chem.*, 183 (1961) 13.
2 H. WENDER, *Gazz. Chim. Ital.*, 19 (1889) 230.

144. *p*-Nitroaniline

(1) Test with ascorbic acid and p-dimethylaminobenzaldehyde [1]

There is no color change, either in the cold or on warming, if the aqueous alcoholic solution of *o*-, *m*-, or *p*-nitroaniline is treated with *p*-dimethylamino-benzaldehyde and ascorbic acid. Under these conditions there is no reduction of NO_2- to NH_2 groups, and therefore no formation of a colored Schiff base by the condensation of the NH_2 group with the aldehyde. However, if such mixtures are taken to dryness, only those containing *p*-nitroaniline leave a red residue. Since a red product is formed by *p*-phenylenediamine and acidic solutions of the aldehyde, the *p*-nitroaniline is obviously reduced by ascorbic acid and the resulting diamine yields a Schiff base:

$$p\text{-}O_2NC_6H_4NH_2 + \text{ascorbic acid} \rightarrow p\text{-}H_2NC_6H_4NH_2 + \text{dehydroascorbic acid}$$

These reactions serve as the basis of a sensitive spot test for *p*-nitroaniline on filter paper. It succeeds well in the presence of the isomeric nitroanilines.

Procedure. A drop of the alcoholic test solution is placed on filter paper (Whatman No. 120) and after drying, a drop of the freshly prepared reagent is applied. The paper is kept at 110° in an oven for 5 to 10 min. The resulting stain is red or pink, depending on the amount of *p*-nitroaniline present. (The color becomes more distinct if the paper is placed in water.)

Reagent: A 16% solution of ascorbic acid in water is saturated with *p*-di-methylaminobenzaldehyde and filtered.

Limit of identification: 0.5 γ *p*-nitroaniline.

The sensitivity is lower in the presence of considerable amounts of *o*- and *m*-nitroaniline because their intense yellow color and adsorption on the paper hide a faint red coloration. Even so, the procedure clearly revealed 8 γ *p*-nitroaniline in the presence of 500 γ of its *o*- and *m*-isomers.

1 F. FEIGL and C. STARK-MAYER, *Chemist-Analyst*, **46** (1957) 37.

(2) Test through conversion into ammonium p-nitrophenylantidiazotate [1]

When *p*-nitroaniline is diazotized (*1*), the resulting diazonium ion is a sensitive spot test reagent [2] for ammonia through the formation (*2*) of red ammonium *p*-nitrophenylantidiazotate:

$$O_2N-\!\!\left\langle\!\!\bigcirc\!\!\right\rangle\!\!-\overset{+}{N}H_3 + HNO_2 \longrightarrow O_2N-\!\!\left\langle\!\!\bigcirc\!\!\right\rangle\!\!-\overset{+}{N}\!\equiv\!N + 2\,H_2O \qquad (1)$$

$$O_2N-\!\!\left\langle\!\!\bigcirc\!\!\right\rangle\!\!-\overset{+}{N}\!\equiv\!N + NH_4OH \longrightarrow O_2N-\!\!\left\langle\!\!\bigcirc\!\!\right\rangle\!\!-\underset{\underset{N-ONH_4}{\|}}{N} + H^+ \qquad (2)$$

The realization of (*1*) and (*2*) allows a test for *p*-nitroaniline and its distinction from the *o*- and *m*-isomers.

When the procedure described here is used, attention must be given to the fact that many diazotized aromatic amines give colored products with ammonia. Most of these are yellow, but pink and violet colorations were observed with benzidine, α- and β-naphthylamine.

Procedure. Minimal quantities of the test material are dissolved in dilute hydrochloric acid, and one drop of the solution (cooled in ice) is placed in the depression of a cooled spot plate. A few crystals of sodium nitrite are added, stirred and treated with 1–2 drops of concentrated ammonia. A positive response is indicated by the appearance of a red color.

Limit of identification: 1 γ *p*-nitroaniline.

1 E. JUNGREIS (Jerusalem), private communication.
2 Comp. F. FEIGL, *Spot Tests in Inorganic Analysis*, 5th ed., Elsevier, Amsterdam, 1958, p. 235.

145. Thiocyano-2,4-dinitrobenzene

Detection through tests for thiocyanate and nitro groups[1]

Though no direct specific test is available for this compound, now widely used as an insecticide, it can be reliably identified through the positive response to the tests for the CNS- and NO_2 groups, and also to the test for *m*-dinitro compounds. The chemistry of these tests and the appropriate procedures and identification limits for thiocyano-2,4-dinitrobenzene are outlined on pp. 302 and 397.

1 Unpublished studies.

146. Phenylisothiocyanate (phenyl mustard oil)

Detection through alkaline saponification[1]

As anil of carbon oxysulfide or thiocarbonic acid, phenylisothiocyanate is saponified (hydrolyzed) when warmed with caustic alkali; aniline and alkali thiocarbonate are formed:

$$C_6H_5N=C=S + 2\ NaOH \rightarrow C_6H_5NH_2 + Na_2CO_2S$$

Aniline vapors can be detected through the color reaction with sodium 1,2-naphthoquinone-4-sulfonate (see p. 153). The formation of thiocarbonate is demonstrated by the evolution of hydrogen sulfide on acidification:

$$Na_2CO_2S + 2\ HCl \rightarrow 2\ NaCl + CO_2 + H_2S$$

The positive response to the test for aniline *during* the test and to the test for hydrogen sulfide *after* the saponification constitute reliable indications of the presence of phenylisothiocyanate.

The alkaline saponification in the water bath, which can be accomplished with a drop of the test solution, will reveal 25 γ phenylisothiocyanate. A more favorable sensitivity is obtained by saponifying in glycerol since a temperature far above 100° is then possible.

Procedure. A drop of the alcoholic test solution is shaken in a micro test tube with 1 drop of a 5% solution of alkali hydroxide and glycerol. The mixture is gently heated initially in a glycerol bath to volatilize the alcohol. The mouth of the tube is then covered with a piece of filter paper moistened with a freshly prepared 2% aqueous solution of sodium 1,2-naphthoquinone-4-sulfonate, and the temperature of the glycerol bath is then raised to 150°. The formation of aniline is indicated by the appearance of a red-violet stain on the reagent paper within 1–2 min. The reaction mixture is then cooled and acidified by adding 1–2 drops of strong hydrochloric acid. The mouth of the tube is covered with a disk of lead acetate paper, and the tube is then placed in boiling water. A black or brown stain on the reagent paper indicates the formation of hydrogen sulfide.

Limits of identification:

   0.5 γ phenylisothiocyanate (through aniline test)
   1 γ ,,   ,, (through hydrogen sulfide test).

1 F. FEIGL and E. LIBERGOTT, *Mikrochim. Acta*, (1964) 259.

147. N-Phenylsulfamic acid

Test through pyrohydrogenolysis[1]

If N-phenylsulfamic acid is heated to 160° with benzoin (m.p. 137°), the benzoin acts as a hydrogen donor and occasions the reaction:

$$C_6H_5NHSO_3H + 2 H^0 \rightarrow C_6H_5NH_2 + H_2O + SO_2$$

Among the products of this redox reaction, aniline can be detected in the gas phase by means of sodium 1,2-naphthoquinone-4-sulfonate (violet color) and sulfur dioxide through the bluing of Congo paper moistened with hydrogen peroxide. The pyrolysis is conducted in a micro test tube whose mouth is covered in turn with disks of the respective reagent papers.

If alkali salts of this acid are presented for examination, they should be fumed off with dilute hydrochloric acid and the evaporation residue brought to 130° to eliminate completely the unconsumed hydrochloric acid. The residue is then fused with benzoin.

The test revealed: 400 γ N-phenylsulfamic acid through the aniline test
                       100 γ ,,   ,,   ,,   ,, SO$_2$,,

1 Unpublished studies.

148. Triphenyl phosphate

Detection through pyroammonolysis [1]

As pointed out on p. 216, the phenol esters of carboxylic acids undergo pyrohydrolysis when they are heated to 160° with hydrates of oxalic acid or of manganous sulfate. The active agent in these cleavages is the water of hydration which is released as superheated steam. The resulting phenol can be readily detected in the gas phase by means of the indophenol reaction (blue color) with 2,6-dichlorobenzoquinone-4-chloroimine and ammonia.

Triphenyl phosphate (m.p. 45°) does not respond to the test; it obviously is resistant to hydrolysis at the specified temperature. However, it does yield phenol if heated with succinic acid, which functions as a water-donor at temperatures above 220° (*Idn. Limit:* 60 γ). The best way to liberate the phenol is by rapid pyrolytic ammonolysis of this ester when heated with guanidine carbonate. The latter acts as ammonia-donor by producing biguanide. The cleavage:

$$(C_6H_5)_3PO_4 + 3\,NH_3 \rightarrow PO(NH_2)_3 + 3\,C_6H_5OH$$

occurs readily with the ammonia formed pyrolytically.

The procedure described here is valid only in the absence of volatile phenols and phenyl esters of carboxylic acids. Such possible interfering compounds should be tested for by a preliminary heating to 160° with hydrated oxalic acid. (Comp. p. 216.)

Procedure.   A drop of the sample, dissolved in ether, chloroform or benzene, is placed in a micro test tube and several cg of guanidine carbonate are added. The solvent is removed. The tube is then placed in a bath which has been preheated to 200°, and a piece of filter paper that has been moistened with a saturated ethereal solution of 2,6-dichloroquinone-4-chloroimine is placed over the mouth of the test tube. The temperature of the bath is then brought to 220–230°. If the response is positive, a blue stain appears on the paper.

Limit of identification: 1 γ triphenyl phosphate.

1 F. FEIGL and E. JUNGREIS, *Anal. Chem.*, 31 (1959) 2102.

Chapter 6

Application of Spot Tests in the Differentiation of Isomers and Homologous Compounds. Determination of Constitutions

General remarks

For a long time the search for new chemical methods of detecting and differentiating isomers and homologs received but little consideration in qualitative organic analysis. The principal reason for this neglect is that such compounds possess characteristic physical constants whose determination by physical methods provides reliable means of identification. Procedures of this kind are adequate for meeting microchemical demands, however they require a knowledge of the pertinent constants which often have not been determined or published. In addition, it is necessary that the isomers or homologs be separated and isolated prior to such physical measurements; the same holds for derivatives prepared from the compounds under study. Chromatographic methods in which the chemical or physical identification is made on adsorptively separated isomeric or homologous compounds likewise assume a knowledge of the pertinent constants *etc.* A compelling reason for the reliance on physical methods (including those that require expensive equipment) was the lack of specific and selective tests whose applicability could be tested with respect to the identification of isomers and homologs. Through the discovery of numerous new reactions of organic compounds that could be utilized for microchemical ends this situation has changed radically, an advance that has been achieved chiefly because of advances in the field of organic spot test analysis. The latter have thus furnished incentives for a more extensive consideration of chemical tests that can be applied directly for the detection and identification of isomers. Studies of reactions that permit the detection of functional groups have shown that their reactivity often is greatly dependent on the remainder of the molecule since the found identification limits for compounds with the same functional groups seldom correspond to equivalent values. The differences are frequently very considerable, and sometimes otherwise very reliable tests will be found to yield no response at all. Hence it was to be expected that the influence of the remainder of the molecule on the reactivity of groups attached to the latter would likewise show up in the chemical behavior of isomeric and homologous compounds and perhaps be of analytical usefulness. The following differentiation possibilities should be kept in mind:

1. The velocities and equilibrium positions of reactions in which the products are directly recognizable by color, solubility behavior, *etc.* differ widely.

2. Functional groups react similarly, but the reaction products behave differently with regard to their direct recognizability or response to supplementary identification reactions.

3. The entire configuration of the molecule permits the occurrence of characteristic reactions only in the case of certain isomers or homologs.

Still other differentiation possibilities arise if one of the isomers or homologs occupies a special position because of volatility (sublimability) or solubility and hence can be detected in a separate phase.

Taking due consideration of the above possibilities, that apply both to reactions in the wet way and pyroreactions, it was found that at the time more than forty spot tests were available through which isomers could be detected without prior separation, and often when present in unfavorable ratios. Since the discovery of more spot tests suited to the resolution of such problems can be awaited with certainty, a new and interesting field for applying spot tests is in the offing.

When isomeric and homologous compounds are to be differentiated the problem invariably arises of coordinating and relating the material being tested with and to compounds of known composition and constitution. Another type of problem is presented when information is desired regarding the constitution of organic compounds whose empirical formulas are known. In such cases it is mostly a matter of identifying certain structural units in the molecule with certainty. At times, this latter step may be of interest in preparative studies concerning the formation, degradation and rearrangement of organic compounds. It has been proven that the responses to spot tests can be of much service when problems of the kinds cited above are presented to the analyst. Pertinent examples will be given in the following sections.

1. Anthracene and phenanthrene

Concentrated benzene solutions of anthracene or phenanthrene react with benzene solutions of picric acid to yield insoluble addition products that differ markedly in color. The anthracene compound is brick red whereas the phenanthrene addition compound is yellow, as is the addition product of naphthalene and picric acid. When benzene or ethereal solutions of their components are evaporated these colored addition compounds are left. They likewise result if the dry components are ground together, thus constituting the products of solid–solid reactions.

Procedure.[1] Filter paper is bathed in a 10% solution of picric acid in benzene and dried in the air. A drop of the ethereal or benzene solution being tested for anthracene is placed on the dry paper. If anthracene is present, a pink or red stain appears on the yellow paper.

Limit of identification: 3 γ anthracene.

Phenanthrene can be identified by the test described on p. 409.

These tests permit a rapid chemical differentiation of the isomers: phenanthrene and anthracene. However, it must be noted that picric acid gives red addition products also with alkyl-substituted naphthalenes, hydrocarbons with condensed benzene rings, and with aromatic amines. The latter compounds can be removed by dissolving the test material in ether or chloroform and extracting the solution with dilute mineral acids.

1 Cl. Costa Neto (Rio de Janeiro), private communication.

2. Phthalic–terephthalic acid. Maleic–fumaric acid

Ordinarily the members of these pairs of isomers are distinguished by determining the respective melting points. However, they can be distinguished by a purely chemical procedure which employs the familiar fact that within these pairs only phthalic acid and maleic acid form anhydrides through the loss of water when subjected to dry heating. This typical loss can be established with even minimal amounts and at temperatures below the melting point of the acids. The resulting superheated steam reacts with thiobarbituric acid or with thio-Michler's ketone to yield hydrogen sulfide. The procedure is given on p. 150.

3. Naphthols, aminophenols and nitroanilines

Selective tests for α- and β-naphthol are described in Chap. 5, Sect. 15 and 16. These tests can be used for the differentiation of the isomers.

Selective tests for *o*-, *m*- and *p*-aminophenol are described in Chap. 5, Sect. 93, 94 and 95. These tests permit the differentiation of the isomers.

Selective tests for *m*- and *p*-nitroanilines are described in Chap. 5, Sect. 143 and 144; these tests are not given by the *o*-isomer.

4. Thiocyanates and isothiocyanates[1]

In contrast to the position-isomers dealt with in Sect. 6, thiocyanates and the isomeric isothiocyanates differ with respect to the —S—C≡N and —N=C=S groups attached to the alkyl or aryl radical. Such isomeric compounds can be detected and differentiated from each other through their positive or negative response to the spot tests cited on pp. 302 and 564 respectively. These tests are based on reductive cleavage or on alkaline hydrolysis. Furthermore, only isothiocyanates respond positively to the test described on p. 84 for thionic compounds based on the realisation of the Wohler's effect.

1 In part private communication from R. POHLOUDEK-FABINI (Greifswald).

5. Eugenol–isoeugenol. Safrole–isosafrole

The differentiation of the members of these two pairs of isomers by fractional distillation is not feasible because the respective boiling points are too close: eugenol (254°) and isoeugenol (266°); safrole (254°) and isosafrole (253°). Eugenol and safrole contain allyl groups, whereas the isomeric propenyl group is present in isoeugenol and isosafrole. As stated on p. 168, the propenyl group is readily iodinated by evaporation with a solution of iodine in carbon disulfide, whereas allyl groups are resistant to iodine. Consequently, a positive response to the iodination tests indicates the presence of the propenyl isomers.

6. Vanillins

The three isomers, namely vanillin (I), isovanillin (II), and orthovanillin (III), can be distinguished from each other through their respective behavior toward reagents employed in tests for aromatic aldehydes.[1]

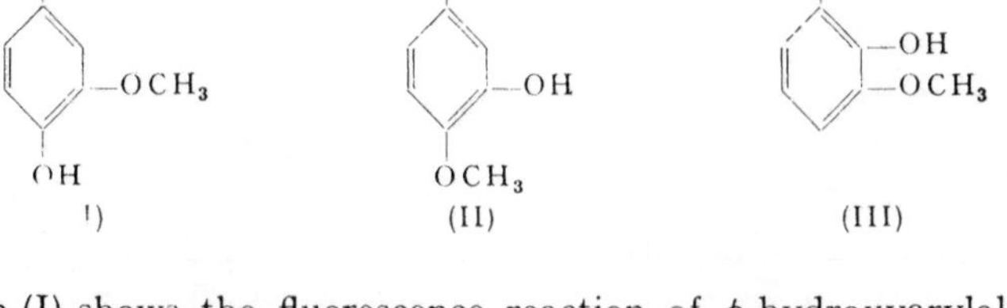

Vanillin (I) shows the fluorescence reaction of p-hydroxyarylaldehydes with barbituric acid (green fluorescence in acid solution, that disappears on the addition of ammonia, see p. 346). Furthermore, it gives a positive response to thiobarbituric acid with yellow fluorescence in acid medium and color change to red without quenching of the fluorescence when alkali is added (see p. 345). It does not produce a fluorescing aldazine with hydrazine sulfate.

Isovanillin (II) yields a yellow fluorescence when it reacts with barbituric acid; the fluorescence is not quenched on the addition of ammonia. It gives no fluorescence reaction with thiobarbituric acid, nor does it yield a fluorescing aldazine on treatment with hydrazine sulfate.

Orthovanillin (III) gives no fluorescence with either barbituric acid or thiobarbituric acid. On the other hand, like other o-hydroxyarylaldehydes, it reacts with hydrazine sulfate (see p. 341) yielding a fluorescing aldazine (orange).

Limits of identification:

    0.01 γ vanillin   0.1 γ isovanillin   0.1 γ orthovanillin

1 V. ANGER and S. OFRI, private communication.

7. Differentiation of dihydroxybenzenes[1]

If solutions of the three isomers in acetone–water (1:1) are adjusted to pH 5 by means of acetic acid (test paper) they show diverse behaviors toward aqueous solutions of potassium bichromate (20 mg/ml), mercurous nitrate (28 mg/ml), and sodium cyanide (38 mg/ml).

o-Dihydroxybenzene	(pyrocatechol)	plus $K_2Cr_2O_7$:	brown-black product	
m-	,,	(resorcinol)	,, $Hg_2(NO_3)_2$:	rust-brown	,,
p-	,,	(hydroquinone)	,, NaCN:	green-brown	,,

The identification of the m- and p-isomers is conducted as spot tests on filter paper; the reaction of the o-isomer is tested on a spot plate as a drop test.

It is possible to detect 25 γ of one of the isomers in the presence of up to 400 times its weight of the other isomers.

1 T. S. MA and A. HIRSCH, *Chemist-Analyst*, 50 (1961) 12.

8. ω-Bromoacetophenone and p-bromoacetophenone[1,2]

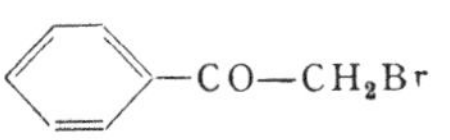

    —CO—CH$_2$Br   Br—   —CO—CH$_3$

      (m.p. 50°)   (m.p. 50°)

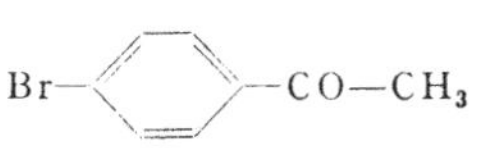

        (I)   (II)

As a primary halogenoalkyl, ω-bromoacetophenone (I) reacts with sodium thiosulfate to yield a so-called Bunte salt:

$$C_6H_5COCH_2Br + Na_2S_2O_3 \rightarrow Na[C_6H_5COCH_2S_2O_3] + NaBr$$

This product decomposes when heated to 180° and sulfur dioxide is given off (see p. 171).

$$2\ Na[C_6H_5COCH_2S_2O_3] \rightarrow C_6H_5COCH_2SSCH_2COC_6H_5 + Na_2SO_4 + SO_2$$

Sulfur dioxide is readily detected in the gas phase by means of Congo paper moistened with hydrogen peroxide.

p-Bromoacetophenone (II) does not show this reaction.

Limit of identification: 10 γ ω-bromoacetophenone.

1 F. FEIGL, V. ANGER and D. GOLDSTEIN, *Mikrochim. Acta*, (1960) 231.
2 F. FEIGL and V. ANGER, *Z. Anal. Chem.*, 181 (1961) 162.

9. Naphthoquinones[1]

Some quinones react with barbituric acid in ammoniacal solution forming colored products. p-Naphthoquinone yields a violet color whereas o-naphthoquinone does not react. The same is true of the homologs of p-naphthoquinone. The chemistry of this color reaction has not been elucidated.

Procedure.   One drop of the test solution is united in a micro test tube with a drop each of a saturated solution of barbituric acid and of conc. ammonia. A positive response is indicated by a violet color.

1 V. ANGER and S. OFRI, *Mikrochim. Acta*, (1964) 917.

10. Isomeric monohalogenofatty acids[1]

The differentiation of α- and β-alanine as described in Sect. 16 combined with the convertibility of halogenated fatty acid into amino fatty acids point to the possibility of learning whether the halogen atom in isomeric monohalogenated fatty acids is in the α- or the β-, γ- *etc.* position with respect to the COOH group. The acid or its alkali salt should be taken to dryness along with ammonium carbonate in order to replace the halogen atom by the NH_2 group. The excess of ammonium carbonate must be expelled by heating (120 min. to 130°). α-Aminocarboxylic acids and their salts give the color reaction with pyridine-2-aldehyde plus Co^{2+} ions (see p. 370). If the response is negative, then the procedure given for the detection of β-alanine should be tried, *i.e.* the residue resulting from the evaporation with ammonium carbonate should be heated to 200° (see p. 576). Immediate evolution of ammonia indicates the presence of β-, γ-, *etc.* halogenated fatty acids.

These tests can be conducted satisfactorily with 0.5–1 mg of the sample.

1 F. FEIGL and S. YARIV, *Anal. Chim. Acta*, 29 (1963) 580.

11. Phenylsulfuric acid and phenolsulfonic acid[1]

In contrast to the phenolsulfonic acids HO—C_6H_4—SO_3H, phenylsulfuric acid C_6H_5—OSO_3H can be readily saponified to yield free phenol by warming with mineral acids. The phenol is volatilized with the water vapor and can be detected in the gas phase by means of the Gibbs reagent (formation of a blue indophenol with 2,6-dichloroquinone-4-chloroimine) as described on p. 185.

p-Phenolsulfonic acid remains completely unaltered under these conditions. However, if heated to 220–250° in the presence of succinic or phthalic acid it is pyrohydrolyzed with production of volatile phenol due to the water formed by anhydration of these acids. The phenol vapors can be detected as mentioned above.

Limits of identification:
    0.3 γ phenylsulfuric acid   0.5 γ phenol-p-sulfonic acid (sodium salt)

1 F. FEIGL and V. ANGER, *Z. Anal. Chem.*, 181 (1961) 164.

12. Naphthylamines

(1) Detection of α-naphthylamine with nitrosoantipyrine (see p. 509)

(2) Detection of β-naphthylamine by condensation with pyridine-2-aldehyde[1]

If an aqueous solution of pyridine-2-aldehyde is treated with β-naphthylamine and ferrous sulfate an intense violet color appears. This color reaction is the result of an initial formation of the Schiff base (I) of the aldehyde which then adds to Fe^{2+} ion to give the complex cation (II) whose chelate ring (III) is responsible for the color.

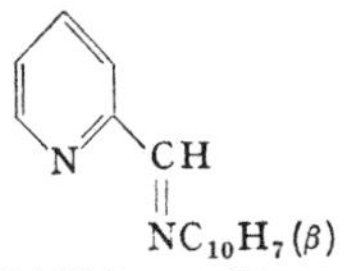

[Fe(Schiff base)$_2$]$^{2+}$
(II)

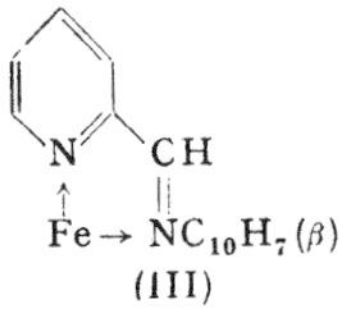

α-Naphthylamine also condenses with pyridine-2-aldehyde but probably because of steric reasons the resulting Schiff base is not capable of coordination on the iron(II) ion. This divergent behavior provides the possibility of detecting β-naphthylamine and of distinguishing it from the α isomer.

Procedure. A drop of the alcoholic test solution is mixed in a micro test tube with a drop of the reagent and then a drop of water is added. A positive response is indicated by the appearance of a blue-violet color. If considerable amounts of α-naphthylamine are present, a white precipitate of this base may be formed, but the latter does not prevent the color reaction.

Limit of identification: 10 γ β-naphthylamine (in the presence of 1000 times as much of the α-isomer).

Reagent: Freshly prepared mixture of 5 ml of 10% aqueous solution of pyridine-2-aldehyde with 10 drops of 0.1 M ferrous sulfate.

1 F. FEIGL, E. JUNGREIS and S. YARIV, *Z. Anal. Chem.*, 200 (1964) 38.

13. α- and β-Naphthylaminesulfonic acids

Isomeric mono- (and di-) sulfonic acids of α- and β-naphthylamine can be differentiated through their ready conversion into the respective amines for which selective tests are available. The conversion is accomplished by warming with Raney nickel alloy and caustic alkali; the aromatically bound SO_3H groups are split off and eventually yield nickel sulfide. The pertinent procedure is given on p. 234. If the desulfurized solution is extracted with ether, the resulting ethereal solution will contain the respective naphthylamine and may be tested for α-naphthylamine or β-naphthylamine by the procedures given on pp. 509 and 573.

14. Benzylamine, monomethylaniline and (o-, m-, p-)toluidine[1]

These isomeric bases boil at 185°, 195° and 200–205° respectively and consequently cannot be separated readily from each other by fractional distillation and identified in this way. They can be distinguished through their diverse behavior toward sodium 1,2-naphthoquinone-4-sulfonate. As was pointed out on p. 153, the aqueous solution of this salt is a sensitive color reagent for reactive CH_2- and NH_2 groups in organic compounds through the production of p-quinoidal condensation products. Accordingly, no reaction occurs with monomethylaniline, whereas benzylamine reacts slowly and the toluidines react very quickly. The resulting condensation products have different colors.

Procedure. Filter paper is moistened with 5% aqueous solution of sodium 1,2-naphthoquinone-4-sulfonate and dried. A drop of the test solution is placed on the reagent paper and after 5 min. the paper is bathed in water to remove the reagent. The results are:

benzylamine	brown-violet stain (*Idn. lim.* 5 γ)			
monomethylaniline	no reaction			
o-toluidine	red stain	(,,	,,	5 γ)
p-toluidine	,, ,,	· (,,	,,	2 γ)

The occurrence of the color reaction with benzylamine can be prevented by adding a drop of dilute formaldehyde. This masking is due to the rapid formation of the Schiff base of this primary amine with formaldehyde. No masking occurs with toluidine. Strangely enough, if p- or o-toluidine is mixed with benzylamine, they too are masked to a distinct extent.

1 F. FEIGL, A. DEL'ACQUA and S. LADEIRA DALTO, unpublished studies.

15. Phenylenediamines

(1) Test for o-phenylenediamine (see p. 510).

(2) Differentiation of phenylenediamines [1]

The solutions of the isomers in acetone–water (1:1), brought to pH 5 with acetic acid (test paper), show diverse behaviors toward solutions of bismuth nitrate (47 mg/ml 6 M acetic acid), sodium nitrite (30 mg/ ml), and potassium permanganate (0.1 mg/ ml).

o-Phenylenediamine plus	$Bi(NO_3)_3$	orange-red brown product		
m-	,,	,,	$NaNO_2$	red ,,
p-	,,	,,	$KMnO_4$	blue-violet ,,

The identification of the o- and the m-isomers is made by spot tests on filter paper; the reaction of the p-isomer is conducted on a spot plate as a drop reaction.

It is possible in this way to detect 10 γ of each isomeric phenylenediamine in the presence of up to 400 times its weight of the other isomers.

1 A. HIRSCH, S. FISCHMAN, M. GOLDBERG, M. OTTENSOSER and M. SCHACHNOW, *Chemist-Analyst*, 50 (1961) 7.

16. α- and β-Alanine

The following tests permit the differentiation between the two isomeric aminopropionic acids $CH_3CH(NH_2)COOH$ and $(NH_2)CH_2CH_2COOH$, frequently called α- and β-alanine, respectively.

(1) Test for α-alanine through condensation with pyridine-2-aldehyde[1]

As pointed out on p. 370, α-aminocarboxylic acids can be detected through the formation of their respective Schiff bases with pyridine-2-aldehyde, followed or accompanied by the production of red water-soluble complex cobalt(III) salts that are resistant to dilute mineral acids.

Procedure.   As given on p. 370.
Limit of identification: 5 γ α-alanine.

(2) Test for β-alanine through pyrolytic production of ammonia[1]

In β-alanine, the NH_2 group is farther from the COOH group than in the α-isomer and consequently the latter exhibits the characteristics of an ammonium salt to a marked degree. Accordingly, the pyrolysis that is characteristic of primary aliphatic amines: $2\ RNH_2 \rightarrow RNHR + NH_3$, occurs more rapidly in the case of the β-isomer than with α-alanine. If the conditions prescribed below are adhered to carefully, the two isomers can be reliably distinguished.

Procedure.   Two micro test tubes are charged with 0.5 mg of α- or β-alanine respectively. A drop of glycerol is added to each tube and the open ends are covered with disks of filter paper moistened with Nessler reagent. The tubes are then immersed in a glycerol bath preheated to 200°. A brown stain due to ammonia appears almost immediately on the reagent paper in the case of the tube containing β-alanine.

1 Unpublished studies, with L. BEN-DOR.

17. Aminobenzoic acids

The procedures given here under *I-III* may be employed to distinguish the three isomeric aminobenzoic acids, either alone or in mixtures.

I. Detection of o-aminobenzoic acid through fluorescence of the alkali salts[1]

If a drop of the alcoholic test solution is placed in a depression of a spot plate or in a micro test tube and observed under ultraviolet light, the following differences will be evident even at high dilutions:

o-Aminobenzoic acid: blue fluorescence (*Idn. lim.* 0.1 γ).
  The fluorescence disappears on addition of mineral acid; it persists on addition of alkali hydroxide or carbonate.
m-Aminobenzoic acid: blue fluorescence (*Idn. lim.* 0.5 γ).
  The fluorescence disappears on addition of acid or alkali.
p-Aminobenzoic acid: no fluorescence.

As shown by the above findings, the isomeric aminobenzoic acids can be identified when alone through the appearance, quenching, or non-appearance of a fluorescence. The persistence of the fluorescence on alkalization is characteristic for *o*-aminobenzoic acid and permits its detection even when mixed with great amounts of the *m*- and *p*-isomers.

II. *Detection of m-aminobenzoic acid through condensation with pyridine-2-aldehyde*[2]

An intense violet color appears when aniline or its salts are treated with a *dilute* solution of pyridine-2-aldehyde containing ferrous sulfate. The color reaction is based on the formation of the anil (I) and the coordination of this product with ferrous ions to yield the complex cation (II), whose color properties can be attributed to the chelate ring (III) it includes:

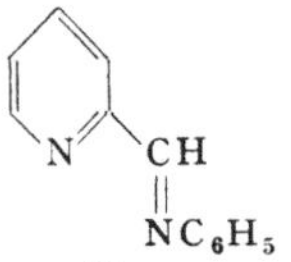

All derivatives of aniline do not show the same behavior. Some do not give the color reaction (*e.g.* benzidine, α-naphthylamine); others yield iron salts whose color is discharged on the addition of alcohol, or the iron salts do not form in an alcoholic milieu. The latter is true of the blue or violet iron compounds normally obtained from *o*- and *p*-aminobenzoic acids. Since the color reaction of *m*-aminobenzoic acid is not impaired by alcohol, this isomer can be detected and distinguished from the other two through this singular behavior.

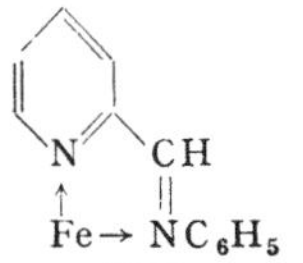

Procedure. A drop of the alcoholic solutions of each of the three amino-benzoic acids is placed in separate micro test tubes and then a drop of the reagent solution is added to each, followed by 2–3 drops of alcohol. The color (violet) persists only in the tube containing the *m*-aminobenzoic acid.

Limit of identification: 25 γ *m*-aminobenzoic acid.

III. *Detection of p-aminobenzoic acid through oxidation with alkali bromate*[1]

A drop of each of the alkaline solutions is placed in micro test tubes and treated with a drop of 1 N potassium bromate and 1 drop of 2 N nitric acid. The following behavior will be seen:

> *o*-aminobenzoic acid: red color   (*Idn. lim.* 50 γ)
> *m*-aminobenzoic acid: yellow color (,,   ,,   50 γ)
> *p*-aminobenzoic acid: ,,   ,,   (,,   ,,   50 γ)

These color reactions permit the identification of the three isomers if they are not mixed with each other; this applies particularly to *p*-aminobenzoic acid in view of the change from violet to yellow.

Because of the lack of validating experimental studies, it is not possible at present to make definite statements concerning the mechanism of the color reaction, though oxidation products of the isomeric aminobenzoic acids are doubtless involved.

1 F. FEIGL and E. JUNGREIS, *Z. Anal. Chem.*, 198 (1963) 419.
2 F. FEIGL and L. BEN-DOR, *Talanta*, 10 (1963) 1111.

18. Metanilic and sulfanilic acid[1]

Dry-heating of metanilic acid with mercuric cyanide yields hydrogen cyanide, a result that is characteristic of acidic compounds (see p. 147). The isomeric sulfanilic acid does not show this reaction because it, as pointed out in Sect. 48, has the constitution of a neutral ammonium salt. This diverse behavior can be used for distinguishing these isomeric acids.

If the absence of metanilic acid has been established, a test for sulfanilic acid can be made by evaporating the sample with formaldehyde in order to methenylate the NH_2 group and thus liberate the acidic SO_3H group. Heating the residue with mercuric cyanide will then result in the evolution of hydrogen cyanide.

Limits of identification:
  5 γ metanilic acid by pyrolysis with $Hg(CN)_2$
  10 γ sulfanilic acid by pyrolysis with $Hg(CN)_2$, follow.ng evaporation with formaldehyde.

The methenylations are best accomplished by heating the test material with solid paraformaldehyde; the excess is driven off completely by heating to 140°.

1 Unpublished studies, with E. LIBERGOTT.

19. H-acid and K-acid[1]

The isomeric 8-amino-1-naphtholdisulfonic acids, namely H-acid (I) and K-acid (II) are important starting materials in the dye industry. Their structures are:

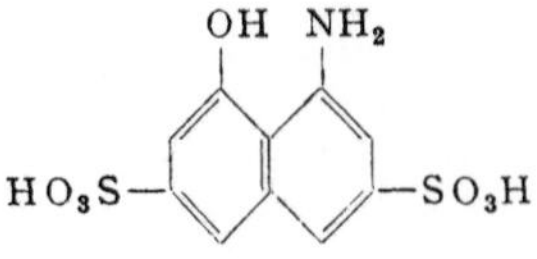

(I) 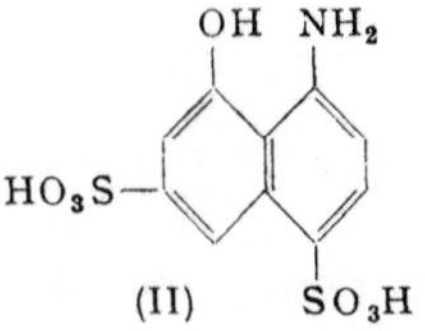   (II) SO_3H

Both acids show the color reaction with sodium 1,2-naphthoquinone-4-sulfonate (see p. 153) but they can be differentiated through their behavior toward nitrosoantipyrine in glacial acetic acid. Only K-acid reacts yielding a blue product. Condensation with loss of sulfuric acid probably occurs with formation of an indamine dye:

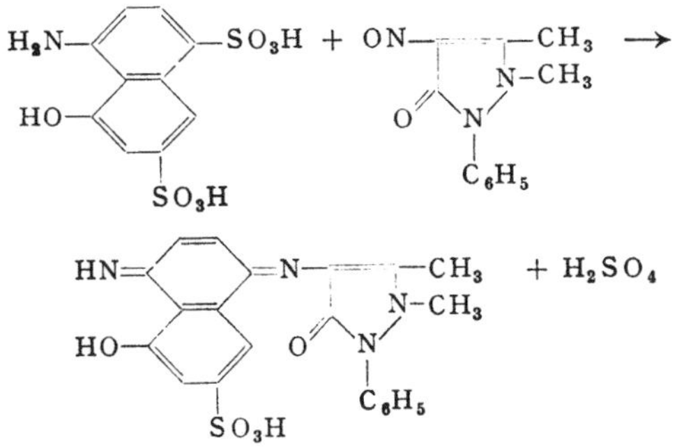

Procedure. A drop of the aqueous test solution or a grain of the sample moistened with water is placed in a micro test tube, or micro crucible, or on a spot plate, and three drops of the reagent solution are added. If K-acid is present, a blue color slowly develops in the cold, and more quickly on warming. A blank test is advisable if small amounts of K-acid are suspected.

Reagent: 2–3 mg of nitrosoantipyrine are dissolved by warming with 10 ml glacial acetic acid.

Limit of identification: 2 γ K-acid.

1 V. ANGER, unpublished studies.

20. Thiouric acids[1]

The structural formulae of the isomers 6-thiouric acid (I) and 8-thiouric acid (II), which have no well-defined melting points,

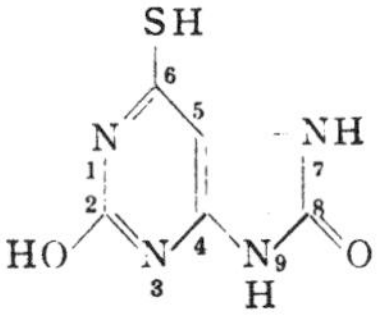

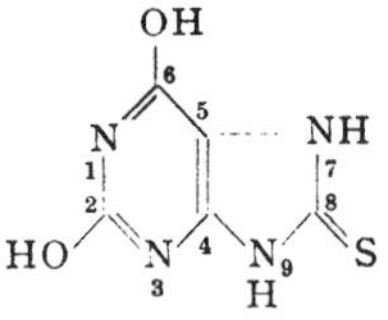

show that (I) is a thiol whereas (II) is a thionic compound. The SH and CS groups, and therefore the isomers, can be differentiated through characteristic spot tests for these groups, described on pp. 222 and 84.

Limit of identification: 5 γ 6-thiouric acid.

For the recognition of the CS group in the acetone-insoluble 8-thiouric acid, it is necessary to heat the sample with conc. $KMnO_4$-solution for 20–30 min. in order to obtain the Wohlers effect. 200 γ of (II) could be identified.

1 Unpublished studies, with S. YARIV.

21. Methylindoles[1]

2-Methylindole, like all indole derivatives with a free β-position, couples in alkaline solution with diazotized p-nitroaniline. The resulting azo dyestuff reacts in its aci-nitro form with magnesium oxide to give a violet lake. 3-Methylindole (skatole) does not react under these conditions.

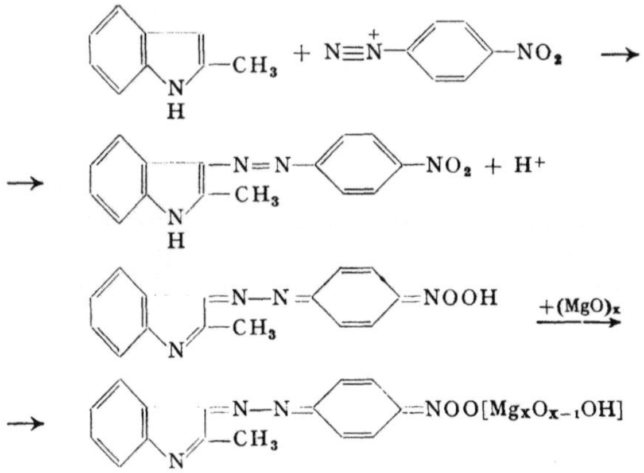

Procedure.   As in test for p-cresol (see p. 420).
Limit of identification: 2 γ 2-methylindole.

1 F. FEIGL and V. ANGER, *Z. Anal. Chem.*, 181 (1961) 165.

22. Quinoline and isoquinoline

These isomers with the boiling points 238° and 243° respectively can be detected and differentiated by the tests outlined and described on pp. 525 and 526 in Chap. 5.

23. Pyridinemonocarboxylic acids

Selective tests are available for each of the three isomeric water-soluble acids and they may be differentiated through their positive response. The tests are:

Picolinic acid: color reaction with iron(II) sulfate (see p. 386).
Identification limit: 5 γ.

Nicotinic acid: a color reaction for pyridine derivatives is described in Chap. 4, Sect. 58. The following modification[1] makes the test selective for nicotinic acid.

The reaction is conducted in a micro test tube with heating in boiling water for about 1 min. followed by cooling. A red or pink color is a positive response.
Identification limit: 0.5 γ.

Isonicotinic acid: color reaction with sodium pentacyanoammineferroate (see p. 291)[2].
Identification limit: 100 γ.

The isomeric acids can be detected even in mixtures with each other by means of the above tests.

1 F. Feigl and A. Del'Acqua, *Z. Anal. Chem.*, 204 (1964) 422.
2 E. F. G. Herington, *Analyst*, 78 (1953) 175.

24. Hydrobenzamide and amarine[1]

Hydrobenzamide (I) and its isomer amarine (II), which is formed from (I) by heating to 120°, are condensation products of 3 moles of benzaldehyde and 2 moles of ammonia:

$$C_6H_5-CH=N \atop C_6H_5-CH=N \Big\rangle CHC_6H_5 \qquad\qquad {C_6H_5-CH-NH \atop C_6H_5-CH-N} {\Big\rangle} CC_6H_5$$

(I)   (m.p. 101°)   (II)   (m.p. 130°; as semihydrate 106°)

As can be seen from the structural formulas, only hydrobenzamide contains a $C_6H_5CH=N-$ group. It was pointed out in Chap. 3 that compounds with this group can be detected through their reaction with thiobarbituric acid, which resembles that of benzaldehyde, namely an orange condensation product results on warming in phosphoric acid solution. Hydrobenzamide can be identified through a positive response to this test and thus distinguished from amarine.

Procedure. As given on p. 205.
Limit of identification: 5 γ hydrobenzamide.

This procedure will reveal hydrobenzamide in the presence of any reasonable amount of amarine.

Another means of differentiating these two water-insoluble isomers is based on the finding that amarine, in line with its constitution, is more basic than hydrobenzamide. Both compounds respond positively to the test which involves the precipitation of red nickel dimethylglyoxime when they are brought into contact with nickel dimethylglyoxime equilibrium solution (see p. 109), a test that reveals organic bases. However, this reaction proceeds much more quickly with amarine than with its isomer. The reaction occurs immediately with 0.4–0.5 mg of amarine, whereas hydrobenzamide if it gives any sign of reaction with the equilibrium solution produces a barely visible precipitate of nickel dimethylglyoxime.

A further differentiation between the isomers may be based on the fact that only the heterocyclic amarine is a tertiary amine showing the test described in Chap. 3, Sect. 57. By means of this test it is possible to show that the transformation of hydrobenzamide into amarine occurs after only 10 min. dry-heating.

1 F. FEIGL and E. LIBERGOTT, *Anal. Chem.*, 36 (1964) 132, see also F. FEIGL and A. DEL'ACQUA, *Z. Anal. Chem.*, 204 (1964) 425.

25. Phenylthiocyanate and phenylisothiocyanate

The isomers C_6H_5SCN (b.p. 231°) and C_6H_5NCS (b.p. 220°) are liquids whose boiling points lie so close together that a precise separation by ordinary fractional distillation is not feasible. However, the thiocyanate gives a positive response to the test described on p. 302 that involves reductive cleavage leading to H_2S. Phenylisothiocyanate can be detected by hydrolytic cleavage, as discussed on p. 564.

26. Phenylacetamide and N-phenylacetamide[1]

Phenylacetamide (m.p. 155°) and N-phenylacetamide (acetanilide, m.p. 115°) can be differentiated through hydrolysis with caustic alkali; they yield ammonia and aniline respectively:

$$C_6H_5CH_2CONH_2 + KOH \rightarrow C_6H_5CH_2COOK + NH_3$$

$$CH_3CONHC_6H_5 + KOH \rightarrow CH_3COOK + C_6H_5NH_2$$

The hydrolysis can be accomplished in a micro test tube, whose mouth is covered with a disk of filter paper moistened with Nessler reagent or an aqueous solution of sodium 1,2-naphthoquinone-4-sulfonate.

A boiling water bath is used. Ammonia vapors produce a brown stain on the Nessler reagent paper; aniline vapors yield a violet stain on the other reagent paper.

The test revealed: 20 γ phenylacetamide   5 γ N-phenylacetamide

1 Unpublished studies.

27. Sulfadiazine and sulfapyrazine[1]

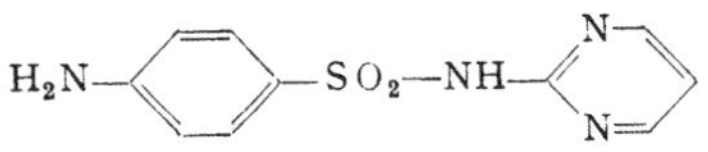

sulfadiazine (I) 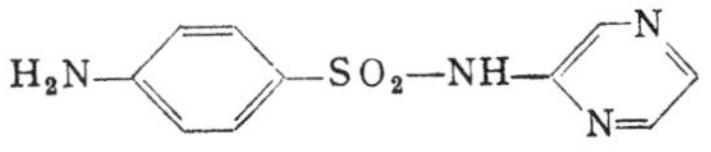   sulfapyrazine (II)

The sulfonamides (I) and (II) differ in the position of the N atoms in the heterocyclic ring. On warming with barbituric or thiobarbituric acid in aqueous solution, only sulfadiazine yields colored (orange or red) condensation products that fluoresce intensely (green or golden yellow) in ultraviolet light. The following condensation seems probable.

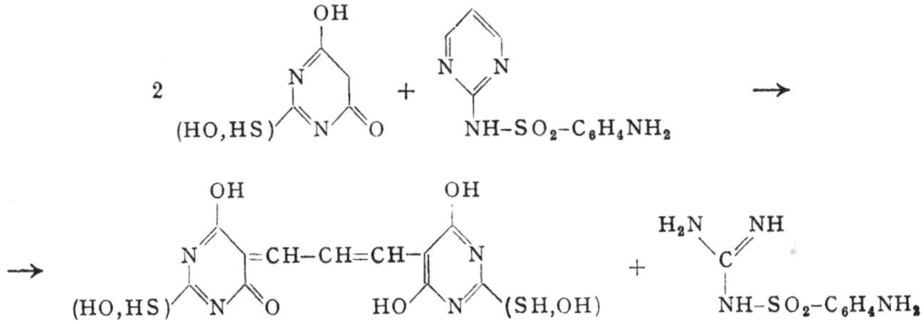

Procedure. As described on p. 645.
Limit of identification: 0.1 γ sulfadiazine.

1 F. FEIGL and V. GENTIL, *Clin. Chim. Acta*, 5 (1960) 153.

28. *p*-Aminoazobenzene and diazoaminobenzene[1]

p-Aminoazobenzene (I) melts at 128° and diazoaminobenzene (II) at 98°. When warmed in aqueous–alcoholic solution with caustic alkali and Devarda alloy, they undergo the following reductive cleavages:

$$C_6H_5—N=N—C_6H_4NH_2 + 4\ H^0 \rightarrow C_6H_5NH_2 + C_6H_4(NH_2)_2 \qquad (1)$$
$$\text{(I)}$$

$$C_6H_5—N=N—NH—C_6H_5 + 6\ H^0 \rightarrow 2\ C_6H_5NH_2 + NH_3 \qquad (2)$$
$$\text{(II)}$$

The above equations show that both of these isomers yield aniline on reduction but ammonia is also produced in the case of diazoaminobenzene. When p-aminoazobenzene, which contains a free NH_2 group, is dry-heated along with hexamine its yields ammonia (see p. 119); this is not true of diazoaminobenzene. All reactions that lead to the release of ammonia (detectable by means of acid–base indicator paper) can be carried out within the technique of spot test analysis, and accordingly a reliable differentiation of these isomers is feasible.

Limits of identification:

10 γ (I) through loss of NH_3 on pyrolysis with hexamine

10 γ (II) through loss of NH_3 on treatment with caustic alkali and Devarda alloy

The aniline produced in the redox reactions (*1*) and (*2*) can be detected in the gas phase through the color reaction with sodium 1,2-naphthoquinone-4-sulfonate (red-violet color). This color reaction also gives a direct indication of the presence of p-aminoazobenzene but not of the isomeric diazoaminobenzene.

If diazoaminobenzene is heated a few degrees above its melting point, it yields nitrogen and aniline.[2] This decomposition can plausibly be represented:

$$4\,C_6H_5—N=N—NH—C_6H_5 \rightarrow 3\,C_6H_5—C_6H_5 + 5\,N_2 + 2\,C_6H_5NH_2$$

If the heating is carried out in a micro test tube immersed in a glycerol bath at 150° and if a disk of filter paper moistened with the sulfonate solution is placed across the mouth of the tube, as little as 5 γ diazoaminobenzene can be detected through the development of a violet stain on the paper. p-Aminoazobenzene does not yield aniline under these conditions.

1 Unpublished studies, with E. LIBERGOTT.
2 F. HEUSLER, *Ann.*, 260 (1890) 231.

29. Benzamidine and benzalhydrazone[1]

The isomers benzamidine (I) and benzalhydrazone (II) yield different products when they undergo acid hydrolysis:

$$C_6H_5C(NH_2)=NH + H_2O + H^+ \rightarrow C_6H_5CONH_2 + NH_4^+$$
$$(I)$$

$$C_6H_5CH=N—NH_2 + H_2O + H^+ \rightarrow C_6H_5CHO + NH_2NH_3^+$$
$$(II)$$

Accordingly, only benzalhydrazone (m.p. 16°), shows the behavior of benzaldehyde on warming with thiobarbituric acid and concentrated phosphoric acid; an orange condensation product results (see p. 205). Benzamidine (m.p. 80°), in contrast to its isomer, is a distinctly basic compound with no reducing properties toward ammoniacal silver solutions.

A good differentiation between benzamidine and benzalhydrazone can be based on the fact that only the first shows the test for amidines, as described in Chap. 3, Sect. 66.

1 F. FEIGL and E. LIBERGOTT, *Anal. Chem.*, 36 (1964) 133.

30. Phenylhydrazine-4-sulfonic acid and phenylhydrazine-N-sulfonic acid[1]

The stable alkali salts of the isomeric phenylhydrazine-4-sulfonic acid (I) and phenylhydrazine-N-sulfonic acid (II):

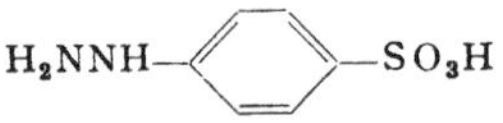

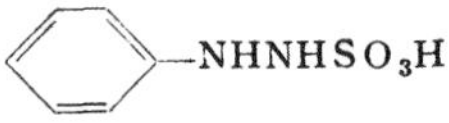

can be distinguished from each other through their behavior when warmed with caustic alkali and Raney nickel alloy. Only (I) is desulfurized with formation of nickel sulfide (comp. test for aromatic sulfonic acids p. 234). The sulfonic acid (II) is cleaved on warming with dilute hydrochloric acid to yield phenylhydrazine chloride and sulfuric acid, so that barium sulfate precipitates at once if barium chloride is present. Within the technique of spot test analysis, it is possible to detect 10 γ of (I) and 25 γ of (II) by means of these procedures.

Hydrazine derivatives of aromatic mono- and disulfonic acids behave in the same manner as (I).

1 Unpublished studies.

31. Ammonium thiocyanate and thiourea

The isomers ammonium thiocyanate and thiourea can be distinguished and also detected in each other's presence through their behavior when warmed with lead hydroxide (or PbO).[1] The reactions are:

$$2\ NH_4CNS + Pb(OH)_2 \rightarrow Pb(CNS)_2 + 2\ NH_3 + 2\ H_2O$$

$$CS(NH_2)_2 + Pb(OH)_2 \rightarrow PbS + NC-NH_2 + 2\ H_2O$$

Accordingly, the evolution of ammonia (detectable with Nessler reagent or acid–base indicator paper) and the leaving of a colorless (yellow) residue are characteristic for ammonium thiocyanate, while the formation of black lead sulfide is characteristic for thiourea. Minimal amounts of the solid specimens or a drop of their solutions are adequate for these identifications. It is only necessary to mix the solid or a drop of the solution with lead hydroxide in a micro test tube and then to heat in boiling water. The occurrence of the reaction will be distinctly visible at once or within 2–3 min.

Limits of identification: 2 γ ammonium thiocyanate   5 γ thiourea

It should be pointed out that the color reaction with iron(III) salts is not reliable for ammonium thiocyanate in the presence of much thiourea because the latter yields a red that gradually fades away. (The reaction may involve formamidine disulfide formed by oxidation of thiourea.) The precipitation of CNS^- ions by Ag^+ ions is not applicable either since white $CS(NH_2)_2.AgNO_3$ is precipitated from acid solutions of thiourea by excess silver nitrate.

The procedure given here will demonstrate that on heating to 160° these isomers undergo a mutual interconversion leading to an equilibrium mixture.[2] This effect is detectable after as little as 15 min. heating.

1 F. FEIGL and E. LIBERGOTT, *Mikrochim. Acta*, (1964) 1111.
2 Comp. BEILSTEIN, *Handbuch Org. Chemie. Vol. 3*, (1921) 181.

32. Nitrosodiphenylamines[1]

The isomers N-nitrosodiphenylamine (I) and *p*-nitrosodiphenylamine (II)

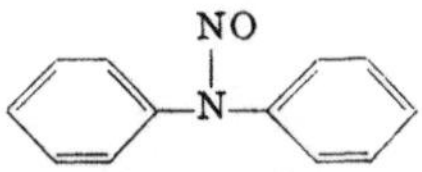

(I) 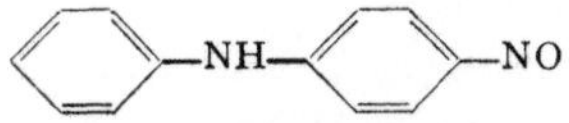   (II)

as well as their nuclear-substituted derivatives give a blue color when treated with concentrated sulfuric acid in the cold or on gentle warming. These color reactions result from the saponification of (I) and (II) by the water in the concentrated sulfuric acid to yield diphenylamine and nitrous acid. The latter oxidizes diphenylamine to blue quinoidal compounds (compare the nitrite and nitrate tests, *Inorganic Spot Tests*, Chap. 4).

The color reaction is specific for nitrosodiphenylamines. No other N- or C-nitroso compound reacts in an analogous manner.

Procedure.   A little of the solid is taken or a drop of the test solution is evaporated to dryness in a micro crucible. A drop of concentrated sulfuric acid is added and the contents of the crucible warmed gently, if necessary. A blue color shows the presence of nitrosodiphenylamine.

Limit of identification: 0.5 γ N- or *p*-nitrosodiphenylamine.

It is possible to differentiate between these two isomeric nitrosodiphenyl-amines by taking advantage of the fact that N-nitroso compounds, in contrast to C-nitroso compounds, are easily denitrosated by hydrazoic acid (see p. 293). If sodium azide and dilute hydrochloric acid are added to N-nitrosodiphenylamine, the reaction

$$(C_6H_5)_2N—NO + HN_3 \rightarrow (C_6H_5)_2NH + N_2 + N_2O$$

occurs almost immediately. Consequently, if the foregoing test gives a positive result, a new portion of the sample is taken to dryness with several cg of sodium azide and a drop of dilute hydrochloric acid. If the residue gives no reaction when treated with concentrated sulfuric acid, or if it yields a distinctly weaker blue color than the same amount of the untreated sample, it may be concluded that the nitrosamine (I) is present.

On the other hand, the C-nitroso compound (II) may be detected with high sensitivity through the test for *p*-nitroso aromatic amines outlined on p. 395.

1 Unpublished studies, with Cl. Costa Neto.

33. Dinitrobenzenes[1]

Compounds that function as hydrogen donors in weak alkaline solution, namely ascorbic acid, sugars, polyphenols, *etc.*, reduce *o*- and *p*-dinitro-benzene to blue-violet and orange water-soluble alkali salts of quinoidal compounds, respectively. (Compare the detection of reductants, p. 133.) No color reaction is given by *m*-dinitrobenzene under these conditions. If the action of the hydrogen donors occurs at particular pH values, the three isomeric dinitrobenzenes differ from each other in characteristic fashion. This is evident in the following compilation:

Dinitrobenzene	0.05 N NaOH	2% NaOH	$MgCO_3$
ortho	blue-violet	blue-violet	—
meta	—	—	—
para	red-orange	yellow	red-orange

It is therefore apparent that the color reactions permit a clear differentia-tion of the *o*- and *p*-isomers. The reaction succeeds with even small amounts and in dilute solutions if phenylhydrazine is employed as the reductant.

Procedure. (*o-Dinitrobenzene*) A spot plate is used. A drop of the alcoholic test solution is treated with 1 drop each of a 2% aqueous solution of phenylhydrazine chloride and 1 drop of a 2% solution of NaOH. The mixture is stirred. Depending on the amount of *o*-dinitrobenzene present, a more or less intense blue-violet color appears.

Limit of identification: 0.2 γ *o*-dinitrobenzene.

When a comparison blank was used, this procedure revealed 0.2 γ *o*-dinitrobenzene in one drop of a saturated alcoholic solution of *p*-dinitrobenzene.

Procedure. (*p-Dinitrobenzene*) 0.01–0.02 g of $MgCO_3$ is placed in a depression of a spot plate and followed (without stirring) by 1 drop each of the alcoholic test solution and 1 drop of a 2% aqueous solution of phenylhydrazine chloride. The mixture is kept at 110–120° for 2–5 min. in a drying oven. A more or less intense orange-colored residue is left, if *p*-dinitrobenzene is present.

Limit of identification: 1 γ *p*-dinitrobenzene.

This identification limit holds likewise for the detection of *p*-dinitrobenzene in a saturated alcoholic solution of *o*-dinitrobenzene.

The isomeric *m*-dinitrobenzene can be detected by the procedure described on p. 397.

1 F. Feigl, J. R. Amaral and V. Gentil, *Mikrochim. Acta*, (1957) 727.

34. Chloronitrobenzenes[1]

Chloronitrobenzenes (*o*- and *p*-) have technical importance. They are obtained by nitrating chlorobenzene and are separated by fractional distillation. Consequently, a rapid method of differentiation is desirable when determining their purity and the procedure given here serves this purpose. The method is based on the ready reduction of these isomeric NO_2 compounds in sulfuric acid solution through Devarda's alloy to the respective NO compounds and the different behavior of the latter. Only the *para*-nitroso compound condenses in strong sulfuric acid to yield 4-nitroso-4'-chlorodiphenyl-hydroxylamine, whereby hypochlorous acid is formed:

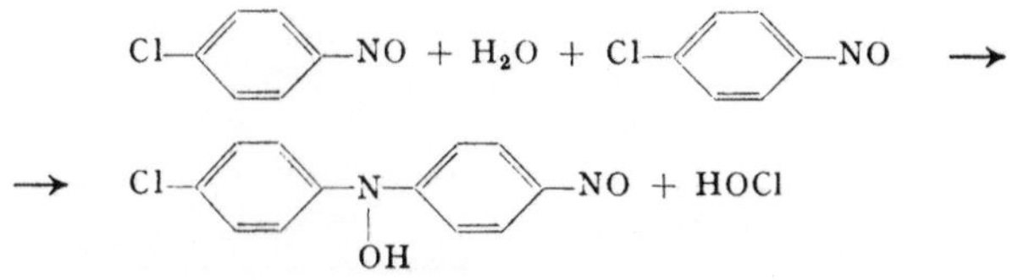

The *ortho*-isomer likewise condenses to yield a Cl and NO substituted hydroxylamine but without the participation of water and with no splitting-out of hypochlorous acid. The occurrence of the above reaction and hence the presence of p-chloronitrobenzene can be recognized through the detection of the resulting hypochlorous acid by means of N,N'-diphenylbenzidine. This base is oxidized in strong sulfuric acid to a blue quinoidal compound (compare p. 301).

Procedure. A micro test tube is used. A drop of the test solution or a tiny amount of the solid material under test is united with 0.5 ml of diphenylbenzidine–sulfuric acid reagent and 1 drop of an aqueous suspension of Devarda's alloy. The mixture is placed in boiling water. If p-chloronitrobenzene is present, a blue color appears.

Reagents: 1) Diphenylbenzidine–sulfuric acid. 1 mg base dissolved in 10 ml 86% sulfuric acid.

2) Reductant. 10 mg Devarda's alloy (finely granulated) suspended in 10 ml of water. The suspension should be shaken before use.

Limit of identification: 5 γ p-chloronitrobenzene.

1 V. ANGER, *Mikrochim. Acta*, (1960) 828.

35. Nitrophenols[1]

o- and p-Nitrophenols can be detected and differentiated through the formation of the corresponding yellow alkali salts under suitable conditions*.

Procedure I. o-Nitrophenol (m.p. 45°, b.p. 214°), when heated to 120° for 5 min. sublimes sufficiently to form vapors which turn filter paper moistened with alkali hydroxide yellow. The m- and p-isomers do not react under these conditions.

Limit of identification: 1.5 γ o-nitrophenol.

Procedure II. p-Nitrophenol (m.p. 113°) has a stronger acidic nature than its isomers. Therefore it reacts nearly instantaneously with fused hydrated sodium thiosulfate at 120°, whereby a yellow melt is formed. When the latter, after cooling, is dissolved in several drops of water and extracted with benzene, a yellow aqueous layer remains. This behavior is not shown by the isomers, even in large amounts.

Limit of identification: 2.5 γ p-nitrophenol.

The above tests can be carried out in a micro test tube with minute quantities of the solid material or with a drop of its solution in alcohol or other organic liquids.

Procedures I and II revealed 2.5 γ o-nitrophenol and 5 γ p-nitrophenol in the presence of 2000 γ of the corresponding isomers.

* See p. 561 regarding the detection of m-nitrophenol.

1 F. FEIGL and D. HAGUENAUER-CASTRO, *Chemist-Analyst*, 49 (1960) 43.

36. Nitrobenzaldehydes[1]

p-Nitrobenzaldehyde. For the detection of this isomer see p. 562.

o-Nitrobenzaldehyde. This compound can be detected selectively in the presence of the other isomeric nitrobenzaldehydes through its reaction with acetone in alkaline solution. Indigo results:

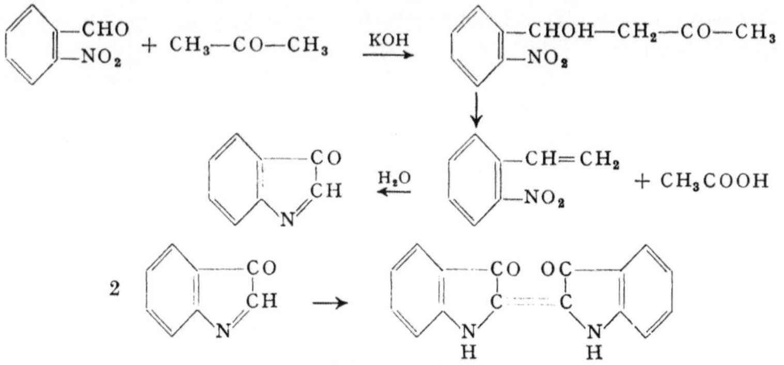

Procedure. A drop of the test solution is united in a micro test tube with 2 drops of a saturated solution of potassium hydroxide in acetone. This mixture is gently heated. A positive response is indicated by a blue color.

Limit of identification: 50 γ o-nitrobenzaldehyde

1 F. FEIGL and V. ANGER, *Z. Anal. Chem.*, 181 (1961) 169.

37. Isomeric methyl derivatives of salicylaldoxime and constitution of copper salicylaldoxime

The isomeric compounds I and II may be differentiated through spot

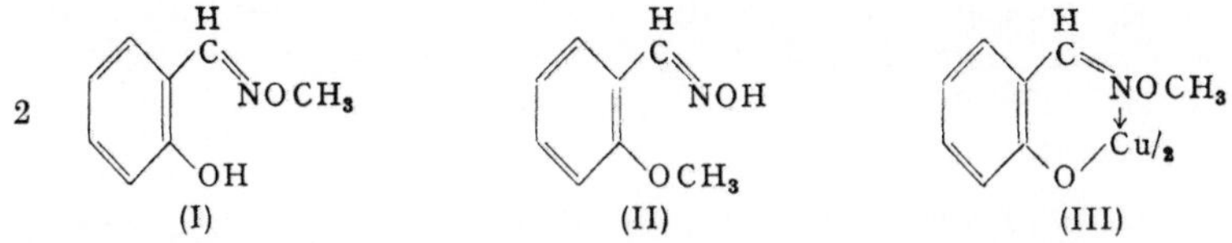

reactions of their alcoholic solutions with copper acetate. Only I gives a brown precipitate due to the formation of the inner-complex salt III, thus indicating that the phenolic OH group and not the NOH group is salified.[1]

The parent compound salicylaldoxime containing acidic OH and NOH groups is a selective precipitant[2] for copper in acid solution, acting here as a monobasic acid. The inner-complex copper salt formed—in accordance with the behavior of I and II—must contain the chelate ring as shown in III.

This assumption was recently confirmed through the isolation of the inner-complex copper salt of the phenylhydrazone of salicylaldehyde, which contains the same chelate ring. The precipitation of this salt from ammoniacal solution permits the gravimetric determination of copper and its separation from cadmium.[3]

1 F. FEIGL and A. BONDI, *Ber.*, **64** (1931) 2819.
2 F. EPHRAIM, *Ber.*, **63** (1930) 1928; **34** (1931) 1210.
3 P. UMPATHY and N. APPALA RAJU, *Ind. J. Chem.*, **2** (1964) 248.

38. Homologous aliphatic aldehydes

Formaldehyde can be detected through the test with chromotropic acid as described on p. 434.

Acetaldehyde, among its homologs, is revealed by means of sodium nitroprusside and piperidine or morpholine (see p. 438).

Propionaldehyde reacts with ninhydrin in concentrated sulfuric to yield a red-brown compound.[1] This interesting and highly selective test will reveal 3 γ propionaldehyde.

1 L. R. JONES and J. R. RIDDICK, *Anal. Chem.*, **26** (1954) 1035.

39. Homologous aliphatic carboxylic acids

Formic acid can be detected in the presence of its homologs by means of the tests given in Chap. 5, Sect. 42.

Among its homologs, the test for acetic acid with *o*-nitrobenzaldehyde is specific (see p. 455).

Propionic acid can be differentiated from acetic and other homologous carboxylic acids by making two tests. While it does not give the reaction with *o*-nitrobenzaldehyde, propionic acid does give a positive response to the test with lanthanum nitrate described on p. 454. If acetic acid is absent, this reaction is characteristic for propionic acid.

40. Differentiation of aniline and benzylamine

The boiling point of these homologous compounds is identical (184°). They can be distinguished from each other through the behavior toward sodium 1,2-naphthoquinone-4-sulfonate. Aniline yields a red condensation product, whereas that obtained with benzylamine is brown-violet.

Procedure. As given in Sect 14.

Limit of identification: 1 γ aniline   5 γ benzylamine

41. Cyanamide and dicyanodiamide[1]

A method for differentiating between cyanamide and its dimer dicyanodiamide is based on their behavior with ammonia. Cyanamide yields guanidine whereas dicyanodiamide produces biguanide:[2]

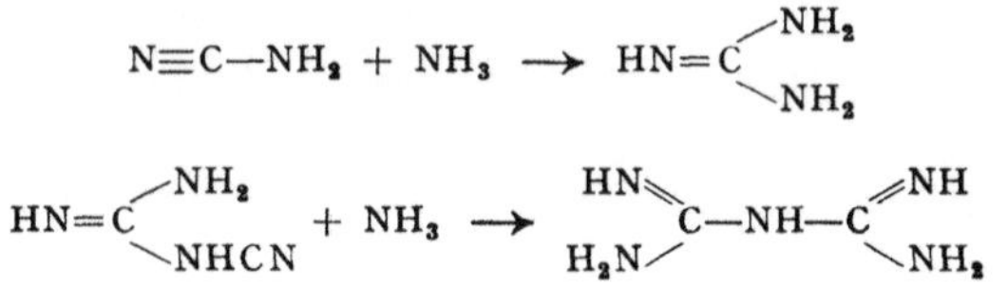

$$N{\equiv}C{-}NH_2 + NH_3 \longrightarrow HN{=}C\diagdown{\diagup}^{NH_2}_{NH_2}$$

$$HN{=}C\diagup^{NH_2}_{NHCN} + NH_3 \longrightarrow \underset{H_2N}{\overset{HN}{\diagdown\diagup}}C{-}NH{-}C\overset{NH}{\underset{NH_2}{\diagup\diagdown}}$$

These addition reactions proceed quantitatively when the respective
amides are evaporated with ammonia water. The resulting products differ
in characteristic fashion when heated to 140–150°. Biguanide remains unaltered, while guanidine loses ammonia and yields biguanide:

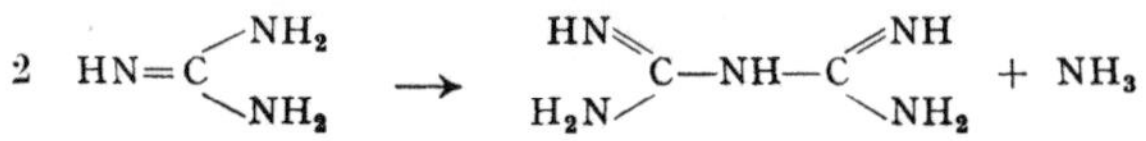

$$2\ \ HN{=}C\diagup^{NH_2}_{NH_2} \longrightarrow \underset{H_2N}{\overset{HN}{\diagdown\diagup}}C{-}NH{-}C\overset{NH}{\underset{NH_2}{\diagup\diagdown}} + NH_3$$

Accordingly, cyanamide can be detected by evaporating with ammonia
followed by dry-heating of the residue. Ammonia is given off and is detected
by means of acid–base indicator paper or with Nessler reagent. Salts of
cyanamide and their N-monosubstituted derivatives behave like the parent
compounds.

Procedure. A small amount of the solid or a drop of its solution is treated
in a micro test tube with a drop of concentrated ammonia water. The mixture is
taken to dryness (complete elimination of the excess ammonia is essential). The
mouth of the tube is covered with a piece of filter paper moistened with Nessler
reagent. The tube is immersed in a glycerol bath preheated to 120° and the
temperature is raised to 150°. The appearance of a brown stain on the reagent
paper indicates the evolution of ammonia. This may be distinct even at 130°. For
the detection of dicyanodiamide by means of diacetyl, comp. p. 263.

1 F. FEIGL and E. LIBERGOTT, *Chemist-Analyst*, 53 (1964) 37.
2 P. KARRER, *Lehrbuch der organischen Chemie,* 13th ed., Georg Thieme Verlag, Stuttgart, 1959, pp. 257, 258.

42. Symmetric and asymmetric diphenylurea

These two isomers (I, II) contain different functional groups for which appropriate tests have been given in Chap. 3.

$$O C \begin{array}{c} \diagup NHC_6H_5 \\ \diagdown NHC_6H_5 \end{array} \qquad\qquad O C \begin{array}{c} \diagup N(C_6H_5)_2 \\ \diagdown NH_2 \end{array}$$

I (m.p. 239°)   II (m.p. 189°)

As shown by (I), *sym*-diphenylurea (carbanilide) is the dianilide of carbonic acid. It was pointed out (p. 261) that anilides of aliphatic carboxylic acids when dry-heated with $MnSO_4.4 H_2O$ undergo pyrohydrolysis and yield aniline that can be detected in the gas phase through the color reaction with *p*-dimethylaminobenzaldehyde, or through the color reaction with sodium 1,2-naphthoquinone-4-sulfonate. (*Idn. Lim.*: 10 γ).

In line with its structure (II), *asym*-diphenylurea is a derivative of diphenylamine. As such it yields a blue color with concentrated sulfuric acid and alkali nitrate or nitrite (*Idn. Lim.*: 10 γ).

The tests mentioned here permit also the differentiation of *sym*- and *asym*-diphenylguanidine.

43. Phenyl- and *sym*-diphenylthiourea[1]

These compounds having the same melting point (154°) can be quickly distinguished from each other by heating above their melting points. Phenylthiourea splits off ammonia as a result of condensation to ω,ω'-diphenyldithiobiuret:

$$2 S C \begin{array}{c} \diagup NH_2 \\ \diagdown NHC_6H_5 \end{array} \longrightarrow \begin{array}{c} S C-NH-C S \\ | \qquad\quad | \\ C_6H_5NH \qquad HNC_6H_5 \end{array} + NH_3$$

Since *sym*-diphenylthiourea has no free NH_2 group it is not capable of splitting off ammonia. The above condensation is accompanied by the loss of hydrogen sulfide that leads to the formation of phenylcyanamide:

$$S C \begin{array}{c} \diagup NH_2 \\ \diagdown NHC_6H_5 \end{array} \longrightarrow H_2S + N\equiv C-NHC_6H_5$$

An analogous loss of hydrogen sulfide leading to diphenylcyanamide occurs likewise with *sym*-diphenylthiourea but only at higher temperatures.

The pyrolytic loss of ammonia and hydrogen sulfide can be accomplished with very small amounts of the sample or with the evaporation residue from a drop of the test solution. The micro test tube is heated in a glycerol bath. The ammonia is revealed by the development of a brown or yellow stain on filter paper moistened with Nessler reagent. Lead acetate paper used in similar fashion will serve to reveal the evolution of hydrogen sulfide.

Limit of identification: 1 γ phenylthiourea.

Phenylurea (m.p. 147°) can be distinguished from *sym*-diphenylurea (m.p. 238°) by pyrolytic splitting off of ammonia.

1 Unpublished studies, with A. DEL'ACQUA.

44. *p*-Phenylenediamine and benzidine[1]

p-Phenylenediamine as well as benzidine form orange-red Schiff bases with *p*-dimethylaminobenzaldehyde. These condensation products are stable toward diluted hydrochloric acid or sulfuric acid, whereas the colored Schiff bases of primary aromatic amines are decomposed. It is highly probable that this peculiar behavior of the diamines is due to the fact that both NH$_2$ groups may condense with *p*-dimethylaminobenzaldehyde. Nevertheless, *p*-phenylenediamine can be detected in the presence of benzidine when a solution of small amounts of *p*-dimethylaminobenzaldehyde in diluted sulfuric acid is used as reagent. Under these conditions insoluble benzidine sulfate is formed, which does not show the color reaction.

Procedure. Filter paper is moistened with a 1% solution of *p*-dimethyl-aminobenzaldehyde in 2N sulfuric acid. If the reagent paper is spotted with a drop of the alcoholic or ethereal solution of *p*-phenylenediamine, a red stain appears within about 1 min.

Limit of identification: 5 γ *p*-phenylenediamine in the presence of 2000 γ benzidine.

1 Unpublished studies, with V. GENTIL.

45. Mono- and di-alkylanilines

A convenient means of distinguishing between all monoalkyl and dialkyl compounds or their anhydrous salts is to heat with hexamine[1] to 100°. The mono compounds yield ammonia whereas the dialkyl compounds are unaltered. It may be assumed that the monoalkylaniline disproportionates:

$$2\ C_6H_5NHR \rightarrow C_6H_5NRR + C_6H_5NH_2$$

and that the resulting aniline condenses with hexamine (see p. 119). Accordingly, when this test is used it is essential to remove any aniline that may be present. This is readily accomplished by warming with nitrous acid (alkali nitrite plus mineral acid). The aniline is thus converted into phenol. If the reaction mixture is then made basic and extracted with ether, the latter takes up the unchanged alkylaniline derivatives. The phenol must be removed because phenols likewise condense with hexamine (see p. 181).

Procedure. A drop of the ethereal solution of the free base is mixed in a micro test tube with excess hexamine and the mouth of the tube is covered with a disk of filter paper moistened with Nessler solution. The mixture is heated in a boiling water bath. A brown-yellow stain indicates the presence of monoalkylaniline.

Anhydrous salts of these bases may likewise be tested. Acid–base indicator papers can be employed instead of Nessler paper.

Limit of identification: 50 γ N-methyl(ethyl)aniline.

1 Unpublished studies.

46. Constitution of the reaction product from thiosemicarbazide and nitrous acid[1]

Thiosemicarbazide and nitrous acid yield a well crystallized compound[2] with the empirical formula CH_2N_4S. Different structures have been suggested, namely aminothiotriazane (I)[3] and thiocarbamic acid azide (II).[4]

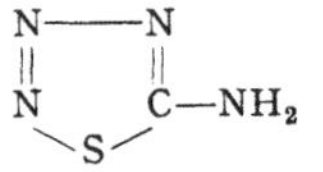

Organic spot test analysis provides a simple and reliable means of determining which of these proposals is correct by applying the iodine–azide reaction (see p. 219). Since the response was negative the compound cannot be thiocarbamic acid azide, and apparently the compound is aminothiotriazane. A further confirmation for the structure (I) is the positive response to the test for tertiary amines described on p. 251.

1 V. ANGER, *Oesterr. Chem. Ztg.*, **62** (1961) 352.
2 M. FREUND and H. SCHINDLER, *Ber.*, **29** (1896) 2502.
3 BEILSTEIN, *Handbuch der organischen Chemie*, 4th ed., Vol. 27 (1937) 781.
4 E. OLIVERIA-MANDALA, *Gazz. Chim. Ital.*, **44** I (1914) 672.

47. Constitution of 6-nitroquinoline

A specific gravimetric method for determining palladium[1] is based on the

precipitation from acid solution of yellow $Pd(C_9H_5N_2O_2)_2$ by 6-nitroquino-
line. According to the composition of this product, 6-nitroquinoline functions
as a monobasic acid, a conclusion that is impossible to justify if 6-nitro-
quinoline is assigned the structure (I). Therefore it was assumed[2] that 6-
nitroquinoline functions as precipitant in the isomeric nitrosophenol
form (II) yielding the phenolic palladium salt (III):

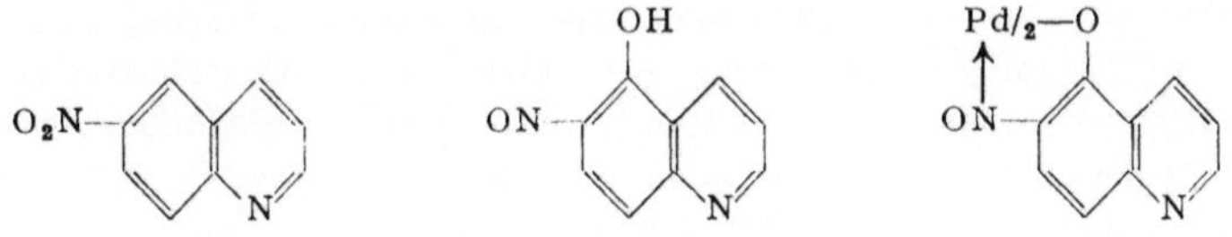

This assumption is supported by the finding that hydrogen cyanide is
evolved if 6-nitroquinoline is dry-heated to 150° with mercuric cyanide. This
effect is characteristic for acidic compounds (see p. 147) including those that
exhibit no acidic character when subjected to other tests. The foregoing
evidence leads to the conclusion that "6-nitroquinoline" has the structure (II) or may react in this form.

1 S. C. OGBURN and A. H. RIESMAYER, *J. Am. Chem. Soc.*, 50 (1928) 3018.
2 F. FEIGL, *Chemistry of Specific, Selective and Sensitive Reactions*, Academic Press,
   New York, 1949, p. 284.

48. Constitution of taurine and sulfanilic acid

In all aminocarboxylic acids the basic character of NH_2 groups and the
acidic character of COOH groups are much less pronounced than in the
parent compounds that lack the NH_2- or COOH groups. However, all
aminocarboxylic acids behave as acids when they are pyrolyzed in mixture
with mercuric cyanide as indicated by the evolution of hydrogen cyanide
(see p. 147). This test evokes no positive response when applied to the
aminosulfonic acids taurine (I) and sulfanilic acid (II). This fact points
to the presence of the respective isomeric forms (Ia) and (IIa) in tautomeric
equilibrium with (I) and (II); these ammonium salt forms are inert toward
mercuric cyanide.

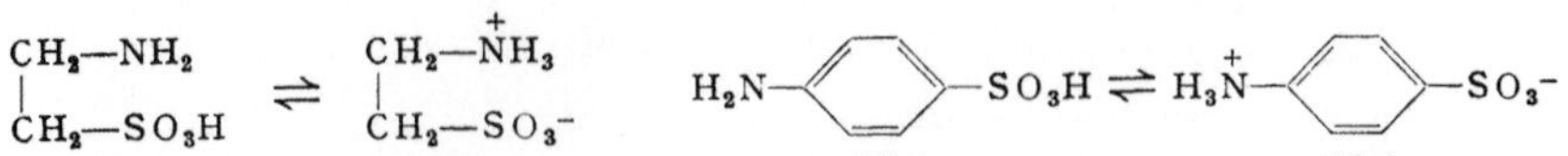

The amount of the amino acid forms (I) and (II) present in the equilibria
is obviously so small that no pyrolytic reaction with the mercuric cyanide

occurs. In this respect, the SO_3H group is therefore masked because of its inner-molecular bonding with the NH_2 group, as shown in the formulations (Ia) and (IIa). Experiences in inorganic analysis have shown that all maskings include the possibility of demasking through disturbance (shifting) and restoration of the equilibria prevailing during the maskings.[1] A demasking of this kind can be accomplished in the cases of taurine and sulfanilic acid by evaporation with formaldehyde (preferably paraformaldehyde). The NH_2 groups in (I) and (II) are methenylated through this action and consequently the now free SO_3H groups are able to enter into the pyrolysis reaction with mercuric cyanide. This is the basis of a test for taurine described on p. 507.

Heating with hexamine also brings about a demasking. In this case the SO_3H group is salified by the decidedly basic hexamine and the freed NH_2 group is methenylated by the hexamine with release of ammonia. The masking and demasking of NH_2- and SO_3H groups discussed above are readily detected in the case of taurine and sulfanilic acid through the negative or positive responses to spot tests.

It should be noted that ammonium chloride and nitrate readily yield hydrogen cyanide when pyrolyzed with mercuric cyanide and accordingly have acidic character. The nonreactivity of taurine and sulfanilic acid demonstrates that these amino acids because of the inner-molecular bonding of the NH_2- and SO_3H groups are more nearly neutral ammonium salts than the salts of ammonia with volatile mineral acids.

1 F. FEIGL, *Chemistry of Specific, Selective and Sensitive Reactions*, Academic Press, New York, 1949, Chap. IV.

49. Constitution of dithiooxalic acid

Dithiooxalic acid has been represented either as the thionic compound (I) or as the isomeric thiolic compound (II):

```
S C—OH            O C—S H
  |                 |
S C—OH            O C—S H
  (I)               (II)
```

These same formulations hold for the alkali salts of this thiocarboxylic acid. If constitution (I) is valid, the alkali salts should respond positively to the test for thionic compounds (containing CS groups) as described on page 84. This test is based on the occurrence of the Wohlers effect when the test material is treated with potassium permanganate containing barium chloride. Hereby violet-tinted $BaSO_4$ is formed. However, the response to this test is negative and consequently there is justification for accepting formulation (II) for dithiooxalic acid and its salts.

50. Constitution of melamine

The tautomeric structures (I) and (II) are possibilities for representing the constitution and modes of reaction of melamine that is employed for the production of synthetic resins with formaldehyde.

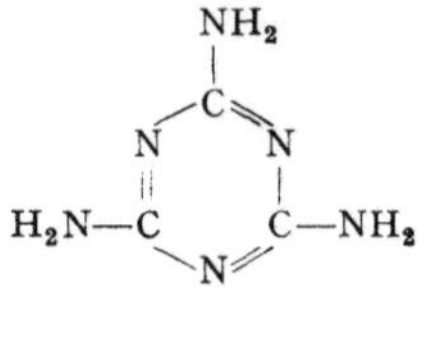

$$
\begin{array}{cc}
\text{NH}_2 & \text{NH} \\
| & \| \\
C & C \\
\diagdown & \diagup \diagdown \\
N \quad\quad N & HN \quad\quad NH \\
\| \quad\quad | & | \quad\quad | \\
H_2N\!-\!C \quad\quad C\!-\!NH_2 & HN\!=\!C \quad\quad C\!=\!NH \\
\diagdown N \diagup & \diagdown N \diagup \\
& H \\
(I) & (II)
\end{array}
$$

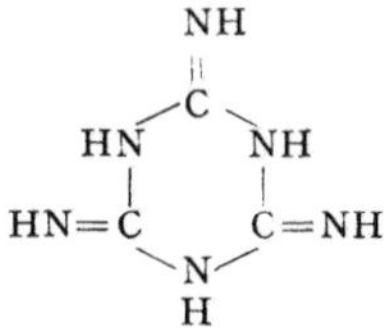

The test for melamine given on p. 532 speaks for (I). Additional supporting evidence for (I) is the positive response obtained with melamine to the test for tertiary amines described on p. 251.

51. Constitution of thiourea

Although thiourea is known only in the thionic form (I) it sometimes reacts in the tautomeric sulfhydryl form (II):

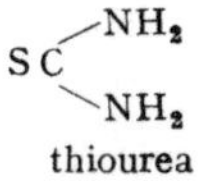

$$
\begin{array}{cc}
\quad\diagup NH_2 & \quad\diagup NH \\
SC \quad\quad (I) & HSC \quad\quad (II) \\
\quad\diagdown NH_2 & \quad\diagdown NH_2 \\
\text{thiourea} & \text{iso(pseudo)thiourea}
\end{array}
$$

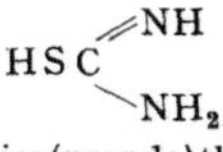

This is exhibited in its behavior toward potassium permanganate. In neutral solution due to the positive response to the test for the thionic group, the oxidation yields urea and sulfuric acid (see p. 84) whereas in acid solution, due to the SH group, the product is the disulfide

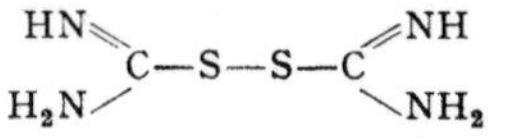

$$
\begin{array}{c}
HN\diagdown \quad\quad\quad\quad\quad \diagup NH \\
C\!-\!S\!-\!S\!-\!C \\
H_2N\diagup \quad\quad\quad\quad\quad \diagdown NH_2
\end{array}
$$

The mode of reaction in the iso- (pseudo-) form of thiourea is also shown by the positive response to the redox reaction with ammonium molybdate in the presence of sulfuric acid. Molybdenum blue results. Furthermore, thiourea yields hydrogen sulfide on fusion with benzoin (comp. p. 146), thus demonstrating its behavior as an aliphatic thiol (II).

52. Reversibility of the benzoin condensation[1]

The familiar conversion of benzaldehyde into its dimer benzoin is brought about by the catalytic action of alkali cyanide:

$$2\ C_6H_5CHO \xrightarrow{KCN} C_6H_5CH(OH)COC_6H_5$$

The fact that this condensation leads to an equilibrium and is reversible can be demonstrated through the finding that heating (160°) of benzoin with syrupy phosphoric acid and thiobarbituric acid leads to a yellow to orange condensation product which is characteristic for aromatic aldehydes (see p. 205). In accord with the fact that a catalyst influences only the speed but not the equilibrium of reversible reactions, the response to the test for benzaldehyde is definitely more pronounced when alkali cyanide is added to the mixture. This effect is easily seen despite the volatility of the hydrogen cyanide formed. Milligram amounts of benzoin are sufficient to detect the release of benzaldehyde from its dimer.

1 Unpublished studies, with A. DEL'ACQUA.

53. Identification of an aminophenol structure in a
new rearrangement product[1]

A product was obtained from a new type of rearrangement for which the isomeric structures (I) and (II) had to be taken into consideration.

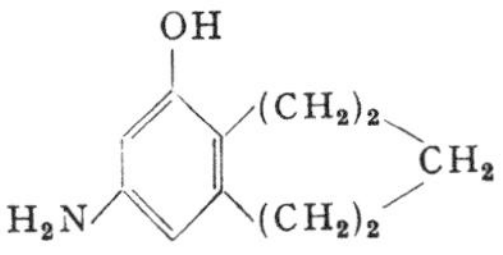

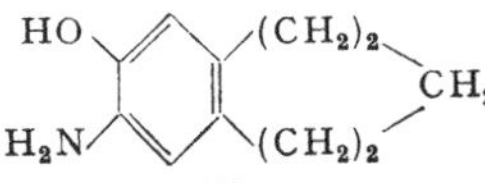

A fast preliminary decision in favor of structure (I) was possible on the basis of a negative response to the test for o-aminophenols described on p. 515. The product gave a weakly but distinctly positive test for m-aminophenols, described on p. 516.

1 M. DVOLAITSKY and A. S. DREIDING, Helv. Chim. Acta, 48 (1965), in press.

54. Differentiation between ar-tetralols and
recognition of a new rearrangement[1]

Among the following ar-tetralols only compounds (I), (III) and (VIII) are phenols with a free para-position and thus show the color reaction with 2,6-dichloroquinone-4-chloroimine described on p. 185.

  (I) 5,6,7,8-Tetrahydronaphthol-1 (m.p. 68°).
 (II) 5,6,7,8-Tetrahydronaphthol-2 (m.p. 61°).
(III) 3-Methyl-5,6,7,8-tetrahydronaphthol-1 (m.p. 98°).
(IV) 4-Methyl-5,6,7,8-tetrahydronaphthol-2 (m.p. 105°).
 (V) 3-Methyl-5,6,7,8-tetrahydronaphthol-2 (m.p. 88°).
(VI) 4-Methyl-5,6,7,8-tetrahydronaphthol-1 (m.p. 88°).
(VII) 1-Methyl-5,6,7,8-tetrahydronaphthol-2 (m.p. 113–114°).
(VIII) 2-Methyl-5,6,7,8-tetrahydronaphthol-1 (m.p. 41°).

In this way it was possible to devise a fast preliminary test for a new rearrangement of compound (II) into (I) and (IV) into (III).

1 W. HOPFF and A. S. DREIDING (Zürich), private communication. See also *Angew. Chem.*, 77 (1965) 717.

Chapter 7

Applications of Spot Reactions in the Testing of Materials, Examination of Purity, Characterization of Pharmaceutical Products, etc.

General remarks

The inspection of materials, control tests during the course of chemical-technological processes, tests of purity of foods and pharmaceutical products, criminal investigations, and researches in the biological sciences often demand a rapid decision as to the presence or absence of a particular material. It has been found that many spot reactions can be usefully applied, either unaltered or with slight modification, for the above purposes and with the consumption of only small amounts of the sample. In addition, it is often of great importance in preparative chemistry, and when studying organic natural products, to be able to determine quickly whether certain organic compounds or members of particular types of compounds have been formed or are present.* The use of spot tests for such special assignments is rather new or at least the literature contains relatively few records along this line. The reason is that spot tests are still far less popular in organic qualitative analysis than in the corresponding inorganic field. However, the future will undoubtedly see a more intensive and extensive employment and development of spot reactions, and it may be safely assumed that the acquired experiences and findings will lead to a further growth of organic spot testing and consequently to its more wide-spread application in the solution of special technical problems. Pertinent examples will be presented in this Chapter, but no claims are made regarding complete coverage of the field.

No discussion of the exceedingly important, sometimes even indispensable applications of organic spot tests in column and paper chromatography has been included. Excellent monographs dealing with this topic are available. These types of chromatography, which are now so important in very many fields, will assuredly make ever-increasing use of sensitive organic spot tests based on the formation of colored reaction products, and especially those

---

* See in this connection H. Lieb and W. Schöniger, *Anleitung zur Darstellung organischer Präparate mit kleinen Substanzmengen*, Springer Verlag, Vienna, 1950.

tests which can be conducted in solutions with an appropriate reagent. To cite a single characteristic example: the iodine–azide reaction (p. 219) was used to detect the sulfur-bearing amino acids separated by paper chromatography.*

Two other remarks are in order here regarding the use of organic spot test analysis in testing materials. The first concerns the fact that sometimes it is possible to make semi-quantitative determinations by means of color reactions carried out in the form of spot tests. Experiences in the inorganic field have demonstrated that fairly exact determinations are possible when spot reactions carried out with the test solutions are compared with the results obtained by the same procedure with standard solutions. This "spot colorimetry", which represents a simple micro or semimicro method, will assuredly also find application in the determination of certain organic compounds or radicals. The second remark relates to the use of spot tests for characterizing pharmaceutical preparations and medicinals, as well as the control examinations of food products. Previous experience leads to the assumption that spot tests are adequate to meet certain requirements of the various pharmacopoeias and accordingly can be considered by the authorities in charge of these important publications. This is all the more justified because spot tests for a considerable number of important organic medicinals and drugs have now been developed for which the pharmacopoeias have given no testing methods**. In addition, certain qualitative procedures, which are prescribed by the pharmacopoeias, may be translated to the technique of spot test analysis. The same is true in the examination of foodstuffs and dyes. In both cases, the resulting economies in time, labor, and material will assuredly prove of advantage. In this connection, attention is called to the application of new tests for nitrogen and volatile phenol in the differentiation of synthetic and fermentation vinegar, and of artificial fibres and resins. These tests require such small amounts of the sample and so little time that they can justly be hailed as veritable triumphs for spot test analysis.

The examples included in this chapter illustrate some of the applications of organic spot test analysis to actual examinations of commercial products and the like. They are based, for the most part, on the preliminary tests and procedures for detecting functional groups and individual compounds, which were discussed in detail in the earlier chapters. With regard to metallo-organic compounds and mixed inorganic–organic systems, the underlying discussions will be found in the volume *Spot Tests in Inorganic Analysis*.

A point that was stressed in the opening chapter needs repeating here where the attention of the reader is centered on "Testing Materials". The

---

* E. CHARGAFF et al., *J. Biol. Chem.*, 175 (1948) 67.
** R. WASICKY and O. FREHDEN, *Mikrochim. Acta*, 1 (1937) 55. See also F. FEIGL and E. SILVA, *Drug Std.*, 23 (1955) 113.

point to be remembered is: No matter what the objective, chemical analysis must never be regarded as a mere means to an end or as a routine task. In view of its scientific foundations, it must always be viewed as experimental chemistry having analytical objectives. Only from this standpoint can the special character of spot test analysis be rightly understood and appreciated with regard to its true position in pure and applied chemistry. In this connection it must be noted that the problems of testing materials often lead to valuable developments in analytical researches, whose results sometimes may have significance far beyond the purely analytical objective.

1. Detection of mineral constituents in papers[1]
(Differentiation of filter papers)

Inner-complex salts are precipitated by the reaction of 8-hydroxyquinoline (oxine) with many metal ions.[2] Some of these products display a yellow-green fluorescence in ultraviolet light. Oxinates are also formed when metal oxides and carbonates come in contact with solutions of oxine in organic liquids, in fact they are formed even by the action of the vapor above solid oxine (see p. 388). Various sorts of papers contain calcium, magnesium, and aluminium, which form fluorescing oxinates. These metals are present in paper as the oxide or carbonate, and also as water-insoluble true or adsorption compounds of cellulose. They are in a high state of dispersion and consequently provide excellent conditions for the action of oxine. If a paper which contains mineral constituents of this kind is spotted with a chloroform solution of oxine, a fluorescent fleck remains after the solvent evaporates. The ash of paper may be tested in the same way.

The fluorescence test will reveal the presence of only the oxinate-forming oxides of calcium, barium, magnesium, aluminium. Titanium, whose dioxide sometimes is used as a filler, and iron, which may be present as an impurity, cannot be detected in this way because their oxinates do not fluoresce. Barium sulfate does not react with oxine, whereas, because of its considerable solubility, calcium sulfate gives a distinct reaction.

The inorganic matter in paper can be converted into fluorescing oxinates by exposure to oxine vapor. This can be accomplished at room temperature by placing the sample over a crucible containing solid oxine. After a few minutes, a weak fluorescence is discernible under ultraviolet light, and the intensity increases with longer exposure. The fluorescence appears sooner if the oxine (m.p. 79°) is warmed. It is remarkable that the low vapor pressure of oxine at even ordinary temperatures is sufficient to produce oxinates. The action of oxine vapor can hardly be explained by assuming a transformation of the inorganic matter into stoichiometric oxinates. It is more probable that the vapor reacts with the metal on the surface of the paper without forming a new phase; in other words it is chemically adsorbed. The same fluorescence occurs, as with inner-complex oxinates (see p. 338), as a result of analogous metal–oxine bondings.[3]

The fluorescence responses observed with various papers and their ashes are given in Table 17.

TABLE 17

FLUORESCENCE TESTS OF PAPERS AND ASH

Paper	Ash(from 15 cm² paper)	Original paper
Filter:		
Qual. 605*	+	—
Qual. 613*	+	+
Qual. 615*	+	+
Quant. 589 Blue**	no ash	—
Quant. 589 Red**	no ash	—
Quant. 589 White**	no ash	—
Quant. 589 Black**	no ash	—
Quant. 589***	no ash	—
News Print	+	+
Bond	+	+
Onion Skin	+	+
Drawing	+	+
Tracing	+	+
Toilet	+	+
Photographic	+	+

* Eaton Dikeman Co., Mt. Holly Springs, Pa.
** Schleicher and Schüll, New York.
*** Schleicher and Schüll, Germany.

The procedure of forming fluorescent flecks on paper makes it possible to examine papers without damage to the specimen; there is no visible wear or tear. It is also possible to gain a clue as to whether the paper contains only calcium or magnesium (possibly along with alumina). In this case, the fluorescent fleck is exposed to acetic acid vapors. This treatment destroys Ca and Mg oxinates, and only the fluorescence due to Al oxinate persists. Consequently, if the fluorescence is quenched completely, or is distinctly weakened, it may be taken as a strong proof of the presence of calcium and/or magnesium.

It is remarkable that, in contrast to filter papers "low in ash", designed for use in qualitative analysis, the so-called "ashless" papers,* prepared for

* The ash content of $5\frac{1}{2}$ cm quantitative papers is below 20 γ and below 100 γ in qualitative papers (C. Schleicher & Schüll Co., *Catalog* 70, New York, 1949).

quantitative work, always contain small amounts of ammonium salts.* This can be established by spotting the specimen with Nessler solution, which yields a brown or yellow stain if the response is positive. The various varieties of filter paper can be easily differentiated by this simple test, for which about 0.5 cm² samples are adequate.

Qualitative and quantitative ashless filter papers can be quickly and surely distinguished by the following procedure: The reagent is an approximately saturated ethereal solution of the acid dye known as "Bleu ciel au chrome, solide B"**. A drop of the reagent solution is placed on the paper and the ether allowed to evaporate. The resulting pink spot is then held over ammonia water. Qualitative paper, which contains mineral matter, immediately changes to blue, whereas with ashless paper the change is to yellow. The blue stain fades rather soon, but the color can be restored by again exposing the fleck to ammonia. The test can also be made with an ammoniacal aqueous solution of the dye. If a micro drop is used, as little as 0.2 cm² of the paper suffices. The basis of the color test probably is that minute amounts of calcium and magnesium compounds highly dispersed in the ordinary filter paper form adsorption complexes or salts with the acid dye, and these in turn produce a blue unstable addition compound with ammonia. This supposition is supported by the finding that a stable blue stain is obtained when qualitative filter paper is spotted with a solution of the dye in ethylenediamine, morpholine, or other non-volatile amines.

An excellent differentiation between all kinds of qualitative and quantitative filter paper may be based on the fact that only the former contains traces of trivalent iron, which can be detected [4] by spotting with a 2% solution of α,α'-dipyridyl in concentrated thioglycolic acid. The latter reduces the trivalent iron, and a red stain of ferrous α,α'-dipyridyl thioglycolate appears at once or within one minute. The intensity of the color is dependent on the amount of iron distributed in the paper. o-Phenanthroline may be used in place of α,α'-dipyridyl.

An advantage of the test prescribed here is that permanent preparations can be made from the tested papers, which can serve as proof of the presence or absence of iron in the specimens. The same is true of the pink spot obtained on ordinary filter papers with "Bleu ciel au chrome" which becomes blue on exposure to ammonia.

The above tests for heat-stable mineral constituents in filter papers can be conducted likewise with tiny amounts of paper ash.

---

* It is likely that small amounts of ammonium salts are left in quantitative filter papers, because the latter are treated with hydrofluoric acid and hydrochloric acid to remove the mineral matter, and then washed also with ammonia.

** The dye is manufactured by Compagnie française des matières colorantes, Paris.

1 F. Feigl and G. B. Heisig, *Anal. Chim. Acta*, 3 (1949) 561; F. Feigl and D. Gold-stein, unpublished studies.
2 Comp. F. J. Welcher, *Organic Analytical Reagents*, Vol. I, Van Nostrand, New York, 1947, p. 263.
3 Comp. F. Feigl, *Chemistry of Specific, Selective and Sensitive Reactions*, Academic Press, New York, 1949, p. 676.
4 F. Feigl and A. Caldas, *Anal. Chem.*, 29 (1957) 580.

2. Detection of organic material in dusts, soils, *etc.*[1]

Dusts, soils, ashes, evaporation residues of water, *etc.* can be tested for organic admixtures or contaminants by the procedure given on p. 62. In this the sample mixed with excess mercuriamido chloride (or with $HgO + NH_4Cl$) is ignited, whereby volatile hydrocyanic acid is formed which can be detected by the bluing of filter paper moistened with a copper acetate–benzidine acetate solution (see p. 546). Due to the fact that the color reaction will reveal 0.25 γ hydrocyanic acid (corresponding to 0.11 γ carbon), only small quantities of the test material are necessary to recognize the presence of organic or metallo-organic substances.

1 A. Caldas and V. Gentil, *Talanta*, 2 (1959) 222.

3. Detection of sulfide sulfur in animal charcoal and dyes[1]

The test for sulfide sulfur, in animal charcoal for instance, using the iodine–azide reaction as described on p. 219, may be carried out on a few mg of sample.

The so-called sulfur dyes are formed by melting aromatic amines or hydroxy compounds with sulfur or alkali polysulfides. Most of them are amorphous insoluble products of unknown constitution. They are probably mixtures of various components.[2] They contain organically bound sulfur, and almost all of them catalyze the iodine–azide reaction.

A little of the solid sample, or else a thread or piece of dyed cotton, is treated with a drop of iodine–azide solution and observed for bubbles of nitrogen. Woollen material cannot be tested in this way, since wool itself gives a positive reaction. The following results were observed:

Primuline yellow*	no reaction
Catigen brown 2 R extra	vigorous evolution of nitrogen
Catigen brilliant green G	evolution of nitrogen
Catigen yellow GG extra	very vigorous evolution of nitrogen
Catigen black SW extra	,, ,,
Catigen indigo CL extra	,, ,,
Catigen violet 3 R	,, ,,

* Primuline yellow (base) is one of the few sulfur dyes whose constitution has been determined. As a mixture of di- and trithiazole compounds, all of its sulfur is in etherlike combination. Its negative response to the iodine–azide test conforms to this structure.

1 Comp. F. FEIGL, *Mikrochemie*, 15 (1934) 1.
2 Comp. P. KARRER, *Organic Chemistry*, 4th Engl. ed., Elsevier, Amsterdam, 1950, p. 626.

4. Detection of free sulfur in solid organic products[1]

When thallous sulfide, which is readily soluble in dilute mineral acids, comes into contact with elementary sulfur dissolved in organic liquids or alkali sulfide it is converted into thallous polysulfide The latter is likewise formed if a filter paper impregnated with Tl_2S is spotted with a drop of a solution of sulfur and then treated with dilute acid. A dark stain remains while the remainder of the black paper is bleached almost instantly. When small amounts of sulfur are involved a layer of thallium polysulfide is produced on the surface of the thallous sulfide, and protects the underlying Tl_2S against being dissolved by the acid.

The rapid production of the polysulfide on thallous sulfide can be used to reveal the presence of uncombined sulfur in insecticides, pharmaceutical materials and the like. A preliminary extraction of the dry sample (finely powdered if possible) with carbon disulfide or pyridine is necessary.* Often it is desirable to subject the sample to a preliminary treatment with hydrochloric acid to remove acid-soluble constituents. The residue is then dried at 105° for several hours. Any amorphous sulfur is thus converted into the crystalline form, which is soluble in carbon disulfide. The extraction can be made in a micro apparatus. One drop of the clear extract is placed on black, freshly prepared Tl_2S paper. The solvent is allowed to evaporate in the air, and the paper is then placed in dilute nitric acid. If sulfur is present, a brown-red or light brown fleck, depending on the quantity, is left at the site of the spotting. The rest of the paper turns perfectly white.

Vulcanized rubber can be tested for uncombined sulfur. A particle of the sample is kept in contact with carbon disulfide for 2 or 3 min. A drop of the liquid is transferred to the reagent paper. A distinct fleck of polysulfide remains after treatment with acid if free sulfur was present.

The following procedure will reveal free sulfur in solid organic materials, gas purifying masses, *etc*. Several mg of the dry, finely pulverized sample are placed on Tl_2S paper and moistened with several drops of carbon disulfide, applied at short intervals. After the solvent has evaporated, the dry powder is brushed off from the paper, which is then placed in dilute acid. A dark fleck develops if free or extractable sulfur was present.

Sulfur dissolved in organic liquids can be detected by placing a drop of

---

* Only the crystalline modifications of sulfur are soluble in carbon disulfide. At room temperature, pyridine dissolves up to 4% of amorphous sulfur[2].

the solution on Tl_2S paper, which is then treated with dilute acid. In this way the completion of a sulfur extraction with organic liquids can be recognized.

Preparation of thallous sulfide paper: Filter paper (S & S 589 or Whatman 42) is bathed in 5% thallous carbonate or acetate solution for several minutes. The excess liquid is drained off and the paper then dried in a blast of heated air. Thallous sulfide is deposited by placing the paper across a beaker containing ammonium sulfide solution warmed to 80°. The conversion to Tl_2S requires not more than several minutes; the side of the paper exposed to the fumes turns perfectly black. It is ready for immediate use and is best cut into strips. Freshly prepared paper should be used when testing for sulfur. On standing, the paper deteriorates because the sulfide is oxidized.

For the sensitive detection of elementary sulphur in solid organic material the fusion reaction with benzoin described in Sect. 5 may also be used.

1 F. FEIGL and N. BRAILE, *Chemist-Analyst*, 32 (1943) 76.
2 R. EDGE, *Ind. Eng. Chem., Anal. Ed.*, 2 (1930) 371.

5. Detection of traces of sulfur in organic liquids[1]

Very small amounts of sulfur dissolved in volatile organic liquids may be detected, after conversion into thiosulfate or mercuric sulfide, by the sensitive iodine–azide reaction described on p. 219. The conversion into thiosulfate is accomplished by heating the evaporation residue of the solution under study with sodium sulfite solution. The conversion into mercuric sulfide can be accomplished by shaking the sample with metallic mercury.

Procedures (I) and (II) may be applied. A further important application is the detection of sulfur in motor fuels, where it exerts a corrosive action on engine parts.[2]

Procedure I. (Formation of $Na_2S_2O_3$)   A drop of the liquid to be tested is allowed to evaporate completely on a watch glass. After evaporation*, two drops of a 5% solution of sodium sulfite are placed on the watch glass. The reaction mixture is then heated 2–3 min. on a water bath and after cooling a few drops of iodine–azide solution (3 g NaN_3 in 100 ml 0.1 N iodine) are added. The evolution of nitrogen bubbles is a positive response.

Limit of identification: 0.5 γ sulfur.

Even smaller amounts of sulfur can be identified by Procedure (II), if larger volumes of the sample are taken for examination.

Procedure II (Formation of HgS)   About 6 to 8 ml of the organic liquid

---

* It is important that any carbon disulfide be completely removed, since it catalyzes the iodine–azide reaction. Compare p. 488.

(preferably CS_2) are shaken vigorously with a drop of mercury in a hard glass test tube or a measuring cylinder. The mercury dropper described in *Inorganic Spot Tests*, Chap. 2, can be used to advantage here. When even very small amounts of free sulfur are present, the surface of the mercury is stained with a black or iridescent film of mercuric sulfide[3] which can be identified by the iodine–azide reaction*. The carbon disulfide is poured off and the mercury is transferred to a watch glass and heated in a current of steam (see *Inorganic Spot Tests*, p. 46) to remove the last traces of carbon disulfide. The mercury is then treated with iodine–azide solution, and a foam of nitrogen bubbles forms around the mercury.

A specific and sensitive test for free sulfur is based on its reaction with melted benzoin (m.p. 133°):

$$C_6H_5COCHOHC_6H_5 + S^0 \rightarrow C_6H_5COCOC_6H_5 + H_2S$$

Benzoin functions as hydrogen donor through the conversion into benzil.[5] The resulting hydrogen sulfide can be detected with lead acetate.

Procedure.[5] Several drops of the liquid or solution are taken to dryness in a micro test tube. A few mg of benzoin are added to the residue. The mouth of the test tube is covered with moist lead acetate paper and 2/3 of the tube is immersed in a glycerol bath heated to 130°. The temperature is gradually brought to 150°. If free sulfur was present the paper acquires a black to brown stain, the color depending on the amount of sulfur present.

Limit of identification: 0.5 γ sulfur.

* The application of this sulfide formation for the detection of small amounts of free sulfur in carbon disulfide, has been recommended.[4] The iodine–azide reaction is positive, even for amounts of HgS too small to be detected visually.

1 F. FEIGL and L. WEIDENFELD, *Mikrochemie (Emich Festschrift)*, (1930) 133.
2 K. WEISSELBERG, *Petroleum Z.*, 31 (1935) No. 10, p. 7.
3 E. OBACH, *J. Prakt. Chem.*, [2], 18 (1878) 258.
4 W. E. ZMACZYNSKI, *Z. Anal. Chem.*, 106 (1936) 32.
5 F. FEIGL and C. STARK, *Anal. Chem.*, 27 (1955) 1838.

6. Detection of carbon disulfide in benzene and carbon tetrachloride[1]

Crude benzene and carbon tetrachloride usually contain a small amount of carbon disulfide; the former about 0.1 to 0.2%; the latter 0.1 to 3.5%. The carbon disulfide enters benzene during its preparation by distillation of hard coals containing sulfur. The carbon disulfide in carbon tetrachloride arises from its preparation from carbon disulfide by chlorination:

$$CS_2 + 2 S_2Cl_2 \rightarrow CCl_4 + 6 S$$

Contamination with carbon disulfide may be detected by the reaction based on the acceleration of the formation of sulfide from CS_2 and alkali in the presence of formaldehyde (p. 489). The sulfide can be detected by the precipitation of PbS from a plumbite solution.

Procedure. Two or three drops of the benzene or carbon tetrachloride are mixed on a spot plate with 1 or 2 drops of a strongly alkaline plumbite solution and 2 drops of 40% formaldehyde solution (formalin). If carbon disulfide is present, a dark ring of lead sulfide is formed at the boundary between the organic solvent and the water layer. Any hydrogen sulfide must be removed before carrying out the test. This removal is described on p. 489.

A reliable test for minimal quantities of carbon disulfide in benzene, carbon tetrachloride *etc.* is based on the formation of yellow-brown copper dimethyldithiocarbamate directly from the components. The details of the procedure are given on p. 489.

1 F. FEIGL and K. WEISSELBERG, *Z. Anal. Chem.*, 83 (1931) 101.

7. Detection of thiophene and thiophene derivatives in benzene and toluene

Benzene and toluene produced by the distillation of coal always contain thiophene and thiophene derivatives, whose quantity varies up to 0.5%. The contaminants cannot be removed by fractional distillation because the respective boiling points lie too close together. Their removal requires treatment with concentrated sulfuric acid; the resulting sulfonic acids remain dissolved in the sulfuric acid.

Benzene or toluene can be tested for thiophene (or its derivatives with an unoccupied position *ortho* to the sulfur atom) by means of the color reaction with 1,2-diketones which is based on the formation of quinoidal indophenine dyes (see pp. 491 and 493). One or two drops of the suspected benzene or toluene is sufficient for the test. The most sensitive reagent for thiophene is ninhydrin. It is advisable to run a comparison test with benzene or toluene known to be free of thiophene and its derivatives. Such pure standards can be prepared by shaking the hydrocarbon with a freshly prepared 1% solution of ninhydrin in concentrated sulfuric acid. The acid layer is poured away, the other layer washed with water, and the hydrocarbon then distilled.

8. Detection of ammonium thiocyanate in thiourea[1]

Thiourea is prepared by prolonged fusing of ammonium thiocyanate. The cooled melt is extracted with strong alcohol to dissolve out the unchanged thiocyanate. The remaining thiourea is purified by recrystallization from hot water. Accordingly, impure thiourea may contain variable amounts of ammonium thiocyanate. Such material can be examined by the procedure given on p. 586, which is based on the diverse behavior of these two isomeric compounds toward lead hydroxide or lead oxide.

1 F. FEIGL and E. LIBERGOTT, *Mikrochim. Acta*, (1964) 1111.

9. Detection of ethanol in motor fuels, benzene, chloroform, *etc.*[1]

A rapid detection of ethanol is of value in the examination of motor fuels, organic liquids, *etc.* The test described on p. 414 can be used to advantage. It is based on the oxidation of ethanol by acidified permanganate solution. The resulting acetaldehyde is easily detected through the color reaction with a solution of sodium nitroprusside containing morpholine. The oxidation proceeds at a satisfactory speed at room temperature and at the comparatively small organic liquid–water interface.

Procedure. A micro test tube is used. A drop or two of the sample and a drop of the acidified solution of permanganate are mixed and the mouth of the tube is covered with filter paper moistened with the reagent solution. If ethanol is present, a blue fleck appears on the paper within 0.5–3 min.

Reagents: 1) 1 N potassium permanganate in sulfuric acid (1 : 1).
　　　　　2) Equal volumes of freshly prepared 5% sodium nitroprusside and 20% morpholine solution.

Technical ether gives a distinct response and even pure ether in large amounts. This is due in part to the alcohol actually present but it is also possible that the permanganate oxidizes the ether:

$$C_2H_5OC_2H_5 + O \rightarrow C_2H_5OH + CH_3CHO$$

followed by oxidation of the ethanol to acetaldehyde.

The test described below will reveal ethanol in motor fuels. It is based on the well-known oxidation:

$$C_2H_5OH + CuO \rightarrow CH_3CHO + Cu^0 + H_2O$$

Procedure. One drop of the sample is placed in a micro test tube. A piece of copper wire is strongly heated and dropped at once into the liquid. The mouth of the tube is immediately covered with filter paper moistened with the reagent solution (2). A blue stain appears at once if the sample contains ethanol.

This test revealed 3 γ ethanol in one drop of benzene.

1 F. FEIGL and C. STARK, *Chemist-Analyst*, 45 (1956) 39.

10. Detection of low-boiling halogenated aliphatic hydrocarbons[1]

Pertinent liquids, such as chloroform, carbon tetrachloride, trichloroethylene, tetrachloroethane, tetrabromoethane, are widely used in industry
as solvents and extracting agents. They can be detected even in mixtures
with benzene or toluene (but not ethers, see p. 188) because of the formation
of halogen when heated with benzoyl peroxide (m.p. 103°) to 100–120°.
The oxidative fission probably may be represented as follows:

$$CCl_4 + 2\ O \rightarrow CO_2 + 4\ Cl^0$$
$$(CH)_2Br_4 + 5\ O \rightarrow 2\ CO_2 + H_2O + 4\ Br^0$$

The liberated halogen may be detected in the gas phase with an appropriate
reagent paper.

Procedure.   A drop of the test solution is united in a micro test tube with
several cg of benzoyl peroxide and the tube is then placed in a glycerol bath
preheated to 60°. The mouth of the tube is covered with a filter paper moistened
with 5% potassium iodide solution and the temperature of the bath is then gradually raised to 110–120°. A positive result is indicated by the development of a
yellow or brown stain (free iodine) on the paper. Amounts in the neighborhood
of 100 γ can be detected in this way. A more sensitive test based on pyrohydrolytic cleavage of hydrogen halides by means of manganous sulfate monohydrate is described on p. 411.

1 F. FEIGL, D. GOLDSTEIN and D. HAGUENAUER-CASTRO, *Rec. Trav. Chim.*, 79 (1960)
531.

11. Differentiation of chloroform and carbon tetrachloride[1]

Chloroform (b.p. 82°) and carbon tetrachloride (b.p. 76°) can be differentiated by the behavior of the vapors derived from one drop of the liquids
when exposed to heated tungstic or chromic acid respectively. If the sample
is heated along with several mg tungstic acid* in a micro test tube, which is
immersed to a depth of 1 cm in a glycerol bath at 150–160°, the carbon
tetrachloride soon undergoes a pyrolytic oxidation:

$$CCl_4 + 2\ O \rightarrow CO_2 + 4\ Cl^0$$

The resulting chlorine can be detected by the blue which it produces on
Congo paper moistened with hydrogen peroxide. Chloroform does not react
under these conditions.

* Commercial tungstic acid often contains small amounts of decomposable chlorides.
The latter can be rendered harmless by heating the preparation for 30 min. at 200°.

If solid chromic acid* is used as oxidant in place of tungstic acid, and the temperature does not exceed 150°, the chloroform is oxidized:

$$2 \, CHCl_3 + 5 \, O \rightarrow 2 \, CO_2 + H_2O + 6 \, Cl^0$$

The resulting chlorine can be detected through the bluing of Congo paper moistened with hydrogen peroxide. Carbon tetrachloride does not react under the conditions prescribed here.

The pyrolytic oxidations are suitable for the differentiation of chloroform and carbon tetrachloride only in the absence of other organic liquids.

* It is advisable to heat several mg of hygroscopic chromic acid for 5 min. at 150°.

1 F. FEIGL, D. GOLDSTEIN and D. HAGUENAUER-CASTRO, *Rec. Trav. Chim.*, 79 (1960) 531.

12. Detection of solvents and products with a glycolic structure

(1) Test through pyrohydrolytic release of acetaldehyde[1]

A number of artificial products manufactured from glycol or ethylene oxide respectively, which may be considered as polymers of ethylene oxide, show the test for the ethylene-group described on p. 166.

The following materials used as solvents and solubilizers give a positive result in mg amounts.

Polyethylene glycol (200, 400); butylpolyglycol; polyethylene glycol methyl ether (400, 1200, 1500); polyethylene glycol propyl ether; "Tween" 20 and 80 (esters of fatty acids with polyethylene glycol sorbitan).

Some waxes have a glycolic structure and therefore show the test for the ethylene group. Pertinent products are polyethylene oxides (water-soluble waxes) and esters of fatty acids with glycol, for instance Lanette wax.

(2) Test through stabilisation of chromium peroxide[2]

A sensitive test for chromic acid is based on the fact that through the addition of hydrogen peroxide a blue color appears due to the formation of perchromic acid or its anhydride CrO_5. These compounds are highly unstable since they are reduced by hydrogen peroxide to green Cr^{3+} ions. If, however, the blue aqueous solution is extracted with ether, the latter turns blue. This color fades only slowly, which effect can be attributed to the formation of an addition compound between CrO_5 and ether. Since polyethylene glycols are high molecular ethers, they too are able to stabilize CrO_5 through addition.[3]

$$HOCH_2\!-\!(CH_2OCH_2)_x\!-\!(CH_2OCH_2)_y\!-\!CH_2OH,$$
$$\downarrow$$
$$CrO_5$$

where x means the number of occupied and y the number of unoccupied oxygen atoms in the polyethylene glycol molecule. This stabilization permits the detection of both liquid and solid polyethylene glycols. Until now, no data about the selectivity of this test are available. It is remarkable that CrO_5 is also stabilized by tributyl phosphate.[4]

Procedure.   To one drop of polyethylene glycol on a spot plate, one drop of 5% potassium chromate followed by a drop of 4% sulfuric acid containing 3% hydrogen peroxide is added. Parallelly a blank test without polyethylene glycol is carried out. In the blank the blue color initially obtained changes to green within a few minutes. The positive response is indicated through the persistence of the blue color during 15 to 20 minutes.

If solid polyethylene glycols have to be tested, solutions in glacial acetic acid must be prepared.

1 F. FEIGL and V. ANGER, unpublished studies.
2 R. SPRINGER and H. ISAK, Z. Anal. Chem., 199 (1964) 363.
3 B. WURZSCHMITT, Z. Anal. Chem., 130 (1949/50) 105.
4 M. N. SASTRI and D. S. SUNDAR, Chemist-Analyst, 50 (1961) 283.

13. Detection of lead tetraethyl or lead tetraphenyl in motor fuels[1]

Certain alkyl, aryl, carbonyl and acetylide metallic compounds are sometimes added to mineral oils, since even small amounts have a decided retarding effect on certain undesirable reactions. These substances are used chiefly as antiknock agents in motor fuels and as antioxidants in hydrocarbons. Since lead tetraethyl (phenyl), which is most frequently used, is very volatile and also extremely toxic, a reliable test for it has important practical significance. The following procedure is easily applied to gasoline and other motor fuels.

Procedure.   One or two drops of the gasoline is placed on filter paper and held in ultraviolet light for about 30 sec. until evaporation is complete.* A drop of a freshly prepared 0.1% solution of dithizone in chloroform is then placed on the paper. If $Pb(C_2H_5)_4$ or $Pb(C_6H_5)_4$ is present, a deep red stain results; otherwise the paper remains green. If the gasoline is highly colored, it should be decolorized beforehand by shaking with activated charcoal.

Oils usually have a deep self color. To test for lead salts, 100 to 150 ml of the oil is mixed with 3% benzene and steam-distilled. The lead test is made on the distillate. Aternatively, a drop of a gasoline possibly containing $Pb(C_2H_5)_4$ is shaken with a 1% aqueous solution of potassium cyanide and

* This treatment may result in the decomposition of organo-lead compounds with deposition of acid-soluble flocks.[2]

exposed to ultraviolet light for a few seconds. A drop of dithizone solution is then placed on the liquid surface.[3]

Traces of nickel carbonyl, sometimes used as antiknock agent, may be detected by an analogous method using ammonia and dimethylglyoxime or rubeanic acid as reagents for nickel. Compare *Inorganic Spot Tests*, Chap. 3.

1 B. STEIGER, *Petroleum Z.*, **33** (1937) No. 27.
2 H. KIEMSTEDT, *Z.Angew. Chem.*, **42** (1929) 1107.
3 B. STEIGER, *Mikrochemie*, **22** (1937) 227.

14. Detection of peroxides in organic liquids[1]

If ether, dioxane, and other oxygen-containing organic liquids are exposed to the air for a considerable time, small amounts of organic peroxides are likely to be formed.[2] Organic liquids contaminated in this fashion are not suitable for extraction purposes in many instances.* They can be tested by placing a drop on potassium iodine–starch paper. After volatilization, ether *etc.* containing peroxide will leave a brown stain, which turns blue when spotted with water. The fleck is produced immediately in the case of impure dioxane containing water.

If filter paper impregnated with *p*-phenylenediamine is spotted with organic liquids containing peroxides, a light blue spot is left.

* A rapid and efficient method for removing peroxides from organic liquids is based on their complete adsorption on activated alumina, such as Grade F–20 chromatographic Alorco activated alumina.[3]

1 Cl. COSTA NETO (Rio de Janeiro), unpublished studies.
2 Comp. A. RIECHE, *Die Bedeutung der organischen Peroxyde für die chemische Wissenschaft und Technik*, Stuttgart, 1936.
3 W. DASLER and C. D. BAUER, *Ind. Eng. Chem., Anal. Ed.*, **18** (1946) 52.

15. Detection of organic stabilizers in hydrogen peroxide[1]

Concentrated (35%) hydrogen peroxide is stabilized usually by acetamide or its N-substituted derivatives. Such stabilizers can be detected even in dilute solutions of hydrogen peroxide by taking the solutions to dryness and then pyrolyzing the residue with mercuric cyanide. Hydrogen cyanide is evolved. (Comp. p. 147).

Procedure. A drop of the hydrogen peroxide being tested is evaporated to dryness in a micro test tube. Several cg of mercuric cyanide are added and mixed with the residue by rubbing with a glass rod. The mouth of the test tube is then covered with a piece of filter paper that has been moistened with reagent for cyanide (see p. 546). The test tube is immersed to about 1/3 its depth in a glycerol bath preheated to 150°. A positive response is indicated by the development of a blue stain on the paper within 1–2 minutes.

The presence of the stabilizers was clearly revealed by this procedure in 1 drop of a 33% solution of hydrogen peroxide after a 1 : 20 dilution. When more dilute solutions are under test, two or more drops must be taken to dryness.

1 Unpublished studies.

16. Detection of morpholine

As shown by its structure, morpholine contains CH_2–CH_2 groups and O- and N atoms bound to them. Such compounds respond positively to the test described on p. 166 based on the formation of acetaldehyde on pyrolysis with zinc chloride (*Idn. Lim.* 2 γ). Furthermore, morpholine gives the color reaction with sodium nitroprusside and acetaldehyde as described on p. 438 (*Idn. Lim.* 5 γ).

Morpholine is a widely used basic solvent that differs from the neutral solvent dioxane in that the latter contains an O atom in place of the NH group of morpholine. These water-soluble compounds can be distinguished in that morpholine responds positively to *both* of the tests cited above, whereas dioxane behaves in the same manner as morpholine with respect to only the pyrolysis with zinc chloride.

17. Detection of pyridine bases in amyl alcohol[1]

Amyl alcohol, prepared by fermentation, always contains distinct amounts of pyridine bases. These can be detected even in commercial amyl alcohol that is represented as 'pure'. The test described on p. 384 serves as a rapid method for detecting pyridine in pure amyl alcohol and in its mixtures with ethyl alcohol. A positive response, namely the production of a violet polymethine dye, is best established, by adding the necessary reagents to the test solution containing amyl alcohol, since the latter dissolves the resulting violet dye to give a brown solution.

1 Unpublished studies, with D. GOLDSTEIN.

18. Detection of pyridine bases in ammonia[1]

Ammonia recovered from the gas liquor of illuminating gas manufactured from coal invariably contains small amounts of pyridine, α- and β-picoline, *etc.* These bases, with the exception of α-picoline, can be detected by suitable modification of the test given on page 384. It is based on the production of a violet polymethine dye with *o*-tolidine.

Procedure.   One drop of the ammonia to be tested is treated on a spot plate with drops of strong bromine water until a permanent yellow is produced. An additional drop of the bromine water is then added. The ammonia is thus completely destroyed. Three drops of 4% potassium cyanide solution are stirred in, and finally three drops of a suspension of *o*-tolidine acetate. Depending on the amount of pyridine bases present, a violet color appears at once or after a few minutes, or the white suspension assumes a blue tint. A blank test is advisable.

Reagents: 1) Saturated solution of bromine in 5% potassium bromide.
2) *o*-Tolidine acetate suspension. 100 ml of a saturated solution of *o*-tolidine chloride plus 1 g of sodium acetate. The resulting suspension is stable if stored in a closed container.

The procedure will reveal 5 γ pyridine.

1 Unpublished studies, with D. GOLDSTEIN.

19. Detection of hydrogen cyanide in illuminating gas[1]

Illuminating gas often contains small amounts of hydrogen cyanide and possibly also dicyanogen, even though it is customary to pass the gas through water to retain ammonia, pyridine bases, hydrogen cyanide, dicyanogen, *etc.* The sensitive test for alkali cyanide (p. 548) can be used to detect traces of hydrogen cyanide and dicyanogen in illuminating gas. It is based on the demasking of dimethylglyoxime from alkaline solutions of complex palladium dimethylglyoxime by cyanide ions.

Procedure.   Filter paper is impregnated with an alkaline solution of palladium dimethylglyoxime (for preparation see p. 548) or a drop of this solution is placed on filter paper. The moist reagent paper or the fleck is held for several seconds or minutes (according to the suspected cyanide content) about a millimeter from the end of a rubber tube (diameter 0.5 mm) connected with the gas supply. The part of the paper which has been exposed to the gas is then spotted with a solution of nickel chloride. A pink to red stain of nickel dimethylglyoxime appears at once, the depth of the color varying with the cyanide content of the gas.

1 F. FEIGL and H. E. FEIGL, *Anal. Chim. Acta*, 3 (1949) 300.

20. Tests for added materials and undesirable admixtures in foods

It is often important to detect certain impurities which may have been added to organic products and also to detect undesirable by-products that have not been removed. This branch of testing is especially necessary in the examination of foods, drugs and cosmetics, when a harmful impurity or unscrupulous adulteration must be detected quickly. If the impurity or adulteration is inorganic, the tests described in *Inorganic Spot Tests* are available. They should be carried out on the ash or acid extract of the sample. In testing for organic substances, some of the spot tests described in Chap. 5 may be applied. Pertinent examples are presented here.

(a) Detection of formic acid

Formic acid should not be present in wood vinegar since it is injurious to health. This acid (which has a definite antiseptic action) is also an inadmissible fermentation product in fruit juices, jams, marmalade, honey, *etc.* If permitted at all, the quantity should be small. Formic acid may be detected as described on p. 451 (conversion to formaldehyde and detection of the latter by the color reaction with a mixture of chromotropic acid and strong sulfuric acid). If the products are colored or have had color added to them, they should be decolorized beforehand (if necessary after solution) with animal charcoal.

(b) Detection of glycerol

The nitroprusside test for glycerol described on p. 416 has many applications. It depends on a color reaction between sodium nitroprusside plus piperidine or morpholine and acrolein, the latter being readily formed by the removal of a molecule of water from glycerol. Since the latter, in the form of glycerol esters of various organic acids, is a constituent of all plant and animal fats, it is possible to test beeswax or lanolin (*adeps lanae anhydr.*) for instance, for any admixture of plant or animal fats, which should not remain in a first-rate product. The glycerol reaction may be used successfully to detect fats in cloudy or dark products. The rapid detection of glycerol is useful in testing cosmetics, and in testing wines and liquors, which may contain glycerol as an illegal sweetening agent.

(c) Detection of citric and malic acid

The detection of citric and malic acid is important in testing fruit vinegar, as this commodity should always contain these acids. Since both citric and malic acid are dicarboxylic acids with the carboxyl groups in the 1,2-positions, the reaction for this group, as described on p. 352, may be applied as a simple and rapid means of testing fruit vinegar.

The fusion reaction with urea (p. 472) which involves the formation of the fluorescent ammonium salt of citrazinic acid, can also be used to detect citric acid and citrates in evaporation residues of vinegars.

(d) Detection of pyridine

A test for pyridine is useful as a rapid means of characterizing commercial alcohol where it is often present as denaturant. The test for tertiary ring bases (see p. 379) and also the test for pyridine (p. 384) may be applied for the rapid detection of small amounts of pyridine.

(e) Detection of saccharin (gluside, crystallose, etc.)

Saccharin, the water-insoluble imide of o-sulfobenzoic acid or its water-soluble sodium salt (crystallose) is sometimes substituted for sugar since it is 500 times as sweet. It may be used illegally in wine, spirits, fruit juices, jams and honey, and hence its detection is very important.

A good test for saccharin (and also crystallose) is based on its saponification when evaporated with hydrochloric acid. The products are ammonium chloride and o-sulfobenzoic acid:

$$C_6H_4{\overset{\displaystyle /SO_2\searrow}{\underset{\displaystyle \searrow CO \nearrow}{}}}NH + HCl + 2\,H_2O \longrightarrow C_6H_4{\overset{\displaystyle /SO_3H}{\underset{\displaystyle \searrow COOH}{}}} + NH_4Cl$$

If the evaporation residue is heated to 120° to remove the excess hydrochloric acid, and then again taken to dryness with a few drops of ammonia, and the residue heated to 250°, the ammonium salt of the sulfobenzoic acid will be left along with the ammonium chloride. The production of these two ammonium salts and hence the presence of saccharin, can be easily established by the Nessler test (*Idn. Limit*: 5 γ saccharin). The test assumes the absence of ammonium salts; therefore a preliminary trial with Nessler reagent is necessary.

If colored solutions are being examined, or if caramelization occurs, it is advisable to run a comparison test. In such cases, the release of ammonia by warming with causic alkali can be detected in the vapor phase with qualitative filter paper moistened with Nessler solution.

21. Detection of salicylic acid in foods, beverages, condiments, etc.[1]

Foods, condiments, beverages, and the like, sometimes contain small amounts of salicylic acid or alkali salicylate which have been added as preservative. Salicylic acid can be detected by the test described on p. 478, which is based on the volatility of this acid and the fluorescence of its magnesium salt. The acidified solution or suspension of the sample in dilute acid must be extracted with ether, which is a good solvent for salicylic acid. The test is carried out with the residue after evaporation of the ether.

Procedure.   About 10 ml of the liquid sample is treated with 10 ml of concentrated sulfuric acid. After cooling, the mixture is shaken with 4 ml of ether. After several minutes, one or two drops of the ethereal solution are evaporated in the bulb of the apparatus shown in Fig. 23, *Inorganic Spot Tests.* The residue is treated with a drop of concentrated sulfuric acid and the apparatus is closed with the stopper, whose knob has been charged with a drop of magnesium hydroxide suspension. The closed apparatus is kept for 10–15 min. in an oven at 130°. The knob is then viewed in ultraviolet light. A blue fluorescence indicates the presence of salicylic acid. A comparison test is advised.

> *Reagent:* Magnesium hydroxide suspension. 30 ml of 10% magnesium sulfate
> solution is mixed with 2 ml of 5 N sodium hydroxide. The suspen
> sion must be shaken before use.

This test will reveal 0.017% salicylic acid in wine or water if one drop of the ether extract is used. Several drops of the extract should be taken when smaller amounts of salicylic acid are suspected.

1 Unpublished studies, with V. GENTIL.

22. Detection of salts of N-cyclohexylsulfamic acid[1]

The water-soluble alkali- and alkaline-earth salts of cyclohexanesulfamic acid (N-cyclohexylsulfamic acid) are used as noncaloric sweetening agents in place of saccharin. The more widely known sodium salt bears the name Cyclamate sodium or Sucaryl sodium; it is 30 to 50 times as sweet as sugar and has no bitter after-taste.

As was pointed out on p. 306, many organic derivatives of sulfamic acid behave like the parent compound in that sulfuric acid results when they are heated with nitrous acid. Accordingly, cyclohexylsulfamic acid can be detected by treatment with an acid solution of barium chloride followed by alkali nitrite; barium sulfate precipitates:

$$\bighexagon\!-\!NHSO_3H + NO_2^- + Ba^{2+} \longrightarrow \bighexagon\!-\!OH + BaSO_4 + N_2 + H^+$$

Procedure.   A tiny portion of the sample or a drop of its aqueous solution is mixed in a micro test tube with one drop of a 1% solution of barium chloride in 2 N hydrochloric acid. Some crystals of sodium nitrite are then added and briefly warmed. A positive response is signalled by a white turbidity or precipitate.

Limit of identification: 5 γ Sucaryl.

If Sucaryl, in mg amounts, is heated to 230° in a glycerol bath, basic vapors are evolved (probably volatile cyclohexylamine) which turn the color of acid–base indicator paper. It is advisable to place a drop of the test solution in a micro test tube along with several mg manganese dioxide and then take the mixture to dryness at 110° in a drying oven. The mouth of the tube is then covered with a piece of indicator paper and the mixture is heated in a glycerol bath to 240°. As little as 25 γ Sucaryl can be detected by the color change of the indicator.

1 F. FEIGL, D. GOLDSTEIN and D. HAGUENAUER-CASTRO, *Z. Anal. Chem.*, 178 (1961) 424.

23. Distinction of fermentation acetic acid from synthetic acetic acid[1]

Fermentation acetic acid and its dilute aqueous solutions (fruit vinegars) invariably contain small amounts of amino acids, and usually also ammonium acetate. In contrast, synthetic acetic acid and the vinegars prepared from it by dilution are free of nitrogen. Accordingly, the detection of nitrogen can serve as the basis for distinguishing between these kinds of acetic acid. Well suited to this purpose is the test for nitrogen (see p. 90), based on the formation of nitrous acid when nitrogenous compounds are pyrolyzed in the presence of oxidants, particularly manganese dioxide. The evaporation residue from 1–2 drops of the acetic acid or vinegar will suffice. Only the residue of fermentation acetic acid yields nitrous acid detectable with the Griess reagent.

1 F. FEIGL and J. R. AMARAL, *Anal. Chem.*, 30 (1958) 1148.

24. Detection of mineral acids and organic hydroxy acids in vinegar[1]

Fruit vinegar, which is prepared by oxidation of dilute alcoholic liquors with the aid of Schizomycetes, contains 3–15% of acetic acid along with small amounts of tartaric, citric, formic and oxalic acid. It contains no mineral acids. On the other hand, synthetic dilute acetic acid is frequently contaminated with mineral acids, such as hydrochloric, sulfuric, phosphoric acid. Consequently, the test for the presence or absence of mineral acids can be based on the fact that after the sample has been treated with excess ammonia and evaporated to dryness, the residue will consist of ammonium acetate (tartrate, *etc.*) plus the ammonium salts of the mineral acids. Only the latter salts can withstand heating to 250°. Ammonium acetate is completely decomposed into acetic acid and ammonia even at 120°, and the ammonium salts of the other organic acids contained in fruit vinegar are decomposed by heating to 250° for ten minutes. The residual ammonium salts of mineral acids are easily detected by the Nessler test. The detection of mineral acids in this way is reliable and requires only a few drops of the vinegar.

If phosphoric acid is suspected, a part of the residue can be tested by the sensitive reaction with ammonium molybdate and benzidine (see p. 97).

Vinegar prepared by diluting pure acetic acid can be identified by evaporating a few drops of the sample with ammonia in a micro crucible and heating the residue to 120° for 5–10 min. If the residue gives a negative response to the Nessler test, no other organic acid is present.

As mentioned above, vinegar which is produced by fermentation of alcoholic liquids, always contains small amounts of organic hydroxy acids. A general test for these acids can be based on the fact that they form soluble complex salts with zirconium in ammoniacal solution, which, even in small amounts, can be detected, after acidification, by a fluorescence reaction with morin (see p. 350). The following steps are necessary: One drop of 1% $ZrCl_4$ solution and a small excess of ammonia are added to 1 or 2 drops of the vinegar. The mixture is warmed and filtered. One drop of 0.05% solution of morin in acetone and one drop of concentrated hydrochloric acid are added to the filtrate. When hydroxy acids are present, a greenish yellow fluorescence appears under ultraviolet light. Vinegar produced by dilution of pure synthetic acetic acid does not show the fluorescence test.

It should be noted that fresh and fermenting fruit juices, which contain sugars, also respond positively to the test described here.

1 Unpublished studies, with J. E. R. MARINS.

25. Detection of vitamin C in vegetable and fruit juices [1]

Many freshly prepared vegetable, and especially fruit juices, contain ascorbic acid, a strong reducing agent. It can be detected by the redox reaction with p-nitrosodimethylaniline (mentioned in Chap. 2), which, carried out in the presence of p-dimethylaminobenzaldehyde, leads to a red Schiff base. The yellow reagent paper, impregnated with both of these amino compounds, is used (for preparation see p. 135). After placing a drop of the juice on the paper and drying at 110°, a red fleck or ring is left. When only small amounts of ascorbic acid are suspected, it is best to bathe the spotted and dried paper in water to wash away the yellow nitroso compound. The red fleck is then seen better.

When examining juices which contain much tannin (which reacts analogously to ascorbic acid), the latter must be precipitated beforehand as the water-insoluble adsorption compound by adding a drop of an aqueous solution of potassium antimonyl tartrate (tartar emetic). The test is then made on a drop of the filtrate or centrifugate. *Idn. Limit:* 1 γ ascorbic acid.

Since the color reaction is so sensitive even dilute vegetable and fruit juices can be tested. The procedure was tried with success on lemon, orange, tomato, pineapple *etc.* juices. The color reaction likewise occurs if a freshly cut surface of fruits and vegetables is pressed against the reagent paper.

1 F. Feigl and D. Goldstein, *Microchem. J.*, 1 (1957) 177.

26. Detection of vitamin B₁[1]

This vitamin, which is also known as thiamine or aneurine, has the structure:

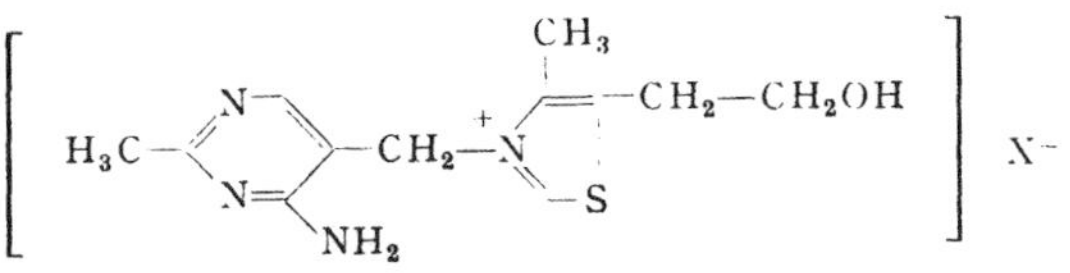

Because of the primary amino group, it undergoes the pyrolytic condensation:

$$2 \ RNH_2 \rightarrow RNHR + NH_3$$

Accordingly, if vitamin B₁ is heated along with dimethyl oxalate, a red product results if thiobarbituric acid is also present. This reaction involves the oxamide that is produced by the action of the pyrolytically formed ammonia on dimethyl oxalate (see p. 150). Consequently, the procedure given on p. 151 for the detection of compounds which yield ammonia on pyrolysis can be applied to the chloride. (*Idn. Lim.*: 2.5 γ)

The detection of vitamin B₁ chloride (nitrate) by pyrolysis in the presence of dimethyl oxalate and thiobarbituric acid may be of value in the investigation of vitamin mixtures, since no other vitamin contains an amino group.

The test may not be applied in the presence of compounds which give off ammonia when pyrolyzed. A list of such compounds is given on p. 150.

The sulfur-bearing vitamin B₁ (chloride, nitrate) can be precipitated from dilute solutions by adding phosphotungstic acid. After centrifuging, the precipitate can be examined for sulfur by applying tests for this element. Appropriate procedures are described in Chap. 2, by using the test for divalent sulfur. Vitamin B₁ salts can be detected in vitamin mixtures in this way.

1 Unpublished studies, with E. Jungreis.

27. Detection of vitamin B₆

(1) Test with 2,6-dichloroquinone-4-chloroimine[1]

Numerous phenols with a free *para* position condense with 2,6-dichloro-quinone-4-chloroimine to give colored indophenols which are acid–base indicators (see p. 185). Vitamin B_6 (pyridoxine chloride), which contains a phenolic OH group with a free *p*-position, reacts:

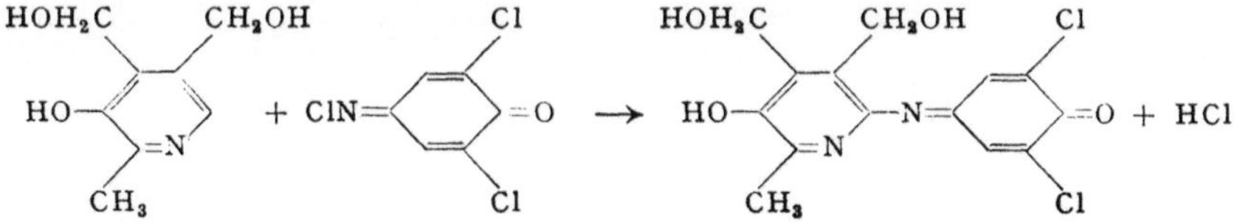

The brown-yellow condensation product becomes blue on exposure to ammonia because of the formation of an unstable ammonium phenolate.

Procedure. Filter paper is bathed in a saturated benzene solution of 2,6-dichloroquinone-4-chloroimine and dried in air. A small amount of the solid sample is rubbed on the paper, or a drop of the aqueous solution is placed on it. A brown-yellow stain appears, which turns deep blue when fumed with ammo-nia. The blue fades rapidly (loss of ammonia from the ammonium ph enolate) but the color is restored by renewed exposure to ammonia.

Limit of identification: 0.5 γ vitamin B_6.

This test cannot be applied directly in the presence of phenols which react with 2,6-dichloroquinone-4-chloroimine, or in the presence of vitamins which are reductants (vitamins B_1, B_2 and C), since they reduce the reagent and thus consume it. In such cases, a drop of the test solution should be treated, in a micro centrifuge tube, with 1–2 drops of a saturated aqueous solution of phosphomolybdic acid, which precipitates vitamin B_6 (a pyridine derivative) as the phosphomolybdate. Any reducing compounds present will be oxidized with production of water-soluble "molybdenum blue" while non-oxidizable phenols and phenolic compounds remain unaltered. The blue suspension is centrifuged, the sediment washed once with water, again centrifuged, and then suspended in 1–2 drops of water. A drop of the suspension is placed on the reagent paper and fumed with ammonia. The phosphomolybdate is decomposed, and the liberated vitamin B_6 gives the indophenol raction. Phosphotungstic acid may also be used as precipitant.

1 F. FEIGL and E. JUNGREIS, *Clin. Chim. Acta*, 3 (1958) 399.

(2) Test with sodium 1,2-naphthoquinone-4-sulfonate and ammonia

Pyridoxine and pyridoxamine give an emerald-green color with 1, 2-

naphthoquinonesulfonic acid and ammonia.[1] A sensitive and selective spot
test for pyridoxine is based on this color reaction. The mechanism may involve the formation of pyridoxamine through reaction with ammonia and
the subsequent reaction of this product with naphthoquinonesulfonic acid,
analogous to the reaction of benzylamine, which likewise gives a green
product (see p. 153). Ammonia is absolutely necessary in this test for pyridoxine and cannot be substituted by any other alkali.

Procedure.[2]   A drop of the test solution is mixed in a depression of a spot
plate with two drops of the reagent and one drop of 2 N ammonia. If pyridoxine
is present, a blue-green to green color appears within one minute. It is advisable
to conduct a blank with the reagent alone since the latter likewise changes color
with ammonia after several minutes and becomes grey-brown. A parallel trial
with 2 N sodium hydroxide instead of 2 N ammonia must not show any blue.
   Reagent: 0.1% solution of sodium 1,2-naphthoquinone-4-sulfonate in meth-
        anol.

Limit of identification: 0.1 γ pyridoxine.

1 L. MAIWALD and H. MASKE, *Z. Physiol. Chem.*, **306** (1956) 143.
2 V. ANGER, unpublished studies.

(3) Identification of pyridoxal by partial tests[1]

   The structural formula shows that pyridoxal is not only a pyridinealde-
hyde, but also an *o*-hydroxyarylaldehyde as well as a phenolic compound
with a free *p*-position.

H_3C
HO— —CH_2OH
CHO

   Appropriate tests are at hand for these parts of the pyridoxal molecule,
and the positive response to them permits a sure identification:
(1) Detection of the pyridine-aldehyde skeleton by condensation with
phenylhydrazine acetate (see p. 386).
   Limit of identification: 0.5 γ pyridoxal.

(2) Detection of the phenolic character of the OH group by means of the
Gibbs test (see p. 185).
   Limit of identification: 0.6 γ pyridoxal.

(3) Detection of the *o*-hydroxyaldehyde group through the formation of the
fluorescing aldazine (see p. 341).
   Limit of identification: 3 γ pyridoxal.

1 L. BEN-DOR, *Chemist-Analyst*, 53 (1964) 8.

28. Tests for rancidity of fats and oils[1]

When fats and oils are stored in contact with air, they deteriorate because of the production of small amounts of hydroxy fatty acids, ketones, aldehydes, and organic peroxides. The resulting alteration in taste and odor, *i.e.* rancidity, is due primarily to organic peroxides, hydroxy fatty acids, and aldehydes. Spot reactions can be employed to reveal the presence of these three types of compounds.

1 J. SCHMIDT and W. HINDERER, *Ber.*, 65 (1932) 87.

(1) Detection of organic peroxides

2,7-Diaminofluorene (I), in acetic acid solution, is converted into the blue quinoneimine compound (II) by the action of many oxidizing agents.[1]

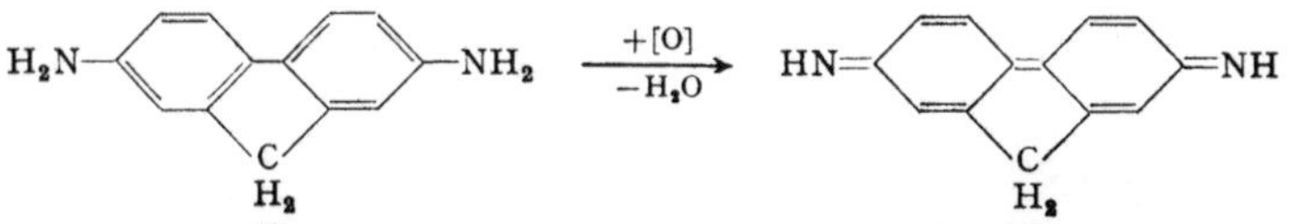

However, organic peroxides and hydrogen peroxide do not have sufficient oxidizing power to convert (I) into (II), but certain ferments, namely the peroxidases, such as the hemin of blood, raise the oxidation power sufficiently so that imine formation results.

Procedure II. A drop of the reagent is placed on filter paper and spotted with a drop of a saturated solution of the fat or oil in peroxide-free ether. A light or deep blue stain appears immediately or within a short time, depending on the peroxide content of the sample. Usually the center of the fleck is green to greenblue, presumably due to a mixture of the blue imine (II) and the yellow to redbrown Schiff base of the reagent with aldehydes in the rancid sample.

Reagent: 100 mg of 2,7-diaminofluorene plus 5 mg hemin dissolved in 5 ml glacial acetic acid. The solution should be freshly prepared.

Another sensitive test for organic peroxides is based on the fact that they, like hydrogen peroxide, convert iron(II) salts into iron(III) salts and the latter in acid solution oxidize N,N-dimethyl-*p*-phenylenediamine to Wurster's red[2] (see p. 513).

Procedure II. A drop of the test solution (glacial acetic acid–chloroform, 1 : 1) is placed on filter paper and treated in succession with a drop of 0.1% ferrous ammonium sulfate and a drop of the reagent. A pink color appears immediately if peroxides are present. The color can be intensified by warming the spotted portion of the paper on a hot plate or in an oven.

Reagent: Freshly prepared 0.1% solution of N,N-dimethyl-*p*-phenylene-diamine in a 5 : 5 : 1 mixture of glacial acetic acid, chloroform, and water.

1 J. Pritzker and R. Jungkunz, *Z. Untersuch. Lebensm.*, 54 (1927) 2425.
2 A. Vioque and E. Vioque, *Grasas y Aceites*, 13 (1962) 211.

(2) *Detection of hydroxyfatty acids* [1]

Hydroxyfatty acids can be detected by the red color which they give with *sym*-diphenylcarbazide. The chemistry of this reaction is not known.

Procedure. One drop of the clear diphenylcarbazide solution is placed on filter paper and, after evaporation of the solvent, is spotted with a drop of a saturated solution of the fat or oil in ether or petroleum ether. The presence of hydroxyfatty acids is indicated by a red fleck.

Reagent: 0.5 g diphenylcarbazide is dissolved in 100 ml of tetrachloroethane, warmed, and filtered after cooling.

Free fatty acids in rancid fats and oils can be detected by the color (fluorescence) reaction with uranyl nitrate and Rhodamine B. The procedure given on p. 704 is suitable.

1 J. Stamm, *Z.Untersuch. Lebensm.*, 62 (1931) 413; St. Korpaczy, *Z. Untersuch. Lebensm.*, 67 (1935) 75.

(3) *Detection of epihydrinaldehyde*

Epihydrinaldehyde is characteristic of the aldehydes in rancid oils and fats. However, there is no proportionality between the degree of rancidity and the content of this rather unstable aldehyde, which is distinctly volatile at room temperature. It may readily be detected by its color reaction with phloroglucinol in the presence of hydrochloric acid.[1] A mechanism for this reaction has been proposed.[2]

Procedure. The fat or oil is thoroughly mixed in a micro crucible with an equal volume of concentrated hydrochloric acid. The crucible is then covered with a piece of filter paper which has been spotted with a 0.1% alcoholic solution of phloroglucinol and several drops of dilute hydrochloric acid. If epihydrinaldehyde is present, a red color appears at once or after gentle warming (40°). Another highly interesting test for rancidity has been described.[3]

1 K. Taeufel and P. Sadler, *Z. Untersuch. Lebensm.*, 67 (1934) 268.
2 J. Pritzker and R. Jungkunz, *Z. Untersuch. Lebensm.*, 54 (1927) 247; J. Pritzker, *Helv. Chim. Acta*, 11 (1928) 445.
3 H. Schmidt, *Naturwiss.*, 46 (1959) 879.

29. Test for pharmaceutical preparations which split off
formaldehyde[1]

Certain important pharmaceutical preparations release formaldehyde when warmed with concentrated sulfuric acid. This hydrolytic splitting is accomplished by the small but definite water content of the concentrated acid. The formaldehyde can be detected by the delicate color reaction with chromotropic acid dissolved in concentrated sulfuric acid (see p. 434).

Procedure.   A little of the solid is placed in a micro test tube or a drop of the test solution is evaporated there. Four drops of concentrated sulfuric acid and 1 or 2 mg of chromotropic acid (as pure as possible) are added. Depending on the amount of formaldehyde released, a more or less intense violet develops at once or within a few minutes.

The hydrolysis may also be conducted in the gas absorbing apparatus (Fig. 23, *Inorganic Spot Tests*). The formaldehyde is revealed in the vapor phase by coming in contact with the suspended drop of chromotropic acid reagent.

The procedure will serve to characterize the preparations listed here under (*a*)–(*g*). The spot reactions used as supplementary identification tests are discussed in detail at other places in this volume.

1 F. FEIGL and E. SILVA, *Drug Std.*, 23 (1955) 113.

(*a*) *Tannaform*

Tannaform, melting at 230° with decomposition, is a water-insoluble condensation product of tannic acid with formaldehyde; it is soluble in ethanol. When dry-heated, it releases formaldehyde.

(*b*) *Methenamine compounds*

Methenamine (hexamethylenetetramine, hexamine) and its salts are cleaved by concentrated sulfuric acid:

$$(CH_2)_6N_4 + 6\ H_2O \rightarrow 6\ CH_2O + 4\ NH_3$$

The procedure was tried with success on very small amounts of the following solids: methenamine, anhydromethylene citrate (Halmitol), methenamine camphorate (Amphotropin), methenamine mandelate, methenamine salicylate (Saliformin).

Methenamine salts may also be detected by the test with β-naphthylamine (p. 498).

(*c*) *Choligen (Bilamid)*

Choligen (Bilamid) is the trade name applied to N-hydroxymethyl-nicotinamide (I). It is chemically related to the widely used nicotinamide (II) and nikethamide (III):

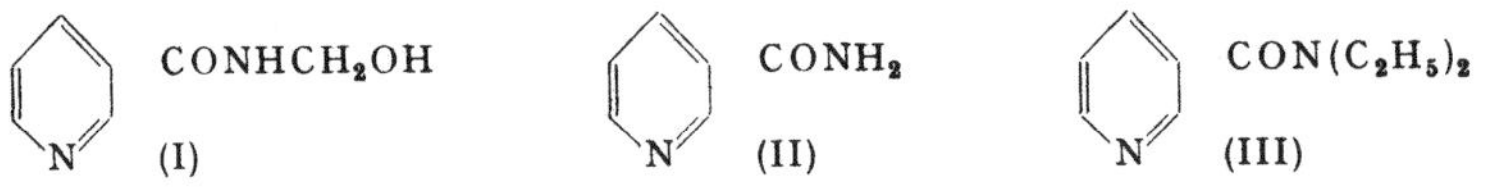

Only (I) yields formaldehyde when warmed with concentrated sulfuric acid.

The detection of the pyridine component can serve as a supplementary identifying test and to differentiate choligen from the other formaldehyde-yielding compounds named here. This test is based on the conversion of pyridine and its derivatives into glutaconic aldehyde (or its derivatives) by means of bromocyanogen. Glutaconic aldehyde and its derivatives form red-violet Schiff bases with o-tolidine (compare the test for pyridine, p. 384).

(d) Monomethylol-dimethylhydantoin

This compound is a white water-soluble crystalline solid recommended as an odorless formaldehyde donor. The release of formaldehyde by cleavage with concentrated sulfuric acid can be represented:

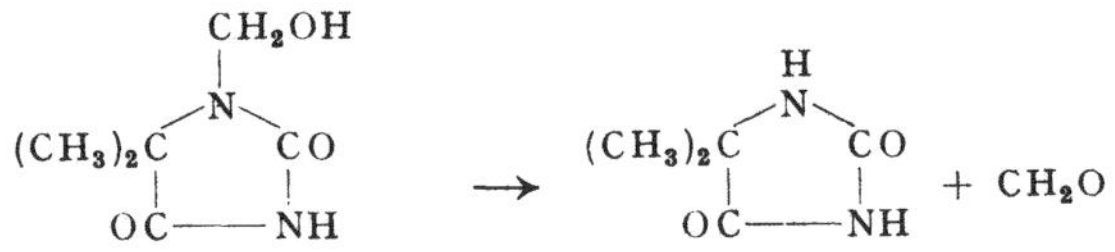

In conformity with this equation, formaldehyde is likewise split off when the material is dry-heated above its melting point (110–117°). The decomposition is noticeable at temperatures as low as 70°.

(e) Sulfoxone sodium (diazone sodium)

This compound has been recommended as a remedy for leprosy.[1] It is split by concentrated sulfuric acid:

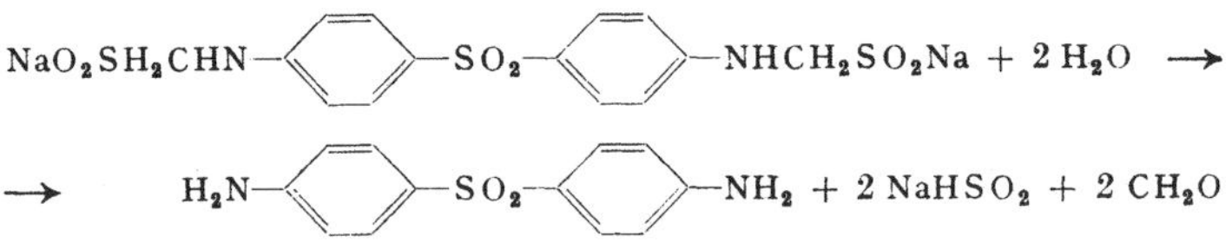

An additional identification test is based on its high reducing power, which is due to its being a derivative of Rongalite (formaldehyde–sodium sulfoxylate) whose detection is outlined on p. 236. Consequently, a water extract of preparations containing sulfoxone sodium yield the color reaction (orange)

on treatment with an ammoniacal alcoholic solution of 1,4-dinitrobenzene.

In addition, the release of hydrogen sulfide when the compound is dry-heated at 250°:

$$2 \text{ NaHSO}_2 \rightarrow \text{Na}_2\text{SO}_4 + \text{H}_2\text{S}$$

can also be utilized to detect the sulfoxylate component[2] (see f).

1 *New and Non-official Remedies*, ed. by the Council on Pharm. and Chem., Am. Med. Ass., Philadelphia, 1954, p. 120.
2 Unpublished studies, with Cl. COSTA NETO.

(f) Neosalvarsan and myosalvarsan

Neosalvarsan (I) and myosalvarsan (II) are derivatives of formaldehyde sulfoxylate and formaldehyde bisulfite respectively. Accordingly, they yield formaldehyde when warmed with concentrated sulfuric acid.

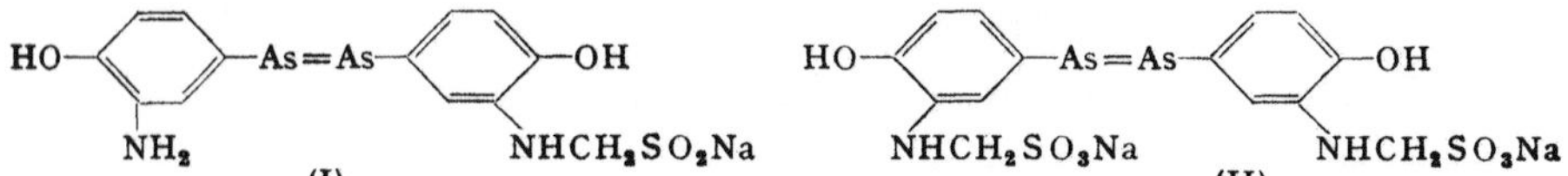

These preparations may be distinguished from each other since only neosalvarsan, as a derivative of Rongalite, gives the color reaction with an ammoniacal solution of 1,4-dinitrobenzene and also it alone responds to the test based on the evolution of hydrogen sulfide described in (e).

As little as 0.1 γ neosalvarsan and 0.5 γ myosalvarsan can be detected by the formaldehyde test. The sulfoxylate test will reveal as little as 15 γ neosalvarsan.

(g) Pyramidone

In the case of pyramidone, the splitting off of formaldehyde is not the result of a hydrolysis but is due to an oxidation by concentrated sulfuric acid (compare p. 635).

30. Detection of preparations containing tannin[1]

Tannin is widely used in pharmaceutical preparations because of its astringent action. Among these are: Albutannin (albumin tannate), Tannaform (tannin formaldehyde), Protan (tannin nucleoprotein).

These products are water-soluble salts or adsorption compounds which taken by mouth yield tannin on hydrolysis in the stomach or intestine. Because of this hydrolysis, tannin preparations show the selective color reaction (p. 135) characteristic of strong reducing agents of which tannin is one. The color reaction is based on the reduction of *p*-nitrosodimethyl-aniline to *p*-aminodimethylaniline, whose condensation with *p*-dimethyl-aminobenzaldehyde leads to a red product.

1 F. Feigl and D. Goldstein, *Microchem. J.*, 1 (1957) 177.

31. Detection of salol

Salol (phenyl salicylate) is used as analgetic, antipyretic and intestinal antiseptic. A rapid test can be based on the fact that the compound, being a phenol ester, may be pyrohydrolyzed, whereby salicylic acid and volatile phenol are formed. The pyrohydrolysis is conducted by heating with oxalic acid dihydrate. The phenol formed is identified in the gas phase by the indophenol color reaction described on p. 185. The sensitive color test for phenoxy compounds described on p. 319 may also be used.

32. Detection of antipyrine[1]

The test given on p. 509 for the detection of α-naphthylamine by means of nitrosoantipyrine is based on the formation of a red-violet azo (pyrazolone) dyestuff. The procedure may be reversed to constitute a specific test for antipyrine (1,5-dimethyl-2-phenyl-3-pyrazolone). It thus becomes possible to distinguish between this analgesic and antipyretic agent and pyramidone (see p. 635) which is used for the same purposes.

The antipyrine is nitrosated in acetic acid[2] as shown in (*1*), the excess nitrous acid is destroyed by sodium azide (*2*), and then α-naphthylamine is introduced to accomplish the condensation to the azo dye (*3*):

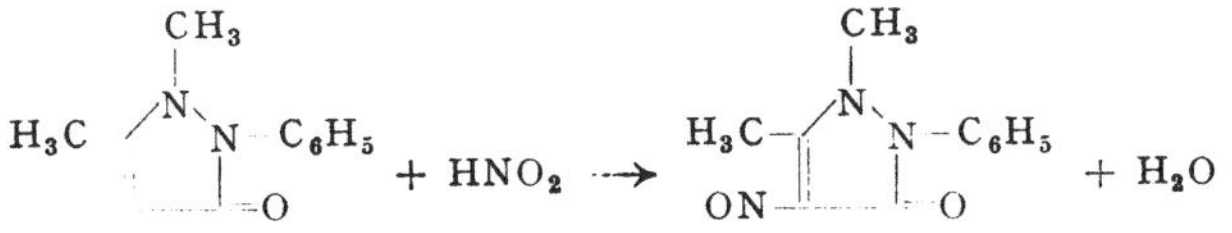

$$HNO_2 + HN_3 \longrightarrow H_2O + N_2O + N_2 \qquad (2)$$

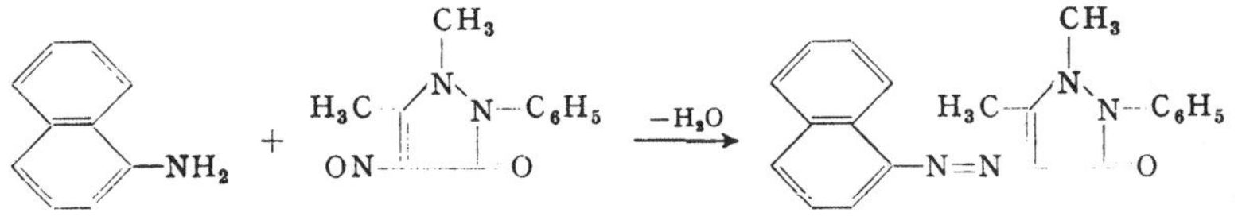

Under the conditions prescribed here, 0.1–0.5 mg of pyramidone gives a faint pink color, which probably arises from a small content of antipyrine, which is the starting material in the preparation of pyramidone.

Procedure.   A drop of the test solution is united in a micro test tube with a drop of glacial acetic acid and a drop of 5% sodium nitrite solution. After 5 min. a pinch of sodium azide is added. When the evolution of gas has stopped, several mg of solid α-naphthylamine are added and the tube is warmed in a water bath. Depending on the amount of antipyrine, a deep or pale violet color appears.

Limit of identification: 2 γ antipyrine.

The structure of antipyrine shows that it contains the NCH$_3$ group, which may be identified through the formation of formaldehyde when fused with benzoyl peroxide (see p. 165). The test will reveal 100 γ antipyrine.

The test is rendered more specific if K-acid is used in place of α-naphthylamine (see p. 579).

In acid solution, antipyrine acquires the character of a phenol with free *ortho*-position and therefore it responds to the test with sodium hexanitritocobaltiate (see p. 183). The test succeeds with antipyrine even in the cold (distinction from phenols).

Limit of identification: 1 γ antipyrine.

1 F. FEIGL, Cl. COSTA NETO and E. SILVA, *Anal. Chem.*, 27 (1955) 1319.
2 L. KNORR, *Ber.*, 17 (1884) 2038.

33. Test for barbiturates [1]

Derivatives of barbituric acid (malonylurea), in which the hydrogen atoms of the CH$_2$ group are replaced by alkyl or aryl groups (or both) are excellent soporifics. It is characteristic of these compounds that they are readily cleaved by warming with caustic alkali solutions:

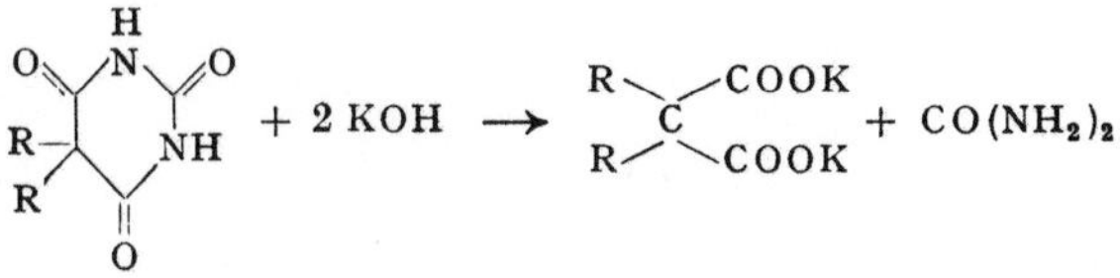

The urea resulting from this hydrolysis can be readily detected in the alkaline solution by means of the test given on p. 529. This test is based on the production of hydrazine from the Hofmann degradation of urea followed by the formation of fluorescing salicylaldazine.

It is notable that barbituric acid, the parent substance of all the barbiturates, does not respond to this test. This failure makes it possible to detect barbiturates in the presence of barbituric acid.

Procedure. A drop of the test solution is placed in a micro test tube along with 2 drops of 5% sodium hydroxide solution. The mixture is kept in a boiling water bath for 3 min. The subsequent treatment of this hydrolyzed material is described on p. 690.

The following were detected: 25 γ luminal (5-ethyl-5-phenylbarbituric acid)

20 γ veronal (5,5-diethylbarbituric acid)

1 Unpublished studies, with V. GENTIL.

34. Detection of novocaine[1]

This compound, widely employed as substitute for cocaine, is the hydrochloride of procaine base, *i.e.* 2-diethylaminoethyl *p*-aminobenzoate

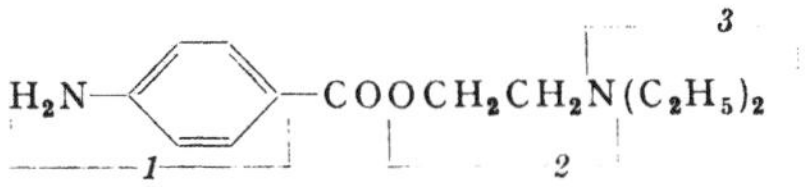

The structural formula shows that novocaine contains three groups that can be detected by spot tests given in Chap. 4:

(*1*) is detectable through the color reaction for primary arylamines (see p. 245);

(*2*) is revealed by the release of acetaldehyde on pyrolysis in the presence of zinc chloride (see p. 166);

(*3*) is detectable through the release of acetaldehyde on heating with benzoyl peroxide (comp. p. 252).

Novocaine can be identified through positive responses to tests *1*), *2*) and *3*) employing micro quantities, and also thus differentiated from the cocaine surrogates: stovaine, panthesine, tutocaine.

1. Unpublished studies.

35. Detection of Dial[1]

A test for the propylene group is described on p. 168; it is based on the fact that compounds of this kind undergo an addition of iodine on the double bond $-CH=CHCH_3$ if they are taken to dryness with a solution of iodine in carbon disulfide. This iodination can be detected through the positive response of the evaporation residue when the latter is subjected to the test for iodine described in Chap. 2. Since allyl compounds isomeric with the propylene compounds are not iodinated by this treatment, Dial, diallyl-barbituric acid (I), remains unaltered. But if Dial is taken to dryness with caustic alkali and then with hydrochloric acid, the resulting residue can be iodinated. Obviously the heating with alkali converts Dial into the isomeric dipropylenebarbituric acid (II):

$$H_2C=CH-CH_2 \diagdown C \diagup CO-NH \diagdown CO \qquad \xrightarrow{KOH} \qquad H_3C-CH=CH \diagdown C \diagup CO-NH \diagdown CO$$
$$H_2C=CH-CH_2 \diagup C \diagdown CO-NH \diagup \qquad\qquad H_3C-CH=CH \diagup C \diagdown CO-NH \diagup$$
$$\text{(I)} \qquad\qquad\qquad\qquad\qquad \text{(II)}$$

Therefore, a test for Dial can be based on the above conversion.

Procedure. A drop of the alcoholic test solution is mixed on a watch glass with a drop of 5% sodium hydroxide. After evaporation to dryness, the residue is kept for about 10 min. in a drying oven at 160–180°. After cooling, the excess alkali is destroyed by adding 2 drops of hydrochloric acid (1 : 1) and taken to dryness. The residue is allowed to stand at room temperature for 5 min. with 6 drops of a 2% solution of iodine in carbon disulfide. The latter is then volatilized and the residue is kept at 140° for 30 min. (drying oven) and cooled. A drop of water is added and the evaporation to dryness is repeated. This latter step is essential to guarantee the complete removal of any iodine incrusted in alkali chloride. The further treatment with bromine water is the same as in the test for propylene compounds.

Limit of identification: 5 γ Dial.

1 F. FEIGL and E. LIBERGOTT, Z. Anal. Chem., 192 (1962) 93.

36. Detection of phenacetin[1]

Phenacetin (acetophenetidine or *p*-ethoxyacetanilide) is often employed medicinally as an analgesic and antipyretic. This compound can be detected through the fact that it yields phenetidine when warmed with caustic alkali. This product in turn is readily revealed by the procedure described on p. 517, which involves oxidation with chloramine T to yield *p*-quinone-4-chloroimine followed by condensation with phenol to give a blue indophenol dye.

Procedure. A drop of the test solution is placed in a micro test tube and warmed with 1 or 2 drops of 2% sodium hydroxide. After cooling, the reaction mass is made acid by adding a drop of concentrated hydrochloric acid, and then a pinch of chloramine T is added. After standing for about 5 min., ether is added and the test conducted from this point on as outlined on p. 517.

Limit of identification: 10 γ phenacetin.

1 F. FEIGL, Cl. COSTA NETO and E. SILVA, Anal. Chem., 27 (1955) 1319.

37. Detection of pyramidone[1]

Pyramidone (4-dimethylamino-1,5-dimethyl-2-phenyl-3-pyrazolone) is a widely used analgesic and antipyretic. It differs from its parent antipyrine not only in the fact that it cannot be nitrosated but also in two characteristics related to its $N(CH_3)_2$ group. In contrast to antipyrine which is neutral, pyramidone is a base and also a reductant.

The reducing action of pyramidone is related to the tendency to tautomerize, a general characteristic of pyrazolones.[2] The oxidation probably begins with the isomeric methoxy form[3] of the pyrazolone:

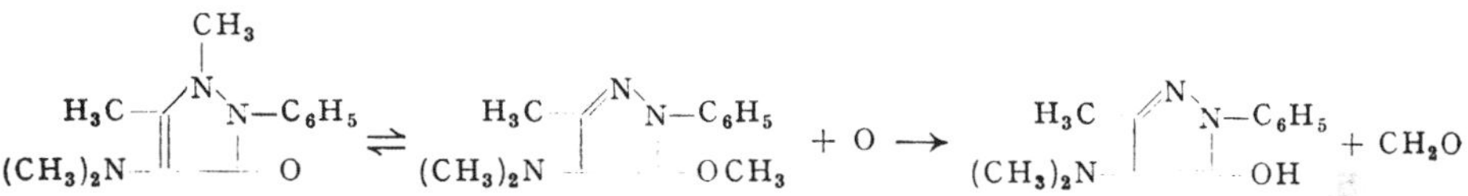

The basic character of pyramidone can be demonstrated by moistening a grain of the solid with a drop of nickel dimethylglyoxime equilibrium solution (p. 109). Red nickel dimethylglyoxime is precipitated.

The reducing action is shown by treating a little of the solid or a drop of its solution in dilute acid with a drop of ferri ferricyanide solution (p. 86). A blue color or precipitate appears (see USA Pharmacopoeia).

Another identification test involves the formaldehyde produced by oxidizing pyramidone. If the solid is gently warmed with a mixture of chromotropic acid and sulfuric acid, the oxidizing action of the latter results in the violet color characteristic for formaldehyde (p. 434). The *limit of identification* is 10 γ pyramidone.

When testing salts of pyramidone, obviously the only tests that may be applied are those based on the reducing action of the parent compound.

1 F. FEIGL and E. SILVA, *Drug Std.*, 23 (1955) 113.
2 P. KARRER, *Organic Chemistry*, 4th Engl. ed., Elsevier, Amsterdam, 1950, p. 798.
3 J. A. SANCHEZ, *Curso de Quimica Analytica Funcional de Medicamentos Organicos*, 2nd ed., Vol. I, Buenos Aires, 1947, p. 438.

38. Detection of morphine and codeine[1]

The opium alkaloids morphine and its monomethyl ether codeine (I) are widely used as narcotics, analgesics, and sedatives. If these compounds are fused at 180° with zinc chloride, there is rapid and extensive conversion into apomorphine and its monomethyl ether apocodeine (II), respectively.[2]

Since these products are phenolic compounds with an unoccupied *para* position, they respond readily to the sensitive indophenol color reaction with 2,6-dichloroquinone-4-chloroimine, described and discussed on p. 185.

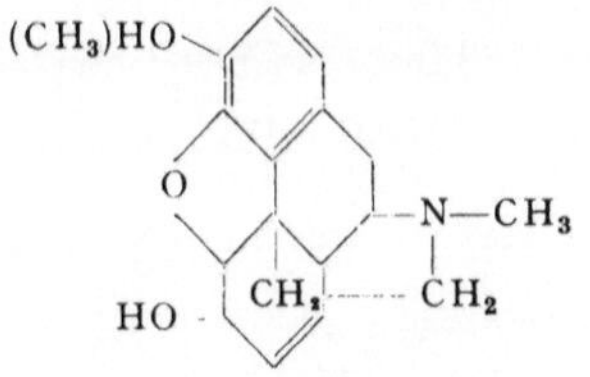

(I) 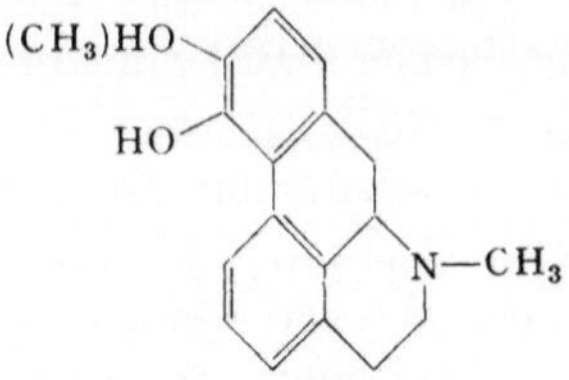   (II)

Accordingly, the zinc chloride fusion together with the indophenol reaction provides a rapid test for morphine and codeine, provided no phenolic compounds with a free *para* position capable of reacting with 2,6-dichloroquinone-4-chloroimine are present. This category of reactive compounds also includes vitamin B_6 (see p. 624)

Procedure. Minimal amounts of the test material or a drop of the test solution is placed in a micro test tube and brought to dryness if need be. A drop of syrupy zinc chloride is added, and the tube is kept for 5 min. in a bath previously heated to 180°. After cooling, the contents of the tube are dropped on filter paper impregnated with a saturated benzene solution of 2,6-dichloroquinone-4-chloroimine. The reagent paper is then fumed over concentrated ammonia. A blue stain indicates a positive response.

Limit of identification: 4 γ morphine and 2 γ codeine.

Morphine and codeine may be differentiated by the test for codeine given in the next section to which morphine does not respond.

1 Unpublished studies, with E. JUNGREIS.
2 E. L. MAYER, *Ber.*, 4 (1871) 121.

39. Detection of codeine

Test through transformation into apocodeine [1]

If codeine or its salts are fused with anhydrous oxalic acid, apocodeine results.[2] The latter as a phenol with a free *para* position shows the indophenol color reaction with 2,6-dichloroquinone-4-chloroimine (see p. 185).

The procedure given here cannot of course be applied directly to samples that also contain phenolic compounds which respond positively to the reagent. Other alkaloids, even the closely related morphine, do not give this test.

Procedure. Small amounts of the test material or a drop of its solution are placed in a micro test tube and taken to dryness if necessary. About 1 cg of anhydrous oxalic acid is added, and the test tube is kept for several minutes in a bath previously heated to 180°. After cooling, the entire contents of the tube are dropped on filter paper impregnated with a saturated benzene solution of 2,6-dichloroquinone-4-chloroimine. The paper is dried and held above concentrated ammonia. A positive reaction is shown by the development of a blue stain.

Limit of identification: 2 γ codeine.

In case interfering phenolic compounds are present, it is necessary to isolate the alkaloid beforehand. This is readily accomplished by making the test solution basic with caustic alkali and then extracting with ether. It is also possible to run the test on the acid-insoluble phosphomolybdate or tungstate of codeine.

1 Unpublished studies, with E. JUNGREIS.
2 L. KNORR and P. ROTH, *Ber.*, **40** (1907) 3357.

40. Detection of ephedrine

(1) *Test through hydramine cleavage*[1]

Compounds which have an OH group in the α-position to an aromatic ring and an amino group in the β-position undergo the so-called hydramine cleavage when subjected to dry heating.[2] The corresponding aromatic-aliphatic ketone results through a change in the position of two hydrogen atoms, and the release of the amine:

$$\text{—CH(OH)—CH—} \rightarrow \text{—CO—CH}_2\text{—} + \text{NH}_2\text{R}$$
$$\qquad\qquad\underset{\text{NHR}}{|}$$

The hydramine cleavage can be accomplished not only with the corresponding salts, but also in the wet way by warming with caustic alkali. Accordingly, ephedrine yields phenyl ethyl ketone and methylamine:

$$C_6H_5CH(OH)-\underset{\underset{NHCH_3}{|}}{C}HCH_3 \rightarrow C_6H_5COCH_2CH_3 + NH_2CH_3$$

Since the hydramine cleavage occurs within a few minutes, and also with small amounts, the detection of methylamine may serve as an indirect test for ephedrine. The condensation reaction of monomethylamine with 2,4-dinitrochlorobenzene can be employed. It leads to a yellow product that is soluble in chloroform (see p. 240).

The detection of the hydramine cleavage, and the associated indirect detection of ephedrine, can be accomplished in two ways. Use can be made of the fact that the methylamine is volatile and hence detectable in the vapor phase (Procedure I). The second method is to warm an ethereal solution of the alkaloid with caustic alkali and an alcoholic solution of 2,4-dinitrochlorobenzene (Procedure II).

Procedure I. The gas absorption apparatus shown in Fig. 23, *Inorganic Spot Tests*, is used. A little of the solid or a drop of its weakly acid solution is placed in the bulb and a drop of 5 *N* alkali added. A drop of a saturated solution of 2,4-dinitrochlorobenzene in alcohol is put on the knob of the stopper. The closed apparatus is kept in a warm water bath for 10 min. If ephedrine is present, the yellow condensation product forms in the hanging drop. It is advisable to rinse the drop into a micro test tube with as little chloroform as possible, and to shake the liquid after adding 1–2 drops of dilute acetic acid. A yellow chloroform layer is characteristic for a positive response.

Limit of identification: 7 γ ephedrine.

This procedure assumes the absence of compounds which yield methylamine when they undergo the hydramine cleavage. Adrenaline is a pertinent instance, also the salts of volatile organic bases.

Procedure II. One drop of the ethereal solution of the test material is mixed in a micro test tube with a drop of 5 *N* alkali and one drop of a 1% ethereal solution of dinitrochlorobenzene. After 5 min. warming in a water bath, the mixture turns yellow-brown. A drop or two of chloroform and several drops of dilute acetic acid are added to the cooled solution and the mixture is shaken. The chloroform layer turns honey yellow to pale yellow.

Limit of identification: 5 γ ephedrine.

Procedure II can be used to detect ephedrine in the presence of adrenaline, since the latter is not soluble in ether. On the other hand, care must be taken to insure the absence of those amines which react similarly with 2,4-dinitrochlorobenzene.

In contrast to adrenaline, ephedrine (after hydramine cleavage) gives the reaction for secondary amines with an ammoniacal solution of copper sulfate and carbon disulfide as described on p. 250; see also Sect. 42 and 43.

1 F. FEIGL and H. E. FEIGL, *Helv. Chim. Acta*, 38 (1955) 459.
2 P. KARRER, *Organic Chemistry*, 4th English ed., Elsevier, Amsterdam, 1950, p. 830.

(2) Test through dry heating with sodium bismuthate [1]

As stated in Test (*1*), alkali readily achieves hydramine cleavage of ephedrine to yield phenyl ethyl ketone and monomethylamine. If heated with sodium bismuthate, ephedrine or its salts likewise undergo hydramine cleavage followed by pyrolytic oxidation of the resulting phenyl ethyl ketone:

$$C_6H_5COCH_2CH_3 + 2 O \rightarrow C_6H_5COOH + CH_3CHO$$

The sodium bismuthate ($NaBiO_3$) functions both as alkalizer and oxidant. The acetaldehyde formed can be detected in the gas phase through the color reaction with sodium nitroprusside and morpholine (see p. 438). Thus there results a characteristic test for ephedrine, which can be employed in the presence of any amount of adrenaline.

Procedure. A tiny amount of the solid or the evaporation residue from a drop of a solution of the alkaloid or its salts is mixed with several cg of sodium bismuthate in a micro test tube. The mouth of the tube is covered with a disk of filter paper moistened with a drop of the reagent for acetaldehyde (for preparation see p. 439). The bottom of the test tube is heated with a micro flame. If the response is positive, a more or less blue stain appears on the yellow paper.

If ephedrine is to be detected in very dilute solution, it is best to place some glass powder in the test tube and to evaporate a drop of the solution on it and then to mix the residue with the sodium bismuthate.

Limit of identification: 5 γ ephedrine.

When using this test, it should be noted that compounds with OC_2H_5 and NC_2H_5 groups likewise yield acetaldehyde when heated with $NaBiO_3$, but to a lesser extent than ephedrine.

1 F. FEIGL and E. SILVA, *J. Am. Pharm. Assoc.*, 47 (1958) 460.

41. Detection of adrenaline

Neither of the tests given here is decisive for adrenaline, but if both give a positive response there is a strong probability that adrenaline is present. The same is true of its salts and derivatives.

(1) Test through hydramine cleavage[1]

When warmed with caustic alkali, adrenaline, like ephedrine, undergoes hydramine cleavage to yield 3,4-dihydroxyacetophenone and methylamine (see p. 637). The latter, and therefore adrenaline indirectly, can be detected by condensation with 2,4-dinitrochlorobenzene to give an orange-yellow compound that is soluble in organic liquids. The method, described as Procedure I for ephedrine (see Sect. 40), should be employed; it involves the detection of methylamine in the vapor phase. The *limit of identification* is 5 γ adrenaline. It must be remembered that ephedrine and volatile aliphatic and aromatic amines react in the same manner.

Procedure II for ephedrine cannot be employed because adrenaline is not soluble in ether. A direct test for adrenaline in aqueous alkaline solutions is uncertain because such solutions turn red on contact with the air and this interferes with the discernment of the condensation product of methylamine with 2,4-dinitrochlorobenzene.

1 F. FEIGL and H. E. FEIGL, *Helv. Chim. Acta*, 38 (1955) 459.

(2) Test by means of 1,2-dinitrobenzene[1]

By virtue of its being an o-diphenol and also because of the CH(OH) group in the α-position, adrenaline is a powerful reducing agent. In alkaline solution it functions as hydrogen donor and thus can be detected through the color reaction with 1,2-dinitrobenzene which is characteristic for such compounds (see p. 133).

Procedure. A micro test tube is used. A drop of the aqueous or alcoholic test solution or a grain of the solid is treated with a drop of a saturated solution of 1,2-dinitrobenzene in alcohol and then with a drop of 0.5 N sodium hydroxide. The mixture is heated not longer than one minute over a free flame. The resulting color is more or less blue according to the amount of adrenaline present.

Limit of identification: 5 γ adrenaline.

This test should not be used in the presence of compounds which act as reductants in alkaline solution.

1 M. PESEZ and P. POIRIER, *Méthodes et Réactions de l'Analyse Organique*, Masson et Cie., Paris, 1954; comp. F. FEIGL and H. E. FEIGL, *Helv. Chim. Acta*, 38 (1955) 459.

42. Detection of emetine and cephaeline

Test through formation of benzene-soluble copper dithiocarbamate[1]

Emetine and the structurally very similar cephaeline are the most important alkaloids in the rhizome of ipecacuanha. The constitutional formula of emetine[2] shows that the molecule contains a heterocyclic NH group (cephaeline has OH in place of one of the OCH₃ groups):

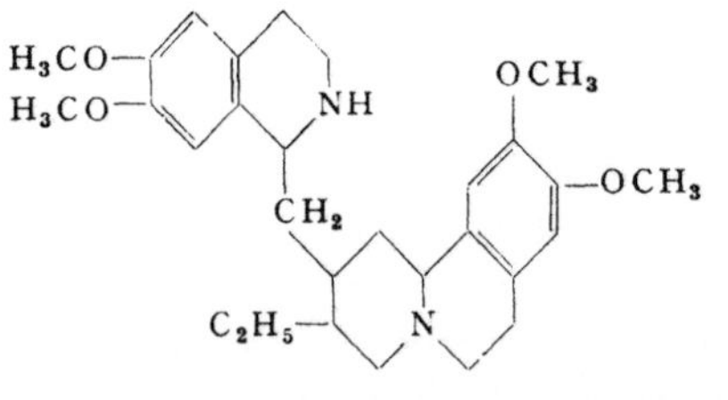

Because of the NH group, the alkaloid behaves analogously to other secondary aliphatic amines toward ammoniacal copper salt solution and carbon disulfide. A brown, benzene-soluble copper salt of a dithiocarbamic acid results (see p. 250).

Procedure. A drop of the acid test solution is treated in a micro test tube with a drop of a 5% solution of copper sulfate and two or three drops of concentrated ammonia. The mixture is shaken with a few drops of a 1 : 3 mixture of carbon disulfide and benzene. A positive response has been obtained if the benzene layer is brown to yellow.

Limit of identification: 2 γ emetine.

This procedure revealed emetine and cephaeline in the hydrochloric acid extract of dried ipecacuanha powder.

Alkaloids that contain an NH group may also be expected to respond to the test for this group. The anhalonium alkaloids (anhalonimide and anhalamine) as well as salsoline belong in this category.

1 Unpublished studies, with J. HAINBERGER.
2 M. PAILER and K. PORSCHINSKY, *Monatsh.*, 80 (1949) 94.

43. Detection of cinchonine and quinine

Test through hydramine fission[1]

If cinchonine (I) is warmed with acetic acid, it undergoes hydramine cleavage (see detection of ephedrine, p. 637). Cinchotoxine (II) results:

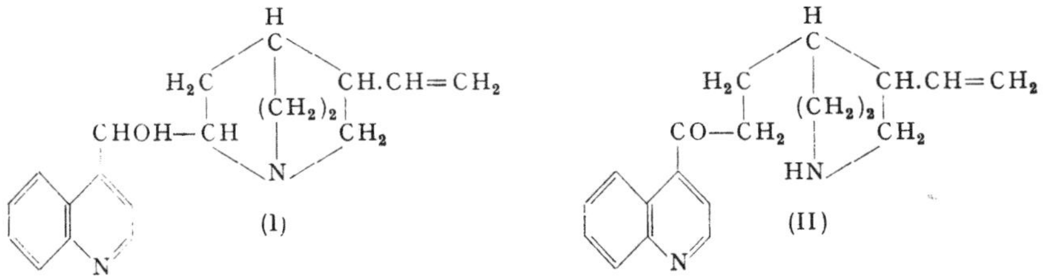

Quinine, which differs from cinchonine merely in a OCH_3 group in the 6'-position, yields the analogous keto compound, quinotoxine, when it undergoes the hydramine cleavage. Both quinotoxine and cinchotoxine are secondary aliphatic bases, which give yellow-brown benzene-soluble copper salts of dithiocarbamic acids on treatment with an ammoniacal copper solution and carbon disulfide (see p. 250).

Procedure. A small quantity of the base and a drop of glacial acetic acid are united in a micro test tube and kept in boiling water for 20 min. The cooled mixture is then subjected to the procedure given for the detection of emetine (see Sect. 42).

If salts of cinchonine or quinine are involved, it is necessary to free the alkaloid prior to the test by evaporating the sample with ammonia.

Limit of identification: 20 γ cinchonine   50 γ quinine hydrochloride

Cinchonidine and hydrocinchonidine, which are isomeric with cinchonine and hydrocinchonine, give the reactions described here, since they likewise yield secondary bases when they undergo the hydramine cleavage.

The test is not appplicable in the presence of emetine and piperine (p. 642) or other secondary aliphatic amines.

Quinine and cinchonine can easily be differentiated through the OCH_3

group contained in quinine. This group is revealed by fusion with benzoyl peroxide to produce formaldehyde (see p. 165).

1 Unpublished studies, with L. HAINBERGER.

44. Detection of piperine

(1) Test through acidic hydrolytic formation of piperidine[1]

The alkaloid piperine (I) is the piperidide of the monobasic unsaturated piperic acid. The acid amide is split by acid hydrolysis into its components, namely piperidine (II) and piperic acid (III):

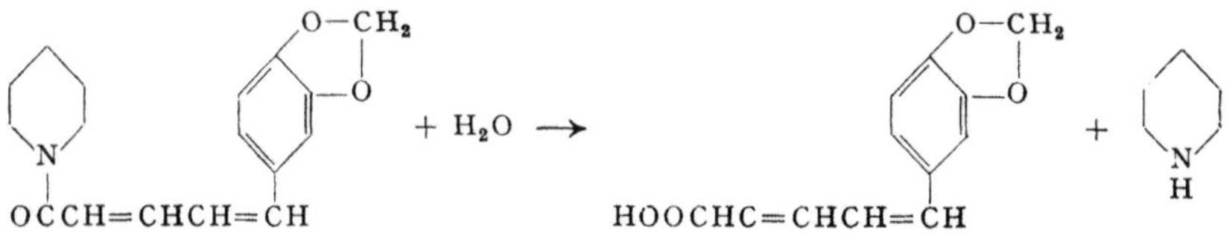

$$+ H_2O \longrightarrow$$

OCCH=CHCH=CH   HOOCHC=CHCH=CH
(I)   (III)   (II)

This hydrolysis can be proved through the detection of the liberated piperidine by the test for secondary amines given on p. 250. Accordingly, the following procedure establishes indirectly the presence of piperine.

The test is specific provided secondary amines are absent. Their absence can be proved prior to hydrolyzing the sample.

Procedure.   A tiny portion of the sample (pulverized pepper will also give satisfactory results) is taken to dryness in a micro test tube along with 2 drops of 1 : 1 hydrochloric acid. The residue is treated with 1 drop of 5% copper sulfate solution and a drop of concentrated ammonia. Two or three drops of a 1 : 3 mixture of carbon disulfide and benzene are added and well shaken. The benzene layer will be brown to yellow if piperidine is present.

Limit of identification: 4 γ piperine.

1 Unpublished studies, with L. HAINBERGER.

(2) Test through alkaline hydrolytic formation of piperidine[1]

The hydrolysis of piperine through the action of aqueous caustic alkali with production of piperidine proceeds slowly and incompletely at water bath temperature. However, due to the excellent solubility of sodium hydroxide and also of piperine in glycerol, the reaction temperature can be raised to 220–240°, with almost immediate liberation of piperidine. The latter (b.p. 106°) volatilizes and can be readily detected in the gas phase by means of the color reaction with sodium 1,2-naphthoquinone-4-sulfonate. (Comp. p. 519).

Procedure. A tiny amount of the solid test material or a drop of its solution is mixed in a micro test tube with 1–2 drops of a saturated glycerol solution of alkali hydroxide and the tube is then immersed in a glycerol bath preheated to 180°. The mouth of the tube is covered with a piece of filter paper moistened with a 0.5% solution of sodium 1,2-naphthoquinone-4-sulfonate. The temperature is brought to 220–230°. A positive response is signalled by the formation of a violet stain on the reagent paper.

Limit of identification: 20 γ piperine.

The method just described is more selective than that given in (*1*) since non-volatile secondary amines do not interfere here. It is imperative that compounds which react in the gas phase with sodium 1,2-naphthoquinone-4-sulfonate be absent (compare p. 153).

1 Unpublished studies, with D. HAGUENAUER-CASTRO.

45. Detection of sulfa drugs

Sulfa drugs of the structure (II) are derivatives of sulfanilamide (I). R usually denotes a nitrogen-bearing heterocyclic, less often guanidine or one of its derivatives.

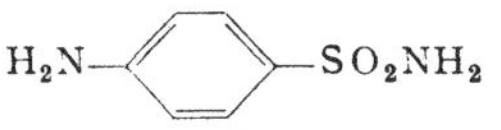

$$H_2N-\langle\ \rangle-SO_2NH_2 \qquad H_2N-\langle\ \rangle-SO_2NHR$$

$$(I) \qquad\qquad\qquad (II)$$

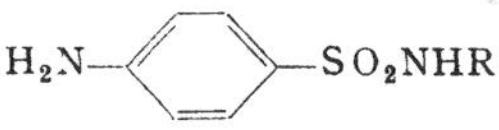

Among these compounds, in which the hydrogen atom of the NH group can be replaced by alkali metal (water-soluble salts result) are: Sulfadiazine, Sulfathiazole, Sulfamerazine, Sulfapyridine, *etc.*

As primary aromatic amines, sulfonamides can be converted to diazonium salts by nitrous acid, which couple with phenols and aromatic amines. If N,N'-dimethyl-α-naphthylamine is used for the coupling, red diazo compounds of the general structure (III) result.

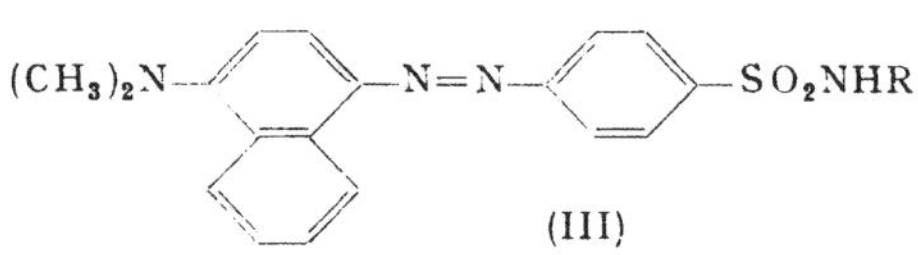

$$(CH_3)_2N-\langle\ \rangle-N=N-\langle\ \rangle-SO_2NHR$$

$$(III)$$

Primary aromatic amines without p-SO$_2$NHR groups yield orange products if diazotized and then coupled with N,N'-dimethyl-σ-naphthylamine.

Procedure.[1]   Filter paper bathed with the reagent and dried in the absence of light and air is spotted with a drop of the test solution and then respotted with a drop of 0.2–0.5% hydrochloric acid. According to the amount of sulfon-amide present, a red stain or ring appears.

Amounts less than 1 γ can be detected.

Reagent: Freshly prepared solution of equal parts of 1% methanolic solution of N,N′-dimethyl-α-naphthylamine and 1% aqueous solution of sodium nitrite.

As aromatic amines the sulfonamides show the color reaction with glu-taconic aldehyde described on p. 245. The test can be conducted on a spot plate. In constrast to other aromatic amines, the sulfonamides do not react with sodium pentacyanoaquoferriate (see p. 248).

Procedure.[2]   A drop of a 1% solution of pyridylpyridinium dichloride is added to a drop of the test solution in a depression of a spot plate. Then a drop of 1 N caustic alkali is added and followed immediately with *one* drop of concen-trated hydrochloric acid. An orange color appears at once if sulfonamides are present. It is advisable to run a blank test.

Limits of identification:

0.05 γ sulfanilamide	0.1 γ sulfamethazine
0.05 γ sulfadiazine	0.2 γ sulfamerazine
0.05 γ sulfaguanidine	

1 C. HACKMANN, *Deut. Med. Wochenschr.*, 72 (1947) 71.
2 V. ANGER and S. OFRI, *Mikrochim. Acta*, (1964) 731.

46. Test for Sulfadiazine[1]

The four sulfa drugs: Sulfadiazine (I), Sulfamerazine (II), Sulfamethazine (III), Sulfapyrazine (IV) have quite similar structures:

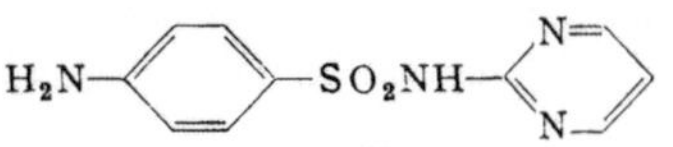

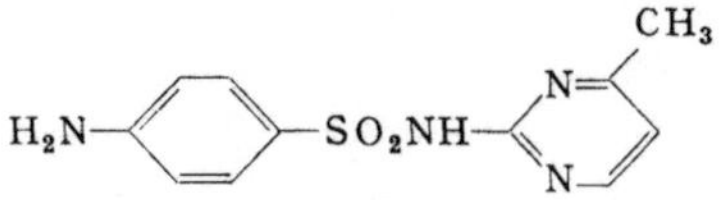

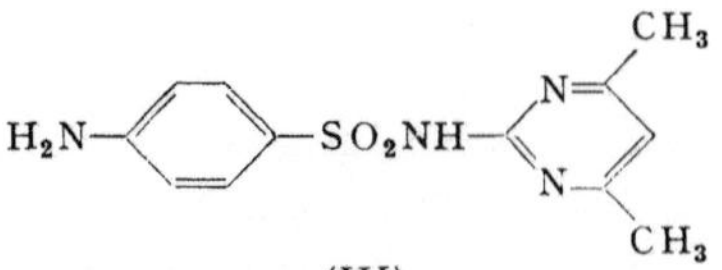

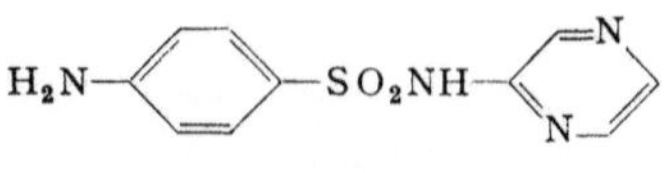

However, only sulfadiazine gives a characteristic reaction when dry-heated with thiobarbituric or barbituric acid. When mixed with either of these reagents and heated to around 190°, a violet or orange product results, which is rather soluble in ethanol. No other sulfa drug tested thus far gives an analogous reaction. Since none of the reactants melts at the specified temperature, the reactions probably fall into the sintering category.

In view of the specificity of the color reaction and the fact that the intact CH_2 group in barbituric and thiobarbituric acid is essential, it may be assumed that a polymethine dye is formed in accord with the equations given on p. 542.

Procedure. A small quantity of the solid is mixed in a micro test tube with 1 mg thiobarbituric acid. The tube is placed in a preheated glycerol bath. The color begins to appear at 140° and reaches its maximum at 190–200°. Thiobarbituric acid yields a violet, barbituric acid an orange product.

Limit of identification: 5 γ sulfadiazine.

1 F. FEIGL, V. ANGER and V. GENTIL, *Clin. Chim. Acta*, 5 (1960) 153.

47. Detection of quaternizable sulfa drugs[1]

The sulfa drugs I–IV cited in Sect. 46, which contain two heterocyclic N atoms, are not converted to the quaternary compounds through evaporation with methyl iodide. However, this does happen with sulfapyridine (a) and sulfathiazole (b) which contain only one N atom in the heterocyclic ring:

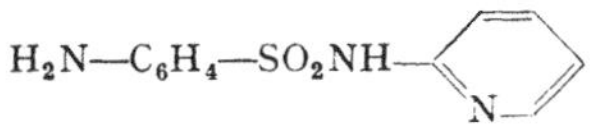

(a) 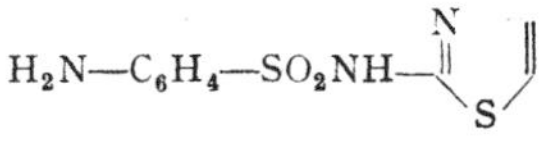   (b)

As stated on p. 377 iodomethylates of this type of tertiary ring bases split off sulfur dioxide when heated with sodium thiosulfate to 180–200°. Therefore quaternizable and non-quaternizable sulfa drugs may be differentiated. Milligram quantities are sufficient to carry out the test.

1 F. FEIGL and V. ANGER, unpublished studies.

48. Tests for sulfa drugs containing urea components[1]

Pertinent examples are: sulfaurea, sulfathiourea and sulfaguanidine. These compounds contain NH_2 groups which condense when dry heated to the respective melting points, whereby ammonia is split off. The latter can be detected by means of indicator paper or Nessler's reagent. In this way the sulfa drugs cited above can be differentiated from all the other sulfa drugs. The identification limits are in the order of 10 γ.

Among the sulfa drugs, whose general detection is given in Sect. 45, sulfathiourea is the only one derived from thiourea. Accordingly, it responds positively to the test that is characteristic for thiourea (and its derivatives) with free NH_2 groups, namely reaction with lead hydroxide or oxide to yield lead sulfide and cyanamide (or N-substituted cyanamide):

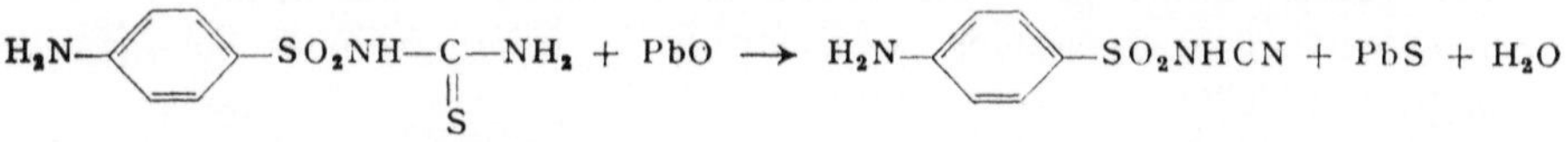

$$H_2N-\langle\ \rangle-SO_2NH-\underset{\underset{S}{\|}}{C}-NH_2 + PbO \longrightarrow H_2N-\langle\ \rangle-SO_2NHCN + PbS + H_2O$$

Procedure. A drop of the test solution is united with several mg of lead hydroxide in a micro test tube and heated in boiling water. A positive response is indicated by the appearance of a black or brown color on the hydroxide.

Limit of identification: 0.6 γ sulfathiourea.

1 Unpublished studies, with W. HOPFF (Zürich).

49. Identification of Entobex[1]

Entobex-Ciba (4,7-phenanthroline-5,6-quinone) is used as an amoebicide and antibacterial drug. As shown by its structural formula

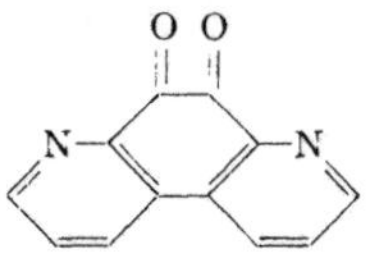

it is an *ortho*-quinoidal compound. Accordingly, when heated at 120–140° with guanidine carbonate, its behavior is analogous to that of phenanthraquinone (see p. 334). A purple condensation product results.

Procedure. A little of the solid test material is mixed in a micro test tube with several cg of guanidine carbonate and 1–2 drops of glycerol (to dissolve the Entobex). The mixture is kept for 1–2 min. in a glycerol bath preheated to 140°. A positive response is indicated by the appearance of a red product on the surface of the guanidine.

Limit of identification: 5 γ Entobex.

1 Unpublished studies, with D. GOLDSTEIN.

50. Tests for isonicotinic acid hydrazide (isoniazide)

This hydrazide has come to the fore as a remedy for tuberculosis. It can be identified in pharmaceutical preparations by condensation with salicylaldehyde[1] or through its redox reaction with phc 'iomolybdate,[2] *i.e.*, by Procedures I and II, respectively.

If a saturated aqueous solution of salicylaldehyde is added to a neutral solution of isonicotinic acid hydrazide (I), the resulting yellow-white crystalline precipitate shows a strong yellow-green fluorescence in ultraviolet light. This product (II) is a hydrazone formed by the condensation:

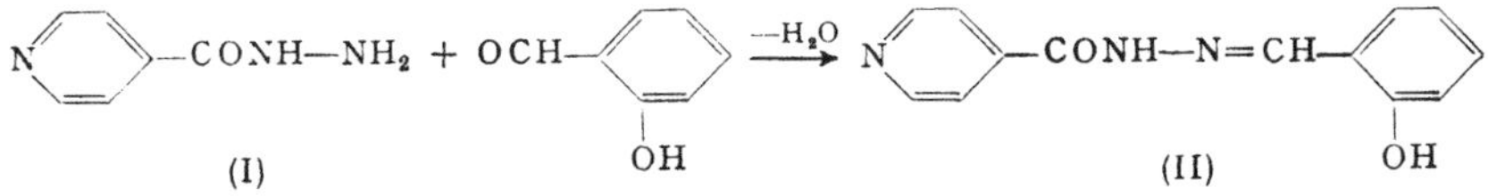

The condensation is analogous to the condensation of hydrazine with salicylaldehyde and other *o*-hydroxyarylaldehydes, which yield water-insoluble acid-resistant aldazines that fluoresce orange-yellow (see p. 275 and 341). However, in contrast to these aldazines, compound (II), by virtue of its pyridine component is soluble in dilute acids and gives a lemon yellow solution, in which the fluorescence has disappeared. The latter is also formed if mineral acid solutions of isonicotinic acid hydrazide (containing colorless cations (III)), are treated with an aqueous solution of salicylaldehyde; the yellow cation (IV) is formed:

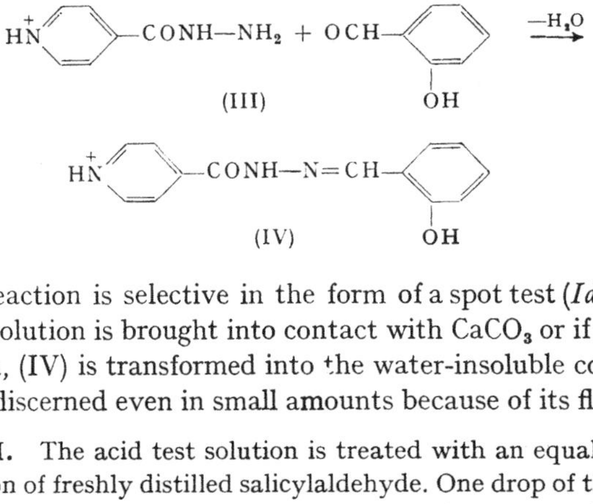

The color reaction is selective in the form of a spot test (*Idn. Lim.* 10 γ). If the yellow solution is brought into contact with $CaCO_3$ or if it is alkalized with ammonia, (IV) is transformed into the water-insoluble compound (II), which can be discerned even in small amounts because of its fluorescence.

Procedure I.   The acid test solution is treated with an equal volume of an aqueous solution of freshly distilled salicylaldehyde. One drop of the light yellow solution, which may appear colorless when only small amounts of the hydrazide are present, is placed on filter paper. After 4 min. the fleck is exposed to ammonia. (The preliminary exposure to air is necessary to volatilize and oxidize the excess salicylaldehyde.) If isonicotinic hydrazide was present, the fleck fluoresces yellow-green in ultraviolet light. The fluorescence disappears when the paper is dried in the air, but reappears if the paper is again held over ammonia. A comparison blank test is advised.

Limit of identification: 0.1 γ isonicotinic acid hydrazide.

The test also succeeds at a like dilution with the volume of a micro drop (0.001 ml), which corresponds to an *identification limit* of 0.002 γ.

Isonicotinic acid hydrazide being a derivative of pyridine can be precipitated from acid solution by phosphomolybdic acid. The crystalline phosphomolybdate is yellow. If the precipitate is treated with excess ammonia, it not only dissolves with production of NH_2NH_2, $(NH_4)_3PO_4$ and $(NH_4)_2MoO_4$, but in addition a redox reaction occurs between the hydrazine and the complexly bound molybdenum. The intensely colored molybdenum blue results. It is not necessary to isolate the phosphomolybdate to accomplish the color reaction. It is sufficient to add phosphomolybdic acid to the acid solution and then alkalize with ammonia. Free hydrazine and oxidizable amines must be absent. The former reduces phosphomolybdic acid to molybdenum blue even in acid solution; the latter are likewise precipitated from acid solution by phosphomolybdate, and molybdenum blue results after addition of ammonia.

Procedure II.   One drop of the neutral or mineral acid test solution is mixed on a spot plate with a drop of a saturated solution of phosphomolybdic acid. After 1–2 min. the mixture is made alkaline with ammonia. Depending on the quantity of isonicotinic acid hydrazide, a blue precipitate or color appears.
  Limit of identification: 0.5 γ isonicotinic acid hydrazide.

1 Unpublished studies, with H. E. FEIGL.
2 Unpublished studies, with W. A. MANNHEIMER.

51. Tests for penicillin G salts[1]

The antibiotics known as penicillin G salts contain not less than 85% of the sodium or potassium salt of the structure:

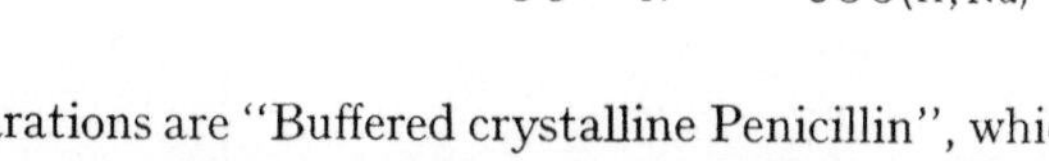

Similar preparations are "Buffered crystalline Penicillin", which is a mixture of the sodium or potassium salt with 4–5% of sodium citrate, and "Penicillin G Procaine", which is the procaine salt of penicillin G[2] (procaine is the 2-diethylaminoethyl ester of *p*-aminobenzoic acid).
  If the samples are pure preparations, the behavior of the ignition residues is characteristic. With the exception of the procaine salt, which burns without leaving a residue, the other preparations yield alkali carbonate. The

ignition tests can be made in a micro crucible with as little as 0.2–0.5 mg of the sample. The test for alkalinity is best made with nickel dimethylglyoxime equilibrium solution as described on p. 108.

A test for penicillin preparations of all kinds consists in treating a tiny portion (fractions of a mg) with a drop of a saturated (room temperature) aqueous solution of phosphomolybdic acid. The test tube containing the mixture is plunged into boiling water for an instant. An intense blue color appears within several seconds if penicillin G compounds are present. The color reaction involves several steps: (a) liberation of the parent acid from its salts by phosphomolybdic acid; (b) hydrolysis of the organic acid to give penicillamine,[3] *i.e.* β,β-dimethylcysteine $(CH_3)_2C(SH)CH(NH_2)CO_2H$. Since the latter is a mercaptan, it is immediately oxidized to the corresponding disulfide by the phosphomolybdic acid with simultaneous formation of molybdenum blue.

Idn. Lim.: 4 γ penicillin G. No color reaction with phosphomolybdic acid is given by streptomycin and dihydrostreptomycin.

The test is not applicable in the presence of strong reducing agents, such as ascorbic acid, phenols, *etc.* because they reduce phosphomolybdic acid even in the cold, whereas the redox reaction with penicillin occurs only on warming.

The hydrolysis of salts of penicillin to penicillamine is so extensive on warming that the color reaction of mercapto compounds with a mixture of p-nitrosodimethylaniline and p-dimethylaminobenzaldehyde occurs (see p. 134). Accordingly, if filter paper impregnated with these two compounds is spotted with a drop of a solution of a penicillin salt and kept at 110° for 2–3 min. in an oven, a red ring appears on the yellow reagent paper. It becomes more distinct if the paper is bathed in water to wash out the yellow p-nitrosodimethylaniline.[4]

Idn. Lim.: 10 γ penicillin.

Penicillin G Procaine can be distinguished from penicillin G not only by the absence of a residue after ignition, but also by the formation of a yellow-orange Schiff base with p-dimethylaminobenzaldehyde, which is a characteristic test for colorless aromatic primary amines (see p. 245). The test is made by placing a pinch of the solid on filter paper impregnated with the aldehyde reagent followed by a drop of dilute acetic acid.

1 Unpublished studies, with D. FONTES (Pernambuco).

2 Comp. the *United States Pharmacopeia*, 15th revision, 1950, p. 427.

3 E. P. ABRAHAM, E. CHAIN, W. BAKER and R. ROBINSON, *Nature*, 151 (1943) 107. Comp. *The Chemistry of Penicillin*, Princeton, 1949, p. 10ff.

4 F. FEIGL and D. GOLDSTEIN, *Microchem. J.*, 1 (1957) 179.

52. Detection of chloromycetin[1]

Chloromycetin (2-dichloroacetamido-1-*p*-nitrophenyl-1,3-propanediol) is the only known antibiotic which has the character of an aromatic nitro compound. The NO_2 group can readily be reduced to the NO group by warming an aqueous solution of chloromycetin with calcium chloride and zinc.[2]

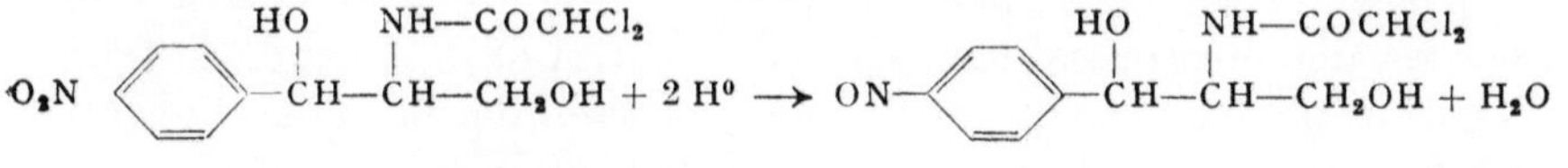

$$O_2N \hspace{1em} \langle\rangle \hspace{-0.5em} -\overset{\displaystyle \underset{|}{HO}}{C}H-\overset{\displaystyle \underset{|}{NH-COCHCl_2}}{C}H-CH_2OH + 2\,H^0 \rightarrow ON- \langle\rangle \hspace{-0.5em} -\overset{\displaystyle \underset{|}{HO}}{C}H-\overset{\displaystyle \underset{|}{NH-COCHCl_2}}{C}H-CH_2OH + H_2O \hspace{1em} (1)$$

The nitroso compound thus produced from chloromycetin condenses, in acetic acid solution, with α-naphthylamine to yield a violet azo dye (compare test for antipyrine, p. 631):

$$C_{10}H_7-NH_2 + ON-C_6H_4-\overset{\displaystyle \underset{|}{}}{C}H-\overset{\displaystyle \underset{|}{}}{C}H-CH_2OH$$
$$OH \quad NH-COCHCl_2$$

$$\downarrow \hspace{6em} (2)$$

$$C_{10}H_7-N{=}N-C_6H_4-\overset{\displaystyle \underset{|}{}}{C}H-\overset{\displaystyle \underset{|}{}}{C}H-CH_2OH + H_2O$$
$$OH \quad NH-COCHCl_2$$

The test given here is based on the realization of (*1*) and (*2*).

Procedure. A little of the solid or a drop of the test solution is treated in a micro test tube with two drops of 10% calcium chloride solution and several mg zinc dust. After warming the mixture in the water bath for two minutes, it is treated with two drops of 5% solution of α-naphthylamine in acetic acid and the warming continued for two minutes longer. In the presence of chloromycetin, a more or less intense violet color appears.

Limit of identification: 10 γ chloromycetin.

Chloromycetin gives a decided reaction for alcohols when treated with a benzene solution of vanadium oxinate (see p. 174) (*Idn. Lim.:* 10 γ).

Chloromycetin can likewise be characterized through tests for NO_2 groups, or for secondary alcohol groups, or for polyhalogen compounds. For details of these tests see Chap. 3.

1 F. FEIGL and E. SILVA, *Drug Std.*, 23 (1955) 113.
2 S. OHKUMA, *J. Japan. Chem. Soc.*, 4 (1950) 622.

53. Detection of aminophylline and euphylline[1]

Aminophylline and euphylline are water-soluble addition products of ethylenediamine with theophylline (1,3-dimethylxanthine), which is only slightly soluble in water. Aminophylline is the 2:1 addition compound of theophylline with ethylenediamine, whereas in euphylline the ratio is 2:3. Both preparations are widely used as diuretics and as myocardial stimulants. Other water-soluble theophylline preparations with the same action include[2] theophylline–diethanolamine, theophylline–calcium salicylate, theophylline–sodium acetate, theophylline–sodium salicylate, and theophylline–methyl glucamine. Aminophylline and euphylline can be identified with certainty through their ethylenediamine component since it reacts with sodium rhodizonate to produce a water-insoluble violet ethylenediamine compound of rhodizonic acid (see p. 497). The test requires the previous transformation of the ethylenediamine contained in the theophylline preparations into its chloride.

Procedure. A little of the test material (fractions of a mg suffice) is placed in a micro crucible along with a drop of hydrochloric acid (1 : 1) and the mixture is taken to dryness on a water bath. The residue is freed completely of any unbound acid by keeping it at 110° for several minutes. After cooling, the residue is treated with a drop of freshly prepared 1% sodium rhodizonate solution. A violet precipitate indicates the presence of aminophylline and/or euphylline. If preferred, a little of the pulverized sample can be placed on filter paper and exposed to the vapors of hydrochloric acid for 1–2 min. and after standing in the air for 1–2 min. the material is spotted with sodium rhodizonate solution.

This test is valid for aminophylline or euphylline only when no theophylline–calcium salicylate is present, since the latter yields violet calcium rhodizonate under the conditions prescribed here. To differentiate between theophylline–calcium salicylate and aminophylline or euphylline, a little of the test material should be stirred in a micro crucible with a drop of ammonia and then treated with a drop of sodium rhodizonate solution. Only calcium rhodizonate but no ethylenediamine rhodizonate is produced (compare p. 659). Consequently, if no reaction is obtained in ammoniacal solution, a positive response to the rhodizonate test after treating the sample with hydrochloric acid indicates the presence of aminophylline or euphylline. If these ethylenediamine compounds of theophylline are to be detected in the presence of theophylline–calcium salicylate, a little of the sample should be digested with sodium carbonate solution whereby $CaCO_3$ is precipitated. The detection of ethylenediamine with rhodizonate can then be conducted by the above procedure using one drop of the clear filtrate or centrifugate.

A rapid but less sensitive test for the ethylenediamine that is loosely bound in aminophylline and euphylline makes use of its condensation with 2,4-dinitrochlorobenzene (compare p. 240); N-alkylated 2,4-dinitroaniline is formed.

1 F. Feigl and E. Silva, *Anais Farm. Quim. Sao Paulo*, 6 (1953) 5.
2 *The Merck Index*, 7th ed., Rahway, N. J., 1960.

54. Detection of piperazine in pharmaceutical preparations[1]

Preparations containing piperazine are widely used as vermifuges. They are mostly salts of this base with organic acids (*e.g.* adipic acid), while in some instances they are liquid mixtures (emulsions) containing water-soluble piperazine hexahydrate. Piperazine can be detected in such preparations or their evaporation residues through the fact that pyrrole is formed by dry heating, whose vapors give a violet color on filter paper moistened with *p*-dimethylaminobenzaldehyde and trichloroacetic acid. See p. 381.

Since this color reaction is shown also by pyrolysis products of proteins, amino acids and sugars (due to furfural and its derivatives), it is advisable to confirm the presence of piperazine by also applying the test for secondary aliphatic amines (see p. 250) in which the brown water-insoluble copper dithiocarbamate is formed.

These tests can be carried out readily on mg quantities of the samples.

1 Unpublished studies, with E. Silva.

55. Detection of guaiacol carbonate[1]

Guaiacol is used for medicinal purposes in the form of its benzoate, cacodylate, cinnamate, salicylate, phosphate, valerate, ethylene ether, and carbonate.[2] The carbonate (I) can be distinguished from the other guaiacol preparations through the fact that it condenses with phenylhydrazine to produce diphenylcarbazide (II)[3]

$$OC \underset{\diagdown OC_6H_4(OCH_3)}{\overset{\diagup OC_6H_4(OCH_3)}{}} + 2\ NH_2NHC_6H_5 \longrightarrow OC \underset{\diagdown NH-NHC_6H_5}{\overset{\diagup NH-NHC_6H_5}{}} + 2\ C_6H_4(OH)(OCH_3)$$

$$\text{(I)} \qquad\qquad\qquad\qquad\qquad \text{(II)}$$

The diphenylcarbazide can be detected by adding an ammoniacal solution of a nickel salt. The blue-violet inner-complex nickel diphenylcarbazide is precipitated and can be extracted by chloroform (comp. p. 390).

Procedure. A little of the solid, or the dry evaporation residue of a solution, is treated in a micro test tube with a drop of phenylhydrazine and heated in a glycerol bath for five minutes at 170°. After cooling, five drops of ammonia (1:1) and five drops of 10% nickel sulfate solution are added. The mixture is then shaken with ten drops of chloroform. A red-violet color in the chloroform layer indicates the presence of guaiacol carbonate.

Limit of identification: 50 γ guaiacol carbonate.

1 Unpublished studies, with E. SILVA.
2 *The Merck Index*, 7th ed, Rahway, N.J., 1960.
3 K. H. SLOTTA and K. R. JACOBY, *Z. Anal. Chem.*, 77 (1929) 344.

56. Tests for drugs containing the dimethylaminoethyl group[1]

Among the modern drugs* containing the dimethylaminoethyl group, —$CH_2CH_2N(CH_3)_2$, bound to N or O are: Pyrilamine maleate (I), an antihistamine also known as Neoantergan or Anthisan, and Tetracaine, a local anesthetic (II), also known as Pantocaine or Anethaine. The structures are:

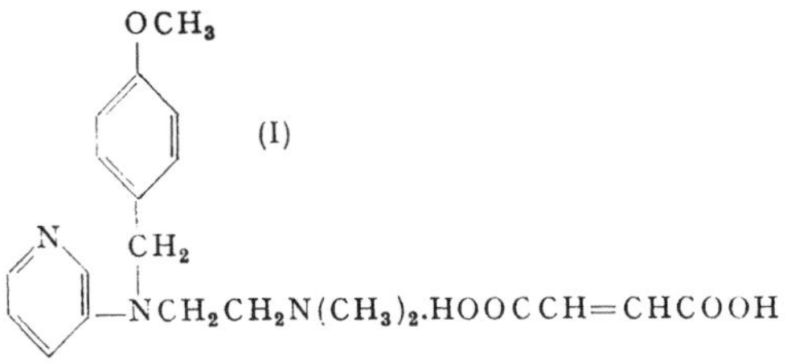

Compounds which contain the dimethylaminoethyl group split off acetaldehyde when heated to 250° with phthalic acid. The latter anhydrizes at this temperature, and the water thus liberated acts as superheated steam on the dimethylaminoethyl group:

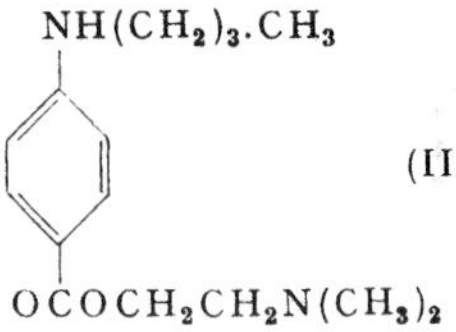

$$RCH_2CH_2N(CH_3)_2 + 2 H_2O \rightarrow RH + HOCH_2CH_2OH + NH(CH_3)_2$$

The resulting glycol loses water and so yields acetaldehyde, whose vapor can be detected by the color reaction with nitroprusside–morpholine mixture as described on p. 438.

The *limits of identification* of the above drugs are less than 0.5 mg, when the test is carried out with the technique of spot test analysis.

* See *New and Nonofficial Drugs*, Philadelphia, 1958.

1 Unpublished studies, with E. SILVA.

57. Detection of plant materials that can be oxidized to mucic acid[1]

A number of plant materials or vegetable products, such as gallactose and pectins, yield mucic acid when warmed with dilute nitric acid. The mucic acid is readily detected (see p. 470) through pyroammonolysis to give volatile pyrrole.

The test for plant materials is conducted in a micro test tube. A little of the sample is taken to dryness with 1–2 drops of dilute nitric acid and the residue is subjected to the procedure given on p. 471.

When only small amounts of mucic acid may be expected, more of the sample should be fumed down with dilute nitric acid in a porcelain crucible. The residue is taken up in the smallest possible volume of water and the resulting solution is then evaporated to dryness and subjected to the test for mucic acid. 1 γ mucic acid was detected in this way.

1 Unpublished studies, with D. GOLDSTEIN.

58. Detection of salicin and populin

Salicin is the glucoside of salicyl alcohol (saligenin), and populin is the benzoate of this glucoside. They occur in the leaves and bark of the poplar and willow.[1] Salicin is used in veterinary practice and in human medicine.[2]

A rapid, sensitive and specific method[3] of detecting salicin and populin is based on the fact that, when pyrolyzed, these compounds yield salicylaldehyde, which can be detected in the gas phase by action with alkali hydroxide or hydrazine to give the alkali salt or the aldazine which respectively exhibit a characteristic blue-violet or yellow-green fluorescence. Comp. p. 344.

The mechanism of this thermal splitting of salicin to give salicylaldehyde probably involves the following stages. When salicin (m.p. 201°) is heated to higher temperatures, some of the sugar caramelizes. The resulting superheated steam saponifies the glucoside as shown in (1) to yield o-hydroxybenzyl alcohol. The vapor of the latter undergoes autoxidation to give salicylaldehyde (2):

$$\text{C}_6\text{H}_4\!\begin{smallmatrix}-\text{OC}_6\text{H}_{11}\text{O}_5\\-\text{CH}_2\text{OH}\end{smallmatrix} + \text{H}_2\text{O} \longrightarrow \text{C}_6\text{H}_4\!\begin{smallmatrix}-\text{OH}\\-\text{CH}_2\text{OH}\end{smallmatrix} + \text{C}_6\text{H}_{12}\text{O}_6 \qquad (1)$$

$$\text{C}_6\text{H}_4\!\begin{smallmatrix}-\text{OH}\\-\text{CH}_2\text{OH}\end{smallmatrix} + \text{O} \longrightarrow \text{C}_6\text{H}_4\!\begin{smallmatrix}-\text{OH}\\-\text{CHO}\end{smallmatrix} + \text{H}_2\text{O} \qquad (2)$$

The occurrence of (2) can be demonstrated by heating a drop of salicyl alcohol in a test tube covered with a piece of filter paper which has been moistened with the reagent for salicylaldehyde (see below).

Procedure. About 1 mg of the sample is placed in a micro test tube or micro crucible covered with a disk of filter paper, which has been moistened with dilute caustic alkali or with a solution of hydrazine sulfate (for preparation see p. 345). The paper is kept in place with a small watch glass. The bottom of the vessel is heated with a micro flame. If salicin is present in not too small amounts, a yellow stain appears on the paper; it is due to the alkali salt or aldazine of salicylaldehyde. Viewed in ultraviolet light, the blue-violet or yellow-green fluorescence of these products is visible.

The test may also be carried out with powder of the sample heated in a capillary tube. When aqueous solutions or extracts are to be examined, a few drops should be taken to dryness beforehand.

The following procedure[4] revealed 5 γ salicin and populin: a drop of the test solution was taken to dryness in a micro test tube along with a little moist manganese dioxide. The residue was then heated to 240°. This treatment saponifies the glucoside and the liberated salicyl alcohol is oxidized to salicylaldehyde, which gives the aldazine reaction in the gas phase.

1 G. Klein, *Handbuch der Pflanzenanalyse*, Vol. 3, Part 2, Hirschwaldse Buchhandlung, Berlin, 1932, p. 815.
2 *The Merck Index*, 7th ed., Rahway, N.J., 1960, p. 916.
3 Unpublished studies.
4 F. Feigl and C. Stark, *Mikrochim. Acta*, (1955) 1000.

59. Detection of helicin[1]

Helicin (m.p. 176°) the glucoside of salicylaldehyde, is saponified by warming its aqueous solution with acids or alkalis:

$$\text{(benzene ring)}\begin{matrix}-OC_6H_{11}O_5\\-CHO\end{matrix} + H_2O \longrightarrow \text{(benzene ring)}\begin{matrix}-OH\\-CHO\end{matrix} + C_6H_{12}O_6$$

The hydrolysis can also be accomplished in the dry way by gentle heating with $MnSO_4.4\ H_2O$, or even better with $H_2C_2O_4.2\ H_2O$. This treatment has the advantage that vapors of salicylaldehyde result, detectable through the formation of fluorescent salicylaldazine (see p. 344).

Procedure. The test is made in a micro test tube. A little of the solid is mixed with finely powdered oxalic acid dihydrate and the mouth of the test tube is covered with a disk of filter paper moistened with the hydrazine reagent solution (for preparation see p. 345). The test tube is kept in boiling water for several minutes. A positive response is indicated by the formation of a yellow stain which fluoresces yellow-green in ultraviolet light.

Salicin and populin are not altered when subjected to the above treatment.

1 Unpublished studies, with E. Jungreis.

60. Detection of coumarin in plant material[1]

As stated on p. 479, coumarin may be detected through the yellow-green fluorescent alkali salt of *trans-o*-hydroxycinnamic acid, which is formed by u.v.-irradiation of the non fluorescing *cis-o*-hydroxycinnamic acid resulting from the cleavage of coumarin by alkali. This characteristic photo-effect together with the volatility of coumarin can be used to reveal this compound in plant materials. The test is made on several cg of the ground moist sample; smaller quantities of the dried sample suffice. The sample is placed in a test tube (3 or 4 cm long) whose mouth is covered with filter paper moistened with dilute sodium hydroxide. The mixture is kept in boiling water for several minutes and the paper is then exposed to ultraviolet light. If coumarin is present, a yellow-green fluorescence appears within a few minutes.

When fruits, Tonka beans, *etc.* are being studied, a freshly cut surface is pressed for several minutes against filter paper moistened with sodium hydroxide solution and then held under the quartz lamp. Impressions of leaves, powders, *etc.* on the paper can likewise be tested in this way. Since many plant materials have a faint self-fluorescence, greater certainty in the test for coumarin can be secured if the stain obtained by sublimation or pressing against hydroxide paper is partly covered with a coin or black paper and then irradiated. If the shielded area is then uncovered under the lamp, there will be distinct differences in the intensity of the fluorescence of the irradiated and unexposed areas when coumarin is present. The difference in the intensity of the fluorescence gradually disappears if the exposure is continued.

1 F. FEIGL, H. E. FEIGL and D. GOLDSTEIN, *J. Am. Chem. Soc.*, 77 (1955) 4162.

61. Detection of antimonial pharmaceutical preparations[1]

A number of organic compounds containing ter- and quinquevalent antimony are used to combat protozoal diseases. The most important of these preparations are:[2]

Tartar emetic (potassium antimonyl tartrate)
Stibamine (sodium *p*-aminophenylantimonate)
Neostibosan (diethylamino-*p*-aminophenylantimonate)
Stibenyl (sodium *p*-acetylaminophenylantimonate)
Antimony thioglycolamide
Antimony sodium thioglycolate
Fuadin (sodium antimony(III) bis-pyrocatechol-3,5-disulfonate)
Glucantime (N-methylglucamine antimonate)
p-Chlorophenylstibonic acid.

The antimony can readily be detected directly in all of the compounds listed here. On treatment with hydrochloric acid, potassium iodide, and sulfurous acid (the latter in the case of quinquevalent antimony) the complex acid $H[SbI_4]$ is produced at once or within a few minutes. Addition of an aqueous solution of the basic dye rhodamine B gives a violet precipitate of the rhodamine B salt. (Comp. p. 100.)

Procedure. A minimal quantity of the solid preparation, or a drop of the solution, is placed in a depression of a spot plate. Drops of potassium iodide solution and dilute hydrochloric acid are added in succession. If the mixture turns brown because of the liberation of iodine, this in itself is an indication of the presence of quinquevalent antimony. In this case, the iodine is removed by adding a drop of sulfurous acid and a drop of the dye solution is introduced.

The separation of a red-violet precipitate proves the presence of antimony.

Reagents: See p. 677.

This test is not directly applicable in the presence of organic bismuth compounds since the latter are converted into $H[BiI_4]$ which behaves like $H[SbI_4]$. Compare Sect. 63, regarding the detection of organic bismuth compounds.

If organic antimony compounds are to be detected in the presence of organic bismuth compounds, the sample should be digested with excess ammonium sulfide. Soluble ammonium sulfoantimonite is formed along with insoluble bismuth sulfide. After filtering (or centrifuging), a drop of the alkaline solution is warmed or taken to dryness with hydrogen peroxide. The resulting residue may then be tested using the above procedure.

Another, though admittedly less sensitive, test for antimony in organic compounds, consists in igniting the sample and treating the residue, which contains Sb_2O_5 and Sb_2O_4, with a drop of a solution of diphenylamine or N,N'-diphenylbenzidine in concentrated sulfuric acid. On stirring, a blue color develops (see p. 100).

If the presence of antimony has been established in pharmaceutical preparations, some of the compounds included in the above list can be characterized by supplementary tests. For instance, Sb-thioglycolamide and -thioglycolate containing divalent sulfur will yield hydrogen sulfide when warmed with hydrochloric acid and Raney nickel alloy (see p. 83). Fuadin, which contains aromatically bonded SO_3Na groups, yields nickel sulfide if warmed with caustic alkali and Raney nickel alloy (see p. 234), and aniline is produced if Stibamine, Neostibosan, or Stibenyl undergo the same treatment (see p. 375). These latter three compounds, as well as *p*-chlorophenylstibonic acid, give a positive response to the test described on p. 309 since they are stibonic acids. In addition, Glucantime can be distinguished from other antimony compounds since as a polyalcohol it shows the test described on p. 315.

1 F. FEIGL and E. SILVA, *Drug Std.*, 23 (1955) 115.
2 Comp. *The Merck Index*, 7th ed., Rahway, N. J., 1960.

62. Detection of arsenic-bearing pharmaceutical preparations

Pharmaceutical preparations containing arsenic yield calcium arsenate if they are ignited with lime. When the residue is fused with sodium formate, hydrogen arsenide is evolved and can be detected by the blackening of filter paper moistened with silver nitrate solution, see p. 98.

The test based on the interaction with phenylhydrazine (formation of phenol) can be recommended for the detection of organic derivatives of arsenic acid. See p. 308 for details.

Arsanilic acid and acetylarsanilic acid or their alkali salts, can be detected through the finding that they behave analogously to sulfanilic acid when warmed with caustic alkali and Raney nickel alloy (see p. 375). Aniline vapors are evolved and can be detected through the violet color they produce on filter paper moistened with sodium 1,2-naphthoquinone-4-sulfonate solution. *Idn. Lim.*: 10 γ.

63. Detection of bismuth-bearing medicinals

Numerous medicinals contain organic compounds in which bismuth is a constituent of normal or complex salts. The most important preparations of this kind are: bismuth subsalicylate, sublactate, subgallate, iodosubgallate, bismuth sodium triglycolamate, basic bismuth pyrogallate, bismuth ethyl camphorate, bismuth glycolyl arsanilate, bismuth tribromophenolate.

A preliminary decomposition of the organic substance is not required for the detection of the bismuth. The small concentration of Bi^{3+}-ions provided by contact of even difficultly soluble compounds with alkaline stannite solution is sufficient to achieve the redox reaction:[1]

$$2\ Bi^{3+} + 3\ SnO_2^{2-} + 6\ OH^- \rightarrow 2\ Bi^0 + 3\ SnO_3^{2-} + 3\ H_2O$$

The elementary bismuth separates in a high state of division and its black color provides a sensitive test (compare *Inorganic Spot Tests*, Chap. 3).

The test may not be applied directly in the presence of organic silver and mercury compounds, since they too are reduced to the respective metals by stannite. In contrast to organic bismuth compounds, which leave Bi_2O_3 on ignition, the ignition residues of organic mercury compounds are free of mercury. Accordingly, when the presence of mercury compounds must be taken into account (for detection see p. 102), the sample must be ignited beforehand. When organic silver compounds are ignited, the metal is left. (Sulfur-bearing compounds yield Ag_2SO_4.) If the presence of a silver-bearing ignition residue is suspected, the latter should be digested with warm diluted hydrochloric acid and the solution then subjected to the test with stannite.

1 L. VANINO and F. TREUBERT, *Ber.*, 31 (1898) 1113.

64. Detection of mercurial pharmaceutical preparations

Most of the mercury compounds used in medicine are salts of phenols, or carboxylic-, sulfonic-, and arsonic acids, or compounds containing HgOH- and HgCN groups attached to carbon. Some are water-soluble, others are only slightly soluble, and in general they dissociate to only a small extent to yield Hg ions. This is also true of mercuric cyanide and oxycyanide, which likewise are used medicinally.

The procedures given in Chap. 2 can be employed to reveal mercury in pharmaceutical preparations. The demasking of $[Fe(CN)_6]^{4-}$ ions to yield Fe^{2+} ions by means of mercury compounds in the presence of α,α'-dipyridyl is recommended particularly. Even small quantities of solid or dissolved mercury preparations give a distinct red color due to $[Fe(\alpha,\alpha'\text{-dip})_3]^{2+}$ ions when warmed with a drop or two of the colorless reagent. This procedure was successful with mercuric cyanide, succinimide, diiodosalicylate, Mercurin (Mercuzan), Mercurosal, Mercurophen.

65. Detection of calcium-bearing pharmaceutical preparations

A rapid and sensitive test for calcium in medicinal preparations can be based on the fact that a yellow aqueous solution of sodium rhodizonate tints calcium oxide violet. Compare *Inorganic Spot Tests*, Chap. 3.

Procedure.[1] A little of the sample is cautiously ignited in a micro crucible. The cold residue is treated with a drop of a 0.2% aqueous solution of sodium rhodizonate. A violet color indicates a positive response.

This procedure was tested successfully with the following calcium preparations:

–glycerophosphate (Neurosin),	–salicylate,
–gluconate,	–dibromobehenate (Sabromin),
–lactate,	–iodobehenate (Calioben),
–levulinate,	–lactophosphate,
–mandelate,	–pantothenate,
–naphtholmonosulfonate (Asaprol),	–creosotate (after ashing).

The foregoing list includes calcium preparations whose organic components can be detected by additional tests. This is true for the detection of glycerol in the glycerophosphate by the procedure given on p. 416. Calcium lactate and lactophosphate respond to the tests described on p. 466. Salicylic acid can be identified in its calcium salts as outlined on p. 478. Calioben is characterized by giving a positive response to the test for iodine described on p. 71. Calcium glycerophosphate and calcium lactophosphate are readily distinguished from other preparations in that phosphate is easily detected in their ignition residues by the procedure given on p. 96.

1 F. FEIGL and E. SILVA, *Drug Std.*, 23 (1955) 114.

66. Detection of iodine-bearing pharmaceutical preparations[1]

Organic iodine compounds are widely used as contrast media in Roentgen diagnosis, as anti-protozoan agents, and in cases of thyroid deficiency. The test for iodine can be executed with minimal quantities of the sample according to the procedure indicated on p. 72, which is based on the direct oxidation of organically bound iodine to iodic acid by bromine.

1 F. FEIGL and E. SILVA, *Drug Std.*, 23 (1955) 114.

67. Detection of hippuran[1]

Hippuran (sodium *o*-iodohippurate) is employed as radiopaque medium in urography. Like all other iodine-containing contrast media used in X-ray diagnosis, this compound gives a positive response to the test for iodine discussed on p. 71. In addition, a reliable identification may be based on the finding that hippuran (in analogy to its iodine-free acidic parent compound) undergoes pyrohydrolysis when heated to 170° with manganese sulfate monohydrate as water donor:

$$IC_6H_4CONHCH_2COONa + H_2O \rightarrow IC_6H_4COOH + H_2NCH_2COONa$$

The resulting glycine may be degraded by means of chloramine T to yield formaldehyde, which can be detected through the color reaction with chromotropic acid (see p. 434).

Procedure.   As given for hippuric acid on p. 503.
Limit of identification: 25 γ hippuran.

1 F. FEIGL, D. GOLDSTEIN and D. HAGUENAUER-CASTRO, unpublished studies.

68. Differentiation of mercuric cyanide and oxycyanide[1]

Mercuric cyanide and oxycyanide, $HgO.Hg(CN)_2$, can be differentiated by the behavior of their solutions toward potassium iodide. The cyanide yields potassium tetraiodomercuriate:

$$Hg(CN)_2 + 4 KI \rightarrow K_2HgI_4 + 2 KCN \tag{1}$$

whereas the oxycyanide gives in addition the reaction:

$$HgO + 4 KI + H_2O \rightarrow K_2HgI_4 + 2 KOH \tag{2}$$

According to (2), only the oxycyanide yields a solution of Nessler reagent, from which the yellow-brown $HgI_2.HgNH_2I$ is precipitated when ammonia is added.

Procedure.[2] A spot plate is used. One drop of the test solution or a tiny portion of the solid is treated in succession with one drop of 0.5% solution of potassium iodide and 1 : 5 ammonia. If mercuric oxycyanide is present, a yellow-brown precipitate or yellow color appears.

Limit of identification: 100 γ mercuric oxycyanide.

Commercial preparations contain about 33% oxycyanide and 67% cyanide. Pure oxycyanide explodes on contact with a flame and also when struck.

1 J. GOISE, *Bull. Soc. Pharm. Bordeaux*, **66** (1928) 209.
2 Unpublished studies, with E. SILVA (Pernambuco).

69. Differentiation of sulfur-bearing and sulfur-free dyes[1]

The procedures given in Chap. 2 can be applied for rapidly detecting sulfur in organic coloring matters. If sufficient amounts of the sample are available and the attainment of the highest sensitivity is not of prime importance, it is advisable to employ the pyrolysis with hexamine leading to ammonium sulfide. The procedure described on p. 82 gave satisfactory results with 100 γ Methylene Blue, 300 γ Methyl Orange, 400 γ Victoria Blue.

1 Unpublished studies.

70. Detection of sulfur dyes containing divalent sulfur

Sulfur dyes, which invariably contain divalent sulfur, can be detected through the fact that they yield hydrogen sulfide when warmed with Raney nickel alloy and hydrochloric acid (blackening of lead acetate paper).

Another test is based on the finding that on pyrolysis with mercuric cyanide these dyes produce vapors which give a blue color when they come into contact with filter paper moistened with copper acetate–benzidine acetate. The chemistry and procedure of this test are discussed on p. 148. The following results were obtained:

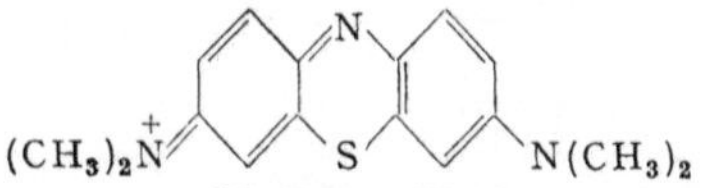

(Methylene Blue)

reduction with Raney nickel alloy: 10 γ
pyrolysis with mercuric cyanide: 10 γ

The detection of sulfur dyes by the wet method with Raney nickel alloy and hydrochloric acid can likewise be used when testing dyed fabrics.

71. Detection of dyes containing di- and tetravalent sulfur

Among these dyes are the thiazine dyes, such as methylene blue and the so-called blue sulfur dyes, the yellow sulfur dyes of the type of benzothiazole, and the thioindigo dyes. It was pointed out in Chap. 2 that organically bound 2- and 4-valent sulfur can be detected through the fact that their compounds undergo reductive cleavage when warmed with Raney nickel alloy and hydrochloric acid with evolution of hydrogen sulfide.

Procedure.   See p. 83.
The test revealed quantities of pertinent dyes below 0.5 mg.

72. Detection of dyes containing hexavalent sulfur

As was pointed out in Chap. 3, aromatic mono- and polysulfonic acids are desulfurized when warmed with Raney nickel alloy and alkali hydroxide. Nickel sulfide is produced. This test can be employed for the detection of sulfonic groups in dyes. Examples are: sulfonated azo dyes, sulfonated alizarin dyes. Alizarin Red S, Indigo Carmine, Sulforhodamine B and also dyes with sulfonamide groups.

Procedure.   See p. 234.
The test revealed γ-amounts of pertinent dyes.

73. Distinguishing between nitrogen-bearing and nitrogen-free dyes[1]

The majority of the dyes used commercially contain nitrogen, but certain important dyestuffs contain only carbon, hydrogen, and oxygen. Pertinent examples are: hydroxy derivatives of anthraquinone and γ-pyrones, benz-anthrone dyes and hydroxytriphenylmethane dyes. Accordingly, a rapid means of distinguishing between nitrogenous and non-nitrogenous dyes may be desirable. The presence of nitrogen can be readily established by the preliminary test based on the formation of nitrous acid when nitrogenous organic or inorganic materials are ignited with manganese dioxide. The nitrous acid is detected in the gas phase by the Griess reaction. The procedure is outlined on p. 91. The test is reliable provided the sample contains no inorganic or organic nitrogenous compounds.

1 Unpublished studies.

74. Distinction between acid and basic dyes

Dyes which are predominantly basic are soluble in melted stearic acid, while predominantly acid dyes dissolve in melted urea. These properties are connected with the electropolar nature of the substances (salt or solvate formation). This behavior may be applied as a quick and simple means of distinguishing between these two classes of dyes.

Basic dyes probably form salts with stearic acid which are readily soluble in molten stearic acid. When urea is melted, it yields biuret (see p. 706) and ammonia is given off. Consequently, there is a possibility that ammonium salts are formed in the case of acid dyes, and the salts then dissolve in molten urea or the biuret–urea mixture.*

Procedure.[1]   Urea and stearic acid are melted in a test tube, a grain of the dye to be tested is added, and the mixture well shaken. When the two layers separate, one will be colored, If the lower layer (urea) is colored, an acid dye is present; if the upper layer (stearic acid) is colored, a basic dye is present.

An excess of dye should be avoided; otherwise both layers are colored.

* Ammonium salts of inorganic and organic acids, and likewise organic carboxylic and sulfonic acids are soluble in molten urea. Furthermore, many primary, secondary, and tertiary amines as well as their salts (including benzidine sulfate) are soluble in fused urea. This is in line with the fact that molten urea is an excellent solvent for many substances.[2]

1 D. REICHINSTEIN, *Helv. Chim. Acta*, 20 (1937) 882.
2 R. E. D. CLARK, *Nature*, 168 (1951) 826.

75. Detection of salts of basic dyes and of dyes with free acidic groups[1]

A procedure is given on p. 117 for the detection of acidic compounds

of all kinds. It is based on the finding that heating of such materials to 160°
along with mercuric cyanide yields hydrogen cyanide which can be identified
in the gas phase by the bluing of a copper–benzidine acetate solution. This
procedure permits the detection of salts (mostly chlorides) of dye bases as
well as of dyes with free COOH-, SO_3H-, and phenolic OH groups. A positive response was given by the following dyes in amounts less than 1 mg:

Neutral Red (chloride)	Victoria Violet 4 B S (SO_3H group)
Malachite Green (chloride)	Naphthol Orange (SO_3H group)
Diazine Green (chloride)	Hematoxylin (polyphenol)
Methyl Red (COOH group)	Quinalizarin (polyphenol)
Fluorescein (COOH group)	Rhodamine B (COOH group)

Dyes containing free basic groups or salified acidic groups do not react
with mercuric cyanide. If such dyestuffs are taken to dryness with dilute
hydrochloric acid and the residue heated to 120° to drive off the excess
hydrochloric acid, the residual products yield hydrogen cyanide when
pyrolyzed with mercuric cyanide.

1 Unpublished studies, with E. Libergott.

76. Detection of dyes containing nitro groups[1]

A test for groups containing N- and O atoms is described in Chap. 2; it is
based on the fact that such compounds yield N_2O_3 or nitrous acid when they
undergo dry heating. The Griess color reaction employed in this test can also
be applied for the detection of nitro groups in dyes, but certain restrictions
must be imposed in this case.

If the dye being examined contains salified hydroxyl- or carboxyl groups,
it is essential to evaporate the sample with hydrochloric acid prior to the
pyrolysis. Otherwise, the Griess reaction will fail.

Special attention must be given to those dyes which contain SO_3H groups
in addition to NO_2 groups. The ratio of these groups is a decisive factor,
since a positive response to the test is given only if the NO_2 group is in excess.

Procedure.   As given on p. 93.
The following amounts of dyes containing nitro groups were detected:

110 γ Alizarin Yellow	50 γ Martius Yellow
14 γ Pigment Orange	100 γ Naphthol Yellow
50 γ C.I. Pigment Red 3	20 γ Alizarin Orange
75 γ C.I. Acid Yellow 63	

1 Unpublished studies.

77. Detection of dyes containing amino groups[1]

A procedure was described in Chap. 2 for the detection of so-called occult NH_2 groups in organic compounds. It is based on the finding that NH_2 groups are methenylated during pyrolysis (160–180°) with hexamine whereby ammonia is produced. The latter is readily detected in the gas phase. Amino groups can be revealed in dyes by this procedure without interference from any acidic groups because the latter are salified by hexamine.

Procedure. As on p. 119.
A positive response was obtained with:

50 γ Victoria Violet	250 γ Acid Green	20 γ Chrysoidine
100 γ Kiton Black HA	10 γ Bismarck Brown	

1 Unpublished studies.

78. Detection of dyestuffs containing N-dialkyl groups[1]

A test for dialkyl groups is given in Chap. 3; it is based on the finding that if compounds in this category are heated to 180° with hexamine the resultant vapors respond to the tests for formaldehyde. Dyestuffs containing one or more N-dialkyl groups can be reliably tested for such groups by the procedure described on p. 254.

The following amounts were detected:

50 γ Methylene Blue	50 γ Malachite Green	500 γ Rhodamine B
80 γ Gallamine Blue	25 γ Neutral Red	

1 Unpublished studies, with E. Libergott and L. Ben-Dor.

79. Differentiation of dyes with N-methyl and N-ethyl groups[1]

Certain dyes differ solely in containing $-N(C_2H_5)_2$ in place of $-N(CH_3)_2$ groups or *vice versa*. In general, the colors of the solid and dissolved compounds are the same, and their dyeing powers and behavior toward reagents are alike. Such pairs include: Methyl Orange and Ethyl Orange; Malachite Green and Brilliant Green; Gallamine Blue and Coelestine Blue.

The presence of a N-methyl compound on one hand, and of a N-ethyl compound on the other, can be revealed by melting the dye with benzoyl peroxide. This treatment (comp. pp. 165 and 252) results in oxidative splitting with formation of aldehydes. The latter can be detected in the gas phase by sensitive color reactions. The methyl compounds yield formaldehyde only, while the ethyl compounds produce acetaldehyde solely. The tests can be made with small quantities of the sample and require not more than a few minutes. The results are reliable. Textiles dyed with these materials can likewise be tested for these pairs.

1 Unpublished studies, with E. Silva.

80. Detection of azo dyes[1]

Azo dyes result from the coupling of diazotized aromatic amines with phenols or amines. They have acidic or basic character according to whether there are acidic or basic groups in the molecule or if one of these groups is in excess. With few exceptions, this characteristic is exhibited through the solubility in dilute alkali or acid. (See also Sect. 74) Azo dyes which contain one or more $N=N$ groups may have this double bond ruptured by reduction with zinc and hydrochloric acid if the dye is acid-soluble:

$$ArN=NAr_1 + 4\,H^0 + 2\,H^+ \rightarrow Ar\overset{+}{N}H_3 + Ar_1\overset{+}{N}H_3$$

The primary aromatic amine produced can be detected with high sensitivity by the condensation with p-dimethylaminobenzaldehyde, which leads to a Schiff base or its colored quinoidal cation.

Procedure. A micro test tube is used. A drop of the test solution is treated with 1 drop of 1 : 10 hydrochloric acid and several pieces of zinc. The reduction is usually complete within 5 min. The reduced solution is placed on filter paper, allowed to soak in, and then dried briefly at 110° in an oven. The spot is treated with a saturated benzene solution of p-dimethylaminobenzaldehyde. A red to yellow fleck indicates a positive response.

The following procedure can be employed for acid-insoluble dyes: The aqueous suspension is warmed with a pinch of sodium hydrosulfite ($Na_2S_2O_4$) whereby the color is discharged. The solution is then acidified with dilute hydrochloric acid, placed on paper, heated briefly in the oven, and the spot then treated with the reagent.

In case the dye contains a free amino group from the start, it is advisable to make a blank test with 1 drop of the acid solution of the unreduced dye by treating it on paper with a drop of the solution of p-dimethylaminobenzaldehyde. Obviously, the reduced dye will give a more intense color with the reagent.

The following amounts of azo dyes gave a positive reaction:

1 γ Victoria Violet	2 γ C.I. Acid Green 20
0.5 γ Congo Red	5 γ C.I. Acid Yellow 23

A positive response was given likewise by C.I. Direct Orange 73, C.I. Direct Orange 1, C.I. Acid Yellow 11.

1 Unpublished studies.

81. Detection of azo dyes derived from aniline[1]

If an alcoholic solution of azobenzene is heated with Devarda alloy and caustic alkali, the reductive cleavage yields aniline:

$$C_6H_5{-}N{=}N{-}C_6H_5 + 4\ H^0 \rightarrow 2\ C_6H_5NH_2$$

Analogously, azo dyes derived from aniline release the latter along with a molecule of a second primary aromatic amine. Aniline and other primary arylamines which are volatalized with steam may be detected in the gas phase through the formation of a yellow Schiff base by reaction with p-dimethylaminobenzaldehyde.

Procedure. A drop of the alcoholic or aqueous test solution is treated in a micro test tube with a few cg of Devarda alloy and a drop of 0.1% sodium hydroxide. The tube is immersed in a heated water bath, and the mouth of the tube is covered with a disk of filter paper impregnated with a saturated benzene solution of p-dimethylaminobenzaldehyde. A yellow stain appears on the paper if the reaction is positive.

The following amounts of dyes were detected:

2 γ Aniline Yellow	10 γ Sudan III
2 γ Chrysoidine	5 γ C.I. Acid Yellow 11
8 γ Kiton Black HA	

1 Unpublished studies, with J. R. AMARAL.

82. Detection and distinction of benzidine dyes[1]

Many substantive (direct) dyes for cotton are prepared from bisdiazotized diamines such as benzidine, o-tolidine, or o-dianisidine by coupling with naphthionic acid, salicylic acid, H-acid, $etc.$ These azo dyes are reductively split by zinc and hydrochloric acid, or by alkaline sodium hydrosulfite ($Na_2S_2O_4$). For example, Congo Red yields benzidine and o-aminonaphthionic acid:

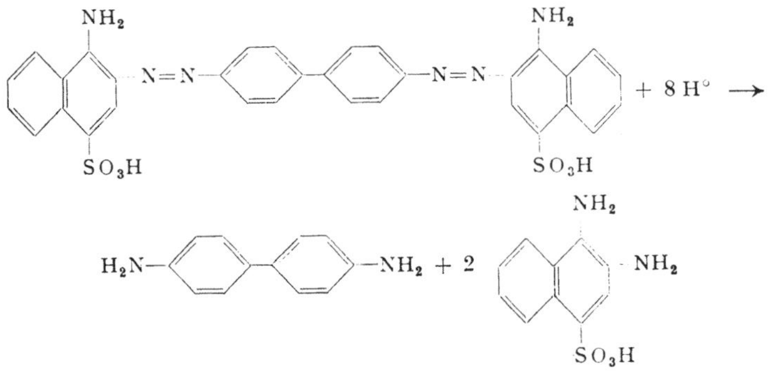

Analogously, the cotton dyes such as Benzopurpurin yield *o*-tolidine or *o*-dianisidine along with *o*-aminonaphthionic acid.

Following reductive cleavage of cotton dyes, a benzidine component can be detected by the following procedure, through the sensitive reaction of benzidine acetate with sodium rhodizonate; a violet precipitate appears (see p. 498). Analogous reductions occur with dyes which contain substituted benzidines but the latter react with rhodizonate only when they are at very high concentrations. Accordingly, cotton dyes containing a non-substituted benzidine component can be identified with certainty.

Procedure. The test is made in a micro test tube. A drop of the test solution is treated with 1 drop of 3% sodium hydroxide followed by about 20 mg of sodium hydrosulfite. The contents of the test tube are kept in a boiling water bath for some minutes. The complete reduction of the dye is indicated by the disappearance of its color. The cooled mixture is shaken with 10 drops of ether, and the ethereal solution is taken from the test tube with a micro pipette. The ether is evaporated on a spot plate and the residue treated with 1–2 drops of freshly prepared 1% solution of sodium rhodizonate. The mixture is stirred by blowing on its surface through a pipette and 1–2 drops of 2 N acetic acid are added. If the response is positive, a violet precipitate appears.

A positive reaction was given by the following amounts:

20 γ Congo Red	50 γ C.I. Direct Blue 6
25 γ C.I. Direct Brown 2	25 γ C.I. Direct Orange 1

If the test for benzidine is negative, the ethereal solution obtained after reduction of the dye may be examined for benzidine derivatives, such as *o*-tolidine, *o*-dianisidine, chlorobenzidine. This examination is based on the production of blue quinoidal compounds when these bases, as well as the parent benzidine, are oxidized by manganese dioxide (Comp. *Inorganic Spot Tests*, Chap. 3).

Procedure. Several drops of the ethereal solution obtained in the preceding procedure are placed in succession on the same area of filter paper impregnated with manganese dioxide (for preparation see p. 474). If the response is positive, a blue color develops when the spotted reagent paper is held in acetic acid vapors.

A positive response was given by the following amounts of cotton dyes derived from *o*-tolidine, *o*-dianisidine, or dichlorobenzidine:

5 γ C.I. Direct Orange 1	5 γ C.I. Direct Blue 14
5 γ C.I. Direct Red 7	25 γ C.I. Direct Blue 22
2.5 γ C.I. Direct Brown 2	5 γ C.I. Direct Red 2
25 γ C.I. Direct Red 61	25 γ C.I. Direct Blue 6

1 Unpublished studies, with V. Gentil.

83. Detection of dyestuffs with *p*-aminophenol(naphthol) structure[1]

Azo dyes which contain a free or alkylated $-\text{N}=\text{N}-\langle\ \rangle-\text{OH}$ group are easily reduced with zinc and hydrochloric acid, whereby *p*-amino-phenol or its alkyl ether are formed. An analogous reduction occurs with dyes derived from *p*-aminonaphthol. A test is described on p. 517 for the detection of these aminophenols; it is based on the fact that when oxidized with chloramine T a *p*-quinoneimine is formed, which condenses with phenol in ammoniacal solution to yield a blue indophenol.

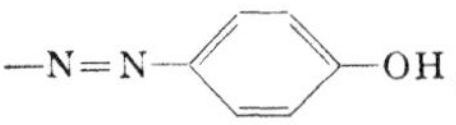

Procedure. A few grains of metallic zinc are added to the hydrochloric acid solution or suspension of the dyestuff and warmed in a water bath. When the color disappears, the solution is decanted from the unused zinc, and an excess of chloramine T added. After about 10 min. the *p*-quinoneimine formed is extracted with ether, and the ethereal solution is brought dropwise on filter paper impregnated with an ethereal solution of phenol. The spot is held over ammonia. A positive response is indicated by the appearance of a blue stain.

The following amounts were detected:

      5 γ α-Naphthol Orange   8 γ Diamine Golden Yellow
      6 γ Chrysophenine

1 Unpublished studies, with V. GENTIL.

84. Detection of dyestuffs with *p*-phenylenediamine and *p*-nitroaniline structure[1]

Dyes, whose production involved *p*-phenylenediamine or *p*-nitroaniline, are reductively cleaved by the nascent hydrogen when they are heated with sodium formate–hydroxide to 210–230° (comp. Sect. 88). *p*-Phenylenediamine is always one of the end products. The net reaction with Aniline Yellow is given as an example:

$$\langle\ \rangle-\text{N}=\text{N}-\langle\ \rangle-\text{NH}_2 + 4\ \text{H}° \longrightarrow \langle\ \rangle-\text{NH}_2 + \text{H}_2\text{N}-\langle\ \rangle-\text{NH}_2$$

The *p*-phenylenediamine yielded by the reductive cleavage sublimes during the heating and can be detected in the vapor phase by means of the phenylene blue reaction (p. 512). The procedure is the same as that described in Sect. 88. An indication of the sensitivity is given by the fact that 5 γ Aniline Yellow can be detected.

The procedure was checked with minute quantities of the dyes shown below. It should be noted that textiles colored with these dyes can likewise be tested in this way. As little as 3–5 mg of the fabrics are sufficient.

Victoria Violet	Diphenyl Brilliant Violet 2R
Kiton Black G	Diphenyl Fast Red 5BLN
Kiton Black HA	Eriochrome Orange R
Carbide Black S	Erio Violet B
Azidine Orange D2R	Alizarin Yellow R

1 F. FEIGL and Cl. COSTA NETO, *J. Soc. Dyers Colourists*, 72 (1956) 239.

85. Detection and differentiation of rhodamine dyes[1]

The following quinoidal structural scheme is characteristic for rhodamine dyestuffs:

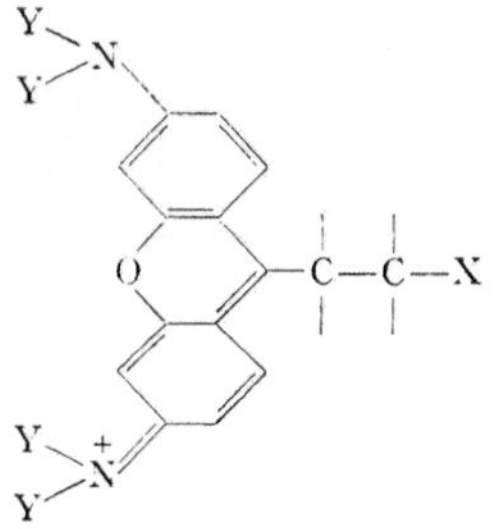

$X = COOH; COOR; SO_3H$

$Y = H; CH_3; C_2H_5; etc.$

Because they are amines, all rhodamines exhibit a decided basic nature. If the molecule contains free SO_3H- or COOH groups, their acidic character becomes evident and the presence of zwitter ions can be expected in the aqueous dyestuff solutions. The neutral, weakly acid or weakly basic aqueous solutions of rhodamines are various shades of red in daylight; in ultraviolet light they exhibit an intense red or orange fluorescence.

Sensitive tests for antimony(V), gold, and thallium(III) and also for antimony(III) by means of Rhodamine B (I) are given in *Inorganic Spot Tests*. They are based on the fact that this basic dye gives violet crystalline precipitates of the complex acids $H[SbCl_6]$, $H[AuBr_4]$, $H[TlBr_4]$ and $H[SbI_4]$. The precipitates are the Rhodamine B salts of the respective acids. The thallium(III) compound dissolves in benzene and the resulting red solution fluoresces intensely orange red in ultraviolet light. In accord with the rule that the analytical effect of organic compounds invariably is related to the presence of certain groups,[2] all rhodamine dyes react analogously to Rhodamine B because of the $\overset{+}{=}NXY$ group they contain. Differences are encountered only with respect to the extractability of the thallium(III) compound with benzene in that the presence of acidic and strongly hydrophilic groups in the dyestuff molecules sometimes impairs the solubility in benzene of the thallium compounds of the respective dyestuffs. Accordingly, the formation of fluorescing thallium compounds that are benzene-soluble can be used to detect strongly basic rhodamine dyestuffs. The following rhodamine dyes were tested:

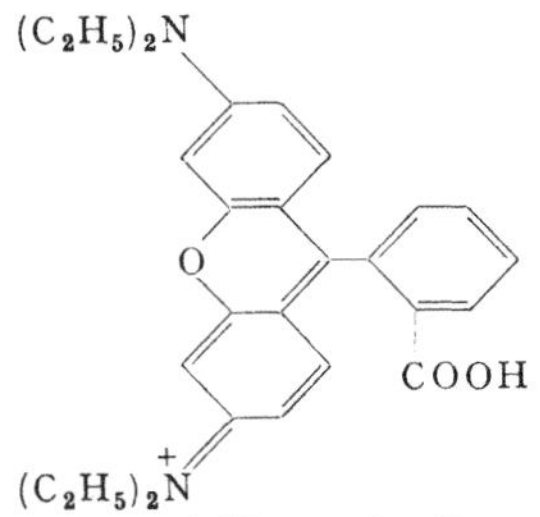

(I) Rhodamine B 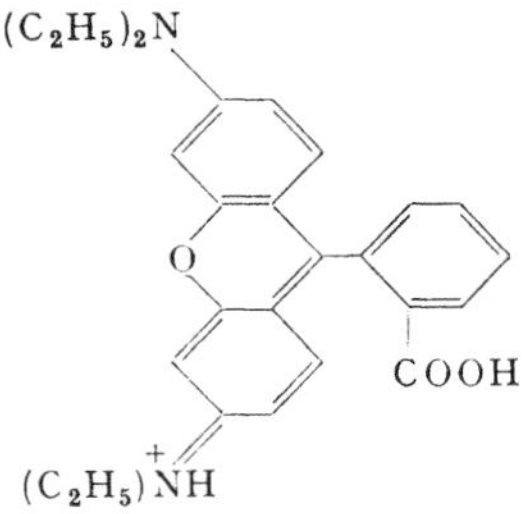   (II) Rhodamine G 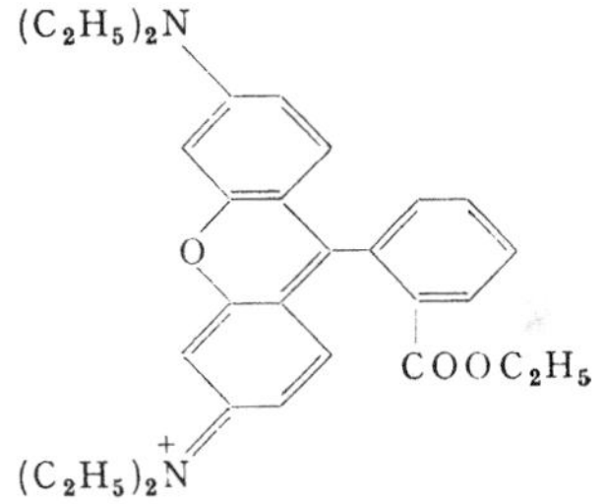   (III) Rhodamine 3 B

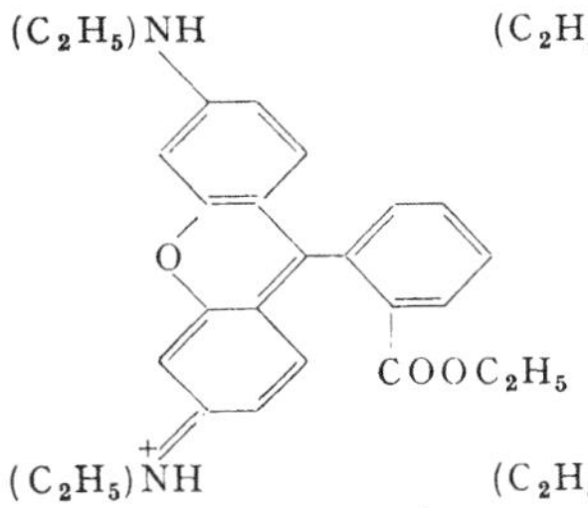

(IV) Rhodamine 6 G 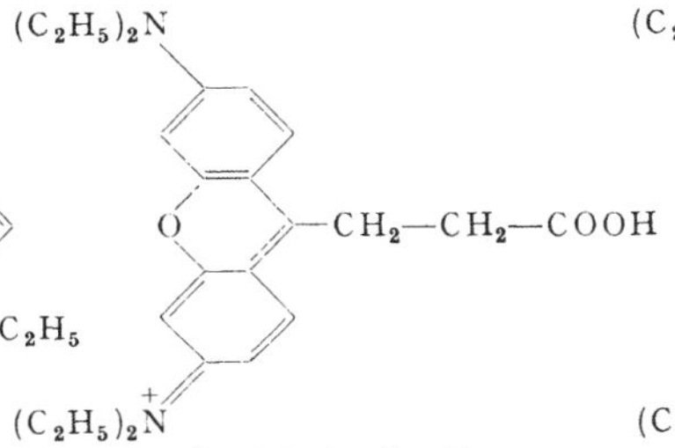   (V) Rhodamine S 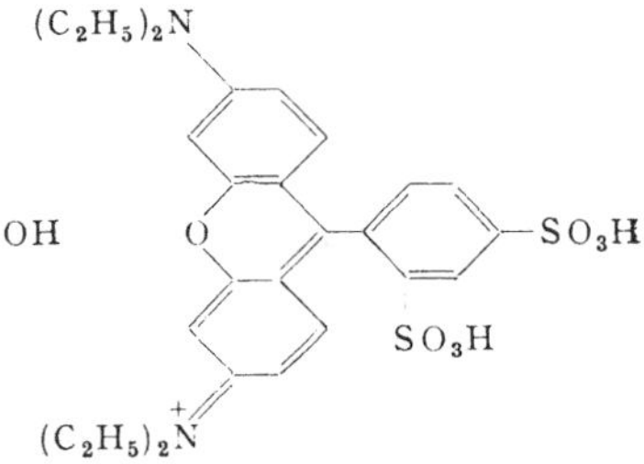   (VI) Sulforhodamine B

In addition, rhodamine dyestuffs whose structure is not given in the Schultz tables were likewise tested: Rhodamine 3R, Rhodamine 6 GH, and Acid Rhodamine R. The findings are included in the compilation given below.

Procedure. One drop of the aqueous test solution is placed in a micro test tube and mixed with a drop of 1 : 1 hydrochloric acid and a drop of a solution of thallic bromide. The mixture is shaken with 5–8 drops of benzene. A red benzene layer, which fluoresces orange in ultraviolet light, indicates the presence of rhodamine dyes.

Reagent: Thallic bromide solution. 2% solution of thallous sulfate is treated with bromine water until a distinct yellow persists. The excess bromine is removed by adding drops of a 10% solution of sulfosalicylic acid. The colorless mixture keeps for several days.

The behavior of rhodamine dyestuffs and the attainable identification limits can be seen in the following compilation; they all yield violet precipitates, most of which are soluble in benzene with fluorescence:

Rhodamine B (I)	0.5 γ	soluble in benzene	orange fluorescence
Rhodamine G (II)	0.25 ,,	,, ,, ,,	,, ,,
Rhodamine 3B (III)	0.25 ,,	,, ,, ,,	,, ,,
Rhodamine 6G (IV)	0.25 ,,	,, ,, ,,	,, ,,
Rhodamine S (V)	—	insoluble in benzene	—
Sulforhodamine B (VI)	—	,, ,, ,,	—
Rhodamine 6GH	0.5 ,,	soluble in benzene	orange fluorescence
Acid Rhodamine R	5 ,,	,, ,, ,,	,, ,,
Rhodamine 3R	0.25 ,,	,, ,, ,,	,, ,,

The above compilation shows that Rhodamine S and Sulforhodamine B are the only rhodamine dyestuffs whose thallium salts are insoluble in benzene. This can evidently be ascribed to the presence of hydrophilic CH_2COOH or SO_3H groups. Differentiation of these two dyes is possible by virtue of the fact that on heating with Raney nickel and alkali hydroxide only Sulforhodamine B gives the desulfonation reaction characteristic for aromatic sulfonic acids, *i.e.* nickel sulfide is formed (comp. p. 234).

The production of red benzene solutions is still visible at dilutions at which the formation of violet thallic precipitates can no longer be seen.

The formation of fluorescent thallic–rhodamine compounds can also be used to detect rhodamine dyes directly in dyed fabrics. Threads of the sample are digested with a drop or two of hydrochloric acid and several drops of the thallic bromide solution. Extraction with benzene follows. A red benzene solution, which fluoresces orange in ultraviolet light, is proof that basic rhodamine dyes were used to color the fabric, provided that a blank containing no $TlBr_3$ shows no fluorescence.

1 Unpublished studies, with J. E. R. MARINS.
2 Comp. F. FEIGL, *Chemistry of Specific, Selective and Sensitive Reactions*, Academic Press, New York, 1949, Chapter 6.

86. Detection of dyestuffs with anthraquinone structure[1]

A test for anthraquinones is given on p. 336. It is based on the formation of a soluble red sodium salt of anthrahydroquinone by the action of alkaline sodium hydrosulfite ($Na_2S_2O_4$). A similar reduction is undergone by acid dyes with anthraquinone structure. However, the latter test is limited in that the reduction is readily observed when the alkaline solution of the dye is not red but blue, violet or yellow.

Procedure. A drop of the alkaline solution of the dyestuff is warmed gently in a micro test tube with a small amount of solid sodium hydrosulfite. If the response is positive, the color turns toward red. If there is a large excess of dyestuff a reoxidation occurs on contact with the air and the original color is restored. In such cases it is essential to add more hydrosulfite.

The test revealed: 5 γ Chrome Blue Black 25 γ Alizarin Saphirol A

       0.5 γ Alizarin Blue ABI 0.5 γ Alizarin Saphirol B

1 Unpublished studies, with D. GOLDSTEIN.

87. Test for hydroxytriphenylmethane dyes[1]

Most of the acid mordant dyes of this category are prepared by condensing benzaldehyde (or its derivatives) with salicylic acid (or its derivatives).[2] When the blue alkaline solutions of these dyes are acidified, the corresponding red or red-brown dye acid is set free. The latter yields a yellow or brown solution in ether. Well known examples of these dyes include: Eriochrome Cyanine (I), Eriochrome Azurol (II), Chrome Fast Pure Blue B (III) and also Chrome Azurol S, Chromate Blue, and aurintricarboxylic acid.

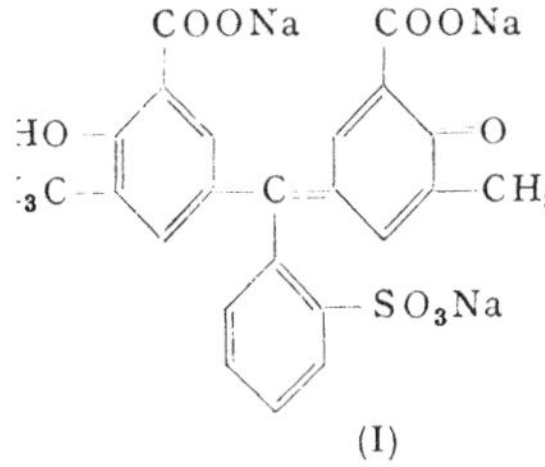

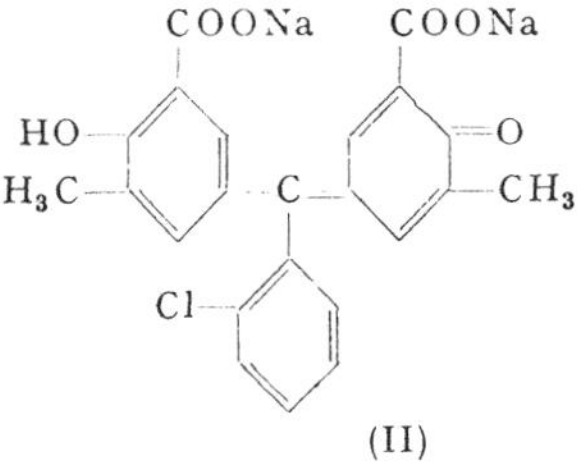

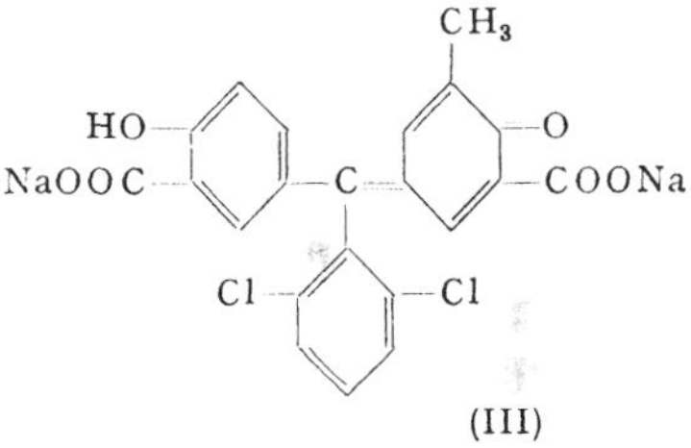

All of these dyes contain the group (IV), which characteristically forms aluminum chelate rings as shown in (V).

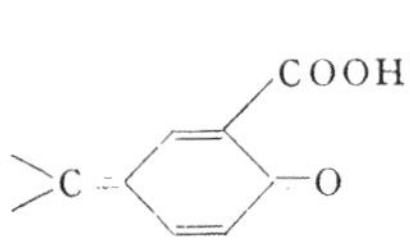

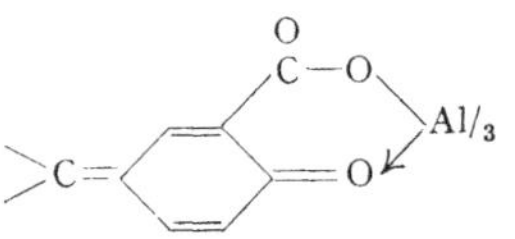

Since the dyes contain acidic groups at other positions in the molecule, the chelate ring and hence the aluminum is a component of the anions. The aluminum chelate compounds of hydroxytriphenylmethane dyes are dark red-violet, sometimes blue. Once formed, they are stable against strong mineral acids. This fact is the basis of a sensitive and specific test for aluminium with Chrome Fast Pure Blue B (see *Inorganic Spot Tests*, Chap. 3).

Procedure.   A drop of the aqueous dye solution (sodium salt) is treated in a micro test tube with a drop of a 5% solution of aluminum sulfate and a drop of dilute ammonia. After shaking, the mixture is made acidic with a drop of 1 : 1 hydrochloric acid. A color results, which persists when the solution is shaken with ether. (The latter extracts the excess dye).

The test revealed:

0.1 γ Chrome Fast Pure Blue B	violet	2.5 γ Naphthochrome Green G	
0.1 γ Eriochrome Cyanine	,,		blue-green
1.0 γ Naphthochrome Violet R	,,	5.0 γ aurintricarboxylic acid	red

When dealing with mixtures of dyes, or when conducting identification tests, it is advisable to suspend the sample in hydrochloric acid and then extract with ether. One drop of the ethereal extract is added to a drop of the $Al_2(SO_4)_3$-solution, the ether evaporated, and the procedure then applied.

1 F. FEIGL and D. GOLDSTEIN, *Anal. Chem.*, 29 (1957) 458.
2 Comp. G. SCHULTZ, *Farbstofftabellen*, 7th ed., Akad. Verlag, Leipzig, 1928-31.

88. Detection of p-phenylenediamine and its oxidation products in hair dyes[1]

Hair, leather and furs are often dyed with p-phenylenediamine. The colorless compound adsorbed by the animal fibers is oxidized to black water-insoluble products through contact with the air or by treatment with hydrogen peroxide. Since this dye and its oxidation products are toxic, their use in cosmetics, *etc.* is legally forbidden.

A rapid test for the free base involves its conversion into phenylene blue by the action of aniline and alkali persulfate in acetic acid solution. The chemistry and procedure for this sensitive test are outlined on p. 512. The test fails when applied to the oxidation products obtained in dyeing of furs with the ursol group of dyes.

Bandrowski's base is the most important oxidation product of p-phenylene-diamine. It is formed, along with p-quinoneimine, by the redox reaction:[2]

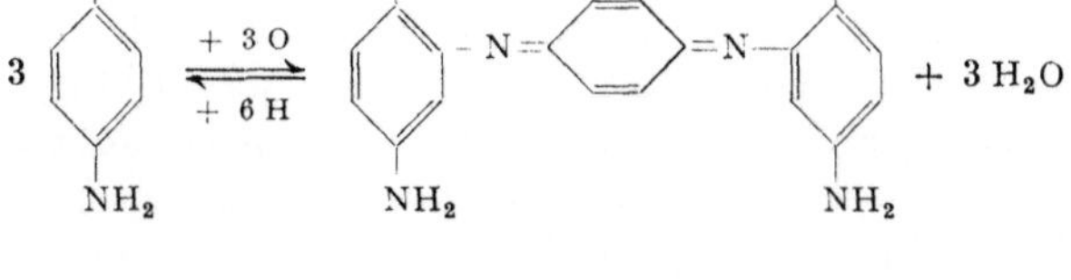

As just indicated, Bandrowski's base can be reconverted to p-phenylene-diamine by reduction with nascent hydrogen. The same is true of p-quinoneimine. This reduction can be accomplished by heating the sample with a mixture of sodium formate and sodium hydroxide. At 210–230°, the mixture decomposes quantitatively[3]:

$$HCOONa + NaOH \rightarrow Na_2CO_3 + 2\ H^0$$

The hydrogen not only converts the Bandrowski's base to phenylene-diamine but it also facilitates the sublimation of this product in the vicinity of its melting point (147°). The sublimed p-phenylenediamine reacts in the vapor phase on contact with aniline and persulfate to yield phenylene blue.

Procedure. A small amount of the solid is mixed in a micro test tube with several cg of the formate–hydroxide mixture, stirred with a drop of water and taken to dryness. The tube is immersed in a glycerol bath previously heated to 205° and the mouth of the tube is covered with filter paper moistened with a drop of the aniline–persulfate solution. The temperature is then raised to 210–230°. Depending on the amount of p-phenylenediamine given off, a light to deep blue stain appears on the paper immediately or within a few minutes.

Reagents: 1) Sodium formate–hydroxide mixture. A solution of 20% sodium formate and 12% sodium hydroxide is evaporated to dryness. It is best to conduct the evaporation in vacuo to avoid the formation of Na_2CO_3.

          2) Solution of 2 drops of aniline in 50 ml 10% acetic acid.

          3) 2% solution of potassium persulfate.

A freshly prepared mixture of equal volumes of 2) and 3) is used.

The test revealed as little as 10 γ Bandrowski's base.

1 F. Feigl and Cl. Costa Neto, *J. Soc. Dyers Colourists*, 72 (1956) 239.
2 J. J. Ritter and G. H. Schmitz, *J. Am. Chem. Soc.*, 51 (1929) 1587.
3 F. Haber and E. Bruner, *Z. Elektrochem.*, 10 (1904) 706.

89. Detection of phthalocyanines[1]

The phthalocyanines used as pigments are inner-complex compounds that have the general structure:

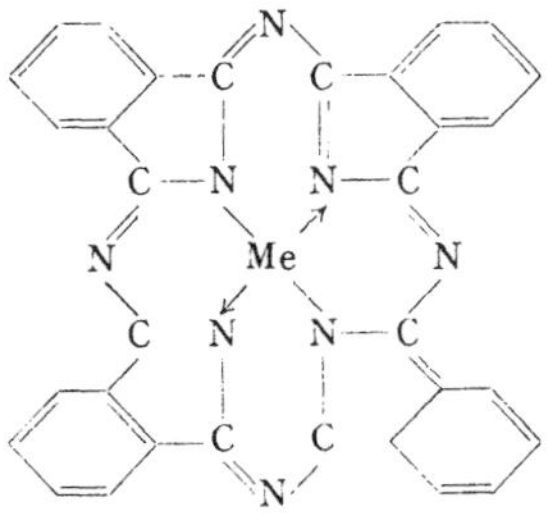

Me = Cu, Fe, Zn, Mn, *etc.*

The phthalocyanines are extraordinarily resistant to concentrated acids and heat (Cu-phthalocyanine is sublimable at 500°). These compounds may be decomposed to learn the identity of the metals included in them by heating mg samples with concentrated perchloric acid to 230° (see p. 101). After dilution with water, the resulting solutions permit the identification of Cu, Fe, Ni, *etc.* ions through the procedures given in *Inorganic Spot Tests*.

The method revealed: copper in Heliogen Blue and Viscotyl Blue; iron in Permanent Red E5B and Viscotyl Green.

1 Unpublished studies, with S. OFRI.

90. Detection of nickel dimethylglyoxime [1]

Scarlet crystalline nickel dimethylglyoxime is well known in inorganic analysis. This inner-complex salt is employed as a sun-fast pigment in paints, lacquers, cellulose compounds, and cosmetics. It can be detected by utilizing its solubility in dilute mineral acids and in alkali cyanide solutions (production of light green Ni^{2+} ions and yellow $[Ni(CN)_4]^{2-}$ ions, respectively). It leaves grey nickel oxide on ignition.

If a little of the sample is treated on a watch glass with several drops of dilute hydrochloric acid (alcoholic, if need be), the color is discharged. The red precipitate is restored by adding concentrated ammonia.

Digestion of Ni dimethylglyoxime with dilute potassium cyanide solution discharges the red color. This is restored on adding mercuric chloride, the nickel dimethylglyoxime being regenerated.

If the sample is ignited, the residue taken to dryness with dilute hydrochloric acid, and then treated with an ammoniacal alcoholic solution of dimethylglyoxime, nickel dimethylglyoxime is formed. It is readily distinguished from other red pigments or admixed organic red materials by the characteristics just cited.

1 Unpublished studies.

91. Detection of antimony and tin mordants in fabrics

A reliable test for antimony in textiles is based on the fact that soluble as well as insoluble compounds of tervalent and quinquevalent antimony are easily transformed into the water-soluble acid $H[SbI_4]$. When an aqueous solution of the basic dyestuff Rhodamine B is added, a red-violet precipitate of the rhodamine salt $[C_{28}H_{31}O_3N_2]$ $[SbI_4]$ is formed (comp. p. 670).

Although the solution of the dye is red, the formation of the red-violet antimony compound, which appears as a fine crystalline precipitate, is readily seen. The *identification limit* is 0.6 γ antimony.

Procedure.[1] Several mm² of the fabric being tested are ashed in a micro crucible. After cooling, the residue, which should be as free as possible from unburned carbon, is treated in succession with single drops of 5% potassium iodide solution, dilute hydrochloric acid, sulfurous acid (to remove the iodine formed by redox reaction with quinquevalent antimony) and 0.5% aqueous dyestuff solution. A violet precipitate indicates the presence of antimony.

When fabrics are to be tested for tin mordants, several mm² of the cloth are ashed; SnO_2 is left. The residue is warmed with concentrated sulfuric acid, to produce $Sn(SO_4)_2$. Then an equal volume of water is added (caution!), and after introducing several crystals of potassium iodide, the mixture is extracted with a benzene solution of iodine. The violet benzene solution, which contains SnI_4, is spotted on filter paper and is held over ammonia to produce $Sn(OH)_4$. The brown fleck is spotted with a drop of 5% sodium sulfite solution to remove the free iodine. Then the fleck is spotted with a drop of 0.05% solution of morin in alcohol or acetone. The paper is bathed briefly in dilute hydrochloric acid (1 : 10) and viewed in ultraviolet light. If tin is present, the fleck exhibits an intense blue-green fluorescence.[2] The latter is due to an acid-resistant fluorescing adsorption compound of $Sn(OH)_4$ with morin. *Inorganic Spot Tests* should be consulted regarding the chemistry of this test, whose *identification limit* is 0.05 γ tin.

Stannic chloride is used to weight silk. The procedure just given can be applied for the detection of small amounts of tin in silk.

1 Unpublished studies, with V. GENTIL.
2 F. FEIGL and V. GENTIL, *Mikrochim. Acta*, (1954) 90.

92. Testing the purity of dimethyl(ethyl)aniline [1]

Since dialkylanilines are prepared from aniline, they may contain the corresponding monoalkylaniline or perhaps aniline itself. Accordingly, when testing the purity of dimethyl(ethyl)aniline, it is well to establish the presence or absence of these contaminants.

As shown on p. 594, aniline and monoalkylanilines yield ammonia when heated to 100° with hexamine. This test, conducted on a drop of the ethereal test solution, therefore gives immediate information since no ammonia is split off by pure dialkylanilines.

To test for aniline, a drop of the ethereal test solution is placed on filter paper which has been impregnated with an ethereal solution of p-dimethylamino-benzaldehyde. The development of a yellow stain, due to the Schiff base, signifies a positive response. If the test for aniline is negative, the liberation of ammonia when the sample is heated with hexamine is a positive proof of the presence of monoalkylanilines. It should also be noted that a dialkyl-aniline which contains free aniline must invariably contain monoalkylaniline.

1 Unpublished studies.

93. Detection of nitrates of organic bases[1]

A reliable test for nitrates of organic bases is based on the fact that they yield nitric as well as nitrous acid when pyrolyzed. The latter arises from an inner-molecular redox reaction between nitric acid and the or-ganic portion of the nitrate. The nitrous acid can be detected in the gas phase through the Griess reaction. If maximum sensitivity is desired, it is well to employ the color reaction with N,N'-diphenylbenzidine,[2] which is given by both the nitrous and nitric acid in the mixture produced by the pyrolysis.

Procedure.   The gas absorption apparatus (Figure 23, *Inorganic Spot Tests*) is used. A little of the solid is placed in the apparatus, or a drop of its aqueous solution is evaporated there. A drop of freshly prepared solution of 10 mg of N,N'-diphenylbenzidine in 50 ml of concentrated sulfuric acid is placed on the knob of the stopper. The apparatus is closed and the bottom is strongly heated with a micro flame. If nitrates of organic bases are present, a blue color appears almost immediately.

The sensitivity of the test is illustrated by the finding that 2.5 γ of caffeine nitrate or guanidine nitrate yields a distinct blue color.

The test cannot be employed in the presence of compounds which contain groups including both N- and O atoms, because such compounds yield nitrous acid when pyrolyzed (see p. 93). In such cases use can be made of the fact that ammonia is produced by warming nitrates with caustic alkali plus Devarda alloy and can be detected with high sensitivity in the gas phase by means of Nessler reagent.

Another test for nitrates of organic bases involves the pyrolysis with benzoin; nitrous acid results. See test for nitrocelluloses, p. 693.

1 F. Feigl, J. R. Amaral and V. Gentil, *Mikrochim. Acta*, (1957) 731.
2 H. Riehm, *Z. Anal. Chem.*, 81 (1930) 439.

94. Detection of naphthol- and naphthylamine-sulfonic acids[1]

A test for naphthol- and naphthylaminesulfonic acids is of interest because many of these compounds are important starting materials in the manufacture of dyes. Likewise, the control of the sulfonation of naphthol and naphthylamine has industrial importance. The two procedures given here can be conducted with small samples. The basis of Procedure I is that aqueous acetic acid solutions of naphthol- and naphthylaminesulfonic acids, which have a free position *ortho* to OH- or NH_2 groups, yield brown water-soluble inner-complex cobalt(III) salts of *o*-nitrosonaphtholsulfonic acids when warmed with sodium cobaltinitrite (see p. 183). Procedure II is based on the fact that many but not all of the sulfonic acids of these classes are oxidatively split when warmed with potassium chlorate and concentrated hydrochloric acid and yield chloranil, which oxidizes tetrabase (see p. 447).

Procedure I.   As described on p. 184.

Procedure II.   A drop of the test solution or a pinch of the solid is warmed in a micro test tube with a drop of a saturated aqueous solution of potassium chlorate and one drop of concentrated hydrochloric acid. Several drops of water are added and the cold solution shaken with 5–10 drops of ether. After the layers have separated, several drops of the ethereal solution are placed on filter paper and the fleck is spotted with a drop of 1% ethereal solution of tetrabase. If the sample contains sulfonic acids, a blue stain remains.

The results of tests for naphthol- and naphthylaminesulfonic acids by Procedures I and II are given in Table 18. + + signifies strongly positive; — negative.

TABLE 18

NAPHTHOL- AND NAPHTHYLAMINESULFONIC ACIDS

Name	Limit of identification or result of test	
	Procedure I	Procedure II
1-Naphthol-4-sulfonic acid	5 γ	5 γ
1- ,, 5 ,, ,,	5 γ	—
2- ,, 6 ,, ,,	2.5 γ	50 γ
2- ,, 7 ,, ,,	+ +	250 γ
2,3-Dihydroxynaphthalene-6-sulfonic acid	5 γ	
1-Naphthylamine-3-sulfonic acid	25 γ	—
2- ,, -6- ,, ,,	2.5 γ	250 γ
Naphthionic acid	2.5 γ	50 γ
2-Naphthol-6,8-disulfonic acid	500 γ	—
Chromotropic acid	5 γ	—
2-Naphthylamine-3,6-disulfonic acid	5 γ	—
H-acid	+ +	—

From the compilation it is evident that, with the exception of 2-naphthol-6,8-disulfonic acid, even small amounts of all naphthol- and naphthylaminesulfonic acids respond to the phenol test. In contrast, the oxidative splitting to produce chloranil is obviously dependent on steric factors. In particular, the chloranil test fails with di- and tri-sulfonic acids.

1 Unpublished studies, with V. GENTIL; see also *Anal. Chim. Acta*, **13** (1955) 210.

95. Differentiation of naphthol- and naphthylamine-sulfonic acids[1]

If great sensitivity is not required, naphtholsulfonic acids and naphthylaminesulfonic acids can be readily distinguished from each other through the fact that only the former yield volatile phenols on pyrolysis. Such phenols are easily detected by the indophenol test (see p. 185).

A more sensitive differentiation is based on the characteristic differences in behavior exhibited by naphthol- and naphthylaminesulfonic acids toward a mixture of formaldehyde and concentrated sulfuric acid. Amost all naphtholsulfonic acids give a characteristic violet-red or yellow color at once, whereas naphthylaminesulfonic acids remain without color change even though warmed with the reagent. The color reaction is probably due to a condensation of the formaldehyde with the particular naphtholsulfonic acid (under the influence of the concentrated sulfuric acid) to yield diarylmethane compounds with OH- and SO_3H groups in the aryl ring. Since such compounds are not likely to be colored, this intial step is doubtless followed by an oxidation to quinoidal colored products, the sulfuric acid functioning as the oxidant.

The most favorable conditions for the realization of the color reaction are provided by employing concentrated sulfuric acid and hexamine (as formaldehyde-donor) and by making the test with solid sulfonic acids or their alkali salts. It should be noted that the reaction products of certain naphtholsulfonic acids exhibit a characteristic fluorescence in ultraviolet light.

Procedure. A granule of the sample or the evaporation residue of a drop of its aqueous solution is treated, in a micro test tube, with several mg of hexamine and a drop or two of concentrated sulfuric acid. Depending on the amount of naphtholsulfonic acid, a more or less intense violet-red or brown color appears. Gentle warming is needed only in the case of chromotropic acid. A quartz lamp should be used to disclose a possible fluorescence of the product.

Table 19 gives the results and *limits of identification* obtained by this procedure. The statements of color and fluorescence refer to 0.25 mg amounts.

TABLE 19

Sulfonic acids	Behavior toward $(CH_2)_6N_4$ + conc. H_2SO_4	
	Color (Fluorescence)	Identification limit
1-Naphthol-4-sulfonic acid	dark brown (lemon)	$1\gamma\,(0.1\gamma)$
1- ,, -5- ,, ,,	orange (lemon)	$1\gamma\,(0.1\gamma)$
2- ,, -6- ,, ,,	red-brown (yellow-green)	2γ (1γ)
2- ,, -7- ,, ,,	orange (no fluorescence)	2γ
2,3-Dihydroxynaphthalene-6-sulfonic acid	red (no fluorescence)	3γ
2-Naphthol-6,8-disulfonic acid	no reaction	
1,8-Dihydroxynaphthalene-3,6-disulfonic acid	violet (no fluorescence)	0.5γ
1-Naphthylamine-3-sulfonic acid	no reaction	
2- ,, -6- ,, ,,	,, ,,	
Naphthionic acid	,, ,,	
2-Naphthylamine-3,6-disulfonic acid	,, ,,	
H-acid	red brown (no fluorescence)	1γ

Naphthylaminesulfonic acids can be detected through their condensation with p-dimethylaminobenzaldehyde to yield orange Schiff bases, provided that other primary aromatic amines are absent. Naphtholsulfonic acids do not react with the aldehyde.

Procedure.   One drop of the test solution and one drop of a saturated solution of p-dimethylaminobenzaldehyde in glacial acetic acid are mixed on a spot plate. Depending on the amount of naphthylaminesulfonic acid present, an orange to light yellow color develops.

This procedure revealed:

0.5 γ 1-naphthylamine-3-sulfonic acid	0.25 γ 2-naphthylamine-3,6-disulfonic
0.1 γ 2- ,, -6- ,, ,,	acid
0.5 γ naphthionic acid	0.5 γ H-acid

Naphthylaminesulfonic acids can also be detected by spot reactions on filter paper impregnated with a saturated benzene solution of p-dimethylaminobenzaldehyde. A yellow fleck appears on the colorless reagent paper because of the formation of the particular Schiff base. The *limits of identification* are somewhat lower than those just mentioned.

Another means of differentiation may be based on the fact that naphthylaminesulfonic acids because of their NH_2 groups give the color reaction (red to violet) with sodium 1,2-naphthoquinone-4-sulfonate (see p. 153).

1 F. FEIGL and L. HAINBERGER, *Mikrochim. Acta*, (1955) 112.

96. Differentiation of naphthylaminemono- and -di-sulfonic acids[1]

As stated on p. 117, acidic compounds may be detected through the evolution of hydrogen cyanide when they are heated to 160° along with mercuric cyanide. This test fails with naphthylaminemonosulfonic acids since they have the character of neutral ammonium salts because of the basic NH_2 group. In contrast, naphthylaminedisulfonic acids, which contain an unsalified SO_3H group, give a positive response to the hydrogen cyanide test and can thus be differentiated from the naphthylaminemonosulfonic acids, provided other acidic materials are absent.

Procedure.   A pinch of the solid test material is placed in a micro test tube and mixed with several mg of mercuric cyanide and a drop of acetone, and taken to dryness. A drop of the aqueous solution of the sample may also be evaporated to dryness. The open end of the tube is covered with a piece of filter paper that has been moistened with a drop of reagent for cyanide (preparation see p. 546). The test tube is then immersed to about half its length in a glycerol bath preheated to 160°. A positive response is signalled by a blue stain.
   Identification limit: 50 γ 2-naphthylamine-3,6-disulfonic acid.

If the response to the above test is negative and naphthylaminemonosulfonic acid is to be detected, the following procedure may be used. The solid sample is heated to 140° with paraformaldehyde; the resulting formaldehyde methenylates the NH_2 group and thus frees the SO_3H group for reaction with the mercuric cyanide.
   Another test for naphthylaminesulfonic acid is based on the finding that the latter is desulfurized when heated with Raney nickel alloy and caustic alkali (see p. 233). Naphthylamine results; it is volatile with steam, and may then be detected in the gas phase by the spot reactions for primary arylamines.

1 Unpublished studies.

97. Examination of gunpowder and explosives[1]

Black gunpowder, which contains large amounts of nitrates, may be tested for nitrate on a drop of the aqueous extract. The color reaction with diphenylamine or diphenylbenzidine, which yields a blue quinoidal oxidation product, may be applied. Similarly, other explosives may be tested for chlorate by the reaction with manganous sulfate and phosphoric acid (formation of red complex Mn(III) phosphate). Azides are revealed by the formation of red ferric azide. The residue from burned black powder always contains thiosulfate, thiocyanate, and sulfide besides some elementary sulfur. Even traces of these sulfur compounds may be detected by the acceleration of the iodine–azide reaction. The test with Nessler reagent or other reagents for free ammonia may be appplied for detecting ammonium salts. Free sulfur can be detected in explosives and gunpowder either directly by fusion with benzoin (see p. 609) or indirectly through the protective layer effect on thallous sulfide (see p. 607). The tests for inorganic ingredients are discussed at length in *Inorganic Spot Tests*.

Aromatic polynitro compounds (picric acid, trinitrotoluene) may be detected by the procedure for *m*-dinitro compounds given on p. 397. The test through pyrolytic release of nitrous acid (see p. 299) is also recommended.

Explosives which contain esters or amides of nitric acid (nitrocellulose nitroglycerol, tetranitropentaerythritol, Tetryl, Hexogen, *etc.*) are readily reduced to nitrous acid which can be identified by the Griess test.

Procedure.   A few mg of the sample are mixed on a spot plate with a drop of sulfanilic acid and a drop of α-naphthylamine solution. A little zinc dust or Devarda alloy is added. If nitro compounds were present, the solution (or solid) is colored bright red by the resulting azo dye. The development of the color may require several minutes.

Reagents: for preparation see p. 92.

Esters of nitric acid may also be detected in threads of textiles. Microscopic examination will reveal admixtures or non-impregnated cotton. When nitrates and water-insoluble nitro compounds are to be detected in the presence of each other, an aqueous extract should be prepared. After centrifuging or filtration, the clear solution is tested for nitrate (see p. 300). The residue, after reduction, is tested for nitrite.

Tests for Tetryl (Tetralyte) and Hexogen (Cyclonite) are described in the following sections.

1 A. A. Azzam, *Mikrochim. Acta*, (1937) 283.

98. Detection of picrylnitromethylamine[1]

Picrylnitromethylamine, also known as Tetryl, Tetralyte, Tityl, is used as primer in explosives. This yellow compound, which is insoluble in water but soluble in benzene, can be detected by heating it to 150°; nitrous acid is split off. If heated to this same temperature with hexamine, vapors of formaldehyde and ammonia are evolved. The former case may involve the innermolecular reaction:

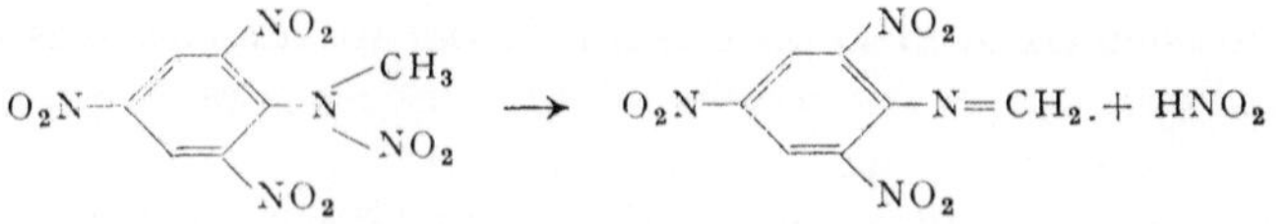

$$O_2N-\underset{NO_2}{\overset{NO_2}{\bigodot}}-N\underset{NO_2}{\overset{CH_3}{<}} \rightarrow O_2N-\underset{NO_2}{\overset{NO_2}{\bigodot}}-N{=}CH_2.+\ HNO_2 \qquad (1)$$

In the second instance, Tetryl functions as oxidant with respect to hexamine and brings about the oxidative decompositions:

$$(CH_2)_6N_4 +\ \ 6\,[O] \rightarrow 6\,CH_2O + 2\,N_2 \qquad (2)$$

$$(CH_2)_6N_4 + 12\,[O] \rightarrow 6\,CO_2 + NH_3 \qquad (3)$$

The pyrolytic reactions (1, 2, 3) can be conducted in micro test tubes whose mouths are covered with pieces of filter paper moistened with Griess or Nessler reagents, respectively. A brown or black stain indicates the occurrence of (2) and (3), a red stain the occurrence of (1).

Identification limits: 8 γ Tetryl through dry-heating
                              10 γ ,,   ,,   pyrolysis with hexamine

The fusion reaction for C-nitro compounds with diphenylamine or tetrabase (see p. 295) can be used as a confirmatory test. The resulting violet addition product will reveal 5 γ Tetryl.

None of the tests given here respond to Cyclonite (see Sect. 99).

1 F. FEIGL and D. HAGUENAUER-CASTRO, *Chemist-Analyst*, 51 (1962) 5.

99. Detection of non-aromatic nitroamine explosives[1]

Hexahydro-1,3,5-trinitrotriazine, also known as Cyclonite, Hexogen, RDX, and T4, is a high explosive. The colorless compound is not soluble in water or benzene but dissolves in acetone. In contrast to Tetryl (*q.v.*) it is not affected by heating to 180°. However, if heated in the presence of manganese dioxide which acts as O donor, Cyclonite yields formaldehyde and nitrous acid:

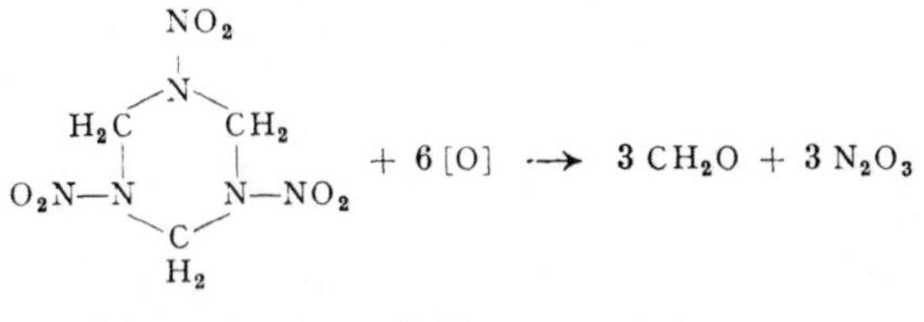

$$\underset{H_2}{\underset{O_2N-N\quad\quad N-NO_2}{\overset{H_2C\quad\quad CH_2}{\overset{\overset{NO_2}{|}}{\overset{N}{\diagup\ \diagdown}}}}}\ +\ 6\,[O]\ \longrightarrow\ 3\,CH_2O + 3\,N_2O_3$$

The pyrolytic oxidation may be conducted in a micro test tube immersed in a glycerol bath at 180°. The evolution of formaldehyde is revealed by the formation of metallic mercury with Nessler reagent, that of the nitrogen oxide by the Griess reaction.

The explosive HMX mentioned in the following section behaves similarly.

Limit of identification: 10 γ Cyclonite (RDX).

1 F. FEIGL and D. HAGUENAUER-CASTRO, *Chemist-Analyst*, 51 (1962) 5.

100. Differentiation of hexahydro-1,3,5-trinitro-*s*-triazine (RDX) and octahydro-1,3,5,7-tetranitro-*s*-tetrazine (HMX)[1]

When dissolved in concentrated sulfuric acid, various nitrate and nitramine explosives react at room temperature with quinalizarin (1,2,5,8-tetrahydroxyanthraquinone) to produce a color change from violet to yellow within 5–60 min. (The color change is due to the oxidative destruction of the dye by the nitric acid split off.) If the same weights of RDX and HMX undergo this treatment, the time required for the color change is not the same for the two compounds since the percentage of nitric acid split off by I is greater than by II.

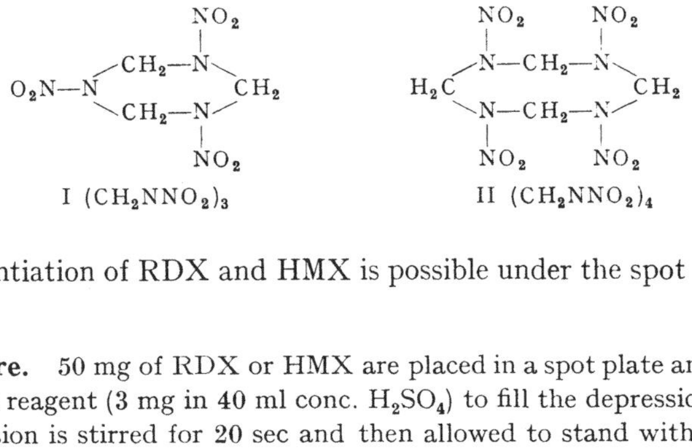

I $(CH_2NNO_2)_3$ II $(CH_2NNO_2)_4$

A differentiation of RDX and HMX is possible under the spot test conditions.

Procedure.   50 mg of RDX or HMX are placed in a spot plate and sufficient quinalizarin reagent (3 mg in 40 ml conc. H_2SO_4) to fill the depression is added. The suspension is stirred for 20 sec and then allowed to stand with occasional stirring. After precisely 20 min. observe the color. At this time, RDX yields a light yellow, HMX a blue color. Again observe the color after 25–30 min; HMX then yields a yellow color. Compare all of these colors with that of the quinalizarin reagent solution.

This procedure serves for the detection and differentiation of RDX and HMX when present alone. Admixtures of these two explosives with each other or with other nitrate and nitramine explosives show different rates of color formation and thus render the results of the test ambiguous.

1 S. SEMEL, *Chemist-Analyst*, 51 (1962) 6.

101. Differentiation of animal and vegetable fibers[1]

Vegetable fibers are free from sulfur, but fibers derived from animal sources contain sulfur compounds in the cysteine components of the protein. These compounds possess an SH group, which accelerates the iodine–azide reaction. Hence, wool and cotton may be differentiated by applying this test (p. 219).

Procedure. The fiber, or small piece of woven material, is moistened with a drop of water or acetone on a watch glass, and treated with 1 or 2 drops of the iodine–azide solution. After a short while, wool is covered with little bubbles, while cotton shows none. Cotton fibers turn dark violet owing to adsorption of iodine, but the color of the bubbling wool remains unchanged. Bleached or unbleached cotton may be used in the test; both are colored violet, with no bubble formation.

Another method of examining textiles for wool is based on the fact that when animal fibers are thermally decomposed, the proteins yield volatile pyrrole and pyrrole derivatives, which can be sensitively detected in the gas phase since, on contact with an acid solution of p-dimethylaminobenzaldehyde, they give violet quinoidal condensation products (see p. 381).

Procedure. Several mg of the test material are pressed to the bottom of a hard glass micro test tube. The mouth of the test tube is covered with a disk of filter paper moistened with a 5% solution of p-dimethylaminobenzaldehyde in concentrated hydrochloric acid. The bottom of the test tube is heated with a micro flame until charring occurs. If animal fibers are present, a red-violet fleck will appear on the paper.

Fabrics containing Nylon, which is a plastic consisting of a superpolymeric amide of protein-like structure, behave like animal fibers when subjected to this test, since the pyrolysis of Nylon obviously produces pyrrole or pyrrole derivatives. On the other hand, Nylon does not accelerate the iodine–azide reaction. Consequently, the latter test can serve to differentiate between Nylon and animal fibers.

Vegetable fibers respond positively to the test for cellulose and cellulose derivatives (p. 691) based on the hydrolysis to furfural, which condenses with aniline to give a red Schiff base.

It may be noted here that animal fibers which contain sulfur and aliphatic bound nitrogen yield H_2S and NH_3 when pyrolyzed with calcium oxalate. Consequently such fibers give a positive response in the preliminary test for pertinent compounds containing nitrogen and sulfur as described on p. 94.

1 F. FEIGL, *Mikrochemie*, 15 (1934) 7.

102. Tests for the identification of artificial fibers[1]

The following tests require no more than 0.5–1 mg of the specimen.

(1) Test for N-bearing fibers by oxidative pyrolytic formation of nitrous acid

A test for nitrogen obviously affords a means of distinguishing synthetic fibers containing nitrogen from those that are free of this element. The nitrogen test discussed and described on p. 90 is suited to this purpose.

A positive response was obtained with the following types of synthetic fibers:

Acrylonitriles	
Orlon 41 (Du Pont)	Dynel
Orlon 42 (Du Pont)	Verel (Acrylic) (Eastman)
PAN fiber (Cassella)	Creslan or X-54
Acrilan	Darlan
Polymers of ε-aminocaproic acid	
Perlon W type (BASF)	Perlon Caprolactam (BASF)
Perlon B type (BASF)	Perlon (Ver. Glanzstoff-Fabr. A.G.)
Hexamethylene adipamides	
Nylon 66	

A negative response was given, as expected, by the following types of fibers:
Polyester fibers: Dacron and Terylene.
Polyvinyl chlorides: PCU and Vynion HH.

(2) Test for Nylon[2]

Since fibers of Nylon and Perlon are polyamides, they give the color test for proteins with tetrabromophenolphthalein ethyl ester described on p. 371.

A specific test for Nylon fibers can be based on the fact that they become blue-violet when brought into contact with a yellow solution of tetrabromophenolphthalein isopropyl ester in diisopropyl ether. Probably a colored adsorption product is formed and it is thought that the diisopropyl ether plays some part in the reaction. The test can be carried out with a small skein of the fiber in a micro test tube. The intensity of the color is not the same in all cases; obviously the degree of polymerization has an influence. However, a positive response is invariably observed if nylon is present.

(3) Test for fibers of the Perlon and Nylon types

All types of nitrogenous fibers cited in (1) yield hydrogen cyanide when pyrolyzed. This product may be readily detected in the gas phase by means of the sensitive color reaction with copper–benzidine acetate (see p. 546). However, if a mixture of the fiber and calcium oxide is ignited, there is no formation of hydrogen cyanide in the case of fibers of the Perlon and Nylon types. This behavior may thus be used to distinguish these types from the other nitrogen-bearing fibers.

(4) Test for Perlon

When fibers of the Perlon type are heated together with sodium thio-sulfate at 220° sulfur dioxide is released which can be detected in the gas phase with appropriate reagent papers. Probably Perlon loses water which acts as superheated steam and then hydrolyzes the thiosulfate according to: $Na_2S_2O_3 + H_2O \rightarrow 2\ NaOH + SO_2 + S$. This behavior appears to be specific for Perlon, provided fibers of the polyvinyl chloride type are absent.

Procedure. A small amount of the sample and several cg of hydrated sodium thiosulfate are put into a micro test tube which is immersed in a glycerol bath previously heated to 180°. The open end of the tube is covered with a piece of Congo paper moistened with 5% hydrogen peroxide. The temperature is raised to 220°. The appearance of a blue stain on the indicator paper is characteristic for Perlon.

Fibers of the polyvinyl chloride type show a remarkable behavior in that they split off hydrochloric acid when heated up to 225°. The acid can be detected in the gas phase with acid–base indicator paper.

(5) Test for fibers with amide structure

Artificial fibers with amide structure such as Nylon, Perlon, and fibers which contain acrylamide as copolymer show a characteristic behavior when heated to 160° with dimethyl oxalate and thiobarbituric acid. A red product is rapidly formed, which is characteristic for oxamide (see p. 254). This is due to the fact that, through pyrolysis, ammonia and/or ammonium salts of aliphatic acids are formed which yield oxamide when heated with dimethyl oxalate. No other fibers cited in (1) show the same behavior.

(6) Test for fibers containing chlorine

Vinyl chloride is a component of Dynel, Verel, PCU, Vynion, DNA, Acrylic, and Phrix. When these fibers are ignited along with manganese dioxide or warmed with a mixture of concentrated sulfuric acid and chromic acid chlorine is liberated. It may be detected through the color reaction with thio-Michler's ketone (see p. 64).

(7) Test for acrylic fibers through pyrolytic formation of thiocyanic acid

The acrylonitrile fibers are polymers of acrylonitrile (vinyl cyanide), either the pure polymer Orlon (E. I. DuPont de Nemours) or copolymers formed with vinyl chloride (Acrilan, Chemstrand Corporation; Dynel, a crimped staple fiber, and Vinyon N, a continuous filament, both manufactured by Union Carbide and Carbon).

The test for the CN groups given on p. 267 which is based on the pyrolytic formation of thiocyanic acid may be employed to characterize the acrylonitrile fibers. All of the acrylonitrile fibers mentioned above give a positive response to this test.

(8) Test for the acrylamide component in acrylonitrile fibers through chemical adsorption of rhodamine B

Among the acrylonitrile fibers and among the synthetic fibers noted above, the fibers containing acrylamide differ in a characteristic manner as regards their behavior toward the red aqueous solution of rhodamine B or toward the colorless benzene solution of this dye. In both cases there is an almost immediate appearance of an intense red coloring which is not removed by washing with water or benzene. The effect is especially striking when the benzene solution of the dye is used because then two colorless components form a red reaction product.

The color reaction results from a chemical adsorption of rhodamine B from its solution. The aqueous solution contains the red quinoid ions, the benzene solution contains the isomeric lactone form of the dyestuff (see p. 704). The chemical adsorption involves a salt-like bonding of the basic rhodamine B on the surface of the fibers through the acidic hydrogen atoms of the acrylic acid amide residing there. The adsorbate exhibits the color and fluorescence of the benzene-soluble salts of rhodamine B with complex metallo-halogeno acids (compare *Inorganic Spot Tests*, Chap. 3) but it differs from these in being insoluble in benzene.

(9) Tests for Dacron and Terylene through pyrolytic formation of volatile phenol or acetaldehyde

Dacron and Terylene are polymers of glycol esters of terephthalic acid. At present they are the only known artificial fibers containing an aromatic component. It was pointed out in Chap. 2 that aromatic compounds which contain oxygen in the nucleus or in an open or closed side chain yield volatile phenols when subjected to dry heating. These phenols may be detected through the color reaction with 2,6-dichloroquinone-4-chloroimine. Accordingly the procedure given on p. 142 may be employed for the detection of Dacron or Terylene.

Another and more sensitive test can be based on the fact that the mentioned fibers split off acetaldehyde when heated with syrupy phosphoric acid or better zinc chloride until carbonization occurs. The reason is that the polyester is first depolymerized and then saponified, whereby glycol is formed. The latter, by loss of water, gives vapors of acetaldehyde (see p. 166). The color reaction with sodium nitroprusside and morpholine or piperidine is used for detecting the aldehyde (see p. 438).

1 F. Feigl, V. Gentil and E. Jungreis, *Textile Research J.*, 28 (1958) 891.
2 V. Anger, unpublished studies.

103. Detection of urea resins[1]

The test for urea, based on the formation of the violet nickel salt of diphenylcarbazide (see p. 390) can be applied for the detection of urea resins, namely polymerized condensation products of urea with formaldehyde *etc.* Lower polymers respond directly; higher polymers require hydrolysis with concentrated hydrochloric acid to liberate the urea.

Procedure.   A few mg of the solid and a drop of concentrated hydrochloric acid are taken to dryness at 110° in a micro test tube. After cooling, the residue is treated with one drop of phenylhydrazine and the mixture is heated to 195° for 5 min. in a glycerol bath. The cooled mixture is stirred with 3 drops of 1:1 ammonia and 5 drops of 10% $NiSO_4$ and then shaken with 10 drops of chloroform. The presence of urea and hence of a urea resin or plastic is indicated by a red-violet to red color in the chloroform.

The test is not sensitive because urea is partly hydrolyzed by acids.

1 Unpublished studies, with V. Gentil.

104. Detection of formol plastics[1]

Plastics of resins produced by condensation of formaldehyde with urea, melamine, or phenols can be detected through the release of formaldehyde on heating with concentrated sulfuric acid. The aldehyde can be identified in the vapor phase by the chromotropic acid color reaction (see p. 434). The splitting of formaldehyde is brought about by the hydrolytic action of the water contained in the concentrated sulfuric acid; it reacts in the form of superheated steam when the acid is heated. The procedure given on p. 324 is recommended.

It is easy to determine whether a formol plastic contains a urea or phenol component. In the first case, pyrolytic oxidation with manganese dioxide will yield nitrous acid (see p. 90). On the other hand, formaldehyde–phenol resins, *e.g.* Bakelite, yield phenol on pyrolysis, which can be detected in the gas phase through the indophenol reaction (see p. 141).

1 F. Feigl and L. Hainberger, *Chemist-Analyst*, 44 (1955) 47.

105. Detection of melamine resins[1]

Melamine resins are hydrolyzed by evaporation with concentrated hydro-chloric acid, whereby a residue of melamine chloride is obtained. The excess hydrochloric acid is completely expelled at 200°. On heating the dry mel-amine chloride with sodium thiosulfate at 140°, sulfur dioxide is formed (see p. 111), which can be detected by the blue color that it produces on Congo red paper moistened with hydrogen peroxide.

All the necessary steps can be carried out with microgram quantities of the sample. The similar urea resins do not interfere, because the urea split-out is hydrolyzed ($CO(NH_2)_2 + H_2O + 2\ HCl \rightarrow CO_2 + 2\ NH_4Cl$) and the ammonium chloride formed does not react with thiosulfate.

1 F. FEIGL and V. ANGER, unpublished studies.

106. Detection of cellulose and cellulose derivatives[1]

When cellulose is heated it decomposes and the resulting superheated steam reacts with unchanged cellulose to produce hexoses, which in turn hydrolyze to give hydroxymethyl-furfural. The latter is volatile with steam and can be detected in the gas phase in the same manner as furfural, *i.e.* by means of the sensitive color test with aniline acetate (see p. 444).

The furfural reaction can be employed to detect cellulose (Procedure I) and, if certain conditions are observed, to reveal the presence of cellulose derivatives such as cellulose acetate (Procedure II). All sugars, starches, and vegetable gums likewise yield furfural or hydroxymethyl-furfural when they are thermally decomposed or hydrolyzed by strong acids.

Procedure I.   A little of the solid, say a thread of the fabric, is placed in a micro crucible. The latter is then covered with filter paper moistened with aniline acetate solution and held in place by a watch glass. The bottom of the crucible is heated for 30 sec. with a micro flame. A red fleck appears on the paper if cotton is present.

Methyl cellulose, ethyl cellulose and acetyl cellulose do not respond to Procedure I. If, however, such cellulose derivatives are heated with concen-trated phosphoric acid, they are saponified with production of cellulose, which, since it is a carbohydrate, does produce furfural when heated. The latter then responds to the test with aniline acetate.

Procedure II.   A little of the solid is placed in a micro crucible along with a drop of syrupy phosphoric acid. The mouth of the crucible is covered with filter paper moistened with aniline acetate solution and weighted down with a watch glass. After cautious heating with a micro flame for 30–60 sec., a red fleck appears on the filter paper if a cellulose derivative is present.

Reagent: 10% solution of aniline in 10% acetic acid

Procedure II, which is more sensitive, may also be applied, in place of Procedure I, for the detection of cellulose.

A good test for cellulose, methyl-, ethyl-, and acetyl cellulose (and carbohydrates) is based on the formation of acetaldehyde when these substances are pyrolyzed. The aldehyde in the vapor phase can be detected through the color reaction with sodium nitroprusside–morpholine reagent (p. 438).

Procedure III. A little of the sample is heated in a micro test tube whose mouth is covered with a disk of filter paper moistened with a drop of the reagent solution (for preparation see p. 439). The heating over a micro flame is continued until charring is seen. A blue stain rapidly appears on the colorless or light yellow reagent paper.

1 Unpublished studies, with J. E. R. MARINS, Cl. COSTA NETO and E. SILVA.

107. Detection and differentiation of methyl- and ethyl cellulose[1]

The tests for methoxy and ethoxy groups (pp. 165, 189) can also be used to distinguish between methyl- and ethyl cellulose. The procedure is based on the production of formaldehyde and acetaldehyde, respectively, when these cellulose esters are cleaved by heating with benzoyl peroxide. The aldehydes are detected in the gas phase. Even mg quantities are adequate.

These esters yield methyl and ethyl alcohol, respectively, when they are saponified by dry heating with $MnSO_4.H_2O$ at 160–200°. The water of crystallization acts as quasi superheated steam at the place of its liberation in contact with the esters. The resulting alcohols undergo significant autoxidation in the gas phase at the temperature of the reaction, and so yield formaldehyde and acetaldehyde, respectively. If the heating is conducted in a micro test tube, whose mouth is covered with a disk of filter paper moistened with Nessler solution, a black stain appears because of the redox reaction:

$$—CHO + [HgI_4]^{2-} + 2\,OH^- \rightarrow —COOH + Hg^0 + H_2O + 4\,I^-$$

The production of acetaldehyde from ethyl cellulose can be detected by the color reaction with sodium nitroprusside described on p. 438.

1 Unpublished studies, with E. SILVA.

108. Plastics and lacquers containing nitrocelluloses [1]

These products, such as celluloid, collodium, zapon varnish, *etc.*, can be identified through the positive response to two tests that are characteristic for nitrocellulose. The $O-NO_2$ group can be detected by fusion with benzoin at around 150°; the pyrohydrogenolysis yields nitrous acid (Procedure I). The basis of the second test is that warming with syrupy phosphoric acid leads to hydrolysis of the nitric acid ester and conversion of the cellulose into methylfurfuraldehyde. The latter shows the characteristic condensation of aromatic aldehydes with thiobarbituric acid (see p. 205) leading to orange products (Procedure II).

Procedure I.   A little of the sample is heated in a micro test tube along with several cg of benzoin by immersing the test tube in a glycerol bath preheated to 130°. The open end of the test tube is covered with a piece of filter paper moistened with Griess reagent (for preparation see p. 92). When the temperature is raised to 150–160° a red stain develops on the reagent paper.

Procedure II.   As described on p. 205.

These two tests can be carried out with 1–2 mg of the sample.

1 F. FEIGL and E. LIBERGOTT, *Chemist-Analyst*, 52 (1963) 47.

109. Test for styrene resins and styrene-containing copolymers [1]

Mononitrobenzene compounds are formed if styrene or its polymers are evaporated with fuming nitric acid. These nitro compounds belong to the group of oxygen-containing aromatics which release phenol when they are subjected to pyrolysis; phenol can be detected in the gas phase by the indophenol reaction, as described on p. 141. Methylstyrene resins and copolymers of styrene and butadiene behave likewise.

Procedure.   A small amount of the sample and 4 drops of nitric acid (d. 1.5) are evaporated to dryness, and then further treated as described on p. 142.

Bakelites and cross-linked epoxy resins, which are oxygen-containing aromatics, also release phenol when pyrolyzed. However, if they are treated with fuming nitric acid, the residue does not yield phenol on pyrolysis.

1 F. FEIGL and V. ANGER, unpublished studies.

110. Test for epoxy resins[1]

Epoxy resins can be cross-linked with aromatic amines, according to the following scheme:

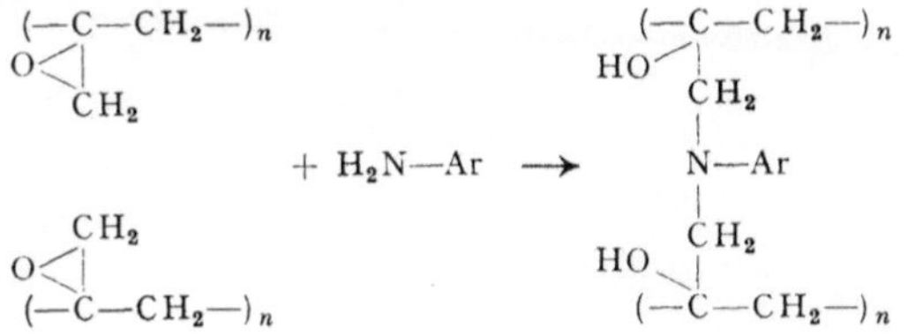

$$(-C-CH_2-)_n \quad\quad\quad (-C-CH_2-)_n$$

Both, untreated and cross-linked epoxy resins, split off acetaldehyde when heated at 240° in a micro test tube; the acetaldehyde thus formed can be detected in the gas phase through the color reaction with sodium nitroprusside and morpholine (see p. 438).

Normal and cross-linked epoxy resins may be differentiated because only the latter yield phenol on hydrolysis. This is due to the fact that, according to the above formulas, the cross-linked epoxy resins are oxygen-containing aromatics for which the pyrolytic release of phenol is characteristic (see p. 141).

1 F. FEIGL and V. ANGER, unpublished studies.

111. Differentiation of chlorinated rubber and neoprene[1]

In the chlorination of rubber, the chlorine is not only added to the double bonds, but it also breaks the long chains of the rubber, producing terminal $-CH_2Cl$ groups. These primary alkyl chlorides can be detected by the production of sulfur dioxide when they are pyrolyzed with sodium thiosulfate (see p. 171). Neoprene, formed by polymerization of 2-chlorobutadiene, does not give this reaction, because it contains only secondary and tertiary bound chlorine.

Another method for differentiation of chlorinated rubber and Neoprene is based on the behavior of these materials when heated to 200–210°.

Procedure.   The sample is heated in a micro test tube immersed in a glycerol bath at 210°. After 10 min. any hydrogen chloride which may have been formed is removed by blowing with a pipet. Then a piece of moistened Congo paper is placed on the mouth of the tube and the temperature of the glycerol bath is lowered to 190–200°. A blue stain on the indicator paper appears if Neoprene was present. This special procedure is necessary because chlorinated rubber also releases a little hydrogen chloride, but only for a short time.

1 F. FEIGL and V. ANGER, unpublished studies.

112. Detection of naphthalene-1,5-diisocyanate[1]

Naphthalene-1,5-diisocyanate, used in manufacturing plastics of the poly-urethane type, condenses with nitrosoantipyrine to a violet-red bisazo dye:

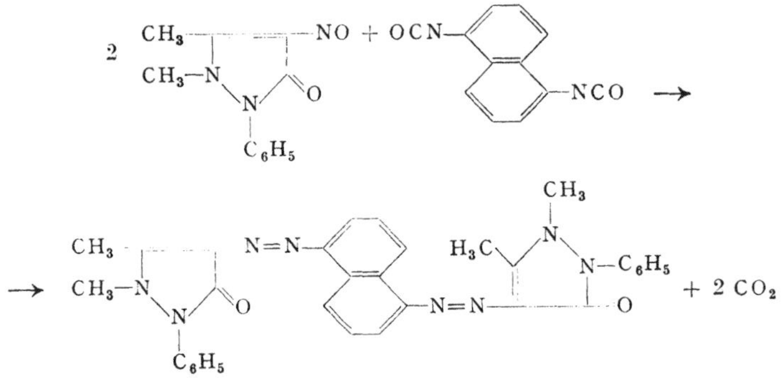

Procedure. Two drops of a solution of 20 mg nitrosoantipyrine in 10 ml 50% acetic acid (compare p. 510) are added to the sample in a micro test tube. The test tube is heated for 10 min. in a boiling water bath. A violet-red color appears, if naphthalene-1,5-diisocyanate is present.

Limit of identification: 5γ naphthalene-1,5-diisocyanate.

1 V. ANGER, unpublished studies.

113. Detection of chromium in leather

Chrome-tanned leather can be recognized by heating with 70% per-chloric acid; the hide substance (collagen) is oxidatively decomposed (see p. 88) and 3-valent chromium is converted to chromic acid. The latter is easily revealed by the sensitive color reaction with diphenylcarbazide.[1]

Procedure.[2] A tiny amount (0.3–0.5 mg) of the leather being tested is mixed in a micro test tube with a drop of 75% perchloric acid and kept at 220° for about 5 min. in a glycerol bath. After cooling, a drop of an alcoholic solution of diphenylcarbazide is added. A positive response is indicated by the appearance of a violet color.

1 P. CAZENEUVE, *Compt. Rend.*, 131 (1900) 346.
2 Unpublished studies.

114. Detection of oxalic acid in leather[1]

Oxalic acid is often used for bleaching vegetable-tanned leathers. It is therefore sometimes desirable to be able to detect this acid in finished leather products. The test by the formation of aniline blue (p. 457) is suitable. It can be carried out directly on a small piece of leather, or on the sole of a shoe, for instance. The intensity of the color reaction indicates whether traces or appreciable amounts of oxalic acid are present.

Procedure. About 0.2 g of diphenylamine is placed on the leather and pressed down with a thick glass rod. The reagent is melted by heating from above with a *very small flame* and carefully kept in fusion for about a minute. Then, 2 or 3 drops of alcohol are gently dropped on, and the test piece exposed to light. The appearance of a blue color indicates oxalic acid. When oxalic acid and diphenylamine are melted together in a crucible, the blue color appears immediately on the addition of alcohol, but on the leather the color develops only after some time. If considerable amounts of oxalic acid are present, the blue color appears after about 1 to 2 h, but for very small amounts, 10 h standing is necessary. The development of the blue color is accelerated by light; samples kept in the dark remain unchanged for days.

1 K. KLANFER and A. LUFT, *Mikrochim. Acta*, 1 (1937) 142.

115. Detection of formaldehyde in leather[1]

Certain tanning operations employ formaldehyde directly or in the form of formaldehyde resins. The formaldehyde, which is chemically bound by the hide substance, can be detected through the color reaction with chromotropic acid and concentrated sulfuric acid (red-violet color). The procedure is that employed to detect formaldehyde-base resins (see p. 690). The test can be made with as little as 0.1–1 mg of leather.

1 Unpublished studies.

116. Differentiation of natural and synthetic tanning agents[1]

I. Many natural tanning agents contain phenolic compounds, whose aqueous–ammoniacal solutions are autoxidizable in contact with the air. Such solutions show a characteristic behavior after the introduction of red ammoniacal solution of iron(II) α,α'-dipyridyl sulfate. A blue-violet precipitate results and its quantity increases when the system is made acidic with acetic acid. Probably the phenolic oxidation products form an adsorption compound with $[Fe(\alpha,\alpha'\text{-dip})_3](OH)_2$. This assumption is supported by the finding that the reaction does not occur in the absence of air or if considerable amounts of alkali sulfite are introduced. No precipitate was obtained with solutions of six synthetic tanning agents, whereas even dilute solutions of natural tanning materials gave a precipitate. Among the materials tested were solutions or extracts of tannic acid, tara powder, gambier, quebracho, acacia, myrobalan, sumac.

Procedure. An Emich centrifuge tube is used. A drop of the aqueous–ammoniacal solution being studied is treated with two drops of reagent and kept in boiling water for several minutes. Then the solution or suspension is acidified with acetic acid and after brief rewarming is centrifuged. A positive response is readily discerned since a violet precipitate collects in the tip of the tube.

Reagent: 0.25 g α, α'-dipyridyl plus 0.14 g $FeSO_4.7 H_2O$ in 50 ml water.

This procedure revealed 1 γ tannic acid.

Natural tanning materials can be detected in tanned hides by this method.

1 F. FEIGL and H. E. FEIGL, *Anais. Assoc. Quim. Brasil*, 6 (1947) 1.

II. Synthetic tanning agents are usually condensation products of formaldehyde and sulfonated phenols (especially crude cresol) or naphtholsulfonic acids[1]. They can be recognized[2] and differentiated from natural (vegetable) tanning materials through the rapid release of formaldehyde on warming with concentrated sulfuric acid. (Procedure as given on p. 324 for detection of formaldehyde). Another general test is the removal of sulfur by warming with Raney nickel alloy and caustic alkali with formation of nickel sulfide (Procedure as for the detection of aromatic sulfonic acids, see p. 234).

Synthetic tanning agents with phenol components can be distinguished from tanning materials with naphthalene components through ignition, since the former evolve phenol vapors which can be detected by means of the indophenol reaction (Procedure see p. 141). Accordingly, if the test for aromatic bound $SO_3H(Na)$ groups is positive and no phenol is evolved on pyrolysis of the sample, it is likely that the sample is a synthetic tanning agent with naphthalene components.

The cited procedures were applied to:

Ignatan T (phenolic)	Carlan O (naphthalenic)
Tamgan LH ,,	Supratan OC ,,

The naphthalenic tanning agents can be distinguished from each other by heating to 160° along with mercuric cyanide. Only Supratan OC yielded hydrogen cyanide indicating that this product contains unsalified SO_3H groups (see p. 117 regarding the detection of acidic compounds by pyrolysis with mercuric cyanide). The synthetic tanning agents with phenolic components behave as neutral products with regard to the mercuric cyanide test; obviously they contained SO_3Na groups.

All of the tests given here can be run with less than 1 mg of the sample.

1 Comp. P. KARRER, *Lehrbuch der organischen Chemie*, 13th ed., Stuttgart, 1959, p. 579.
2 Unpublished studies.

117. Detection and differentiation of organic tanning agents in leather[1]

Leather made with vegetable tanning agents contains polyphenols. In alkaline solution, the latter reduce 1,2-dinitrobenzene to the violet salt of the *aci*-form of *o*-nitrophenylhydroxylamine (p. 133).

The sample (less then 1 mg) is treated in a micro test tube with a drop of 25% Na_2CO_3 solution and one drop of 5% benzene solution of 1,2-dinitrobenzene. When warmed in a water bath, a deep violet color appears within a minute or two. The reaction is not given by leather prepared with synthetic tanning agents, which are condensation products of sulfonated phenols (particularly crude cresols) or naphthalenesulfonic acids with formaldehyde.[2] Such materials (Neradole, Supratan) when warmed with Raney nickel alloy and caustic alkali produce nickel sulfide, a result which is characteristic for aromatic sulfonic acids. The procedure described on p. 234 can be used for the detection of the synthetic tanning agents of the type under consideration here.

Synthetic tanning agents without phenolic character can be differentiated from vegetable tanning materials through the fact that only the latter release phenol when ignited. See p. 185 for the detection of phenol vapors.

1 Unpublished studies, with Cl. COSTA NETO.
2 Comp. P. KARRER, *Lehrbuch der organischen Chemie*, 13th ed., Stuttgart, 1959, p. 579.

118. Detection of the contamination of fabrics by mammalian urine[1]

Urea which is characteristic for urine may be sensitively detected by the evolution of ammonia in the presence of urease. The ammonia can be detected by the sensitive test with manganese nitrate–silver nitrate solution (p. 89) which yields manganese dioxide and elementary silver.

Procedure. The fabric suspected of being soiled with urine is viewed (if necessary under ultraviolet light) and the stain is outlined with a pencil. The stained area is treated with 2–4 drops of urease solution, which is allowed to soak in for 5–10 sec. The cloth is then placed on a heated steam bath and covered at once with a piece of freshly prepared reagent paper so that the paper is wetted by the damp area of the cloth. A firm contact should be maintained by pressing across the stained area. If the stain was due to urine, a black spot appears on the filter paper in about 30 sec.

Reagents: 1) Urease solution. A 10% slurry of jack bean meal in water is allowed to settle for a few minutes; the supernatant liquid is used as urease solution. (The slurry may be stored as long as a week in a refrigerator without apparent loss of activity.)

2) Mangenese–silver reagent paper. Coarse filter paper is soaked in the neutralized $Mn(NO_3)_2$–$AgNO_3$ solution (p. 90) and quickly dried on a steam bath at about 100°.

1 N. H. ISHLER, K. SLOMAN and M. E. WALKER, *J. Assoc. Offic. Agr. Chemists*, 30 (1947) 670.

119. Test for alkali alkyl sulfates used as wetting agents[1]

The esters of alkali bisulfates with fatty alcohols are widely used as wetting agents. Being alkyl esters of non-carboxylic acids they form sulfur dioxide when pyrolyzed with sodium thiosulfate (comp. p. 311). A differentiation between this kind of wetting agent and others is thus possible.

The procedure is the same as described on p. 171.

The following products give a positive response:

Sodium lauryl sulfate, sodium octadecyl sulfate, sodium heptadecyl sulfate, sodium methylhexadecyl sulfate, sodium methyloctadecyl sulfate. The test for esters of sulfuric acid described on p. 235, may also be used for the examination of pertinent wetting agents.

1 F. FEIGL and V. ANGER, *Chemist-Analyst*, 49 (1960) 13.

120. Detection of Chloramine T[1]
(Differentiation from alkali hypochlorite)

Chloramine T is used as an antiseptic, and because of its powerful oxidizing action is also employed as a bleaching agent. Therefore, it sometimes is desirable to distinguish it from alkali hypochlorites (or chlorinated lime). The sensitive color reaction (blue coloration) produced by thio-Michler's ketone can be applied (see p. 65). A preliminary decomposition of hypochlorite ions is essential since they too oxidize the reagent. Hydroxyl ions must also be removed. Both of these objectives can be attained by adding hydrogen peroxide and zinc chloride. The reactions:

$$ClO^- + H_2O_2 \rightarrow Cl^- + H_2O + 2\,O \quad \text{and} \quad 2\,OH^- + Zn^{2+} \rightarrow Zn(OH)_2$$

produce a practically neutral solution of Chloramine T and hence provide the best conditions for the color reaction since excess peroxide and Chloramine T react but slowly.[2]

Procedure. A drop of the test solution is shaken in a micro test tube with 1 drop of 10% hydrogen peroxide. A drop of a concentrated solution of zinc chloride is then added and the mixture again shaken. A blue color appears on adding a drop of a saturated alcoholic solution of thio-Michler's ketone if chloramine T is present. A comparison blank with chloramine T solution is advisable if small amounts of the latter are suspected.

Limit of identification: 1 γ chloramine T.

A sensitive test for chloramine T (even in the presence of hypochlorite) is based on the color reaction with thiodiphenylamine described on p. 521.

If hypochlorite is to be detected in the presence of chloramine T, a drop of the test solution should be treated on a spot plate with a 4% alkaline solution of thallous sulfate (see p. 127). If hypochlorite is present, a brown precipitate appears at once or within 5 minutes.

Limit of identification: 2.5 γ alkali hypochlorite.

Freshly prepared solutions of chloramine T do not show this reaction; they respond only on warming. At room temperature they undergo slow hydrolysis with production of sodium hypochlorite as can be recognized through the redox reaction with thallous hydroxide.

1 F. FEIGL and R. A. ROSELL, *Z. Anal. Chem.*, 159 (1958) 335.
2 J. W. FENTON and G. K. INGOLD, *J. Chem. Soc.*, 131 (1928) 3295.

121. Detection of Rongalite[1]

Rongalite is the stable addition product – or more correctly condensation product – of sodium sulfoxylate and formaldehyde. Its formula is $NaHSO_2$. $CH_2O.2\ H_2O$. This colorless crystalline compound is a powerful reductant which, in the dry state, like the equally stable sodium hydrosulfite or hyposulfite ($Na_2S_2O_4$) is used in the textile industry as a printing reducing agent. The reducing power of Rongalite is due to the ready oxidation of its divalent sulfur to sulfate:

$$H_2C\begin{matrix} \diagup OH \\ \diagdown O-S-ONa \end{matrix} + 2\ O \longrightarrow CH_2O + NaHSO_4$$

Rongalite may be identified by sensitive tests for its formaldehyde and sulfoxylate components. The former can be detected by the color reaction with chromotropic acid and sulfuric acid (p. 434), using the fact that Rongalite loses formaldehyde above 125°.[2] The formaldehyde vapor is brought into contact with the reagent and a violet color appears (Procedure I). The sulfoxylate can be detected through its redox reaction with an alcoholic ammoniacal solution of *p*-dinitrobenzene (Procedure II). The product is the violet ammonium salt of a *para*-quinoidal nitrogen acid (see p. 133).

The test for Rongalite with p-dinitrobenzene is not impaired by the sulfite and thiosulfate always present in the commercial preparations. However, any sulfide will interfere since it yields an orange color with the reagent. This interference can be averted by shaking the Rongalite solution with a little lead carbonate (formation of PbS), and conducting the test with a drop of the filtrate or centrifugate.

Procedure I (formaldehyde detection). The apparatus shown in Figure 23, *Inorganic Spot Tests*, is used. A little of the solid is placed in the bulb or a drop of the test solution is evaporated there. A drop of a solution of chromotropic acid in concentrated sulfuric acid is suspended on the knob of the stopper. The evaporation residue or the sample is cautiously warmed. If Rongalite was present, the hanging drop turns red-violet.

Limit of identification: 0.02 γ Rongalite.

If solid Rongalite or the evaporation residue is treated directly with the reagent solution, as little as 0.5 γ Rongalite will give a violet color.

Procedure II (hydrosulfite detection). A micro test tube is used. A little of the solid or a drop of a solution is treated in succession with one drop of a saturated solution of p-dinitrobenzene in alcohol and one drop of concentrated ammonium hydroxide. The mixture is warmed above an open flame for 30 to 60 sec. If Procedure I gave a positive result for formaldehyde, a violet color in the present test indicates the presence of Rongalite.

Limit of identification: 3 γ Rongalite.

1 F. FEIGL and L. HAINBERGER, *Mikrochim. Acta*, (1955) 119.
2 G. PANIZZON, *Melliand Textilber.*, 12 (1931) 119.

122. Identification of binders used in mineral coatings on papers[1]

The identification of important binders generally used in mineral coatings on papers and paperboards is possible by means of characteristic reactions carried out in greater part as spot tests with appropriate reagents. Pertinent instances are:

Tetrabromophenolphthalein ethyl ester for protein and butyl rubber;
Iodine (0.1 N) for starch and polyvinyl acetate;
Ninhydrin for protein;
Acetic anhydride–66% sulfuric acid for styrene–butadiene and polyvinyl resins;
Morpholine for polyvinylidene chloride;
Heated copper wire for polyvinylidene chloride;
Nitric acid (digestion) for styrene–butadiene;
Mercuric sulphate for butyl rubbers;
Iodine–boric acid for polyvinyl alcohol.

The original paper should be consulted for the identification scheme and details of the test procedures.

1 W. M. Boast and S. W. Trosset, Jr., *Tech. Assoc. Pulp & Paper Industry (TAPPI)*, 45 (1962) 873.

123. Detection of adhesives containing proteins[1]

Adhesives of animal origin (glue, gelatin, casein *etc.*) contain proteins and therefore nitrogen. In contrast, adhesives derived from plant sources (gum arabic, pectins, starches, dextrin, *etc.*) are polysaccharides or their closely related products, and hence are free of nitrogen. Accordingly, these two classes of adhesives can be distinguished from each other by the test for nitrogen. This can be done by the procedure described on p. 90 in which the sample is pyrolyzed in the presence of MnO_2 or Mn_2O_3 to yield nitrous acid. The directions given there should be followed.

The sensitivity of the test for the present purpose is indicated by the finding that a positive response to the Griess color reaction in the gas phase was given by 0.1 γ gelatin. Because of this high sensitivity, traces of adhesives containing protein can be readily detected on paper, wood, cotton, *etc.* Obviously, the absence of nitrogenous dyes, nitro compounds and amines, as well as inorganic nitrogenous materials is essential.

1 F. Feigl and J. R. Amaral, *Anal. Chem.*, 30 (1958) 1148.

124. Detection of acetic acid (acetate) in formic acid (formate)[1]

Commercial formic acid and alkali formates often contain notable quantities of acetic acid or acetate even though they are designated as "pure". To test for these admixtures or impurities, use can be made of the fact that formic acid containing acetic acid yields a mixture of the calcium salts when taken to dryness with an excess of calcium carbonate. If the residue is then strongly heated, acetaldehyde results:

$$(HCOO)_2 Ca + (CH_3COO)_2Ca \rightarrow 2\ CaCO_3 + 2\ CH_3CHO$$

An analogous reaction occurs with a mixture of the alkali salts.

The acetaldehyde resulting from either of these reactions can be detected in the vapor phase by the color reaction with a solution of sodium nitroprusside containing morpholine (see p. 438). Analogous reactions are given by lactic, butyric, and propionic acid, but the test is far less sensitive in the two latter cases than for acetic acid.[2]

Procedure. A micro test tube is used. One drop of the suspected sample is stirred with a slight excess of powdered calcium carbonate and the suspension brought to dryness in an oven or by cautious direct heating. The mouth of the tube is covered with filter paper moistened with the reagent solution and the bottom of the tube is then heated over a micro flame. If acetic acid was present, a more or less intense blue appears on the paper.

To test for acetate in formates, several mg of the solid or the evaporation residue of a solution is heated as just prescribed, but without addition of calcium carbonate.

This procedure was applied to samples of concentrated formic acid from various sources, and whose acetic acid content was given as below 0.4%. In practically all cases, the reaction for acetic acid was very marked, and there was a positive response in some instances after the sample had been diluted with 10–20 volumes of water. Similarly, many samples of sodium and calcium formate gave an acetaldehyde reaction, which sometimes was remarkably distinct even though only mg quantities were taken for the test.

1 F. Feigl and C. Stark, *Chemist-Analyst*, 45 (1956) 46.
2 J. V. Sanchez, *Chem. Abstr.*, 30 (1936) 4432.

125. Test for free acid in succinic and phthalic anhydrides[1]

It is stated on p. 149 that compounds which lose water when heated up to 190° can be detected through the fact that the water released during pyrolysis reacts with thio-Michler's ketone or thiobarbituric acid. The products of this pyrohydrolysis are Michler's ketone or barbituric acid and hydrogen sulfide. The latter can be detected through the blackening of lead acetate paper. Since succinic acid and phthalic acid form anhydrides at the given temperature and hereby lose water, these acids may be detected in the presence of their respective anhydrides by the pyrohydrolysis of thio-Michler's ketone or thiobarbituric acid.

Procedure. As given on p. 150.

1 Unpublished studies.

126. Detection of higher fatty acids in paraffin wax and vaseline[1]

Paraffin waxes and vaseline are natural or prepared mixtures of saturated hydrocarbons. Paraffin consists wholly of solid hydrocarbons, whereas the oily consistency of vaseline is due to hydrocarbons which are liquid at ordinary temperatures. Pure paraffin wax and vaseline ought not to contain any admixed higher fatty acids with low melting points. A content of fatty acids may also come from saponified oils and fats.

A rapid test for higher fatty acids in paraffins and the like is based on a reversal of a sensitive test for uranium. (Compare *Inorganic Spot Tests*, Chap. 3.) If a neutral uranyl solution is shaken with a colorless saturated solution of rhodamine B in benzene, the benzene layer turns red and fluoresces with an intense orange color in ultraviolet light if carboxylic acids, soluble in benzene, are present. This effect may be explained as follows: The colorless benzene solution of rhodamine B contains the lacto-form (I) of the dyestuff in equilibrium (*1*) with minimal quantities of the quinone-form (II). When the uranyl salt solution is added and the mixture shaken, this equilibrium is not disturbed. If, however, the benzene solution of the dyestuff contains carboxylic acids, the latter react with UO_2^{2-} ions to form small quantities of the complex uranyl carboxylic acids (*2*) which in turn react with (II) to yield (*3*) benzene-soluble red salts of the dye:

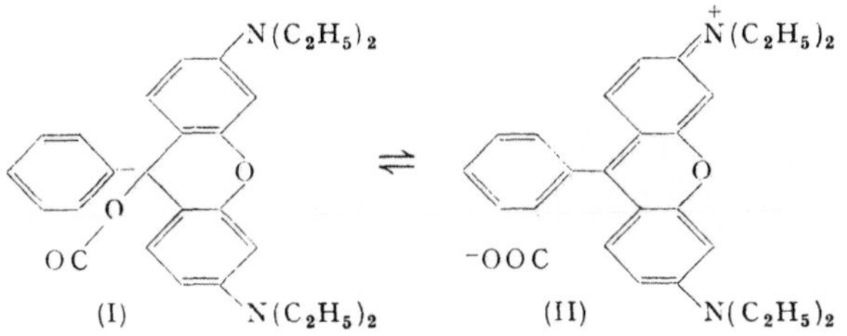

$$UO_2^{2+} + 3\,RCOOH \rightleftharpoons H[UO_2(RCOO)_3] + 2\,H^+ \qquad (2)$$

Hence this procedure involves the small amounts of the compounds furnished by the equilibria (*1*) and (*2*), and after they have been consumed they are continuously delivered again. The amounts thus made available are adequate for the color reaction (*3*) to proceed to a visible extent.

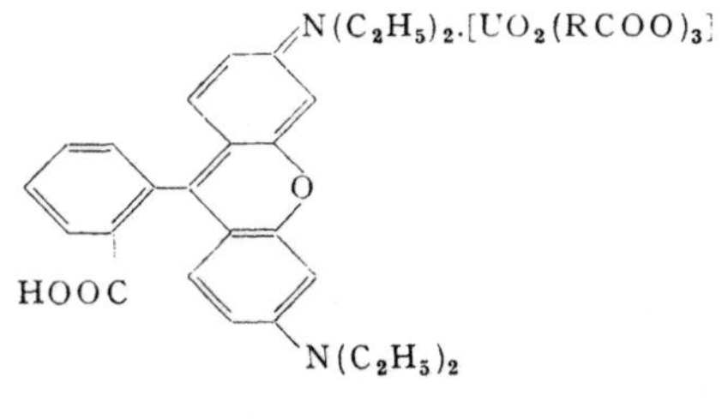

Procedure.   A micro test tube is used. A drop of a benzene solution of the sample and a drop of saturated solution of rhodamine B in benzene are mixed and then treated with 1–2 drops of 1% aqueous solution of uranyl nitrate or acetate. The mixture is shaken. If fatty acids are present, the benzene layer turns red to pink, depending on the quantity of acid. When examined in ultra-violet light, the benzene layer exhibits an orange fluorescence. If an emulsion forms, the benzene and water layers can be separated by centrifuging.

This procedure revealed 5 γ palmitic or stearic acid.

1 F. FEIGL, D. GOLDSTEIN and V. GENTIL, unpublished studies.

127. Detection of fats and waxes[1]

Fats and fatty oils, which are esters of glycerol with higher and middle fatty acids, may be detected by the glycerol reaction given on p. 416, which is based on the dehydration of glycerol to acrolein. The latter may be identified through the color reaction with sodium nitroprusside solution containing piperidine. The various kinds of waxes consist predominantly of esters of higher monobasic carboxylic acids with higher monohydric alcohols. They may be detected by the ester test described on p. 214, which involves conversion to hydroxamic acids followed by the formation of colored inner-complex ferric salts. The presence of waxes and resins in mineral oils may be detected through this reaction.

The hydroxamic acid reaction gives the following colors with the waxes and resins:

Beeswax	violet-brown	Tolu Balsam	brown-green
Carnauba wax	dark brown-violet	Colophony	,, ,,
Candelilla wax	lilac	Congo copal	brown
Montan wax	green-yellow	"Soromin" (synthetic ester)	violet

A rapid and sensitive test for fats, fatty oils, and waxes can be based on the production of ammonia when these materials are heated to 130–160° with biuret. The ammonia can be readily detected in the gas phase by means of acid–base indicator papers or Nessler reagent. The test is based on the ammonolysis which occurs when these esters are heated with biuret. Amides of the respective fatty acids are formed with liberation of glycerol or higher monohydric alcohols. These alcohols then condense with biuret to yield urethanes, and ammonia is produced concurrently.

1 F. FEIGL, V. ANGER and O. FREHDEN, *Mikrochemie*, 15 (1934) 18.

128. Detection of calcium cyanamide and urea in fertilizers and soils[1]

Calcium cyanamide (Nitrolime) and urea (Floramide) are often incor-porated in fertilizers as sources of nitrogen.[2] They may be detected by the release of ammonia when the sample is taken to dryness along with dilute hydrochloric acid and the evaporation residue then heated to 250°.

When calcium cyanamide is boiled down with hydrochloric acid it is converted to the chloride of dicyanodiamide:

$$2\ N\equiv C\!-\!NCa + 5\,HCl \longrightarrow 2\ CaCl_2 + HN\!=\!C\diagup^{NH_2.HCl}_{\diagdown NHCN}$$

This product, like many other guanidine derivatives, yields ammonia on heating (see p. 262). When urea is heated above its melting point (133°), it yields ammonia because of the formation of biuret. Since no fertilizers or soils contain materials which yield ammonia under these conditions, the following procedure can be used to reveal cyanamide salts and urea.

Procedure. Several mg of the solid (0.5 mg suffices for pure compounds) along with 1–3 drops of dilute hydrochloric acid are evaporated in a micro test tube. The tube is then immersed in a glycerol bath previously heated to 150–200°, and the mouth of the tube is covered with filter paper moistened with Nessler solution. The temperature of the bath is then raised to 230–240°. A brown or yellow stain appears on the paper.

Calcium cyanamide can be easily detected, even in the presence of urea, by warming with zinc and sulfuric acid; hydrocyanic acid results. The chemical background and the procedure of the HCN-test are outlined on p. 546.

1 Unpublished studies, with C. STARK.
2 Comp. G. H. COLLINGS, *Commercial Fertilizers*, 2nd Ed., Blakiston, Philadelphia, 1938, pp. 81, 91; as well as W. B. ANDREWS, *The Response of Crops and Soils to Fertilizers*, State College, Mississippi, 1947, pp. 72, 103.

129. Detection of alkyl phosphates used as insecticides[1]

As outlined in Chap. 3, alkyl esters of non-carboxylic acids yield easily detectable sulfur dioxide when heated with sodium thiosulfate at 160–180°. Accordingly, this behavior is shown by alkyl esters of phosphoric and thio-phosphoric acid which are used as insecticides. The procedure described on p. 171 was applied. The following compounds in amounts less than a mg gave a positive response:

Para-oxon

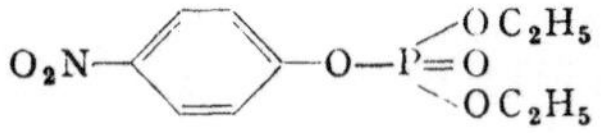

$$O_2N\!-\!\!\left\langle\bigcirc\right\rangle\!-\!O\!-\!P\!\!\diagup^{OC_2H_5}_{\diagdown OC_2H_5}\,\,^{\Vert}_{O}$$

Parathion

$$O_2N\!-\!\!\left\langle\bigcirc\right\rangle\!-\!O\!-\!P\!\!\diagup^{OC_2H_5}_{\diagdown OC_2H_5}\,\,^{\Vert}_{S}$$

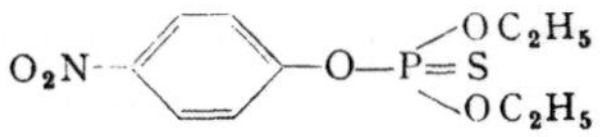

Methyl-parathion
$$O_2N-\!\!\left\langle \bigcirc \right\rangle\!\!-O-P{=}S\genfrac{}{}{0pt}{}{OCH_3}{OCH_3}$$

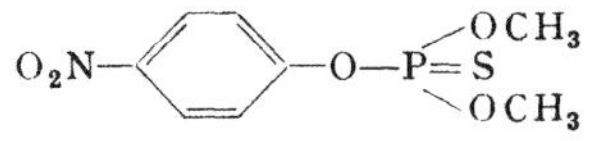

Chlorthion
$$\underset{O_2N-\!\!\left\langle \bigcirc \right\rangle\!\!-O-P{=}S}{\overset{Cl}{}}\genfrac{}{}{0pt}{}{OC_2H_5}{OC_2H_5}$$

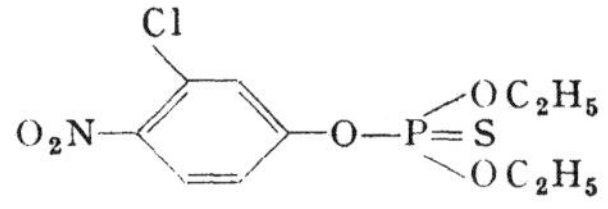

Diazinon
$$\underset{\quad CH_3}{\overset{(CH_3)_2CH}{\left\langle \bigcirc \right\rangle}}-O-P{=}S\genfrac{}{}{0pt}{}{OC_2H_5}{OC_2H_5}$$

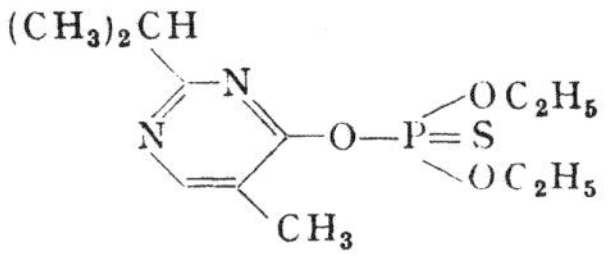

Malathion
$$\begin{array}{l}C_2H_5OOC-CH-S-P{=}S\genfrac{}{}{0pt}{}{OC_2H_5}{OC_2H_5}\\[2pt] \quad\quad\quad\;\;| \\[2pt] C_2H_5OOC-CH_2\end{array}$$

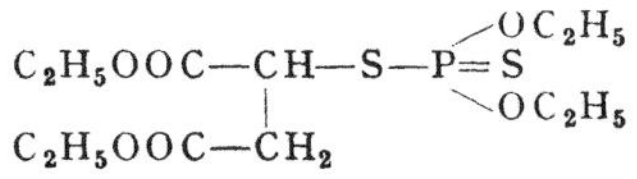

Diptorex
$$Cl_3C-CHOH-P{=}O\genfrac{}{}{0pt}{}{OCH_3}{OCH_3}$$

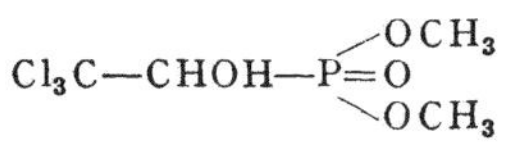

Among the insecticides included in the above list, those containing sulfur respond positively to the iodine–azide test (see p. 219). The test for thionic compounds based on the realization of the Wohlers effect (comp. p. 84) can also be used. The chlorine-bearing compounds give the Beilstein test (see p. 63) and those containing NO_2 groups show the color reaction on fusion with tetrabase or diphenylamine (see p. 295).

1 F. FEIGL and V. ANGER, *Chemist-Analyst*, **49** (1960) 13; and unpublished studies.

130. Detection of arylmercury nitrates[1]

If arylmercury nitrates (seed disinfectants) are heated to 140–150° with benzoin (m.p. 137°), nitrous acid is split off during the following partial reactions, in which benzoin in its reductone form functions as acid and as hydrogen-donor:

$$2\,ArHgNO_3 + C_6H_5C(OH){=}C(OH)C_6H_5 \rightarrow (ArHg)_2[C_6H_5COCOC_6H_5] + 2\,HNO_3 \quad (1)$$

$$HNO_3 + C_6H_5C(OH){=}C(OH)C_6H_5 \rightarrow C_6H_5COCOC_6H_5 + H_2O + HNO_2 \quad (2)$$

Since nitrous acid is readily detected in the gas phase by the Griess test, the realization of reactions (1) and (2) permits the selective detection of arylmercury nitrates, provided the absence of nitrates of organic bases, which behave in a similar manner toward benzoin through occurrence of reaction (2). It is also assumed that the presence of mercury has been established (see p. 102).

Procedure. A little of the solid sample or a drop of its aqueous solution is united in a micro test tube with several cg of benzoin and taken to dryness if necessary. The mouth of the test tube is then covered with a disk of filter paper moistened with Griess reagent and the test tube is immersed in a glycerol bath preheated to 130°. The temperature is then raised to 150–160°. A positive response is signalled by the development of a red fleck on the reagent paper.

Limit of identification: 8 γ phenyl (tolyl)mercuric nitrate.

1 F. FEIGL, *Chemist-Analyst*, 50 (1962) 4.

131. Detection of some enzymatic activities

Peroxidases, acid phosphatase, polyphenoloxidases and unspecific esterases can be detected through characteristic and sensitive color reactions.

Stable reagent papers which can be applied for spot tests were recently developed. They are of value for documentation in the control of manufacturing processes in the food and canning industry and also in crime investigations.

Information in detail is available from the manufacturer Atesmo AG, Zürich, Switzerland.

Individual Compounds and Products Examined

The compounds are listed alphabetically under their most well-known trivial names. Under each compound the Chapter, Section and Test number are given, followed by the limit of identification or other indication of the result of the reaction, whereby (+) stands for "positive reaction", (±) for "non-characteristic reaction" and (—) for "no reaction".

Acacia extract
7/116　　　(+)
acenapthoquinone
4/14(1)　　(—)
4/14(2)　　(—)
4/15　　　(+)
5/32(2)　　(—)
acetaldehyde
3/18(1)　　0.5 γ
3/18(2)　　0.25 γ
3/20(1)　　4 γ
3/20(2)　　5 γ
3/20(3)　　30 γ
3/20(4)　　0.01 γ
3/20(5)　　1 γ
3/20(6)　　23 γ
3/21(2)　　(—)
3/24(2)　　100 γ
3/25　　　(—)
5/9　　　(—)
5/28(1)　　(—)
5/28(2)　　(+); (—)
5/29　　　1 γ
5/42(4)　　(—)
5/52(1)　　(+)
5/53(1)　　(+)
5/53(2)　　(+)
acetaldehyde ammonia
3/20(4)　　0.06 γ
acetamide
2/26　　　3 γ
2/40　　　0.3 γ
3/62(2)　　8 γ
acetamidine chloride
3/66　　　7.5 γ

m-acetamidobenzo-
trifluoride
2/10(2)　　25 γ
3-acetamino-4-methoxy-
benzenesulfinic acid
2/12(4)　　2.5 γ
2/12(5)　　5 γ
2/12(8)　　2.5 γ
3/38(1)　　20 γ
3/38(3)　　2.5 γ
acetanilide (N-phenyl-
acetamide)
2/34　　　(+)
2/36　　　(+)
3/48　　　4 γ
3/52　　　(—)
3/63　　　0.5 γ
3/65　　　10 γ
6/26　　　5 γ
acetic acid
2/42(6)　　(—)
5/42(2)　　(—)
5/42(4)　　(—)
5/43(1)　　50 γ
5/43(2)　　60 γ
5/43(3)　　10 γ
5/45(2)　　(—)
5/52(1)　　(—)
5/55　　　(—)
7/124　　　(+)
acetic anhydride
3/32　　　5 γ
acetindoxyl
3/48　　　20 γ

acetoacetic acid ethyl ester
2/24(7)　　25 γ
2/44　　　1.2 γ
3/13　　　60 γ
3/24(1)　　4 γ
3/24(2)　　300 γ
5/65　　　10 γ
aceto-(2-chloromercury-
ethyl)-mesidide
2/22(2)　　2.25 γ
acetofluoroglucose
2/10(1)　　200 γ
acetoguanamine
3/67(2)　　(—)
acetohydroxamic acid
3/87(2)　　5 γ
acetoin
3/23(2)　　0.5 γ
acetone
3/18(1)　　50 γ
3/18(2)　　40 γ
3/23(1)　　20 γ
3/23(2)　　50 γ
3/24(1)　　10 γ
3/24(2)　　100 γ
3/25　　　1 γ
5/9　　　(—)
5/28(1)　　(—)
5/32(2)　　(—)
5/35　　　0.2 γ
acetone oxime
3/85　　　0.08 γ
acetophenone
2/34　　　(+)
3/18(1)　　20 γ

2/32   (−)

2/35   5 γ

2/44   0.12 γ

3/48   5 γ

3/49(1)   1.6 γ

3/49(2)   50 γ

3/50   0.2 γ

3/52   0.05 γ

3/54(1)   0.5 γ

6/40   1 γ

aniline sulfate

2/27   5 γ

3/50   2.2 γ

3/53   10 γ

Aniline Yellow, see also
    p-aminoazobenzene

7/81   2 γ

7/84   5 γ

anisalacetone

5/32(2)   (−)

anisaldehyde

3/18(1)   15 γ

3/18(2)   0.35 γ

3/19   0.01 γ

3/20(3)   2 γ

3/20(4)   5 γ

3/20(5)   0.1 γ

3/21(1)   2 γ

3/21(2)   0.1 γ

3/21(3)   0.3 γ

3/21(4)   0.1 γ

3/22(1)   0.1 γ

3/22(2)   0.2 γ

4/27   5 γ

5/52(1)   (+)

anisaldehyde bisulfite

3/20(4)   6.5 γ

anisaldehyde phenyl-
    hydrazone

3/80(2)   (+)

anisaldehyde-2-sulfonic
    acid

2/34   (+)

anisic acid

2/34   (+)

anisic acid hydrazide

3/78(4)   80 γ

p-anisidine

3/54(1)   0.5 γ

anisoin

4/29(1)   5 γ

4/29(2)   30 γ

anisole

2/32   (+)

3/18   1 γ

4/3   0.5 γ

anisyl alcohol

5/42(4)   (−)

anol

3/4(1)   40 γ

3/4(2)   4 γ

Anthisan, see Pyrilamine
    maleate

anthracene

2/5   0.5 γ

2/32   25 γ; (+)

2/33   (+)

2/36   (−)

5/3   2 γ

5/4   3 γ

6/1   3 γ

anthragallol (1,2,3-tri-
    hydroxyanthraquinone)

4/28   2.5 γ

anthranilic acid (o-amino-
    benzoic acid)

3/26   11 γ

3/52   0.5 γ

3/54(1)   10 γ

4/44   25 γ

6/17-I   0.1 γ

6/17-II   50 γ

anthraquinone

2/32   (−)

2/33   (−)

2/34   (+)

2/36   (−)

4/13   0.05 γ

4/14(1)   (−)

4/14(2)   (−)

4/18   0.5 γ

5/32(2)   (−)

anthraquinone-1,5-
    disulfonic acid

3/27   (−)

3/43   1 γ

4/38(2)   30 γ

4/38(3)   2 γ

anthraquinone-2,6-
    disulfonic acid

3/27   (−)

3/43   1 γ

4/38(3)   2 γ

anthraquinone-1-
    sulfonic acid

3/27   (−)

3/43   2.5 γ

(K salt)

4/38(3)   1 γ

anthraquinone-2-
    sulfonic acid

2/12(4)   5 γ

2/12(5)   2.5 γ

2/28   15 γ

4/13   0.5 γ

4/38(1)   2 γ

4/38(2)   0.4 γ

4/38(3)   1 γ

anthrone

2/34   (+)

antimony sodium thio-
    glycolate

7/61   (+)

antimony thioglycol-
    amide

7/61   (+)

antipyrine (phenazone)

3/1   100 γ

3/14(5)   (+)

3/14(7)   (−)

3/57   (+)

7/32   2 γ; 1 γ

arabinose

2/31(7)   (+)

4/1(2a)   25 γ

4/1(2b)   5 γ

4/19   2.5 γ

4/20(1)   0.2 γ

5/9   (±)

5/28(1)   (±)

5/42(1)   (−)

5/63(1)   (−)

arabinose osazone

3/74   1 γ

3/77   1 γ

arbutin

3/14(5)   (+)

5/21(1)   (+)

5/21(2)   (−)

arginine

3/53   10 γ

3/66   10 γ

3/67(1)   5 γ

3/67(2)   1 γ

4/46   25 γ

4/47(1)   60 γ

4/47(2)   (+)

arginylglycine

3/51   25 γ; 30 γ

arsanilic acid

3/54(1)   (−)

4/50   10 γ

4/51(1)   6 γ

4/51(2)   5 γ

7/62   10 γ

ascorbic acid (Vitamin C)

2/31(4)   5 γ

2/31(5)   0.2 γ

2/31(7)   10 γ

2/31(8)   1 γ

camphoric anhydride
  3/32   10 γ
camphor-10-mercury
  iodide
  2/22(2)   2.5 γ
camphor oxime
  2/16   (+)
  3/86   40 γ
camphorquinone
  4/12   1 γ
camphorsulfonic acid
  2/13(2)   5 γ
  3/43   (−)
Candelia wax
  7/127   (+)
capric acid
  2/24(4)   2 γ; 5 γ
  2/24(5)   (+)
  3/27   (−)
caproic acid
  2/24(5)   (+)
caprylic acid
  2/24(5)   (+)
  3/27   (−)
carbanilide (sym-
  diphenylurea)
  3/64   0.6 γ
  3/65   10 γ
  6/42   10 γ
carbazole
  2/32   (+)
  2/33   (+)
  3/48   30 γ
  4/41(1)   0.5 γ
  4/57   1 γ
  5/99   0.2 γ
Carbide Black S
  7/84   (+)
carbon
  7/2   0.11 γ
carbon disulfide
  5/68(1)   0.14 γ
  5/68(2)   10 γ
  5/68(3)   3 γ
  7/6   (+)
carbon tetrachloride
  3/8   (−)
  5/8(1)   5 γ
  5/8(2)   200 γ
Carlan O
  7/116   (+)
Carnauba wax
  7/127   (+)
carvacrol
  3/14(7)   0.5 γ
  3/14(8)   1 γ

casein
  4/48(1)   0.5 γ
  4/48(2)   5 γ
Catigen Black SW extra
  7/3   (+)
Catigen Brilliant
  Green G
  7/3   (+)
Catigen Brown 2R extra
  7/3   (+)
Catigen Indigo CL extra
  7/3   (+)
Catigen Violet 3R
  7/3   (+)
Catigen Yellow GG extra
  7/3   (+)
cellulose
  4/21   0.3 γ
  7/106   (+)
cephaeline
  7/42   (+)
cetyl alcohol
  3/10(1)   6.5 γ
cetyl palmitate
  3/28   10 γ
cetylphenylamine
  3/55   10 γ
chelidonic acid
  4/54   25 γ
chelidonine
  4/9   0.1 γ
cherry gum
  4/1(2a)   (+)
chloral-2-ethylquinoline
  2/6(4)   30 γ
chloral hydrate
  2/7(1)   16 γ
  2/7(2)   10 γ
  3/8   0.5 γ
  3/20(1)   (−)
  3/20(5)   10 γ
  3/21(2)   (−)
  5/28(1)   (−)
  5/30(1)   50 γ
  5/30(2)   10 γ
chloramine B
  2/31(1)   0.04 γ
  3/102   1 γ
chloramine T
  2/7(2)   10 γ
  2/12(7)   400 γ
  2/17   100 γ; 250 γ
  2/31(1)   0.04 γ
  2/31(2)   2.5 γ
  3/102   1 γ
  4/64(1)   (−)
  7/120   1 γ

chloranil
  2/6(3)   5 γ
  2/31(1)   0.12 γ
  4/13   (+)
  4/14(1)   2 γ
  4/14(2)   5 γ
  5/39(1)   0.25 γ
  5/39(2)   2 γ
chloranilic acid
  4/14(1)   (−)
  4/14(2)   (−)
chloroacetamide
  3/7   10 γ
  3/62(1)   10 γ
ω-chloroacetophenone
  3/19   0.1 γ
chloroacetyl chloride
  3/33   0.1 γ
4-chloro-2-amino-
  anisole
  2/6(3)   100 γ
o-chloroaniline
  2/7(1)   20 γ
  2/7(2)   30 γ
p-chloroaniline
  2/7(1)   16 γ
  2/7(2)   30 γ
  2/36   0.5 γ
  3/48   4 γ
p-chloroaniline
  biguanide
  3/67(2)   (+)
chloroarsanilic acid
  2/19   5 γ
p-chlorobenzaldehyde
  2/34   (+)
  3/22(1)   25 γ
chlorobenzene
  2/7(1)   20 γ
  2/32   (+)
p-chlorobenzenestibonic
  acid
  3/105   1 mg
p-chlorobenzoic acid
  hydrazide
  3/78(4)   (+)
4-chlorobenzonitrile
  3/70   20 γ
p-chlorobenzophenone
  2/7(1)   16 γ
1-chloro-4-bromo-
  dinitrobenzene
  3/94   (+)
2-chloro-m-cresol
  3/14(8)   0.5 γ
4-chloro-m-cresol
  3/14(8)   1.6 γ

cinnamic acid
2/32 (+)
2/33 (+)
2/34 (+)
3/6 100 γ
3/26 33 γ
3/27 (+)
5/44 12 γ
5/55 (−)
5/56(1) (−)
5/56(2) (−)
cinnamic acid N-cyano-
methylamide
5/44 15 γ
cinnamic alcohol
3/5 (+)
cinnamoyl chloride
3/33 0.2 γ
cinnamoylformic acid
phenylhydrazone
3/76 2 γ
C.I. Pigment Red 3
7/76 50 γ
C.I. Solvent Red 23, see
Sudan III
citral
3/20(3) 0.1 γ
3/21(2) (−)
3/21(4) (±)
3/22(1) 50 γ
3/22(2) 6 γ
5/31 130 γ
citric acid
2/4(1) 8 γ
2/4(3) 2 γ
2/24(6) 10 γ
2/27 2.5 γ
2/40 10 γ
3/26 (+)
3/27 (+)
4/1(2) (−)
4/31 15 γ; 1 γ
5/9 (−)
5/27 10 γ
5/42(1) (−)
5/42(2) (−)
5/42(4) (−)
5/45(2) (−)
5/52(1) (−)
5/54 (−)
5/56(1) (−)
5/56(2) (−)
5/58(1) 6 γ
5/58(2) 2 γ
5/59(1) (+)
5/59(2) (−)
5/59(4) (−)

5/63(1) (−)
5/63(3) (−)
citronellal
3/20(3) 10 γ
citrulline
4/47(2) (+)
clupeine
4/48(1) 0.5 γ
codeine
2/15(1) 3 γ
3/1 40 γ
3/11 2 γ
3/57 (+)
7/38 2 γ
7/39 2 γ
Coelestine Blue
3/58 (+)
7/79 (+)
colophony
7/127 (+)
Congo copal
7/127 (+)
Congo red
2/36 (+)
3/41(2) 20 γ
7/80 0.5 γ
7/82 20 γ
coniferyl alcohol
3/5 10 γ
4/8 1 γ
4/25 5 γ
coniferylaldehyde
3/22(2) 0.1 γ
4/25 0.1 γ
4/26 0.01 γ
coniine
3/56(1) 10 γ
copper benzoinoxime
2/16 (+)
copper salicylaldoxime
2/16 (+)
coramine (N,N-diethyl-
nicotinamide)
3/58 (+)
4/53(2) (+)
4/58(1) 6 γ
coumarin
2/36 (+)
3/28 5 γ
5/62(1) 0.5 γ
5/62(2) 1 γ
7/60 (+)
creatine
3/57 (+)
creatine hydrate
3/67(2) 1 γ

creatinine
3/67(2) (−)
creosol (2-methoxy-4-
methylphenol)
3/14(8) 1.8 γ
Creslan
7/102(1) (+)
o-cresol
2/32 4 γ
3/14(2) 2 γ
3/14(5) (+)
3/14(6) 2 γ
3/14(8) 0.5 γ
4/3 (+)
5/21(1) (±)
m-cresol
3/14(5) (+)
3/14(6) 2 γ
3/14(7) 1 γ
3/14(8) 0.6 γ
5/21(1) (±)
p-cresol
3/14(5) (+)
3/14(7) (−)
3/14(8) 1.2 γ
4/3 (−)
4/14 0.2 γ
5/21(1) (±)
Crotamiton (N-ethyl-o-
crotono-toluidide)
3/4(1) 50 γ
3/4(2) 12 γ
crotonaldehyde
3/18(2) 0.1 γ
3/19 0.1 γ
3/20(3) 2 γ
3/20(4) 0.12 γ
3/20(5) 0.05 γ
3/21(2) (−)
3/21(4) (±)
3/22(1) 10 γ
3/22(2) 0.5 γ
4/26 (−)
5/12(2) (+)
5/28(1) (−)
5/29 (+)
crotonic acid
3/26 33 γ
cumene
2/32 (+)
cumic aldehyde
3/20(3) 3 γ
curcumin
2/33 (+)
2/36 (−)
cyanamide
3/71 80 γ

3/130   5 γ
6/41   (+)
o-cyanoacetanilide
   3/70   1 γ
cyanoacetic acid
   2/6(1)   0.3 γ
   3/69(1)   25 γ
   3/70   2 γ
cyanoacetic acid amide
   5/129   0.1 γ
cyanoacetic acid anilide
   3/64   1.7 γ
   5/129   0.2 γ
cyanoacetic acid
   N-methylamide
   5/129   0.2 γ
cyanogen bromide
   5/128   0.3 γ
α-cyano-N-methylamino-
   cinnamic acid
   3/69(1)   50 γ
   3/70   2 γ
3-cyanoquinoline
   3/70   20 γ
   5/104   (−)
cyanuric chloride
   3/33   0.05 γ
cyclobutanone
   3/25   4 γ
cycloheptane-1,2-dione
   dioxime
   2/16   (+)
   3/86   20 γ
   4/65   25 γ
cyclohexane-1,2-dione
   4/11   5 γ
cyclohexane-1,2-dione
   dioxime
   3/85   0.1 γ
   3/86   (+)
   4/65   25 γ
cyclohexanol
   3/10(1)   10 γ
   3/11   500 γ
   3/25   (−)
   5/11(1)   6 γ
   5/11(2)   (+)
cyclohexanone
   3/18(2)   25 γ
   3/23(1)   10 γ
   3/23(2)   2.5 γ
   3/25   5 γ
   5/11(1)   6 γ
cyclohexylamine
   2/40   10 γ
cyclohexyl oxalate
   5/46   2.5 γ

N-cyclohexylsulfamic
   acid, Na salt
   3/44   20 γ
   7/22   5 γ
Cyclonite (RDX)
   7/98   (−)
   7/99   10 γ
   7/100   (±)
cyclopentanone
   3/23(2)   5 γ
   3/25   0.2 γ
cysteine
   2/17   25 γ
   2/24(2)   2.5 γ
   2/31(8)   2 γ
   3/35(2)   100 γ
   3/35(3)   100 γ
   4/47(2)   3 γ
   5/59(5)   (+)
cystine
   2/12(8)   2.5 γ
   2/17   300 γ; 50 γ
   3/36(1)   0.005 γ
   3/36(2)   5 γ
   4/47(2)   3 γ

Dacron
   7/102(1)   (−)
   7/102(9)   (+)
Darlan
   7/102(1)   (+)
DDT
   2/6(4)   40 γ
   2/33   (−)
   3/8   (−)
1,10-decanedisulfonic
   acid
   3/43   (−)
decyl aldehyde
   3/20(3)   200 γ
dehydroascorbic acid
   4/13   (+)
7-dehydrocholesterol
   4/2   1 γ
desoxybenzoin
   3/19   0.05 γ
desoxybenzoin oxime
   3/85   0.4 γ
dextrin
   4/1(2a)   12.5 γ
   4/1(2b)   6 γ
   4/21   0.2 γ
diacetyl
   2/31(4)   5 γ
   3/24(1)   10 γ
   3/24(2)   40 γ

3/25   (−)
4/10(1)   5 γ
4/10(2)   (−)
4/10(3)   0.05 γ
4/10(4)   2 γ
4/11   0.5 γ
4/13   0.05 γ
5/32(1)   (−)
5/32(2)   (−)
5/36   2 γ
5/42(4)   (−)
diacetyl monoxime
   2/16   (+)
   3/85   8 γ
   4/10(1)   (+)
   4/10(3)   0.05 γ
5,5-diallylbarbituric acid
   (Dial)
   3/5   (−)
   7/35   5 γ
diallyl ether
   3/5   (−)
diallyl thioether
   3/5   (−)
Diamine Golden Yellow
   7/83   8 γ
2,5-diaminoanisole
   3/54(1)   1 γ
4,4′-diaminodiphenyl
   sulfone
   3/39   10 γ
   3/40   2.5 γ
4,4′-diaminodiphenyl-2-
   sulfonic acid, Na salt
   2/12(4)   4 γ
   2/12(5)   1 γ
1,2-diaminonaphthalene
   5/88(3)   5 γ
1,8-diaminonaphthalene,
   see 1,8-naphthylene-
   diamine
2,3-diaminonaphthalene
   5/88(3)   1 γ
2,4-diaminophenol
   3/54(1)   1 γ
   4/52   0.1 γ
   5/95(1)   (−)
2,4-diaminopyridine
   3/52   0.05 γ
3,4-diaminopyridine
   4/58(2)   (−)
Diazine Green
   2/26   20 γ
   7/75   (+)
Diazinon
   7/129   (+)

diazoaminobenzene
  2/25   25 γ
  6/28   5 γ
diazone sodium, see
  sulfoxone sodium
dibenzalacetone
  3/6   50 γ
  5/32(2)   (−)
dibenzoylmethane
  2/44   12 γ
dibenzyl
  2/32   (+)
dibenzylcyanamide
  3/71   200 γ
dibenzyldiselenide
  3/46   1 γ
dibenzyl ketone
  3/25   (−)
1,2-dibromoethane
  5/8(1)   (+)
  5/8(2)   (+)
5,7-dibromoisatin
  4/10(3)   0.0001 γ
2,3-dibromonaphtho-
  quinone
  4/17   0.5 γ
2,4-dibromophenoxy-
  acetic acid
  4/35   (+)
5,7-dibromoquinaldine
  4/61   1 γ
2,6-dibromoquinone-4-
  chloroimine
  2/31(1)   2 γ; 0.5 γ
  4/64(1)   0.5 γ
  4/64(2)   0.5 γ
di-n-butylamine
  3/55   10 γ
N,N-di-n-butylcarbamate
  4/62   (−)
dibutylthiourea
  2/12(9)   25 γ
dicetyl sulfone
  3/39   20 γ
dichloramine (N,N'-
  dichloro-p-toluene-
  sulfonamide)
  3/102   15 γ
dichloroacetic acid
  3/26   11 γ
  5/50   50 γ
dichloroacetyl chloride
  3/33   0.5 γ
3,4-dichloroaniline
  2/7(1)   8 γ
3,3'- and 4,4'-dichloro-
  azoxybenzene,
  3/83   (+)

2,6-dichlorobenzaldehyde
  2/6(3)   50 γ
3,4-dichlorobenzaldehyde
  2/34   (+)
6,8-dichlorobenzoylene-
  urea
  4/62   (−)
2,6-dichlorobenzyl
  cyanide
  3/69(1)   5 γ
  3/70   3 γ
2,3-dichloro-5,6-dicyano-
  quinone
  4/14(1)   1 γ
  4/14(2)   1 γ
β,β'-dichlorodiethylmethyl-
  amine
  3/2   (+)
β,β'-dichlorodiethyl sulfide
  3/2   (+)
1,2-dichloroethane
  5/8(1)   (+)
  5/8(2)   (+)
dichlorofluorescein
  2/6(3)   25 γ
5,7-dichloro-8-hydroxy-
  quinaldine
  2/7(2)   20 γ
2,3-dichloronaphtho-
  quinone
  4/17   0.5 γ
2,4-dichlorophenoxyacetic
  acid
  4/35   0.05 γ
2,4-dichlorophenoxy-
  ethanol
  3/10(1)   20 γ
5,7-dichloroquinaldine
  4/61   0.8 γ
2,6-dichloroquinone-4-
  chloroimine
  2/31(1)   1 γ; 0.25 γ
  4/64(1)   0.1 γ
  4/64(2)   0.5 γ
dichlorothiourea
  2/12(8)   2.5 γ
N,N'-dichloro-p-toluene-
  sulfonamide, see di-
  chloramine
dicyanodiamide
  3/66   5 γ
  3/67(1)   0.3 γ
  3/67(2)   10 γ
  3/71   100 γ
  5/131(1)   25 γ
  5/131(2)   200 γ

dicyanodiamidine sulfate
  3/67(1)   1 γ
  3/67(2)   50 γ
dicyanogen
  5/127   1 γ
N,N-dicyclohexylsulfamic
  acid, Na salt
  3/44   25 γ
  3/100   25 γ
4,4'-di-(dimethylamino)-
  diphenylmethane, see
  tetrabase
diethanolamine
  2/24(1)   20 γ; 15 γ
  2/27   5 γ
  3/2   (+)
  3/48   4 γ
  3/51   1 γ; 0.6 γ
  3/54   10 γ
  3/55   1 γ
  3/56(2)   100 γ
  4/43   50 γ
diethyl allylmalonate
  3/29   (+)
diethylamine
  3/56(2)   4 γ
N,N-diethylaniline
  2/35   2 γ
  3/49(1)   1.7 γ
  3/58   (+)
  4/40(1)   3 γ
  4/40(2)   1 γ
diethylethanolamine
  4/43   75 γ
diethyl ether
  3/15   20 γ
1,1-diethylhydrazine
  3/73(1)   0.6 γ
diethyl ketone
  3/25   (−)
diethyl malonate, see
  malonic acid diethyl ester
N,N-diethylnicotin-
  amide, see coramine
diethyl oxalate
  2/16   20 γ
  3/28   3 γ
  3/29   (+)
diethyl phthalate
  3/29   (+)
diethyl sulfate
  3/44   10 γ
sym-diethylthiourea
  2/12(9)   200 γ
  2/31(9)   15 γ
4,4'-difluorobenzophenone
  2/10(2)   50 γ

dihydroxyacetone
 5/42(4)   (−)
2,4-dihydroxyaceto-
 phenone
 2/34   (+)
 4/23   0.1 γ
1,2-dihydroxyanthra-
 quinone, see alizarin
2,3-dihydroxyanthra-
 quinone, see hystazarin
1,8-dihydroxyanthra-
 quinone-3-carboxylic
 acid
 4/28   (+); (−)
2,4-dihydroxybenz-
 aldehyde
 2/34   (+)
 3/14(5)   0.5 γ
 3/20(1)   (−)
 5/18   (±)
2,5-dihydroxybenz-
 aldehyde
 5/21(1)   (−)
 5/21(2)   (±)
3,4-dihydroxybenzaldehyde
 see protocatechualdehyde
2,3-dihydroxybenzoic acid
 4/1(1)   (+)
2,4-dihydroxybenzoic acid,
 see β-resorcylic acid
2,5-dihydroxybenzoic acid,
 see hydroquinone carbo-
 xylic acid
2,6-dihydroxybenzoic acid,
 see γ-resorcylic acid
3,4-dihydroxybenzoic acid,
 see protocatechuic acid
3,5-dihydroxybenzoic acid,
 see α-resorcylic acid
3,4-dihydroxycinnam-
 aldehyde
 5/17(1)   (−)
 5/17(2)   (±)
 5/23   (±)
o,o'-dihydroxydiphenyl
 5/18   (−)
 5/21(1)   (±)
p,p'-dihydroxydiphenyl
 5/18   (−)
 5/21(1)   (±)
1,8-dihydroxy-3-hydroxy-
 methyl-anthraquinone
 4/28   (−); (+)
dihydroxymaleic acid
 4/31   40 γ; 15 γ
 5/45(2)   (−)

dihydroxy-2-methylanthra-
 quinone, 1,3-, 1,4- and
 1,8-
 4/28   (−); (+)
1,8-dihydroxy-3-methyl-
 anthraquinone
 4/28   (−); (+)
1,8-dihydroxy-3-methyl-
 6-methoxyanthraquinone
 4/28   (−); (+)
 5/17(1)   (−)
 5/17(2)   (−)
1,3-dihydroxynaphthalene,
 see naphthoresorcinol
1,4-dihydroxynaphthalene
 5/21(1)   (±)
 5/21(2)   (±)
1,5-dihydroxynaphthalene
 5/21(1)   (+)
1,8-dihydroxynaphthalene
 5/17(1)   (−)
 5/17(2)   (−)
 5/18   (−)
 5/21(1)   (±)
 5/22   (±)
2,3-dihydroxynaphthalene
 5/17(1)   (−)
 5/17(2)   (−)
 5/18   (−)
 5/22   (−)
2,6-dihydroxynaphthalene
 5/17(2)   (±)
 5/21(1)   (±)
 5/22   (±)
2,7-dihydroxynaphthalene
 2/34   (+)
 3/14(4)   50 γ; 15 γ
 5/17(1)   (±)
 5/17(2)   (−)
 5/21(1)   (±)
 5/22   (−)
2,3-dihydroxynaphthalene-
 6-sulfonic acid
 2/13(1)   5 γ
 5/17(1)   (−)
 5/17(2)   (−)
 7/94   5 γ
3,4-dihydroxyphenyl-
 alanine
 5/17(2)   (±)
2,6-dihydroxypyridine-
 4-carboxylic acid
 4/54   25 γ
dihydroxytartaric acid
 4/10(1)   1 γ
 4/10(4)   5 γ
 5/54   (−)

5/56(2)   (±)
dihydroxytartaric acid
 osazone
 3/72   (−)
 3/74   30 γ
 3/76   0.7 γ
 3/80(2)   (−)
diiodotyrosine
 2/9(1)   0.5 γ
 3/9   0.002 γ
 4/47(1)   30 γ
diisoamylamine
 3/51   100 γ; 115 γ
 3/56(2)   2 γ
diisopropyl ether
 3/15   (−)
diisopropylthiourea
 2/12(9)   50 γ
dimethoxyazoxybenzene,
 2,2'- and 4,4'-
 3/83   (+)
4,4'-dimethoxydiphenyl
 sulfone
 3/40   25 γ
p,p'-dimethoxythiobenzo-
 phenone
 3/34   0.1 γ
dimethylamine
 3/55   0.5 γ
4'-dimethylamino-
 azobenzene-4-arsonic acid
 3/104   10 γ
p,p'-dimethylamino-
 benzalazine
 3/59   100 γ
p-dimethylaminobenz-
 aldehyde
 3/18(2)   1.2 γ
 3/19   0.0001 γ
 3/20(1)   (−)
 3/20(2)   (−)
 3/20(3)   0.2 γ
 3/20(5)   0.5 γ
 3/21(2)   0.2 γ
 3/21(3)   0.3 γ
 3/21(4)   2 γ
 3/22(1)   25 γ
 3/22(2)   0.5 γ
 3/48   4 γ
 3/49(1)   0.4 γ
 3/57   4 γ
 3/59   100 γ
 4/26   0.5 γ
p-dimethylamino-
 benzylidenerhodanine
 2/12(9)   5 γ
 3/59   100 γ

dioxan
2/2 (+)
3/15 (+)
α,β-diphenoxyethane
3/17 10 γ
1,3-diphenoxy-2-propanol
4/3 2.5 γ
diphenyl
2/32 (+)
2/33 15 γ
diphenylamine
2/25 50 γ
2/31(5) 0.1 γ
2/32 (−)
2/33 (+)
2/36 0.25 γ
3/48 8 γ
3/49(1) 1.6 γ
3/49(2) 50 γ
3/50 0.6 γ
3/52 0.05 γ
3/55 10 γ
4/41(1) 0.5 γ
4/41(2) 0.1 γ
4/41(3) 0.25 γ
5/63(1) (−)
diphenylamine-2,2'-
dicarboxylic acid
4/41(3) 0.4 γ
diphenylamine-4-
sulfonic-acid
2/17 (−)
4/41(1) 2 γ
4/41(3) 0.3 γ
N,N-diphenylanthranilic
acid
4/41(3) 4 γ
5/96 4 γ; 6 γ
1,4-diphenylbenzene
2/33 (+)
N,N'-diphenylbenzidine
2/32 (+)
3/52 0.05 γ
4/41(1) 5 γ
4/41(2) 0.1 γ
4/41(3) 0.17 γ
Diphenyl Brilliant
Violet 2R
7/84 (+)
N,N'-diphenylcarbamate
4/62 (−)
diphenylcarbazide
2/25 10 γ
3/72 0.05 γ
3/81 5 γ
diphenylcarbazone
2/25 15 γ

3/72 0.01 γ
3/79 10 γ
3/81 5 γ
diphenyl carbonate
3/30 10 γ
diphenyl disulfide
3/36(2) 5 γ
diphenyl ether
2/32 (+)
4/3 2.5 γ
Diphenyl Fast Red 5 BLN
7/84 (+)
N,N-diphenylformamide
4/41(1) 2.5 γ
asym-diphenylguanidine
4/41(3) 100 γ
sym-diphenylguanidine
(melaniline)
2/41 25 γ
3/67(2) (−)
1,1-diphenylhydrazine
3/73(1) 7 γ
3/73(2) 1 γ
3/74 5 γ
4/41(1) 5 γ
1,2-diphenylhydrazine,
see hydrazobenzene
diphenylmercury
3/107 60 γ
diphenylmethane
2/32 (+)
diphenylmethylcarbinol
2/34 (+)
diphenylnitrosamine, see
N-nitrosodiphenylamine
diphenyl oxalate
5/46 3.5 γ
diphenyl phthalate
3/30 20 γ
4/3 2 γ
1,4-diphenylsemi-
carbazide
3/79 15 γ
4,4-diphenylsemi-
carbazide
4/41(1) 3 γ
4/41(3) 5 γ; 2.5 γ
diphenylthiocarbazone
(dithizone)
2/12(4) 8 γ
2/12(5) 5 γ
2/12(6) 50 γ
2/17 1000 γ; 500 γ
3/34 2.5 γ
3/79 10 γ
3/81 5 γ

diphenyl thiocarbonate
3/34 1 γ
asym-diphenylthiourea
2/31(9) 20 γ
sym-diphenylthiourea,
see thiocarbanilide
asym-diphenylurea
2/41 17 γ
4/41(1) 3 γ
4/41(3) 110 γ
4/62 150 γ
5/108(1) (−)
6/42 10 γ
sym-diphenylurea, see
carbanilide
dipropylamine
3/51 140 γ; 125 γ
di-n-propyl ketone
3/25 (−)
diptorex
7/129 (+)
α,α'-dipyridyl
2/24(2) (−)
2/32 (−)
4/53(1) 10 γ
4/53(2) 5 γ
4/58(2) (−)
dithiooxamide (rubeanic
acid)
2/12(8) 5 γ
2/12(9) 1.8 γ
2/17 200 γ; 1 mg
2/31(5) 0.5 γ
2/31(9) 2.5 γ
3/34 0.03 γ
3/35(1) 0.5 γ
dithiotartaric acid
2/31(9) 5 γ
dithizone, see diphenyl-
thiocarbazone
di-o-tolylmercury
3/107 100 γ
di-p-tolylmercury
3/107 100 γ
di-p-tolyl trithio-
carbonate
3/34 0.1 γ
sym-di-m-tolylurea:
4/62 150 γ
5/108 (−)
dixanthogen
2/12(9) 10 γ
3/36(2) 5 γ
5/71 20 γ; 5 γ
dulcitol
5/9 (−)

duroquinone
4/14(1)   (−)
4/14(2)   (−)
Dynel
7/102(1)   (+)
7/102(6)   (+)
7/102(7)   (+)

Edestin
4/48(1)   5 γ
4/48(2)   10 γ
EDTA (ethylenediamine-
tetraacetic acid)
3/2   (+)
4/44   (+); (−)
5/84(1)   1.7 γ
5/84(2)   8 γ
5/84(3)   10 γ
egg albumin
4/48(1)   0.5 γ
4/48(2)   30 γ
elaidic acid
2/24(5)   (+)
ellagic acid
2/34   (+)
2/36   (−)
3/14(5)   2 γ
emetine
2/34   (+)
2/36   (+)
3/56(1)   (+)
7/42   2 γ
Endoxan
3/2   5 γ
Entobex (4,7-phenanthro-
line-5,6-quinone)
4/14(1)   2 γ
4/14(2)   2 γ
4/15   5 γ
7/49   5 γ
eosin
2/6(1)   0.25 γ
ephedrine
2/36   (+)
3/12   200 γ
3/56(1)   2 γ
7/40(1)   7 γ; 5 γ
7/40(2)   5 γ
epichlorohydrin
3/7   100 γ
3/18(2)   55 γ
epihydrinaldehyde
7/28(c)   (+)
epoxy resin
7/110   (+)

ergosterol
4/2   2 γ; 0.1 γ
ergosteryl acetate
4/2   1 γ
ergosteryl butyrate
4/2   1 γ
Eriochrome Cyanine
7/87   0.1 γ
Eriochrome Orange R
7/84   (+)
Erio Violet R
7/84   (+)
erucic acid
2/24(5)   (+)
erythritol
4/1(2a)   (+)
5/9   (−)
erythronic acid
5/54   (−)
5/56(2)   (−)
estragol
3/5   10 γ
4/8   10 γ
ethanol
3/10(1)   20 γ
3/11   1 mg
5/9   (−)
5/10   3 γ
7/9   3 γ
ethanolamine
2/24(3a)   10 γ
2/24(3b)   10 γ
4/1(1)   (+)
ethanol-2-sulfonic acid,
Na salt
3/43   1 mg
2-ethoxybenzoyl chloride
3/33   0.1 γ
3-ethoxy-4-methoxybenzyl
cyanide
3/69(1)   2 γ
3/70   1 γ
m-ethoxyphenoxyacetic
acid
3/17   10 γ
ethyl acetate
3/28   2 γ
3/29   5 γ
ethyl acetoacetate, see
acetoacetic acid ethyl
ester
ethylacetoacetic ethyl
ester
5/65   (−)
ethylamine
2/15(2)   0.2 γ
3/48   10 γ
3/51   0.5 γ; 3 γ

ethylamine chloride
3/53   10 γ
ethyl α-aminoacetate (glycine ethyl ester)
3/28   10 γ
3/48   20 γ
3/51   100 γ; 70 γ
N-ethyl-4-aminocarb-
azole
3/58   (+)
4/57   2.5 γ
N-ethylaniline
3/55   2 γ
3/58   (+)
6/45   50 γ
ethylbenzene
2/32   (+)
ethyl benzoate
3/29   5 γ
ethyl n-butyl ketone
3/25   (−)
ethyl carbamate
2/15(2)   0.4 γ
N-ethylcarbazole
3/49(1)   5 γ
4/41(1)   1 γ
4/57   0.5 γ
ethyl cellulose
4/19   (+)
7/106   (+)
7/107   (+)
ethyl cinnamate
3/29   (+)
N-ethyl-o-crotono-
toluidide, see Crotamiton
N-ethyl-3-diethylamino-
carbazole
4/57   0.5 γ
N-ethyl-4-diethyl-amino
carbazole
4/41(1)   1 γ
ethyl N,N′-diphenyl-
carbamate
4/41(1)   5 γ
4/41(3)   15 γ
ethylene chlorohydrin
3/2   (+)
ethylenediamine
2/15(2)   0.1 γ
2/24(1)   5 γ; 10 γ
2/24(2)   2 γ
2/24(3a)   100 γ
2/24(3b)   2.5 γ
3/49(1)   6 γ
3/49(2)   15 γ
5/77   0.3 γ

ethylenediamine dichloride
2/27   5 γ
2/41   15 γ
3/50   1 γ
3/53   10 γ
ethylenediaminetetraacetic
acid, see EDTA
ethylene dithiocyanate
3/97(1)   2.5 γ; 5 γ
ethylene glycol, see glycol
ethylenethiourea
2/12(9)   50 γ
ethyl formate
3/28   10 γ
5/42(4)   (−)
ethyl γ-methyl-γ-nitro-
valerate
2/16   (+)
ethylmorphine chloride
dihydrate (Dionine)
3/16   200 γ
Ethyl Orange
3/58   15 γ
7/79   (+)
ethyl orthoformate
3/28   10 γ
o-ethylphenol
3/14(8)   0.4 γ
ethyl phenylacetate
3/29   (+)
5-ethyl-5-phenyl-
barbituric acid,
see luminal
ethylphenylhydrazine
3/72   0.05 γ
N-ethylpiperidine
3/57   7 γ
ethyl stearate
3/28   10 γ
ethyl p-stibonobenzoate
3/105   1 mg
ethylsulfuric acid, Na salt
3/44   10 γ
3/108   5 γ
ethyl thiocyanate
3/97(2)   110 γ
ethyl p-toluenesulfonate
3/108   20 γ
ethyl urethan
3/28   10 γ
4/62   100 γ
ethylvanillin
3/20(1)   (−)
eugenol
3/14(5)   (+)
3/14(8)   1.5 γ
4/8   1 γ

4/25   5 γ
5/21(1)   (±)
eugenol methyl ether
3/5   10 γ
4/8   10 γ
euphylline
7/53   (+)

Fluorene
2/32   (+)
fluorescein
2/34   (+)
7/75   (+)
o-fluorobenzoic acid
2/10(1)   100 γ
β-fluoronaphthalene
2/10(1)   100 γ
formaldehyde
2/31(6)   0.1 γ
3/18(1)   0.05 γ
3/18(2)   0.25 γ
3/20(1)   1 γ
3/20(2)   2 γ
3/20(3)   50 γ
3/20(4)   0.02 γ
3/20(5)   0.2 γ
3/20(6)   0.2 γ
3/25   (−)
5/28(1)   0.14 γ
5/28(2)   0.5 γ; 3.5 γ
5/28(3)   0.01 γ
5/32(2)   (−)
5/42(4)   (−)
5/52(1)   (+)
5/53(1)   (±)
5/53(2)   (+)
5/56(1)   (±)
5/56(2)   (−)
7/115   (+)
formaldoxime
5/28(1)   (+)
formamide
3/62(2)   10 γ
5/112   10 γ
formamidine acetate
3/66   5 γ
formamidine disulfide
3/88(2)   (+)
formanilide
3/64   1 γ
3/65   10 γ
5/114   50 γ
formic acid
2/24(6)   (−)
2/31(6)   1 γ
3/20(5)   0.005 γ

5/42(1)   1.4 γ
5/42(2)   5 γ
5/42(3)   2.5 γ
5/43(4)   0.04 γ
5/45(2)   (−)
5/52(1)   (−)
5/55   (−)
5/59(4)   (−)
fructose
3/18(1)   500 γ
4/1(2a)   12.5 γ
4/1(2b)   5 γ
4/19   2.5 γ
4/20(1)   0.2 γ
4/21   0.2 γ
4/22   8 γ
5/9   (±)
5/28(1)   (±)
5/42(1)   (±)
5/63(1)   (±)
fuadin
7/61   (+)
2-furaldoxime
(furfuraldoxime)
2/16   (+)
3/86   (+)
furfural
3/18(1)   5 γ
3/18(2)   0.35 γ
3/19   0.001 γ
3/20(1)   20 γ
3/20(2)   12 γ
3/20(3)   0.02 γ
3/20(5)   0.05 γ
3/20(6)   15 γ
3/21(1)   1 γ
3/21(2)   0.05 γ
3/21(3)   0.05 γ
3/21(4)   0.01 γ
3/22   0.01 γ
4/26   (−)
4/27   (±)
5/9   (±)
5/28(1)   (±)
5/34(1)   0.05 γ
5/34(2)   0.01 γ
furfural bisulfite
3/20(4)   0.7 γ
furfuraldoxime, see
2-furaldoxime
furfural phenylhydrazone
3/80(2)   (+)
furil
4/13   2.5 γ
furil dioxime
2/32   (+)

5/42(4)	(−)
5/45(2)	(−)
5/54	(−)
5/56(1)	(±)
5/56(2)	(±)
5/63(1)	1 γ
5/63(2)	0.5 γ
5/63(3)	1 γ

glyoxime
5/116(3)	(+)

gramine
4/56	(−)

guaiacol
3/14(7)	1 γ
3/14(8)	0.2 γ
5/17(1)	(±)
5/17(2)	(−)
5/21(1)	(±)

guaiacolaldehyde
5/17(2)	(±)

guaiacol benzoate
3/17	2.5 γ

guaiacol carbonate
7/55	50 γ

guanidine (salts)
2/27	50 γ
3/61 (1)	5 γ
3/66	7.5 γ
3/67(1)	2 γ
3/67(2)	(−)
5/6(2)	30 γ
5/109	5 γ

guanidine nitrate
7/93	2.5 γ

guanine chloride
3/67(2)	(−)

gum arabic
4/1(2a)	(+)

gum tragacanth
4/19	(−)

H-acid
2/12(1)	1.2 γ
3/14(4)	(−)
3/14(8)	(−)
3/41(1)	1.5 γ
3/43	2 γ
4/44	(+)
4/52	1 γ
5/14	(+)
5/87	(±)
7/94	(+); (−)
7/95	1 γ; 0.5 γ

helicin
7/59	(−)

Heliogen Blue
7/89	(+)

hematommic acid methyl ester monobenzyl ether
4/23	1.8 γ

hematoxylin
7/75	(+)

hemipinic acid
3/27	(+)

hemipinic anhydride
3/32	5 γ
4/31	(+)

hemoglobin
4/48(1)	0.5 γ

heptylamine
3/51	12 γ; 7 γ

2,4,6,2′,4′,6′-hexabromo-azoxybenzene
3/83	(+)

hexaethylbenzene
2/32	(+)

hexachlorobenzene
2/6(3)	0.5 γ
2/33	(−)
2/36	(−)

hexachlorocyclohexane
2/6(4)	30 γ

hexahydro-1,3,5-trinitro-s-triazine, see Cyclonite

1,2,3,5,6,7-hexahydroxy-anthraquinone, see rufigallic acid

hexamethyleneimine
3/55	1 γ

hexamethylenetetramine, see hexamine

hexamine (hexamethylene-tetramine, urotropine, methamine)
3/20(4)	50 γ
3/20(6)	1 γ
5/28(1)	0.03 γ; 0.003γ
5/78(1)	2.5 γ
5/78(2)	5 γ
5/78(3)	20 γ
5/116(1)	(+)
5/116(2)	(+)
7/29(b)	(+)

hexanitrodiphenylamine
3/94	0.5 γ

hexanol
3/10(1)	(+)

4-(n-hexyl)-resorcinol
5/18	(±)

hide powder
4/48(2)	100 γ

hippuran
7/67	25 γ

hippuric acid
2/34	(+)
5/63(1)	(−)
5/81	12 γ

histidine chloride
3/49(2)	50 γ
3/57	(+)

HMX
7/100	(+)

homocysteine
3/36(2)	5 γ

hordenine
3/57	(+)

hydantoin
5/125	20 γ

hydrastine
4/9	0.2 γ

hydrazobenzene
2/31(7)	(+)
3/48	4 γ
3/72	(−)
5/134	15 γ

α-hydrindone
3/23(2)	10 γ

hydrobenzamide
3/19	0.01 γ
6/24	5 γ

hydrocyanic acid (prussic acid)
5/126(1)	0.25 γ
5/126(2)	0.25 γ
7/119	(+)

hydroquinone (quinol)
2/31(5)	0.05 γ
2/31(6)	2.5 γ
2/31(7)	10 γ
2/32	(+)
2/33	(+)
3/14(1)	10 γ
3/14(3)	10 γ
3/14(5)	(+)
3/14(7)	(−)
3/14(8)	1.1 γ
5/17(1)	(±)
5/17(2)	(±)
5/21(1)	7 γ
5/21(2)	0.5 γ
5/21(3)	0.2 γ
5/21(4)	0.2 γ
5/22	(±)
5/23	(±)
6/7	25 γ

hydroquinone carboxylic acid
5/21(1)	(±)

hydroquinone monomethyl
ether (*p*-hydroxyanisole)
   3/14(7)   (−)
o-hydroxyacetophenone
   3/14(5)   (+)
   3/14(8)   1.1 γ
p-hydroxyacetophenone
   3/14(8)   8 γ
o-hydroxyanisole, see
   guaiacol
p-hydroxyanisole, see hydroquinone monomethyl
   ether
p-hydroxybenzaldazine
   3/20(5)   0.005 γ
o-hydroxybenzaldehyde,
   see salicylaldehyde
m-hydroxybenzaldehyde
   2/31(5)   10 γ
   2/31(6)   5 γ; 10 γ
   2/35   0.1 γ
   2/36   (+)
   3/14(3)   10 γ
   3/14(7)   (−)
   3/14(8)   1 γ
   3/18(1)   1 γ
   3/18(2)   0.5 γ
   3/19   0.01 γ
   3/20(1)   50 γ
   3/20(2)   50 γ
   3/20(3)   4 γ
   3/20(4)   0.5 γ
   3/21(1)   1 γ
   3/22(1)   5 γ
p-hydroxybenzaldehyde
   2/31(5)   10 γ
   2/31(6)   500 γ
   2/34   (+)
   2/35   50 γ
   2/36   0.25 γ
   3/14(1)   (−)
   3/14(3)   1 γ
   3/14(7)   (−)
   3/14(8)   1.9 γ
   3/18(1)   10 γ
   3/20(1)   1 mg
   3/20(2)   (−)
   3/20(3)   5 γ
   3/21(1)   1 γ
   3/21(3)   0.3 γ
   3/22(1)   0.5 γ
o-hydroxybenzoic acid,
   see salicylic acid
m-hydroxybenzoic acid
   3/14(4)   10 γ; 1 γ
   3/14(8)   1 γ
   5/17   (−)
   5/21(1)   (±)

   5/56(2)   (−)
   5/64(1)   (−)
p-hydroxybenzoic acid
   3/14(1)   (−)
   3/14(3)   2 γ
   3/14(8)   2 γ
   5/17   (−)
   5/21(1)   (−)
   5/23   (−)
   5/56(2)   (−)
   5/63(1)   (−)
p-hydroxybenzoic acid
   hydrazide
   3/14(8)   1.8 γ
4-hydroxybenzophenone
   2/34   (+)
α-hydroxybutyric acid
   5/53(1)   (+)
   5/54   (−)
   5/56(1)   (−)
   5/63(1)   (−)
β-hydroxybutyric acid
   5/54   (−)
   5/56(2)   (−)
   5/63(2)   (−)
4-hydroxy-7-chloro-
   quinoline
   5/104   (−)
4-hydroxy-7-chloro-
   quinoline-
   3-carboxylic acid
   5/104   (−)
m-hydroxycinnamic acid
   3/14(3)   5 γ
p-hydroxycinnamic acid
   3/6   100 γ
o-hydroxydiphenyl
   (*o*-phenylphenol)
   2/32   (+)
   2/34   (+)
   3/14(4)   20 γ; 5 γ
   3/14(7)   2 γ
   4/3   (+)
   5/21(1)   (±)
m-hydroxydiphenyl
   5/21(1)   (±)
p-hydroxydiphenyl
   (*p*-phenylphenol)
   2/32   (+)
   2/34   (+)
   2/35   5 γ
   2/36   (+)
   3/14(4)   2.5 γ; 1 γ
   3/14(5)   (+)
   3/14(7)   (−)
   3/14(8)   1.9 γ
   4/3   (+)

1-β-hydroxyethyl-2-hepta-
   decylglyoxalidine
   4/43   120 γ
N-β-hydroxyethyl-
   morpholine
   4/43   25 γ
N-β-hydroxyethyl-
   propylenediamine
   4/43   30 γ
hydroxyhydroquinone
   2/34   (+)
   3/14(7)   (+)
   5/17(1)   (±)
   5/18   (±)
   5/21(1)   (+)
   5/21(2)   (±)
   5/22   (±)
   5/23   5 γ
2-hydroxy-3-methoxy-
   benzaldehyde, see orthovanillin
4-hydroxy-3-methoxy-
   benzaldehyde, see vanillin
1-hydroxy-2-methyl-
   anthraquinone
   4/28   (−); (+)
2-hydroxy-3-methyl-
   benzoic acid
   3/14(8)   (−)
N-hydroxymethyl-nicotin-
   amide, see choligen
6-hydroxymethyl-
   picolinic acid
   4/53(2)   (−)
   4/60   7 γ
2-hydroxy-1-naphth-
   aldehyde
   3/18(2)   0.5 γ
   4/24(2)   (+)
1-hydroxy-2-naphthoic
   acid
   5/18   (−)
   5/21(1)   (±)
2-hydroxy-3-naphthoic
   acid
   5/21(1)   (±)
2-hydroxy-3-naphthoic
   acid hydrazide
   3/78(4)   (+)
8-hydroxynaphthyl-1-
   hydrazine-
   3,6-disulfonic acid
   3/73(1)   3 γ
4-hydroxy-3-nitrobenzoic
   acid
   2/16   (+)
12-hydroxyoleic acid
   2/24(5)   (+)

menthol
3/10(1) 5 γ
3/12 200 γ
5/21(1) (±)
2-mercaptobenzothiazole
2/25 20 γ
2/31(5) 10 γ
2/31(8) 5 γ
2/32 (−)
2/36 10 γ
3/35(1) 5 γ
mercaptophenylthia-
diazolone
3/35(1) 2.5 γ
mercuric cyanide
7/64 (+)
mercuric diiodo-
salicylate
7/64 (+)
mercuric oxycyanide
7/68 100 γ
mercuric succinimide
7/64 (+)
Mercurin
7/64 (+)
Mercurophen
7/64 (+)
Mercusal
7/64 (+)
mesitylene
2/32 (+)
mesoxalic acid
4/10(2) (+)
5/54 (−)
5/56(2) (±)
metaldehyde
5/53(1) (-)
5/53(2) (-)
metanilic acid
3/43 5 γ
6/18 5 γ
methamine, see hexamine
methanol
3/10(1) 20 γ
3/11 1 mg
5/9 25 γ
methionine
4/44 (-)
4/47(2) 2.5 γ
methoxybenzaldehyde-
sulfonic acids
4/71 5 γ; 10 γ
o-methoxybenzaldoxime
3/85 6 γ
p-methoxybenzohydr-
oxamic acid
2/16 0.2 γ

3/87(1) 40 γ
3/87(2) 0.25 γ
o-methoxybenzoic acid
3/17 5 γ
p-methoxycinnamaldehyde
4/26 0.5 γ
2-methoxy-m-cresol
3/14(8) 0.4 γ
2-methoxy-4-methyl-
phenol, see creosol
6-methoxy-8-nitro-
quinoline
3/93(2) (+)
(o-methoxyphenoxy)-ethyl
bromide
2/6(4) 5 γ
5-methoxysalicylic acid
hydrazide
3/78(4) 80 γ
methyl acetylsalicylate
3/31 100 γ
methylamine
2/15(2) 0.1 γ
2/24(3a) 1 γ
2/24(3b) 0.1 γ
3/48 4 γ
p-methylaminophenol
sulfate
2/31(5) 0.5 γ
5/95(1) 5 γ
1-(methylamino)-3-phenyl-
propane
3/56(2) 10 γ
β-(methylamino)-pyridine
4/58(1) 0.1 γ
N-methylaniline
3/49(2) 50 γ
3/55 0.5 γ
6/14 (−)
6/45 50 γ
N-(α-methylbenzyl)-
diethanolamine
4/42 5 γ
methyl benzyl ketone
3/25 100 γ
3-methylbutane-1-sulfonic
acid, Na salt
3/42 15 γ
3/43 (−)
methyl n-butyl ketone
3/25 0.2 γ
methyl cellulose
4/19 (+)
7/106 (+)
7/107 (+)
methyl cinnamate
3/6 100 γ

methylcyclohexanol
3/10(1) (+)
methyl n-decyl ketone
3/25 1 γ
6-methyl-4,5-dihydroxy-
salicylaldehyde
4/23 0.8 γ
N-methyldiphenylamine
4/41(1) 0.5 γ
methylene blue
2/12(4) 5 γ
2/12(5) 2.5 γ
2/12(7) 100 γ
2/12(8) 2.5 γ
2/36 1 γ
7/69 100 γ
7/70 10 γ
7/78 50 γ
methylephedrine
3/57 (+)
methylethylglyoxime
4/65 5 γ
methyl ethyl ketone
3/18(1) 20 γ
3/23(1) 20 γ
3/23(2) 25 γ
3/24(1) 10 γ
3/24(2) 150 γ
3/25 0.3 γ
4-methyl-5-ethylpyridine-
2,3-dicarboxylic acid
4/60 (+)
methylglyoxal
(pyruvaldehyde)
5/9 (−)
5/32(1) ?
5/32(2) (−)
4-methylhematommic acid
4/23 1.1 γ
methyl n-heptadecyl
ketone
3/25 1 γ
2-methylheptan-6-one
3/24(1) 10 γ
3/24(2) 150 γ
methyl n-hexyl ketone
3/25 0.2 γ
methyl p-hydroxybenzoate
3/14(1) (−)
3/14(3) 1 γ
3/14(8) 8.4 γ
2-methylindole
4/55 (+)
4/56 0.1 γ
6/21 2 γ
3-methylindole, see skatole

3/20(1)	40 γ	p-nitrobenzoic acid		2/24(7)	200 γ
3/20(2)	10 γ	3/93(1)	2.5 γ	2/44	(+)
3/20(3)	5 γ	3/93(2)	(+)	3/91	50 γ
3/20(4)	0.6 γ	m-nitrobenzoic acid		3/92(1)	0.5 γ
3/20(5)	0.5 γ	hydrazide		5/32(2)	(−)
3/21(1)	4 γ	3/78(4)	60 γ	5/139(1)	0.6 γ
3/22(1)	5 γ	p-nitrobenzoic acid		5/139(2)	2.5 γ
3/22(2)	0.5 γ	hydrazide		2-nitro-4-methylbenzene-	
3/91	1 γ	3/78(1)	1 γ	seleninic acid	
3/93(1)	4 γ	3/78(2)	(−)	2/14	5 γ
4/26	(−)	3/78(3)	0.1 γ	2-nitro-4-methylphenyl-	
4/74	5 γ; 1 γ	3/78(4)	(+)	selenocyanate	
6/36	50 γ	p-nitrobenzonitrile		2/14	5 γ
m-nitrobenzaldehyde		3/91	0.3 γ	1-nitronaphthalene	
3/18(1)	1 γ	4-nitrobenzoyl chloride		2/16	0.2 γ
3/20(1)	40 γ	3/33	0.5 γ	2/34	(+)
3/20(2)	20 γ	p-nitrobenzyl chloride		3/91	1 γ
3/21(1)	1 γ	2/6(3)	5 γ	1-nitro-2-naphthol	
3/21(2)	0.2 γ	2/6(4)	100 γ	3/93(2)	(+)
3/93(1)	4 γ	p-nitrobenzyl cyanide		1-nitro-5-naphthylamine	
4/74	5 γ; 1 γ	2/16	(+)	3/93(1)	0.3 γ
p-nitrobenzaldehyde		3/69(1)	2 γ	3/93(2)	(+)
3/18(2)	0.2 γ	3/70	3 γ	3-nitrophenanthra-	
3/19	0.0001 γ	3-nitrocarbazole		quinone	
3/20(3)	1 γ	4/57	10 γ	4/13	0.002 γ
3/20(4)	0.8 γ	nitrocellulose		p-nitrophenetole	
3/20(5)	0.1 γ	7/108	(+)	3/91	0.2 γ
3/21(1)	2 γ	o-nitrocinnamic acid		o-nitrophenol	
3/21(2)	0.2 γ	3/93(1)	3 γ	2/32	(−)
3/21(4)	(−)	3/93(2)	(+)	3/14(2)	5 γ
3/22(2)	0.5 γ	m-nitrocinnamic acid		3/14(7)	(−)
3/93(1)	4 γ	3/93(1)	3 γ	3/14(8)	(−)
4/74	5 γ; 1 γ	3/93(2)	(+)	3/93(1)	0.4 γ
5/142	1 γ	p-nitrocinnamic acid		3/93(2)	(+)
nitrobenzene		3/93(1)	3 γ	4/3	(−)
2/34	(+)	3/93(2)	(+)	6/35	1.5 γ
2/36	(+)	nitroethane		m-nitrophenol	
3/91	0.8 γ	2/15(2)	0.3 γ	2/24(7)	5 γ
3/93(1)	1.5 γ	2/16	(+)	3/14(7)	(−)
3/93(2)	(+)	3/92(1)	0.5 γ	3/14(8)	(−)
4/70	(−)	5/140	7 γ	3/91	2 γ
p-nitrobenzenearsonic acid		5-nitrofurfural semi-		4/3	(−)
4/51(1)	6 γ	carbazone		5/141	3 γ
2-nitrobenzene-		3/93(2)	(+)	p-nitrophenol	
seleninic acid		nitrofurylglyoxal		2/16	(+)
2/14	5 γ	4/10(4)	5 γ	2/24(7)	6 γ
p-nitrobenzenestibonic acid		nitroguanidine		2/36	0.5 γ
4/51	5 γ	2/16	0.1 γ	3/14(1)	(−)
o-nitrobenzoic acid		3/67(1)	5 γ	3/14(3)	2 γ
2/16	(+)	3-nitro-4-hydroxybenzyl		3/14(7)	(−)
2/32	(−)	alcohol		3/14(8)	(−)
2/34	(+)	3/10(1)	10 γ	3/91	2 γ
2/36	(+)	3/91	2.5 γ	3/93(1)	0.4 γ
3/91	3 γ	nitromethane		3/93(2)	(+)
m-nitrobenzoic acid		2/15(2)	0.3 γ	4/3	(−)
3/93(1)	8 γ	2/16	(+)	6/35	2.5 γ
3/93(2)	(+)				

7/102(4)   (+)
7/102(5)   (+)
Perlon W
  7/102(1)   (+)
Permanent Red E5B
  7/89   (+)
pervitin
  3/56(1)   5 γ
phenacetin
  3/16   50 γ
  7/36   10 γ
phenanthraquinone
  2/5   0.05 γ
  2/33   (+)
  2/34   (+)
  4/10(1)   0.5 γ
  4/10(2)   10 γ
  4/12   0.5 γ
  4/13   0.002 γ
  4/14(1)   (−)
  4/14(2)   (−)
  4/15   1.5 γ
  5/32(2)   (−)
  5/40   3 γ
phenanthrene
  2/32   (+)
  2/33   15 γ
  5/3   3 γ
  5/5   20 γ
o-phenanthroline
  2/24(2)   25 γ
  4/53(1)   10 γ
4,7-phenanthroline-5,6-
  quinone, see Entobex
phenazone, see antipyrine
m-phenetidine
  2/34   (+)
  3/17   20 γ
phenetole
  2/34   (+)
  3/16   30 γ
  4/3   0.8 γ
phenobarbital, see
  luminal
phenol
  2/5   0.1 γ
  2/31(5)   0.05 γ
  2/32   2 γ; (+)
  2/35   0.2 γ
  3/14(1)   1 γ
  3/14(2)   1 γ
  3/14(3)   1 γ
  3/14(4)   12 γ; 0.25 γ
  3/14(6)   0.3 γ
  3/14(7)   5 γ
  3/14(8)   0.1 γ
  4/3   0.25 γ

5/17(1)   (±)
5/17(2)   (−)
5/18   (−)
5/21(1)   (±)
5/21(2)   (−)
5/22   (−)
5/23   (−)
phenolphthalein
  2/33   (−)
  2/34   (+)
  2/36   1 γ
phenol-o-sulfonic acid
  3/14(8)   0.9 γ
phenol-m-sulfonic acid
  3/14(8)   1 γ
phenol-p-sulfonic acid
  3/14(7)   (−)
  3/14(8)   2 γ
  4/36   0.5 γ
  6/11   0.5 γ
phenoxyacetic acid
  2/34   (+)
  3/17   10 γ
  4/3   20 γ
  4/35   (+)
  5/61   50 γ
γ-phenoxybutyric acid
  2/34   (+)
phenylacetaldehyde
  3/18(2)   2 γ
  3/20(4)   10 γ; 5 γ
  3/23(2)   5 γ
phenylacetamide
  3/62(2)   5 γ
  6/26   20 γ
N-phenylacetamide, see
  acetanilide
phenyl acetate
  3/28   2 γ
  3/30   10 γ
phenylacetic acid
  2/34   (+)
  3/26   11 γ
  3/27   (+)
phenylacetohydroxamic
  acid
  2/16   0.2 γ
  3/87(1)   40 γ
  3/87(2)   0.1 γ
phenylacetyl chloride
  3/33   0.1 γ
phenyl acetylsalicylate
  4/3   2.5 γ
phenylalanine
  2/34   (+)
  4/44   (+)
  4/47(2)   (+)

N-phenylanthranilic acid
  2/34   (+)
  3/30   20 γ
  4/41(1)   5 γ
phenylarsonic acid
  2/5   0.2 γ
  2/19   1.4 γ
  2/32   (+)
  2/34   (+)
  3/104   5 γ
4-phenylazobenzoyl
  chloride
  3/33   0.2 γ
phenyl benzoate
  3/30   20 γ
  4/3   5 γ
phenylbiguanide
  2/41   25 γ
  3/67(2)   (+)
1-phenyl-1-chloro-
  ethylene-2-phosphonic
  acid
  2/18   1 γ
phenyl (2,4-dinitro-
  phenyl) diselenide
  2/14   5 γ
o-phenylenediamine
  3/50   0.5 γ
  3/54(1)   1 γ
  5/88(1)   5 γ
  5/88(2)   2 γ
  5/88(3)   1 γ
  6/15   10 γ
m-phenylenediamine
  3/54(1)   2 γ
  6/15   10 γ
p-phenylenediamine
  2/24(1)   15 γ
  2/31(5)   0.25 γ
  3/48   1 γ
  3/50   0.2 γ
  3/52   0.1 γ
  3/54(2)   0.1 γ
  5/89   0.5 γ
  6/15   10 γ
  6/44   5 γ
phenylethyl alcohol
  3/11   100 γ
phenylglyoxal
  2/24(7)   20 γ
phenylglyoxylic acid
  4/10(2)   (+)
phenylhydrazine
  2/31(4)   (+)
  2/31(6)   2 γ
  2/31(7)   5 γ
  3/51   250 γ; 200 γ

picramine
  4/70   2 γ
picric acid
  2/36   5 γ
  3/93(1)   8 γ
  3/93(2)   (+)
  3/94   0.1 γ
  4/70   2 γ
picrolonic acid
  2/24(7)   0.1 γ
  2/32   (−)
picryl chloride
  4/70   2 γ
picrylnitromethyl-
  amine (Tetryl)
  7/98   8 γ; 10 γ; 5 γ
Pigment Orange
  7/76   14 γ
pilocarpine
  3/57   (+)
pimelic acid
  3/27   10 γ
pinacol hydrate
  5/21(1)   (±)
piperazine
  3/55   0.5 γ
  5/98   20 γ
  7/54   (+)
piperidine
  2/44   0.6 γ
  3/48   4 γ
  3/49(1)   3.3 γ
  3/51   4.5 γ
  3/55   10 γ
  3/56(1)   0.2 γ
  3/56(2)   5 γ
  5/97   12.5 γ
piperidine N-oxide
  2/16   (+)
piperidinium piperidyl-
  dithiocarbamate
  3/99   1.8 γ
piperine
  4/9   0.1 γ
  7/44(1)   4 γ
  7/44(2)   20 γ
piperonal
  2/34   (+)
  3/18(2)   1 γ
  3/20(4)   4 γ
  3/21(1)   1 γ
  3/21(2)   0.1 γ
  3/21(3)   0.3 γ
  3/21(4)   0.1 γ
  3/22(2)   0.5 γ
  4/26   0.2 γ
  4/27   5 γ

5/17(1)   (±)
5/17(2)   (−)
piperonal phenyl-
  hydrazone
  3/80(2)   (+)
piperonylic acid
  5/42(4)   (−)
polyethylene glycol
  7/12(1)   (+)
  7/12(2)   (+)
polyethylene glycol
  methyl ether
  7/12(1)   (+)
polyethylene glycol
  propyl ether
  7/12(1)   (+)
polyethylene oxide
  7/12(1)   (+)
polyvinyl acetate
  7/122   (+)
polyvinyl alcohol
  7/122   (+)
polyvinyl chloride
  (PCU, Vynion HH)
  7/102(1)   (−)
  7/102(6)   (+)
polyvinylidene chloride
  7/122   (+)
populin
  7/58   5 γ
potassium cyanurea
  3/71   200 γ
potassium ethylxanthate
  2/31(5)   0.5 γ
  2/31(7)   (+)
  3/28   10 γ
  3/34   0.04 γ
  3/35(1)   1 γ
  3/37   1 γ
  5/68(3)   (+)
potassium isatinsulfonate
  2/12(1)   1.2 γ
potassium methionate
  2/13(2)   6 γ
Primuline Yellow
  7/3   (−)
procaine
  3/57   (+)
  3/58   (+)
procaine hydrochloride,
  see novocaine
procaine penicillin G
  3/58   (+)
proline
  3/56(2)   1 γ
n-propanol
  3/11   1 mg

5/9   (−)
5/10   (−)
propane-2-sulfonic acid,
  Na salt
  3/43   (−)
propanone-1,1-dicarboxylic
  acid
  3/24(1)   15 γ
2-propene-1-sulfonic acid,
  Na salt
  3/43   (+)
propionaldehyde
  3/20(3)   20 γ
  3/20(4)   20 γ
  5/28(1)   (−)
  5/42(4)   (−)
  5/53(1)   (+)
  6/38   3 γ
propionaldehyde
  cyanohydrin
  3/20(4)   34 γ; 10 γ
propionaldehyde
  phenylhydrazone
  3/76   0.1 γ
propionanilide
  3/65   25 γ
propionic acid
  2/24(6)   (−)
  5/43(1)   50 γ
  5/43(2)   (−)
  5/45(2)   (−)
propyl acetate
  3/29   (+)
propylamine
  3/51   13 γ; 10 γ
propyl benzoate
  3/28   3 γ
propylenediamine
  5/77   0.3 γ
propylene glycol
  4/1(1)   (+)
  5/9   (−)
protein
  7/122   (+)
protocatechualdehyde
  3/14(1)   (−)
  3/14(3)   4 γ
  3/18(2)   1 γ
  3/20(4)   1.2 γ
  3/21(1)   1 γ
  5/17(1)   (±)
  5/17(2)   (±)
protocatechuic acid
  5/17(1)   (−)
  5/17(2)   (±)
  5/18   (−)
  5/21(1)   (−)

5/22 (±)
5/23 (±)
5/42(1) (−)
5/56(2) (−)
5/63(1) (−)
protocetraric acid
4/23 2.2 γ
prussic acid, see hydro-
cyanic acid
psoromic acid
4/23 3.5 γ
purpurin (1,2,4-trihydroxy-
anthraquinone)
4/28 0.1 γ
Pyramidone
3/1 20 γ
7/29(g) (+)
7/37 10 γ
pyrazinecarboxylic acid
4/60 (+)
pyrazine-2,3-di-
carboxylic acid
4/60 (+)
pyrazine-2,5-di-
carboxylic acid
4/60 (+)
pyrazinetetracarboxylic
acid
4/60 (+)
pyrazinetricarboxylic acid
4/60 (+)
pyrazole
3/70 40 γ
pyridazine-3-carboxylic
acid
4/60 (+)
pyridazinetetra-
carboxylic acid
4/60 (+)
pyridine
2/24(2) (−)
2/24(3) 15 γ; 5 γ
3/49(2) 2 mg
3/57 2 γ
4/53(2) 30 γ
4/54 12 γ
4/58(1) 0.2 γ
4/58(2) 0.01 γ
5/101(1) 2.5 γ
5/101(2) 2.5 γ
7/17 (+)
7/18 5 γ
pyridine-2-aldehyde
3/19 0.05 γ
3/21(2) 1 γ
3/21(3) 100 γ
3/22(2) 2 γ

4/59 0.5 γ
5/102 4 γ
pyridine-3-aldehyde
3/21(2) 0.5 γ
3/21(4) 50 γ
3/22(2) 100 γ
4/58(2) 0.5 γ
4/59 0.5 γ
pyridine-4-aldehyde
3/19 0.01 γ
3/21(4) (−)
3/22(2) 1 γ
4/58(1) 1 γ
4/58(2) 0.5 γ
4/59 0.5 γ
β-pyridinecarbinol
5/101(1) (−)
pyridine-2-carboxylic-5-
acetic acid
4/60 (+)
pyridine-2,3-dicarboxylic
acid
4/53(2) (−)
4/60 10 γ
pyridine-2,4-dicarboxylic
acid
4/60 (+)
pyridine-2,5-dicarboxylic
acid
4/60 (+)
pyridine-2,6-dicarboxylic
acid
3/57 (+)
4/60 (+)
pyridine-3,4-dicarboxylic
acid
4/53(2) (−)
pyridine-3,5-dicarboxylic
acid
4/58(2) (−)
pyridinepentacarboxylic
acid
4/60 (+)
pyridine-3-sulfonic acid
5/103 30 γ; 10 γ
pyridine-2,3,4,6-tetra-
carboxylic acid
4/60 (+)
pyridine-2,3,5,6-tetra-
carboxylic acid
4/60 (+)
pyridine-2,3,5-tricarboxylic
acid
4/60 (+)
pyridine-2,3,6-tricarboxylic
acid
4/60 (+)

pyridine-2,4,5-tricarboxylic
acid
4/60 (+)
pyridine-2,4,6-tricarboxylic
acid
4/60 (+)
pyridoxal
4/23 (+)
7/27(3) 0.5 γ; 0.6 γ;
3 γ
pyridoxine
3/14(7) (−)
4/53(2) (−)
4/58(2) (−)
7/27(1) 0.5 γ
7/27(2) 0.1 γ
α-pyridylbenzoyl phenyl-
hydrazone
3/76 0.1 γ
β-pyridylcarbinol tartrate
4/53(2) 40 γ
4/58(1) 0.5 γ
Pyrilamine maleate
(Neoantergan, Anthisan)
7/56 (+)
pyrimidine-4-carboxylic
acid
4/60 (+)
pyrocatechol
2/31(7) 2.5 γ
2/32 (+)
2/33 (+)
2/34 (+)
2/36 (−)
3/14(1) 5 γ
3/14(2) 4 γ
3/14(3) 5 γ
3/14(6) 1 γ
3/14(7) 0.5 γ
3/14(8) 2.3 γ
4/1(1) (+)
4/4 5 γ
5/17(1) 4 γ
5/17(2) 0.5 γ
5/17(3) 1 γ
5/21(1) (±)
5/21(2) (±)
5/21(4) (−)
5/22 (±)
5/23 (±)
6/7 25 γ
pyrocatecholdisulfonic
acid
2/24(6) 4 γ
2/28 25 γ
3/14(5) (+)

pyrogallic acid
| 5/17(1) | (−) |

pyrogallol
2/31(5)	0.05 γ
2/31(6)	5 γ
2/31(7)	5 γ
2/32	(+)
2/33	(+)
2/36	(−)
3/14(1)	10 γ
3/14(2)	7 γ
3/14(3)	5 γ
3/14(7)	2.5 γ
3/14(8)	(−)
4/4	25 γ
4/7	0.1 γ
5/17(2)	(±)
5/18	(±)
5/21(1)	(±)
5/21(2)	(±)
5/21(3)	(−)
5/21(4)	(±)
5/22	1 γ
5/23	(±)
5/29(9)	(+)

pyrogallolcarboxylic acid
5/21(1)	(−)
5/23	(±)
5/56(2)	(−)

pyrrole
2/44	0.6 γ
3/48	40 γ
4/55	0.04 γ
5/14	(+)

pyrrolidine
| 3/55 | 5 γ |
| 3/56(2) | 0.5 γ |

pyruvaldehyde, see
methylglyoxal

pyruvic acid
5/32(3)	(+)
5/42(1)	(±)
5/53(1)	(+)
5/53(2)	(−)
5/54	(−)
5/56(2)	(−)
5/63(1)	(−)
5/64(1)	3 γ
5/64(2)	0.2 γ

pyruvic ester
| 2/44 | (+) |
| 3/24(1) | 15 γ |

Quebracho extract
| 7/116 | (+) |

quercetin
| 4/30 | (+) |

quinacrine chloride
| 3/58 | (+) |

quinaldine
2/33	(+)
4/53(2)	50 γ
4/54	12 γ
5/104	(+)

quinalizarin
2/36	(−)
4/28	0.8 γ
7/75	(+)

quinhydrone
| 5/21(1) | (+) |
| 5/21(2) | (+) |

quinic acid
5/17(1)	(−)
5/17(2)	(−)
5/21(1)	(+)
5/22	(−)
5/23	(±)
5/54(2)	(±)
5/56(2)	(−)

quinine
2/33	(+)
2/36	(−)
3/3	10 γ
3/57	(+)
4/9	(−)
4/53(1)	10 γ
4/53(2)	50 γ
7/43	50 γ

quinoline
2/33	(+)
2/36	(+)
3/57	(+)
4/53(2)	2 γ
4/54	25 γ
4/58(2)	(−)
5/104	20 γ; 45 γ

quinoline-2-carboxylic
acid
| 4/60 | (+) |

quinoline-2,3-dicarboxylic
acid
| 4/60 | (+) |

quinoline iodomethylate
| 5/104 | (+) |

quinolinic acid, see
pyridine-2,3-dicarboxylic
acid

quinolinic acid 3-
methyl ester
| 4/60 | (+) |

quinone (p-benzoquinone)
2/34	(+)
2/36	(+)
4/13	(+)

4/14(1)	0.2 γ
4/14(2)	0.5 γ
5/21(1)	(+)
5/21(2)	(+)
5/21(3)	(+)
5/38	2 γ

quinone dioxime
| 3/88(2) | 0.8 γ |
| 3/88(3) | 30 γ |

quinoxaline-2,3-di-
carboxylic acid
| 4/60 | 2 γ |

Raffinose
| 5/42(1) | (±) |

RDX, see cyclonite

resorcinol
2/31(5)	0.01 γ
2/32	(+)
2/33	(−)
2/34	(+)
2/35	0.8 γ
2/36	(−)
2/44	(+)
3/14(1)	5 γ
3/14(2)	2 γ
3/14(3)	0.5 γ
3/14(4)	3 γ; 0.25 γ
3/14(5)	0.5 γ
3/14(7)	0.5 γ
3/14(8)	0.6 γ
4/4	25 γ
4/5	1 γ
4/6	1 γ
5/14	(+)
5/18	1 γ
5/21(1)	(±)
5/21(2)	(−)
5/22	(±)
5/23	(±)
6/7	25 γ

α-resorcylic acid
5/18	(±)
5/21(1)	(±)
5/23	(−)
5/56(2)	(−)

β-resorcylic acid
5/18	(±)
5/21(1)	(±)
5/23	(−)
5/56(2)	(−)

γ-resorcylic acid
| 5/56(2) | (−) |

rhamnose
| 2/31(7) | (+) |
| 5/42(1) | (±) |

Rhodamine B
3/58   50 γ
7/76   (+)
7/78   500 γ
7/85   0.5 γ
Rhodamine 3B
7/85   0.25 γ
Rhodamine G
7/85   0.25 γ
Rhodamine 6G
7/85   0.25 γ
Rhodamine 6GH
7/85   0.5 γ
Rhodamine 3R
7/85   0.25 γ
Rhodamine S
7/85   (−)
rhodanine
2/12(9)   2.5 γ
2/44   0.6 γ
3/34   0.003 γ
rhodizonic acid
2/31(5)   0.5 γ
4/10(1)   5 γ
4/10(2)   4 γ
4/13   0.5 γ
4/14(1)   (−)
4/14(2)   (−)
5/41   0.1 γ
icinoleic acid
3/26   15 γ
rongalite
3/45   (+)
7/121   0.2 γ; 0.5 γ;
        3 γ
Rose Bengal
2/9(2)   0.005 γ
rotenone
2/34   (+)
rubber, chlorinated
7/111   (+)
rubeanic acid, see
  dithiooxamide
rufigallic acid
4/28   1 γ
rutin
3/14(8)   (−)

Saccharic acid
5/54   (−)
5/56(2)   (−)
saccharin
2/4(2)   5 γ
2/12(4)   6 γ
2/12(5)   2 γ
2/17   (+)

2/36   (−)
3/43   1 γ
3/61   10 γ
4/31   10 γ; 5 γ
7/20(e)   5 γ
saccharose (sucrose,
  cane sugar)
2/4(1)   5 γ
2/4(3)   1.5 γ
3/18(1)   (−)
4/1(2a)   25 γ
4/1(2b)   7 γ
4/19   25 γ
4/21   0.25 γ
4/22   15 γ
5/9   (±)
5/28(1)   (±)
5/42(1)   (±)
5/63(1)   (−)
safrole
2/34   (+)
3/5   (+)
4/9   0.4 γ
salazinic acid
4/23   (+)
salicin
2/36   2.5 γ
7/59   5 γ
salicyl alcohol (saligenin)
3/10(1)   3 γ
4/3   (−)
5/26   2 γ
salicylaldazine
3/19   0.05 γ
3/21(4)   1 γ
salicylaldehyde
2/31(5)   0.1 γ
2/31(6)   1 mg
2/32   (−)
3/14(1)   2 γ
3/14(3)   5 γ
3/14(7)   (−)
3/14(8)   1.2 γ
3/18(1)   4 γ
3/18(2)   0.35 γ
3/19   0.001 γ
3/20(1)   100 γ
3/20(2)   100 γ
3/20(3)   5 γ
3/20(4)   0.7 γ
3/20(5)   0.1 γ
3/21(1)   1 γ
3/21(2)   1 γ
3/21(3)   0.3 γ
3/21(4)   0.5 γ
3/22(1)   2.5 γ
3/22(2)   0.5 γ

4/3   (−)
4/23   1 γ
4/24(1)   0.05 γ
4/24(2)   0.05 γ
4/26   (−)
5/52(1)   (+)
salicylaldehyde bisulfite
3/20(4)   0.8 γ
salicylaldoxime
2/16   5 γ
2/32   (−)
2/34   (+)
3/19   0.01 γ
3/86   (−)
6/37   (+)
salicylalhydrazone
3/14(8)   0.5 γ
3/19   0.01 γ
3/21(2)   1 γ
3/21(4)   0.5 γ
3/22(2)   0.5 γ
3/72   0.01 γ
3/80(1)   5 γ
3/80(2)   (−)
salicylalphenylhydrazone
3/72   0.05 γ
3/80(1)   0.5 γ
3/80(2)   1 γ
salicylamide
2/24(7)   1 γ
3/14(7)   (−)
3/14(8)   0.4 γ
3/60(1)   5 γ
3/60(2)   25 γ
3/61(1)   5 γ
4/34   (+)
salicylanilide
2/24(7)   5 γ
3/64   2 γ
4/34   (+)
salicylhydroxamic acid
3/87(1)   40 γ
salicylic acid
2/4(1)   3 γ
2/4(3)   3 γ
2/24(7)   1 γ
2/32   (+)
2/36   1 γ
3/14(4)   5 γ; 2 γ
3/14(5)   (+)
3/14(8)   4 γ
3/27   (+)
4/3   (−)
5/17(1)   (−)
5/17(2)   (−)
5/18   (−)
5/21(1)   (±)

5/22	(−)
5/42(1)	(−)
5/42(4)	(−)
5/52(1)	(−)
5/55	(−)
5/56(1)	(−)
5/56(2)	(−)
5/60	5 γ
7/21	(+)

salicylic acid alkyl esters
5/60	(+)

salicylic acid hydrazide
3/14(8)	1.9 γ
3/72	1 γ
3/73(1)	0.1 γ
3/78(1)	5 γ
3/78(2)	0.5 γ
3/78(5)	1 γ
4/34	(+)

salicylideneaniline
3/21(2)	5 γ
3/21(4)	1 γ
3/22(2)	1 γ

saligenin, see salicyl
alcohol

salmine
4/48(1)	0.5 γ

salol, see phenyl
salicylate

salvarsan
2/19	2 γ; 20 γ

scopolamine
3/57	(+)

sebacic acid
2/4(3)	0.5 γ
2/24(4)	2 γ; 5 γ

sedulone
4/53(2)	(−)

semicarbazide
2/31(8)	2 γ
2/44	0.6 γ
3/72	1 γ
3/73(1)	0.3 γ
3/78(1)	2.5 γ
3/78(2)	1 γ
3/78(3)	0.5 γ
5/108(1)	(−)

serine
4/1(2)	(+)

serum albumin (blood
albumin)
4/48(1)	0.35 γ
4/48(2)	20 γ

sitosterol
3/10(1)	(+)

skatole
4/55	(+)
4/56	0.1 γ

sodium acetate
3/26	100 γ

sodium acetylarsanilate
2/19	5 γ

sodium diethyldithio-
carbamate
2/12(9)	10 γ
2/31(5)	1 γ
3/99	2.5 γ

sodium heptadecyl
sulfate
7/119	(+)

sodium p-hydroxyphenyl-
arsonate
2/19	4 γ

sodium lauryl sulfate
7/119	(+)

sodium methylhexyl
sulfate
7/119	(+)

sodium methyloctadecyl
sulfate
7/119	(+)

sodium octadecyl sulfate
7/119	(+)

l-sorbose
2/31(7)	5 γ
4/19	2.5 γ
4/22	8 γ

Soromin
7/127	(+)

spermidine phosphate
2/40	50 γ
3/56(2)	70 γ

spermine phosphate
2/40	50 γ
3/56(2)	80 γ

starch
4/1(2a)	50 γ
4/1(2b)	20 γ
4/19	3 γ
4/21	0.2 γ
7/122	(+)

stearic acid
2/24(4)	5 γ; 50 γ
2/24(5)	(+)
2/24(7)	2.5 γ
3/26	20 γ
3/27	(−)
7/126	5 γ

Stibamine
7/61	(+)

stibanilic acid
3/105	1 mg
7/61	(+)

Stibenyl
7/61	(+)

p-stibonobenzamide
3/105	1 mg

p-stibonobenzoic acid
3/105	1 mg

stilbene
2/32	(+)
3/6	50 γ

Stovarsol
3/14(5)	(+)
3/63	(+)

streptidine
3/67(1)	(+)

streptomycin
3/67(1)	100 γ
3/67(2)	(+)

strychnine
3/57	(+)

styrene
3/6	500 γ

styrene resin
7/109	(+)
7/122	(+)

succinic acid
2/5	0.1 γ
2/24(6)	10 γ
2/40	5 γ
3/26	11 γ
3/27	(+)
4/31	5 γ
5/45(2)	(−)
5/48	5 γ
5/52(1)	(−)
5/55	(−)
5/56(1)	(−)
5/57	(+)

succinic anhydride
3/32	5 γ
4/31	5 γ
5/48	(+)

succinimide
3/61	0.5 γ
4/31	5 γ
5/48	(+)
5/118	2.5 γ

succinyl chloride
3/33	0.2 γ

sucrose, see saccharose

Sudan III
2/36	(−)
7/81	3 γ

sugar, reducing
4/20(2)	0.1–10 γ

sulfadiazine
4/44	(+)
6/27	0.1 γ
7/45	0.05 γ
7/46	5 γ

sulfaguanidine
  3/67(2)   (−)
  7/45   0.05 γ
  7/48   10 γ
sulfamerazine
  7/45   0.2 γ
sulfamethazine
  2/36   (+)
  7/45   0.1 γ
sulfanilamide
  2/12(4)   10 γ
  2/17   400 γ; 100 γ
  2/24(2)   25 γ
  3/52   0.5 γ
  3/60(1)   20 γ
  4/44   (+)
  7/45   0.05 γ
sulfanilic acid
  2/12(1)   1.2 γ
  2/12(4)   20 γ
  2/12(5)   5 γ
  2/12(6)   50 γ
  2/13(1)   1 γ
  2/13(2)   12 γ
  2/15(1)   2 γ
  2/17   (−)
  2/24(6)   10 γ
  2/26   10 γ
  2/27   5 γ
  2/28   30 γ
  2/36   0.5 γ
  3/27   (−)
  3/41(1)   25 γ
  3/43   2.5 γ
  3/52   0.1 γ
  3/54(1)   (−)
  4/44   50 γ
  4/49   1 γ
  4/50   0.5 γ
  6/18   10 γ
sulfanilurea
  3/60(1)   10 γ
  7/48   10 γ
sulfapyridine
  2/17   (−)
  2/36   (+)
  3/60(1)   30 γ
  7/47   (+)
sulfathiazole
  2/24(2)   5 γ
  2/36   (+)
  3/70   25 γ
  4/53(1)   10 γ
  7/47   (+)
sulfathiourea
  7/48   0.6 γ

sulfonal
  2/13(2)   6 γ
  3/39   5 γ
Sulforhodamine B
  7/72   (+)
  7/85   (−)
sulfosalicylic acid
  2/4(3)   2 γ
  2/12(1)   1 γ
  2/12(5)   2.5 γ
  2/24(6)   5 γ
  2/28   25 γ
  2/32   (+)
  2/34   (+)
  2/36   1 γ
  2/40   2.5 γ
  3/14(5)   1 γ
  3/41(2)   10 γ
  3/43   5 γ
  4/3   (−)
  4/49   500 γ
  5/74(1)   3.5 γ
  5/74(2)   1.5 γ
Sulfotepp, see tetraethyl
  dithiopyrophosphate
Sulfoxone sodium (Diazone
  sodium)
  3/45   (+)
  7/29(e)   (+)
sumac extract
  7/116   (+)
Supratan OC
  7/116   (+)
  7/117   (+)
synthalin
  3/67(1)   (+)

Tangan LH
  7/116   (+)
tannaform
  7/29(a)   (+)
tannic acid
  2/31(7)   10 γ
  2/31(8)   5 γ
  4/7   2 γ
  5/23   (±)
  7/30   (+)
  7/116   (+)
tara powder
  7/116   (+)
tartaric acid
  2/24(6)   5 γ
  2/27   5 γ
  2/31(4)   50 γ
  2/40   1 γ
  3/27   (+)

  4/1(2a)   100 γ
  4/31   25 γ
  5/9   (−)
  5/27   10 γ
  5/42(2)   (−)
  5/45(2)   (−)
  5/52(1)   (±)
  5/54   (±)
  5/56(1)   2 γ
  5/56(2)   10 γ
  5/59(4)   (−)
  5/63(1)   (−)
  5/63(3)   (−)
tartrazine
  2/13(2)   12 γ
tartronic acid
  5/54   (±)
  5/55   (±)
  5/56(1)   (±)
  5/56(2)   (±)
  5/63(1)   (−)
taurine
  2/12(4)   5 γ
  2/12(5)   0.5 γ
  2/13(1)   0.25 γ
  2/17   100 γ; 50 γ
  2/26   25 γ
  2/27   10 γ
  2/28   30 γ
  2/40   5 γ
  3/41(2)   1 γ
  3/43   (−)
  5/85   10 γ
terephthalaldehyde
  5/32(2)   (−)
terephthalic acid
  2/33   (−)
terpenes
  3/10(1)   (+)
terpineol
  3/10(1)   (+)
terpinol hydrate
  5/21(1)   (±)
Terylene
  7/102(2)   (−)
  7/102(9)   (+)
tetrabase (4,4′-di-
  (dimethylamino)-
  diphenylmethane)
  2/5   0.2 γ
  2/24(1)   15 γ; 10 γ
  2/31(5)   0.05 γ
  2/36   (+)
  3/49(1)   0.2 γ
  3/49(2)   50 γ
  5/91   0.5 γ; 2.5 γ

thymine
2/35   (+)
thymol
3/14(1)   5 γ
3/14(3)   (−)
3/14(5)   (+)
3/14(6)   0.5 γ
3/14(7)   0.5 γ
3/14(8)   1 γ
4/3   (−)
5/23   (±)
thyroxine
3/9   0.002 γ
tiglic aldehyde
5/29   (+)
Tolu balsam
7/127   (+)
toluene
2/32   1.5 γ
2/33   (+)
2/36   (+)
toluene-2,3-dithiol
2/31(8)   10 γ
tolueneseleninic acid
3/47   120 γ
toluenestibonic acid, m-
and p-
3/105   1 mg
toluenesulfinic acid
3/38(1)   5 γ
3/38(2)   5 γ
3/38(3)   2.5 γ
N-p-toluenesulfonyl-
carbazole
4/41(1)   1.5 γ
4/57   2.5 γ
toluhydroquinone
3/14(8)   1 γ
o-toluidine
3/54(1)   0.2 γ
6/14   5 γ
p-toluidine
6/14   2 γ
p-toluquinone
4/14(1)   1 γ
4/14(2)   1 γ
5/38   100 γ
3-tolu-o-quinone diacetate
4/14(1)   5 γ
4/14(2)   1 γ
m-toluylenediamine
3/54(1)   1 γ
tolylmercuric chloride
3/106   30 γ
tolylmercuric nitrate
7/130   8 γ

m-tolylstibonic acid
2/34   (+)
tolylurea
3/63   (+)
m-tolylurea
4/62   150 γ
5/108(1)   (−)
p-tolylurea
3/64   2 γ
tosyl-m-hydroxybenzo-
trifluoride
2/10(2)   40 γ
tribenzylamine
3/57   (+)
tribromoaniline
2/26   20 γ
2/33   (−)
4/42   10 γ
tribromoethanol
3/8   (+)
tricarballylic acid
3/26   11 γ
4/31   5 γ
5/42(2)   (−)
trichloroacetic acid
3/8   0.5 γ
5/51   50 γ
1,1,1-trichloro-2,2-bis-
(p-chlorophenyl)-ethane,
see DDT
trichloroethylene
3/8   (+)
1,1,1-trichloro-2-hydroxy-
3-phenylaminopropane
2/6(4)   20 γ
N-trichloromethyl-
mercapto-tetrahydro-
phthalimide
2/6(4)   20 γ
triethanolamine
3/2   (+)
3/57   (+)
4/42   30 γ
triethylbenzene
2/32   (+)
2,4,6-triethyleneamino-
1,3,5-triazine
3/2   4 γ
N-trifluoroacetyl-p-tolu-
idine
2/10(2)   50 γ
m-trifluoromethyl-benzoic
acid
2/10(2)   20 γ
1,2,3-trihydroxyanthra-
quinone, see anthragallol
1,2,4-trihydroxyanthra-
quinone, see purpurin

2,3,4-trihydroxybenz-
aldehyde
5/17(1)   (±)
5/17(2)   (±)
5/18   (±)
2,4,6-trihydroxybenz-
aldehyde
5/18   (±)
2,3,4-trihydroxybenzoic
acid, see pyrogallol-
carboxylic acid
2,4,6-trihydroxybenzoic
acid
5/18   (±)
5/23   (±)
3,4,5-trihydroxybenzoic
acid, see gallic acid
1,6,8-trihydroxy-3-methyl-
anthraquinone
4/28   (−); (+)
triiodothyronine
3/9   0.002 γ
trimellitic acid trimethyl
ester
4/31   2.5 γ
trimethylamine
3/57   (+)
trimethylene-α-mannitol
5/28(1)   (+)
2,4,6-trinitroaniline
2/26   20 γ
1,3,5-trinitrobenzene
3/93(2)   (+)
3/94   (+)
2,4,6-trinitro-m-cresol
3/94   (+)
2,4,6-trinitrotoluene
2/44   (+)
3/93(2)   (+)
3/94   (+)
trional
2/13(2)   10 γ
triphenylarsine
3/103   2.5 γ
triphenylarsinic oxide
2/34   (+)
1,3,5-triphenylbenzene
2/32   (+)
triphenylbismuthine
3/103   5 γ
triphenylmethane
2/32   (+)
triphenyl phosphate
5/148   1 γ
triphenylstibine
3/103   5 γ

Author Index

Subject Index

(For tests for individual compounds, see
Individual Compounds and Products Examined, p. 709)